Economics of Space and Time
The Measure-Theoretic Foundations of Social Science

Arnold M. Faden

The Iowa State University Press,
Ames, Iowa, U.S.A.

ARNOLD M. FADEN, Professor of Economics at Iowa State University, holds the Ph.D. degree from Columbia University. He has originated the post-Bayesian approach to statistical inference and is also doing work on quantitative history.

© 1977 The Iowa State University Press
Ames, Iowa 50010. All rights reserved

Composed and printed by Science Press, Ephrata, Pa. 17522

First edition, 1977

Library of Congress Cataloging in Publication Data

Faden, Arnold M   1934–
  Economics of space and time.

  Includes bibliographical references.
  1. Space in economics.   2. Measure theory.
3. Social sciences—Methodology.   I. Title.
HB199.F23   330.1   77-4471
ISBN 0-8138-0500-7

To Karl A. Fox

CONTENTS

FOREWORD

THE LITERATURE of location theory and spatial economics has long had a life of its own apart from the mainstream of economic theory. This is regrettable but understandable as long as economic theory rests on the assumption that the spatial dimension can be suppressed. However, space is reasserting itself in economics through the emergence of regional and urban economics as areas of current concern.

At this critical juncture, a book of the format of A. M. Faden's is doubly welcome. It supplies a novel and thoroughly modern approach to the economics of space, incorporating much of the classical economics of location in its modern framework. This is the framework of measure theory, which proves indeed to be a natural one for many aspects of spatial economics. Looking back, it now appears that Launhardt and Lösch, to name but two of the great classics, were groping for this approach. It should now be possible to restate much of classical location theory in more general and more operational terms. Such is the power of the new approach that the present book though covering much ground has not exhausted its possibilities. It will thus give a stimulus to all of us to take a new look at the problems of spatial economics.

Martin J. Beckmann

München, Oktober 1976

PREFACE

IN THE COURSE of working on various problems in locational economics, it gradually dawned on me that much deeper results could be obtained if these problems were reformulated in terms of measure theory. I pursued this line of inquiry for several years, and found not only that most of classical location theory could be reformulated with profit in terms of measures but also that the scope of the approach was much broader than I had anticipated. In short, it turns out that measure theory is a vehicle of universal descriptive power that provides a framework into which all of social science can fit.

It is easiest to summarize the book from back to front. Chapters 8 and 9 contain the reformulation of most of classical location theory, and a good deal more. Chapters 5 and 7, on the allocation-of-effort and transportation problems respectively, provide results used in many of the models in Chapters 8 and 9 and are of considerable interest in their own right. Chapter 6 covers markets in general, in particular the real-estate market. Chapters 2 and 4 provide the overall descriptive framework. Chapter 3 develops the deepest mathematical innovation in the book, the concept of pseudomeasure. Finally, Chapter 1 elaborates the quick summary of this Preface, discusses what remains to be done, and provides a detailed synopsis and reader's guide.

This work was done in splendid isolation—perhaps too splendid. But I must mention Karl Fox, to whom this book is dedicated. Fox is a man of vision who is launching what amounts to a second career with his work on social indicators and historical interpretation. He was Head of the Iowa State University Department of Economics during the book's gestation period, and his encouragement was a necessary input to the sustained effort that went into creating it. It must not have been easy to keep the faith when year after year went by without anything palpable to show for it. I trust that the present book justifies the wait. Finally, a note of thanks to my editor, Nancy Bohlen, who handled a difficult manuscript and temperamental author with grace.

Economics of Space and Time

*The Measure-Theoretic Foundations
of Social Science*

1

Introduction

1.1. AIMS, ACHIEVEMENTS, AND SHORTCOMINGS

It is the thesis of this book that measure theory is the natural language for spatial economics and, indeed, for all social science.

As I carried forward the work started in my dissertation, *Essays in Spatial Economics,*[1] it gradually became apparent that most of the discussion could be made deeper and more general if it were reformulated in measure-theoretic terms rather than in the elementary algebraic and geometric language commonly used in all standard treatises on location theory. I resolved to restructure my planned book along these lines after learning enough of the mathematics of measure theory to do so. (At this time I had only the most limited acquaintance with measure theory. It was mentioned once or twice in my dissertation, but only peripherally.) It eventually became clear that even this fairly ambitious goal did not go far enough. To deal with the problems that arose in location theory it was necessary to go beyond the existing corpus of mathematical knowledge. As has happened so often in the past, especially with physics, the demands of an applied field stimulated new discoveries in pure mathematics. This book, taken as a whole, may fairly be said to found a new branch of mathematics— one that might be called the *measure theory of optimization;*[2] and not only to found it, but to bring it to a fairly high state of development.

At the same time, the applications appeared to be broadening: first within location theory, where old results (including my own) were gen-

---

1. A. M. Faden, *Essays in Spatial Economics,* Ph.D. Diss., Dept. Econ., Columbia Univ., 1967.
2. Certain existing disciplines fall under this rubric, notably parts of the theory of games and parts of mathematical statistics. But this book, with the exception of Chapter 5, is independent of these. It is also independent of the one existing direct application of measures in economic theory, the "continuum of traders" model. For comments on the latter see Section 6.7.

eralized and put on a more rigorous basis and where new problems were formulated and solved. Theory and applications developed hand-in-hand, each stimulating the other. I came to realize that the framework I had constructed for dealing with locational problems was in a way universal and could in principle handle all problems of social science.

Thus one starts with a limited problem and ends up with a world view. This broadest of extensions is not developed in any detail in the present work, but only sketched (mainly in Chapters 2 and 4). Nonetheless, on the basis of these results it might be predicted that much of the social science of the future will be written in the language of measure theory—but not all: Here the analogy with probability theory is instructive. Since the work of Kolmogorov in 1933, probability theory has come to be viewed as a branch of measure theory. This does not mean that it is worthwhile translating every probability problem into measure-theoretic language. But the possibility of such a translation helps one to think clearly and to generalize and makes available a powerful body of mathematical results if they are needed. Similarly, I do not mean that social science will be straitjacketed into a limited number of categories or that it will lose its richness of detail or become (further) dehumanized. Rather, the fact that its statements can be translated into the framework of this book makes for a certain unity of vision lacking at present.

Let me now turn to a more detailed discussion. I shall take up the nature of measure theory, of location theory, and of the interpretation of the latter in terms of the former; the problems that arise; and what this book accomplishes and does not accomplish.

Measure theory will be presented *analytically* in Chapter 2; here, however, the subject is treated *historically*. This theory originated around 1900 in the work of Borel and Lebesgue and was first directed toward two related problems: to extend the notion of length from intervals to rather general subsets of real numbers in a reasonable way, and to extend the notion of integral to a rather general class of real functions. This program was carried out successfully in the first decade of this century, mainly by Lebesgue, and the new theory found a number of unexpected applications. The developments that are really of interest to us occurred, however, in the next decade. First, the new theory was connected with an old idea due to Stieltjes, who in 1894 defined the concept of integration with respect to an arbitrary *distribution of mass* on the real line.[3] From this more general point of view, the integral of Lebesgue is merely the special case in which mass is uniformly distributed over the line. Second, the work of Cara-

---

3. T. Hawkins, *Lebesgue's Theory of Integration: Its Origins and Development* (Univ. Wis. Press, Madison, 1970), pp. 180–81.

théodory and Fréchet generalized the concepts of measure and integral from the real line (and its easy extension to *n*-space) to *abstract* spaces.

This generality is the key to the applications of measure theory in this book. For the points of the measure space could be, e.g., points of (physical) space, space-time, personality types, resource types in general, technical processes, time paths of development, etc., or even more complicated entities built up from these. The measures themselves could represent physical mass, numbers of events, dollar values, man-hours, volumes, durations, degrees of belief, etc. In Chapter 2 it will be shown in detail that most statistical data may be construed directly as measures, and that by suitable construction, this representation may be extended (in principle) to literally all facts relevant to social science.

I skip over the further historical development of measure theory and conclude with a discussion of the work of Kolmogorov.[4] This consisted of interpreting the probability calculus as a branch of measure theory. That is, there is a way of translating the vocabulary of probability theory into measure-theoretic terms so that true propositions in the former discipline are translated into true propositions in the latter. (Some examples of such a translation are: "Event" becomes "measurable set"; "random variable" becomes "measurable function"; "expectation" becomes "integral.") From this perspective Kolmogorov was able to place probability theory on a rigorous axiomatic basis and to clarify and develop certain portions (most notably the foundations of stochastic processes).

Some intriguing issues arise in comparing Kolmogorov's work with this book. Let me note some contrasts. First, the probability calculus already had a large and impressive body of doctrine; by contrast social science is in a more primitive state. Second, the probability calculus was a branch of pure mathematics (though its foundations were somewhat vague); Kolmogorov's work amounted to an *intra*mathematical reduction of one branch to another. By contrast, the present work has the task of representing real-world situations in measure-theoretic terms. This requires a much longer and more open-ended discussion. Third, the path leading from probability calculus to measure theory is rather short and direct. By contrast, the path from social science to measure theory is tortuous, involving the construction of a complex conceptual apparatus and considerable development of measure theory itself. For all these reasons Kolmogorov's work could be accomplished quite briefly, while the present work had to be much longer.

---

4. A. N. Kolmogorov, *Grundbegriffe der Wahrscheinlichkeitsrechnung* (Springer, Berlin, 1933), trans. N. Morrison as *Foundations of the Theory of Probability,* 2nd ed. (Chelsea, New York, 1956).

The most intriguing point of comparison, however, is that both books attempt to reduce their respective subjects to the same mathematical discipline, measure theory. This is not a coincidence. To unearth the connection, one has to examine the historical roots of probability theory. For, although it is now a self-contained mathematical theory, its basic concepts were designed for the analysis of certain real-world situations, so that it was and still is very much an applied theory. Specifically, the work of Pascal and Fermat was directed toward the analysis of gambling games. From this start the applications were gradually broadened to include insurance; errors of observation; and certain parts of physics, genetics, survey sampling, etc.

To what, then, do probability statements refer? The classical answer is that probability is the number of favorable equally likely elementary events divided by the total number of such events. This makes probability an additive set function, in fact, a measure. This additivity property was then carried over to various generalizations: to unequal elementary probabilities, to continuous distributions, etc. Indeed, it is additivity, together with the multiplication rule for combining "independent" events, that enabled one to speak of a "calculus" of probabilities.

This rather crude, almost circular, definition of probability seemed to need clarification. Many different interpretations have been offered, and the question of which is "correct" remains a matter of lively controversy to the present day. The major dichotomy is between those who take probability statements to represent "degrees of belief" and those who take them to represent "relative frequencies." A careful discussion would further distinguish several schools within each of these camps; e.g., "relative frequency" can refer to a series of observations already made (or which will be made, say, in the next hundred years) or to the limit, if it exists, resulting from the indefinite repetition of an experiment; "degree of belief" can refer to what one *does* believe or to what one should rationally believe on the evidence available.

The relative merits of these approaches will not be reviewed here. Let me, however, indicate their relation to the measure-theoretic translation of probability. Consider first the "relative frequency" interpretation (the "finite" version, not the "limiting" version). Relative frequencies are a special case of the "physical measures" that are the subject matter of this book. One goes to the general case in two steps. First, it is usually more convenient to use absolute rather than relative frequencies; second, one often deals with measures in which there are no "natural units," hence no frequencies per se—e.g., the spatial distribution of water: one is "measuring" rather than "counting."

On the other hand, Jeffreys, de Finetti, and Savage have argued convincingly that for applications of probability some version of the "degree of belief" interpretation must take primacy.[5] Yet despite much axiomatic work, one is not quite convinced that the probability calculus as it stands today is the proper vehicle for representing degrees of belief. The representability of physical measures is much more firmly established. The framework of this book is broad enough to take account simultaneously of both physical measures (including relative frequencies) and degrees of belief. A brief sketch of how to do this is presented in Section 2.8.

I turn now to the nature of the real-world interpretations, the "co-ordinating definitions," of measure-theoretic categories. My concern here is not with details, which will be amply covered in Chapter 2 and throughout the book, but with broad ideas and motivations.

The basic observation is that the world can be described by stating how much matter is embodied in what forms at what places at what dates—or, more generally, how much matter is undergoing such-and-such transformations at what places at what dates.

The language of measure theory is ready-made for descriptions of this sort: Quantities of matter are represented by the values of measures, while the underlying universe set whose subsets are given these values is built up from the fundamental categories of Resources, Space, and Time, describing forms, places, and dates respectively. (Methods of dealing with transformations and other more complex descriptions are spelled out in Chapter 2.) Measurement need not be in terms of physical mass, but may be in terms of numbers of units, volumes, money values, etc., as befits the particular entities under discussion.

But as soon as one realizes it is possible to *describe* the world in measure-theoretic terms, it becomes clear that one can also *analyze* the world in these terms. Laws of development can be formulated in measure-theoretic language, problems can be posed and solved in it, etc.

The question remains, If measure-theoretic language *can* be used in this way, is it worthwhile to do so? Why should it be more useful to translate into measure theory than to translate into, say, Esperanto? The only answer is to make the translation and see what ensues. This book shows that one can obtain a massive and rather imposing body of results, and it seems clear that these are just a small part of the results that await further research along these lines.

We all have the experience of working fruitlessly on a problem for a time, only to see the solution clearly when we reformulate the problem in the right way. Call this the "natural" formulation for the problem in

---

5. H. Jeffreys, *Theory of Probability,* 3rd ed. (Clarendon Press, Oxford, 1961), esp. Ch. 7; H. E. Kyburg, Jr., and H. E. Smokler, eds., *Studies in Subjective Probability* (Wiley, New York, 1964).

question. Now think, not of one problem, but of a whole range of problems constituting some discipline. And think, not of a single formulation, but of a range of interrelated formulations constituting a consistent point of view, a method, a body of doctrine, a theory, or whatever you wish to call it. If it often happens that problems from some discipline yield when formulated in terms of a certain theory, one may speak of that theory as the "natural language" for that discipline.

Measure theory is the natural language for spatial economics in this sense. Let me examine the status of this latter discipline. If one compares the standard treatises of the 1930s on location with those of the present,[6] one's general impression is that not much progress has been made—at least relative to the rest of economic theory, much of which has been transformed almost beyond recognition over this period. Why should this be? Certainly not from lack of interest. Indeed, regional and urban economics, which are the applied parts of the discipline, have grown enormously. Nor is it clear that location theory is so much more difficult than other branches of economics. Indeed, "space" is really a very simple category. Transportation is much simpler conceptually than most manufacturing processes. A location can be characterized by three (or two) numbers, while the state of a person or a commodity requires an indefinitely long description. So, with such simple subject matter, why is the field so backward?[7]

The answer is that location theorists have up to now simply not found the proper language in which to formulate their problems. Measure theory is that language. This entire book is an argument for this statement. Let me indicate the kinds of snarls that are encountered if one does *not* use measure theory.

Consider the theory of Thünen systems (Chapter 8). Von Thünen's work dates from 1826 and contains the oldest formal model in location theory. The problem is to determine the spatial distribution of agricultural land uses, given that all deliveries are to be made to a single isolated city (thought of as a geometrical point). There are several basic conceptual difficulties. First, what is a "land use?" An indefinitely large va-

---

6. For the 1930s see T. Palander, *Beiträge zur Standortstheorie* (Almqvist och Wiksell, Uppsala, 1935); E. M. Hoover, Jr., *Location Theory and the Shoe and Leather Industries* (Harvard Univ. Press, Cambridge, 1937). For the present there are several choices, e.g., H. W. Richardson, *Regional Economics: Location Theory, Urban Structure, and Regional Change* (Praeger, New York, 1969). An overview may be found in the articles on Spatial Economics by E. M. Hoover and L. N. Moses in *International Encyclopedia of the Social Sciences* (Macmillan, New York, 1968). Recently the "new urban economics" has developed rapidly, in a direction complementary to this book. See Chapter 8, note 15.

7. Of course, a number of disciplines have evolved to deal with spatial problems: network flow theory, traffic flow theory, the linear programming transportation problem, information diffusion models, etc. But none of these directly touches the traditional problems of location theory. Chapter 7 generalizes the transportation problem to measure-theoretic form, whereupon it *does* become directly relevant to several of these problems.

riety of commodities exist that may enter as inputs or outputs in a land use (e.g., the different possible qualities of corn or fertilizer); the proportions in which these commodities enter may vary continuously; finally, there may be indefinite variations in the time distribution of these inputs and outputs (sequencing, seasonal cycles, etc.). Clearly, "land use" is no simple concept. The complications become even worse with the modern observation that the Thünen analysis is relevant to the spatial distribution of *urban* land uses, so that it becomes a theory of the internal structure of the city. For this application the concept of land use must be extended to include various kinds of residential processes, manufacturing, trade, services, office activities, etc. Furthermore, it is desirable to distinguish various time patterns within each of these uses as well as the succession of different uses on the same site.

Existing studies deal with these possibilities by drastic simplification, either by assuming there are just a finite number of possible land uses or by assuming they may be described by the variation of a few parameters, e.g., "intensity." By contrast, the measure-theoretic approach easily accommodates all the complications mentioned above.

A second difficulty involves the concept of "spatial distribution" of land uses. There is an infinite variety of possible distributions even for the case of just two land uses: Corn and oats may be mixed in proportions that vary with location, or they may be segregated (in which case the region assigned to each must be specified). For the general set of land uses as discussed above, the possibilities are of course much richer. Existing theories usually cope with this difficulty by restricting a priori the set of distributions, e.g., assuming that land uses are segregated into geometrically simple regions. The measure-theoretic approach, however, accommodates the general case.

A third difficulty involves the allocation of "continuous" space to "discrete" decision makers, e.g., farmers. The solution offered in Section 8.7 seems conceptually superior to that of preceding writers,[8] though the assumptions concerning preferences may well be criticized as unrealistic.

The "snarls" in traditional approaches involve inadequacies of descriptive power. More serious are their inadequacies of analytical power. Consider the normative analysis of Thünen systems, which is the main focus of Chapter 8. The question of whether, or in what sense, the Thünen arrangement of land uses is optimal is unresolved in the literature.[9] It turns out that the Thünen solution *is* optimal for a certain simple

---

8. E. S. Dunn, Jr., *The Location of Agricultural Production* (Univ. Florida Press, Gainesville, 1954); W. Isard, *Location and Space-Economy* (MIT Press, Cambridge, 1956), Ch. 7; W. Alonso, *Location and Land-Use* (Harvard Univ. Press, Cambridge, 1964).

9. See the discussion in Alonso, *Location and Land-Use*, Ch. 1 and pp. 101–5. The first semirigorous argument for the affirmative appears to be Faden, *Essays*, pp. 170–88, though it is couched in equilibrium rather than optimizing terms. This argument is essentially repeated on pp. 500–504.

linear-programming transportation problem! This is not your ordinary transportation problem, however, but a measure-theoretic generalization of it (see Chapter 7). Indeed, applying the ordinary transportation problem in this way requires a most unnatural "discretization" of space as well as a restriction to a mere finite number of land uses; thus the measure-theoretic generalization is needed to deal adequately with the Thünen problem.

The Thünen model arises from the interplay of a single "discrete" city and a "continuous" hinterland. More generally, this interplay of discrete and continuous pervades location theory: A discrete plant pollutes a continuous environment; a discrete policeman patrols a continuous neighborhood; a continuously dispersed rural population condenses to a discrete city.

This interplay is another source for the descriptive and analytical superiority of measure-theoretic over traditional approaches. Measure-theoretic description allows for "discrete," "continuous," and "mixed" distributions over Space (e.g., urban-rural population), over Time (e.g., "lumpy" and "continuous" production), and over Resources (e.g., positive amounts of some resource together with a "continuous" distribution over others) as well as over more complex spaces. As for analytic power, note first that many problems can scarcely be formulated without measure-theoretic language; e.g., is it optimal for a certain industry to be spatially distributed into discrete plants, or continuously, or with a mixture of these, etc.?[10] Most of the models of Chapter 9 involve this discrete-continuous interplay.[11]

These examples should give some inkling of the sense in which measure theory is the natural language for spatial economics. It lends the subject generality, unity, and mathematical power; above all, it is an aid to clear thinking.

Spatial economics is not just another branch of learning.[12] That Space (and Time) are fundamental categories in any attempt to understand the world is an old idea, going back at least to the Greek atomists. If one tries to specify the somewhat elusive reasons for this notion, one

---

10. A very simple model exists on the uniform plane with uniform costs in which the optimal distribution appears to be "mixed"; see A. M. Faden, Inefficiency of the regular hexagon in industrial location, *Geographical Analysis* 1(Oct. 1969):321–28; see also pp. 667–68.

11. The terms "discrete" and "continuous" are used throughout this discussion in an intuitive sense. The technically correct terms are "atomic" and "nonatomic" respectively.

12. A semantic note: The disciplines called regional science, human ecology, theoretical geography, and ekistics are in my view essentially the same as spatial economics or location theory in the sense that they share a common core of central ideas, though each retains its particular flavor. There is product differentiation in the marketplace of ideas as well as that of commodities.

arrives at the following intuitive arguments: First, their *universality:*
Everything has a location and a date.   Second, their *universal required-
ness:* Every activity needs room in which to operate, and a certain duration in which to be consummated. (Thus the classical economists spoke of
"land" and the "period of production.")   Third, their connection with
causation: things must be in contact to interact.   More generally, they
may interact indirectly, e.g., by sending signals through the communications system; more generally still, the intensity of interaction depends on
distance and relative position.

This book continues this "spatiotemporal" tradition, the detailed
constructions occurring mainly in Chapters 2 and 4. To assure that it does
not become just another abortive attempt at system building, to be for-
gotten in its turn, I have concentrated on turning out a body of results too
large and impressive to be ignored. This again explains the size of the
book.

Let me now try to summarize what this book does and does not accomplish. First, the book takes a quantum jump over existing location
theory in generalizing existing results, in putting them on a more rigorous
basis, and in striking out in new directions. I have not attempted to write
a comprehensive treatise on location theory but only to cover the parts
where I have succeeded in making substantial progress via measure-
theoretic formulation. These, however, do constitute the bulk of mainstream tradition of von Thünen, Launhardt, Weber, Palander, Hoover,
Lösch, Ponsard, Isard, Alonso, and Beckmann.[13] Second, the book develops the mathematics of measure theory itself.   The development is
ultimately directed toward applications, but much of it should be of considerable interest even to the "pure" mathematician. Third, the book develops a framework of ideas that seems to be adequate for all social
science, not just location theory.

It is important to take full note of the things this book does not accomplish, to gain perspective and to answer the question, Where do we go
from here? I divide these "nonaccomplishments" into four groups: theoretical, foundational, empirical, and practical.

Despite its bulk, the book discusses only a fraction of the relevant
theoretical models. I have concentrated on optimization models and have
slighted, though not ignored, equilibrium models.[14] The latter are partial,
not general (in the sense in which economists use these terms; each disci-

---

13. Examples of parts not covered in this book are the control-theoretic models in
E. S. Mills, *Studies in the Structure of the Urban Economy* (Johns Hopkins Press, Baltimore,
1972) and the Löschian comparative statics, the homogeneous Thünen, and the commuting
models in Faden, *Essays.*
14. Equilibrium models may be found throughout Chapter 6 and in Sections 5.8, 8.7,
8.8, 9.2, 9.5, 9.6.

pline has its own particular notion of what is "general"). Almost all the models are deterministic, though some could be interpreted as applying to expected values.

The sense in which these models are or are not dynamic requires some discussion. The overall framework of the book has Time as one of its basic categories and allows one to describe change and development without trouble. But a dynamic model in addition puts restrictions on the kinds of changes that can occur. There are several ways of specifying these restrictions. One is to state "laws of development," say in the form of differential equations. In this sense very few of the models are dynamic. Another approach is to specify a set of feasible "activities," each activity being defined by a complete input-output structure over Time. Here change is determined implicitly by the choice of activity. This latter approach is the one used in Chapter 8 and to some extent in Chapter 9. Still another approach is to consider Time as simply a fourth spatial dimension. This is the approach used in the real-estate model of Chapter 6. These and other approaches are considered in Chapter 4.

One would like to introduce nonlinear features into several of the models; e.g., preferences are usually expressed by integrals and are thereby linear in the corresponding measures and integrands. But even in the realm of the linear there is plenty of room for generalization. One expects that linear programs in general can be put into measure-theoretic form along the lines of Chapter 7, which is restricted to transportation and transhipment problems.

The "foundational gap" is the one between the conceptual framework of this book and the world of experience. The claim is that one can describe the world in terms of measures; but the description is highly formalized, several steps removed from everyday perception. Specifying the values of a *measure* involves a process of *measurement* that needs elucidation. How does one recognize the same location at different times? the same physical object persisting through time? the same resource type at different times or places? What is a resource type anyway? What is the significance of countable additivity? of measurability?

All these questions deserve extended discussion. Some are discussed in Chapter 2, but only cursorily. I have concentrated instead on setting up the structure, not worrying overmuch about the foundations. One must first attain a body of results and only then inquire as to what they mean.

As for the "empirical gaps," remember that this book is intended as a work in theory: It is not concerned directly with particular real-world situations. On the other hand, it is important to have a large number of links between the theory and the real world, both to clarify the concepts and to indicate directions of application. Accordingly, literally scores of illustrations of theoretical models are scattered throughout the text. These illustrations should be understood as hypotheses that the suggested interpretations are reasonable approximations to the real-world situation. For example, in Chapter 8 I suggest that the observed tendency for population density to be higher along the fringes of continents than in their interiors is explainable along von Thünen lines. In effect, this draws attention to a small number of facts (e.g., water vs. land transport cost differentials), which with some simplifying assumptions yield the general configuration observed; other relevant facts such as climatic differentials are ignored and would be incorporated in a more complete theory.

Lacking here is a "spatial econometrics," a body of doctrine to guide statistical inference for the models of location theory. In addition, of course, in any particular application a detailed factual investigation would be needed to assess the "goodness of fit" of the models, and not merely the "casual empiricism" appropriate for this book.

Finally, there are the "practical gaps," the absence in most cases of feasible computational schemes for obtaining solutions to the various models. It is one thing to show that a certain equilibrium exists or that a certain problem has an optimal solution, and quite another to calculate these solutions with a reasonably small amount of computational effort. I have indicated in some cases how this might be done—usually successive approximation schemes via price adjustments—but these have not been carried out in detail.[15]

In this connection, there is the practical-minded man's objection to the whole enterprise of this book. The argument might go as follows: "You have this (allegedly) wonderful new theory. But how does it help me meet a payroll? Specifically, a measure is, in general, a very complicated object. To reduce information handling costs, you have to aggregate it to a discrete distribution. But if you do this, doesn't your theory collapse to something we all know already—ordinary linear programming, for instance?"

I will answer this objection in three stages. First, for practical work one must certainly approximate by a family of objects that can be indexed by a small number of parameters. But there are many ways to do this. Population distribution over a region may be represented by partitioning into subregions and giving the population of each (discrete aggregation). It might be more convenient, however, to approximate the distribution of a city population by a circular normal or exponential distribution, and the population of a region by a sum of such measures. How best to approximate is itself a genuine problem, which can be solved only by taking

---

15. A brief survey of algorithms for location theory may be found in A. J. Scott, *Combinatorial Programming, Spatial Analysis and Planning* (Methuen, London, 1971).

the "raw" measure in all its complexity as a starting point. Second, for many of the problems in this book, though they involve unrestricted measures, it can be shown that they may be indexed by a small number of parameters. The Weber problem on the plane (Section 9.4) has two parameters; the market-region problem with n plants (Section 9.5) has n or $n - 1$ parameters, viz., prices; the Löschian one-industry problem with fixed deliveries (Section 9.6) has two parameters, scale and spacing; the interplant and interindustry problems of Section 9.3 can be reduced to combinatorial problems. Furthermore, several of the problems have solutions that have very simple characterizations in terms of the given data; these include the allotment-assignment problem (Section 8.5) and some of the police-criminal-victim models of Section 5.8. Thus all these models are computable. What should be done with the remaining models, such as the transportation problem of Chapter 7, from the practical point of view remains an open question. It is hardly justified, however, to replace these problems a priori by discrete aggregations (several of the "computable" problems above are in fact special cases of the transportation problem).

The third stage of the rebuttal takes a different tack. Suppose for the sake of argument that none of the models in this book were computable. Would it still have been worth writing? My answer is yes. The theory lends a certain power and clarity to thought, which must issue indirectly in a firmer grasp of practical problems. And the overall view of the world it contains has a certain grandeur.

1.2. SYNOPSIS OF CHAPTERS

CHAPTER 2

Chapter 2 has two objectives: to serve as an exposition and reference to the measure theory used here and to lay out the basic conceptual framework of the book. Sections 2.1, 2.4, and 2.6 are devoted to measure theory. Measures are defined in Section 2.1, and Section 2.2 then introduces the discussion of how real-world data may be represented as measures. Sections 2.3 and 2.4 are preliminary to Section 2.5, which sets out the general theory of the measure-theoretic representation of the real world. Section 2.3 introduces the fundamental categories of Space, Time, and Resources, while Section 2.4 introduces the measure-theoretic concepts and theorems needed to understand Section 2.5. The major exposition of measure theory occurs in Section 2.6. Section 2.7 introduces activities and related concepts; these are essentially aids to picking out interesting patterns from the all-embracing flux of Section 2.5. Finally,

Section 2.8 notes several real-world phenomena whose representation requires multilayer measures, i.e., measures over a space whose points are themselves measures.

How does the exposition of measure theory here compare with that in standard treatises? I have stressed measures over *abstract* spaces rather than over the real line or *n*-space. This is not just idle generality but is needed because many of the spaces dealt with are rather complicated entities, not easily reducible to the real numbers; these include the space of Resources, of Histories, of Activities, multilayer spaces, etc.

Most of the material covered in standard treatises appears here, and conversely most of the mathematical material in this chapter appears in standard treatises. But the overlap is not complete in either direction. For example, I have omitted any discussion of measures on groups, differential measure theory, vector measures, Hausdorff measures, complete measures, outer measures, Daniell integrals, and Lebesgue decompositions simply because these concepts are not used in the remainder of the book. Furthermore, to save space I have omitted all proofs of standard theorems (except occasionally when these are very short). Hence the reader who wants a deep understanding of the subject should by all means consult the standard treatises even for the topics covered.

Some results that appear to be new are covered in Section 2.6, in particular in the subsection "Abcont and Product Measures" (where full proofs are provided) and also in the subsection "Extension of Set Functions."

As for the "world-representation" whose exposition culminates in Section 2.5, the reader acquainted with the theory of stochastic processes will note the resemblance to the concepts used in that subject. But note also the differences: Measures are not normalized and may be infinite; more significantly, the "realizations" (here called "histories") do not all have the same time domain; finally, of course, the interpretations of the concepts used are completely different from those common in stochastic processes.[16]

CHAPTER 3

Mathematically, Chapter 3 is the deepest in the book. The theory generalizes the concept of signed measure. Section 3.1 works with measures in general. Section 3.2 specializes to σ-finite measures, leading to

---

16. This framework is adequate for descriptive purposes in all social science. What about natural science? On the whole it remains adequate; but the framework is incompatible with relativity theory, since it assumes "absolute" Space and Time. And it is (probably) incompatible with quantum theory, since it allows for complete state descriptions.

the fundamental concept of *pseudomeasure*.   Algebraic operations and integration with pseudomeasures are defined in this section, while various order relations are defined in Section 3.3.   There are two basic types of applications of pseudomeasures in this book.   First, pseudomeasures may be used to represent "net" values (e.g., net production, net migration) even in cases where both "gross" measures are infinite, as might occur in models with infinite horizons.   The main application of this type is to the transhipment problem, Sections 7.7–7.11 (see also Section 6.9).   Much more important is the use of pseudomeasures to represent preferences. Pseudomeasures are used throughout the book in this way, and this application allows the formulation of models that are both more general and simpler than would otherwise be the case.[17]   The use of pseudomeasures also generalizes neatly such apparently unrelated subjects as the Ramsey-Weizsäcker "overtaking" approach and the Bernoulli-von Neumann–Morgenstern expected utility approach, the former by going from Time to more general spaces, the latter by allowing unbounded utility functions.   These topics are discussed in Section 3.3.

Pseudomeasure theory is quite elegant from a purely mathematical point of view, and Chapter 3 is just a bare introduction to it.   The subject is developed further in Sections 6.9 and 7.11.

CHAPTER 4

While Chapter 2 contains a descriptive framework valid in "all possible worlds," so to speak, Chapter 4 is devoted to various approaches to the representation of "attainable worlds," i.e., to the representation of an agent's power.   As introduced in Section 4.1, power is limited in many different ways: by lack of knowledge, by natural law, by resource availability, by limited authority, by legal constraints, by budget constraints, etc.   These are all discussed in Chapter 4 (except for budget constraints, discussed in Section 6.1).   Section 4.2 elucidates the concept of uncontrollability, e.g., the sense in which the past is uncontrollable.   Section 4.3 discusses cross-sectional constraints, i.e., constraints on the possible distributions over Space and Resources at a moment of Time.   Section 4.4 explores a few simple cases of the potentially much richer class of constraints involving several moments of Time; it ends with an example involving pollution, essentially a simple diffusion model.   Section 4.5 builds on the activity concepts of Section 2.7.   It first specializes to the case of

---

17. These virtues arise from the fact that it is unnecessary to impose artificial restrictions to guarantee that utilities are real valued.   Of course, simplicity is in the eye of the beholder.   If English spelling were made phonetic, it would be objectively simpler, but not necessarily to someone who had learned to spell the old way.

"simply located sedentary" activities, i.e., activities located at a single fixed point. The construction that follows is essentially a measure-theoretic generalization of the standard activity analysis approach. Sections 4.6 and 4.7 examine certain assumptions implicit in this approach. In Section 4.6 the assumption that there are no "neighborhood effects" is explicated and then criticized as to its realism. In Section 4.7 the same is done for the assumption of "constant returns to scale." Three different meanings of this concept are distinguished, stemming from three different scale concepts discussed in Section 2.7. Section 4.8 briefly discusses the Knightian contention that "indivisibility" is responsible for "nonconstant returns to scale." Six different meanings of indivisibility are distinguished, none of which support Knight's contention. The discussion in Sections 4.7 and 4.8, which is by no means definitive, is an example of how explicit consideration of the spatial variable can clarify certain unsettled issues in economic theory. Finally, Section 4.9 is an informal discussion of the modes by which one controls the location of things (and thereby their interaction): bringing things together, keeping them apart, maintaining relative positions, etc., by means of barriers, walls, bindings, transportation construction, etc.

Chapter 4 is the most unfinished of all the chapters. The discussion is at a rudimentary level, and there is no overall point of view uniting the sections as in most other chapters.

CHAPTER 5

Chapter 5 is devoted mainly to the study of one very general and pervasive type of optimization problem, an example of which is: Given alternative projects, returns from each depending on the amounts of effort invested, with constraints on the total effort available and on the amount that may be invested in each project, allocate effort among projects to maximize overall return. Section 5.1 discusses various interpretations: Effort may be money, time, resources, personnel, etc. The formal problem is in one sense a special case of the Neyman-Pearson problem, special in that it has just one allocative constraint; but it is extremely general in all other respects. The problem is formally set up in Section 5.2, and analyzed in Sections 5.3-5.7. In particular, in Section 5.4 a shadow-price condition associated with optimal solutions is derived. Finally, in Section 5.8 these results are applied to the problem of explaining the spatial distribution of crime. This is determined by a game between Criminals and Victims on the one hand, and between Criminals and Police on the other. In these games the participants are faced with allocation-of-effort problems. The results predict a positive association between crime rates and population (or wealth) density, and also predict a tendency for crime to suburbanize with increased numbers of police. The allocation-of-effort problem finds further applications in Chapters 8 and 9.

CHAPTER 6

Sections 6.1–6.3 focus on the budget constraints facing a single agent. Section 6.1 stresses basic accounting identities in perfectly competitive markets, spread over a measurable space of commodities at various times and places. Section 6.2 extends the analysis to rental markets, discussing ownership vs. control, bailments, and service activities. Section 6.3 indicates how the measure-theoretic approach extends to encompass imperfections in the commodity and/or capital markets. The discussion is rudimentary throughout.

Section 6.4 introduces the main topic of this chapter—the real-estate market, which allocates the ownership and control of Space-Time. Discussion of the organization of this market carries over into Section 6.5, which stresses the question of how best to represent the preferences of the participants in the market, balancing simplicity against realism. Assuming these simplifications, Section 6.6 proves the existence of equilibrium in the real-estate market and explores certain of its properties. Section 6.7 indicates how joint control of real estate among several agents might be represented, in contrast to the exclusive control approach of Section 6.6. Section 6.8 is a critique of a leading alternative theory of the real-estate market, that of Alonso. Finally, Section 6.9 generalizes the analysis of Section 6.6 in a direction that might be relevant in unbounded Space or Time, where the number of agents might be infinite, rents might be infinite, preferences no longer representable by real numbers, etc. These complications are all handled by introducing pseudomeasures, and the analysis closely parallels that of Section 6.6. An appendix to Section 6.9 develops certain order properties of the space of pseudomeasures that grow out of this analysis.

An alternative approach to the real-estate market occurs in Section 8.7. This latter model is deeper than the one in this chapter in that preferences over regions are derived from the uses one is going to run there, rather than being assumed directly. However, the Section 8.7 model operates in a Thünen context, while the one in Chapter 6 has no such restriction.

CHAPTER 7

Chapter 7 develops the measure-theoretic generalization of the transportation (Sections 7.1–7.5) and transhipment (Sections 7.6–7.11) problems of linear programming. Capacities and requirements are represented as measures, and cost as an integral (in general, a pseudomeasure). Topological concepts are introduced as needed. The section headings are mostly self-explanatory. "Potentials" in Section 7.5 refer to the system of shadow-prices associated with optimal solutions. Section 7.6 discusses the relation between the two problems and the realism of their assumptions. Further connections between the two problems are obtained in the latter part of Section 7.7 and in Section 7.10. The "skew" problem of Section 7.11 is the transhipment problem in terms of *net* shipments. It is interesting that in one formalization the problem is to find an optimal pseudomeasure. All other problems in this book involve finding unknown measures, point functions, individual points, or sets.

The transportation problem is the foundation of Chapter 8, and both transportation and transhipment appear in various sections of Chapter 9.

CHAPTER 8

Chapters 8 and 9 (and Chapter 6 to a lesser extent) deal with the traditional models of location theory, though they go far beyond the standard agenda of topics. Chapter 8 deals with Thünen models, introduced in Section 8.1. Stress is placed on the optimization aspects of these models, though equilibrium is covered as well. Section 8.2 introduces "ideal" weights and distances, which are needed because real-world transport costs are not proportional to ton-miles of shipments. The special modifications of these concepts needed in the Thünen context are discussed in Section 8.3. Section 8.4 takes off from the activity analysis framework of Chapter 4. It indicates how "amounts" of land uses may be measured in "ideal" acres. ("Ideal" area is the third ideal concept used; it allows for "fertility" differentials and land availability not proportional to physical area; "land uses" are simply space-using activities.) Also it indicates how resource-time ideal weights translate into land-use weights. These formalities allow much simplification in the expressions for transport costs and areal capacity constraints. Interpretations of land uses in terms of time patterns, multiple stories, and ideal weight assessments are discussed.

Section 8.5 gives the basic Thünen optimization model: to minimize overall transport costs subject to areal capacity and land-use allotment constraints. Formally, this becomes a special case of the measure-theoretic transportation problem. A detailed analysis follows, the special features allowing sharper results than in Chapter 7. Section 8.6 then applies these abstract results to concrete Thünen systems and also gives several other applications, some far removed from the original spatial context.

Section 8.7 gives the basic Thünen equilibrium model. Each agent must decide which land to acquire and what land uses to run there. Under fairly weak assumptions concerning preferences, etc., it is shown that any equilibrium will satisfy the Thünen pattern in the overall spatial distribution of land uses and land values. In view of the very complicated arguments needed for rigor, a simplified heuristic approach is also given (taken from my *Essays*).

Section 8.8 discusses many interpretations, predictions, and informal extensions of this very flexible model: the spatial distribution of multiple-story structures, land speculation, people by income, family size, and car ownership; the influence of pollution, crime, and danger; the effects of taxation; and the welfare implications of the model. Section 8.9 discusses a variety of real-world Thünen systems classified in several ways: first, by spatial scale—from workbenches to theaters, farms, villages, city neighborhoods, cities, metropolitan regions, and subcontinental systems, up to the entire world (the latter, coupled with ideal distance distortions, explains the tendency for continents to have relatively empty interiors) and second, dynamically—in particular whether they develop by inward migration (e.g., rural-urban, dual economies) or by outward migration (e.g., suburbanization, pioneering). Thünen systems fall into a hierarchical structure, and this yields an explanation of the "central place" arrangement. Some indices for assessing Thünen fits are noted, such as transport nets, Stewart potentials, and percentage of nonlocal flows.

CHAPTER 9

Section 9.1 discusses informally a variety of preference patterns in a spatial context, involving pollution, layout problems, attraction and repulsion among persons and groups, density preferences, visits to facilities, etc. Section 9.2 discusses a number of informal models involving interacting agents having some of these preference patterns. Specifically, agents are assumed to be attracted to or repelled from various others and can act by moving themselves or by inducing others to move (or to stay put). Another group of models involves attraction to facilities or services for which the agent has a taste, leading to generalizations of the Tiebout model, urban neighborhood formation, segregation by income, etc. Changes in preference are briefly discussed, such as fatigue, deprivation, extinction, and habituation of tastes.

The rest of Chapter 9 is more formal in tone. Section 9.3 deals with certain interplant and interindustry optimization problems, in which flows between all pairs of plants or industries are given, along with the locations of some plants or industries, and the others are to be located to minimize overall transport costs. Under certain conditions there is a tendency for optimal locations to coincide, a clue to the observed tendency toward agglomeration in the real world. The concept of "orientation" of an industry is clarified, and an explanation is given for the tendency of industries to dichotomize into "resource" and "market" orientation. A different model deals with the "localization" of industries into small regions.

Section 9.4 deals with the classical Weber problem—to choose a plant location minimizing transport costs given the spatial distribution of shipments to and from the plant. A multiplant generalization is also considered, in which interplant flows are also given, and all plants must be located. Many interpretations are offered, along with a discussion of subproblems in general. The connection with "nodality" is discussed. Results concerning existence and bounds on solutions are obtained, as well as the use of symmetry and convexity in finding solutions. Finally, the multiplant Weber problem in a Thünen context is examined with some surprising results.

Section 9.5 deals with a class of problems of which the following is typical: Given a system of plants with locations and capacities specified, and given a spatial distribution of demand, find a shipping pattern that satisfies these constraints at minimal transport cost. This is a special case of the transportation problem of Chapter 7, special in that the origin space is countable. Just as with the allotment-assignment problem (Section 8.5), this special feature leads to sharper results. Specifically, optimal flows are associated with a system of shadow mill prices that determine market regions, each plant shipping exclusively to its own region. The implications between these prices, various market region concepts, and optimal flows are explored in detail.

In the "service systems" of Section 9.6, finally, one is to find the number, location, and output of plants as well as the spatial distribution of outputs, both from the viewpoint of optimization (maximizing total benefits minus transport and production costs) and for various social equilibria. A classification scheme is set up and the goodness of fit of various real-world systems to the model is discussed. The case of a single plant with fixed location and output is taken up first; this reduces to an allocation-of-effort problem (Chapter 5). Then the discussion moves on to many plants, again with all locations and outputs fixed; the market region problem (Section 9.5) is a subproblem of this. The analysis turns on the relations between optimal solutions and the shadow-price conditions of the market region and allocation-of-effort subproblems. The next topic is the relation between spatial prices and shipment patterns, in particular the Mills-Lav demonstration that parts of the field may go unserved. Finally, the difficult subject of variable plant locations and numbers is broached. The need (for the first and only time in this book) for introducing *extended* ordering of pseudomeasures is noted. The inefficiency of the honeycomb lattice arrangement of plants is demonstrated.[18] This is contrary to many statements in the literature and indicates the unsettled nature of the subject.

1.3. READER'S GUIDE

This is a difficult book for several reasons: its length; its novelty; the inherent complexity of its arguments; and the fact that its central mathematical apparatus, measure theory, is at present unfamiliar to most economists and other social scientists. Hence the need for a "reader's guide."

One pervasive problem in organizing the book was the double task of developing the formal theory on the one hand, and connecting it with real-world concepts on the other. The first task called for a rigorous self-contained approach; the second, for an open-ended intuitive approach. I have tried to accomplish both tasks by segregating the formal and intuitive portions into separate sections; this has been only partially successful. My main device has been to signal formal discussion by using *Definition, Theorem, Proof,* and ∎ at the end of a proof. The text thus blocked out is meant to be self-contained and to meet full standards of mathematical rigor (in contrast to much contemporary literature that is studded with pseudo-"proofs" of "theorems").

The only mathematical prerequisite for this book is elementary calculus (e.g., real numbers, limits, mean-value theorem, absolute convergence, uniform continuity, Riemann integral), though the reader whose knowledge stops at this point will not have an easy time. All other concepts and theorems are developed as needed.

It should be possible to understand the text by skipping all proofs, reading only the definitions and theorems, thereby avoiding the most difficult parts of the book. Reading the proofs, of course, is needed for a really deep understanding of the theories developed (as well as for verifying results).

The diagram below indicates the precedence relations among the chapters—the solid arrow indicating strong dependence and the broken arrow indicating weak dependence. Getting down to the section level,

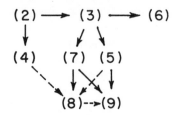

---

18. Following Faden, Inefficiency of the regular hexagon, pp. 321–28.

Chapter 4 is needed for Section 8.4, Chapter 5 is needed for Sections 8.6 and 9.6, Chapter 7 (transportation half) is needed for Sections 8.5 and 9.5, Chapter 7 (transhipment half) is needed for Section 9.3, Chapter 8 is needed for Section 9.4.

Chapter 2 is precedent to all other chapters. Perhaps Sections 2.1–2.5 should be read straight through and Section 2.6 used as a general reference on measure theory.

Chapter 3 presents a special problem because of its novelty. It is precedent to all other chapters except 2 and 4; but the reader who does not feel at home with pseudomeasures may circumvent this dependence, at least in part. First, many sections in Chapters 5–9 do not use pseudomeasures. Second, the Standard Integral Theorem (p. 161) allows one in some cases to translate a utility function expressed in pseudomeasures into more familiar terms (this works in Chapter 5, for example, but not in most of Chapter 7). Third, one may add extra conditions such as boundedness to various premises to insure that utilities are expressible as real numbers, thus circumventing Chapter 3 at the cost of some generality. All this still leaves a residue of results that require a working knowledge of pseudomeasures, which are really not all that difficult to master. Chapter 3 has to be read only through standard order; extended order is used only in Section 9.6 except for passing references.

Section Y of Chapter X is indicated by X.Y. Formula Z of Section X.Y is indicated by (X.Y.Z). It is referred to as (Z) within Section X.Y, referred to as (Y.Z) elsewhere in Chapter X, and referred to as (X.Y.Z) in other chapters.

2

The Problem of Description

Consider the following data on the catch of fishing craft landing at Boston, Gloucester, and New Bedford, 1948–50.

	Boston			Gloucester			New Bedford		
Variety	1948	1949	1950	1948	1949	1950	1948	1949	1950
					mil lb				
Cod	34.7	28.6	24.4	7.7	7.0	6.0	6.3	4.1	5.1
Haddock	105.3	90.1	107.4	11.2	8.5	10.0	11.4	9.9	11.5
Mackerel	3.5	0.7	1.1	19.9	6.9	5.7	1.7	0.3	0.3
Flounder	9.2	8.1	10.0	2.7	7.6	7.4	41.0	33.9	29.4

Source: U.S. Bureau of the Census, *Statistical Abstract of the United States: 1956* (Washington, 1956), p. 720.

A listing of the ingredients of this table might run as follows: (i) a "universe of discourse" consisting of certain *regions* (ports), at certain *time periods* (1948–50), at which certain *resources* (types of fish) appear; and (ii) a *measure* of the quantities involved for each possible combination of place, time, and resource.

It turns out that these data can, in fact, be represented by a *measure* in the technical mathematical sense of the term. Furthermore, this is true not only for the present example but for most of statistical data in general—population, births, migration, marriages, production, transportation, income, wealth. Even this rather sweeping statement understates the possibilities of describing the world in terms of measures. Most kinds of data that cannot be represented directly as measures (such as prices, population densities, per capita incomes) are derived from underlying measures in ways that themselves are well-known parts of the theory.

This chapter will carry out this descriptive program. It will expound the concepts, terminology, and basic theorems of measure theory as used in this book,[1] and it will build a unified apparatus for describing the world in terms of these concepts. The unity arises from the fact that the universes of discourse over which the various measures are defined are always built up from the three basic sets of Space, Time, and Resources, just as in the fish example above (but not always in so simple a fashion).

Of course, the fact that this apparatus *can* be constructed does not mean it is *useful* to do so. The rest of this book may be considered an argument for the proposition that it *is* useful.

2.1. MEASURE THEORY, I

This section is a purely formal exposition; no real-world interpretations are offered. We follow in general the practice of omitting proofs of theorems when these are available in standard treatises.[2] The concept of "measure" rests on three others: "sigma-field," "the extended real numbers," and "countable additivity." We discuss these in turn.

SIGMA-FIELDS

Recall some notations from elementary set theory. Write $x \in A$ to indicate that x *belongs to* or is a *member, element,* or *point* of the set A, $A \subseteq B$ (or $B \supseteq A$) indicates that A is a *subset* of B (i.e., every element of A is an element of B). $A = B$ signifies that $A \subseteq B$ and $B \subseteq A$, so that sets are considered equal iff they have the same members (we use "iff" to abbreviate "if and only if" throughout this work). Also, $\{x_1, x_2, \ldots\}$ is the set whose elements are $x_1, x_2, \ldots$; $\{x\}$ is the set with the single member x. $\emptyset$ (the Norwegian letter O) stands for the *empty set,* the set with no members.

Let P be a certain property, and let the symbol $\{x \mid x$ has the property $P\}$ stand for the set of all objects having the property P. Given two sets

---

1. Certain new mathematical results will be given later. Most mathematical material in this chapter is standard except possibly for terminology.

2. The following books are recommended for readers wishing to go beyond the necessarily sketchy outline of measure theory presented in this chapter. They are roughly in decreasing order of difficulty: N. Dunford and J. T. Schwartz, *Linear Operators,* Vol. 1 (Wiley-Interscience, New York, 1958), Ch. 3; S. Saks, *Theory of the Integral* (Stechert-Hafner, New York, 1937); H. Hahn and A. Rosenthal, *Set Functions* (Univ. New Mexico Press, Albuquerque, 1948); P. R. Halmos, *Measure Theory* (Van Nostrand, Princeton, 1950); S. K. Berberian, *Measure and Integration* (Chelsea, New York, 1970); M. E. Munroe, *Measure and Integration,* 2nd ed. (Addison-Wesley, Reading, Mass., 1970); A. E. Taylor, *General Theory of Functions and Integration* (Blaisdell, New York, 1965), second half of book.

Books on the theory of functions of a real variable will generally have pertinent material. But the books above have been chosen for their abstract approach to the subject, which is suitable for the applications we wish to make. Terminology has not been fully standardized, so that these books differ among themselves and with this book.

A and *B*, their *intersection,* written *A* ∩ *B*, is defined as the set of elements they have in common; i.e., *A* ∩ *B* = {*x* | *x* ∈ *A* and *x* ∈ *B*}. The *union* of *A* and *B*, written *A* ∪ *B*, is the set of elements in at least one of the two sets; i.e., *A* ∪ *B* = {*x* | *x* ∈ *A* or *x* ∈ *B* (or both)}. The *complement of B with respect to A,* written *A**B* (A slash B) is the set of elements in *A* but not in *B*: *A**B* = {*x* | *x* ∈ *A* and *x* ∉ *B*}. (A line drawn through any relation signifies that the corresponding proposition is not true: thus *A* ⊄ *B*, *A* ≠ *B*, as well as *x* ∉ *B*.)

Now consider a set $\mathcal{G}$ whose elements are themselves sets. For euphony, $\mathcal{G}$ will be called a *class* or a *collection* of sets, rather than a set of sets. (We follow the convention of using small letters for points, capital letters for sets of points, and script letters for classes of sets.)

The *intersection* of $\mathcal{G}$, written ∩$\mathcal{G}$, is defined as the set of points common to all the members of $\mathcal{G}$: ∩$\mathcal{G}$ = {*x* | *x* ∈ *G* for all *G* ∈ $\mathcal{G}$}. If $\mathcal{G}$ has as members just the two sets *A* and *B*, this reduces to *A* ∩ *B*. Similarly, we define the *union* of $\mathcal{G}$, written ∪$\mathcal{G}$, as {*x* | *x* ∈ *G* for at least one *G* ∈ $\mathcal{G}$}.

There is a basic distinction between sets with a finite and an infinite number of members (finite and infinite sets, for short). Another distinction of great importance is that between *countable* and *uncountable* sets. A set is *countably infinite* iff its members can be ticked off in an infinite sequence: x_1, x_2, x_3, ... (i.e., the set can be placed in 1–1 correspondence with the positive integers). Examples of countably infinite sets are the set of positive integers, the set of all integers, the set of rational numbers, and the set of lattice points in the plane (i.e., the set of all points (m, n) where m, n are both integers). On the other hand, the set of real numbers is not countably infinite. A set is *countable* iff it is either finite or countably infinite. The following result is used repeatedly, generally without explicit mention.

Theorem: Let $\mathcal{G}$ be a countable collection of sets, each of which is itself countable; then ∪$\mathcal{G}$ is countable.

We are now ready to define "sigma-field" (sometimes called "sigma-algebra"). Suppose we are given *A*, a set, and Σ, a collection of subsets of *A*.

Definition: Σ is a *sigma-field* (σ-field) (with *universe set A*) iff

(i) ∅ ∈ Σ and *A* ∈ Σ;
(ii) if *E* ∈ Σ and *F* ∈ Σ, then *E**F* ∈ Σ;
(iii) if $\mathcal{G}$ is a countable collection of members of Σ, then ∪$\mathcal{G}$ and ∩$\mathcal{G}$ are both members of Σ.

We express (ii) by saying Σ is closed under differences, and (iii) by saying Σ is closed under countable unions and intersections.

Definition: If Σ is a σ-field with universe set A, the pair (A, Σ) is called a *measurable space;* the members of Σ are called *measurable sets.*

The definition of σ-field given above is redundant in the sense that some of the conditions follow from the others. To verify that a given collection of sets is a σ-field, it is then useful to have a stripped-down criterion. First, if the universe set A is given, $A \backslash B$ is simply called the *complement* of B.

Theorem: Let Σ be a collection of subsets of A; Σ is a σ-field (with universe set A) iff $\emptyset \in \Sigma$, and Σ is closed under complementation and countable unions (i.e., $E \in \Sigma$ implies that $A \backslash E \in \Sigma$, and $\cup \mathcal{G} \in \Sigma$ for any countable collection $\mathcal{G}$ of sets belonging to Σ).

Thus we need to verify only half of conditions (i) and (iii) and a weakened form of (ii).

We now give some examples of σ-fields:

1. The collection of the two sets A, $\emptyset$ by themselves constitute a rather trivial σ-field.
2. The collection of all subsets of A (including A and $\emptyset$) is a σ-field.
3. For this example we first give some definitions that are quite important in their own right. Let $\mathcal{G}$ be a collection of subsets of A.

Definition: $\mathcal{G}$ is a *covering* (of A) iff $A \subseteq \cup \mathcal{G}$; $\mathcal{G}$ is a *packing* iff no two members of $\mathcal{G}$ have a point in common (i.e., if G_1 and G_2 belong to $\mathcal{G}$, then $G_1 \cap G_2 = \emptyset$); $\mathcal{G}$ is a *partition* iff it is both a packing and a covering.

These definitions may also be expressed as follows: $\mathcal{G}$ is a covering iff every point of A belongs to *at least one* member of $\mathcal{G}$; $\mathcal{G}$ is a packing iff every point of A belongs to *at most one* member of $\mathcal{G}$; $\mathcal{G}$ is a partition iff every point of A belongs to *exactly one* member of $\mathcal{G}$.

Now let $\mathcal{G}$ be a given partition of A and let Σ consist of all sets of the form $\cup \mathcal{F}$, where $\mathcal{F}$ ranges over all possible subclasses of $\mathcal{G}$. (Since $\cup \emptyset = \emptyset$, the empty set belongs to Σ.) One may verify that Σ is then a σ-field. An important special case arises when $\mathcal{G}$ is a *finite* partition. If $\mathcal{G}$ has N nonempty member sets, the σ-field Σ has 2^N member sets. It may be shown that *all finite* σ-fields are of this form.

4. In the three examples above, we could give a simple property characterizing the measurable sets (i.e., the members of Σ). Usually, how-

ever, this is not possible. Instead, we typically characterize σ-fields as being *generated* from a class of sets given in advance. We now turn to this important concept.

Let **S** be a nonempty collection of σ-fields, all relative to the same universe set A. One may verify that $\cap$**S** is then itself a σ-field relative to A. (Notice that **S** is a "third-order" construct: it is a set whose members are sets, the members of these in turn being sets; $\cap$**S** is then a "second-order" construct, a class of sets—which turns out to be a σ-field.)

Definition: Let $\mathcal{G}$ be a collection of subsets of A; the *σ-field generated by* $\mathcal{G}$ is $\cap$**S**, where **S** is the collection of all σ-fields (relative to A) containing $\mathcal{G}$ as a subclass: **S** = $\{\Sigma \mid \Sigma$ is a σ-field and $\mathcal{G} \subseteq \Sigma\}$.

As an example, let $\mathcal{G}$ be a countable partition; then the generated σ-field is precisely Σ as constructed above under example 3. To prove this, one shows first that Σ is indeed a σ-field, and second that every member of Σ must belong to every σ-field containing $\mathcal{G}$; this second statement follows at once from the fact that σ-fields are closed under countable unions.

We may say for short that the σ-field generated by a collection $\mathcal{G}$ is the "smallest" σ-field containing $\mathcal{G}$. The discussion above explicates this concept and shows that indeed there is such a smallest σ-field.

With the aid of this concept we may now define what is historically the granddaddy of all σ-fields: *the Borel field on the real line.* Here the universe set is the real line, and the Borel field is simply the σ-field generated by the class of finite open intervals, i.e., all sets of the form $\{x \mid a < x < b\}$, where a and b range over the real numbers.

This σ-field is quite important, and it is useful to note that it may be generated by a variety of different collections. Besides the finite open intervals, it is generated by the class of sets of the form $\{x \mid x < b\}$, b ranging over the real numbers; and also by the class of sets $\{x \mid x > a\}$, a ranging over the reals. Furthermore, it is generated by any of these classes when a or b ranges merely over the rational numbers instead of the reals.[3] Finally, closed sets could be used in any of these cases instead of open sets (just substitute the weak inequality $\leq$ for the strict inequality $<$).

There is no direct way of characterizing the real Borel field, so one must be satisfied to define it in terms of "generation." This situation is the rule, not the exception.

The Extended Real Numbers

We now turn to the second concept needed in the definition of measure: the extended real number system. The ordinary real number

---

3. Any dense subset of the reals could be used.

system is augmented by two "points at infinity," and the relation of order and the operations of arithmetic are then extended to these ideal points.

Definition: *The extended real number system* consists of the real numbers together with two new points, written $+\infty$ and $-\infty$; $a \geq b, a + b$, $a \cdot b$, etc., retain their usual meanings when a and b are both real numbers; when one or both of these is $\pm\infty$, the order relation is extended as follows: $+\infty > a$; $a > -\infty$; $+\infty > -\infty$, for any real number a. The operations of arithmetic are extended as follows.

Addition: $a + (+\infty) = (+\infty) + a = +\infty$; $a + (-\infty) = (-\infty) + a = -\infty$, for any real number a. Also, $(+\infty) + (+\infty) = (+\infty)$; $(-\infty) + (-\infty) = -\infty$. (The expressions $(+\infty) + (-\infty)$ and $(-\infty) + (+\infty)$ are not defined and are to be considered meaningless.)

Negation: $-(+\infty) = -\infty$; $-(-\infty) = +\infty$.

Subtraction: The rule: $a - b = a + (-b)$ defines subtraction in terms of addition and negation.

Multiplication: If a is an extended real number >0, then $a \cdot (+\infty) = (+\infty) \cdot a = +\infty$; $a \cdot (-\infty) = (-\infty) \cdot a = -\infty$. If a is an extended real number <0, then $a \cdot (+\infty) = (+\infty) \cdot a = -\infty$; $a \cdot (-\infty) = (-\infty) \cdot a = +\infty$. And $0 \cdot a = a \cdot 0 = 0$ for any extended real number a.

All these extensions are "natural," except perhaps the rule that zero times any extended real number, including $\pm\infty$, yields zero (which is useful in integration theory). The fact that addition (hence subtraction as well) is not always defined reveals a basic disparity between real, and extended real, arithmetic. (Division by $\pm\infty$ has also not been defined, but this is of no importance since no occasion arises in this book where one would want such an operation.)

The number $+\infty$ will usually be abbreviated simply to ∞.[4] The Borel field of the *extended* reals (or the extended real line in geometric terminology) may be defined as the σ-field generated by the class of sets of the form $\{x \mid x > a\}$, where a ranges over the real numbers. These sets are now subsets of the extended reals, so that the number $+\infty$ belongs to all.

It may be verified that this σ-field consists precisely of all sets having any of the following four forms: $E, E \cup \{\infty\}, E \cup \{-\infty\}, E \cup \{\infty\} \cup \{-\infty\}$, where E ranges over the Borel field of the *real* numbers.

COUNTABLE ADDITIVITY

First we recall the definition of function. A *function f* with *domain A* and *values in B,* written $f{:}A \rightarrow B$, is a set of ordered pairs (a, b), where

---

4. The numbers $\pm\infty$ in the extended real number system should not be confused either with "indeterminate forms" in calculus, which are just abbreviations for certain limit operations, or with infinite cardinal numbers in set theory.

$a \in A$ and $b \in B$, each point of A being the first component of exactly one such pair. For each $a \in A$ the point of B thus associated with it is called the *value* of f at a, and is written $f(a)$. The set $\{b \mid b = f(a)$ for at least one $a \in A\}$ is called the *range* of f. It need not include all of B.

Given a measurable space (A, Σ), consider a function $\mu{:}\Sigma \rightarrow$ non-negative extended reals; i.e., μ assigns to each measurable set a value that is either a nonnegative real number or $+\infty$.

Definition: μ is *finitely additive* iff for every pair of measurable sets E, F that are disjoint (i.e., $E \cap F = \emptyset$), we have $\mu(E \cup F) = \mu(E) + \mu(F)$.

Theorem: Let $\mu{:}\Sigma \rightarrow$ nonnegative extended reals be finitely additive. If $E \subseteq F$, where E, F are measurable sets, then $\mu(E) \leq \mu(F)$.

Proof: Since E, F are measurable, so is $F \backslash E$. Also $E, F \backslash E$ are disjoint, and $E \cup (F \backslash E) = F$. Hence $\mu(E) + \mu(F \backslash E) = \mu(F)$. Since $\mu(F \backslash E) \geq 0$, it follows that $\mu(E) \leq \mu(F)$. ∎

The property expressed by this theorem is called *monotonicity* and implies that μ takes on its maximum value for the universe set A. If μ is finitely additive, and E, F, G are three measurable sets no pair of which have a point in common (i.e., the class consisting of these three sets is a packing), then

$$\mu(E \cup F \cup G) = \mu(E) + \mu(F \cup G) = \mu(E) + \mu(F) + \mu(G)$$

By induction a similar rule extends to any finite packing. However, we want to go further and define such an additivity rule for all *countable* packings of measurable sets, not just for all finite packings.

Definition: Let μ take values in the nonnegative extended real numbers, its domain being σ-field Σ. μ is *countably additive* iff, for any countable packing of measurable sets $\mathcal{G}$, we have

$$\mu(\cup \mathcal{G}) = \mu(G_1) + \mu(G_2) + \mu(G_3) + \cdots \qquad (2.1.1)$$

Here $G_1, G_2, G_3, \ldots$ is any enumeration of the members of $\mathcal{G}$ in a sequence, and the right-hand side of (1) is to be understood as the ordinary sum of an infinite series.

The possibility that μ may take on the value ∞ causes no problems. If $\mu(G_n) = \infty$ for some n, the right side of (1), hence the left side, equals ∞; if the partial sums on the right increase beyond any finite bound, both sides again must equal ∞.

MEASURES

We now put all these concepts together.

Definition: A *measure* μ is a function

   (i) whose domain is a σ-field Σ,
   (ii) which takes values in the nonnegative extended real numbers,
   (iii) which is countably additive,
   (iv) and for which $\mu(\phi) = 0$.

(Only one function satisfies (i), (ii), and (iii) but not (iv), viz., the function assigning the value ∞ to all sets. Thus (iv) serves only to exclude this rather trivial case; cf. the first example below.)

Definition: The triple (A, Σ, μ), where A is the universe set, Σ a σ-field (relative to A), and μ a measure with domain Σ, is called a *measure space* (whereas the first two alone, without μ, constitute a *measurable space*).

The following result establishes a "continuity" property of sorts for measures.

Theorem: Let (A, Σ, μ) be a measure space, and let $G_1, G_2, \ldots$ be a sequence of measurable sets that is *increasing;* i.e., $G_1 \subseteq G_2 \subseteq G_3 \subseteq \cdots$. Then limit $\mu(G_n)$ as $n \to \infty$ is $\mu(\cup \mathcal{G})$, ($\mathcal{G}$ being the collection of all the G's). If instead the sequence is *decreasing* $(G_1 \supseteq G_2 \supseteq \cdots)$, *and* $\mu(G_k) < \infty$ for some k, then $\lim \mu(G_n) = \mu(\cap \mathcal{G})$.

To prove the first part, partition $\cup \mathcal{G}$ into sets $G_n \backslash G_{n-1}$ and apply countable additivity; to prove the second part, take complements with respect to G_1 and apply the first part.

We complete this section by giving some examples of measures:

1. Let (A, Σ) be any measurable space; the function assigning the value zero to every member of Σ is a measure (the *identically zero* measure, which we write simply as 0); so is the function assigning the value ∞ to every nonempty member of Σ (the *identically infinite* measure, written ∞).

2. Let the σ-field Σ be *finite*. Then Σ is generated by a partition $\mathcal{G}$, and we may assume $\phi \notin \mathcal{G}$. Assign nonnegative numbers arbitrarily to the members of $\mathcal{G}$; any member of Σ has a unique representation $\cup \mathcal{F}$, where $\mathcal{F} \subseteq \mathcal{G}$; assign to this set the value equal to the sum of the numbers assigned to members of $\mathcal{F}$. The result is a measure.

3. Let (A, Σ) again be any measurable space. Define μ by: $\mu(E) =$ num-

ber of points in E if E is finite and $E \in \Sigma$; $\mu(E) = \infty$ if E is infinite
and $E \in \Sigma$. Then μ is a measure, the *counting* measure.
4. Again let (A, Σ) be arbitrary. Choose a fixed point $a_o \in A$, and define
μ by: $\mu(E) = 1$ if $a_o \in E$ and $E \in \Sigma$; $\mu(E) = 0$ if $a_o \notin E$ and $E \in \Sigma$.
μ is a special kind of measure, an *atomic* measure (we discuss this more
fully later). Note that we need not assume that the singleton set $\{a_o\}$ is
itself measurable.
5. Our last example is the most famous of all measures. Let A be the real
line and let Σ be the Borel field on it. It may be shown there is exactly
one measure μ having the property that $\mu\{x \mid a < x < b\} = b - a$ for
all pairs of real numbers such that $a < b$; i.e., μ assigns to any finite
open interval its length. μ is known as *(1-dimensional) Lebesgue
measure*.

2.2. REPRESENTATION OF THE REAL WORLD BY
MEASURES: PRELIMINARIES

Having at last defined the concept of measure, we now go to real-
world interpretations. (In this chapter we follow the practice of alternating formal development of measure theory with interpretations.)

First, we make some general philosophical comments. We can start
with some portion of the real world and *represent* it in the language of
some formal system. Or we can start with some formal system and *interpret* (or "apply") the statements in it to refer to some part of the world.
The first process clearly involves a severe abstraction (only a small fraction of facts about the world can be translated into the formal system).[5]
The second process also involves an abstraction; it is not always possible
to find a "fact" corresponding to every valid statement in the formal
theory. One must then be satisfied with a partial interpretation of the
formal system.

As an example, consider the representation of *time* by the real numbers. It is easy to interpret statements like "$t_1 > t_2$," or "$t_1 - t_2 = t_3 - t_4$."
But what facts correspond to the statements "t is irrational," or "every
Cauchy sequence $\{t_n\}$ has a limit?" As far as facts are concerned, one
could do just as well representing time by the rational numbers; however,
the real numbers are more convenient.

In exactly the same way, while measure theory is a remarkably flexible and natural instrument for describing the world, we cannot expect
every statement in it to correspond to a fact. The formal apparatus of the

---

5. It also often involves a *distortion*, to squeeze the facts into the categories of the
formal system, e.g., the assumption of perfect vacuums, ideal gases, and pure substances
and, in social science, of perfect competition, lightning calculation, ideal types, rationality,
economic man, political man, libidinal man, etc.

theory is designed for mathematical power and elegance, and as a result we might say the theory *outruns* what can be observed or measured in the real world.

What kinds of real-world facts are representable by measures? As an example think of the countries of the world as being identified with their territories. Consider the set of all locations on the surface of the earth (we idealize by thinking of each location as an extensionless point). The United States is a certain subset of these points, Switzerland is another subset, etc. Furthermore, no two of these subsets have a point in common.[6] Thus the collection of all countries is a packing. If we add to this collection the set consisting of the rest of the world (it will include the high seas, Antarctica, etc.), we have a partition of the surface of the earth. This partition generates a σ-field, viz., all sets of the form $\cup \mathfrak{F}$, where $\mathfrak{F}$ ranges over all subcollections of this partition.

Now choose any fixed date, and to the set $\cup \mathfrak{F}$ assign the number that is the total population of the territory $\cup \mathfrak{F}$ at this date. The function thus defined on the σ-field is a measure. For, obviously, the total population of the union of two disjoint regions is the sum of the populations of the respective regions, so that the function is *finitely additive*. Furthermore, since we have a finite σ-field, *finite* additivity is equivalent to *countable* additivity in this case. This proves we have a measure.

Any territorial magnitudes having the finite additivity property can be represented as measures in exactly the same way as population. This includes territorial area, wealth, coal reserves, miles of highways, and (for a fixed time interval) steel production, steel consumption, births, deaths, marriages, divorces, murders, Ph.D.'s granted, and innumerable others.[7] Measurement units can be quite varied—numbers of objects, mass, dollar value, acres.

Statistical tables presenting data of this sort will not typically write out the entire measure. If the surface of the earth is partitioned into, say, 130 nations plus the rest of the world, a complete table must assign a value to each of the 2^{131} members of the σ-field. This is impossible in practice and also unnecessary, since if we are given the values for the 131 partition elements, the value for any other measurable set is given by the addition of the values of the appropriate subclass of partition elements. Thus in practice, tables will just give values for the generating partition, plus perhaps a few other "marginal subtotals." Any table of statistical data (if it can be represented by a measure at all) can be represented in the foregoing simple form, with the σ-field generated by a finite partition.

---

6. Common borderlines may be thought of as "no-man's land" which belongs to none of the abutting countries.

7. These correspond roughly to Norman Campbell's "fundamental magnitudes." Cf. N. R. Campbell, *Foundations of Science* (Dover, New York, 1957), Ch. 10.

For a second example, suppose one wanted to represent the concept "quantity of time." We represent time itself as usual by the real numbers, each number representing an "instant." We want to assign numbers to various sets of time instants to measure the quantity of time embodied in that set. To an interval $\{t \mid a < t < b\}$ (where a, b are real numbers with $a < b$) is assigned the value $b - a$. Also, it seems reasonable that quantity of time should be at least finitely additive. This suggests that Lebesgue measure, on the Borel field of the real line, is an appropriate mathematical representation of this intuitive concept.

This example is quite instructive in illustrating how the requirements of intuition and considerations of mathematical power combine to suggest the appropriate representation. First, what is the appropriate class of subsets of the real line for which the assignment of a number representing quantity of time is to be considered meaningful? Intuition demands that it include all subsets for which, conceptually, an observation could be made (say by observing the angle through which a clock hand turns), hence certainly all finite intervals should be included. Mathematical elegance demands that it be a σ-field. To satisfy both these demands, it must include at least the entire Borel field.[8] This requirement leads again to the outrunning of the facts by the theory. For what observation could confirm the statement: "The quantity of time embodied in the set of all rational time instants is zero?"

Consider next the various additivity conditions. As suggested above, finite additivity has strong intuitive appeal; but measure theory comes into its own with the stronger requirement of countable additivity.[9] It may perhaps be contended that if a real-world magnitude is already finitely additive, it is intuitively plausible, or even demanded by intuition, that it be in fact countably additive. We know of no philosophical discussion of this issue.[10] Since it seems to be of some importance, we offer some reflections in an appendix to this section.

---

8. It could be more inclusive than the Borel field. But the attempt to extend Lebesgue measure to the class of *all* subsets of the real line runs into counterintuitive paradoxes—at least in the realm of "standard" measure theory, the kind used in this book and everywhere else until very recently. But cf. A. R. Bernstein and F. Wattenberg, Nonstandard Measure Theory, pp. 171–85, in *Applications of Model Theory to Algebra, Analysis and Probability*, W. A. J. Luxemburg, ed. (Holt, Rinehart and Winston, New York, 1969).

9. In fact, the great contribution of Lebesgue consists in sensing and systematically developing the consequences of countable additivity as opposed to the earlier finitely additive "Jordan content."

10. De Finetti has argued that subjective probabilities should be only finitely additive. But the measures we are discussing represent physical magnitudes, not degrees of belief (with the exception of Section 2.8), and his strictures do not apply to them. B. de Finetti, *Probability, Induction and Statistics* (Wiley, New York, 1972).

APPENDIX ON ADDITIVITY

Let μ be a function defined on a σ-field with range in the nonnegative extended reals, representing some real-world data. We suppose μ is finitely additive. There is then some plausibility for the view that it should be countably additive. But why stop at countable additivity? Does not intuition demand that a real-world-representing μ be uncountably additive? Here we have run ahead of ourselves, since we have yet to give a definition of "uncountable additivity." First, we need another concept.

Definition: Given a set of extended real numbers E, the *supremum* of E is the smallest extended real number at least as large as every member of E.

For example, the supremum of the real numbers is ∞; the supremum of the negative numbers is 0. One of the advantages of the *extended* reals is that *every* subset of them has a supremum.

Now let I be an arbitrary nonempty set—finite or infinite, countable or uncountable. Let f be a function with domain I and range in the nonnegative extended reals. Take any finite subset of I, say $\{i_1, i_2, \ldots, i_n\}$, and form the sum $f(i_1) + f(i_2) + \cdots + f(i_n)$. The *summation* of f is now defined as the supremum of the set of these sums formed by ranging over all possible finite subsets of I.

It is easy to verify that if I is finite, the summation of f is its summation over I itself; if I is countable, summation of f is the limit of the series formed when the elements of I are arranged in any sequence. The generalization arises when I is uncountable.[11] With the aid of these concepts we may now formulate a definition.

Definition: μ is *uncountably additive* iff, for any packing of measurable sets $\mathcal{G}$ such that $\cup \mathcal{G}$ is measurable, the summation of μ on $\mathcal{G}$ equals $\mu(\cup \mathcal{G})$.

Here $\mathcal{G}$ plays the role of the index set I above, and $\mu(G)$ with $G \in \mathcal{G}$ corresponds to $f(i)$ with $i \in I$; $\mathcal{G}$ need not be countable.

Uncountable implies countable additivity. Reverting to our argument above, it is hard to see why intuition should swallow countable additivity but strain at uncountable additivity. But uncountable additivity appears to be too strong a condition for many purposes. For example,

---

11. This reduces to the countable case as follows. If the set $I' \subseteq I$ on which f is positive is uncountable, the summation of f equals ∞; if I' is countable, the summation of f is the same as that of f restricted to I'. But this reduction does not detract from the intuitive appeal of a single definition covering all cases.

Lebesgue measure does not satisfy it. To see this, note that if the function f on the index set I is identically zero, its summation equals zero. Since any set is the union of a packing of singleton sets, it follows that if μ is uncountably additive and $\mu\{x\} = 0$ for all singleton sets $\{x\}$, μ must be identically zero. But Lebesgue measure assigns value zero to each singleton set and is not identically zero, hence it is not uncountably additive.

Possibly the difficulty can be resolved by going beyond the (standard) extended real-number system. In any case, we shall assume our representations are countably, but not necessarily uncountably, additive; i.e., we assume they are measures in the ordinary sense of the term. This allows us to apply the great resources of measure theory to real-world problems. In this respect we merely follow in the footsteps of the great probabilists and statisticians.

2.3. SPACE, TIME, AND RESOURCES

We can give many more examples of the representation of facts by measures. But we want to do more than this. We want a unified way of looking at the world, so that all these examples fall out naturally as specializations and we do not need an ad hoc argument for each new case. The claim is that this unified view can be built up from three basic sets: "Resources," the set of resource types; "Space," the set of locations; and "Time," the set of instants. We abbreviate these here, and throughout the book, as R, S, and T, respectively. In this section we describe these three sets. In later sections we put them together.

Time is well represented by the real numbers (smaller numbers being prior to larger numbers in the temporal sense). The measurable subsets of T will always be taken to be the Borel field on the real line.

Definition: A subset of Time is called a *period* iff it is measurable.

Space is thought of most naturally as a 3-dimensional continuum. Sometimes it is more useful to identify S with the surface of the earth, since almost all human activity takes place in a thin film at the surface (even in the space age). In this connection one often makes two further idealizations. First, the earth is taken to be a perfect sphere. Second, the spherical surface itself is flattened into a planar region, or even an infinite plane; thus three dimensions collapse to two.

When dealing with S and T jointly, we need a convention about what locations at two different times are to be considered identical. We always assume the earth is at rest. This geocentric convention would not be made by an astronomer, but for social science purposes it is by far the most convenient. Thus Portugal can be identified with the same subset of S at different times, whereas it would be wandering about under any other convention.

Let us turn to the problem of defining an appropriate σ-field on S. By "appropriate" we refer to the following somewhat vague desiderata. Any subset of S on which an observation could be made conceptually, should be included. In particular, the various simple geometric figures (cubes and spheres in 3-space, squares and circles in the plane, etc.) should be included. But one should not go much beyond the σ-field generated by these because both mathematical and conceptual difficulties arise in defining measures on these very rich classes. On the real line the Borel field fits these specifications fairly well. A natural generalization to higher dimensional sets serves much the same purpose.

Definition: On the plane the *(2-dimensional) Borel field* is the σ-field generated by the class of open rectangles, i.e., by the sets of the form $\{(x, y) \mid a < x < b, \ c < y < d\}$, with a, b, c, d being real numbers.

Just as on the real line this σ-field is generated by many other simple classes, e.g., by the open discs, the sets of the form $\{(x, y) \mid (x - a)^2 + (y - b)^2 < c^2\}$ with a, b, c being real numbers; also by the closed rectangles or discs obtained by substituting $\leq$ for $<$. All ordinary geometric plane figures (thought of as including their boundaries) belong to this σ-field.

Similarly, in 3-space the *3-dimensional Borel field* is that generated by the open prisms $\{(x_1, x_2, x_3) \mid a_i < x_i < b_i, i = 1, 2, 3\}$, and this σ-field is also generated by the open discs $\{(x_1, x_2, x_3) \mid (x_1 - a_1)^2 + (x_2 - a_2)^2 + (x_3 - a_3)^2 < b^2\}$.

Finally, given a subset E of 3-space, such as the (idealized) surface of the earth, the *relative Borel field of E* is the class of sets of the form $\{E \cap F \mid F \in 3\text{-dimensional Borel field}\}$. For example, if we take a plane or a line embedded in 3-space, its relative Borel field according to this definition may be shown to coincide exactly with the 2- and 1-dimensional Borel fields respectively.

These constructions give reasonably good solutions to the problem of the appropriate σ-field on S. In some cases another choice may be better. In particular, it is often sufficient to work with a small subclass of the complete Borel field (as in our example of population distribution by country).[12]

---

12. How do we know that the territories occupied by countries *are* in fact Borel sets (i.e., members of the Borel field)? An element of convention enters here. Without attempting a rigorous discussion, it may be said that any real-world region presenting itself as an observational unit is empirically indistinguishable from some Borel set. Thus taking it to be a Borel set is a mathematical convenience that does no violence to the facts.

In much of this book the particular structure of Space (dimensionality, shape, etc.) is irrelevant. In this case it suffices to think of S as just some arbitrary measurable space. This gain in generality is important, because we often want to deal with Space of a highly non-Euclidean character, its structure determined by the irregularities of transportation cost and land quality.

From now on we use the term "region" in the following technical sense.

Definition: A subset of Space is called a *region* iff it is measurable.

We now turn to Resources. This has a much more complicated structure than S or T. Fortunately, a great many results do not depend on a detailed knowledge of this structure. Also, certain conceptual problems are tied up with R. We accordingly present here a "naive" description of R, reserving the discussion of difficulties for an appendix.

An object is identified by specifying *where* it is, *when* it is, and *what kind of thing* it is. *Space* is the set of possible answers to the first question, *Time* to the second, and we mean *Resources* to be the set of possible answers to the third.

The elements of R are *types* of things rather than specific entities. Thus "water" is a resource (or rather, a set of resources, since it can be differentiated by temperature, pressure, purity, etc.), but any specific drop of water must be identified further by its position in S and T.

What types of things are included in R? All possible types that are relevant and with as much fineness of distinction as is useful for the problem in hand. This will include natural resource types: soils, minerals, water, air, vegetation, animals. It will include manufactured commodities, crops, machinery, and structures. It will include sewage, garbage, trash, and junk. It will include all types of people, distinguished by sex, age, race, skills, beliefs, attitudes, tastes, personality, and any other relevant traits. It may even include such intangibles as light, sound, electricity, and gravity.[13]

Two apparent difficulties may be cleared up at once. The first refers to nonexistent resource types. Should "unicorn" be included in R? Actually, it does no harm to include nonexistents; as we shall see, existence is described by a measure placed on R, not by R itself. Second, can uniqueness or individuality be represented by a model that deals only in types? The answer is yes, if the distinctions made in R are sufficiently

---

13. It is clear from this list that the term "resource" is misleading. Other possible terms, such as "substance," "essence," "quiddity," or "quality," seem even worse. One should keep in mind, then, that "resource" is a general neutral term embracing people types as well as goods types, and "illth" as well as "wealth."

fine. If one gives a very detailed description of a certain type of person, there will be at most one person at one time fitting that description—say George Washington at noon, July 4, 1776.

One sometimes distinguishes between different resource types and different varieties or qualities of the same resource type, e.g., minor variations of brand-name goods. From our present point of view, different varieties are also simply different types. We do not take account of the fact that a codfish is somehow more similar to a mackerel than it is to a cabbage.

We turn to the problem of finding an appropriate σ-field for R. Following our previous approach, we should include all subsets on which, conceptually, a measurement could be taken. Thus "man," "fish," "water," "glove," and "car" determine sets (the set of resource types that are men, fish, etc.) that should be measurable. We could systematically go through the dictionary, and most nouns and adjectives would determine measurable subsets of R in the same way. However, most English words are more or less vague, and borderline cases arise: "Is this creature to be considered a fish or not?" Once the class of conceptually observable subsets of R is determined, the σ-field generated by them would be the one recommended.

Unlike the case of S and T where the Borel fields are the natural choices, the proper choice of σ-field for R is still unclear, as indicated above. Fortunately, nothing here hinges on a detailed specification, and it is sufficient to suppose that R comes supplied with some σ-field, making it a measurable space.

APPENDIX ON RESOURCES

We discuss certain additional problems concerning the set of Resources. First is the problem of self-reference. Among the attributes of people will be their mental states—their beliefs, perceptions, thoughts. But to describe these, we must refer back to R (and to S and T and measures over these sets). Further complications arise if these mental states refer to still other mental states, and we can even get an infinite regress of the kind sometimes discussed in connection with strategy and games: "He thinks that I think that he thinks. . . ." We take up this point again in the more general context of multilayered theories (see Section 2.8).

Second is the problem of inclusiveness. When we take account of our limited information and the fact that there are "more things . . . than are dreamed of in our philosophy," anything smaller than the set of all possible resource types may be descriptively inadequate. But the concept of an all-inclusive R is not very clear and may even entail a logical contradiction.

Third is the problem of complex resources. It will generally not do, for example, to think of an entire river valley as a single resource type. Instead, we think of it as a spatial configuration of resources: water, soil, trees, roads, houses, people.[14] But the same reasoning applies to each of these smaller units. An automobile of a certain specific type is just a spatial configuration of steel, rubber, glass, and paint, and we could continue down to the molecular-atomic level.

From the practical point of view, where we stop in this analysis depends on the size of the unit of interest. The physiologist may take a person to be a configuration of tissues; the biochemist may view him as a configuration of molecules. The social scientist rarely has occasion to split people up spatially in this way.

For the ways we are using the set R, to call something a resource type, when it is actually a heterogeneous spatial configuration, involves distortion. The distortion is greater the larger the object is because we use the concept of resource-location pair (r, s), referring to a resource type r located at a point s. This is somewhat ill-defined if the resource type is by its nature spread over a region of greater or lesser extent.

For the scale on which social science models typically operate, the distortion involved in treating people as resource types is negligible for the most part. The same is probably true for most ordinary commodities, although, as stated, we draw the line at resources as big as river valleys.

This discussion raises the question, If we steadfastly refuse to admit spatial configurations as resource types, what becomes of R? We are driven to resolve one configuration into its components, and these into further components until, presumably, we arrive at a small number of elementary particles out of which everything else is built. From the practical point of view this procedure is absurd. We hardly expect atomic physics to be a prerequisite for social science. Nonetheless, the chance exists that a mathematically convenient theory can be built up by following this route, and we offer a few speculations as to what it would look like.

There is a general tendency for the number of kinds of things to become less as we descend the spatial hierarchy (a great many different types of houses can be constructed by arranging one type of brick in different ways). Suppose, to make the theory as simple as possible, that everything ultimately reduces to just *one* kind of thing: "matter." The objects of everyday life would then be identified with certain distributions of matter over S (or possibly S and T), i.e., with measures assigning to each region the quantity of matter in it and all these regions forming a σ-field relative to the universe set, which is the region the object in question actually occupies.

---

14. Spatial "configurations" as we shall see can be represented by measures.

Such a theory has two great virtues. First, it avoids the distortions that arise when treating people and commodity types as members of R. Second, it eliminates R itself (by reducing it to a single point) and thus simplifies the situation and avoids all the other difficulties connected with R we have been discussing.

In the remainder of this book we usually ignore the issues raised in this appendix, and interpret R as outlined in the main body of this section. This is done partly for pedagogic reasons (descriptively, the theory runs closer to intuition) but mainly because we have not yet arrived at satisfactory answers to the issues raised.

2.4. MEASURE THEORY, II

In this section we return to pure mathematics to define some concepts needed for further developments.

RESTRICTED MEASURES

Let (A, Σ, μ) be a measure space. Just as with any other function we may consider the *restriction* of μ to a subdomain of its domain Σ; i.e., we take a subclass $\Sigma' \subseteq \Sigma$ and define a function μ' with domain Σ' by the rule

$$\mu'(E) = \mu(E), \qquad \text{all } E \in \Sigma' \tag{2.4.1}$$

The only special condition we insist on is that Σ' itself be a σ-field (not necessarily relative to the original universe set A). One says that Σ' is a *sub-σ-field* of Σ.

It is then immediate that μ' is a measure; the facts that it takes values in the nonnegative extended real numbers and is countably additive follow at once from its definition, (1). Two special cases deserve mention.

Definition: μ' is an *aggregation* of μ iff $A \in \Sigma'$.

That is, A still remains the universe set, although Σ' is a "thinning-out" of the original σ-field. As an example, let A be the surface of a sphere and Σ the Borel field on A. Let $\mathcal{G}$ be a finite partition of A into Borel sets, and let Σ' be the σ-field generated by $\mathcal{G}$. The distribution of population by countries fits this model (see Section 2.2). It is clear why the term "aggregation" is used for this relation. While μ gives, say, the complete distribution of population, μ' gives only the distribution for entire countries.

Definition: Given measure space (A, Σ, μ) and $B \in \Sigma$, μ' is the *restriction* of μ to B iff μ' is the restriction of μ whose domain Σ' is the class of all measurable subsets of B.

We easily verify that this Σ' is indeed a σ-field, whose universe set, however, is B, not A. As an example, again take the case of population distribution over the surface of the earth. One may be interested only in the distribution within some particular region B, in which case one studies the restriction of μ to B. In general, the notion of restriction to B enables one to isolate particular objects, activities, or situations within the overall description. Each different measurable subset B yields a different restriction.

Sometimes one is given, not the entire measure μ, but pieces or *patches,* each defined on the measurable subsets of some set B. If these patches cover the entire universe set A, the question arises: Can these patches be put together to yield a single measure on the entire measurable space? The following theorem gives the answer.

Theorem: Patching Theorem. Given a measurable space (A, Σ) and, for each $n = 1, 2, \ldots,$ a measure space (or "patch") (B_n, Σ_n, μ_n), satisfying the conditions:

(i) $B_n \in \Sigma$ for all n, and $\mathcal{B}$, the collection of all the B_n's, is a covering of A (i.e., $\cup\mathcal{B} = A$);

(ii) $\Sigma_n = \{E \mid E \subseteq B_n \text{ and } E \in \Sigma\}$, for all n;

(iii) the μ_n's are compatible in the sense that if $E \in (\Sigma_{n_1} \cap \Sigma_{n_2})$, then $\mu_{n_1}(E) = \mu_{n_2}(E)$, for all n_1, n_2, and E.

Then there is exactly one measure μ on (A, Σ) such that μ_n is the restriction of μ to B_n for all n.

Proof: First we prove there is at most one such μ. Let $B_1' = B_1, B_2' = B_2 \backslash B_1$, and in general $B_n' = B_n \backslash (B_1 \cup \cdots \cup B_{n-1})$. Now suppose μ satisfies the conclusion of the theorem. For any $G \in \Sigma$,

$$\mu(G) = \sum_{n=1}^{\infty} \mu(G \cap B_n') = \sum_{n=1}^{\infty} \mu_n(G \cap B_n') \qquad (2.4.2)$$

(The first equality in (2) arises from the facts that the sets $\{G \cap B_n' \mid n = 1, 2, \ldots\}$ are a packing whose union is G and μ is countably additive; the second equality arises from the facts that $G \cap B_n' \in \Sigma_n$ and μ_n is the restriction of μ to Σ_n.) Since μ is explicitly determined by the μ_n's in (2), it is unique.

It remains to show that the μ defined by (2) does actually satisfy the theorem. First we show that for each n, μ_n is the restriction of μ to Σ_n. Let $G \in \Sigma_n$. For all $k > n$, $G \cap B_k' = \emptyset$. For all $k \leq n$, $\mu_k(G \cap B_k') = \mu_n(G \cap B_k')$, by conditions (ii) and (iii). Hence, by (2),

$$\mu(G) = \sum_{k=1}^{n} \mu_n(G \cap B'_k) = \mu_n(G \cap B_n) = \mu_n(G)$$

proving that μ_n is the restriction of μ to B_n.

Next, $\mu(\phi) = 0$, since $\mu_n(\phi) = 0$ for all n. It remains to show only that μ given by (2) is countably additive. Let $\mathcal{G} = \{G_m \mid m = 1, 2, \ldots\}$ be a measurable packing. Then

$$\mu(\cup \mathcal{G}) = \sum_{n=1}^{\infty} \mu_n[(\cup \mathcal{G}) \cap B'_n] = \sum_{n=1}^{\infty} \left[\sum_{m=1}^{\infty} \mu_n(G_m \cap B'_n) \right]$$

$$= \sum_{m=1}^{\infty} \left[\sum_{n=1}^{\infty} \mu_n(G_m \cap B'_n) \right] = \sum_{m=1}^{\infty} \mu(G_m)$$

(Here the first and last equalities come from (2), the second from the countable additivity of each μ_n, and the reversal of summation order in the third equality from the fact that all summands are nonnegative.) This proves that μ is countably additive. ∎

The most important case of the patching theorem is where the class $\mathcal{B}$ is a *partition* of A. Here we say that measure space (A, Σ, μ) is the *direct sum* of the measure spaces (B_n, Σ_n, μ_n) and write

$$A = B_1 \oplus B_2 \oplus \cdots, \qquad \Sigma = \Sigma_1 \oplus \Sigma_2 \oplus \cdots, \qquad \mu = \mu_1 \oplus \mu_2 \oplus \cdots,$$

or, for short, $\mu \doteq \oplus_n \mu_n$, etc.

Given *any* countable collection of measure spaces with disjoint universe sets, there is a unique way of combining them into a direct sum (see Section 4.6 for full definition). The direct sum should not be confused with the (ordinary) sum of measures, which is defined only for identical σ-fields; see below.

PRODUCT SPACES

Consider any function with domain I and range in Y. If one is mainly interested in the range space, such a function may be written as (Y_i), $i \in I$, and is referred to as *family of elements of Y, indexed by set I*. This is more general than the notion of "subset of elements of Y" since it allows for repetitions; the same Y-element may be assigned to more than one I-element. We distinguish families from subsets by using parentheses ( ) or brackets [ ] instead of braces { }. The elements of Y may be of any nature, e.g., numbers, functions, or sets.

Consider the case where Y is a collection of sets. We rewrite it as $\mathcal{G}$

to conform to our customary notation. We then have a family of sets indexed by I, viz., (G_i), $i \in I$, where each $G_i \in \mathcal{G}$.

Definition: The *cartesian product* of the family (G_i), $i \in I$, is the set $\{(g_i), i \in I \mid g_i \in G_i \text{ for all } i \in I\}$.

That is, each member of the cartesian product is itself a family of elements of $\bigcup \mathcal{G}$ indexed by I; specifically, a family having the property that the element assigned to index i, g_i, always belongs to the set G_i assigned to index i in the original family of sets (G_i), $i \in I$. The cartesian product will be denoted by $\Pi_{i \in I} G_i$, or even by ΠG_i if no confusion is possible. If all the G_i's are identical ($= G$, say), the cartesian product is written G^I.

If the index set I is finite, the cartesian product assumes a fairly simple form; e.g., let I contain just two elements, say $I = \{1, 2\}$, and let the family be (A_1, B_2) (i.e., set A is assigned to $i = 1$, and B to $i = 2$). The cartesian product of this family may be identified with the class of *ordered pairs* $\{(a, b) \mid a \in A, b \in B\}$. Here A and B may overlap, or even be identical. The cartesian product in this case is written $A \times B$.

Similarly, if the family of sets is (A_i), $i = 1, \ldots, n$, the cartesian product may be identified with the set of *ordered n-tuples* $(a_1, \ldots, a_n)$, where $a_i \in A_i$, $i = 1, \ldots, n$, and these choices are made in all possible ways. This may be written $A_1 \times A_2 \times \cdots \times A_n$. Again, some or all of the A_i's may be identical. If all $A_i = A$, this may be written A^n. As an example, let each of the A_i's be the real line. Then $A_1 \times \cdots \times A_n$ is simply n-space, the set of all n-tuples of real numbers.

We now introduce measure-theoretic concepts. Suppose we have a family of measurable spaces $[(A_i, \Sigma_i)]$, $i \in I$; i.e., for each index i there is given a σ-field Σ_i with universe set A_i. There is a standard method for defining a σ-field on the cartesian product ΠA_i. First, we introduce a preliminary concept.

Definition: Subset E of the cartesian product ΠA_i is a *rectangle* iff there is a family of sets (E_i), $i \in I$, such that $E_i \subseteq A_i$ for all $i \in I$, and $E = \Pi E_i$.

As an example let A be the real line, and let E_1, E_2, be two finite intervals of real numbers. Then $E_1 \times E_2$ is literally a rectangle in the plane—the plane being, of course, the cartesian product $A \times A$. This is the origin of the abstract concept "rectangle."

Definition: Given the family $[(A_i, \Sigma_i)]$, $i \in I$, the *product σ-field* is the σ-field on the cartesian product ΠA_i generated by the class of all rectangles $E = \Pi E_i$ having the following properties:

(i) $E_i \in \Sigma_i$ for all i,

(ii) $E_i = A_i$ for all i except for at most one index i_o.

In (ii) the phrase "at most one" could be replaced by "at most a finite number of" or "at most a countable number of"; i.e., we can show that all three of these classes of rectangles generate the same σ-field. It follows that if the index set I is countable, (ii) is trivial and may be dropped from the definition. Rectangles satisfying (i) are called *measurable*.

The product σ-field is denoted by $\Pi_{i \in I} \Sigma_i$ or $\Pi \Sigma_i$, and the resulting product measurable space is then $(\Pi A_i, \Pi \Sigma_i)$. If the (A_i, Σ_i) are identical for all I ($= (A, \Sigma)$, say) this may be written (A^I, Σ^I) or perhaps $(A, \Sigma)^I$. When I is finite, so that the family may be written $[(A_i, \Sigma_i)]$, $i = 1, \ldots, n$, the product σ-field is written $\Sigma_1 \times \Sigma_2 \times \cdots \times \Sigma_n$; and the resulting product space is then $(A_1 \times \cdots \times A_n, \Sigma_1 \times \cdots \times \Sigma_n)$.[15]

As an example let Σ be the Borel field on the real line. Then one may verify that $\Sigma \times \Sigma$ is simply the Borel field on the plane; in fact this provides an alternative definition for that σ-field. Similarly, $\Sigma \times \Sigma \times \Sigma$ is the Borel field in 3-space, and we may define the Borel field in n-space (or even in arbitrary cartesian products of the real line with itself) in an analogous way.

Suppose one is given a measure space of the form $(A \times B, \Sigma' \times \Sigma'', \mu)$. That is, the product measurable space is built up from the two components (A, Σ') and (B, Σ'') in the manner just described, and a measure μ is given whose domain is the product σ-field $\Sigma' \times \Sigma''$.

Definition: μ', the *left marginal measure* of μ, has domain Σ' and is given by $\mu'(E) = \mu(E \times B)$ for all $E \in \Sigma'$.

It is easily verified that μ' is indeed a measure. We may think of μ' as being constructed in two steps. First, consider those members of $\Sigma' \times \Sigma''$ of the form $E \times B$, where $E \in \Sigma'$. These form a sub-σ-field, and μ restricted to this subdomain is an *aggregation*, as defined earlier. Second, since the "right side" of all the rectangles $E \times B$ is the same, we may regard μ as a function of its "left side" only; this yields μ'. In the same way, the *right marginal measure* μ'', with domain Σ'', is given by $\mu''(F) = \mu(A \times F)$ for all $F \in \Sigma''$.

As examples, take any cross-classification, e.g., population classified by location and hair color or shipments by origin and destination. If μ is the total distribution by numbers or mass, the left marginal μ' will give the distribution of population by location alone or of shipments by origin

---

15. Note that $\Pi \Sigma_i$ and $\Sigma_1 \times \cdots \times \Sigma_n$ are not the cartesian products of the family (Σ_i), $i \in I$. No confusion should result from this ambiguous notation.

only. The right marginal will give population by hair color or shipments by destination.

Statistical tables frequently give data for product spaces, and it is customary to give the marginal measures in addition to the original measure. (More accurately, it is customary to give the data for the generating partitions of the component measurable spaces. These are just the "marginal subtotals.") The "marginal" terminology appears in particular in *probability theory*.

Definition: A *probability* is a measure that assigns the value 1 to the universe set.

If μ in $(A \times B, \Sigma' \times \Sigma'', \mu)$ is a probability, we verify immediately that μ' and μ'' are also probabilities—the *left* and *right marginal probabilities* respectively.

Suppose one has an arbitrary product measurable space $(\Pi_{i \in I} A_i, \Pi_{i \in I} \Sigma_i)$. Let $\{I', I''\}$ be a partition of the index set into two nonempty pieces. One may verify that the product space is the same as

$$[(\Pi_{i \in I'} A_i) \times (\Pi_{i \in I''} A_i), (\Pi_{i \in I'} \Sigma_i) \times (\Pi_{i \in I''} \Sigma_i)]$$

That is, we arrive at the same result by first taking the products over I' and I'' respectively, and then taking the product of these products. Thus an arbitrary product space can be expressed as the product of two spaces in many ways. For any such factoring one can define left and right marginals exactly as above.

MEASURABLE FUNCTIONS

Let (A, Σ') and (B, Σ'') be two measurable spaces, and let f be a function with domain A and values in B. Unlike measures, which assign values to *subsets* of A, f assigns values to individual *points* of A. It is customary to refer to the former type as *set* functions and the latter as *point* functions.

Definition: f is a *measurable function* (with respect to Σ', Σ'') iff $\{a \mid f(a) \in E\} \in \Sigma'$ for all $E \in \Sigma''$.

The set $\{a \mid f(a) \in E\}$ is called the *inverse image* of E, so that the definition may be paraphrased: f is measurable iff the inverse image of every Σ''-measurable set is a Σ'-measurable set. If there is no ambiguity, the reference to Σ', Σ'' may be omitted, and one simply writes: f is measurable (or not). We give some examples:

1. Let Σ' = all subsets of A. Then *any* function is measurable.

2. Let Σ'' consist of the two sets ϕ, B. Then again any function is measurable, since the inverse image of B is A, and of ϕ is ϕ.

3. Let f be a constant; i.e., there is a $b_o \in B$ such that $f(a) = b_o$ for all $a \in A$. Then f is measurable. (Proof: If $b_o \in E$, the inverse image of E is A; if $b_o \notin E$, the inverse image is ϕ.)

4. Let $A = B$, and $\Sigma'' \subseteq \Sigma'$. Then the *identity* function, given by $f(a) = a$, is measurable. (Proof: The inverse image of any set is itself.)

5. Let (A, Σ') and (B, Σ'') both be the real line with Borel field. It may be shown that any *continuous* function is measurable.

6. Let $(A, \Sigma') = (B \times C, \Sigma'' \times \Sigma''')$. Let $f(b, c) = b$. Then f is measurable. (Proof: Let $E \in \Sigma''$; the inverse image of E is the set $E \times C$, which always belongs to $\Sigma'' \times \Sigma'''$.)

Example 6 has an important generalization; f is an example of a projection.

Definition: The *projection* from the cartesian product $\Pi_{i \in I} A_i$ to the i_oth component space A_{i_o} is the function that assigns the value a_{i_o} to the point $(a_i), i \in I$.

That is, it picks out the i_oth "coordinate" of any element of the cartesian product. This function is written π_{i_o}. These projections are always measurable, the proof of this fact being a minor elaboration of that given under example 6.

The following theorem gives a very useful criterion for the measurability of a function.

Theorem: Given measurable spaces (A, Σ), (B, Σ'), and $f:A \to B$, let $\mathcal{G}$ be a collection of sets that generate Σ'. Then f is measurable iff $\{a \mid f(a) \in G\} \in \Sigma$ for all $G \in \mathcal{G}$.

Proof: The "only if" statement is trivial. Conversely, let $\mathcal{F}$ be the class of all subsets of B whose inverse images are Σ-measurable. By assumption, $\mathcal{G} \subseteq \mathcal{F}$. If $E \in \mathcal{F}$, then $B \backslash E \in \mathcal{F}$; this follows from the fact that since $\{a \mid f(a) \in E\}$ belongs to Σ, so does its complement $A \backslash \{a \mid f(a) \in E\} = \{a \mid f(a) \in B \backslash E\}$. Similarly, if $\mathcal{H} \subseteq \mathcal{F}$ and $\mathcal{H}$ is countable, then $\cup \mathcal{H} \in \mathcal{F}$. To see this, note that $\{a \mid f(a) \in H\} \in \Sigma$ for all $H \in \mathcal{H}$; hence, the union of these sets over $H \in \mathcal{H}$ belongs to Σ; but this union is $\{a \mid f(a) \in \cup \mathcal{H}\}$, the inverse image of $\cup \mathcal{H}$. It follows that $\mathcal{F}$ is a σ-field. Since it contains $\mathcal{G}$ which generates Σ', we must have $\mathcal{F} \supseteq \Sigma'$. Hence f is measurable. ∎

Let us apply this theorem to the case where B is the real or extended real numbers and Σ' is the corresponding Borel field. Then Σ' is generated by the class of sets $\{x \mid x > b\}$ where b ranges over the real numbers, and the same is true if $>$ is replaced by any of the three signs $<$, $\geq$, $\leq$. Hence to verify that some function f is measurable, it suffices to check that $\{a \mid f(a) > b\} \in \Sigma$ for all real b, or to do this with any of the other three signs in place of $>$. (In fact, checking this for b rational is sufficient.)

The proof above gives a paradigm for proving general statements about all the members of a σ-field: prove the property for a generating class, and prove the class possessing this property is closed under complements and countable unions. Another useful theorem proved in exactly this way is the following.

Theorem: Measurable Section Theorem. Let $(A \times B, \Sigma' \times \Sigma'')$ be a product space. For all $E \in (\Sigma' \times \Sigma'')$ and for all $b \in B$, $\{a \mid (a,b) \in E\} \in \Sigma'$.

Proof: Let $\mathfrak{F}$ be the class of subsets E of $A \times B$ having the property that $\{a \mid (a,b) \in E\} \in \Sigma'$ for all $b \in B$. We verify routinely that $\mathfrak{F}$ is closed under complementation and countable unions. Next, consider the measurable rectangle $E' \times E''$. If $b \in E''$, then $\{a \mid (a,b) \in (E' \times E'')\} = E'$; and if $b \notin E''$, this set $= \emptyset$. Hence all such rectangles belong to $\mathfrak{F}$. But these generate $\Sigma' \times \Sigma''$. ∎

Suppose we are given a function $f: A \times B \to C$. For a point $a_o \in A$ we define $f(a_o, \cdot)$ to be the function with domain B and range in C whose value at $b \in B$ is $f(a_o, b)$. This is the *right a_o-section* of f. Similarly, for $b_o \in B$, the *left b_o-section* of f, written $f(\cdot, b_o)$, is the function with domain A whose value at $a \in A$ is $f(a, b_o)$. Here the sets A and B may themselves be cartesian products.

Theorem: Given $(A \times B, \Sigma' \times \Sigma'')$ and (C, Σ), suppose $f: A \times B \to C$ is measurable. Then all its left and right sections are measurable.

Proof: Consider any left section $f(\cdot, b_o)$. For any $E \in \Sigma$ the set $\{(a, b) \mid f(a, b) \in E\}$ belongs to $\Sigma' \times \Sigma''$, since f is measurable. Hence $\{a \mid f(a, b_o) \in E\} \in \Sigma'$ by the measurable section theorem above. But this set is the inverse image of E under $f(\cdot, b_o)$, hence the latter is measurable. The proof for right sections is similar. ∎

Suppose one is given a measurable space (A, Σ) and a function $f: A \to B$. We use f and Σ to define a certain σ-field on B.

Definition: The class Σ' of all subsets $E \subseteq B$ having the property that $\{a \mid f(a) \in E\} \in \Sigma$ is called the *σ-field induced by f on B*.

It can be verified that Σ' *is* a σ-field and f is measurable with respect to Σ, Σ'. In fact, Σ' may be characterized as the *largest* σ-field on B such that f remains measurable with respect to Σ, Σ'. This approach also works in reverse. This time suppose (B, Σ') is the measurable space and again $f:A \rightarrow B$ is given.

Definition: The class Σ of all subsets of A of the form $\{a \mid f(a) \in E\}$, where E ranges over Σ', is called the *σ-field inversely induced by f on A.*

Again, one verifies routinely that Σ *is* a σ-field and f is measurable with respect to Σ, Σ'. In fact, Σ may be characterized as the *smallest* σ-field on A such that f remains measurable with respect to Σ, Σ'.

Induction applies to measures as well as to σ-fields.

Definition: Given measure space (A, Σ, μ), measurable space (B, Σ'), and measurable function $f:A \rightarrow B$, the *measure μ' induced by f on Σ'* is given by

$$\mu'(E) = \mu\{a \mid f(a) \in E\} \qquad (2.4.3)$$

for all $E \in \Sigma'$.

One easily verifies that μ' *is,* in fact, a measure. As an example take the product space $(A \times B, \Sigma \times \Sigma', \mu)$ and the component space (B, Σ'), and let $f:A \times B \rightarrow B$ be the projection, given by $f(a,b) = b$. Then for any $E \in \Sigma'$, we have, by (3), $\mu'(E) = \mu\{(a,b) \mid f(a,b) \in E\} = \mu(A \times E)$. But this is precisely the definition of the right marginal measure, so this concept could have been defined as the measure induced on (B, Σ') by the projection of $A \times B$ on B. Of course, the left marginal measure is that induced on (A, Σ) by the projection of $A \times B$ on A. These inductions can be combined: Starting with a measure space (A, Σ, μ) and a function $f:A \rightarrow B$, one may first induce the σ-field Σ' on B, and then the measure μ' on (B, Σ').

Measurability of functions is preserved under a great variety of operations. We conclude by listing a few results of this type. The operations themselves are quite useful apart from any question of measurability.

Definition: Given sets A, B, C and functions $f:A \rightarrow B$ and $g:B \rightarrow C$, the *composition* of f and g, written $g \circ f$, is the function with domain A and range in C given by $(g \circ f)(a) = g(f(a))$.

Theorem: Given measurable spaces $(A, \Sigma), (B, \Sigma'), (C, \Sigma'')$; if $f:A \rightarrow B$ and $g:B \rightarrow C$ are measurable, so is $g \circ f$.

To prove this, note that the inverse image under $g \circ f$ is the inverse image under f of the inverse image under g.

We have defined the *supremum* of a set of extended real numbers E as the smallest number not less than any $x \in E$ (notation: sup E). Now suppose we have a collection of functions $\mathfrak{F}$, all extended real-valued with common domain A.

Definition: The *supremum* of $\mathfrak{F}$, written sup $\mathfrak{F}$, is the function with domain A whose value at $a \in A$ is sup$\{f(a) \mid f \in \mathfrak{F}\}$.

Definition: Similarly, the *infimum* of a set of extended real numbers E, written inf E, is the largest number not greater than any $x \in E$. The *infimum* of collection $\mathfrak{F}$ is the function whose value at $a \in A$ is inf$\{f(a) \mid f \in \mathfrak{F}\}$.

Theorem: Let (A, Σ) be a measurable space, and $\mathfrak{F}$ a *countable* collection of extended real-valued functions with common domain A. If each $f \in \mathfrak{F}$ is measurable, then sup $\mathfrak{F}$ and inf $\mathfrak{F}$ are measurable.

Given two functions $f, g : A \rightarrow$ reals, the *sum, difference,* and *product* may be defined in the usual pointwise manner; e.g., $(f + g)(a) = f(a) + g(a)$. If f and g are measurable, so are their sum, difference, and product.

Similarly, given a sequence of functions $f_n : A \rightarrow$ reals, $n = 1, 2, \ldots$, their *pointwise limit* (if it exists) is the function f whose value at $a \in A$ is $\lim_{n \to \infty} f_n(a)$. If all f_n are measurable, f is measurable.

Given $f : A \rightarrow B$ and $E \subseteq A$, the *restriction* of f to E, written $f \mid E$, is the function with domain E that coincides with f there. Given (A, Σ) and (B, Σ'), if f is measurable and $E \in \Sigma$, then $f \mid E$ is measurable.

Theorem: Given a measurable space (A, Σ), a family of measurable spaces $(B_i, \Sigma_i), i \in I$, and a corresponding family of functions (f_i), $i \in I$, where $f_i : A \rightarrow B_i$ for all i; let $g : A \rightarrow \Pi B_i$ be given by $g(a) = (f_i(a))$, $i \in I$. Then g is measurable iff all the functions $f_i, i \in I$, are measurable.

2.5. REPRESENTATION OF THE REAL WORLD BY MEASURES: GENERAL THEORY

We now have the tools to build a unified framework from the three basic sets: Resources, Space, and Time.

HISTORIES

Suppose we had a complete description of a person at some instant of his life. This will include his *state* (height, weight, blood pressure, skills, attitudes, thoughts). From our discussion of the set of Resources, this state may be identified with a point in R. It will also include his *location*, which is a point in S. We may identify this complete description, then, with a pair of points, one in R and one in S, and thus with a point in the cartesian product $R \times S$.

This is for a single instant in T. To give a complete lifetime picture of a person, we must repeat this procedure for each such instant of his life. Suppose a person is born at time t_1 and dies at time t_2, so that he is alive in the interval $\{t \mid t_1 \leq t \leq t_2\}$.[16] A complete description would then be represented by a function whose domain is $\{t \mid t_1 \leq t \leq t_2\}$ and whose range is in $R \times S$. Equivalently, it is represented by a pair of functions, both with domain $\{t \mid t_2 \leq t \leq t_2\}$, one with values in R and giving the person's state at each moment of his life, the other with values in S and giving his location at each moment.

Next consider a machine. It is "born" in some factory, is transported to another, lives out a productive life there, moves into semiretirement, and finally "dies" on the scrap heap. All this can again be described by a function whose domain is an interval of T and whose range is in $R \times S$.

An apple is "born" on the branch of a tree, is harvested, moves through the channels of trade to a household, and ends its existence in somebody's stomach. A certain rock was formed when the earth was created and will persist until the earth is destroyed. And so it goes. One can think of the world as a concatenation of processes of this type, each representable as a function whose domain is an interval of T and whose range is in $R \times S$.

Certain problems are connected with this point of view. First, by the law of the conservation of mass, "births" and "deaths" are transformations from one form of matter to another. In principle this can be handled by our apparatus: The apple disappears, but the person who has eaten it becomes slightly different in state: less hungry, better nourished, heavier.

Second, there are ambiguities in the description. Suppose a handle and a blade are combined to make a knife. One possible description of this event is that the handle and blade both cease to exist at this instant, and the knife begins to exist. Another possibility is to have the blade and handle maintain their separate existences, merely being combined from then on in a certain spatial configuration. In this second approach "knife" is not a resource type at all but the name for certain spatial configurations of other resources. The difficulty here is similar to that discussed in the Appendix on Resources of Section 2.3. If the program sug-

---

16. There is a strong element of convention in this statement. One can hardly identify a unique moment at which a person pops into or out of existence (e.g., we might start with conception rather than birth; there are problems with resuscitation, suspended animation, etc.) The inclusion of the endpoints t_1, t_2 in the interval is also obviously a pure convention.

gested there could be carried out, it would avoid this problem as well. The solution lies in making conventions as to what is to be considered a resource, as opposed to a spatial configuration of other resources.

A third problem concerns resources that are "continuously" spread over space. The precise meaning of this term will be taken up later, but for the present we may take it to refer to such resources as air, water, soil, wheat, cement, and steel, as opposed to people, animals, cars, and machines which are more naturally thought of as "discrete" particles.[17] All the examples given are of the "discrete" type, and the question arises, Can we describe the continuous resources in the same terms? The answer is yes. One of the great advantages of the measure-theoretic approach is that it can handle discrete, continuous, and mixed distributions with equal facility. We take up this point later when the approach has been more fully expounded.

We now return to the main line of argument: the world is being viewed as a collection of processes, each of which can be represented as a function whose domain is an interval of Time, and whose range is in $R \times S$.

We need not exclude the possibility that a given process has no birth, so that its existence stretches indefinitely into the past; or that it has no death, so that its existence stretches indefinitely into the future; or that it has neither birth nor death. The following definition formalizes these considerations.

Definition: A *history* is a function taking values in $R \times S$ and whose domain is a closed T-interval, i.e., a subset of T (the real line) of one of the following four types: either $\{t \mid t_1 \leq t \leq t_2\}$, where $t_1 < t_2$ are real numbers; $\{t \mid t \leq t_2\}$; $\{t \mid t \geq t_1\}$; or T itself.

The history of a person, i.e., a history taking on only person types as values in R, may be referred to as a *biography*.

Definition: For a given history h, the function taking on the value s ($\in S$) when h takes on the value (r, s) will be called the *itinerary* of that history.

Definition: The function taking on the value r ($\in R$) when h takes on the value (r, s) will be called the *transmutation path* of that history.

---

17. This distinction hinges on the scale of observation. All the "continuous" resources mentioned reveal a "granular" structure under the microscope. Conversely, from a large-scale point of view it may be useful to think of people, say, as being continuously distributed, as when one speaks of "population density" or "migration flow."

Thus the itinerary traces out the locations occupied by a history in the course of its existence; the transmutation path traces out the *states in the Resources set* through which the history passes. For example, if we take the biography of a person, his itinerary will trace out all his movements, trips, visits, migrations, and commuting patterns over his lifetime. His transmutation path will trace out his progress from infancy to childhood to adulthood to old age, with the accompanying moods, experiences, activities, speech, etc. We denote the itinerary of history h by h_s, and its transmutation path by h_r. Thus h_r takes values in R, and h_s takes values in S.

Let Ω be the *set of all possible histories,* i.e., the set of all functions from closed T-intervals to $R \times S$ (not merely those histories realized by an actual "particle"). We now show how the world may be described as a measure space (Ω, Σ, μ) with universe set Ω.

The measure μ has the following intuitive interpretation. For a set of histories $E \in \Sigma$, $\mu(E)$ is the total "mass" flowing through the locations and forms at the times indicated by the various histories of E. This vague characterization will be elucidated in the next few pages.

Along with the measure space (Ω, Σ, μ) we consider certain families of functions, all with domain a subset of Ω, and with values in various product spaces built up from R, S, and T. Each such function corresponds to the asking of a question, the answer to which appears as a measure on the space in which its values lie.

Formally, let f be one such function; f takes values in a set A that is typically of the form $R^a \times S^b \times T^c$ for some nonnegative integers a, b, c, though it may be more complex. We assume R, S, and T come supplied with appropriate σ-fields Σ_r, Σ_s, Σ_t respectively, and these determine a product σ-field on A. Then for any measurable subset $G \subseteq A$, we assume the set of histories $\{h \mid f(h) \in G\}$ belongs to Σ; i.e., we assume f is measurable. Then f induces the measure μ onto the measurable space A, and this induced measure is intuitively the answer to the question embodied in the function f. We now illustrate, beginning with specifics and then generalizing.

CROSS-SECTIONAL MEASURES

Consider the question, What is the total quantity of water in Lake Erie at noon, January 26, 1970 (in tons)? The answer is given by the μ-value of a certain set of histories E. Specifically, E is the set of histories whose transmutation paths at the moment noon, January 26, 1970, lie in the subset of R labeled "water," and whose itineraries at that instant are located in the region "Lake Erie." Set E can be written symbolically as

$$\{h \mid h(\text{noon, January 26, 1970}) \in (\text{water} \times \text{Lake Erie})\} \quad (2.5.1)$$

Questions of the general form, What is the total quantity of resources of types F in region G at time t? (of which the above is an example) may be called *cross-sectional* questions. It should be clear that any cross-sectional question (with $F \in \Sigma_r$, $G \in \Sigma_s$, $t \in T$) has as its answer the μ-value of a certain set of histories, viz.,

$$\{h \mid h(t) \in F \times G\} \qquad\qquad (2.5.2)$$

Consider the logic of the situation. The various drops or molecules of water in Lake Erie at the moment in question had a variety of past histories: some fell directly as rain, some flowed in from Lake Huron, some entered as sewage, some as industrial effluent. And they will have a variety of future histories: some evaporating, some flowing out to sea, some entering samples taken by pollution researchers. All these combinations and more will be in the set of histories (1). But cross-sectional questions are not aimed at eliciting this detail; instead they lump together all such histories, "from whatever source derived" and "to whatever destiny aimed," and this is just what sets of the form (2) do.

Before giving further examples, we examine the assumptions behind this whole approach. At first glance, a conservation of mass assumption seems to be involved. The same total "quantity of matter" is carried through time along the paths traced out by the histories, merely changing its form and location, i.e., redistributing itself over $R \times S$. And, indeed, this literal interpretation of "mass" is perfectly adequate for many kinds of histories.

Trouble arises when we consider biographies of persons. Mass in the literal sense changes as one advances from infancy to adulthood to dotage. But we have a certain freedom in choosing measurement units. In the case of resources that come in natural units (such as people, cars, or cattle) it is common to measure in terms of number of entities rather than in terms of number of pounds.[18]

Which measurement units should one choose? We enunciate the principle: choose measurement units in such a way that the resulting "mass" *is* conserved as one traces out the path of histories through time. Thus for most social science purposes the "number of persons" measurement approach is the correct one because it gives each person the constant "mass" of 1 over his lifetime.

From now on we drop the quotation marks around "mass," it being understood that the appropriate units are used for the various histories (pounds, numbers, acres, yards, board-feet). Two points should be noted. First, there is no theoretical objection to adding together measurements

---

18. For people, the "number of entities" approach is almost universal. No political system is organized on the principle, "one pound, one vote!"

using different units for the different components of the sum. As long as
the various units are known, the measure μ carries the information without loss. Furthermore, if we switch from one system of measurement
units to a completely different system, a simple formula enables us to
translate the old measure μ into a new measure μ' in terms of the new
units.[19] Second, it is not clear a priori that one *can* define measurement
units in such a way that the desired goal of mass conservation is attainable, even approximately. For further discussion, see the appendix following this section.

To generalize, consider the function f_t (t being a fixed real number)
that assigns to history h the value $h(t)$, which is a point in $R \times S$. Here
the domain of f_t consists of all histories in existence at instant t; f_t is assumed to be measurable (with respect to Σ restricted to the domain of f_t
and with respect to $\Sigma_r \times \Sigma_s$, the σ-field of $R \times S$) and induces the measure μ onto the space $(R \times S, \Sigma_r \times \Sigma_s)$.

What value is assigned to the measurable rectangle $F \times G$ ($\subseteq R \times S$)? The value that μ assigns to the inverse image $\{h \mid f_t(h) \in F \times G\}$.
But this is the same as the set (2). The measure induced by f_t is the *cross-
sectional measure* giving the distribution of mass over $R \times S$ at time t.
This measure provides the answer to any question concerning the world
at time t_2, i.e., to all possible cross-sectional questions. All this information is contained in the original measure μ over the space of possible histories (Ω, Σ) and is extracted from that measure by means of the mapping
f_t. Let μ_t be the cross-sectional measure for time t, so that $\mu_t(E) = \mu\{h \mid h(t) \in E\}$ for all $E \in \Sigma_r \times \Sigma_s$.

In general, one will not be interested in the entire realm $R \times S$. A
regional geographer who wants to know everything about Austria at time
t, for example, will restrict μ_t to $R \times$ Austria. But someone who wants
to know everything about steelmaking at t_o, wherever it exists, will restrict μ_{t_o} to $F \times S$, where F is the set of resource types having to do with
steelmaking (ore, coke, slag, blast furnaces, steelworkers). Generally, one
makes both restrictions, narrowing attention to a subset of resources in
some region.

This is perhaps the time to bring up the question of *practice*. Even
after restricting one's attention, the resulting measure is a very complicated business. In practice, must we not simplify drastically to say anything at all? There are three answers to this question. First, we indeed
simplify in practice. The most common method is to *aggregate* into a simpler sub-σ-field, usually one generated by a finite partition. The result, of
course, is still a measure.

The second answer is that it is possible to simplify *without* aggregat-

---

19. This point will be elaborated on after we have defined the concept of "integral."

ing. Practice demands that a description be specifiable by a small number of numerical parameters. Aggregation does this. But it can also be accomplished by having a stock of standard measures available, indexed by a small number of parameters. If the stock is well chosen and versatile, we can find an element that is a good approximation (or "fit") to the actual measure. Examples are the Pearson family of distributions in statistics and the approximation of functions by polynomials or trigonometric sums. Indeed, the general practice of approximating things by other things in a smaller, simpler family is a universal principle of scientific work; much literature exists on how to find the best approximation or test for goodness of fit.

The third answer refers to the division of labor between practical and theoretical work. Consider numerical calculation. In practice we need only the rational numbers (or even less—say rationals of the form $N \cdot 10^{-20}$, N an integer). But for theoretical work this would be a crippling restriction. The real numbers are needed even for evolving and justifying practical procedures of numerical calculation itself. In the same way, even if the only measures ever to be used in a practical way are the aggregations into finite σ-fields (a premise we do not grant), we would still want to use measure theory to gain theoretical insight.

We now return to cross-sectional measures. It has been mentioned that "complex" resources may be thought of as *spatial configurations* of simpler resources. We are now in a position to pin down the concept of "configuration." Consider a certain building at time t, for example, which is a configuration of bricks, wood, plaster, and glass. Let E be the region occupied by this building.[20] The configuration that is this building is then simply the cross-sectional measure μ_t restricted to $R \times E$. This restriction tells us how much of every kind of material is present in each part of E, which is just the information we need to describe the building completely. In general, any "spread-out" entity at a time t may be identified with the cross-sectional measure μ_t restricted to $R \times E$, E being the region occupied by the entity in question.

This more or less takes care of a specific entity at a specific time. It is also of interest to define the concept of a *type* of configuration, not tied down to any specific region or time instant. We delay giving such a definition until further mathematical concepts have been introduced.[21] Different configurations with the same R-marginal may be referred to as *isomers*, to borrow a term from chemistry.

We have been dealing with facts involving one point in T. We now

---

20. An element of convention is involved in defining E; e.g., Is E to include the entire volume enclosed by the building, or just the shell? We suppose this has been decided.

21. In particular, the concept of "congruent measures." See Section 2.7.

go on to facts involving two points in T, which introduces transformations, transportation, and storage. For example, how many people alive at time t_1 have died by time $t_2 > t_1$? The answer is

$$\mu\{h \mid h_r(t_1) \in \text{person, and } t_2 \text{ is not in the domain of } h\} \quad (2.5.3)$$

The set of histories in (3) is exactly that called for in the question. (It is assumed here that for histories of this type the measurement units are "numbers of entities."

If instead, t_2 precedes t_1, then (3) gives the number of people *born* between t_2 and t_1 and still alive at t_1. How many people migrated from region F_1 at time t_1 to region F_2 at time t_2 $(t_2 > t_1)$? The answer is

$$\mu\{h \mid h(t_1) \in (\text{person} \times F_1), \text{and } h(t_2) \in (\text{person} \times F_2)\} \quad (2.5.4)$$

We should qualify this statement. Expression (4) gives the number of people in region F_1 at t_1 and in region F_2 at t_2. Hence, first, it says nothing about their itineraries within the interval; these may involve all sorts of spatial maneuvers. Second, it gives the number of people physically present in these regions rather than resident in them, change in residence being the usual definition of migration. Residential location could be represented, but it is a more complicated concept than physical location, involving mental states and legal documents.

As a special case of (4) we could have $F_1 = F_2 = F$. Then (4) would count the number of people who stayed in region F throughout the interval, but it would also count those who wandered out of the region after time t_1 but returned by time t_2.

How much cotton yarn at time t_1 has been converted into shirts at t_2 $(t_2 > t_1)$? The answer is

$$\mu\{h \mid h_r(t_1) \in \text{cotton yarn, and } h_r(t_2) \in \text{shirts}\} \quad (2.5.5)$$

Expression (5) gives the mass of the set of histories whose transmutation path was in the resource set "cotton yarn" at instant t_1 and in the resource set "shirts" at instant t_2. Again some qualifications are in order. As above, there is no restriction on what these histories do in the interim. More serious is the fact that, depending on how histories are defined, (5) may give a "wrong" answer. Recall the discussion of knives, blades, and handles, where it is pointed out that when a given history is "born" or "dies" is partly a matter of convention. If things are defined so that cotton yarn ends its existence when converted into shirts, (5) gives the answer zero. The difficulty harks back to the problem of defining the set R in a satisfactory manner.

Instead of considering questions piecemeal, let us set up a measure that answers all such questions systematically. We have considered the case of a single moment t_o and the resulting cross-sectional measure on universe set $R \times S$. Now we consider two moments and get a measure over $(R \times S)^2$. Such measures are called *two-timing* (or perhaps *double-cross-sectional*).

Given two moments t_1 and t_2 ($t_1 < t_2$), define the function f_{t_1,t_2} by

$$f_{t_1,t_2(h)} = (h(t_1), h(t_2)) \tag{2.5.6}$$

The domain of f_{t_1,t_2} is the subset of histories in existence at both times t_1 and t_2. It is assumed that f_{t_1,t_2} is measurable. Hence from μ it induces a measure μ_{t_1,t_2} on the range space $(R \times S \times R \times S)$. The intuitive meaning of μ_{t_1,t_2} is as follows. Let E and F be measurable subsets of $R \times S$. Then $\mu_{t_1,t_2}(E \times F)$ is the total mass of all histories having a value in E at moment t_1 and in F at moment t_2.

Again we may let the sets E and F themselves be rectangles in $R \times S$: Let $G_1, G_2 \in \Sigma_r$ and $H_1, H_2 \in \Sigma_s$. Then $\mu_{t_1,t_2}[(G_1 \times H_1) \times (G_2 \times H_2)]$ is the mass embodied in the histories in resource set G_1 and region H_1 at time t_1 and in resource set G_2 and region H_2 at time t_2.

This measure gives no information concerning histories that are "born" or "die" between t_1 and t_2. To answer such questions systematically, we proceed as follows (details concerning measurability are omitted). We add an artificial point z_o, signifying nonexistence, to the set $R \times S$. Again f_{t_1,t_2} is defined as in (6), but its domain is now all of Ω; if history h is not in existence at time t_i, then $h(t_i)$ is to be understood as z_o. This extended function induces a measure μ_{t_1,t_2} onto the space $[(R \times S) \cup \{z_o\}]^2$; e.g., $\mu_{t_1,t_2}[(\text{person} \times S) \times \{z_o\}]$ would be the same as (3), i.e., the total number of persons alive at time t_1 who have died by t_2.

Having gone from one to two time points, it is simple to go to three, to a finite, or even to a countably infinite number. For example, choose a measurable set $E_t \subseteq [(R \times S) \cup \{z_o\}]$ for each *integer* t $(0, \pm 1, \pm 2, \ldots)$. The measure of the set $\{h \mid h(t) \in E_t, t = 0, \pm1, \pm2, \ldots\}$ gives the mass of all histories passing through each of the sets E_t at the respective integer times. One could even do this for all the *rational* instants t, since these are countable—as good a monitoring system as one could hope for.

PRODUCTION AND CONSUMPTION

A broad category of questions concerns births or production over time, such as, How much corn was grown in Iowa in 1948? or, How many people were born in New York in 1934? To give a general method for drawing such descriptions out of the measure space of histories (Ω, Σ, μ), we first restrict Ω to the subset Ω_-, consisting of all histories with a date of birth, i.e., which do not exist indefinitely far back into the past. Then define the function $f:\Omega_- \to R \times S \times T$ by

$$f(h) = (h(t_1), t_1) \tag{2.5.7}$$

where t_1 is the moment of birth of the history h. That is, f assigns a pair, the second component of which is the moment of birth (= the earliest time at which h takes a value in $R \times S$), and the first is the value h takes at that time.

In terms of f the amount of corn grown in Iowa in 1948 is the measure of the set of histories

$$\{h \mid f(h) \in \text{corn} \times \text{Iowa} \times 1948\} \qquad (2.5.8)$$

since (8) is precisely the set of histories "born" in the time interval 1948, whose transmutation path starts in the resource set "corn" and whose itinerary starts in the region "Iowa."

Assuming f to be measurable, it induces the measure μ (restricted to Ω_-) onto $(R \times S \times T, \Sigma_r \times \Sigma_s \times \Sigma_t)$. We call this induced measure λ_1. On rectangles, λ_1 can be given a simple intuitive interpretation. Let $E \in \Sigma_r, F \in \Sigma_s, G \in \Sigma_t$; then $\lambda_1(E \times F \times G)$ = total mass of all histories starting at some instant in period G in region F in resource set E. Thus λ_1 gives the distribution of "births" or "production" over R, S, and T.

By an argument exactly parallel to the one just given we can describe the distribution of "deaths" or "consumption." Omitting details, we restrict Ω to histories having an end in T, then take a function g having as its value the pair consisting of the date of death and the point in $R \times S$ occupied by the history at that moment. Assuming g is measurable, it induces a measure λ_2 onto $(R \times S \times T, \Sigma_r \times \Sigma_s \times \Sigma_t)$. The interpretation of λ_2 on rectangles is $\lambda_2(E \times F \times G)$ = total mass of all histories ending at some instant in period G in region F in resource set E.

Finally, we consider the joint pattern of production and consumption. First, restrict Ω to the set of histories having both a beginning and an end; call this Ω_o. Now define the function

$$k:\Omega_o \to (R \times S \times T \times R \times S \times T)$$

(= $(R \times S \times T)^2$, for short) given by $k(h) = (f(h), g(h))$, where f is defined by (7), and g is the similar function defined above. That is, $k(h)$ is a quadruple giving the point in $R \times S$ at which h starts, the time it starts, the point in $R \times S$ at which it ends, and the time it ends.

On $(R \times S \times T)^2$ we take the sextuple product field $(\Sigma_r \times \Sigma_s \times \Sigma_t)^2$. Since f and g are measurable, so is k. Let ν be the measure it induces on $((R \times S \times T)^2, (\Sigma_r \times \Sigma_s \times \Sigma_t)^2)$. On rectangles the interpretation of ν is as follows. Let $E_1, E_2 \in \Sigma_r; F_1, F_2 \in \Sigma_s; G_1, G_2 \in \Sigma_t$; then $\nu(E_1 \times F_1 \times G_1 \times E_2 \times F_2 \times G_2)$ = total mass of all histories starting at some instant in period G_1 in region F_1 in resource set E_1 and ending at some instant in period G_2 in region F_2 in resource set E_2. If we think of $(R \times S \times T)^2$ as the product of $R \times S \times T$ by itself, λ_1 and λ_2 are precisely the left and right marginal measures of ν respectively.

The main lines of development of our descriptive program should now be clear. Extensive magnitudes in general may be represented as measures, and a large variety of these may be derived from one underlying measure μ on the space of histories. We could extend this section indefinitely, systematically deriving more and more complex varieties of data from the underlying measure space (Ω, Σ, μ). As an exercise, the reader is invited to puzzle out how the fishing example that begins this chapter may be derived from (Ω, Σ, μ). This is more complex than our previous cases because, e.g., codfish are not produced in the port of Boston but arrive there from elsewhere. The solution, as a restriction of a measure over $R \times S \times T$, involves counting each history the number of times it enters a given subset of $R \times S$.

We make one final comment on the scope of this program. Our examples have been drawn exclusively from statistical data, i.e., the kind of data appearing in tabular numerical form in census reports, etc. These data have a certain precision that makes them easy to discuss. However, since our model deals with the redistribution of matter in the most general sense of the term, both in location and in form (= resource state), in principle it should be able to handle "literary" data as well (history, travel, biography, belles lettres). "He swept her up in a passionate embrace" *could* be translated into the language of measures; the only conceptual difficulty lies in the vagueness of the description.

Appendix on Histories

One disconcerting feature of our model is the extreme generality of the concept of "history." Between birth and death any function with values in $R \times S$, however erratic, is an admissible history. This in itself is not disqualifying. If in the real world frogs do not turn into princes and Dr. Jekyll does not become Mr. Hyde, this is indicated by assigning the measure zero to the appropriate set of histories. But difficulties remain.

Trouble arises from the diversity of measurement units. The more a given history wanders over the set of Resources, the harder it becomes to assign units in such a way that "mass" is preserved over time. This suggests the following kind of modification (or rather, restriction) on the set of histories Ω. The set of Resources is given a partition $\mathcal{R}$ into measurable subsets, such that the elements of any set $E \in \mathcal{R}$ are similar to each other in some sense. In particular, they are similar in the sense that the same kinds of measurement units are applicable to all the elements of any given set of the partition. Thus we would not put into the same set resource types that come in natural units and resource types that lend themselves to measurement by weight.

Having set up the partition $\mathcal{R}$ of R into fairly homogeneous subsets, we now admit only histories whose transmutation paths stay entirely within some one set $E \in \mathcal{R}$. This restriction on the set of histories alleviates the measurement unit problem. Each $E \in \mathcal{R}$ may now be tagged with its "natural" unit (pounds, number of entities, acres, etc.). In setting up a σ-field on the restricted Ω, one may begin by taking the set of all histories with R-values in set E to be measurable for each $E \in \mathcal{R}$. (In contrast to R, Space requires no such breaking up. Because of its homogeneous nature, the wandering of itineraries over S creates no measurement unit problems.)

The restriction just discussed is a kind of "boundedness" constraint, limiting the "distance" over which any transmutation path is allowed to wander in R. A different kind of restriction also suggests itself—one prohibiting "discontinuous jumps." The quotation marks are used in this paragraph because we have not yet defined any structure on R or S that would give meaning to them.[22]

Without going into details here, suppose the concept of continuity for histories has been defined, and in such a way that the maxim *natura non facit saltum* is valid.[23] This does not by itself disqualify our original scheme. It just means that measurable sets of discontinuous histories are assigned the value zero. However, in this case there may be some advantage to restricting our original Ω to the simpler subset of continuous histories.

Finally, we mention the measurability problem on the space of histories. This is the problem of identifying which sets of histories correspond to observations that might be made, at least conceptually. The criterion of "conceptual observability" is itself vague; but even if it were pinned down one would still have to classify systematically the possible kinds of data and find the subsets of Ω corresponding to each.

2.6. MEASURE THEORY, III

This section differs in style from Sections 2.1 and 2.4 in two respects. First, the ratio of theorems to definitions is higher: We are more concerned with stating the results of the theory, and less with merely outlining the concepts. Second, we give illustrations not only from pure mathematics but also from the applied concepts we have been building (R, S, T,

---

22. We have deliberately refrained from introducing the metrical or topological concepts that would be needed for this. These notions play a decidedly secondary role in this book, and we have therefore concentrated on building up the theory of measure per se, which does not depend on them.

23. A history is said to be continuous iff it is continuous at all instants of time except for the moments of birth and death, and continuous from the (future, past) at (birth, death) respectively.

histories, etc.) No confusion should result from this mixture. As before, we omit proofs unless they are very short or instructive or not readily available.

Finite and Sigma-finite Measures

A first distinction is between *finite* and *infinite* measures, the latter being those taking on the value ∞ at least once. Since a measure attains its maximum value on the universe set A, a measure μ is finite iff $\mu(A)$ is finite and is infinite iff $\mu(A) = \infty$.

Definition: Consider any function f whose range is in the extended real numbers. Function f is *finite above* iff it never takes on the value $+\infty$, *finite below* iff it never takes on the value $-\infty$, *finite* or *real valued* iff it is finite both above and below. On the other hand, f is *bounded above (bounded below)* iff there is a real number L such that $f(x) \leq L$ ($f(x) \geq L$) for all x in the domain of f; f is *bounded* iff it is bounded both above and below.

Equivalently, f is bounded iff there is a real number L such that $-L \leq f(x) \leq L$ for all x in the domain of f, which can also be expressed by writing $|f(x)| \leq L$ for all x, vertical bars indicating the absolute value of a number ($|a| = a$ if $a \geq 0$; $|a| = -a$ if $a < 0$). Note that f bounded above, below, or both implies f finite above, below, or both respectively, but the converse is not necessarily true. For example, the identity function $f(x) = x$ on the real line is finite but not bounded.

However, for *measures* the properties "finite," "finite above," "bounded," and "bounded above" are all equivalent. To see this, first note that measures are automatically bounded below, since they are non-negative; second, if μ is finite, the real number $\mu(A)$ provided an upper bound: $\mu(E) \leq \mu(A)$ for all measurable E. The terms "bounded measure" and "finite measure" will be used interchangeably.

A related very important concept is σ-finiteness.

Definition: Let (A, Σ, μ) be a measure space. Measure μ is *σ-finite* iff there is a countable partition $\mathcal{G}$ of A into measurable sets such that for each $G \in \mathcal{G}$ the restriction of μ to G is finite.

Consider the following examples:

1. Any finite measure is σ-finite. (Proof: Let the partition $\mathcal{G}$ have as its only member the universe set A itself.)
2. Consider Lebesgue measure on the real line. This is certainly not finite, since it assigns the value ∞ to the entire line. But it *is* σ-finite.

(Proof: Take the countable measurable partition consisting of the sets $\{x \mid n \leq x < n + 1\}$, where n runs through the integers $0, \pm1, \pm2, \ldots$; μ is finite on each piece—in fact, $\mu\{x \mid n \leq x < n + 1\} = 1$ for all n.)

3. Any measure that assigns the value ∞ to some singleton set $\{x\}$ is *not* σ-finite. (Proof: For any partition $\mathcal{G}$, if $x \in G \in \mathcal{G}$, the restriction of μ to G remains infinite.)

4. As a less trivial example of a non-σ-finite measure, let A be the real line, Σ any σ-field on A, and μ the counting measure ($\mu(E)$ = number of points in E). (Proof: Let $\mathcal{G}$ be any countable packing of measurable sets such that $\mu(G)$ is finite for all $G \in \mathcal{G}$; then each G is finite so $\cup\mathcal{G}$ is a countable set; since A is uncountable, $\mathcal{G}$ cannot be a partition.)

The importance of σ-finite measures stems from two facts: they have many useful properties not shared by measures in general; and most measures that come up, even in theoretical investigations, are σ-finite.

ATOMIC AND NONATOMIC MEASURES

We have mentioned the distinction between resources typically distributed "discretely" over Space and those distributed "continuously." We now define these concepts rigorously and abstractly. Actually, two entirely different concepts explicate the notion of "continuous distribution," one of them involving a single measure, the other a certain relation between two measures. We now give the first.

Definition: A measure μ is *nonatomic* iff, for any measurable set E for which $\mu(E) > 0$, there is a pair of measurable sets F, G such that $F \cap G = \emptyset$, $F \cup G = E$, $\mu(F) > 0$, and $\mu(G) > 0$.

Briefly, μ is nonatomic iff any set of positive measure can be split into two pieces, each of positive measure. Consider the following examples:

1. It may be shown that Lebesgue measure on the real line is nonatomic.

2. Let $\mu(\{x\}) > 0$ for some measurable singleton set $\{x\}$; then μ cannot be nonatomic (since $\{x\}$ cannot be split). (The converse of this statement is not true. There are measures for which $\mu(\{x\}) = 0$ for all x, yet they are not nonatomic.)

Definition: Given a measure space (A, Σ, μ), a set $E \in \Sigma$ is called an *atom* for μ iff $\mu(E) > 0$; and, however E is split into two measurable sets F, G ($F \cap G = \emptyset$, $F \cup G = E$), either $\mu(F) = 0$ or $\mu(G) = 0$. Thus a measure is nonatomic iff Σ contains no atoms.

At the other extreme we have the following definition.

Definition: Let (A, Σ, μ) be a measure space. Measure μ is *atomic* iff A itself is an atom.

That is, $\mu(A) > 0$, and for any $E \in \Sigma$, either $\mu(E) = 0$ or $\mu(A \backslash E) = 0$. If μ is finite, then μ is atomic iff its range consists of exactly two values, 0 and $\mu(A)$. (This follows at once from $\mu(A) = \mu(E) + \mu(A \backslash E)$ and the definition.)

Definition: μ is *simply concentrated* iff there is a point $a_o \in A$ having the property

$$\mu(E) = 0 \text{ if } a_o \notin E, \qquad \mu(E) = \mu(A) > 0 \text{ if } a_o \in E \qquad (2.6.1)$$

for all measurable E.

A simply concentrated measure is atomic. Although not all atomic measures are simply concentrated, the latter is the most important kind found in practice. Choosing an arbitrary point a_o, an arbitrary positive number for $\mu(A)$, and μ according to (1) gives a simple recipe for constructing atomic measures (the point a_o need not be unique, in general).

Definition: μ is a *σ-atomic* measure iff there is a countable measurable partition $\mathcal{G}$ of A such that G is an atom for all $G \in \mathcal{G}$.

Otherwise expressed, μ is σ-atomic iff the universe set A can be split into a countable number of measurable pieces such that μ restricted to each piece is an atomic measure.

Theorem: Atomic Decomposition Theorem. Given measure space (A, Σ, μ), with μ σ-finite, there is a set $E \in \Sigma$ such that

  (i) μ restricted to E is nonatomic,
  (ii) μ restricted to $A \backslash E$ is σ-atomic.

If E' is another set satisfying this theorem, then $\mu(E \backslash E') = 0$ and $\mu(E' \backslash E) = 0$: E is "almost" unique. Furthermore, the atoms in any two such decompositions may be paired off such that a similar relation holds between each pair of atoms. This is the first of several basic decomposition theorems whose aim is to represent measures as built up in one way or another from simpler measures.

We illustrate with a real-world example. Consider the rural-urban distribution of population over the surface of the earth. It is a useful approximation to think of the urban population as being concentrated in cities, each located at a single point on the earth's surface (say at $a_1, a_2, \ldots$), while the rural population is "smeared" over the surface.

If Σ for this example is the Borel field, so that all singleton sets $\{a_i\} \in \Sigma$, then an atomic decomposition is given by $A \backslash E = \{a_1, a_2, \ldots\}$. That is, on this "urban set" population distribution is σ-atomic (each singleton set $\{a_i\}$ being an atom), and on the complementary "rural set" it is nonatomic.[24]

Having decomposed μ atomically, one is then in a position to take advantage of the special properties of each part. For nonatomic measures the following property is very useful.

Theorem: Given (A, Σ, μ), with μ a nonatomic measure, let $\mu(E)$ be finite. Then for any real number x such that $0 \leq x \leq \mu(E)$, a set $F \in \Sigma$ exists such that $\mu(F) = x$.

That is, if μ is finite and nonatomic, it takes on every single real value in the interval from 0 to $\mu(A)$. Next suppose μ is infinite but σ-finite. It follows easily from the definition that for any real number x, μ takes on a *real* value greater than x. Combining this observation with the theorem just stated, we conclude that an infinite, σ-finite, nonatomic measure takes on all positive real numbers as values.

INTEGRATION

We start with a measurable space (A, Σ). The integral will be a certain function that assigns an extended real number to every pair consisting of (i) a measure μ on (A, Σ) and (ii) a measurable function f with domain A, and with values in the nonnegative extended real numbers.[25] Our notation for the integral is

$$\int_A f \, d\mu, \quad \text{or} \quad \int_A f(x) \, \mu(dx) \tag{2.6.2}$$

We start with a certain special kind of function f.

Definition: Function f (with domain A) is *simple* iff it is measurable, real valued, nonnegative, and takes on only a finite number of values.

As an example, the constant function $f(x) = c$ (where $\infty > c \geq 0$) is sim-

---

24. Can one go further and decompose the rural population into a "2-dimensional" part (e.g., the farm population) and a "1-dimensional" part (e.g., population living along roads, rails, or rivers), and perhaps a residual? The answer is yes, with the aid of the *Lebesgue decomposition theorem,* which we shall not cover in this book. For this "dimensional decomposition" see Hahn and Rosenthal, *Set Functions,* pp. 106–9.

25. Later we will allow both μ and f to take on negative values. Unless otherwise noted, all functions from now on will have their range in the extended real numbers. (This, of course, does not mean they *must* take on infinite values, only that they *may* do so.)

ple. Another example, which merits a definition of its own, is the following.

Definition: The *indicator function* of set E (notation I_E) is given by $I_E(a) = 1$ if $a \in E$; $I_E(a) = 0$ if $a \notin E$.

The indicator function of any measurable set E is simple.

Now let f be simple, and let $\{x_1, \ldots, x_n\}$ be its range. Since f is measurable, each set $\{a \mid f(a) = x_i\}$ is measurable, and the collection of these, for $i = 1, \ldots, n$, constitutes a finite partition of A. We now define $\int_A f \, d\mu$ to equal

$$x_1 \mu \{a \mid f(a) = x_1\} + \cdots + x_n \mu \{a \mid f(a) = x_n\} \qquad (2.6.3)$$

In evaluating (3), recall the rules of arithmetic in the extended real number system. In particular, $x \cdot \infty = \infty$ if $x > 0$, and $0 \cdot \infty = 0$.

Examples: (i) For the constant function $f(x) = c$, (3) consists of just one term, $c \cdot \mu(A)$. (ii) For indicator functions, $\int_A I_E \, d\mu = 1 \cdot \mu(E) = \mu(E)$.

We now define the integral in general in terms of its value for simple functions. We use the notation $f \geq g$ for two functions on A to indicate that $f(a) \geq g(a)$ for all $a \in A$. Also, sup abbreviates supremum.

Definition: Given measure space (A, Σ, μ) and nonnegative measurable function f on A,

$$\int_A f \, d\mu = \sup \left\{ \int_A g \, d\mu \,\middle|\, g \leq f, g \text{ simple} \right\} \qquad (2.6.4)$$

That is, we consider the set of all simple functions bounded above by f; for each of these we form its integral, and the integral of f is defined as the supremum of the resulting set of extended real numbers.[26]

Note that (4) is not circular, since the integral of a simple function has already been defined by (3). Definition (4) is also consistent in the sense that if f itself is a simple function, then (4) gives the same answer as (3).

A useful extension of this definition is to integration over a measurable subset E of A. This is denoted $\int_E f \, d\mu$ and is simply the ordinary integral (4) when f and μ are both restricted to E. (For $E = \emptyset$, we set it equal to zero.) This may also be written as an integral over A. In fact, $\int_E f \, d\mu = \int_A I_E \cdot f \, d\mu$ for all $E \in \Sigma$. (The function being integrated on the right is the product of f and the indicator function of E, so that it coincides with f for points of E and is identically zero off E.)

---

26. There are a large number of seemingly different definitions of the integral in the literature. Most are either equivalent to (4) or minor variants.

Compare this with the ordinary Riemann integral. Let f be real valued, continuous, and nonnegative on the closed interval $\{x \mid a \le x \le b\}$ (a, b real numbers). Then

$$\int_a^b f(x)\, dx = \int_{\{x \mid a \le x \le b\}} f\, d\mu \qquad (2.6.5)$$

Here the left expression is the Riemann integral in its usual notation, μ on the right is Lebesgue measure, and (5) shows how to translate the Riemann integral into the form (2). Equation (5) is valid for any function f having a Riemann integral. (Actually, for some "pathological" functions one needs a richer σ-field—the "Lebesgue completion" of the Borel field—for (5) to be valid for any such f. This concept is unimportant for our purposes and we pass over it.)

The integral (2)–(4) constitutes a triple generalization of the Riemann integral. First, the class of functions f possessing an integral is broadened. Second, the integral is defined not only for Lebesgue measure but for measures in general. Third, the integral is defined for any abstract measurable space, not just the real line. In view of this enormous generality the following theorem is surprising because it shows that the general integral can be expressed in terms of the Riemann integral, in fact, as the Riemann integral of a monotone nonincreasing function.

Theorem: Let (A, Σ, μ) be a measure space and f a measurable nonnegative function on A. Then

$$\int_A f\, d\mu = \int_0^\infty \mu\{a \mid f(a) > t\}\, dt = \int_0^\infty \mu\{a \mid f(a) \ge t\}\, dt \qquad (2.6.6)$$

Here the middle and right expressions are improper Riemann integrals defined by the usual limiting processes. Since the integrands are monotonic, there is no problem of existence, though $+\infty$ is a possible value. Equation (6) is proved by comparing the Riemann sums approximating the middle and right integrals with the integrals of simple functions approximating the left expression. The middle or right-hand form in (6) will be referred to as the *Young integral*.[27]

To illustrate (6), take the constant function $f(x) = c$ ($c > 0$). Then $\{a \mid f(a) > t\} = A$ if $t < c$, and $= \phi$ if $t \ge c$; hence the middle integrand equals $\mu(A)$ up to $t = c$, and equals 0 beyond that point; the right-hand integrand is identical except at the single point $t = c$; hence both integrals equal $c \cdot \mu(A)$, which we have already verified to be the value of $\int_A c\, d\mu$.

---

27. After W. H. Young, 1905. The general integral (4) is essentially due to M. Fréchet, 1915.

We list some standard properties of the integral; f and g are assumed to be measurable nonnegative extended real-valued functions on A.

$$\int_A f \, d\mu \geq 0 \tag{2.6.7}$$

If $f > 0$ (i.e., $f(a) > 0$ for all a), and $\mu(A) > 0$, then

$$\int_A f \, d\mu > 0 \tag{2.6.8}$$

If f and μ are both bounded, then

$$\int_A f \, d\mu \text{ is finite} \tag{2.6.9}$$

If c is a positive number, then

$$c \int_A f \, d\mu = \int_A cf \, d\mu \tag{2.6.10}$$

$$\int_A f \, d\mu + \int_A g \, d\mu = \int_A (f + g) \, d\mu \tag{2.6.11}$$

INDEFINITE INTEGRALS

We have defined $\int_E f \, d\mu$, where E is a measurable set. Now for fixed f and μ (where these are defined as above), consider the function ν with domain Σ which is given by $\nu(E) = \int_E f \, d\mu$.

Definition: ν is known as the *indefinite integral* of f with respect to μ. We shall use the notation $\int f \, d\mu$ for the indefinite integral.

Theorem: $\int f \, d\mu$ is a measure. If f and μ are both bounded, then $\int f \, d\mu$ is bounded. If f is finite and μ is σ-finite, then $\int f \, d\mu$ is σ-finite.

The first statement may be proved by aid of the monotone convergence theorem, which we come to later. We prove the last two statements. If f and μ are both bounded, the boundedness of $\int f \, d\mu$ is immediate from (9). Let f be finite and μ σ-finite, and let $\{G_m\}$, $m = 1, 2, \ldots$, be a measurable partition of A such that $\mu(G_m)$ is finite for all m. Let $E_{mn} = \{a \mid a \in G_m \text{ and } n \leq f(a) < n + 1\}$, where $m = 1, 2, \ldots,$ and $n = 0, 1, 2, \ldots$. The class of sets E_{mn} is a countable measurable partition of A, and f and μ are both bounded on each piece. Hence by (9), $\int_{E_{mn}} f \, d\mu$ is finite for all m, n, so that $\int f \, d\mu$ is σ-finite, completing the proof.

One application of indefinite integrals is to the problem of change in measurement units, which we left hanging in Section 2.5. Let (A, Σ, μ) be a measure space representing some real-world data. Measurement units need not be homogeneous; they may be "acres" in one portion of A, "pounds" in another, "numbers of entities" in a third.[28] Suppose the measurement units are changed in some arbitrary manner. The same data will now be represented by a new measure ν in terms of the new measurement units. For example, if everything was previously measured in kilograms and we convert to grams, obviously μ is blown up by a factor of 1000:

$$\nu(E) = 1000 \, \mu(E) \qquad \text{for all } E \in \Sigma \qquad (2.6.12)$$

But what is the general relation between μ and ν?

The change in measurement units can be represented by a function f on A: $f(a)$ = number of new units equivalent to one old unit at point a; f is real valued and positive. The only other restriction we impose is that it be measurable. The relation between μ, ν, and f is then

$$\nu = \int f \, d\mu \qquad (2.6.13)$$

As an example, take the conversion above of kilograms into grams. In this case $f(a) = 1000$ for all $a \in A$. Checking the formula for the integral of a constant, we see that (13) does indeed reduce to (12) in this case.

Let μ be a measure, and f, g two nonnegative measurable functions on (A, Σ). Since the indefinite integral $\int g \, d\mu$ is a measure, one can integrate f with respect to it.

Theorem: Let (A, Σ, μ), f, and g be as stated. Then

$$\int f \, d\left[\int g \, d\mu\right] = \int fg \, d\mu \qquad (2.6.14)$$

That is, the indefinite integral of f with respect to the measure $\int g \, d\mu$ is the same as the indefinite integral of fg with respect to μ.

---

28. For readers troubled by this cavalier addition of heterogeneous units—often said to be "invalid"—it should be mentioned that it is clearer to think of the measurement unit as being part of the definition of the concept, the measurement number itself being "pure." Thus "the length of this bar in meters: 3.7," rather than "the length of this bar: 3.7 meters." Cf. R. Carnap, *Introduction to Symbolic Logic and Its Applications*, W. H. Meyer and J. Wilkinson, trans. (Dover, New York, 1958), p. 169. In any case, the treatment here shows how to work with heterogeneous units with safety and convenience.

As an illustration take the measurement-unit transformation discussed above. Suppose one changes measurement units according to function g and then changes them again according to function f. The composite result is the left expression in (14), and the theorem states that this yields the same transformation as a single change in units represented by $h(a) = f(a) \cdot g(a)$.

In particular, if we simply invert the previous change, then $f = 1/g$, and by (14),

$$\int \frac{1}{g} \, d\left[\int g \, d\mu \right] = \int 1 \, d\mu = \mu$$

as it should.

<small>DENSITIES</small>

We pull together two ideas in this section. First, we have mentioned that the intuitive concept of "continuous distribution" has two quite distinct explications. One is the concept of "nonatomic" measure, which we have discussed. The second, "absolute continuity," will be taken up here.

Second, we mentioned at the beginning of this chapter that certain kinds of data (such as prices or population densities), which could not themselves be represented as measures, could be derived from measures in a certain way. The same circle of ideas serves to accomplish this.

Definition: Let μ, ν be two measures on the same measurable space (A, Σ). ν is *absolutely continuous* with respect to μ iff, whenever $\mu(E) = 0$, then $\nu(E) = 0$. The notation for this state of affairs is $\nu \ll \mu$.

It follows at once that absolute continuity is transitive: if λ, μ, ν are three measures on (A, Σ) such that $\nu \ll \mu$ and $\mu \ll \lambda$, then $\nu \ll \lambda$. Also, of course, $\mu \ll \mu$.

Theorem: Let ν be the indefinite integral $\int f \, d\mu$. Then $\nu \ll \mu$.

Proof: Suppose $\mu(E) = 0$ for $E \in \Sigma$ so that μ restricted to E is identically zero. If g is a simple function, it follows from (3) that $\int_E g \, d\mu = 0$. Then from (4) this must be true for any nonnegative measurable function g, in particular for f. Hence $\nu(E) = 0$. ∎

The basic result concerning absolute continuity is that under slight restriction the converse of this statement is true.

Theorem: Radon-Nikodym Theorem. Let μ, ν be two measures over (A, Σ) such that μ is σ-finite and $\nu \ll \mu$. Then a nonnegative measurable function f exists such that

$$\nu = \int f \, d\mu \qquad (2.6.15)$$

(f is known as the *density*, or *Radon-Nikodym derivative*, of ν with respect to μ and is sometimes written $d\nu/d\mu$). To make a statement concerning the extent to which the density f is uniquely determined by ν and μ, we need the following concepts.

Definition: Let (A, Σ, μ) be a measure space; a property P is said to hold μ-*almost everywhere*, or for μ-*almost all points*, iff a set $E \in \Sigma$ exists such that $\mu(E) = 0$, and P is true for all points of $A \backslash E$.

Another way of expressing the same thing is in terms of null sets. A set F is μ-*null* iff there is a set $E \in \Sigma$ such that $F \subseteq E$ and $\mu(E) = 0$. Then P holds μ-almost everywhere iff the set $\{a \mid P$ is not true for $a\}$ is μ-null. In all this, if μ is understood it may be omitted; thus one says simply "almost everywhere," etc.

Definition: Two functions f, g are μ-*equivalent* (or μ-*almost identical*) iff the set $\{a \mid f(a) \neq g(a)\}$ is a μ-null set.

Theorem: Let f, g be two nonnegative measurable functions and μ a σ-finite measure on (A, Σ). Then for the indefinite integrals we have $\int f \, d\mu = \int g \, d\mu$ iff f and g are μ-equivalent.

This answers the uniqueness question concerning the Radon-Nikodym derivative. In that theorem only μ is required to be σ-finite. If ν is also σ-finite, we can state the stronger conclusion that there is a *finite* density f satisfying (15).

We give some possible real-world examples and compare all this with the intuitive concept of "density." Take 3-dimensional space with the Borel field. Just as the ordinary concept of length extends to Lebesgue measure on the real line, the ordinary concept of volume extends to a measure known as 3-dimensional Lebesgue measure in 3-space.[29] For simplicity we continue to refer to this extension as "volume" and denote it by μ. Let ν be the mass distribution of some resource type over Space. The *average density* of this resource in a region E of positive volume is given by $\nu(E)/\mu(E)$. (Average density is not defined if $\mu(E) = 0$.) Average density is thus a set function, whose domain is a certain subclass

---

29. We discuss higher dimensional Lebesgue measure below.

of the Borel field Σ. This is rather unwieldy, and one would like to go from average density to density at a point. A rough analogy is the process of going from average slopes to the more useful derivatives.

The Radon-Nikodym theorem pins down these vague notions. We assume that ν is absolutely continuous with respect to volume μ; i.e., for any region E if E has no volume, E has no resource content. Since μ is σ-finite, it follows that a point function f exists satisfying (15), and this is exactly the property one would want a point density to have.

Now the foregoing analysis did not depend in any way on the particular natures of the two measures involved. This raises the possibility of thinking of a great many other types of data as being "densities" derived in this way from two measures. We give several examples:

1. Let μ be population distribution over the surface of the earth, by place of residence (measurement unit = number of people). Let ν be the distribution of total income, again attributed by residence. We see that $\nu \ll \mu$, since no income accrues to unpopulated regions. The "density" $d\nu/d\mu$ in this case is simply per capita income.
2. Let μ be the distribution of economic commodities over S, measured in *mass* units and perhaps quite heterogeneous. Let ν be the same distribution measured in *value* or *wealth* terms (unit = dollars). Again $\nu \ll \mu$. The "density" $d\nu/d\mu$ in this case may be interpreted as *prices*. In somewhat more detail, the universe set is a subset of $R \times S$; the density $p(r, s)$ is then the price of resource type r at location s. The units in which $p(r, s)$ is measured will be dollars per acre, or gram, or litre, corresponding to whatever units μ was measured in at point (r, s).
3. Consider the concept of the "quality" of resources: gold has higher quality than brass, etc. One explication of this somewhat elusive concept is to define quality as the ratio of value to weight. Thus if we let μ be the distribution of resources by weight and ν their distribution by value, "quality" comes out as the density $d\nu/d\mu$.
4. This and the next example show that *index numbers* may be construed as densities. We are given two times t_0, t_1, with $t_0 < t_1$. The universe set A is some appropriate subset of R or $R \times S$. p_0, p_1 exist at times t_0, t_1 respectively. Formally, p_0 and p_1 are measurable positive functions on the universe set. Prices may be given directly or may themselves be derived as densities, as under example 2. Also, there are two quantity measures μ_0, μ_1 referred to the respective times. These may represent stocks, production, consumption, exports. We suppose each of these measures is absolutely continuous with respect to the other; i.e., $\mu_0(E) = 0$ iff $\mu_1(E) = 0$ for all measurable sets $E \subseteq A$.

   The *Laspeyres price index* for measurable set $E \subseteq A$ is now defined as

$$\int_E p_1 \, d\mu_0 \Big/ \int_E p_0 \, d\mu_0 \qquad (2.6.16)$$

The *Paasche price index* substitutes μ_1 for μ_0 in (16). As E varies, the numerator and denominator of (16) define indefinite integrals, and the price index comes out as an average density of these. The point density, or Radon-Nikodym derivative of $\int p_1 \, d\mu_0$ with respect to $\int p_0 \, d\mu_0$, is simply $f(a) = p_1(a)/p_0(a)$, since by (14),[30]

$$\int f \, d\Big[\int p_0 \, d\mu_0 \Big] = \int f p_0 \, d\mu_0 = \int p_1 \, d\mu_0$$

5. The *Laspeyres quantity index* for measurable E is defined as

$$\int_E p_0 \, d\mu_1 \Big/ \int_E p_0 \, d\mu_0 \qquad (2.6.17)$$

The *Paasche quantity index* substitutes p_1 for p_0 in (17). Again this is the average density derived from two indefinite integrals. The point density of $\int p_0 \, d\mu_1$ with respect to $\int p_0 \, d\mu_0$ is simply $d\mu_1/d\mu_0$, since for $f = d\mu_1/d\mu_0$ we obtain, by (14),

$$\int p_0 \, d\mu_1 = \int p_0 \, d\Big[\int f \, d\mu_0 \Big] = \int p_0 f \, d\mu_0 = \int f \, d\Big[\int p_0 \, d\mu_0 \Big]$$

These examples should illustrate the variety of data that can be brought under the rubric "density." Examination of statistical compilations would show that, of the data that cannot be represented directly as measures, the great bulk can be represented as densities with respect to some pair of measures.[31]

In our examples we have presented the pair of measures first and derived the density from them. However, in some cases the density is more readily observable than one of the measures (in which case the measure may be constructed as an indefinite integral).

Consider the standard exercise in capital theory of converting from current to discounted dollars. Let μ be a measure with universe set T, having the interpretation $\mu(E)$ = value in current dollars of that portion of an income stream arriving in time period E. (The use of measure language enables us to cover the cases of lump-sum accruals, continuous accruals, and mixtures of the two all in a single notation.) Assuming for simplicity a constant discount rate i and discounting to moment t_o, the

---

30. By the uniqueness theorem, any function identical to p_1/p_0 except for a set of μ_0-measure zero is also a Radon-Nikodym derivative. In this case, as in many others, the derivative p_1/p_0 is more "natural" than the other functions μ_0-equivalent to it.

31. Densities correspond roughly to Campbell's "derived magnitudes." See Campbell, *Foundations of Science*, Ch. 10, and note 7 above.

income stream expressed in discounted dollars is simply the indefinite integral

$$\int e^{-i(t-t_o)} \mu(dt) \tag{2.6.18}$$

Here the density $f(t) = e^{-i(t-t_o)}$ is more or less directly observable, and the discounted income stream is constructed from it.[32]

Again consider prices. One can observe list prices if these exist. Or one can take the ratio of money passing in one direction to goods passing in the opposite direction. The first gives a direct observation of a density (perhaps misleading if there are trade discounts, etc.) The second derives price as an average density of two measures.

Before dropping this topic, consider the concept of "uniformity." One speaks, especially in spatial economics, of uniform population distribution, uniform resources, uniform planes. A moment's reflection indicates that these terms are expressing the proportionality of the measure in question to some other implicit measure, usually surface area. Thus if μ is areal measure and ν is population distribution, the assertion is that there is a number c such that $\nu(E) = c\mu(E)$ for all measurable sets E ($0 < c < \infty$). This in turn may be abbreviated $\nu = c\mu$.

Definition: Let μ and ν be two measures over the space (A, Σ); ν *is uniform with respect to* μ iff there is a positive real number c such that $\nu = c\mu$.

This is an equivalence relation among measures and implies that each is absolutely continuous with respect to the other. An equivalent way of stating the relation is that the density $d\nu/d\mu$ is equal to a positive real constant (μ-almost everywhere). We recognize, of course, that any such relation between disparate measures is at best an approximation. In general, it should not be taken literally at the microscopic level: A literally uniform distribution of land and water would just yield mud everywhere.

INDUCED INTEGRALS

Let (A, Σ, μ), (B, Σ', μ') be two measure spaces. Let $f{:}A \to B$ be measurable such that for all $E \in \Sigma'$,

$$\mu'(E) = \mu\{a \mid f(a) \in E\} \tag{2.6.19}$$

---

32. In rate-of-return calculations the density is unknown, but its derivation is not comparable to our procedure. Incidentally, (18) illustrates the alternative notation for the integral given in (2): When several letters are floating around, it clarifies which measure and integrand we are referring to.

i.e., μ' is induced by f from μ. Finally, let g be a measurable, nonnegative extended real-valued function on B.

Theorem: Induced Integrals Theorem. Under the conditions stated,

$$\int_B g \, d\mu' = \int_A (g \circ f) \, d\mu \qquad (2.6.20)$$

Proof: A quick proof may be obtained by using the Young integral (6). In fact, from (19), $\mu'\{b \mid g(b) > t\} = \mu\{a \mid (g \circ f)(a) > t\}$ so that

$$\int_B g \, d\mu' = \int_0^\infty \mu'\{b \mid g(b) > t\} \, dt = \int_0^\infty \mu\{a \mid (g \circ f)(a) > t\} \, dt$$

$$= \int_A (g \circ f) \, d\mu \quad \blacksquare$$

Here $g \circ f$ is the composition of f and g. If g can take on negative values, neither integral in (20) has yet been defined. However, to anticipate, the theorem is still true in this case, in the sense that if either integral in (20) is well defined, then so is the other, and they are equal.

The elementary rules concerning substitution of variables in integration may be derived from (20).

There is a more general way of looking at relation (20). If we consider the *indefinite* integrals $\int (g \circ f) \, d\mu$ and $\int g \, d\mu'$, then the latter is the measure induced on (B, Σ') by f from the former on (A, Σ).

CONVERGENCE THEOREMS

The following theorems are among the most useful in measure theory and will be used repeatedly. It is convenient to state them for integrands that are unrestricted in sign, even though we have so far only defined integration for nonnegative integrands. For the more general definition, see the subsection on signed measures below.

We distinguish formally between the *sequence* of extended real numbers $(x_1, x_2, \ldots)$, which is a family of numbers indexed by the integers $1, 2, \ldots$, and the *set* $\{x_1, x_2, \ldots\}$, which is the range of this family. We have defined the concepts of the supremum and infimum of a set of numbers. The sup and inf of the sequence $(x_1, x_2, \ldots)$ are defined simply as the sup and inf respectively of the set $\{x_1, x_2, \ldots\}$. Two slightly more complicated operations on sequences are needed here: *lim sup* and *lim inf* (limit superior and inferior).

Let $(x_1, x_2, \ldots)$ be a sequence of extended real numbers. Let $y_n =$

$\sup\{x_n, x_{n+1}, \ldots\}$, $n = 1, 2, \ldots$; i.e., y_n is the supremum of the numbers left in the sequence after deleting the first $n - 1$ in order.

Definition: *Lim sup* of the sequence $(x_1, x_2, \ldots)$ is defined as the inf of the set $\{y_1, y_2, \ldots\}$.

A similar construction reverses the roles of inf and sup.

Definition: Let $z_n = \inf\{x_n, x_{n+1}, \ldots\}$, $n = 1, 2, \ldots$. Then $\sup\{z_1, z_2, \ldots\}$ is known as the *lim inf* of the sequence $(x_1, x_2, \ldots)$.

Definition: Sequence $(x_1, x_2, \ldots)$ *converges* to x_o iff the lim sup and lim inf of the sequence both equal x_o. The number x_o is the *limit* of the sequence and we write $x_n \rightarrow x_o$.

Some examples follow:

1. Let $(x_1, x_2, \ldots)$ be a sequence of real numbers. This converges to the real number x_o in the sense of this definition iff it converges to x_o in the ordinary sense of the term "converge."
2. The sequence $(1, 0, 1, 0, 1, \ldots)$ has a lim sup of 1 and a lim inf of 0. (Proof: $y_n = 1$ and $z_n = 0$ for all n.)
3. Let $(x_1, x_2, \ldots)$ be a nondecreasing sequence, and let $x_o = \sup\{x_1, x_2, \ldots\}$; then $(x_1, x_2, \ldots)$ converges to x_o. (Proof: $y_n = x_o$ for all n, hence $\inf\{y_1, y_2, \ldots\} = x_o$; $z_n = x_n$ for all n, hence $\sup\{z_1, z_2, \ldots\} = x_o$.) Note, e.g., that the sequence $(1, 2, 3, \ldots)$ converges to $+\infty$.

Now let $(f_1, f_2, \ldots)$ be a sequence of extended real-valued functions, with a common domain A.

Definition: *Lim inf* $(f_1, f_2, \ldots)$ is the function with domain A whose value at $a \in A$ equals lim $\inf(f_1(a), f_2(a), \ldots)$. *Lim sup* $(f_1, f_2, \ldots)$ is defined analogously. If these two values are the same for all $a \in A$, the common function f thus determined is called the *limit* of the sequence $(f_1, f_2, \ldots)$, and we write $f_n \rightarrow f$.

One special case in which the limit exists is when the sequence is nondecreasing; i.e., $f_n(a) \leq f_{n+1}(a)$, $n = 1, 2, \ldots$, and all $a \in A$. This follows from example 3, which also shows that for the limiting function f, $f(a)$ is the sup of $\{f_n(a)\}$, $n = 1, 2, \ldots$, for all $a \in A$.

Theorem: Monotone Convergence Theorem. Let (A, Σ, μ) be a measure space; let (f_n), $n = 1, 2, \ldots$, be a nondecreasing sequence of measurable functions on A, with limit f. If

$$\int_A f_1 \, d\mu > -\infty \tag{2.6.21}$$

then

$$\int_A f_n \, d\mu \rightarrow \int_A f \, d\mu \tag{2.6.22}$$

Here we may note that the sup of a sequence of measurable functions is measurable. Also, condition (21), together with nondecreasingness, guarantees that all the integrals appearing in (22) are well defined. It frequently happens that all the f_n's are nonnegative, in which case (21) is automatically fulfilled. Another version of the monotone convergence theorem uses infinite *series* rather than sequences.

Theorem: Let (A, Σ, μ) be a measure space; let $(f_n), n = 1, 2, \ldots$, be a sequence of nonnegative measurable functions. Then

$$\int_A (f_1 + f_2 + \cdots) \, d\mu = \int_A f_1 \, d\mu + \int_A f_2 \, d\mu + \cdots \tag{2.6.23}$$

Here on the right we have an ordinary infinite series whose sum is defined as usual as the limit of the partial sums. On the left the integrand is expressed as an infinite series of functions. This is to be understood pointwise: the value at point $a \in A$ is $f_1(a) + f_2(a) + \cdots$. Convergence is assured on both sides of (23) by nonnegativity. Equation (23) follows at once from the application of the monotone convergence theorem to the partial sums $f_1 + \cdots + f_n, \; n = 1, 2, \ldots$.

A closely related result involves an infinite series of measures. Let μ_1, $\mu_2, \ldots$ all be measures on (A, Σ). The sum of the series $\mu_1 + \mu_2 + \cdots$ is defined as the set function μ whose value at $E \in \Sigma$ is $\mu_1(E) + \mu_2(E) + \cdots$. One easily verifies that μ is a measure. We then have the following theorem.

Theorem: Let f be a nonnegative measurable function in (A, Σ), and let $\mu_1 + \mu_2 + \cdots$ be a series of measures with sum μ. Then

$$\int_A f \, d\mu = \int_A f \, d\mu_1 + \int_A f \, d\mu_2 + \cdots$$

Proof: Apply monotone convergence to the Young integral:

$$\int_A f \, d\mu = \int_0^\infty \mu\{a \mid f(a) > t\} \, dt = \int_0^\infty (\mu_1 + \mu_2 + \cdots)\{a \mid f(a) > t\} \, dt$$

$$= \int_0^\infty \mu_1\{a \mid f(a) > t\} \, dt + \cdots = \int_A f \, d\mu_1 + \cdots \quad \blacksquare$$

These two preceding theorems are used in the subsection on product measures.

Theorem: Fatou's Lemma. Let (A, Σ, μ) be a measure space; let (f_n), $n = 1, 2, \ldots$, be a sequence of measurable functions on A such that $f_n \geq g$ for all n, g being another measurable function on A. If

$$\int_A g \, d\mu > -\infty \tag{2.6.24}$$

then

$$\int_A (\liminf f_n) \, d\mu \leq \liminf \int_A f_n \, d\mu \tag{2.6.25}$$

In (25), the lim inf on the left defines a function, which is to be integrated; the lim inf on the right applies to the ordinary sequence of extended real numbers whose nth term is $\int_A f_n \, d\mu$. Note that the lim inf of any sequence of measurable functions is measurable. Also condition (24) guarantees that all integrals appearing in (25) are well defined. If all f_n's are nonnegative, as is common, we may take $g = 0$, and (24) is automatically fulfilled.

An example will clarify the meaning of Fatou's lemma. Choose two sets E_1, $E_2 \in \Sigma$. Let f_n be the indicator function of E_1 if n is odd, and of E_2 if n is even. Lim inf f_n then equals $I_{E_1 \cap E_2}$; the sequence on the right of (25) is $\mu(E_1)$, $\mu(E_2)$, $\mu(E_1)$, $\ldots$; finally, (25) states that $\mu(E_1 \cap E_2) \leq$ minimum of $(\mu(E_1), \mu(E_2))$.

Theorem: Dominated Convergence Theorem. Let (A, Σ, μ) be a measure space; let (f_n), $n = 1, 2, \ldots$, be a sequence of measurable functions on A, with limit f; let $|f_n| \leq g$ for all n, g being another measurable function on A such that

$$\int_A g \, d\mu < \infty \tag{2.6.26}$$

Then

$$\int_A f_n \, d\mu \rightarrow \int_A f \, d\mu \tag{2.6.27}$$

The condition $|f_n| \leq g$, which states that the absolute value of f_n is dominated by g, may also be written as $-g(a) \leq f_n(a) \leq g(a)$ for all $a \in A$, and all $n = 1, 2, \ldots$. As above, the condition (26) guarantees that all integrals appearing in (27) are well defined. But even if all the integrals in (22), (25), or (27) are well defined, we still cannot drop the conditions (21), (24), or (26), respectively, with impunity; this may be shown by counterexamples.

EXTENSION OF SET FUNCTIONS

A σ-field is usually specified by mentioning a class of sets that generates it; e.g., the Borel field on the line is generated by the class of intervals. Similarly, a measure is often specified by stating its values just on some of the sets of its domain; e.g., Lebesgue measure is the one that assigns to each interval its ordinary length.

Let (A, Σ) be a measurable space, let $\mathfrak{R}$ be a subclass of Σ, and let $\mu : \mathfrak{R} \to$ extended reals be a set function defined on this subclass. The question arises, Does there exist a measure $\nu : \Sigma \to$ extended reals that coincides with μ on the latter's domain: $\nu(E) = \mu(E)$ for all $E \in \mathfrak{R}$? In other words, can μ be *extended* to a measure on Σ? Furthermore, are there several such extensions or at most one? We now specify certain conditions on μ and $\mathfrak{R}$ that enable us to answer such questions.

Definition: Set function $\mu : \mathfrak{R} \to$ nonnegative extended reals is *countably additive* iff, for any countable packing $E_1, E_2, \ldots$ of $\mathfrak{R}$-sets whose union E is also an $\mathfrak{R}$-set, we have $\mu(E) = \mu(E_1) + \mu(E_2) + \cdots$.

This is a slight generalization of the concept of countable additivity on a σ-field: the condition that $E \in \mathfrak{R}$ must be stated explicitly, since $\mathfrak{R}$ is not necessarily closed under countable unions.

Definition: Collection $\mathfrak{R}$ is a *semiring* iff

(i) $\emptyset \in \mathfrak{R}$;
(ii) if $E, F \in \mathfrak{R}$, then $E \cap F \in \mathfrak{R}$;
(iii) if $E, F \in \mathfrak{R}$, and $E \subseteq F$, then there is a finite sequence $G_0, \ldots, G_n$ of $\mathfrak{R}$-sets such that $G_0 = E$, $G_n = F$, $G_{i-1} \subseteq G_i$, and $G_i \backslash G_{i-1} \in \mathfrak{R}$, $i = 1, \ldots, n$.[33]

For example, the collection of all intervals on the real line (together with $\emptyset$)

33. Halmos, *Measure Theory*, p. 22. The proof of the following theorem may be found in Halmos, Ch. 3, or in J. von Neumann, *Functional Operators*, Vol. 1, *Measures and Integrals*, Annals of Mathematics Studies #21 (Princeton Univ. Press, Princeton, N.J., 1950).

is a semiring. The collection of measurable rectangles in a product space (and ϕ) is a semiring.

Theorem: Let (A, Σ) be a measurable space, let $\Re$ be a semiring that generates Σ, and let nonnegative $\mu:\Re \to$ extended reals be countably additive, with $\mu(\phi) = 0$. Then there is a measure ν with domain Σ which extends μ.

If, in addition, a countable collection $\mathcal{G} \subseteq \Re$ covers A such that $\mu(G) < \infty$ for all $G \in \mathcal{G}$, then there is exactly one such extension.

The premise that $\Re$ is a semiring can be weakened considerably without invalidating these conclusions.

Definition: Collection $\Re$ is a *weak semi-σ-ring* iff

(i) $\phi \in \Re$;
(ii) for all E, $F \in \Re$, $E \backslash F$ can be partitioned into a countable number of $\Re$-sets.

Collection $\Re$ is a *semi-σ-ring* iff it is a weak semi-σ-ring, and for all E, $F \in \Re$, $E \cap F$ can also be partitioned into a countable number of $\Re$-sets.

Any semiring is a semi-σ-ring. For let $E, F \in \Re$; then $E \cap F \in \Re$, so the collection consisting of $E \cap F$ alone is a countable partition of $E \cap F$ into $\Re$-sets; furthermore, the collection $\{G_1 \backslash G_0,\ G_2 \backslash G_1, \ldots, G_n \backslash G_{n-1}\}$, where $G_0 = E \cap F$, $G_n = E$, $G_{i-1} \subseteq G_i, i = 1, \ldots, n$, is a finite, hence countable, partition of $E \backslash F$ into $\Re$-sets. This shows that the following theorem is a generalization of the preceding one.

Theorem: Let (A, Σ) be a measurable space, let $\Re$ be a *semi-σ-ring* that generates Σ, and let nonnegative $\mu:\Re \to$ extended reals be countably additive with $\mu(\phi) = 0$. Then there is a measure ν with domain Σ that extends μ.

Instead, let $\Re$ be a *weak semi-σ-ring* that generates Σ; let ν_1, ν_2 be two measures on Σ that coincide on $\Re$-sets; and let there be a countable collection $\mathcal{G} \subseteq \Re$ that covers A such that $\nu_1(G) = \nu_2(G) < \infty$ for all $G \in \mathcal{G}$. Then $\nu_1 = \nu_2$ throughout Σ.

The proof of this theorem is very long and would take us too far afield to set down here.

ABCONT AND PRODUCT MEASURES

Let (A, Σ, μ) and (B, Σ', ν) be two measure spaces. We have defined the concept of the product $(A \times B, \Sigma \times \Sigma')$ of the two measurable spaces

(A, Σ) and (B, Σ'). We now define the notion of a measure λ on the product space which is in a sense the product of the measures μ and ν.

Definition: Measure λ on $(A \times B, \Sigma \times \Sigma')$ is a *generalized product of μ and ν* iff

$$\lambda(E \times F) = \mu(E) \cdot \nu(F) \qquad (2.6.28)$$

for all $E \in \Sigma$ and $F \in \Sigma'$.

Intuitively, (28) says that the λ-measure of any "rectangle" $E \times F$ is the ordinary product of the measures of its "sides." (In evaluating the right-hand side, remember that $0 \cdot \infty = 0$.) As an example, let μ and ν both be Lebesgue measure (= length) on the real line, and let λ be (2-dimensional) Lebesgue measure (= area) on the plane. Then (28) is satisfied: The area of a rectangle is the product of its sides.

Does a generalized product exist for any pair of measures? Yes it does. This may be proved via the monotone convergence theorem coupled with the extension theory just discussed. Is it unique? No, not always; but if μ and ν are both σ-finite, uniqueness is guaranteed. Thus 2-dimensional Lebesgue measure is the only possible product of Lebesgue measure with itself.

We are interested here only in product measures that can be expressed as integrals. Choose $G \in (\Sigma \times \Sigma')$ and consider the following expression:

$$\lambda(G) = \int_A \nu\{b \mid (a,b) \in G\} \, \mu(da) \qquad (2.6.29)$$

For a given point $a_o \in A$, the set $\{b \mid (a_o, b) \in G\}$ is the right-hand a_o-section of G. We know this is a measurable subset of B, hence it has a ν-value. We associate this value with point a_o. We now have a well-defined function with domain A, taking values in the nonnegative extended reals; this function is precisely the integrand in (29). The integral with respect to μ may now be taken, *provided* the integrand just defined is measurable with respect to (A, Σ). Suppose this to be the case for each $G \in (\Sigma \times \Sigma')$. Then (29) defines a set function λ.

It can be shown that λ is a measure (use monotone convergence in its additive form). Furthermore, letting $G = E \times F$, we obtain $\lambda(E \times F) = \int_E \nu(F) \, d\mu = \mu(E) \cdot \nu(F)$, so that λ is a generalized product of μ and ν.

Thus the expression (29) yields a generalized product, provided the sticky issue of measurability is resolved. If ν is finite, it is known that the integrand in (29) is measurable for all $G \in (\Sigma \times \Sigma')$. To generalize this result, we now introduce a new class of measures, the *abcont* measures.

Theorem: Let (A, Σ, μ) be a measure space. If μ has *any* of the following four properties, then it has *all* of them:

(i) there is a series of *finite* measures ν_n, $n = 1, 2, \ldots$, on (A, Σ) such that

$$\mu = \nu_1 + \nu_2 + \cdots \tag{2.6.30}$$

(ii) there is a measure space (B, Σ', ν), ν *σ-finite*, and a measurable function $f : B \to A$ such that μ is the measure induced by f from ν;

(iii) there is a *finite* measure ν on (A, Σ) with respect to which μ is absolutely continuous $(\mu \ll \nu)$;

(iv) there is a *finite* measure ν on (A, Σ), and a measurable function, $f : A \to$ nonnegative extended reals such that $\mu = \int f \, d\nu$.

Proof: (i) *implies* (ii): Let $N = \{1, 2, \ldots\}$, let Σ'' be the class of all subsets of N, let $(B, \Sigma') = (A \times N, \Sigma \times \Sigma'')$, let $f : B \to A$ be the projection, $f(a, n) = a$, and define ν as follows. For any set of the form $E \times \{n\}$, where $E \in \Sigma$, $n = 1, 2, \ldots$, let $\nu(E \times \{n\}) = \nu_n(E)$. Any set $G \in \Sigma'$ is a countable disjoint union of sets of this form, so measure ν is fully determined on Σ'; ν is σ-finite, since $\nu(A \times \{n\}) = \nu_n(A) < \infty$, $n = 1, 2, \ldots$, and the sets $A \times \{n\}$ partition B. Also, for any $E \in \Sigma$,

$$\nu\{b \mid f(b) \in E\} = \nu(E \times N) = \nu(E \times \{1\}) + \nu(E \times \{2\}) + \cdots$$

$$= \nu_1(E) + \nu_2(E) + \cdots = \mu(E)$$

Thus μ is indeed the measure induced by f from ν.

(ii) *implies* (iii): If $\mu = 0$, this is trivial; hence we may assume $\mu \neq 0$, so that $\nu \neq 0$. Let $\{B_1, B_2, \ldots\}$ be a countable measurable partition of B such that $0 < \nu(B_n) < \infty$ for all n, and define the set function λ on (A, Σ) as follows. For $E \in \Sigma$, $\lambda(E)$ is the summation of

$$2^{-n} \nu[B_n \cap \{b \mid f(b) \in E\}] / \nu(B_n) \tag{2.6.31}$$

over $n = 1, 2, \ldots$. Each of these terms defines a measure on (A, Σ), hence λ itself is such a measure. Furthermore, $\lambda(A) = 2^{-1} + 2^{-2} + \cdots \leq 1$, so λ is finite. It remains to show that $\mu \ll \lambda$. Suppose $\lambda(E) = 0$; then each of the terms (31) vanishes so that $\nu[B_n \cap \{b \mid f(b) \in E\}] = 0$, $n = 1, 2, \ldots$. Adding over n, we obtain $\nu\{b \mid f(b) \in E\} = 0$; since μ is induced by f from ν, it follows that $\mu(E) = 0$; hence $\mu \ll \lambda$.

(iii) *implies* (iv): Apply the Radon-Nikodym theorem.

(iv) *implies* (i): For each $n = 1, 2, \ldots$ define $f_n : A \to$ reals by

$$f_n(a) = 0 \qquad\qquad \text{if } f(a) \le n - 1$$
$$f_n(a) = f(a) - (n - 1) \qquad \text{if } n - 1 < f(a) < n$$
$$f_n(a) = 1 \qquad\qquad \text{if } f(a) \ge n$$

Each f_n is nonnegative and measurable. Define ν_n by $\nu_n = \int f_n \, d\nu$. Since ν and f_n are bounded, each ν_n is bounded. Finally,

$$\mu = \int f \, d\nu = \int [f_1 + f_2 + \cdots] \, d\nu$$
$$= \int f_1 \, d\nu + \int f_2 \, d\nu + \cdots = \nu_1 + \nu_2 + \cdots$$

establishing (30). (The third equality follows from monotone convergence.)

We now have a closed circle of implications, hence these four properties are logically equivalent. ∎

Definition: Measure μ is *abcont* iff it satisfies any (hence all) of the foregoing properties.

Abcont is an acronym for "absolutely continuous" and is suggested by property (iii). But keep in mind that absolute continuity is a relation between *two* measures, while abcontness is a property of a *single* measure.

Any σ-finite measure is abcont. This can be seen from property (ii); for if μ is σ-finite, we may take $(B, \Sigma', \nu) = (A, \Sigma, \mu)$ and let f be the identity mapping. Also, in property (iv) if we impose the additional condition that f be finite, we have a characterization of σ-finite measures. This also yields a decomposition for abcont measures. When restricted to the set $\{a \mid f(a) < \infty\}$, μ is σ-finite; and when restricted to the set $\{a \mid f(a) = \infty\}$, μ takes on just two possible values, 0 and ∞.

Abcont measures exist that are not σ-finite. In fact, to produce such a measure, take any finite measure $\nu \ne 0$ and set $\mu(E) = \infty$ whenever $\nu(E) > 0$ and set $\mu(E) = 0$ whenever $\nu(E) = 0$. It can be seen that μ is a measure; that it is abcont follows from the observation that $\mu = \nu + \nu + \nu + \cdots$ (property (i)), that $\mu \ll \nu$ (property (iii)), or that $\mu = \int (\infty) \, d\nu$ (property (iv)).

Abcont measures are of much less importance to us than the narrower class of σ-finite measures. There are two reasons for introducing them. First, even if one is interested only in σ-finite measures, abcont non-σ-finite measures may appear as the result of perfectly straightforward operations, e.g., induction as in property (ii). For example, one or both marginals of a σ-finite measure ν on a product space $A \times B$ may have this property. (Exercise: Verify that the marginals of 2-dimensional Lebesgue measure on the plane are abcont non-σ-finite.)

Second, in many cases abcontness is a more natural assumption than σ-finiteness in that it yields stronger and more transparent results, with clearer proofs. This is especially the case in this section on product measures and happens from time to time throughout the book.

Each of the four characterizations of abcont measures yields a "closure" theorem. We gather these results below.

Theorem:

(i)   The sum of a series of abcont measures on (A, Σ) is abcont;

(ii) let (B, Σ', ν) and (A, Σ, μ) be measure spaces, and let measurable $f : B \rightarrow A$ induce μ from ν; if ν is abcont, μ is abcont;

(iii) if $\mu \ll \nu$ and ν is abcont, then μ is abcont;

(iv) let measure ν on (A, Σ) be abcont, and let $f : A \rightarrow$ extended reals be nonnegative measurable; then $\mu = \int f \, d\nu$ is abcont.

Proof:

(i)   Let ν_m be abcont, $m = 1, 2, \ldots$, so that $\nu_m = \lambda_{m1} + \lambda_{m2} + \cdots$ for some finite measures $\lambda_{mn}, m, n = 1, 2, \ldots$. The sum of the ν_m's is then the sum of the double series λ_{mn}. Since the λ's are countable and nonnegative, they may be rearranged in a single series. Hence summation ν_m is abcont by property (i).

(ii) Since ν is abcont, there is a measure space (C, Σ'', λ), λ σ-finite, and a measurable function $g : C \rightarrow B$ inducing ν from λ. But then $(f \circ g) : C \rightarrow A$ induces μ from λ, so that μ is abcont by property (ii).

(iii) Since ν is abcont, there is a finite measure λ for which $\nu \ll \lambda$. But then $\mu \ll \lambda$, so μ is abcont by property (iii).

(iv) Since ν is abcont, there is a finite measure λ and a nonnegative measurable $g : A \rightarrow$ extended reals such that $\nu = \int g \, d\lambda$. But then $\mu = \int f \, d[\int g \, d\lambda] = \int (fg) \, d\lambda$, so μ is abcont by property (iv).   ∎

We now return to the problem of constructing product measures. Consider again the integral expression (29), and suppose now that ν is abcont. The first property of abcont measures—that they are the sums of series of finite measures—is the key to the discussion from here on. Thus we may write $\nu = \nu_1 + \nu_2 + \cdots$, where each ν_n is finite. Now, from what was said above, the function on A given by $f_n(a) = \nu_n\{b \mid (a, b) \in G\}$ is measurable for each $n = 1, 2, \ldots$; hence the sum of all of them is a measurable function. That is, the abcontness assumption on ν guarantees that the integrand in (29) is measurable, so that the integral expression (29) yields a well-defined product measure. Note that no restriction on μ need be imposed.

Next, suppose μ and ν are both abcont. In this case it is possible to reverse their roles, yielding the expression

$$\tilde{\lambda}(G) = \int_B \mu\{a \mid (a,b) \in G\} \, \nu(db) \qquad (2.6.32)$$

We now have two product measures λ and $\tilde{\lambda}$. However, these are *identical*. To show this, we start from the observation made above that two finite (in fact, σ-finite) measures have a *unique* product measure. Let ν be decomposed as above and, similarly, write $\mu = \mu_1 + \mu_2 + \cdots$, where all measures μ_m are finite. Substituting μ_m and ν_n for μ and ν respectively in (29) and (32), we obtain two measures that may be written λ_{mn} and $\tilde{\lambda}_{nm}$. We must have $\lambda_{mn} = \tilde{\lambda}_{nm}$, since both are products of the finite measures μ_m and ν_n. Taking the double summation over $m, n = 1, 2, \ldots$ and applying two versions of monotone convergence, we arrive finally at $\lambda = \tilde{\lambda}$.

Definition: Let (A, Σ, μ) and (B, Σ', ν) be two measure spaces, at least one of which is abcont; $(\mu \times \nu)$ is the product measure obtained from (29) (if ν is abcont) or from (32) (if μ is abcont).

If both are abcont then of course either integral formula may be used, yielding the same result. From here on we refer to $\mu \times \nu$ simply as *the product* of μ and ν. Note, however, that generalized products of μ and ν may exist other than $\mu \times \nu$. Indeed, one such case occurs with μ abcont and ν bounded, but we shall not examine this counterexample.

If μ and ν are both finite, then $\mu \times \nu$ is finite (since $\mu(A) \cdot \nu(B) < \infty$). If μ and ν are both σ-finite, then $\mu \times \nu$ is σ-finite (countably partition A and B so that μ and ν are finite on each respective piece; $\mu \times \nu$ is then finite on each $A_m \times B_n$). Finally, if μ and ν are both abcont, then $\mu \times \nu$ is abcont (to see this, note that each λ_{mn} above is finite). However, in this case other generalized products of μ and ν sometimes exist which are not abcont.

Since $\mu \times \nu$ is a measure, we may integrate with respect to it. The following result is important.

Theorem: Let (A, Σ, μ) and (B, Σ', ν) be measure spaces, with ν abcont; let $f : A \times B \rightarrow$ extended reals be measurable and nonnegative. Then

$$\int_{A \times B} f \, d(\mu \times \nu) = \int_A \mu(da) \int_B \nu(db) \, f(a,b) \qquad (2.6.33)$$

The right-hand side of (33) is to be interpreted as follows. For fixed $a_o \in A$, $f(a_o, \cdot)$ has domain B. Integrating with respect to ν, we obtain a number that depends on the point a_o. The resulting function with domain

A is measurable and may be integrated with respect to μ. Thus (33) expresses a *single* integral with respect to *product* measure in terms of an *iterated* integral with respect to the *component* measures. Results of this kind go under the name of *Fubini's theorem* (or Fubini-Tonelli's theorem).

Two observations are worth making. First, let $G \in \Sigma \times \Sigma'$, and let $f = I_G$, the indicator function of set *G*. Then one easily verifies that (33) reduces to (29), the defining equation for product measure. Second, suppose μ and ν are both abcont. Then the roles of μ and ν may be reversed. Combining this version of Fubini's theorem with the one above, we see that the same result is obtained independent of the order in which an iterated integration is carried out (provided the integrand is nonnegative).

The proof of Fubini's theorem may be outlined. For *f* an indicator function the result is immediate, as noted above. A simple function is a weighted sum of indicators, and the result then follows for *f* simple. Finally, noting that any nonnegative measurable function is the pointwise limit of a nondecreasing sequence of simple functions, we apply monotone convergence to the result just obtained, yielding the general theorem.

We generalize these results in two directions: first, to the product of more than two spaces; second, from ordinary to conditional measures. The second direction is of greater importance for us, and we start with it.

Definition: Let (A, Σ) and (B, Σ') be measurable spaces; a *conditional measure* is a function $\nu : A \times \Sigma' \to$ extended reals such that

   (i) for each $a_o \in A$ the right section $\nu(a_o, \cdot)$ is a measure on (B, Σ');
   (ii) for each $E \in \Sigma'$ the left section $\nu(\cdot, E)$ is a measurable function on (A, Σ).[34]

Note the peculiar domain of ν: the cartesian product of the universe set *A* of one space and the σ-field Σ' of the other space. Thus ν assigns numbers to pairs (a, E) consisting of a point and a set.

Suppose we have a conditional measure ν as above and an ordinary measure μ on (A, Σ). Let $G \in \Sigma \times \Sigma'$ (this is the product σ-field, not the cartesian product), and consider the expression

$$\lambda(G) = \int_A \nu[a, \{b \mid (a, b) \in G\}] \; \mu(da) \qquad (2.6.34)$$

This is the same as (29) except for the extra *a* in the argument of ν. This causes no complications. As before, for given *G* the expression

---

34. In the literature, conditional measures arise mainly in probability theory, e.g., in J. L. Doob, *Stochastic Processes* (Wiley, New York, 1953), App. "Conditional probability" is often used in a quite different sense than the one employed here.

$\nu(a, \{b \mid (a, b) \in G\})$ defines a function with domain A, which is to be integrated with respect to μ. *If* the integrand is measurable for each $G \in \Sigma \times \Sigma'$, then λ is a well-defined set function; in fact, λ is a measure, as one verifies by applying monotone convergence.

A known sufficient condition for the integrand in (34) to be measurable is that ν be a *finite* conditional measure. To generalize, we must extend the abcontness concept.

Definition: Conditional measure $\nu: A \times \Sigma' \to$ extended reals is *abcont* iff a series of finite conditional measures $\nu_n, n = 1, 2, \ldots$, exists such that

$$\nu = \nu_1 + \nu_2 + \cdots \qquad (2.6.35)$$

Here (35) is to be understood as usual in the pointwise sense: For the pair $(a, E) \in (A \times \Sigma')$, the values $\nu_n(a, E)$ constitute a numerical series whose sum is $\nu(a, E)$ (both sides may equal $+\infty$).

We now proceed as above. For given $G \in \Sigma \times \Sigma'$, the expression $\nu_n(a, \{b \mid (a, b) \in G\})$ defines a measurable function on A for each n; hence the sum of all these functions, which is the integrand in (34), is measurable. Thus $\lambda(G)$ is a well-defined measure. We refer to this as the *product* of μ and ν and denote it as above by $\mu \times \nu$. Here μ may be any measure on (A, Σ), while ν is an abcont conditional measure with domain $A \times \Sigma'$.

This generalizes the preceding construction if we identify the ordinary measure ν on (B, Σ') with the conditional measure $\tilde{\nu}$ given by

$$\tilde{\nu}(a, G) = \nu(G) \qquad (2.6.36)$$

In other words, a conditional measure that is independent of its first argument may be thought of as an ordinary measure with domain Σ', in which case the formula (34) reduces to (29). However, by taking ν conditional, we have lost the symmetry between μ and ν. No reversal of roles is possible even if μ is abcont. Also, with the identification (36), conditional abcontness reduces to ordinary abcontness.

Fubini's theorem remains valid if ν is taken to be abcont conditional; simply insert the extra argument a in (33).

We now generalize to more than two spaces. Let measurable spaces $(A_1, \Sigma_1), \ldots, (A_n, \Sigma_n)$ be given. We are also given functions $\mu_1, \ldots, \mu_n$; μ_i has domain $(A_1 \times \cdots \times A_{i-1}) \times \Sigma_i$ and is a conditional measure. That is, for a given point $(a_1, \ldots, a_{i-1})$ in the product space $A_1 \times \cdots \times A_{i-1}$, the right section $\mu_i(a_1, \ldots, a_{i-1}, \cdot)$ is a measure on (A_i, Σ_i); and for a given set $E \in \Sigma_i$ the left section $\mu(\cdot, \cdot, \ldots, \cdot, E)$ is a measurable function on $(A_1 \times \cdots \times A_{i-1}, \Sigma_1 \times \cdots \times \Sigma_{i-1})$. ($\mu_1$ is just an ordinary measure with domain Σ_1.) Finally, let a nonnegative measurable function $f: A_1 \times \cdots \times A_n \to$ extended reals be given, and consider the expression

$$\int_{A_1} \mu_1(da_1) \int_{A_2} \mu_2(a_1, da_2) \int_{A_3} \cdots \int_{A_n} \mu_n(a_1, \ldots, a_{n-1}, da_n) f(a_1, \ldots, a_n)$$

$$(2.6.37)$$

These iterated integrals are to be evaluated from right to left; i.e., first fix $a_1, \ldots, a_{n-1}$ and integrate the right section of f with respect to the right section of μ_n over A_n. This yields a function with domain $A_1 \times \cdots \times A_{n-1}$. Next, fix $a_1, \ldots, a_{n-2}$ and integrate this function with respect to μ_{n-1} over A_{n-1} to obtain a function with domain $A_1 \times \cdots \times A_{n-2}$. Continue, finishing with an integration over A_1 with respect to μ_1.

For this process to be well defined we must obtain a measurable integrand at each stage. What conditions guarantee this? We start with the case where f is an indicator function. Let $G \in (\Sigma_1 \times \cdots \times \Sigma_n)$ (this is the product σ-field, not the cartesian product), and let $f = I_G$.

Theorem: If $\mu_2, \mu_3, \ldots, \mu_n$ are all abcont conditional measures, then the iterated integral is well defined for $f = I_G$, for any $G \in (\Sigma_1 \times \cdots \times \Sigma_n)$. The resulting set function $\lambda(G)$ is a measure on $(A_1 \times \cdots \times A_n, \Sigma_1 \times \cdots \times \Sigma_n)$.

Definition: $\lambda(G)$ is called the *product measure* of $\mu_1, \ldots, \mu_n$ and is written $\mu_1 \times \cdots \times \mu_n$.

No restriction need be placed on μ_1. Also, the conditioning structure permits no role reversals; the successive integrations must be performed in the specified order. A special case arises when the μ's are actually independent of their point arguments; if so, each μ_i may be thought of as an ordinary measure on (A_i, Σ_i). If we let G be the measurable rectangle $E_1 \times \cdots \times E_n$, we obtain in this case

$$(\mu_1 \times \cdots \times \mu_n)(E_1 \times \cdots \times E_n) = \mu_1(E_1) \cdot \mu_2(E_2) \cdots \mu_n(E_n)$$

generalizing the product measure relation to n components. For example, n-dimensional Lebesgue measure is the product of n 1-dimensional Lebesgue measures.

We next generalize Fubini's theorem.

Theorem: Let $\mu_2, \ldots, \mu_n$ and f be as above. Then (37) is well defined and equals

$$\int_{A_1 \times \cdots \times A_n} f \, d(\mu_1 \times \cdots \times \mu_n)$$

This and the preceding theorem are best proved simultaneously by induction on n, the number of components. The case $n = 2$ has already been discussed, and these results may be used to go from n to $n + 1$. We omit details.

Finally, we comment on some properties of product measures. If μ_1 is abcont (as well as $\mu_2, \ldots, \mu_n$ being conditionally abcont), $\mu_1 \times \cdots \times \mu_n$ is abcont. This may be proved by induction on n. If $\mu_1, \mu_2, \ldots, \mu_n$ are all *bounded*, then $\mu_1 \times \cdots \times \mu_n$ is bounded. (Note that boundedness is a stronger condition than finiteness for conditional measures; for ordinary measures the two concepts of course coincide.) What about the intermediate case of σ-finiteness? For this we need one more concept.

Definition: Given (A, Σ) and (B, Σ'), let $\nu : A \times \Sigma' \to$ extended reals be a conditional measure; ν is *uniformly σ-finite* iff there is a countable collection $\mathcal{G} \subseteq \Sigma'$ such that, for all $(a, b) \in A \times B$, there is a set $E \in \mathcal{G}$ such that $b \in E$ and $\nu(a, E) < \infty$.

This property is a little stronger than mere σ-finiteness of each right section $\nu(a, \cdot)$. It implies conditional abcontness. As an example, let ν be independent of its point argument; then ν is uniformly σ-finite iff it is σ-finite when thought of as an ordinary measure.

Theorem: Let μ_1 be σ-finite, and let $\mu_2, \ldots, \mu_n$ be uniformly σ-finite; then $\mu_1 \times \cdots \times \mu_n$ is σ-finite.

Proof: First take the case $n = 2$: μ_2 is uniformly σ-finite, so Σ_2-sets $F_1, F_2, \ldots$ exist such that for any (a_1, a_2) there is an F_i for which $a_2 \in F_i$ and $\mu_2(a_1, F_i) < \infty$. Also a covering $E_1, E_2, \ldots$ of A_1 exists such that $\mu_1(E_j) < \infty$ for all j, by μ_1 σ-finite. Define

$$G_{ijk} = \{a_1 \mid a_1 \in E_j \text{ and } \mu_2(a_1, F_i) < k\} \times F_i$$

for all $i, j, k = 1, 2, 3, 4, \ldots$. These sets G_{ijk} form a countable measurable covering of $A_1 \times A_2$. Measurability follows from the fact that $\mu_2(\cdot, F_i)$ is measurable; also, any $(a_1, a_2) \in (E_j \times F_i)$ with $\mu_2(a_1, F_i) < \infty$ for some i, j, so $(a_1, a_2) \in G_{ijk}$ for some $k = 1, 2, \ldots$, proving the covering property. Finally, $(\mu_1 \times \mu_2)(G_{ijk}) \leq k \cdot \mu_1(E_j) < \infty$, since the integrand has k for an upper bound and is zero outside E_j. This proves $\mu_1 \times \mu_2$ is σ-finite.

For the general case proceed by induction on n: assume true for $n - 1$, so that $\mu_1 \times \cdots \times \mu_{n-1}$ is σ-finite. From Fubini's theorem we obtain

$$\mu_1 \times \cdots \times \mu_n = (\mu_1 \times \cdots \times \mu_{n-1}) \times \mu_n$$

But this expresses the n-fold product as a 2-fold product, of which the left component is σ-finite by induction hypothesis, and the right component

μ_n is uniformly σ-finite. Hence $\mu_1 \times \cdots \times \mu_n$ is σ-finite, completing the induction. ∎

DISTRIBUTION FUNCTIONS

Our entire discussion of measure theory has been framed to apply to measurable spaces in general. Distribution functions[35] will be the one exception in that the concepts apply only to finite products of the real line, i.e., to n-space, the set of all n-tuples of real numbers, with the corresponding n-dimensional Borel field. This measurable space will be denoted (A^n, Σ^n) in this discussion.

Definition: Let μ be a measure on (A^n, Σ^n), and let f be a real-valued function with domain A^n; f is a *distribution function* for μ (in the *narrow* sense) iff for every n-tuple of real numbers $(b_1, \ldots, b_n)$ we have

$$f(b_1, \ldots, b_n) = \mu\{(x_1, \ldots, x_n) \mid x_i < b_i, i = 1, \ldots, n\} \quad (2.6.38)$$

For $n = 1$ the indicated set on the right is the ray $\{x \mid -\infty < x < b_1\}$. For $n = 2$ it is a "southwest" quadrant of the plane.

Definition: Let μ be a measure on (A^n, Σ^n), and let f be a real-valued function on A^n; f is a *distribution function* for μ (in the *wide* sense) iff for every n pairs of real numbers $(a_i, b_i), i = 1, \ldots, n$, with $a_i < b_i$, the value

$$\mu\{(x_1, \ldots, x_n) \mid a_i \leq x_i < b_i, i = 1, \ldots, n\} \quad (2.6.39)$$

is equal to the following sum of 2^n terms:

$$f(b_1, \ldots, b_n) - f(a_1, b_2, \ldots, b_n) - f(b_1, a_2, b_3, \ldots, b_n) - \cdots$$
$$+ f(a_1, a_2, b_2, \ldots, b_n) + \cdots + (-1)^n f(a_1, \ldots, a_n) \quad (2.6.40)$$

In (40) we run over terms of the form $f(y_1, \ldots, y_n)$, where $y_i = a_i$ or b_i, and all 2^n possible selection patterns are used. If an even number of a_i's appear, the term enters with a $+$ sign; if an odd number, with a $-$ sign.

For $n = 1$ the set appearing in (39) is an interval, including its left but not its right endpoint. For $n = 2$ the set is an ordinary rectangle, including two of its four sides and one of its four corners. It is convenient to refer to sets in general of the form (39) as *bounded intervals*. Every bounded interval in n-space has 2^n corners, and f is a distribution function for μ iff the measure of any interval is equal to the sums and differences of the values of f on these corners, according to the sign rule stated above.

---

35. On distribution functions see von Neumann, *Functional Operators, I, Measures and Integrals*, pp. 160–72; H. Cramér, *Mathematical Methods of Statistics* (Princeton Univ. Press, Princeton, N.J., 1946), pp. 77–82. Definitions vary from one author to another.

Some examples follow:

1. $\mu = 0$ and $f = 0$ identically; then f is a distribution function for μ in both senses.
2. $\mu = 0$, and f is constant $\neq 0$; then f is a distribution function in the wide but not the narrow sense. (Proof: There are an equal number of $+$ and $-$ terms in (40), so the sum is zero.)
3. $\mu(E) = 1$ if $(0, \ldots, 0) \in E$ and $= 0$ otherwise; the function f for which $f(b_1, \ldots, b_n) = 1$ if all b_i's are positive and $= 0$ otherwise is a distribution function in both senses.
4. Let μ be n-dimensional Lebesgue measure; this has no distribution function in the narrow sense, but the function $f(b_1, \ldots, b_n) = b_1 b_2 \ldots b_n$ is one in the wide sense. (Proof: Lebesgue measure in (39) is the product $(b_1 - a_1)(b_2 - a_2) \ldots (b_n - a_n)$; when multiplied out we get 2^n terms, which are exactly the terms of (40) with the proper signs.)

We want answers to the following questions. Given μ, when does it have a distribution function in either sense, and are these unique? Given f, when is it the distribution function of some measure, in either sense, and is this measure unique?

A partial answer can be given immediately. From (38) it can be seen that a measure has at most one distribution function in the narrow sense and that it *does* have such a function iff it is finite on every set of the type appearing in (38). The corresponding result for wide-sense distribution functions is a little more difficult.

Theorem: Measure μ has a distribution function in the wide sense iff it is finite on every bounded interval. If f is a wide-sense distribution function for μ, then g is another one iff $g - f$ is of the form $h_1 + \cdots + h_n$, where $h_i : A^n \to$ reals does not depend on its ith coordinate.

That is, a change in the ith coordinate produces no change in $h_i(x_1, \ldots, x_n)$, whatever the values of the other $n - 1$ coordinates.

Thus if f is a wide-sense distribution function for μ, one can add any real-valued function of $n - 1$ variables to f and still retain that property. For $n = 1$, a constant may be added. For $n = 2$, $g(x_1, x_2) = f(x_1, x_2) + b_1(x_1) + b_2(x_2)$ is a wide-sense distribution function for μ if f is, because any such additions cancel out in the differencing process (40).

If μ is finite on every bounded interval, one can write an explicit formula for f, viz.,

$$f(b_1, \ldots, b_n) = \pm \mu\{(x_1, \ldots, x_n) \mid 0 \le x_i < b_i \text{ or } b_i \le x_i < 0, i = 1, \ldots, n\}$$

$$(2.6.41)$$

Here the condition $0 \le x_i < b_i$ is to be imposed if b_i is positive, the condition $b_i \le x_i < 0$ if b_i is negative; the $+$ is to be taken if the number of negative b_i's is even, the $-$ sign if that number is odd; finally, $f(b_1, \ldots, b_n) = 0$ if any of the b_i's equal zero. As an example, for Lebesgue measure, (41) comes out to $f(b_1, \ldots, b_n) = b_1 b_2 \ldots b_n$, a distribution function already referred to. The general wide-sense distribution function is then obtained by adding arbitrary functions $h_1 + \cdots + h_n$ to (41), as in the theorem above.

The various conditions imposed on μ have the following relations. If μ is finite on every set of the form (38), then μ is finite on every bounded interval; this in turn implies that μ is σ-finite. However, as one may show by examples, neither of these two implications can be reversed.

Next we come to the relation between wide and narrow sense.

Theorem: f is a distribution function for μ in the *narrow* sense iff it is a distribution function for μ in the *wide* sense and, for any $i_o = 1, \ldots, n$ and for any $n - 1$ fixed real numbers b_i $(i = 1, \ldots, n; i \ne i_o)$, the limit of $f(b_1, \ldots, b_n)$ as $b_{i_o} \to -\infty$ exists and equals zero.

Much deeper and more important are the converse results, giving conditions on f that make it a distribution function. We need the following concept.

Definition: $f: A^n \to$ reals is *continuous from below* iff, for any n-tuple of real numbers, $(x_1, \ldots, x_n)$ and any real $\epsilon > 0$, a real $\delta > 0$ exists such that $|f(y_1, \ldots, y_n) - f(x_1, \ldots, x_n)| < \epsilon$ for any $(y_1, \ldots, y_n)$ satisfying $x_i \ge y_i \ge (x_i - \delta)$, $i = 1, \ldots, n$.

Theorem: Let f be a real-valued function with domain A^n such that

    (i) f is continuous from below;
    (ii) for every n pairs of real numbers (a_i, b_i), $i = 1, \ldots, n$, with $a_i < b_i$, the expression (40) is nonnegative.

Then there is exactly one measure μ such that f is a distribution function for μ in the wide sense.

By combining the last two theorems, we get a sufficient condition for f to be a distribution function in the *narrow* sense; viz., (i) and (ii) of the preceding theorem, with (iii) $f(b_1, \ldots, b_n) \to 0$ as $b_{i_o} \to -\infty$ for any $i_o = 1, \ldots, n$, the other b_i's being held fixed. (Conditions (i), (ii), and (iii) are also necessary for f to be a distribution function in the narrow sense.)

SIGNED MEASURES

In Section 2.5 we discussed the measure λ_1 on the measurable space $(R \times S \times T, \Sigma_r \times \Sigma_s \times \Sigma_t)$, representing "production" or "births," and also the measure λ_2 on the same space, representing "consumption" or "deaths." One wonders if there is any way of representing "*net* production," or "natural increase," the difference between these two measures.

Such "netting out" procedures are very common in practice. Thus one subtracts imports from exports to get net exports, in-migration from out-migration to get net migration, debts from claims to get net creditor position.

Formally, one has two measures, say μ and ν, over the same measurable space (A, Σ) and one wants to attach a meaning to the subtraction operation $\mu - \nu$. (In the examples mentioned—exports, migration, and debts—the universe set may be taken to be Space or perhaps, if one has full "origin-destination" data, $S \times S$.)

The obvious way to define $\mu - \nu$ is as the set function with domain Σ, whose value at $E \in \Sigma$ is equal to $\mu(E) - \nu(E)$. There are two difficulties with this procedure. First, $\mu - \nu$ will in general take on negative values and is therefore no longer a measure. Second, if μ and ν are both *infinite* measures, the meaningless expression $\infty - \infty$ is indicated as the value of $\mu - \nu$ for certain sets (e.g., for the universe set A itself); thus things are not even well defined in this case.

We avoid the second difficulty for the time being by assuming that at least one of the two measures μ, ν is finite; $\mu - \nu$ as defined above is then a well-defined set function on (A, Σ). The important point is that this set function has all the defining characteristics of a measure, with the single exception that it can take on negative values.

This suggests the following definition.

Definition: The set function μ is a *signed measure* iff

  (i) its domain is a σ-field,
  (ii) its range lies in the extended real numbers,
  (iii) it is countably additive,
  (iv) $\mu(\emptyset) = 0$.

The only property that needs explaining is (iii). Let $\mathcal{G}$ be a countable packing of measurable sets, and let $G_1, G_2, G_3, \ldots$ be an enumeration of the members of $\mathcal{G}$; then we must have

$$\mu(\cup \mathcal{G}) = \mu(G_1) + \mu(G_2) + \cdots \qquad (2.6.42)$$

in the sense that the limit of the right-hand series exists, and equals $\mu(\cup \mathcal{G})$. Furthermore, this equality is required to hold no matter how the terms of the right-hand series are rearranged. For measures where all terms are nonnegative, this imposes no additional restriction, since the sum of an infinite series of nonnegative terms is invariant under rearrangement of terms. But it *is* an additional restriction when terms of opposite sign occur in (42). The total requirement may be cast in the following convenient form.

Consider just the *positive* terms among the $\mu(G_n)$; let the sum of these terms alone be P (as stated, P does not depend on the order of arrangement of these terms; if there are no positive terms, set $P = 0$). Similarly, consider just the *negative* terms among the $\mu(G_n)$, and let their sum be N (if there are no negative terms, set $N = 0$). Then it is required that, first, at least one of P, N be finite and, second, that $P + N = \mu(\cup \mathcal{G})$.

Consider some examples of signed measures:

1. Any measure is a signed measure; this follows at once from the definitions (and of course any nonnegative signed measure is a measure).
2. Let Σ be a finite σ-field generated by partition $\mathcal{G}$, where $\emptyset \notin \mathcal{G}$; assign extended real numbers arbitrarily to the members of $\mathcal{G}$, with the restriction that at most one of the two numbers $\{+\infty, -\infty\}$ is used; for every $\mathcal{F} \subseteq \mathcal{G}$, assign to the set $\cup \mathcal{F}$ the sum of the numbers assigned to the elements of $\mathcal{F}$; finally, assign 0 to the empty set $\emptyset$. The result is a signed measure; in fact, all signed measures defined on finite σ-fields are of this form.

It is trivial that if a measure is finite above (i.e., does not take on the value $+\infty$), it is bounded above. The same property holds for signed measures in general (the proof in this case requiring some effort). We have the following theorem.

Theorem: If a signed measure is finite above (finite below), then it is bounded above (bounded below).

Thus finiteness and boundedness are synonymous properties for signed measures. From this theorem we obtain the following important property.

Theorem: A signed measure is either bounded above or bounded below (or both).

Proof: Let μ be a signed measure. We show that it cannot take on both values $+\infty$ and $-\infty$. For suppose $\mu(E) = \infty$, $\mu(F) = -\infty$ for some measurable sets E, F; since $\mu(E \cap F) + \mu(F \setminus E) = \mu(F)$, we must have $\mu(E \cap F) \neq \infty$; then, since $\mu(E \cap F) + \mu(E \setminus F) = \mu(E)$, it follows that $\mu(E \setminus F) = \infty$; $E \setminus F$ and F are disjoint; but $\mu(E \setminus F) + \mu(F)$ is undefined, contradicting countable additivity.

Thus a signed measure is finite above, or below, or both; by the theorem above, it is then bounded above, or below, or both. ∎

Definition: Let μ, ν be two signed measures over the same measurable space (A, Σ), such that $\mu(A)$, $\nu(A)$ are not infinite of opposite sign. The sum $\mu + \nu$ is the set function with domain Σ whose value at E is given by $\mu(E) + \nu(E)$.

The restriction on $\mu(A)$, $\nu(A)$ ensures that the meaningless expression $\infty - \infty$ does not arise.

Definition: Let μ be a signed measure over (A, Σ), and let c be a real number. The *scalar product* $c\mu$ is the set function with domain Σ whose value at E is given by $c \cdot \mu(E)$.

In particular, $(-1)\mu$ is written simply as $-\mu$.

Theorem: The sum $\mu + \nu$ and the scalar product $c\mu$ are signed measures, where μ, ν, and c are restricted as indicated in their respective definitions.

This theorem contains the statement made above concerning the difference of two measures, for $\mu - \nu$ is the same as $\mu + (-\nu)$ and is therefore a signed measure. Again, if μ and ν are measures, then $\mu + \nu$ is always well defined and is a signed measure (in fact, a measure, since $\mu + \nu$ is obviously nonnegative).

Let **M** be the set of all *finite* signed measures over space (A, Σ); **M** is closed under addition and scalar multiplication. That is, if $\mu \in$ **M**, $\nu \in$ **M**, and c is real, then $\mu + \nu \in$ **M**, and $c\mu \in$ **M**. We make the important observation that **M** with these two operations is a vector space. Let us define this abstractly.

Definition: Let **M** be a set, $+$ a function with domain **M** $\times$ **M** and range in **M**, and $\cdot$ a function with domain (real numbers $\times$ **M**) and range in **M**. (These are called *addition* and *scalar multiplication* respectively; we use the notation $\mu + \nu$ and $c\mu$ instead of the clumsy $+ (\mu, \nu)$ and $\cdot (c, \mu)$.) Then **M** with these two operations is a *(real) vector space* iff the following eight conditions are fulfilled:

(i) $(\lambda + \mu) + \nu = \lambda + (\mu + \nu)$ for all $\lambda, \mu, \nu \in$ **M**;
(ii) $\mu + \nu = \nu + \mu$ for all $\mu, \nu \in$ **M**;
(iii) there is an element of **M**, denoted 0, such that $\mu + 0 = \mu$ for all $\mu \in$ **M**;
(iv) there is an element $\nu \in$ **M** such that $\mu + \nu = 0$ for all $\mu \in$ **M**;
(v) $b(\mu + \nu) = b\mu + b\nu$ for all real b, all $\mu, \nu \in$ **M**;
(vi) $(b + c)\mu = b\mu + c\mu$ for all real b, c, all $\mu \in$ **M**;
(vii) $b(c\mu) = (bc)\mu$ for all real b, c, all $\mu \in$ **M**;
(viii) $1 \cdot \mu = \mu$ for all $\mu \in$ **M**.

The element 0 in (iii) is unique and known as the zero or neutral element of **M**. The ν in (iv) is also uniquely determined by μ and is equal to $(-1)\mu = -\mu$.

This definition has nothing in particular to do with measures, although we have kept the measure notation. The most familiar example of a vector space is the set of all n-tuples of real numbers under the operations + and · given by

$$(x_1, \ldots, x_n) + (y_1, \ldots, y_n) = (x_1 + y_1, \ldots, x_n + y_n)$$

$$c(x_1, \ldots, x_n) = (cx_1, \ldots, cx_n)$$

We now state formally the following.

Theorem: Let (A, Σ) be a measurable space, and **M** the set of all finite signed measures on it. Then **M**, with the operations $\mu + \nu$ and $c\mu$ defined above, is a vector space.

Proof: We have stated that **M** is closed under these two operations. The zero element 0 is the identically zero measure; the element ν in (iv) is $-\mu$; (i)–(viii) are then immediate consequences of the definitions. ∎

We now turn to the difficulties raised by *infinite* measures, signed or unsigned. Throughout this discussion we have made qualifications to avoid the meaningless expression $\infty - \infty$. Any attempt to make a vector space out of a larger set of signed measures than **M** seems to fail because there is no signed measure ν satisfying $\mu + \nu = 0$ if μ takes on infinite values, so that condition (iv) of the definition of vector space breaks down.

This is unfortunate because many situations of theoretical interest appear to call for a concept that is, in effect, the difference of two infinite measures. Take *net production,* which is the signed measure over $(R \times S \times T, \Sigma_r \times \Sigma_s \times \Sigma_t)$ obtained from the difference, production minus consumption, as discussed above. There is no a priori reason why the two measures, production and consumption, should not both be infinite. Indeed, in problems with an unlimited time horizon or an "endless plane," the presumption is that both *will* be infinite.

Again suppose we are evaluating economic development policies by comparing costs and benefits against some benchmark. Benefits and costs can be represented as measures on the Time axis. What if both are infinite, a not implausible situation if the horizon is unlimited? We would still like to evaluate benefits minus costs and if possible to compare two such evaluations.[36]

---

36. We shall see later that the method for solving these problems is closely related to work of Ramsey, Weizsäcker, and others on the evaluation of infinite development programs. In fact, it incorporates these "overtaking criteria" as special cases. See Sections 3.3 and 3.4.

We have developed the concept of "pseudomeasure" to overcome these difficulties. This concept, being outside the corpus of present-day measure theory, deserves full-scale treatment in a chapter of its own, so we do not define it here. But we do wish to indicate how it jibes with the more familiar concepts presently under discussion. Pseudomeasures are generalizations of σ-finite signed measures, just as the latter are generalizations of σ-finite measures. With their aid, σ-finite signed measures can be added freely, even when infinite of opposite sign. When one extends **M** to all σ-finite signed measures, it ceases to be a vector space; but extending it still further to all pseudomeasures restores this property. Furthermore, one can order pseudomeasures in ways that are elegant and intuitively appealing with respect to the problems mentioned above. This discussion of signed measures may be viewed as a halfway point, to be suitably generalized when we come to pseudomeasures in Chapter 3.

We have seen that for any pair of measures μ, ν, not both infinite, the difference $\mu - \nu$ is a signed measure. A basic result is that the converse is also true: any signed measure can be expressed as the difference of two measures.

Theorem: Jordan Decomposition Theorem. Let μ be a signed measure on space (A, Σ), and consider the following two set functions, both with domain Σ.

$$\mu^+(E) = \sup\{\mu(F) \mid F \subseteq E, F \in \Sigma\} \qquad (2.6.43)$$

$$\mu^-(E) = \sup\{-\mu(F) \mid F \subseteq E, F \in \Sigma\} \qquad (2.6.44)$$

Then μ^+ and μ^- are measures, not both infinite, and $\mu = \mu^+ - \mu^-$. (In (43) and (44) sup abbreviates supremum, which is taken over the set of all values assumed by μ, respectively $-\mu$, on measurable subsets of E.)

Definition: μ^+ in (43) is known as the *upper variation* of μ, μ^- in (44) as the *lower variation* of μ. The sum $\mu^+ + \mu^-$ is known as the *total variation* of μ and is denoted $|\mu|$. The pair (μ^+, μ^-) is the *Jordan decomposition* of μ.

For example, let μ be a measure; then $\mu^+ = \mu$ and $\mu^- = 0$. (Proof: By monotonicity, $\mu(E) \geq \mu(F)$ if $F \subseteq E$; hence $\mu^+(E) = \mu(E)$. Since μ is nonnegative, the supremum of $-\mu$ is attained on the empty set ϕ; hence $\mu^-(E) = 0$.) Similarly, if μ is *nonpositive* (i.e., μ is the negative of a measure), then $\mu^+ = 0$, and $\mu^- = -\mu$.

Suppose we start with a pair of measures (μ_1, μ_2) (not both infinite), form their difference $\mu = \mu_1 - \mu_2$, and then take the Jordan decomposition (μ^+, μ^-). What is the relation between these two pairs? The answer is given by the following.

Theorem: Let μ_1, μ_2 be two measures over (A, Σ), not both infinite, and let $\mu_1 - \mu_2 = \mu$. Then there is a finite measure ν such that

$$\mu_1 = \mu^+ + \nu, \qquad \mu_2 = \mu^- + \nu \qquad (2.6.45)$$

Proof: If E, $F \in \Sigma$ and $F \subseteq E$, then $\mu(F) \leq \mu_1(F) \leq \mu_1(E)$. It follows from (43) that $\mu^+(E) \leq \mu_1(E)$ for all $E \in \Sigma$. Similarly, $-\mu(F) \leq \mu_2(F) \leq \mu_2(E)$, so $\mu^-(E) \leq \mu_2(E)$ for all $E \in \Sigma$, from (44).

If μ_2 is finite, so is μ^-; in this case set $\mu_2 - \mu^- = \nu$; ν is a finite measure, and the relation $\mu_1 - \mu_2 = \mu^+ - \mu^-$ yields (45).

If μ_1 is finite, so is μ^+; set $\mu_1 - \mu^+ = \nu$, and we again get (45). ■

Thus for given signed measure μ the Jordan decomposition is the *smallest* pair of measures whose difference is μ. Any other pair of measures having this property is obtained by adding the same finite measure ν to both halves of (μ^+, μ^-). The Jordan decomposition also has the following deeper property.

Definition: Measures μ, ν over space (A, Σ) are *mutually singular* iff there is a partition of A into two measurable sets, P, N (i.e., $P \cup N = A$, $P \cap N = \phi$) such that $\mu(N) = 0$ and $\nu(P) = 0$.

Theorem: Hahn Decomposition Theorem. The upper and lower variations μ^+ and μ^- of any signed measure μ are mutually singular.

This may be stated in the following slightly different form.

Theorem: Hahn Decomposition Theorem, Second Version. For any signed measure μ on space (A, Σ), A can be split into two measurable sets P, N such that for all $E \in \Sigma$,

   (i) if $E \subseteq P$, then $\mu(E) \geq 0$;
   (ii) if $E \subseteq N$, then $\mu(E) \leq 0$ $\qquad\qquad (2.6.46)$

We prove that these two versions imply each other. Assume the first, and let P, N be a measurable partition of A such that $\mu^+(N) = 0$ and $\mu^-(P) = 0$. If measurable $E \subseteq P$, then $\mu(E) = \mu^+(E) - \mu^-(E) = \mu^+(E) \geq 0$. If $E \subseteq N$, then $\mu(E) = \mu^+(E) - \mu^-(E) = -\mu^-(E) \leq 0$. This proves version two.

Conversely, assume the second version. Then $\mu(E) \leq 0$ for every measurable subset of N; hence $\mu^+(N) = 0$ by (43). Similarly, $-\mu(E) \leq 0$ for $E \subseteq P$, $E \in \Sigma$; hence $\mu^-(P) = 0$. Thus μ^+, μ^- are mutually singular. This completes the proof.

Definition: Given signed measure μ on (A, Σ), any ordered measurable partition (P, N) of A satisfying (46) will be called a *Hahn decomposition* for μ; P is the *positive half* and N the *negative half* of this decomposition.

Equivalently, this could have been defined as a measurable partition satisfying $\mu^-(P) = 0, \mu^+(N) = 0$.

Thus, while the Jordan decomposition is a pair of measures, the Hahn decomposition is a pair of sets. The Jordan decomposition is unique; the Hahn decomposition is "almost" unique in the sense that if (P', N') is another Hahn decomposition, then $|\mu|(P \cap N') = 0$ and $|\mu|(P' \cap N) = 0$, $|\mu|$ being the total variation of μ. Given P, $\mu^+(E) = \mu(E \cap P)$ since, within E, μ takes on its supremum at $E \cap P$; similarly, $\mu^-(E) = -\mu(E \cap N)$.

Among all pairs of measures whose difference is μ, the Jordan decomposition is the only mutually singular pair. As an example, let μ be a measure. Then $P = A$, $N = \phi$ is a Hahn decomposition for μ. (Proof: immediate from (46).)

The Hahn and Jordan decompositions of real-world signed measures have simple intuitive interpretations. Suppose, for example, that the universe set is Space, and let signed measure μ be net exports of some commodity: $\mu(E)$ = exports from E minus imports to E, for every region E. The Hahn decomposition theorem shows that S can be split into two regions P and N such that every subregion of P (N) is a net exporter (importer); μ^+ and μ^- are then the export and import measures *exclusive* of transhipment.[37]

Most of the theorems discussed in Section 2.6 have generalizations from measures to signed measures. Because of the Hahn and Jordan decomposition theorems, it is unnecessary for the most part to state these separately: we simply perform the appropriate decomposition, apply the theorem in question to each piece separately, and then put the pieces together to get the generalization to signed measures. There are just two cautions to be observed: (i) theorems involving inequalities do not always generalize, thus the simple statement $\int_A f \, d\mu \geq 0$ need not hold if f or μ can assume negative values (see below for definition) and (ii) we sometimes need additional assumptions to guarantee that the meaningless expression $\infty - \infty$ does not arise.[38]

There are a few concepts whose generalization to signed measures

---

37. $\mu^+(E)$ must include exports from E to itself, not merely to $A \backslash E$, for otherwise μ^+ would not be an additive set function; similarly for $\mu^-(E)$. These sketchy statements will be elaborated when we discuss transportation and transhipment in Chapter 7.

38. When generalizing even further to pseudomeasures, we do not have to do this. See Chapter 3.

deserves explicit mention. We have already discussed σ-finiteness, whose definition carries over without change.

Definition: Let μ be a signed measure on (A, Σ). μ is *σ-finite* iff there is a countable partition $\mathcal{G}$ of A into measurable sets such that $\mu(G)$ is finite for all $G \in \mathcal{G}$.

We now come to integration. Given measurable space (A, Σ), the integral $\int_A f \, d\mu$ has been defined in the case where μ is a measure and f a nonnegative measurable function. We now remove both of these sign restrictions.

Definition: Let f be an extended real-valued function on domain A. Functions f^+ and f^- on domain A are given by

$$f^+(a) = \max(f(a), 0), \qquad f^-(a) = \max(-f(a), 0)$$

(max abbreviates maximum; i.e., f^+ coincides with f where the latter is positive and equals zero elsewhere; f^- coincides with $-f$ where f is negative and equals zero elsewhere); f^+ is known as the *positive part* of f, f^- as the *negative part* of f.

Note that both f^+ and f^- are nonnegative functions. Also that $f^+ - f^- = f$, while $f^+ + f^- = |f|$, the absolute value of f.

Definition: Let μ be a signed measure, and f a measurable function on the measurable space (A, Σ). The *integral* of f *with respect to* μ is given by:

$$\int_A f \, d\mu = \left[\int_A f^+ \, d\mu^+ \right] + \left[\int_A f^- \, d\mu^- \right] - \left[\int_A f^+ \, d\mu^- \right] - \left[\int_A f^- \, d\mu^+ \right]$$

$$(2.6.47)$$

provided the right-hand expression is not of the form $\infty - \infty$. (If it *is* of this form, then $\int_A f \, d\mu$ is considered to be meaningless.)

In each of the four integrals on the right of (47), the integrands f^+ and f^- are nonnegative, and μ^+ and μ^- (the upper and lower variations of μ) are measures. Thus these integrals have already been defined, and (47) gives the integral for signed measures and "signed functions" in terms of these already defined integrals.[39]

---

39. One of the advantages of pseudomeasures is that the proviso concerning expressions of the form $\infty - \infty$ may be dropped. When (47) is suitably generalized, the (indefinite) integral of *any* measurable real-valued function with respect to *any* pseudomeasure is well defined. See Chapter 3.

Definition (47) is consistent in the following sense. If f is non-negative and μ is a measure, the definition in (47) coincides with the original. This follows from the fact that, in this case, $f^+ = f$, $f^- = 0$, $\mu^+ = \mu$, $\mu^- = 0$; thus three of the four integrals are zero, and the last gives $\int_A f^+ \, d\mu^+ = \int_A f \, d\mu$, where the right-hand expression has the original definition (4).

Two "halfway" cases arise. If μ is a measure, (47) reduces to $\int_A f^+ \, d\mu - \int_A f^- \, d\mu$; and if f is nonnegative, (47) reduces to

$$\int_A f \, d\mu^+ - \int_A f \, d\mu^-$$

As an example, let $f(a) = c$ for some fixed real number c. Then (47) yields $\int_A c \, d\mu = c\mu(A)$, which is exactly the same formula as when $c \geq 0$ and μ is a measure. (Proof: Consider separately the two cases $c \geq 0$, $c < 0$.)

We conclude this discussion with a few integration formulas. On measurable space (A, Σ), μ and ν are signed measures, f and g are measurable extended real-valued functions, and c is a real number.

If f and μ are both bounded, then

$$\int_A f \, d\mu \qquad\qquad (2.6.48)$$

is well defined and is finite.

$$c \int_A f \, d\mu = \int_A cf \, d\mu \qquad\qquad (2.6.49)$$

$$c \int_A f \, d\mu = \int_A f \, d(c\mu) \qquad\qquad (2.6.50)$$

$$\int_A f \, d\mu + \int_A g \, d\mu = \int_A (f + g) \, d\mu \qquad\qquad (2.6.51)$$

$$\int_A f \, d\mu + \int_A f \, d\nu = \int_A f \, d(\mu + \nu) \qquad\qquad (2.6.52)$$

Equations (49)–(52) are to be read as follows. If both sides are well defined, they are equal; (48), (49), and (51) thus generalize (9), (10), and (11) respectively. In (50) $c\mu$ is the scalar product; in (52) $\mu + \nu$ is the sum of two signed measures.

2.7. ACTIVITIES

The measure space of histories (Ω, Σ, μ), which in principle gives a complete description of the world, is rather unwieldy as a whole; we often want to focus attention on one or another aspect of special interest. We have discussed how certain data may be extracted (e.g., cross-sectional and double cross-sectional measures, production and consumption). Here we continue this discussion, concentrating not so much on material typically appearing in statistical tabulations but on less sharply defined categories: "situations," "events," "processes," "activities."

Thus out of the flux one distinguishes a house, a crowd of people, a town, a seacoast; or a driver traveling along the highway, a sugar refinery in operation, a farmer plowing his field, an army on the march. The first four items are "cross-sectional configurations"; i.e., they describe a part of the world at an instant of time. The last four refer to something going on over an interval of time.

One can distinguish situations in an indefinitely large variety of ways. Out of these possibilities a much smaller, but still enormous, number are actually distinguished and named in the words of some language. Why some possibilities are selected and others are not is itself an interesting question, to be answered on the one hand by relations of similarity, contiguity, contrast, closure, and other characteristics of "good gestalt," and on the other by causal relations. Situations tend to be selected so that their parts are mutually interdependent and relatively independent of the rest of the world.

Causation will be discussed in Chapter 4. Here, however, we are concerned only with problems of description. We want a framework adequate for describing a variety of situations or processes, whether they make "causal sense" or not. If one takes the stork population of Sweden and the human birth rate of that country as constituents of a single activity, this activity can be perfectly well defined (although not very useful perhaps).

Cross-sectional configurations have been discussed briefly in Section 2.5. It was mentioned that a house, for example, could be represented as a measure over universe set $R \times F$, R being the set of resource types and F the region occupied by the house. This measure gives the spatial distribution of the resources constituting the house. One point was left hanging in this discussion. A measure over $R \times F$ represents only a particular house, viz., the house occupying the particular region F (at the moment for which the cross-section was defined). We have no way as yet for representing a house *type*, or configuration type in general, as opposed to any particular specimen of that type. The following construction fills this gap.

METRIC SPACES AND CONGRUENT MEASURES

Definition: A *metric space* consists of a set A and a real-valued function d with domain $A \times A$, satisfying

(i) $d(x, x) = 0$ for all $x \in A$;

(ii) $d(x, y) > 0$ if $x \neq y$ for all $x, y \in A$;

(iii) $d(x, y) = d(y, x)$ for all $x, y \in A$;

(iv) $d(x, y) + d(y, z) \geq d(x, z)$ for all $x, y, z \in A$.

Condition (iii) is called *symmetry,* condition (iv) the *triangle inequality.*
The metric space itself is written as the pair (A, d); d is called the *metric*
or the *distance function.* If d is understood, A itself may be referred to as
a metric space.[40]

Our first example is the most familiar case.

Definition: Let A be n-space. The *Euclidean* metric gives the distance
between $x = (x_1, \ldots, x_n)$ and $y = (y_1, \ldots, y_n)$ as

$$d(x, y) = [(x_1 - y_1)^2 + \cdots + (x_n - y_n)^2]^{1/2} \qquad (2.7.1)$$

Conditions (i), (ii), and (iii) on d are verified immediately. Condition (iv),
which is a little harder to verify, states exactly that in the triangle with
vertices $x, y,$ and z the length of the side from x to z does not exceed the
sum of the lengths of the other two sides.

Definition: Let A again be n-space. The *city-block* metric (also
known variously as the *rectangular, metropolitan, manhattan,* or *mid-
western* metric) gives the distance between $x = (x_1, \ldots, x_n)$ and $y =
(y_1, \ldots, y_n)$ as

$$d(x, y) = |x_1 - y_1| + \cdots + |x_n - y_n| \qquad (2.7.2)$$

Conditions (i)–(iv) on the d defined by (2) are immediate consequences of
the properties of absolute values. When $n = 1$, we have the real line,
and in this case (1) and (2) both reduce to the same function, viz.,
$d(x, y) = |x - y|$. But for $n > 1$ the two metrics are distinct. The most
important case for our later work will be $n = 2$, the resulting metric
spaces being called the *Euclidean plane* and the *city-block plane* respectively.

   These two metrics are examples of *normed* metrics; i.e., A is a vector
space, and there exists a function $\| \cdot \|$ on A (called a *norm*) satisfying
$\|x\| > 0$ for all $x \neq 0$, $\|x + y\| \leq \|x\| + \|y\|$, $\|px\| = |p| \cdot \|x\|$ for p real,
and such that $d(x, y) = \|x - y\|$ for all $x, y \in A$.

   As a third example let A be the surface of a sphere of radius r and
center $c = (c_1, c_2, c_3)$. The *great-circle* metric gives the distance between $x, y \in A$ as $r \cdot$angle (xcy), the angle being measured in radians.

   As a final example let A be any nonempty set, and define d by

---

40. This is the first occasion where non-measure-theoretic (specifically, metric) concepts
are being used.

$d(x, y) = 1$ if $x \neq y$; $d(x, x) = 0$.   Since this satisfies (i)–(iv) it is a bona fide distance function, known as the *discrete* metric.

Let (A, d) be a metric space, and B a subset of A; B can be considered a metric space in its own right by taking $d(x, y)$ for x, $y \in B$, to be the distance from x to y in metric space B. This amounts to defining the metric on B to be the restriction of d to the subdomain $(B \times B) \subseteq (A \times A)$. We always consider any subset of a metric space to be itself a metric space in this way.

In a metric space, the sets $\{y \mid d(x, y) < r\}$, $\{y \mid d(x, y) \leq r\}$ are known as the *open, closed discs* of center x and radius r, respectively.

Definition: Let (A, d), (B, d') be two metric spaces.   Mapping $f : A \to B$ is a *congruence* or an *isometry* from A to B iff

  (i) f is *onto* (i.e., for all $b \in B$, an $a \in A$ exists for which $f(a) = b$);
  (ii) $d(x, y) = d'(f(x), f(y))$ for all $x, y \in A$.   (2.7.3)

Equation (3) states that f is distance preserving; A and B may be overlapping or even identical, and in the latter case d' may or may not be the same as d.

Definition: Metric spaces (A, d) and (B, d') are *congruent*, or *isometric*, iff there exists an isometry $f : A \to B$.

If (A, d) is congruent to (B, d'), which in turn is congruent to (C, d''), then (A, d) is congruent to (C, d''). This follows from the fact that if $f : A \to B$ and $g : B \to C$ are isometries, the composition $g \circ f : A \to C$ is an isometry. Thus the relation of congruence between metric spaces is *transitive*. Furthermore, if (A, d) is congruent to (B, d'), then (B, d') is congruent to (A, d). For if $f : A \to B$ is an isometry, it has an *inverse* function $g : B \to A$ (i.e., $g(f(a)) = a$ for all $a \in A$, and $f(g(b)) = b$ for all $b \in B$), and g is also an isometry.   Thus congruence is a symmetric relation. Finally, (A, d) is obviously congruent to itself, the identity map $f(a) = a$ being an isometry; thus congruence is *reflexive*. In short, congruence is an *equivalence* relation between metric spaces.

Next, let (A, Σ) and (B, Σ') be two measurable spaces.

Definition: $f : A \to B$ is *measurability preserving* (from A to B) iff

  (i) f has an inverse function $g : B \to A$;
  (ii) both f and g are measurable.

Measurability of f means, of course, that $\{a \mid f(a) \in E\}$ belongs to Σ whenever $E \in \Sigma'$; measurability of g reverses these roles, so that

$\{b \mid g(b) \in F\}$ belongs to Σ' whenever $F \in \Sigma$. Condition (i) holds iff f is 1-1 and onto.

Let (A, Σ, μ) and (B, Σ', μ') be two measure spaces.

Definition: $f: A \rightarrow B$ is *measure preserving* iff

(i) f is measurability preserving,
(ii) $\mu'(E) = \mu\{a \mid f(a) \in E\}$ for all $E \in \Sigma'$.

Condition (ii) is merely the statement that μ' is the measure induced by f from μ. This is all completely symmetric because μ is also the measure induced from μ' by the inverse mapping g.

These concepts may now be combined. Suppose A is provided with a metric d and with a σ-field Σ and measure μ; thus A is both a metric space and a measure space, and we write this as a quadruple (A, d, Σ, μ).[41]

Definition: Let (A, d, Σ, μ) and (B, d', Σ', μ') be two metric-measure spaces. These are *measure congruent* iff an isometry $f: A \rightarrow B$ exists (for the pair (A, d) and (B, d')), which is also *measure preserving* (for the pair (A, Σ, μ) and (B, Σ', μ')).

There must be a *single* function that simultaneously preserves distance and measure. If the metrics are understood, one says that μ and μ' are *congruent* measures.

To illustrate these concepts, we take the Euclidean plane with the 2-dimensional Borel field and Lebesgue measure. Let E, F be two measurable sets congruent in the sense of plane geometry; they have the same "size" and "shape." That is, there is a function $f: E \rightarrow F$ whose range is all of F and for which $d(x, y) = d(f(x), f(y))$ for all x, $y \in E$, where d is Euclidean distance. Letting Σ_E, Σ_F be the Borel field restricted to subsets of E, F respectively, one can show that f is measurability preserving. In fact, f is measure preserving with respect to Lebesgue measure restricted to (E, Σ_E) and (F, Σ_F) respectively. Hence E and F are measure congruent. Measure congruence is an equivalence relation among metric-measure spaces, just as congruence per se is an equivalence relation among metric spaces.

CONFIGURATION TYPES

We now apply these concepts. Suppose that Space has a metric d and a σ-field Σ (e.g., S is 3-space, d the Euclidean metric, Σ the Borel field). Let E and F be two regions. These may be thought of as metric spaces and also as measurable spaces in their own right.

---

41. One typically postulates certain further relations between d and Σ. However, this is not necessary for this discussion.

Let μ, ν be two distributions of mass over universe sets E, F respectively. (We suppress discussion of the Resource set for the moment; μ and ν may be thought of as the marginals of measures over $R \times E$ and $R \times F$ respectively.) It is intuitively appealing to explicate the vague statement, "μ and ν are the same *type of configuration*" by the precise statement, "μ and ν are *measure congruent*." The latter requires that E and F have the same size and shape and that the distributions over E and F have the same "pattern," so that ν is in a sense just a rigid displacement of μ.[42]

We now bring R into the discussion. Let μ, ν be mass distributions over spaces $(R \times E, \Sigma_r \times \Sigma_E)$ and $(R \times F, \Sigma_r \times \Sigma_F)$ respectively. Here E and F are regions as above; Σ_E and Σ_F are the restrictions of the σ-field Σ_s of S to measurable subsets of E, F respectively; and Σ_r is the σ-field of R; regions E and F are provided with the metric of S, but R is not assumed to have any metric.

Definition: We say that μ and ν *represent the same configuration-type* iff

(i) there is an isometry $f : E \rightarrow F$ that is also measurability-preserving, with respect to (E, Σ_E) and (F, Σ_F);

(ii) for all $G \in \Sigma_r \times \Sigma_F$,

$$\nu(G) = \mu\{(r,s) \mid (r, f(s)) \in G\} \qquad (2.7.4)$$

Note that f is a mapping between *regions* and is not involved directly with R at all. However, f determines a certain mapping from $R \times E$ onto $R \times F$, viz., the one carrying the point (r,s) to $(r, f(s))$: Location shifts, but resource type is held fixed. Equation (4) then asserts that ν is the measure induced by this mapping from μ.

Indeed, it can be seen that the mapping $(r,s) \rightarrow (r, f(s))$ is measure preserving. First, it is measurable, since the two functions $(r,s) \rightarrow r$ and $(r,s) \rightarrow f(s)$ are both measurable. Second, since f has an inverse function g, the mapping $(r,s) \rightarrow (r, g(s))$ from $R \times F$ onto $R \times E$, is the inverse of $(r,s) \rightarrow (r, f(s))$; and it is measurable since g is.

If μ and ν represent two houses, say, then condition (i) requires that they have the same size and shape, while condition (ii) requires that the same materials (wood, glass, brick) be arranged in the same relative positions in both houses. Thus this definition appears to capture quite well

---

42. One fine point deserves footnote mention. The distributions representing the left and right gloves of a pair are measure congruent but cannot be transformed into one another by a rigid motion: one must be turned inside-out. We could therefore insist that sameness of type requires not only measure congruence, but preservation of "parity" or "orientation." Since this condition does not seem important for social science problems, we pass over it without further discussion.

the intuitive notion of "two instances of the same kind of thing." Resources and Space are not treated symmetrically in this definition. The reason lies in the "heterogeneous" nature of R as opposed to the "homogeneous" nature of S. There is no analog in R to the congruence relation among regions of S, at least not in general.[43]

If μ and ν represent the same configuration type, then their right marginals μ' and ν', on spaces (E, Σ_E) and (F, Σ_F) respectively, are congruent measures. This follows from (4) upon taking $G = R \times F'$, where $F' \in \Sigma_F$:

$$\nu'(F') = \nu(R \times F') = \mu\{(r, s) \mid f(s) \in F'\} = \mu'\{s \mid f(s) \in F'\}$$

We have defined the condition under which two specific configurations are to be considered of the same type. But we have not yet defined the notion of configuration type abstractly and divorced from any particular specimen of that type. For example, we can say that this house in Lisbon is of the same type as that one in Hong Kong, but cannot yet speak of this house type not in any particular place. Recall that we conceive of S in real terms, its regions being actual geographical places, so that our definition above is "tied" to the real world.

This difficulty is easily overcome. We suppose R and its σ-field Σ_r to be given. A *configuration type* is then defined as a quadruple, consisting of a set E, a metric d on E, a σ-field Σ_E on E, and a measure μ on $(R \times E, \Sigma_r \times \Sigma_E)$. This configuration type is *exemplified* on region F at moment t iff there is a measurability-preserving isometry $f: E \rightarrow F$ such that the mass distribution ν, over $R \times F$ at time t, satisfies (4) with μ.

In other words, the definition of "exemplification" is formally identical to that of "same configuration type," and merely differs in interpretation. In the latter, E and F are both regions, and we get a relation between two *particulars;* in the former, E is an abstract set endowed with a metric and σ-field, and we get a relation between a *universal* and a *particular*. Furthermore, we can run this definition in reverse and identify the abstract configuration type of any particular: merely strip the region it occupies of all properties except its metric and σ-field. There is no need to abstract from R as we did from S because R is already the set of *types* of resources.

We make one comment on the relation between these concepts and ordinary language. Clearly "house," "forest," "crowd of people," and the like refer, not to a single configuration type, but instead roughly delimit a set of such types. The more detailed the description, the smaller the set of configurations satisfying it; but even an encyclopedic descrip-

---

43. The "same" melody in two different keys, or the replacement of pine by spruce in a house, are examples involving shifts among resource types analogous to interregional shifts. But these apply only within small "homogeneous" subsets of R.

tion does not narrow the range down to one. Furthermore, and logically distinct from the ambiguity just mentioned, there is a penumbra of vagueness about ordinary language, so that there are "borderline cases" and "twilight zones" where it is uncertain whether a given configuration type satisfies a given description.

Suppose we have a description of the world at some moment t in the form of a list of configuration types exemplified in various regions, $E_1, E_2, \ldots$. Some of these regions may overlap: If $F \subseteq (E_i \cap E_j)$ where F is another region, this means that the mass distributed over F participates simultaneously in two configurations; e.g., the ceiling of one apartment may be the floor of another. If the descriptions are accurate, the two measures μ_i and μ_j will then be identical over F. We know from the patching theorem that this is a sufficient condition for the various μ_i's to determine a unique measure μ on the universe set $\bigcup\{E_n \mid n = 1, 2, \ldots\}$, μ_n being the restriction of μ to E_n. Thus the complete cross-sectional measure may be "patched" together.

In particular, if $E_i \subseteq E_j$, we have the relation of *part* and *whole* (e.g., a neighborhood in a city, a family in a nation); i.e., the relation of whole to part is represented formally by the measure μ_j over $R \times E_j$ and its restriction to $R \times E_i$, which is μ_i. One frequently deals with a whole hierarchy of parts and wholes (e.g., items into packages into cartons into carloads into trainloads).

ACTIVITIES

Our discussion thus far has been confined to cross-sectional measures; we now turn to "dynamic configurations." "Activity" will be used as a generic term for such processes. There seem to be several related but distinct concepts here.

We start with the measure space of histories (Ω, Σ, μ). Just as a configuration is a restriction of a cross-sectional measure, one may think of an activity as a restriction of μ itself; i.e., one takes a measurable subset of histories, $H \in \Sigma$, and refers to μ restricted to H as an activity.[44] For example, H may be the set of all histories whose itineraries lie in region F and whose transmutation paths lie in resource set E at moment t or, more generally, lie in $E_i \times F_i$ at moment t_i, $i = 1, \ldots, n$. Or H can be the set of histories lying in $E \times F$ ($\in \Sigma_r \times \Sigma_s$) at least once during time interval G, or lying in $E \times F$ throughout time interval G.

A different approach starts, not with (Ω, Σ, μ), but with the production and consumption measures λ_1 and λ_2 derived from it. Recall that the

---

44. We have mentioned some problems concerning which subsets are to be considered measurable in the set of histories. Here we assume implicitly that all sets mentioned are in fact measurable.

universe set here is $R \times S \times T$, with the interpretation: $\lambda_1(E \times F \times G) =$ mass of all histories "produced" or "born" during period G and starting in resource set E and region F. And λ_2 is the same, with "consumed" or "dying" in place of "produced" or "born." An activity in this sense is a restriction of λ_1 or λ_2 (or both, or net production $\lambda_1 - \lambda_2$) to a subset of $R \times S \times T$ of the form $R \times E$, E being a measurable subset of $S \times T$.

It is very common and useful to combine these approaches. Consider, for example, the description of the operation of a certain shoe factory from time t_1 to t_2. The appropriate set of histories H consists of all that act as "factors of production" for some time during the interval $[t_1, t_2]$—the factory building, the land on which it sits, the tools, the machinery, the workers and management personnel, even the air and the gravitational field at the site. For production and consumption, take the restricted set $R \times F \times [t_1, t_2]$, F being the region occupied by the factory and grounds. Measure λ_1 will include the production of shoes, but also the production of smoke, noise, odor, leather scraps, etc; λ_2 will include the consumption of rubber, leather, nails, glue, water, electricity, fuel, etc. The activity in question may now be defined as the triple of measures $(\mu, \lambda_1, \lambda_2)$, μ being restricted to H and λ_1, λ_2 to $R \times F \times [t_1, t_2]$, or as the pair $(\mu, \lambda_1 - \lambda_2)$, $\lambda_1 - \lambda_2$ being a signed measure or, more generally, a pseudomeasure.

This is by no means the only possible representation of this process. For one thing, it omits inventories of materials that are raw, in-process, and finished. These may be incorporated if desired by expanding the set of histories H to include them. The distinction between materials and factors is in any case not a sharp one. Materials change form (i.e., position in R) relatively fast, factors relatively slowly. But it is sometimes convenient to treat the commuting work force as if it were consumed upon arriving each morning and produced anew upon leaving. In this representation, labor would be recorded in the λ_1, λ_2 accounts, each worker being counted once for each separate commuting trip he makes.

Let us try to classify various activities. A first distinction is between activities that "stay put" and those that change location over time. Letting H be the set of histories over which a given activity q is defined, the *location* of q at time t is the region

$$F_t = \{x \mid h_s(t) = x \text{ for some } h \in H\} \qquad (2.7.5)$$

(Recall that h_s is the itinerary of history h. The fictitious point z_o, indicating that a given history is not in existence at instant t, is excluded from this set.) If F_t is constant over the interval $[t_1, t_2]$, the activity is said to be *sedentary* in that interval.

We can also define the location of an activity in terms of the production-consumption measures λ_1 and λ_2. Their universe set is $R \times E$, and the spatial cross-section of E at time t, $\{s \mid (s, t) \in E\}$, is the *location* of the activity at t. If an activity is defined having both μ (stock) and λ (flow) components, it is generally convenient to make this set coincide with F_t of (5).

Is the shoe factory activity discussed above sedentary in the interval $[t_1, t_2]$? Not according to the original definition, because the work force commutes in and out of the plant site. But if we transfer labor (and any other factors entering or leaving the plant site during the interval in question) to the production-consumption account, as suggested above, then it becomes sedentary. This shows that whether a certain process is to be considered sedentary is at least partly a matter of convention.

A *simply located* activity[45] is one whose location at any instant is a single point of S (or the empty set). If this single point is the same over the interval $[t_1, t_2]$, it is also a sedentary activity. The concept of simple location for an activity is of course an idealization, but a very useful one, as Chapter 8 will demonstrate.

Among sedentary activities we distinguish those that are *stationary* or *steady state*. First, look at activities defined in terms of production and consumption over universe set $R \times F \times T$, F being a region. As usual T is represented by the real numbers. λ, over the measurable space $(R \times F \times T, \Sigma_r \times \Sigma_F \times \Sigma_t)$, is *stationary* iff for all $E \in \Sigma_r \times \Sigma_F \times \Sigma_t$ and for all real numbers c,

$$\lambda(E) = \lambda\{(r, s, t) \mid (r, s, t + c) \in E\} \qquad (2.7.6)$$

Here λ can stand for production or consumption or their difference, net production. Equation (6) states that, say, production is invariant under displacement in time, and this captures the intuitive notion of a "steady rate of production."

The concept of "steady state" for activities defined in terms of histories is a bit more complicated.

Definition: Let h be a function whose domain is the real numbers with values in a set B, and let c be a real number; the *c-displacement* of h is the function h^c:reals $\rightarrow B$ given by

$$h^c(x) = h(x - c) \qquad (2.7.7)$$

all real x.

Interpreting the domain as Time, (7) states that wherever h goes, h^c goes c time units later ($-c$ time units earlier if c is negative). If h is a history, then h^c is also a history for all real c.

Now let H be the set of histories over which an activity q, with re-

---

45. Also called a Weberian activity, after Alfred Weber. See Section 9.4.

stricted measure μ, is defined. Activity q is *stationary* iff

(i) H is closed under displacement (i.e., if $h \in H$, then the c-dis-
placement of h also belongs to H for all real c);
(ii) if $E \subseteq H$ is measurable and c is real, then $\mu(E) = \mu\{h \mid h^c \in E\}$.

Condition (i) implies that q is sedentary; condition (ii) implies that
the cross-sectional distribution of mass for the histories H at time t is independent of t, that the double cross-sectional distribution at times t_1 and
t_2 depends only on the difference $t_2 - t_1$, and, in general, that the entire
process "looks the same" if shifted arbitrarily through time. Finally, if an
activity is given by a pair $(\mu, \lambda_1 - \lambda_2)$ or triple $(\mu, \lambda_1, \lambda_2)$, it is to be
considered stationary iff all its components are stationary according to the
respective definitions above.

Stationarity is a severe requirement. Under it there can be no batch
production, only a continuous flow; no shifts, only a continuous arrival
and departure of workers; no daily, weekly, or seasonal cycles. It is the
ultimate in uneventfulness.

ACTIVITY TYPES

Just as with configurations one distinguishes *particular* activities, located in specific portions of Space-Time, and activity *types*. We follow
the same procedure as above; viz., first to determine when two specific
activities are considered to be of the same type, and second to define
activity type abstractly.

Let f be a measurability-preserving isometry of S onto itself. We
consider two activities to be of the same type iff there is such an f that,
together with a time displacement, transforms one of these activities into
the other.

Consider two activities defined in terms of production or consump-
tion: say λ on the universe set $R \times G$, and λ' on the universe set $R \times G'$
(where G and G' are measurable subsets of $S \times T$). Then λ and λ' are
said to be of the *same activity type* iff there is an $f : S \to S$ as above and a
real number c such that

$$G = \{(s, t) \mid (f(s), t + c) \in G'\} \qquad (2.7.8)$$

and such that

$$\lambda'(E) = \lambda\{(r, s, t) \mid (r, f(s), t + c) \in E\} \qquad (2.7.9)$$

for all measurable $E \subseteq R \times G'$.

Equation (8) states that the two Space-Time "regions" G and G' have
the same "size" and "shape," while (9) states that the relative patterns of
production (or consumption) within these regions are the same. These
activities need not be sedentary.

Next take two activities defined in terms of histories: say μ on universe set H, and μ' on H' (H and H' being measurable subsets of Ω). With $f:S \rightarrow S$ as above, c a real number, and h a history, the c, f-*transformation* of h is the history $h^{c,f}$ given by the following rules:

For the transmutation path: $h_r^{c,f}(t) = h_r(t - c)$, all $t \in T$.

For the itinerary: $h_s^{c,f}(t) = f(h_s(t - c))$, all $t \in T$.

If $h(t - c) = z_o$, it is understood that $h^{c,f}(t) = z_o$.

That is, the transmutation path of $h^{c,f}$ is simply the c-displacement of the transmutation path of h; this is unaffected by f. The itinerary of $h^{c,f}$ is the f-transformation of the c-displacement of the itinerary of h. (If f is the identity, this simply reduces to the c-displacement of h.)

Measures μ and μ' are then said to be of the *same activity type* iff there is a measurability-preserving isometry $f:S \rightarrow S$ and a real number c such that

$$H = \{h \mid h^{c,f} \in H'\} \tag{2.7.10}$$

and such that

$$\mu'(E) = \mu\{h \mid h^{c,f} \in E\} \tag{2.7.11}$$

for all measurable $E \subseteq H'$.

Finally, if two activities q and q' are both triples $(\mu, \lambda_1, \lambda_2)$ and $(\mu', \lambda_1', \lambda_2')$ respectively, or pairs (μ, λ) and (μ', λ'), they are said to be of the *same activity type* iff each of the components is of the same activity type according to the above respective definitions, with the *same* f and c satisfying all simultaneously.

The similarity between these definitions and that of stationarity is patent. In fact, an activity is stationary iff it is of the same type as any time displacement of itself, S being held fixed.

We now define the concept of activity type in the abstract, first for activities of the production-consumption type. We are given the measurable space of Resources (R, Σ_r). Let (T', Σ_t') be the real line and its Borel field. An *activity type* is then defined as a quintuple, consisting of a set S', a metric d' on S', a σ-field Σ_s' on S', a subset G' of $S' \times T'$ which is measurable: $G' \in \Sigma_s' \times \Sigma_t'$; and a measure λ' on universe set $R \times G'$.

Here the notation S', T' is meant to suggest "abstract" Space and Time; there is no need to abstract from R, since this is already a set of resource types. Measure λ' then gives the production or consumption pattern over the abstract space $R \times G'$; if we are dealing with *net* production, λ' is a signed measure or, more generally, a pseudomeasure.

This activity type is *exemplified* on the set $R \times G$ (where G is now a subset of "real" Space-Time $S \times T$) iff there is a measurability-preserving isometry $f:S \rightarrow S'$ and a real number c such that (8) and (9) are satisfied.

The construction of activity types in the sense of histories is similar. With S', T' as above, an *abstract history* is a function with domain T' and range in $(R \times S') \cup \{z_o\}$ satisfying the requirements for being a history in the ordinary sense. An *activity type* is then a measure μ' whose universe set is a measurable set of abstract histories H'. This is *exemplified* on the set of "real" histories H iff there is an $f:S \rightarrow S'$ and a number c, as above, such that (10) and (11) are satisfied.

Finally, we may have an activity type having both a "histories" component and a production and/or consumption component, both structures being superimposed on abstract Space-Time $S' \times T'$. This complex activity type is then *exemplified* on "real" sets H and $R \times G$ iff there is an f and c, as above that satisfy (8), (9), (10), and (11) *simultaneously*.

SCALE OF ACTIVITIES

The question of whether there are "constant returns to scale" remains a vexing one in the economics literature. This is properly a question of technology, not description, and we therefore do not discuss it here. We do suggest, however, that much of the disagreement arises from the fact that "scale" is an ambiguous concept. In this section we spell out several of its possible meanings.

Given two activities q and q', when is q' a *k-fold expansion* of q, where k is a positive real number? All the following answers have this in common: when $k = 1$, they reduce to the concept defined above of q and q' being the same activity (or activity type). Thus we are really seeking to generalize to the case where q and q' are somehow "similar" but unequal in "size."

Specifically, let q be the complex activity consisting of the measure μ over the measurable subset of histories H, and λ over $R \times G$ (G being a measurable subset of $S \times T$). Measure λ represents the production-consumption components and may be a signed measure, a pair of measures, etc. (the following discussion is valid for all these possibilities). Similarly, q' consists of the measure μ' over H', and λ' over $R \times G'$ (λ' being of the same character as λ; $G' \subseteq S \times T$ and measurable).

Definition: q' is a *k-fold expansion* of q in the *intensive sense* iff there is a measurability-preserving isometry $f:S \rightarrow S$ and a real number c such that

(i) equations (8) and (10) are valid;

(ii) equation (9) is replaced by

$$\lambda'(E) = k \cdot \lambda\{(r, s, t) \mid (r, f(s), t + c) \in E\} \qquad (2.7.12)$$

for all measurable $E \subseteq R \times G'$;

(iii) equation (11) is replaced by

$$\mu'(E) = k \cdot \mu\{h \mid h^{c,f} \in E\} \qquad (2.7.13)$$

for all measurable $E \subseteq H'$.

That is, the "locations" of the two activities in Space-Time have the same "size" and "shape"; the relative distribution of mass over these locations is also the same, but the absolute levels on corresponding sets are k times greater for q'. If these represent two shoe factories, we would find k times as much machinery, inventories, workers, etc., crowded into the same area, turning out shoes and consuming materials at k times the rate in one of these factories as compared to the other.

This is a rather unusual conception of "scale," and we list it first only because it is the simplest of the possibilities. Indeed, one is tempted to say, "This is not a scale expansion at all: *all* factors, including land, must be multiplied in proportion, while the 'intensive' concept leaves the quantity of land unchanged!" We now try to pin down the alternative notion of scale that underlies this expostulation.

The first difficulty revolves about the concept "quantity of land." "Land" is an ambiguous term, sometimes referring to a certain class of resources that includes dirt, minerals, and trees; and sometimes it is a synonym for Space. Now, in the intensive scale concept, land in the first sense *has* been multiplied by k: the soil is k times more densely packed, etc. The protest above must therefore refer to the second meaning of the term "land." But what then is the "quantity of Space," and how does one multiply it by k?

The simplest approach is to identify "quantity of Space" with *volume* in the case of 3-space, and with *area* in 2-dimensional cases such as the plane or the surface of a sphere.[46] We assume that S is endowed with such a quantity measure α and refer to it generically as "area."

Next we need to generalize the concept of isometry.

Definition: Let (A, d) and (B, d') be two metric spaces. Mapping $f : A \rightarrow B$ is a *similarity* iff

---

46. How to define these measures in the case of more complicated manifolds is itself a rather difficult problem into which we do not delve. See L. Cesari, *Surface Area*, Annals of Mathematics Studies, #35, (Princeton Univ. Press, Princeton, N.J., 1956); or T. Radó, *Length and Area*, Vol. 30 (American Mathematical Society Colloquium Publ., Providence, R. I., 1948).

(i) f is onto;

(ii) there is a positive real number m such that

$$d'(f(x), f(y)) = m \cdot d(x, y) \text{ for all } x, y \in A$$

Number m is called the *dilatation* of f.

Thus f has the effect of stretching all distances by the factor m. (If $m < 1$, this is of course a shrinkage; f is an isometry iff $m = 1$.) If f is a similarity with dilatation m, it has an inverse $g: B \rightarrow A$, which is a similarity with dilatation $1/m$. Roughly speaking, a similarity preserves "shape" but not "size."

Let A be 3-space, and d the Euclidean metric on A; let f be a similarity from (A, d) to itself, with dilatation m. Then it may be shown that f is measurability preserving and that

$$\mu(E) = m^3 \mu\{a \mid f(a) \in E\} \tag{2.7.14}$$

for all Borel sets $E \subset A$, where μ is volume (i.e., 3-dimensional Lebesgue measure). Thus f expands volume by the cube of the dilatation. Similarly, if (A, d) is the Euclidean plane then (14), with m^2 substituted for m^3 and μ being 2-dimensional Lebesgue measure, is valid. We confine attention to these two cases.

Again let q be the activity given by measure μ on the set of histories H and λ on $R \times G$; similarly, q' is given by μ' on H' and λ' on $R \times G'$.

Definition: q' is a *k-fold expansion* of q in the *extensive sense* iff there is a real number c and a similarity $f: S \rightarrow S$ with dilatation $k^{1/D}$ such that (8), (10), (12), and (13) are valid. (Here D is the dimensionality of Space: $D = 2$ for the plane, and 3 for 3-space.)

Thus the single difference between the intensive and extensive scale concepts is that the spatial transformation has a dilatation of 1 in the former case, and of $k^{1/D}$ in the latter. The reason for this latter choice is that area (or volume for $D = 3$) expands in exactly the same ratio as μ and λ expand on corresponding sets. Thus the *average density* of all measures with respect to the "quantity of Space" is the same for q and q' for corresponding sets. This presumably is the meaning of a "proportional expansion of all factors, including land."

If q is the normal shoe factory and q' is a k-fold expansion of q in the extensive sense, the workers in q' would be Brobdingnagians (if $k > 1$) or Lilliputians (if $k < 1$); all machinery and plant would expand or shrink in the same proportion. Stocks of resources and rates of output and inflow would expand by the factor k, but per unit area (or volume) would remain the same as before.

We merely mention here another class of "scale" concepts, those involving *time dilatation*. In all cases discussed the only transformation to which T was subjected was a simple translation $t \rightarrow t + c$. But there could also be a scale factor $t \rightarrow kt + c$, where k is a positive real number other than 1. The effect of this is to change the *speed* at which processes occur, the rate at which "particles" fulfill their histories. There are numerous possibilities, depending on whether S is also subjected to a dilatation, on the factors multiplying μ and λ and on the relations among these four magnitudes. An example of a time dilatation is the relation between a film run at normal speed and the same in slow motion. One could even have $k < 0$, which corresponds to running the film backward.

Another concept rather different from any of the foregoing is that of scale in the *duplicative* sense. Here the expansion factor k must be a positive integer. Again let q be the activity given by μ on H and λ on $R \times G$, and q' the activity given by μ' on H' and λ' on $R \times G'$.

Definition: q' is a *k-fold expansion* of q in the *duplicative sense* iff there is a measurable partition of H' into k pieces (say $H_1, \ldots, H_k$) and of G' into k pieces (say $G_1, \ldots, G_k$) such that the activity q_i', defined by μ' restricted to H_i and λ' restricted to $R \times G_i$, is the same activity as q for all $i = 1, \ldots, k$, with the same time translation c serving for all $i = 1, \ldots, k$.

This definition captures the notion of the same processes running "side by side," as in row housing, banks of machines, or the plants of a perfectly competitive industry. The stipulation on c requires simultaneous acting out by the k-fold duplicates; this could be relaxed to allow for staggered timing or even for duplication by a k-fold repetition in T.

This completes our short survey of some meanings of "scale." We shall expand on it later with a discussion of "returns to scale" (Section 4.7).

SOME EVERYDAY ACTIVITIES

We examine here how various broadly defined activity categories—such as mining, transportation, and services—fit into the present framework. Since these categories were not designed to be so fitted and their definitions involve many ad hoc elements, we can hope at best for a broad-brush characterization, with many errors in detail.

One can classify activities from many points of view, e.g., by the number of persons participating. Thus one may distinguish *natural* activities (no participants), *private* activities (one participant), and *shared* activities (more than one). Of the latter, one may distinguish various authority structures, cooperative vs. conflictive aspects, who performs what services for whom. At the moment, however, we are mainly interested in the physical structure of activities. We take activities in the histories sense and pose the problem as follows: What characterizes the defining set of histories H of, say, an activity classified as "construction?"

Consider transportation, for example. An *ideal transportation activity* is one in which all histories $h \in H$ have constant transmutation paths, at least over the interval $[t_1, t_2]$ to which one is referring. That is, a typical "particle" may change its location in S but not its resource form $r \in R$. This is of course an approximation: travelers get fatigued, cargo spoils, vehicles suffer wear and tear.

The foregoing approximate description applies not only to the activities customarily called "transportation" but to several others as well: utilities such as water, gas, electricity, and sewage disposal; communications such as telephone and broadcasting. The postal system consists mainly of transportation activities. Most everyday activities, in fact, will have a transportation subactivity in them.

For some transportation activities it is useful to idealize even further and take H so that all its members have the same itinerary over the relevant time interval. This makes it simply located, and the activity may be represented by a resource bundle (i.e., a measure over a subset of R) traveling over the "track" determined by the common itinerary. This approximation is good for transportation that goes in channels (roads, pipes, wires) but poor for broadcasting.

A special case of transportation is storage, the simplest of all activities. An *ideal storage activity* is one in which all histories are constants, at least over the relevant time interval; i.e., the "particles" change neither their resource forms $r \in R$ nor their locations $s \in S$. This approximates processes in the real world that change "slowly." What is to be considered "slow" depends on one's focus of attention and scale of observation. To the historical geologist the earth has undergone great changes, but on the human scale it has a certain massive sameness except for changes in the weather and "minor" fluctuations such as earthquakes and floods. Again, an economist interested in short-term business fluctuations can treat the stock of capital goods and population as constants, but this is not true for one studying economic development.

Note that when one compares ideal transportation or storage with the bundle of processes labeled "transportation" or "storage" in the real world, one must not only approximate but also abstract from certain aspects. The fuel consumed in refrigerating a warehouse does not itself satisfy the conditions for ideal storage even approximately. In simply located transportation one focuses on the bundle being moved (including the vehicle, if any) and abstracts from the fixed plant of the transportation system: the train, but not the rails; the electric current, but not the wires.

Trade, retail and wholesale, is largely a matter of transportation and storage.

Motion in S means motion relative to the earth, since we have conventionally taken the earth to be fixed in S. There is another class of activities, however, in which the essential feature resides, not in motion relative to the earth, but in the motion of the "particles" relative to each other. In particular, a *fission activity* is one in which the itineraries diverge from each other over time, and a *fusion activity* is one in which they converge toward each other over time.

These characterizations are rather vague. One could distinguish further according to whether the divergence of itineraries did or did not depend on the resource states of the histories, giving us *segregating activities* or simple *scattering activities* respectively. Also, one could go into the various ways of measuring dispersion and association of spatial distributions.[47] But this is unnecessary for the present discussion, which is impressionistic in any case.

Going to *manufacturing,* it appears (very roughly, and with many exceptions) that one can classify manufacturing processes into fission activities, in which things are taken apart or separated into component substances, and fusion activities, in which they are put together or assembled into larger units. First-stage processing of raw materials is generally of the fission type because natural products are unwieldy in size or ingredients are combined in nonuseful forms: crude oil is refined, ores are beneficiated, crops are winnowed, logs and carcasses are chopped up. Later stages tend to be of the fusion type: cars are assembled; cotton is spun, woven, and sewn into clothing; drugs are blended.[48]

Construction is a kind of fusion process distinguished by the nature of its product. This is not so much a question of size (supertankers and jumbo jets are larger than most buildings) but rather that the product is attached to the earth; it is "real" rather than "movable" property. There are cases where it is not clear whether a given item is real or movable (e.g., fixtures, "mobile" homes), but generally, buildings, dams, bridges, roads, railway tracks, airports, docks, and pipelines all belong to the former category, while vehicles, machines, and consumer goods belong to the latter. In summary, construction is a fusion to the earth.

Mining is the reverse of construction. It consists of detaching pieces

47. D. S. Neft, *Statistical Analysis for Areal Distributions,* Monogr. Ser. #2 (Reg. Sci. Res. Inst., Philadelphia, 1966).

48. Fission and fusion activities correspond roughly to Beverly Duncan's "processing" and "fabricating" industries, and even more roughly to Alfred Weber's "material-oriented" and "market-oriented" industries. See. O. D. Duncan, W. R. Scott, S. Lieberson, B. Duncan, and H. H. Winsborough, *Metropolis and Region* (Johns Hopkins Press, Baltimore, 1960), pp. 57–58 and Ch. 7; and E. M. Hoover, *The Location of Economic Activity* (McGraw-Hill, New York, 1948), pp. 31–38, respectively.

from the earth. This will include not only the extraction of minerals in the conventional sense but also the undoing of previous construction in demolition work. Tunneling also would be considered a form of mining, just as land filling would be considered a form of construction. In summary, mining is a fission from the earth.

What about *agriculture, forestry, hunting,* and *fishing?* These are generally classified as extractive, and indeed they have a strong mining component, as defined above. (A certain style of agriculture is known pejoratively as "soil mining.") But these increasingly tend to be run as self-sustaining processes by reseeding, restocking, and fertilization, so that the "construction" aspect is becoming as important as the "mining" aspect.

This brings us to *services.* At first glance, this seems to be such a heterogeneous category (embracing repairs; business, personal, and professional services; entertainment; education) that no succinct property could begin to approximate it. Indeed, this will be our contention as far as the *physical* structure of these processes is concerned.

Adam Smith divided workers into productive and unproductive, and it is clear from his examples of the latter (servants, lawyers, musicians) that he had in mind more or less the present-day distinction between those engaged in the production of goods vs. services.[49] Although service workers have long since been admitted as contributors to the national product,[50] the tradition lingers that they produce an "intangible." This is clearly wrong in detail: laundering is a service and there is nothing intangible about dirty laundry (or, rather, the transformation from dirty to clean laundry). What makes laundering a service is that the laundry does not own the item it is cleaning. We claim that this characteristic, rather than any intangibility, is what distinguishes the bulk of the activities known as services.

If true, this means that services are such, not because of any physical property of the activity, but because of the ownership relations among the interested parties. Hence the same activity may be either a service or not, depending on the organization of the industry. Suppose laundries operated as used car dealers do, buying dirty shirts, cleaning them, and reselling them on the second-hand market. This may well be considered goods production. Conversely, suppose copper refineries operated as follows. Miners ship their ore to the refineries without relinquishing ownership; the refined copper is then returned to the owners, who pay a fee for the service. This is completely analogous to the organization of laundries. Would not copper refining then be considered a service industry?

We quickly run through the major service categories to indicate how

---

49. Adam Smith, *Wealth of Nations,* (Modern Library, New York, 1937), Book 2, Ch. 3.
50. In the West, not in the Communist world.

well this characterization applies. There is no problem with repair services in general, e.g., watches, shoes, cars, radios. In all cases owner *A* relinquishes possession of the item to repairman *B*, who fixes and returns the item to *A*.

What about *rentals,* e.g., of houses, hotel rooms, or cars? Let us look at the servicing relation a little more closely. Person *A* owns some items α; *B* owns some items β; α and β are brought together with the result that α is benefited, in return for which *A* pays a fee to *B*. For repairs, β typically consists of the repairman himself and his tools; for laundering it consists of cleaning equipment, etc. Now the fee can be described as a rental payment for the services of β; rentals and services are two ways of looking·at the same transaction. When gardener *B* trims *A*'s rosebush, we may say either that *A* rents *B*'s labor services or that *A* relinquishes possession of his rosebush to *B*, who returns it to *A* in improved condition. In the case of house rental, β is the house itself. What is α? It is *A* himself and his possessions, which are provided with shelter services.

Since a person always owns his own body (in a nonslave society) any benefits to *A*'s body (including his mind) made by another person *B* automatically fall into the category of service activity, according to our ownership criterion. This includes the services of physicians, dentists, barbers, sex partners, and, ultimately, morticians. It includes the services of clergymen, entertainers, and all who provide information: teachers, lawyers, physicians again, consultants, employment agencies, private detectives, credit bureaus, telephone answering services, and astrologers. Perhaps this group explains the connection of services with intangibles: one cannot see directly the changes in a person's information state or welfare level.

Most *government* activities would be services as here defined because they provide benefits to persons and goods not owned by government.

This brief survey appears to cover most activities customarily classified as "services." Of the remainder, a number seem simply to be misclassified. (We are of course now turning the tables and using our ownership criterion to determine what "should be" considered a service activity.) From our point of view, photographers, duplicating services, and sign painters are goods producers. It is true that their products are closely tailored to individual clients, but the same is true of much housebuilding, job shop work, printing, and other activities classified as goods production. Similarly, a lawyer writing up a will or a contract is engaged in goods production. The most important misclassified industry is advertising, whose product (again tailored to individual clients) is advertising copy, consisting of jingles, skits, blurbs, etc.

Services are discussed further in connection with rental markets and bailments in Section 6.2 and with "service systems" in Section 9.6.

2.8. MULTILAYER MEASURES

For a general measure space (A, Σ, μ) nothing specific needs to be assumed about the nature of the points of A. In our applications A has variously been a product space built up from R, S, and T, or a space of functions (histories) whose domain and ranges are built up from these, or subsets of the foregoing, etc. We now discuss some cases in which the points of universe set A are themselves measures over some other measurable space.

Let us spell this out. Given a fixed measurable space (B, Σ), let $\mathbf{M}$ be the set of all measures over it ($\mathbf{M}$ could also be the set of all signed measures or all pseudomeasures; the discussion would be unaffected). Now consider a measure space $(\mathbf{M}, \Sigma', \mu)$, in which $\mathbf{M}$ itself plays the role of universe set. We shall refer to this as a *2-layer measure*. Next, suppose B itself is a set of measures over still another space, so that each member of $\mathbf{M}$ is itself a 2-layer measure; then $(\mathbf{M}, \Sigma', \mu)$ will be referred to as a *3-layer measure*. This clearly extends to any finite $n = 1, 2, \ldots$.

We consider some ways in which such multilayer structures arise in applications. Let us first bring in the factor of *uncertainty*. We have noted that in principle the measure space of histories (Ω, Σ, μ) provides a complete description of the world for social science purposes. But one never knows exactly what the measure μ is. It is desirable, then, to try to represent states of relative ignorance or degrees of belief concerning the true measure μ.

We adopt a Bayesian point of view, according to which "state of belief" is representable as a probability measure over the universe set of "possible worlds."[51] Specifically, a state of belief is given by $(\mathbf{M}, \Sigma', \pi)$, where $\mathbf{M}$ is the set of all measures over the measurable space of histories (Ω, Σ), Σ' is a σ-field on $\mathbf{M}$, and π is a probability measure with domain Σ'. For any $E \in \Sigma'$, $\pi(E)$ is the probability ($=$ "degree of belief") that the true mass distribution μ, over the space of histories, belongs to the set of measures E. This is a 2-layer measure.

How is Σ' determined? The heuristic principle we have used before states that all sets that are "conceptually observable" should be considered measurable. Here the equivalent principle would seem to be: any set of measures that is "sufficiently simple" so that a mind could, conceptually, hold a degree of belief concerning it should be considered measurable. This is rather vague and is best explained by examples. If F is a measurable set of histories and c is a number, the event "the total mass concentrated on the histories in F exceeds c" would appear to be one to which a degree of belief could be attached. This means that the set of

---

51. See H. E. Kyburg, Jr., and H. E. Smokler, eds., *Studies in Subjective Probability* (Wiley, New York, 1964), especially the essays by B. de Finetti and B. O. Koopman.

measures

$$\{\mu \mid \mu(F) > c\} \tag{2.8.1}$$

is to be considered measurable (i.e., belongs to Σ') for all $F \in \Sigma$, all real c. In particular, suppose F is the set of histories originating in subset G of $R \times S \times T$; then the probability π attached to the set (1) gives the degree of belief that total births or production in G exceeds the value c.

This example provides an illustration of how the probability π can be induced onto simpler spaces. Let F retain its meaning of the set of histories originating in fixed set G, and consider the function with domain **M** that assigns $\mu(F)$ to measure μ. This induces a probability measure on the real line, which is exactly the state of belief concerning production in set G. This induction process is completely analogous to the many examples in Section 2.5 of the induction of μ on the space of histories (Ω, Σ) onto simpler spaces.[52]

The case of perfect certainty, with a known measure μ_o over (Ω, Σ), may be identified with the special case of the probability measure $(\mathbf{M}, \Sigma', \pi)$ in which π is *simply concentrated* with all mass at the "point" $\mu_o \in \mathbf{M}$.

For a second example, consider the structure of the Resources set R. Taking people types as points of R, a complete specification of a person $r \in R$ will include his mental state, in particular his state of knowledge. Assume for the momemt that r describes a person in a state of complete certainty. His state of knowledge will then include a description of the world, which is represented as a measure over the space of histories Ω. This leads again to something resembling a 2-layer measure, for some of the points of R have an internal structure involving measures, while the overall descriptive measure is on a universe set built up in part from R.

Once we admit structures into R that involve measures over universe sets involving R, we appear to be led to infinite-layered structures. The reason is that a person's state of knowledge will itself be (at least) 2-layered, since it involves knowledge of other states of knowledge; and one cannot stop at any finite number of layers.

Whether we can build a useful (or even consistent) theory from such an infinite regress remains to be seen. One consideration simplifies things, however. The measure space (Ω, Σ, μ) gives a complete description of the world. But any person, even in a state of perfect certainty, will have a

---

52. For the theory of such "random measures" see D. J. Daley and D. Vere-Jones, A summary of the theory of point processes, in P. A. W. Lewis, ed., *Stochastic Point Processes: Statistical Analysis, Theory, and Applications* (Wiley-Interscience, New York, 1972), pp. 299–383, esp. pp. 316–18. Much of this survey applies to general random measures, not just to integer-valued measures.

limited capacity to assimilate information. This limitation may be represented formally by replacing Σ by a small sub-σ-field $\Sigma' \subseteq \Sigma$, yielding an *aggregation* of the original measure and losing detail. Knowledge of other people's states of knowledge (and of one's own past and future states) would be in terms of an even smaller σ-field $\Sigma'' \subseteq \Sigma'$, hence even more aggregative and sketchy, etc.

Finally, mental states of uncertainty can be represented as above by a probability measure over a universe set **M** of physical measures. We again get an infinite regress in the form: A's degree of belief concerning B's degree of belief. . . .[53]

In this book we shall not have occasion to use multilayer measures in any of the interpretations just discussed. We do, however, use them in another way, as a representation of technology. Here **M** will be the set of "basic feasible activities," and feasible activities in general will be measures over **M**. For detailed discussion see Section 4.5.

---

53. Even an omniscient Deity would have need for probability concepts to represent the states of mind of the less than omniscient creatures inhabiting the world.

3

The Comparison of Infinite Measures

This chapter develops the theory of pseudomeasures. These are extensions of signed measures that enable one to carry out, for example, the operation of subtraction even for infinite measures. Much of standard measure theory generalizes to pseudomeasures, and many theorems can be stated without qualifying conditions as to finiteness, integrability, etc. Thus the theory should interest even "pure" mathematicians.

The theory also has numerous applications. First, it enables one to "net" freely, even when both "grosses" are infinite. The subtraction of consumption from production has already been discussed. Another example is migration: One would like to get net migration by subtracting gross out-migration from in-migration, even when the latter two measures are infinite (e.g., this might occur on an infinite plane or with an infinite time horizon). In general, it enables one to perform arithmetical accounting operations freely on measures without worrying about the appearance of the meaningless expression $\infty - \infty$.

Second, it allows one to compare different "infinite utility streams" such as arise in the evaluation of economic development programs. The "overtaking" and similar criteria developed to deal with these problems find their natural place within the theory and emerge as special cases of a general approach.

Even more generally, pseudomeasures are in many cases a natural way of representing preference orderings, i.e., instead of representing preferences by real-valued utility functions, one uses pseudomeasure-valued utility functions with various natural orderings on the space of pseudomeasures. This arises for infinite-horizon development programs, for problems of location theory on the infinite plane, and for preferences among uncertain situations.

3.1. JORDAN DECOMPOSITION THEORY

The formal development of pseudomeasure theory goes through two stages. The first involves a generalization of the concept of Jordan decomposition. This operation, which applies to any pair of measures, has an interesting and elegant theory by itself. In this section we develop only that portion of the theory which leads directly to pseudomeasures or has direct applications elsewhere in this book. Other results will be presented as exercises (which are generally fairly difficult to prove).

Pseudomeasures per se arise from the application of the Jordan decomposition to σ-finite measures. This enables one to define algebraic operations, integration, and ordering relations in a manner that is fruitful for applications and also of considerable mathematical interest in itself.

Let (A, Σ) be a measurable space. All sets referred to below are assumed to belong to Σ, and all measures and other set functions are assumed to have Σ as their domain.

Definition: Let (μ, ν) be an ordered pair of measures; the *upper variation* of (μ, ν) is the set function λ^+ given by

$$\lambda^+(E) = \sup\{\mu(F) - \nu(F) \mid F \subseteq E, \nu(F) < \infty\} \qquad (3.1.1)$$

for all $E \in \Sigma$.

That is, to find $\lambda^+(E)$, form the difference $\mu(F) - \nu(F)$ for each measurable subset F of E for which $\nu(F)$ is finite, and then take the supremum over these numbers. Similarly, consider the following.

Definition: The *lower variation* of (μ, ν) is the set function λ^- given by

$$\lambda^-(E) = \sup\{\nu(F) - \mu(F) \mid F \subseteq E, \mu(F) < \infty\} \qquad (3.1.2)$$

for all $E \in \Sigma$.

Definition: The *Jordan decomposition* of the ordered pair of measures (μ, ν) is the ordered pair of set functions (λ^+, λ^-) given by (1), (2).[1]

Theorem: The set functions λ^+, λ^- are actually measures.

Proof: λ^+ and λ^- are nonnegative (since $\phi \subseteq E$ for all $E \in \Sigma$), and $\lambda^+(\phi) = \lambda^-(\phi) = 0$. It remains only to prove countable additivity.

Let (E_n), $n = 1, 2, \ldots$, be a countable measurable packing with $E = \cup\{E_n\}$. Let F satisfy $F \subseteq E$, $F \in \Sigma$, $\nu(F) < \infty$; then $F \cap E_n$ ful-

---

1. C. Carathéodory defines a similar operation in his *Algebraic Theory of Measure and Integration*, F. E. J. Linton, trans., P. Finsler, A. Rosenthal, R. Steuerwald, eds. (Chelsea, New York, 1969), pp. 299–304, in the context of "regular outer measures" on a "σ-ring of somas." But he then takes the difference of the two variations (on the somas where this is well defined) and thereby misses the following theory, which depends on retaining λ^+, λ^- as separate entities.

fills the same conditions with E_n in place of E for all $n = 1, 2, \ldots$. Hence

$$\lambda^+(E_1) + \lambda^+(E_2) + \cdots \geq [\mu(F \cap E_1) - \nu(F \cap E_1)]$$

$$+ [\mu(F \cap E_2) - \nu(F \cap E_2)] + \cdots = \mu(F) - \nu(F) \qquad (3.1.3)$$

Taking the supremum over all such sets F, we obtain

$$\lambda^+(E_1) + \lambda^+(E_2) + \cdots \geq \lambda^+(E) \qquad (3.1.4)$$

It remains to establish the opposite inequality. If $\lambda^+(E_n) = \infty$ for any n, then $\lambda^+(E) = \infty$, since λ^+ is clearly monotone nondecreasing; in this case we get equality in (4). The remaining case is where $\lambda^+(E_n)$ is finite for all n. Choose a real number $\epsilon > 0$ and, for each $n = 1, 2, \ldots$, choose $F_n \subseteq E_n$ such that

$$\mu(F_n) - \nu(F_n) \geq \lambda^+(E_n) - \epsilon \cdot 2^{-n} \qquad (3.1.5)$$

Noting that $(F_1 \cup \cdots \cup F_N) \subseteq E$, and adding (5) over $n = 1, \ldots, N$, we obtain

$$\lambda^+(E) \geq [\mu(F_1) - \nu(F_1)] + \cdots + [\mu(F_N) - \nu(F_N)]$$

$$\geq \lambda^+(E_1) + \cdots + \lambda^+(E_N) - \epsilon(2^{-1} + \cdots + 2^{-N}) \qquad (3.1.6)$$

Letting $N \to \infty$ in (6), we obtain $\lambda^+(E) \geq -\epsilon + \lambda^+(E_1) + \lambda^+(E_2) + \cdots$. Since $\epsilon > 0$ is arbitrary, we obtain (4) with inequality sign reversed. Hence λ^+ is countably additive. By symmetry, so is λ^-. ∎

How does this operation compare with the ordinary Jordan decomposition of a signed measure λ? We know that λ can be expressed as the difference of two measures (say $\lambda = \lambda_1 - \lambda_2$) where λ_1 or λ_2 is finite. Let (λ^+, λ^-) be the (generalized) Jordan decomposition of the pair (λ_1, λ_2). It is then easily verified that λ^+, λ^- coincide with the ordinary upper and lower variations of λ respectively. Note that the operation above is well defined even if μ and ν are both infinite. In this sense it represents a true generalization of the ordinary Jordan decomposition.

Write $J(\mu, \nu)$ for the Jordan decomposition of (μ, ν). Let ρ_1, ρ_2 be set functions. We write $\rho_1 \leq \rho_2$ to indicate that $\rho_1(E) \leq \rho_2(E)$ for all sets $E \in \Sigma$. Then we have the following.

Theorem: Let $(\lambda^+, \lambda^-) = J(\mu, \nu)$; then

$$\lambda^+ \leq \mu \text{ and } \lambda^- \leq \nu \qquad (3.1.7)$$

Proof: Choose $E \in \Sigma$. For any $F \subseteq E$ with $\nu(F) < \infty$, we have $\mu(E) \geq \mu(F) \geq \mu(F) - \nu(F)$. Taking the supremum over all such F, we obtain $\mu(E) \geq \lambda^+(E)$. Thus $\mu \geq \lambda^+$. The proof that $\nu \geq \lambda^-$ is similar. ∎

Thus J is a "shrinking" operation. We see below that J in effect removes the common part of μ, ν from each of them.

Let μ, ν be measures, with $\mu \geq \nu$. We want to define the operation of *subtracting* ν from μ in a reasonable way. One's first impulse is to take $\mu(E) - \nu(E)$, but this introduces the meaningless operation $\infty - \infty$ if μ and ν are both infinite measures.

Definition: Let $\mu \geq \nu$ be measures; $\mu - \nu$ is defined as the upper variation of the pair (μ, ν).

Note that $\mu - \nu$ is only defined for the case $\mu \geq \nu$. One easily sees, incidentally, that the *lower* variation of (μ, ν) is 0, the identically zero measure.

The following theorem shows that "minus" has at least some of the properties of ordinary subtraction.

Theorem:

(i) Let $\mu \geq \nu$ be measures; then

$$\mu = (\mu - \nu) + \nu \qquad (3.1.8)$$

(ii) Let μ, ν, θ be measures, with $\mu = \nu + \theta$; then

$$\theta \geq (\mu - \nu) \qquad (3.1.9)$$

If ν is σ-finite, (9) is an equality.

Proof: (i) If $\nu(E) = \infty$, then $\mu(E) = \infty$ and (8) is satisfied at E. If $\nu(E) < \infty$, then for any $F \subseteq E$,

$$\mu(E) - \nu(E) = [\mu(F) - \nu(F)] + [\mu(E\backslash F) - \nu(E\backslash F)] \geq \mu(F) - \nu(F)$$

so that $\mu(F) - \nu(F)$ attains its supremum at $F = E$; hence

$$(\mu - \nu)(E) = \mu(E) - \nu(E) \qquad (3.1.10)$$

which again verifies (8) at E; this proves (i).

(ii) Choose $E \in \Sigma$. For any $F \subseteq E$ such that $\nu(F) < \infty$, we have $\theta(E) \geq \theta(F) = \mu(F) - \nu(F)$. Taking the supremum over all such F, we obtain (9). Finally, let ν be σ-finite, so that there is a measurable partition $(A_n), n = 1, 2, \ldots,$ of A such that $\nu(A_n) < \infty$, all n. Then

$$\theta(E \cap A_n) = \mu(E \cap A_n) - \nu(E \cap A_n) = (\mu - \nu)(E \cap A_n) \qquad (3.1.11)$$

from (10). Summing (11) over n, we obtain (9) with equality. ∎

Note that the inequality (9) is sometimes strict; e.g., let A consist of one point, and let $\mu(A) = \nu(A) = \infty$, $\theta(A) = 1$.

Exercises:

1. Let $\mu_1 \geq \mu_2 \geq \cdots \geq \mu_n$ be measures. Show that $(\mu_1 - \mu_n) = (\mu_1 - \mu_2) + (\mu_2 - \mu_3) + \cdots + (\mu_{n-1} - \mu_n)$. (Hint: Prove for $n = 3$ and use induction; consider separately the two cases $(\mu_{n-1} - \mu_n)(E) = \infty, < \infty$.)
2. Let $\mu \geq \nu \geq \theta$ be measures. Show that $(\mu - \nu) = (\mu - \theta) - (\nu - \theta)$. (Hint: Use the result of the preceding exercise for a start.)

We now define two further operations on a pair of measures (μ, ν). Consider measures θ satisfying $\theta \leq \mu$, $\theta \leq \nu$. Is there a largest among them? That is, is there a measure $\tilde{\theta}$ satisfying these conditions and $\geq \theta$ for any θ satisfying them? There is, and it is known as the *infimum* of μ and ν, written $\inf(\mu, \nu)$. One can give an explicit formula for this measure:

$$\inf(\mu, \nu)(E) = \inf\{\mu(F) + \nu(E \backslash F) \mid F \subseteq E\} \qquad (3.1.12)$$

for all $E \in \Sigma$. Note the distinction between the two infs in (12). The one on the right is the ordinary infimum of a set of extended real numbers, viz., $\mu(F) + \nu(E \backslash F)$ for all measurable subsets F of E.

Similarly, the *supremum* of μ and ν, written $\sup(\mu, \nu)$, is the smallest measure $\geq \mu, \nu$. The formula for this is the same as (12), with (ordinary) sup in place of inf on the right-hand side.[2]

Theorem: Let μ, ν be measures, and let $(\lambda^+, \lambda^-) = J(\mu, \nu)$; then

$$\mu + \lambda^- = \nu + \lambda^+ = \sup(\mu, \nu) \qquad (3.1.13)$$

Proof: First we prove the right-hand equality in (13). Choose $E \in \Sigma$. If $\nu(E) = \infty$, this equality is satisfied at E. If $\nu(E) < \infty$, then

$$\nu(E) + \lambda^+(E) = \nu(E) + \sup\{\mu(F) - \nu(F) \mid F \subseteq E\}$$

$$= \sup\{\mu(F) + [\nu(E) - \nu(F)] \mid F \subseteq E\}$$

$$= \sup\{\mu(F) + \nu(E \backslash F) \mid F \subseteq E\} = \sup(\mu, \nu)(E)$$

so the right-hand equality again holds at E. In a similar manner, with μ in place of ν, we prove that $\mu + \lambda^- = \sup(\mu, \nu)$, which establishes (13). ∎

This result has several applications. As a first, we show that the Jordan decomposition operator is *idempotent*. That is, since $J(\mu, \nu)$ is an ordered pair of measures, we may apply the J operator again; but nothing new arises: $J^2(\mu, \nu) = J(J(\mu, \nu)) = J(\mu, \nu)$.

---

2. For discussion of infima and suprema of measures, see N. Dunford and J. T. Schwartz, *Linear Operators*, Vol. 1, pp. 162–63.

Theorem: The Jordan decomposition operator satisfies $J^2 = J$.

Proof: Let μ_0, ν_0 be measures, let $(\mu_1, \nu_1) = J(\mu_0, \nu_0)$, and let $(\mu_2, \nu_2) = J(\mu_1, \nu_1)$. We must show that $\mu_2 = \mu_1$, $\nu_2 = \nu_1$. It suffices to show that $\mu_2 \geq \mu_1$, $\nu_2 \geq \nu_1$, since the opposite inequalities are already known by (7). Choose $E \in \Sigma$. We then have

$$\mu_1(E) = \sup\{\mu_1(F) - \nu_1(F) \mid F \subseteq E, \nu_0(F) < \infty\} \qquad (3.1.14)$$

To see this, note that

$$\mu_1(F) + \nu_0(F) = \mu_0(F) + \nu_1(F) \qquad (3.1.15)$$

by (13). Also $\nu_0(F) < \infty$ for the sets F in (14), and $\nu_1(F) \leq \nu_0(F)$, by (7). Hence we may transpose the ν-terms in (15) to obtain $\mu_1(F) - \nu_1(F) = \mu_0(F) - \nu_0(F)$, from which (14) follows. In turn, (14) implies that

$$\mu_1(E) \leq \sup\{\mu_1(F) - \nu_1(F) \mid F \subseteq E, \nu_1(F) < \infty\} \qquad (3.1.16)$$

For the set of numbers over which the sup is taken in (16) is at least as large as the set in (14), since $\nu_1 \leq \nu_0$. But (16) states that $\mu_1(E) \leq \mu_2(E)$; hence $\mu_1 = \mu_2$. The proof that $\nu_1 = \nu_2$ is similar. ∎

The following important result is a second application.

Theorem: Let μ_1, ν_1, μ_2, ν_2 be measures such that

$$J(\mu_1, \nu_1) = J(\mu_2, \nu_2) \qquad (3.1.17)$$

Then

$$\mu_1 + \nu_2 = \nu_1 + \mu_2 \qquad (3.1.18)$$

Proof: By contradiction. Suppose (λ^+, λ^-) is the common Jordan decomposition of (μ_1, ν_1) and (μ_2, ν_2), and let (18) be false, so that there is an $E \in \Sigma$ for which, say,

$$\mu_1(E) + \nu_2(E) < \nu_1(E) + \mu_2(E) \qquad (3.1.19)$$

Hence $\mu_1(E) < \infty$, so that $\lambda^+(E) < \infty$; also, $\nu_2(E) < \infty$, so that $\lambda^-(E) < \infty$. Now we have $\mu_i(E) + \lambda^-(E) = \nu_i(E) + \lambda^+(E)$, $i = 1, 2$, by (13). Adding, we obtain

$$\mu_1(E) + \lambda^-(E) + \nu_2(E) + \lambda^+(E) = \nu_1(E) + \lambda^+(E) + \mu_2(E) + \lambda^-(E)$$

Being finite, the λ-terms drop out and we are left with a contradiction of (19). If the inequality in (19) is reversed, the same argument again leads to a contradiction. Hence (18) is true. ∎

It will turn out, under σ-finiteness assumptions, that (17) and (18) are actually equivalent, a basic result for pseudomeasures.

Theorem: Let μ, ν be measures, let $(\lambda^+, \lambda^-) = J(\mu, \nu)$, and let $\theta = \inf(\mu, \nu)$; then

$$\lambda^+ = \mu - \theta, \qquad \lambda^- = \nu - \theta \qquad (3.1.20)$$

$$\mu = \lambda^+ + \theta, \qquad \nu = \lambda^- + \theta \qquad (3.1.21)$$

Proof: $(\mu - \theta)$ is the upper variation of (μ, θ), while λ^+ is the upper variation of (μ, ν). Since $\nu \geq \theta$, and the upper variation is a nonincreasing function of the right-hand component of the pair $(\mu, \cdot)$, it follows that

$$\lambda^+ \leq (\mu - \theta) \qquad (3.1.22)$$

To prove the converse inequality, choose $E \in \Sigma$; for any $F \subseteq E$ such that $\theta(F) < \infty$ and for any finite $\epsilon > 0$, there is a measurable $G \subseteq F$ such that

$$\theta(F) \geq \mu(F \backslash G) + \nu(G) - \epsilon \qquad (3.1.23)$$

by definition (12). We then have

$$\mu(F) - \theta(F) \leq \mu(G) - \nu(G) + \epsilon \leq \lambda^+(E) + \epsilon \qquad (3.1.24)$$

The left inequality in (24) follows from (23); the right inequality is true by definition of λ^+, on noting that $G \subseteq E$ and $\nu(G) \leq \theta(F) + \epsilon < \infty$, by (23).

Taking the supremum in (24) over such sets F, we obtain $(\mu - \theta)(E) \leq \lambda^+(E) + \epsilon$. Since ϵ is arbitrary, we obtain (22) with inequality sign reversed. This establishes $\lambda^+ = \mu - \theta$. The proof that $\lambda^- = \nu - \theta$ is similar. Hence (20) is established.

Finally, (21) follows from (20), e.g., $\lambda^+ + \theta = (\mu - \theta) + \theta = \mu$, by (8). ∎

The results (20) are intuitively appealing. $\text{Inf}(\mu, \nu)$ may be thought of as the mass distribution that μ and ν share in common. Then (20) states that the Jordan decomposition operator subtracts out this common part from μ and ν respectively. (One should not jump to the conclusion, however, that λ^+, λ^- have nothing in common: $\inf(\lambda^+, \lambda^-)$ is not always 0. See below.)

It follows from (21) and (9) that

$$\theta \geq (\mu - \lambda^+), \qquad \theta \geq (\nu - \lambda^-) \qquad (3.1.25)$$

with equality if λ^+ or λ^- is σ-finite respectively. (In general, (25) cannot be strengthened to equality: take $A = \{a\}$, $\mu(A) = \infty$, $\nu(A) = 1$.)

Results (21) furnish alternative proofs for two preceding theorems of importance: the idempotency of J, and the equality (18) for two pairs with the same Jordan decomposition. Taking the latter first, assume (17) with (λ^+, λ^-) the common decomposition, and let $\theta_i = \inf(\mu_i, \nu_i)$, $i = 1, 2$. Then

$$\mu_1 + \nu_2 = (\lambda^+ + \theta_1) + (\lambda^- + \theta_2) = (\lambda^- + \theta_1) + (\lambda^+ + \theta_2) = \nu_1 + \mu_2$$
$$(3.1.26)$$

by (21), which yields (18). As for idempotency, let $(\mu_1, \nu_1) \doteq J(\mu_0, \nu_0)$ and $(\mu_2, \nu_2) = J(\mu_1, \nu_1)$. The hard thing to prove is that $\mu_2 \geq \mu_1$, $\nu_2 \geq \nu_1$; but by (21) we have

$$(\mu_1, \nu_1) = J(\mu_1 + \theta, \nu_1 + \theta) \qquad (3.1.27)$$

where $\theta = \inf(\mu_0, \nu_0)$. Considering θ in (27) as a variable measure, we verify that upper and lower variations are both nonincreasing functions of θ, so that, indeed, $\mu_1 \leq \mu_2$, $\nu_1 \leq \nu_2$, implying $J^2 = J$.

Exercises:

1. Let $(\lambda^+, \lambda^-) = J(\mu, \nu)$. Show that $\lambda^+ = \sup(\mu, \nu) - \nu$, that $\lambda^- = \sup(\mu, \nu) - \mu$, and that $\lambda^+ + \lambda^- = \sup(\mu, \nu) - \inf(\mu, \nu)$. (Hint: Start with (13).)
2. Let $J(\mu_1, \nu_1) = J(\mu_2, \nu_2)$. Show that the common value in (18) is $\sup(\mu_1, \nu_1) + \inf(\mu_2, \nu_2) = \inf(\mu_1, \nu_1) + \sup(\mu_2, \nu_2)$. (Hint: This follows easily from (21) and (13).)

A final cluster of ideas centers on the concepts of Hahn decomposability and mutual singularity. These have already been discussed in connection with signed measures, and we repeat the definitions here for convenience.

Definition: An ordered pair of measures (μ, ν) is *Hahn decomposable* iff universe set A can be split into two measurable pieces P and N such that $\mu(E) \geq \nu(E)$ for all measurable $E \subseteq P$, and $\mu(F) \leq \nu(F)$ for all measurable $F \subseteq N$. The ordered pair (P, N) is a *Hahn decomposition* for (μ, ν).

Definition: An ordered pair of measures (μ, ν) is *mutually singular* iff A can be split into two measurable pieces P and N such that $\nu(P) = 0$ and $\mu(N) = 0$.

Mutual singularity implies Hahn decomposability, since the pair (P, N) is a Hahn decomposition.

Theorem: Let μ, ν be measures, with $(\lambda^+, \lambda^-) = J(\mu, \nu)$. Each of the following conditions implies the other two:

   (i) (μ, ν) is Hahn decomposable;
   (ii) (λ^+, λ^-) is Hahn decomposable;
   (iii) (λ^+, λ^-) is mutually singular.

Proof: (i) *implies* (iii): Let (P, N) be a Hahn decomposition for (μ, ν). Then $\lambda^-(P) = 0$, since $\nu \leq \mu$ on subsets of P; similarly, $\lambda^+(N) = 0$.

(iii) *implies* (ii): Clear.

(ii) *implies* (i): Since J is idempotent, (λ^+, λ^-) is its own Jordan decomposition; hence (λ^+, λ^-) is mutually singular, by the argument showing that (i) implies (iii). Let (P, N) split A so that $\lambda^-(P) = 0$, $\lambda^+(N) = 0$. For any $E \subseteq P$ such that $\mu(E) < \infty$, we have $\lambda^-(P) \geq \nu(E) - \mu(E)$. Hence $\nu \leq \mu$ on the subsets of P. A similar argument yields $\mu \leq \nu$ on N.

We now have a closed circle of implications, so these three conditions are equivalent. ∎

We are interested in conditions that guarantee Hahn decomposability.

Theorem: Let μ, ν be measures. Any of the following three conditions implies that (μ, ν) is Hahn decomposable:

   (i) μ is abcont;
   (ii) ν is abcont;
   (iii) $\inf(\mu, \nu)$ is σ-finite.

Proof: (i) Consider packings of sets $E \in \Sigma$ satisfying $\mu(E) > \nu(E)$. A maximal packing $\mathcal{G}$ of this sort exists, i.e., a packing not properly contained in any larger such packing. (This inference requires the axiom of choice, say in the form of Zorn's lemma.) We show that $\mathcal{G}$ must be countable. Since μ is abcont, a finite measure ρ exists with $\mu \ll \rho$. For each $E \in \mathcal{G}$, $\mu(E) > 0$, hence $\rho(E) > 0$. The class of $\mathcal{G}$-sets E on which $\rho(E) > 1/n$ must be finite for each $n = 1, 2, \ldots$, since ρ is finite. Then $\mathcal{G}$ itself, as the union of these classes, must be countable.

We may then write $\mathcal{G} = \{E_1, E_2, \ldots\}$. For each m, $\nu(E_m) < \infty$; hence, restricting everything to E_m, $\mu - \nu$ is an ordinary signed measure and so has a Hahn decomposition $P_m \cup N_m = E_m$. We claim that $(P, A \backslash P)$ is a Hahn decomposition for (μ, ν), where $P = P_1 \cup P_2 \cup \cdots$. To verify this, let $F \subseteq P$; we have

$$\mu(F \cap P_m) \geq \nu(F \cap P_m) \tag{3.1.28}$$

for each m; by summation, $\mu(F) \geq \nu(F)$; thus $\mu \geq \nu$ on P. Conversely, let $F \subseteq A \backslash P$; F may be written in the form

$$F = [F \backslash (\cup \mathcal{G})] \cup (F \cap N_1) \cup (F \cap N_2) \cup \cdots \qquad (3.1.29)$$

We have

$$\nu(F \cap N_m) \geq \mu(F \cap N_m) \qquad (3.1.30)$$

for each m. Furthermore, we have

$$\nu(F \backslash \cup \mathcal{G}) \geq \mu(F \backslash \cup \mathcal{G}) \qquad (3.1.31)$$

For if (31) were false, we could form a larger packing on which $\mu(E) > \nu(E)$ by including the set $F \backslash \cup \mathcal{G}$; this contradicts the maximality of $\mathcal{G}$. Adding (31) to the sum of (30), we obtain $\nu(F) \geq \mu(F)$, so that $\nu \geq \mu$ on $A \backslash P$. This concludes the proof.

(ii) Same as (i), with roles of μ, ν interchanged.

(iii) Let (A_n), $n = 1, 2, \ldots$, be a partition of universe set A such that $\inf(\mu, \nu)(A_n) < \infty$, all n. For each n there is then a set $F_n \subseteq A_n$ such that

$$\mu(F_n) < \infty, \qquad \nu(A_n \backslash F_n) < \infty \qquad (3.1.32)$$

When restricted to F_n or to $A_n \backslash F_n$, (32) shows that $\mu - \nu$ is a signed measure and thus has a Hahn decomposition. Let P be the union of the pieces of the decompositions on which $\mu \geq \nu$, and let N be the union of the pieces on which $\nu \geq \mu$. Then (P, N) is a Hahn decomposition for (μ, ν) by the argument of (i). ∎

Exercise: Show that each of the three premises in the preceding theorem implies the following condition and that this condition in turn implies that (μ, ν) is Hahn decomposable. A set $E \in \Sigma$ exists such that μ restricted to E is abcont and ν restricted to $A \backslash E$ is abcont.

For all we know to this point, any pair of measures might be Hahn decomposable. The following counterexample scotches this possibility.

Theorem: A pair of measures (μ, ν) exists that is *not* Hahn decomposable and that, furthermore, is its own Jordan decomposition.

Proof: Let A be uncountable, and let Σ consist of all countable subsets of A and their complements; split A into two uncountable pieces P and N (note that P, N are *not* measurable), and let μ, ν be counting measure restricted to P, N respectively. That is, for $E \in \Sigma$, if $E \cap P$ is finite, then $\mu(E)$ = number of points in $E \cap P$; otherwise, $\mu(E) = \infty$; ν is defined similarly, with N in place of P. One easily checks that these are bona fide measures.

Now suppose $(E, A\backslash E)$ were a Hahn decomposition for (μ, ν). Either E or $A\backslash E$ must be countable. If E is countable, then $P\backslash E$ is nonempty; choosing $a_o \in P\backslash E$, we have $\nu\{a_o\} = 0 < 1 = \mu\{a_o\}$, so $\nu \geq \mu$ is false on $A\backslash E$, contradiction. If $A\backslash E$ is countable, then $N \cap E$ is nonempty; choosing $a_o \in N \cap E$, we have $\mu\{a_o\} = 0 < 1 = \nu\{a_o\}$, so $\mu \geq \nu$ is false on E, contradiction. Hence there is no Hahn decomposition.

Next let $(\lambda^+, \lambda^-) = J(\mu, \nu)$, and let E be a countable set. $\nu(E \cap P) = 0$, hence

$$\lambda^+(E) \geq \mu(E \cap P) - \nu(E \cap P) = \mu(E) \qquad (3.1.33)$$

Also, $P\backslash E$ is infinite and hence contains an infinite countable set F; $\nu(F) = 0$, so that

$$\lambda^+(A\backslash E) \geq \mu(F) - \nu(F) = \infty \qquad (3.1.34)$$

Relations (33) and (34) show that $\lambda^+ \geq \mu$, so these measures are equal. A similar argument yields $\lambda^- = \nu$. Thus (μ, ν) is its own Jordan decomposition. ∎

For this counterexample, one easily verifies that $\inf(\mu, \nu)$ takes the value 0 on countable sets and takes the value ∞ on their complements. Thus $\inf(\lambda^+, \lambda^-)$ is not always 0.

Theorem: Let μ, ν be measures. If the pair (μ, ν) is mutually singular, then (μ, ν) is its own Jordan decomposition.

Proof: Let (P, N) be a measurable partition of A such that $\nu(P) = \mu(N) = 0$. Let $(\lambda^+, \lambda^-) = J(\mu, \nu)$. For any $E \in \Sigma$, we have $\nu(E \cap P) = 0$, so

$$\lambda^+(E) \geq \mu(E \cap P) - \nu(E \cap P) = \mu(E \cap P) = \mu(E)$$

Hence $\lambda^+ \geq \mu$, so these are equal. A similar argument yields $\lambda^- = \nu$. ∎

The counterexample above shows that the converse of this theorem does not always hold.

Exercises:

1. Show that the following condition is necessary and sufficient for (μ, ν) to be its own Jordan decomposition. For any $E \in \Sigma$, if (μ, ν) restricted to E is Hahn decomposable, then (μ, ν) restricted to E is mutually singular.

2. Let (μ, ν) be its own Jordan decomposition. Show that $\theta = \inf(\mu, \nu)$ can take only the values 0 and ∞. (Hint: Use (21) to deduce that $2\theta \leq$

θ.) This last exercise may be compared with the result: $\inf(\mu, \nu) = 0$ iff (μ, ν) is mutually singular. For a proof see p. 379.

3.2. PSEUDOMEASURES

From now on all measures will be σ-finite, unless explicitly noted otherwise. All measures are on the same space (A, Σ). We are concerned with ordered pairs (μ, ν) of such measures. Among these pairs the mutually singular ones will play a key role. The following theorem gives some characterizations of these pairs.

Theorem: Let (μ, ν) be a pair of σ-finite measures with Jordan decomposition (λ^+, λ^-). If (μ, ν) has any of the following properties, it has all of them:

(i) (μ, ν) is mutually singular;
(ii) (μ, ν) is its own Jordan decomposition;
(iii) $\inf(\mu, \nu) = 0$;
(iv) $\lambda^+ = \mu$;
(v) $\lambda^- = \nu$;
(vi) $\lambda^+ + \lambda^- = \mu + \nu$;
(vii) $\lambda^+ + \lambda^- = \sup(\mu, \nu)$;
(viii) $\mu + \nu = \sup(\mu, \nu)$.

Proof: (i) *implies* (ii): Already proved.

(ii) *implies* (iv) *and* (v). Obvious.

(iv) *implies* (vi): $\lambda^+ + \lambda^- = \mu + \lambda^- = \nu + \lambda^+ = \nu + \mu$. (The middle equality is from (1.13).)

(v) *implies* (vi): $\lambda^+ + \lambda^- = \lambda^+ + \nu = \lambda^- + \mu = \nu + \mu$.

(vi) *implies* (vii) *and* (viii):

$$\lambda^+ + \lambda^- \leq \mu + \lambda^- = \sup(\mu, \nu) \leq \mu + \nu \qquad (3.2.1)$$

(The equality is from (1.13); the right-hand inequality follows from the definition of $\sup(\mu, \nu)$.) Since the two extreme expressions in (1) are equal, the middle expressions must equal each of them; this yields (vii) and (viii).

(vii) *implies* (vi): As is easily verified, the following equality holds in general:

$$\mu + \nu = \sup(\mu, \nu) + \inf(\mu, \nu) \qquad (3.2.2)$$

By assumption the right side of (2) equals

$$\lambda^+ + \lambda^- + \inf(\mu, \nu) = \lambda^+ + \nu = \sup(\mu, \nu) = \lambda^+ + \lambda^- \quad (3.2.3)$$

The left equality in (3) arises from (1.21) and the middle equality from (1.13); (2) and (3) together yield (vi).

(i) *implies* (iii): A may be split into P, N such that $\mu(N) = 0$, $\nu(P) = 0$; hence $\inf(\mu, \nu)(A) \leq \mu(N) + \nu(P) = 0$.

(iii) *implies* (viii): This follows at once from (2).

So far we have made no use of the σ-finiteness assumption. We now use it to show that (viii) implies (i). This establishes a closed circle of implications and shows that all eight properties are logically equivalent.

(viii) *implies* (i): By σ-finiteness, (μ, ν) has a Hahn decomposition (P, N). Let $\mathcal{G}$ be a countable measurable partition of P such that $\mu(G) < \infty$ for each $G \in \mathcal{G}$. Since $\mu \geq \nu$ on every subset of G, we have $\mu(G) = \sup(\mu, \nu)(G) = \mu(G) + \nu(G)$, implying $\nu(G) = 0$. This is true for each such G, so that $\nu(P) = 0$. A similar argument yields $\mu(N) = 0$. Thus (μ, ν) is mutually singular. ∎

Now consider the set **M** of all ordered pairs of σ-finite measures (μ, ν) on (A, Σ). Two such pairs are said to be *equivalent* iff they have the same Jordan decomposition. This equivalence relation determines a partition Ψ of **M**; viz., each element ψ of Ψ is the set consisting of all pairs having some particular (λ^+, λ^-) as their common Jordan decomposition.

Definition: Each element $\psi \in \Psi$ is called a *pseudomeasure*. The common Jordan decomposition of all members of ψ is called the *Jordan form* of pseudomeasure ψ and will usually be written as (ψ^+, ψ^-); the measures ψ^+, ψ^- are called respectively the *upper* and *lower variations* of ψ; Ψ itself is the *space of pseudomeasures* over (A, Σ).

Theorem: Let ψ be a pseudomeasure, and let A be split into two measurable sets P, N. Each of the following conditions implies the other two:

(i) (P, N) is a Hahn decomposition for every pair of measures (μ, ν) belonging to ψ;

(ii) (P, N) is a Hahn decomposition for at least one pair (μ, ν) belonging to ψ;

(iii) $\psi^-(P) = \psi^+(N) = 0$.

Proof: (i) *implies* (ii): Obvious, since ψ is not empty.

(ii) *implies* (iii): $\mu \geq \nu$ on subsets of P, hence $\psi^-(P) = 0$; $\nu \geq \mu$ on subsets of N, hence $\psi^+(N) = 0$.

(iii) *implies* (i): Suppose (i) is false, so that there is a pair $(\mu, \nu) \in \psi$ and a set E such that, say, $E \subseteq P$ and $\mu(E) < \nu(E)$. But then $\psi^-(P) \geq \nu(E) - \mu(E) > 0$, so that (iii) is false. If, instead, $E \subseteq N$ and $\nu(E) < \mu(E)$, then $\psi^+(N) \geq \mu(E) - \nu(E) > 0$, so that (iii) is again false.

This establishes a closed circle of implications, so the three conditions are logically equivalent. ∎

Definition: (P, N) is a *Hahn decomposition* for pseudomeasure ψ iff any (hence all) of the conditions above are satisfied.

Every pseudomeasure has a Hahn decomposition, since any pair of σ-finite measures is Hahn decomposable.

The basic relations between pseudomeasures and their Jordan forms are spelled out in the following results.

Theorem: The mapping that associates each pseudomeasure with its Jordan form establishes a 1–1 correspondence between Ψ and the set of *mutually singular* σ-finite pairs (μ, ν). The Jordan form of ψ belongs to ψ. In fact, ψ consists of all pairs of measures of the form $(\psi^+ + \theta, \psi^- + \theta)$, where θ ranges over the set of σ-finite measures.

Proof: Different pseudomeasures have different Jordan forms, and vice versa, so we must show that the set of Jordan forms (ψ^+, ψ^-) coincides with the set of mutually singular measures. If (μ, ν) is a Jordan form, then by a preceding theorem, (μ, ν) is mutually singular. Conversely, if (μ, ν) is mutually singular, it is its own Jordan decomposition. Hence it is the Jordan form of the pseudomeasure to which it itself belongs.

Let (μ, ν) belong to pseudomeasure ψ, so that $(\psi^+, \psi^-) = J(\mu, \nu)$. But then, by (1.21), we have $\mu = \psi^+ + \inf(\mu, \nu)$, $\nu = \psi^- + \inf(\mu, \nu)$, so that (μ, ν) is indeed of the form $(\psi^+ + \theta, \psi^- + \theta)$. Conversely, let (μ, ν) be of this form, and let $(\lambda^+, \lambda^-) = J(\mu, \nu)$. Choose $E \in \Sigma$, and let $F \subseteq E$ satisfy: $\psi^-(F) + \theta(F) < \infty$; then

$$\psi^+(E) \geq \psi^+(F) \geq [\psi^+(F) + \theta(F)] - [\psi^-(F) + \theta(F)]$$

Taking the supremum over such sets F, we obtain $\psi^+(E) \geq \lambda^+(E)$.

To prove the reverse inequality, let (P, N) split A so that $\psi^-(P) =$

$\psi^+(N) = 0$, and let $\mathcal{G}$ be a countable partition of A such that $\theta(G) < \infty$, all $G \in \mathcal{G}$. For any such G we have

$$\psi^-(E \cap G \cap P) = 0 \quad \text{and} \quad \theta(E \cap G \cap P) < \infty$$

so that

$$\lambda^+(E \cap G) \geq [\psi^+(E \cap G \cap P) + \theta(E \cap G \cap P)]$$
$$- [\psi^-(E \cap G \cap P) + \theta(E \cap G \cap P)]$$
$$= \psi^+(E \cap G \cap P) = \psi^+(E \cap G)$$

Adding these inequalities over all $G \in \mathcal{G}$, we obtain $\lambda^+(E) \geq \psi^+(E)$. Thus $\lambda^+ = \psi^+$. A similar argument establishes $\lambda^- = \psi^-$. It follows that (μ, ν) belongs to ψ. ∎

Thus a pseudomeasure is a collection of pairs of measures, among which is one special "canonical" pair, the Jordan form. This is the unique pair that is mutually singular, is its own decomposition, and has the smallest left component among all the pairs as well as the smallest right component. (Proof: Let (μ, ν) not be the Jordan form, hence not mutually singular, hence $\lambda^+ \neq \mu$, $\lambda^- \neq \nu$.) The Jordan form may be recovered from any pair (μ, ν) by subtracting out their "common part" $\inf(\mu, \nu)$ from each of them.

The following result establishes a very useful criterion for equivalence.

Theorem: Equivalence Theorem. Let μ_1, ν_1, μ_2, and ν_2 be σ-finite measures. The pair (μ_1, ν_1) is equivalent to (μ_2, ν_2) iff

$$\mu_1 + \nu_2 = \nu_1 + \mu_2 \tag{3.2.4}$$

Proof: Half of this theorem has already been proved: (4) is implied by $J(\mu_1, \nu_1) = J(\mu_2, \nu_2)$. See (1.17)–(1.18).

Conversely, let (4) hold. Let $\mathcal{G}_i$ be a countable partition of A such that $\nu_i(G_i) < \infty$ for all $G_i \in \mathcal{G}_i$, $i = 1, 2$. Letting G_1, G_2 be sets from these respective partitions, note that ν_1 and ν_2 are both finite on subsets of $G_1 \cap G_2$, hence may be subtracted from both sides of (4) on such sets. This justifies the middle equality in the following chain. Let $(\lambda_i^+, \lambda_i^-) = J(\mu_i, \nu_i)$, $i = 1, 2$, and choose $E \in \Sigma$. Then

$$\lambda_1^+(E \cap G_1 \cap G_2) = \sup\{\mu_1(F) - \nu_1(F) \mid F \subseteq (E \cap G_1 \cap G_2)\}$$
$$= \sup\{\mu_2(F) - \nu_2(F) \mid F \subseteq (E \cap G_1 \cap G_2)\}$$
$$= \lambda_2^+(E \cap G_1 \cap G_2)$$

Adding over all $G_1 \in \mathcal{G}_1$, all $G_2 \in \mathcal{G}_2$, we obtain $\lambda_1^+(E) = \lambda_2^+(E)$. A similar argument yields $\lambda_1^- = \lambda_2^-$. Hence (μ_1, ν_1) and (μ_2, ν_2) are equivalent. ∎

Exercise: Show that this result remains true if the σ-finiteness assumption is weakened to: $\inf(\mu_i, \nu_i)$ is σ-finite for $i = 1, 2$.

We now make a few notational conventions. Pairs (μ, ν) will generally be used to denote the pseudomeasures to which they belong. Equivalence between pairs will be denoted by the equality sign. Thus $(\mu_1, \nu_1) = (\mu_2, \nu_2)$ does *not* mean that $\mu_1 = \mu_2$, $\nu_1 = \nu_2$; it means that these pairs belong to the same pseudomeasure, so that only (4) is true. Similarly, we write $(\mu, \nu) = \psi$ to indicate that (μ, ν) belongs to pseudomeasure ψ.

Sigma-finite measures and signed measures may now be thought of as special kinds of pseudomeasures. Specifically, the measure μ may be identified with the pseudomeasure $(\mu, 0)$. (Here 0 is the identically zero measure.) If μ is a σ-finite signed measure let (μ^+, μ^-) be its ordinary Jordan decomposition. We now identify μ with the *pseudomeasure* (μ^+, μ^-).

Pseudomeasure ψ is *bounded* iff both ψ^+ and ψ^- are bounded measures. The class of bounded pseudomeasures may be identified with the class of bounded signed measures. Next consider the case where *exactly one* of ψ^+ and ψ^- is infinite. The class of these pseudomeasures may be identified with the class of infinite (σ-finite) signed measures. Finally, we have the case where *both* ψ^+ and ψ^- are infinite measures. These "proper" pseudomeasures are new kinds of entities and provide the rationale for this whole development.

The Algebra of Pseudomeasures

We now define various algebraic operations on pseudomeasures. The result we are aiming at is that, under various natural definitions, the set of all pseudomeasures Ψ becomes a (real) vector space. First we define addition.

Definition: The *sum* of the two pseudomeasures (μ_1, ν_1) and (μ_2, ν_2) is the pseudomeasure $(\mu_1 + \mu_2, \nu_1 + \nu_2)$.

This definition is not quite as straightforward as it appears because the pairs (μ, ν) stand, not for themselves, but for the pseudomeasures to which they belong. For this definition to be consistent, the pseudomeasure represented by the sum must not depend on the particular pairs chosen for the summands; i.e., if another two pairs (μ_1', ν_1') and (μ_2', ν_2') are respectively equivalent to (μ_1, ν_1) and (μ_2, ν_2), then $(\mu_1' + \mu_2', \nu_1' + \nu_2')$

must be equivalent to $(\mu_1 + \mu_2, \nu_1 + \nu_2)$. This fact is, however, an easy consequence of the equivalence criterion just proved and is left as an exercise. Note also that the sum of two σ-finite measures is σ-finite.

We now want to verify that the properties of vector spaces, insofar as they refer to addition, are satisfied by this definition. It is obvious that $\psi_1 + \psi_2 = \psi_2 + \psi_1$ and that $\psi_1 + (\psi_2 + \psi_3) = (\psi_1 + \psi_2) + \psi_3$. We now need the concepts of zero and negation.

Definition: The *zero pseudomeasure* is the one whose Jordan form is $(0, 0)$.

This is the pair both of whose members are the identically zero measure. We denote this pseudomeasure simply by 0 if no confusion is possible. From the equivalence criterion it is immediate that the σ-finite pair (μ, ν) belongs to this pseudomeasure iff $\mu = \nu$.

Definition: The *negation* of pseudomeasure (μ, ν) is pseudomeasure (ν, μ).

Once again this definition must be checked for consistency: The negation of an equivalent pair must be equivalent to the negation of the original pair. This follows immediately from the equivalence criterion. Negation will be denoted as usual by a minus sign. *Subtraction* is defined as follows.

Definition: $\psi_1 - \psi_2 = \psi_1 + (-\psi_2)$.

These definitions again satisfy the conditions for a vector space: $\psi + 0 = \psi$ for any pseudomeasure ψ, and $-\psi$ is the unique additive inverse of ψ: $\psi + (-\psi) = 0$.

Next we define *scalar multiplication*.

Definition: The *product* of the real number b and the pseudomeasure (μ, ν) is the pseudomeasure $(b\mu, b\nu)$ if $b \geq 0$ and is the pseudomeasure $((-b)\nu, (-b)\mu)$ if $b < 0$.

Here $b\mu$ is the measure that assigns the value $b \cdot \mu(E)$ to the measurable set E. Note that measures are always multiplied by nonnegative numbers, so that they remain measures. Again, a proof of consistency is required for this operation, and the proof is trivial. The second part of this definition could have been framed in terms of the first part as follows: if $b < 0$, then $b\psi = (-b)(-\psi)$.

The remaining axioms for a vector space may now be verified routinely. For real numbers b_1, b_2 and pseudomeasures ψ_1, ψ_2, we have

$$b_1(\psi_1 + \psi_2) = b_1\psi_1 + b_1\psi_2; \qquad (b_1 + b_2)\psi_1 = b_1\psi_1 + b_2\psi_1$$
$$b_1(b_2\psi_1) = (b_1b_2)\psi_1; \qquad\qquad 1 \cdot \psi_1 = \psi_1$$

The only minor complications arise with the second equality, where the various sign combinations for b_1, b_2 and $b_1 + b_2$ must be examined. Details are omitted. We summarize as follows.

Theorem: Under the foregoing definitions the set of all pseudomeasures Ψ is a (real) vector space.

As discussed in Section 2.6, the set of bounded signed measures is a vector space. This property is lost for the larger set of σ-finite signed measures because addition sometimes leads to the meaningless expression $\infty - \infty$ and is therefore not well defined for certain pairs. What we have done, in effect, is to embed this set in a still larger set and to extend the domains of addition and scalar multiplication in such a way that the vector space property is restored.

Note that subtraction of *measures* is compatible with subtraction of *pseudomeasures* wherever both operations apply. To see this, let $\mu \geq \nu$ be σ-finite measures. The difference $\mu - \nu$ was defined in the preceding section by the relation

$$(\mu - \nu, 0) = J(\mu, \nu) \tag{3.2.5}$$

Identifying μ and ν with the pseudomeasures $(\mu, 0)$, $(\nu, 0)$ respectively and using the new definition of subtraction, we obtain

$$(\mu, 0) - (\nu, 0) = (\mu, 0) + (0, \nu) = (\mu, \nu) = (\mu - \nu, 0)$$

This last equality follows from (5), or from the equivalence theorem (4) upon noting that $\mu + 0 = (\mu - \nu) + \nu$. Thus $\mu - \nu$ in the pseudomeasure sense equals the pseudomeasure $(\mu - \nu, 0)$, which may be identified with the measure $\mu - \nu$, subtraction being defined as in (5). Neither subtraction concept may be subsumed under the other, however, since their domains of definition differ.

To illustrate these concepts, consider the case of a *finite* σ-field Σ. Except for the trivial case $A = \phi$, Σ is generated by a finite partition of universe set A into nonempty sets, say $A_1, \ldots, A_n$. We claim that the space of pseudomeasures Ψ is here isomorphic to n-space, the set of all real n-tuples, with the usual definitions of addition and scalar multiplication. To see this, note first that a measure on (A, Σ) is completely determined by its values on the partition elements $A_1, \ldots, A_n$. Thus measures may be "coded" by n-tuples of nonnegative numbers. This establishes a

1–1 correspondence between the set of (σ-)finite measures on (A, Σ) and the nonnegative orthant of n-space. Furthermore, this correspondence extends in an obvious way to a 1–1 relation between the set of (σ-)finite *signed* measures and all of n-space. If we identify the finite signed measure λ with the set of all pairs of finite measures (μ, ν) such that $\mu - \nu = \lambda$, we can check that this is precisely the operation of gathering these pairs into pseudomeasure classes. The correspondence, pseudomeasures $\leftrightarrow$ signed measures $\leftrightarrow$ n-tuples, is then verified to be an isomorphism among vector spaces in the sense that it is preserved under addition and scalar multiplication in the respective systems.

There would be little point in constructing the elaborate machinery of pseudomeasure theory if we were dealing only with finite σ-fields. The point is, of course, that these concepts carry over to arbitrary measurable spaces (A, Σ), yielding results that are far from trivial.

INTEGRATION WITH PSEUDOMEASURES

Just as the concept of addition was extended above with the use of pseudomeasures, so will the concept of integration now be extended. Recall that everything is defined on the fixed measurable space (A, Σ).

Definition: Let f be a real-valued measurable function. The *(indefinite) integral* of f with respect to the pseudomeasure (μ, ν) is the pseudomeasure

$$[\int f^+ \, d\mu + \int f^- \, d\nu, \ \int f^- \, d\mu + \int f^+ \, d\nu] \tag{3.2.6}$$

Expression (6) is to be understood as follows. First, f^+ and f^- are the nonnegative functions given by

$$f^+(a) = \max(f(a), 0), \qquad f^-(a) = \max(-f(a), 0)$$

Next the four integrals in (6) are ordinary indefinite integrals. The indefinite integral of a nonnegative real-valued function with respect to a σ-finite measure is itself a σ-finite measure. Hence (6) is a pair of σ-finite measures, and as such it represents a pseudomeasure.

A consistency question again arises with respect to this definition. Namely, if a pair (μ', ν') equivalent to (μ, ν) is substituted in (6), will the resulting pair be equivalent to the original (6)? The answer is yes, and the proof is again an easy consequence of the equivalence criterion together with the elementary integration rule,

$$\int g \, d\lambda_1 + \int g \, d\lambda_2 = \int g \, d(\lambda_1 + \lambda_2) \tag{3.2.7}$$

Details are left as an exercise.

Note that (6) is well defined for *any* real-valued measurable function and *any* pseudomeasure. In particular, it is valid for any (σ-finite) signed measure interpreted as a pseudomeasure. This contrasts with the usual definition, which sometimes leads to the meaningless expression $\infty - \infty$. From our point of view, what happens is that the operation of indefinite integration sometimes leads out of the class of signed measures into the essentially wider realm of pseudomeasures, just as addition sometimes does. It can be seen that, when the ordinary indefinite integral is well defined (and the integrating signed measure is σ-finite), it yields a signed measure equivalent to (6). Thus our definition does indeed extend the ordinary definition.

We use the notation $\int f \, d(\mu, \nu)$ or $\int f \, d\psi$ for integration with respect to a pseudomeasure.

Most of the elementary theorems concerning integrals generalize to pseudomeasure integrals. We consider in this section a few of these theorems involving equalities. Theorems involving inequalities will be discussed later.

Theorem: Let f be a real-valued measurable function, and let ψ_1 and ψ_2 be pseudomeasures. Then

$$\int f \, d\psi_1 + \int f \, d\psi_2 = \int f \, d(\psi_1 + \psi_2) \tag{3.2.8}$$

Proof: Choose arbitrary members (μ_i, ν_i) of ψ_i, $i = 1, 2$; then $(\mu_1 + \mu_2, \nu_1 + \nu_2)$ belongs to $\psi_1 + \psi_2$. Expanding the two sides of (8) with these according to the rule (6), the left (right) side becomes a pair, each measure of which is the sum of four (two) indefinite integrals. Equality of these pairs is established by applying the rule (7) four times for the nonnegative functions $g = f^+$ or $g = f^-$, combined with the measures $\lambda_i = \mu_i, i = 1, 2$, or $\lambda_i = \nu_i, i = 1, 2$. ∎

Theorem: Let f and g be real-valued measurable functions, and ψ a pseudomeasure. Then

$$\int f \, d\psi + \int g \, d\psi = \int (f + g) d\psi \tag{3.2.9}$$

Proof: A rule of the same form as (9) holds for ordinary indefinite integrals with two nonnegative functions and a measure. Let (μ, ν) be an arbitrary member of ψ, and expand both sides of (9) by the rule (6). The left side of (9) becomes

$$\left[\int (f^+ + g^+) \, d\mu + \int (f^- + g^-) \, d\nu, \int (f^- + g^-) \, d\mu + \int (f^+ + g^+) d\nu \right]$$

while the right side becomes a similar pair with $(f + g)^+$ in place of $(f^+ + g^+)$ and with $(f + g)^-$ in place of $(f^- + g^-)$. Testing by the equivalence criterion (4), we find that these two pairs are equivalent if the following equation holds.

$$f^+ + g^+ + (f + g)^- = f^- + g^- + (f + g)^+ \qquad (3.2.10)$$

But the validity of (10) follows at once from the fact that

$$(f^+ - f^-) + (g^+ - g^-) = f + g = (f + g)^+ - (f + g)^- \quad \blacksquare$$

Theorem: Let f be a real-valued measurable function, ψ a pseudomeasure, and b, c real numbers. Then

$$\int (bf)\, d(c\psi) = bc \int f\, d\psi \qquad (3.2.11)$$

Proof: A rule of the same form as (11) holds for ordinary integrals, and choosing an arbitrary member (μ, ν) of ψ, expansion of both sides of (11) by rule (6) yields a routine verification. (The four possible sign combinations of b, c must be dealt with separately.)   ■

In explanation of the following theorem, note that since $\int g\, d\psi$ is a pseudomeasure, it makes perfectly good sense to integrate another function f with respect to *it*. The left side of (12) represents the resulting iterated integral, and (12) states that this can actually be expressed by a single integral.

Theorem: Iterated Integral Theorem.   Let f and g be real-valued measurable functions, and ψ a pseudomeasure.   Then

$$\int f\, d\left[\int g\, d\psi\right] = \int fg\, d\psi \qquad (3.2.12)$$

Proof: A rule of the same form holds for ordinary integrals for two nonnegative functions and a measure. Choosing an arbitrary pair (μ, ν) belonging to ψ, we first expand $\int g\, d\psi$ by (6), and then expand the integral of f by *this* pair, again by (6). The result is a pair, the left measure of which is

$$\int f^+g^+\, d\mu + \int f^+g^-\, d\nu + \int f^-g^+\, d\nu + \int f^-g^-\, d\mu \qquad (3.2.13)$$

and the right measure of which is obtained from (13) by switching μ and ν. The equality of this pair with the expansion of $\int fg\, d(\mu, \nu)$ follows from the fact that

$$(fg)^+ = f^+g^+ + f^-g^-, \qquad \text{and} \qquad (fg)^- = f^+g^- + f^-g^+ \qquad (3.2.14)$$

(The validity of (14) is established by considering the four possible sign combinations of f, g separately.)   ■

These four theorems all follow the same pattern. The equalities (8), (9),

(11), (12) are already known to hold for ordinary integrals with non-negative functions and measures, and this fact is used to show that the expansions of the corresponding pseudomeasure integrals are equivalent.

Let 1 be the function everywhere equal to 1. The following is easily verified:

$$\int 1 \, d\psi = \psi \qquad (3.2.15)$$

These results concerning integrals—(8), (9), (12), and (15)—may be summarized in algebraic terms. Let $\mathfrak{F}$ be the set of all real-valued measurable functions on (A, Σ). Set $\mathfrak{F}$ is a *ring* in the algebraic sense under pointwise addition and multiplication. In fact it is a commutative ring with unity, the unit element being 1. Define addition on the space of pseudomeasures Ψ as above; define "scalar multiplication" $\circ$ as a mapping from $\mathfrak{F} \times \Psi$ to Ψ, viz., $f \circ \psi = \int f \, d\psi$. Then these results, along with the preceding ones, state that Ψ is a (unitary) *module* over the ring $\mathfrak{F}$ with respect to these operations.[3] Then Ψ as a real vector space may be thought of as a module over the subring of constant functions if we identify the real number c with the constant function $f = c$.

We will sometimes need the Jordan form of an integral. This is easily found.

Theorem: Let (ψ^+, ψ^-) be the Jordan form of pseudomeasure ψ, and let $f : A \to$ reals be measurable. Then

$$[\int f^+ \, d\psi^+ + \int f^- \, d\psi^-, \ \int f^- \, d\psi^+ + \int f^+ \, d\psi^-] \qquad (3.2.16)$$

is the Jordan form of $\int f \, d\psi$.

Proof: It is clear that the pair (16) belongs to pseudomeasure $\int f \, d\psi$. The only thing left to prove is that (16) is mutually singular. Split A into P, N so that $\psi^-(P) = \psi^+(N) = 0$. Then the two left integrals in (16) are zero on the set

$$[N \cap \{a \mid f(a) \geq 0\}] \cup [P \cap \{a \mid f(a) < 0\}]$$

while the two right integrals are zero on the complementary set

$$[P \cap \{a \mid f(a) \geq 0\}] \cup [N \cap \{a \mid f(a) < 0\}] \qquad \blacksquare$$

Definition: The *total variation* of pseudomeasure ψ is the measure $\psi^+ + \psi^-$.

This is a direct generalization of the same concept for signed measures. We shall denote the total variation by $|\psi|$.

---

3. For the concepts involved see N. Jacobson, *Lectures in Abstract Algebra,* Vol. 1 (Van Nostrand, Princeton, 1951), pp. 162–67.

Next recall that if μ and ν are two measures, μ is said to be absolutely continuous with respect to ν iff, for any measurable set E, if $\nu(E) = 0$, then $\mu(E) = 0$. The notation for this is $\mu \ll \nu$. We now extend this concept to pseudomeasures.

Definition: Let ψ_1 and ψ_2 be pseudomeasures; ψ_1 is *absolutely continuous* with respect to ψ_2 iff $|\psi_1| \ll |\psi_2|$.

This is well defined and not circular, because $|\psi_1|$ and $|\psi_2|$ are ordinary measures. The same notation will be used: $\psi_1 \ll \psi_2$.

We end this subsection with two generalizations of well-known theorems.

Theorem: Let f be a real-valued measurable function, and ψ a pseudomeasure. Then $\left| \int f \, d\psi \right| = \int |f| \, d|\psi|$.

Here $|f| = \max(f, -f)$, the absolute value of f. The expression on the right is an ordinary indefinite integral, and the claim is that it equals the total variation of the pseudomeasure $\int f \, d\psi$.

Proof: Since (16) is the Jordan form of $\int f \, d\psi$, the total variation $\left| \int f \, d\psi \right|$ is the sum of the four integrals in (16), which is

$$\int (f^+ + f^-) \, d(\psi^+ + \psi^-) = \int |f| \, d|\psi| \qquad \blacksquare$$

Theorem: Radon-Nikodym Theorem for Pseudomeasures. Let ψ_1 and ψ_2 be pseudomeasures. A real-valued measurable function f exists such that $\psi_1 = \int f \, d\psi_2$ iff $\psi_1 \ll \psi_2$.

Proof: The "only if" part is simple. Let $\psi_1 = \int f \, d\psi_2$, and let E be a measurable set such that $|\psi_2|(E) = 0$. Hence $\psi_2^+(E) = 0$ and $\psi_2^-(E) = 0$. We have proved that (16) is the Jordan form for an integral $\int f \, d\psi$. Since both measures in (16) clearly equal zero at E for $\psi = \psi_2$, we have $\psi_1^+(E) = \psi_1^-(E) = 0$, so that $|\psi_1|(E) = 0$. This proves that $\psi_1 \ll \psi_2$.

Conversely, assume $\psi_1 \ll \psi_2$, so that $|\psi_1| \ll |\psi_2|$. These are both σ-finite measures, hence by the ordinary Radon-Nikodym theorem a nonnegative real-valued measurable function g exists such that $|\psi_1| = \int g \, d|\psi_2|$. Now let (P_i, N_i) be a Hahn decomposition for ψ_i, $i = 1, 2$, and define the function f as follows:

$$f(a) = g(a) \qquad \text{if } a \in (P_1 \cap P_2) \cup (N_1 \cap N_2)$$

$$f(a) = -g(a) \qquad \text{if } a \in (P_1 \cap N_2) \cup (N_1 \cap P_2)$$

We now show that f is the required function: $\psi_1 = \int f \, d\psi_2$. Expanding the integral in the form (16) for $\psi = \psi_2$, we show that the pair of measures in (16) is, in fact, (ψ_1^+, ψ_1^-). It suffices to prove this equality for measurable subsets of each of the four sets $(P_1 \cap P_2)$, $(N_1 \cap N_2)$, $(P_1 \cap N_2)$, and $(N_1 \cap P_2)$ because, since these partition A, equality for any measurable set follows by summation. We will carry out the analysis for $P_1 \cap N_2$, the argument for the other three sets being similar. Since $\psi_1^-(P_1) = 0$ and $\psi_2^+(N_2) = 0$, it follows that $|\psi_1| = \psi_1^+$ and $|\psi_2| = \psi_2^-$ when all measures are restricted to $P_1 \cap N_2$. Also, f is nonpositive on $P_1 \cap N_2$, so that $f^+ = 0$ and $f^- = g$ on this set. Hence, restricted to $P_1 \cap N_2$, the four integrals in (16) (with $\psi = \psi_2$) reduce to $(0 + \int g \, d\psi_2^-, 0 + 0)$. Now $\psi_1^- = 0$ on $P_1 \cap N_2$, while $\psi_1^+ = |\psi_1| = \int g \, d|\psi_2| = \int g \, d\psi_2^-$ on $P_1 \cap N_2$, proving equality. ∎

There is also a theory of product and conditional pseudomeasures. In particular, there is a "Fubini theorem" for pseudomeasures. We shall not discuss this.

APPLICATIONS OF PSEUDOMEASURES

Having formally introduced pseudomeasures, we consider some of the ways they can be used. A number of the following examples have already been mentioned, but we can now discuss them more coherently. We assume that all measures discussed are σ-finite.

Let μ and ν be measures over Space as universe set, with the interpretation that $\mu(E) =$ gross production of a certain resource in region E and $\nu(E) =$ gross consumption of that resource in E. If both measures are infinite, they cannot be subtracted to yield net production. We can, however, represent net production by the *pseudomeasure* (μ, ν). What can be done with this representation?

Consider first the Jordan form (λ^+, λ^-) of this pseudomeasure, with a Hahn decomposition (P, N). We know that $\mu \geq \nu$ when both are restricted to P and that $\nu \geq \mu$ when both are restricted to N. Thus (P, N) splits S into the region of net production and the region of net consumption, and λ^+, λ^- give these respective "net" measures. When these pseudomeasures reduce to ordinary measures or signed measures (which occurs when λ^+ and λ^- are not both infinite), they do so in an intuitively appealing way. For example, suppose production is everywhere 3 times consumption. The pseudomeasure $(3\nu, \nu)$ has the Jordan form $(2\nu, 0)$, which is the ordinary measure 2ν, and states that net production is 2 times consumption.

Problems involving infinite "gross" measures often arise when the horizon is unlimited: the infinite plane of location theory with unlimited space horizon, or economic development programs with unlimited time horizon. In such situations it is convenient (though not usually essential) to frame the problem in the form "find the optimal pseudomeasure such that...."

We have just discussed one broad category of application for pseudomeasures: the representation of physical situations. Another perhaps more important application is to the representation of preferences. Consider, e.g., an economic development program with infinite horizon. Typically, one represents the "payoff" from a policy p by an integral of the form

$$\int_0^\infty f(p, t) \, dt \qquad\qquad (3.2.17)$$

where $f(p, t)$, for example, may be determined by total consumption under policy p at time t. One chooses the attainable policy that maximizes (17). There are two difficulties with an objective function of the form (17). First, suppose the value $+\infty$ can be attained with several policies. Are these to be considered equally good? Simple examples suggest otherwise. Suppose policies p' and p'' are such that $f(p', t) > f(p'', t)$ for all t, but both policies give the value $+\infty$ in (17). Intuitively, we would be inclined to say that p' is the better policy. This means that (17) does not properly represent the structure of preferences.

The second difficulty appears to be even more troublesome. What about feasible policies for which (17) is not well defined, i.e., where its evaluation leads to the meaningless expression $\infty - \infty$. These policies would simply be incomparable with others under the objective function (17). Yet in many cases simple intuition does suggest that some of these policies are better than others, e.g., when they are related as p' and p'' above. Thus again (17) does not properly represent the structure of preferences. Function (17) is an integral over Time. But the same problems can arise with integrals over Space, Space-Time, or abstract spaces.

Are these difficulties serious? One can of course frame models that avoid them and insure that all integrals (17) that arise are well defined and finite. This is done in practice by truncating at a finite horizon, introducing time discounts, etc. But these restrictions prevent one from coming to grips with many significant problems. Several of these arise in location theory and will be taken up later. We mention one or two others here.

Consider the problem of global welfare maximization. We adopt a terminology and point of view that is currently out of fashion. Suppose we wish to maximize the balance of total "pleasure" over total "pain" in the world. Both the foregoing difficulties may arise. Because the time horizon is infinite, all integrals may diverge to $+\infty$ no matter what policy is followed. (A pessimist would maintain that all integrals diverge to $-\infty$, which creates the same difficulty.) And, if the total amount of both "pleasure" and "pain" is infinite, none of the integrals will be well defined.

A rather different example is that of Bernoullian (or von Neumann-Morgenstern) utility.[4] Abstractly, one is given a measurable space (A, Σ), and the problem is to characterize the preference orderings that a "rational" man might entertain over the set of all possible probability measures on (A, Σ). The main result is the "expected utility" principle: for a rational man there is a measurable function $u : A \to$ reals, such that he prefers probability μ_1 over μ_2 iff

$$\int_A u \, d\mu_1 > \int_A u \, d\mu_2 \qquad (3.2.18)$$

Now there is no difficulty if u is a bounded function, for then the integrals in (18) are always finite. If u is unbounded above or below, however, one can show there are probability measures for which the integrals (18) are both infinite. And if u is unbounded both above and below, there are probabilities for which the integrals (18) are not well defined.

We shall argue below that there is no compelling reason why u *should* be bounded. There are perfectly reasonable preference orderings calling for an unbounded utility function u. But in this case what are we to make of the integrals in (18), and how are we to compare them? One possibility is to restrict comparisons to probability measures concentrated on a finite number of points, for with these the integrals (18) are finite even if u is unbounded. This unduly restrictive solution may be avoided, however, *if* we interpret these integrals as pseudomeasures and the order relation as standard ordering of pseudomeasures. All this will be fully explained below, and an axiomatic justification for this procedure will be given for the case of a countable universe set.

We also discuss the ideas of Ramsey, and of succeeding writers such as Weizsäcker and Gale, on how to deal with unbounded sums and integrals. These ideas—the so-called "overtaking" criteria—drop out as special cases of the development below. Thus the use of pseudomeasure-valued utility indicators leads to a unified theory that includes not only 1-dimensional unbounded integrals (the "overtaking" case) but also higher dimensional cases (such as spatial integrals in location theory) and, at the same time, incorporates Bernoullian utility theory with unbounded utility functions u.

Starting from ordinary integrals like (17) or (18), the first step is to go from the definite to the indefinite integral. In comparing unbounded integrals, merely taking note of the value $+\infty$ loses essential information. We should take into account the entire distribution pattern.

---

4. A good general reference is R. Duncan Luce and H. Raiffa, *Games and Decisions* (Wiley, New York, 1957).

The second step is to note that an indefinite integral is a signed measure. The fact that it appears in the form of an indefinite integral is irrelevant for the following analysis. The problem has become one of comparing signed measures.

The third step is to allow for integrals that are not well defined in the ordinary sense by interpreting them as pseudomeasures. We recall that $\int f \, d\mu$ is always a well-defined pseudomeasure for any real-valued measurable f and any pseudomeasure μ. (In the examples above, μ is just an ordinary measure—Lebesgue measure for the development problem and a probability measure for the Bernoulli problem.) Thus we allow pseudomeasure-valued objective functions.

We now have a problem that embraces all the others as special cases: develop a plausible criterion for deciding when one pseudomeasure is larger, or "better," than another. Our investigation will be guided partly by intuition and also by the requirement that, when the pseudomeasures reduce to bounded signed measures, their ordering should be compatible with that induced by the comparison of finite definite integrals.

We have introduced pseudomeasures in connection with the difficulty of ill-defined integrals. However, pseudomeasures are also essentially involved in the difficulty of comparing unbounded integrals. From our point of view, both these difficulties are the same, and insofar as our program is successful, both are resolved in one stroke.

3.3. NARROW AND STANDARD ORDERING OF PSEUDOMEASURES

The problem we have just formulated is: given pseudomeasures ψ_1 and ψ_2, to give a rule for deciding when ψ_1 is to be considered larger than ψ_2. Our discussion will proceed through stages of increasing concreteness. First we take up comparisons in general, with a discussion of partial orderings. Our development makes essential use of the fact that the pseudomeasures are a vector space, and, accordingly, we next discuss partial orderings on vector spaces. We then come to the space of pseudomeasures itself, and the discussion goes through several more stages.

Partial Orderings in General

Let H be a set.[5] A *relation* on H is a subset of the cartesian product $H \times H$. The particular kinds of relations we are interested in are called

---

5. The material in this subsection and the next is well known, but terminology is not completely standardized, and we have selected those aspects that are relevant for our particular purposes.

partial orders, and will be denoted by $\geq$ or $\succcurlyeq$. If x and y are members of H, the notation $x \geq y$ will indicate that the ordered pair (x, y) is in the relation $\geq$; $y \leq x$ means the same thing.

Definition: The relation $\geq$ is a *partial order* iff

(i) for all $x, y, z \in H$, if $x \geq y$ and $y \geq z$, then $x \geq z$ (*transitivity*);
(ii) for all $x \in H$, $x \geq x$ (*reflexivity*).

The interpretations we have in mind for $\geq$ refer to "size" or to "preferredness," and the statement $x \geq y$ may then be read: x is at least as big as, or at least as good as, y, depending on context. Transitivity and reflexivity are conditions with obvious intuitive appeal under such interpretations.

Given the partial order $\geq$ on H, we now define two further relations. Let $x, y \in H$.

Definition: $x > y$ iff $x \geq y$ but not $y \geq x$ (*strict order*); $x \sim y$ iff $x \geq y$ and $y \geq x$ (*indifference*).

Relation $x > y$ may be read: x is greater than, or better than, y, $x \sim y$ may be read: x is as big as, or indifferent to, y.

For any pair of elements x, y there are now four possibilities, exactly one of which must hold: (i) $x \sim y$; (ii) $x > y$; (iii) $x < y$ (i.e., $y > x$); (iv) none of these, which occurs when $x \geq y$ and $y \geq x$ are both false. In the first three cases we say that x and y are *comparable* (under the relation $\geq$), in the last case, *incomparable*.

Definition: A partial order $\geq$ on H is *complete* iff any pair of elements of H are comparable; $\geq$ is *antisymmetric* iff $x \sim y$ implies $x = y$ for all $x, y \in H$.

Thus a partial order is complete iff for any pair $x, y \in H$ either (x, y) or (y, x) (or both) stand in the relation $\geq$. A partial order is antisymmetric iff no two *distinct* elements are indifferent ($x \sim x$ is always true, owing to reflexivity).

Here are some examples. The natural ordering on the real numbers is both complete and antisymmetric. In utility theory one customarily assumes that a decision maker's preference ordering is complete, but not necessarily antisymmetric. Suppose each of a set I of different people has a preference ordering $\succcurlyeq_i$ over a set of alternatives H ($i \in I$). The *Pareto ordering* $\succcurlyeq$ determined by these is given by: $x \succcurlyeq y$ iff $x \succcurlyeq_i y$ for all $i \in I$. This need not be either complete or antisymmetric. In what follows we make no assumptions concerning completeness or antisymmetry.

Definition: Given partial order $\geq$ on H, a point $x^o \in H$ is *greatest*, or *best*, iff $x^o \geq x$ for all $x \in H$. Point $x^o \in H$ is *unsurpassed*[6] iff there is no point $x \in H$ such that $x > x^o$.

The following result is immediate.

Theorem: In partial order $\geq$ over H, x^o is greatest iff x^o is unsurpassed and comparable to all other $x \in H$.

Thus any greatest element is unsurpassed, and if $\geq$ is complete, the two concepts coincide. There may not be a greatest, or even an unsurpassed, element; or there may be several. Any two greatest elements must be indifferent; any two unsurpassed elements must be either indifferent or incomparable.[7]

Definition: Let $\geq_1$ and $\geq_2$ be two partial orderings on set H. Relation $\geq_2$ *extends* $\geq_1$ iff for all $x, y \in H$,

(i) $x >_1 y$ implies $x >_2 y$,
(ii) $x \sim_1 y$ implies $x \sim_2 y$.

That is, $\geq_2$ extends $\geq_1$ iff whenever two elements are comparable under $\geq_1$ that order relation is retained under $\geq_2$. In our ensuing discussion we place a number of partial orders on the space of all pseudomeasures, each an extension of the preceding.

Definition: Let $f: H_1 \rightarrow H_2$ be a function, and let $\geq_2$ be a partial order on H_2. The *partial order induced on H_1 by f* is the relation $\geq_1$ on H_1 satisfying $x \geq_1 y$ iff $f(x) \geq_2 f(y)$ for all $x, y \in H_1$.

One easily verifies that $\geq_1$ is indeed a partial order. Note that the induction here is backward, from the range space to the domain.

Definition: Let $(H_1, \geq_1)$ and $(H_2, \geq_2)$ be two partially ordered spaces. $(H_1, \geq_1)$ is *representable* by $(H_2, \geq_2)$ iff there is a function $f: H_1 \rightarrow H_2$ such that $x \geq_1 y$ iff $f(x) \geq_2 f(y)$ for all $x, y \in H_1$.

These two definitions underlie the representation of preferences by utility functions, for example. Here the space H_2 is usually the real numbers, and $\geq_2$ their natural ordering; H_1 is the space of possible alternatives for

---

6. The customary mathematical term for this concept is "maximal."
7. A point is said to be *Pareto efficient* for the family of partial orders ($\geq_i$), $i \in I$, over H iff it is unsurpassed in their Pareto ordering.

the problem in hand, and f the utility function. In our discussion H_2 will be the space of pseudomeasures Ψ, and $\geq_2$ will be one or another of the partial orders to be specified. Our claim is that this space provides a convenient representation for some problems in which preferences are not conveniently representable, or not representable at all, by the real numbers.

PARTIAL ORDERINGS ON VECTOR SPACES

Let V be a vector space, so that there is an operation of addition (from $V \times V$ to V) and of scalar multiplication (from the real numbers $\times$ V to V). We are interested in a certain restricted class of partial orders on V.

Definition: A relation $\geq$ on a vector space V is a *vector partial order* iff

(i) $\geq$ is a partial order in the ordinary sense;
(ii) if $x \geq y$, then $x + z \geq y + z$, for all $x, y, z \in V$;
(iii) if $x \geq y$ and b is a positive real number, then $bx \geq by$.

As an example, take n-space with the definition $(x_1, \ldots, x_n) \geq (y_1, \ldots, y_n)$ iff $x_i \geq y_i$ for all $i = 1, \ldots, n$ (the second $\geq$ referring to the natural ordering of the real numbers). Vector partial orders may be characterized in a very simple and useful fashion.[8] First, we need one more concept.

Definition: Subset P of a vector space V is a *convex cone* iff

(i) $0 \in P$;
(ii) if $x \in P$ and $y \in P$, then $x + y \in P$;
(iii) if b is a positive real number and $x \in P$, then $bx \in P$.

Theorem: Relation $\geq$ is a vector partial order on the vector space V iff there is a convex cone P such that, for all $x, y \in V$,

$$x \geq y \quad \text{iff} \quad x - y \in P \tag{3.3.1}$$

Letting $y = 0$ in (1), it is clear that $P = \{x \mid x \geq 0\}$. This is called the *positive cone* of the ordering $\geq$. If P is an arbitrary convex cone, and we use (1) to *define* the relation $\geq$, it follows that this relation is a vector partial order. This is generally the most convenient way to specify vector partial orderings.

---

8. See, e.g., J. L. Kelley, I. Namioka et al., *Linear Topological Spaces* (Van Nostrand, Princeton, 1963), p. 16.

One easily verifies that $x > y$ iff $(x - y) > 0$ and that $x \sim y$ iff $(x - y) \sim 0$, for vector partial orders.

NARROW ORDERING OF PSEUDOMEASURES

We now come to the vector space Ψ of all pseudomeasures over a fixed measurable space (A, Σ). Here we define a vector partial ordering called *narrow order;* in the next subsection we define another called *standard order,* which extends narrow order; and finally we define a variety of *extended orderings,* which all extend standard order.

If μ, ν are a pair of measures, or signed measures, we have already used the notation $\mu \geq \nu$ to abbreviate the condition: $\mu(E) \geq \nu(E)$ for all measurable sets E. We now define an ordering on pseudomeasures that generalizes this relation and reduces to it when both pseudomeasures are signed measures.

Definition: The *narrow order* $\geq$ on the space of pseudomeasures is the vector partial order whose positive cone is $\{\psi \mid \psi^- = 0\}$.

The pseudomeasures whose lower variation is zero are precisely those which are measures, so that the narrow order is the one whose positive cone is the set of (σ-finite) measures. (One verifies immediately that this set is, in fact, a convex cone, so that the definition is consistent.)

The following theorem gives several necessary and sufficient conditions for two pseudomeasures to be related by the narrow ordering $\geq$. These conditions are all in the form $\mu \geq \nu$, where μ and ν are *measures*, and this is to be interpreted in the ordinary sense that $\mu(E) \geq \nu(E)$ for all $E \in \Sigma$.

Theorem:

   (i) Let (μ, ν) be a pseudomeasure; $(\mu, \nu) \geq 0$ iff $\mu \geq \nu$;
  (ii) let (μ_1, ν_1) and (μ_2, ν_2) be pseudomeasures; $(\mu_1, \nu_1) \geq (\mu_2, \nu_2)$ iff $\mu_1 + \nu_2 \geq \nu_1 + \mu_2$;
 (iii) let ψ_1, ψ_2 be pseudomeasures; $\psi_1 \geq \psi_2$ iff $\psi_1^+ \geq \psi_2^+$ and $\psi_1^- \leq \psi_2^-$.

Proof: (i) Let (λ^+, λ^-) be the Jordan form of (μ, ν). $(\mu, \nu) \geq 0$ iff $\lambda^- = 0$. It is immediate from the definition of λ^- that $\mu \geq \nu$ implies $\lambda^- = 0$. Conversely, if $\nu(E) > \mu(E)$ for some measurable E, then $\lambda^-(E) \geq \nu(E) - \mu(E) > 0$; hence $(\mu, \nu) \not\geq 0$.

(ii) Since $\geq$ is a vector partial order, $(\mu_1, \nu_1) \geq (\mu_2, \nu_2)$ iff

$$(\mu_1 + \nu_2, \nu_1 + \mu_2) = (\mu_1, \nu_1) - (\mu_2, \nu_2) \geq 0$$

(the equality comes from the equivalence theorem). The result now follows from (i).

(iii) If $\psi_1^+ \geq \psi_2^+$ and $\psi_1^- \leq \psi_2^-$, then $\psi_1^+ + \psi_2^- \geq \psi_1^- + \psi_2^+$, so that $\psi_1 \geq \psi_2$ from (ii).

Conversely, suppose $\psi_1^+ \geq \psi_2^+$ is false, so that $\psi_1^+(E) < \psi_2^+(E)$ for some measurable set E. Let (P_2, N_2) be a Hahn decomposition for ψ_2. Then

$$\psi_1^+(E \cap P_2) + \psi_2^-(E \cap P_2) = \psi_1^+(E \cap P_2)$$

(since $\psi_2^-(P_2) = 0$)

$$\leq \psi_1^+(E) < \psi_2^+(E) = \psi_2^+(E \cap P_2)$$

(since $\psi_2^+(N_2) = 0$)

$$\leq \psi_2^+(E \cap P_2) + \psi_1^-(E \cap P_2)$$

Hence $\psi_1^+ + \psi_2^- \not\geq \psi_1^- + \psi_2^+$, so that $\psi_1 \not\geq \psi_2$ from (ii).

Finally, if $\psi_1^-(E) > \psi_2^-(E)$ for some measurable E, let (P_1, N_1) be a Hahn decomposition for ψ_1. An argument similar to the one just given, but with $E \cap N_1$ in place of $E \cap P_2$, again shows that $\psi_1 \geq \psi_2$ is false. ∎

It now follows that the narrow ordering $\geq$ reduces to the ordinary $\geq$ when the pseudomeasures are ordinary signed measures. For, letting μ and ν be signed measures and identifying them with the pseudomeasures $(\mu^+, \mu^-), (\nu^+, \nu^-)$ respectively, we have that $(\mu^+, \mu^-) \geq (\nu^+, \nu^-)$ iff $\mu^+ \geq \nu^+$ and $\mu^- \leq \nu^-$, which is necessary and sufficient for $\mu \geq \nu$ in the ordinary sense. Our notation is therefore consistent.

Narrow order is *antisymmetric*. For if $\psi_1 \geq \psi_2$ and $\psi_2 \geq \psi_1$ are both true, then $\psi_1^+ \geq \psi_2^+ \geq \psi_1^+$ and $\psi_1^- \leq \psi_2^- \leq \psi_1^-$, so that $\psi_1^+ = \psi_2^+$ and $\psi_1^- = \psi_2^-$; that is, $\psi_1 = \psi_2$.

It follows that $\psi_1 > \psi_2$ iff $\psi_1 \geq \psi_2$ and $\psi_1 \neq \psi_2$. Applied to the theorem above, this yields criteria for one pseudomeasure being bigger than another. For example, the pseudomeasure $(\mu, \nu) > 0$ iff $\mu \geq \nu$ and $\mu(E) > \nu(E)$ for at least one measurable set E.[9]

Theorem: Narrow order is incomplete, except when Σ is the trivial σ-field $\{A, \emptyset\}$.

---

9. This relation may be written $\mu > \nu$. This notation is consistent with the corresponding pseudomeasure relation $(\mu, 0) > (\nu, 0)$. Note that $\mu > \nu$ does not mean that $\mu(E) > \nu(E)$ for *all* $E \in \Sigma$. In fact, the latter condition never holds, since all measures are zero at $E = \emptyset$.

Proof: The case $\Sigma = \{A, \phi\}$ is left as an exercise. (The space of pseudomeasures is isomorphic to the real numbers in this case, if $A \neq \phi$.)

If Σ is not trivial, there is a measurable E^o such that neither E^o nor $A\backslash E^o$ is empty. Choose points $a \in E^o$, $b \in A\backslash E^o$, and define the measures μ, ν by $\mu(F) = 1$ if $a \in F$, $\mu(F) = 0$ otherwise; and $\nu(F) = 1$ if $b \in F$, $\nu(F) = 0$ otherwise, for all $F \in \Sigma$. Then $\mu(E^o) = \nu(A\backslash E^o) = 1$, and $\nu(E^o) = \mu(A\backslash E^o) = 0$, so that μ and ν are not zero and they are mutually singular. Hence the pseudomeasure (μ, ν) is not comparable to 0, and the narrow order $\geq$ is incomplete. ∎

With the aid of narrow order, we can generalize the standard inequality theorems for integrals to pseudomeasure integrals. In the following, $\geq$ or $>$ when used between pseudomeasures refers to narrow order, while the expression $f \geq g$ between point functions f and g means that $f(a) \geq g(a)$ for all $a \in A$.

Theorem: Inequalities for Pseudomeasure Integrals. Let ψ_1 and ψ_2 be pseudomeasures, and let f and g be real-valued measurable functions, all on measurable space (A, Σ).

   (i) If $\psi_1 \geq 0$ and $f \geq 0$, then $\int f \, d\psi_1 \geq 0$.

   (ii) If $\psi_1 \geq 0$ and $f \geq g$, then $\int f \, d\psi_1 \geq \int g \, d\psi_1$.

   (iii) If $\psi_1 \geq \psi_2$ and $f \geq 0$, then $\int f \, d\psi_1 \geq \int f \, d\psi_2$.

   (iv) If $\psi_1 > 0$ and $f(a) > 0$ for all $a \in A$, then $\int f \, d\psi_1 > 0$.

   (v) If $\psi_1 > 0$ and $f(a) > g(a)$ for all $a \in A$, then $\int f \, d\psi_1 > \int g \, d\psi_1$.

   (vi) If $\psi_1 > \psi_2$ and $f(a) > 0$ for all $a \in A$, then $\int f \, d\psi_1 > \int f \, d\psi_2$.

Proof: (i) By assumption, $\psi_1^- = 0$ and $f^- = 0$, so the expansion of $\int f \, d\psi_1$ by (2.16) reduces to $(\int f^+ \, d\psi_1^+, 0)$. Hence $(\int f \, d\psi_1)^- = 0$, so that $\int f \, d\psi_1 \geq 0$.

   (ii) $\int f \, d\psi_1 - \int g \, d\psi_1 = \int (f - g) \, d\psi_1 \geq 0$ from (i). Hence

$$\int f \, d\psi_1 \geq \int g \, d\psi_1$$

   (iii) Since $\psi_1 \geq \psi_2$, $(\psi_1 - \psi_2) \geq 0$;

$$\int f \, d\psi_1 - \int f \, d\psi_2 = \int f \, d(\psi_1 - \psi_2) \geq 0$$

from (i). Hence $\int f \, d\psi_1 \geq \int f \, d\psi_2$.

   (iv) Since $\psi_1 > 0$, $\psi_1^+(A) > 0$; also f is positive, so by a standard integration theorem, $\int_A f \, d\psi_1^+ > 0$. Hence, $\int f \, d\psi_1 = (\int f^+ \, d\psi_1^+, 0) > 0$.

(v) $\int f \, d\psi_1 - \int g \, d\psi_1 = \int (f - g) \, d\psi_1 > 0$ from (iv).   Hence $\int f \, d\psi_1 > \int g \, d\psi_1$.

(vi) Since $\psi_1 > \psi_2$, $(\psi_1 - \psi_2) > 0$.   Hence $\int f \, d\psi_1 - \int f \, d\psi_2 = \int f \, d(\psi_1 - \psi_2) > 0$ from (iv), so that $\int f \, d\psi_1 > \int f \, d\psi_2$.   ∎

STANDARD ORDERING OF PSEUDOMEASURES

While narrow order is quite useful, it is literally too narrow to represent preferability relations among pseudomeasures in an intuitively plausible way.   Consider, for example, the representation of preferences by definite integrals over Time:

$$U(p) = \int_0^\infty f(p, t) \, dt \qquad (3.3.2)$$

p being a policy.   Going from definite to indefinite integrals and thence to signed measures, we can translate the criterion (2) as follows.   With each policy p is associated a signed measure μ_p.   Policy p_1 is at least as good as policy p_2 iff $\mu_{p_1}(A) \geq \mu_{p_2}(A)$.   Note that the value of μ_p on the universe set A is all that counts; this corresponds precisely to the use of the *definite* integral in (2).   We have argued that this sort of criterion is counterintuitive when both signed measures are infinite (of the same sign), but it is perfectly adequate for comparing policies if their corresponding signed measures are both *finite*.

We would like, then, an ordering on the space of pseudomeasures such that when two pseudomeasures are finite signed measures μ, ν, the relation between these agrees with the criterion above: μ is at least as good as ν iff $\mu(A) \geq \nu(A)$.   Narrow order does not accomplish this: take any two mutually singular finite nonzero measures.   (These always exist if the σ-field Σ is not trivial.)   These are not comparable under narrow ordering but are comparable under the criterion just stated.

Standard order, which we are about to define, meets these desiderata.   We first need a preliminary result.

Lemma: Given measurable space (A, Σ), the set of pseudomeasures ψ satisfying

$$\psi^+(A) \geq \psi^-(A), \quad \text{and} \quad \psi^-(A) < \infty \qquad (3.3.3)$$

is a convex cone.

Proof: Clearly, 0 belongs to this set.   If ψ is a pseudomeasure and b a positive real number, then $(b\psi)^+ = b \cdot \psi^+$, and $(b\psi)^- = b \cdot \psi^-$.   Hence if ψ belongs, then $b\psi$ belongs.

Finally, suppose ψ_1, ψ_2 both belong; we must show that $\psi_1 + \psi_2$ belongs. First, from (3), $\psi_i^-(A) < \infty$, $i = 1, 2$. Also $(\psi_1 + \psi_2)^- \leq \psi_1^- + \psi_2^-$ (from the minimizing property of the Jordan form). Hence

$$(\psi_1 + \psi_2)^-(A) < \infty \qquad (3.3.4)$$

Also

$$(\psi_1 + \psi_2)^+(A) + \psi_1^-(A) + \psi_2^-(A) = (\psi_1 + \psi_2)^-(A) + \psi_1^+(A) + \psi_2^+(A)$$

$$(3.3.5)$$

by the equivalence criterion (2.4). Since $\psi_1^-(A)$ and $\psi_2^-(A)$ are both finite, we may subtract them from both sides of (5) to obtain

$$(\psi_1 + \psi_2)^+(A) =$$

$$(\psi_1 + \psi_2)^-(A) + [\psi_1^+(A) - \psi_1^-(A) + \psi_2^+(A) - \psi_2^-(A)]$$

$$\geq (\psi_1 + \psi_2)^-(A) \qquad (3.3.6)$$

since (3) applies to ψ_1 and ψ_2. Inequalities (4) and (6) show that (3) is satisfied by $\psi_1 + \psi_2$. ∎

We are now assured that the following definition is consistent.

Definition: *Standard order* $\geqslant$ on the space of pseudomeasures is the vector partial order whose positive cone is

$$\{\psi \mid \psi^+(A) \geq \psi^-(A), \text{ and } \psi^-(A) < \infty\} \qquad (3.3.7)$$

We use the notations $\geqslant$ and $\geq$ to distinguish standard from narrow order respectively; $\succ$ and $>$ are the corresponding strict inequality signs; the indifference sign $\sim$ will refer to indifference under *standard* order only. (It is not needed for narrow order, since indifference coincides with equality there.)

The positive cone (7) consists precisely of those pseudomeasures which are signed measures μ satisfying $\mu(A) \geq 0$. It follows that a pseudomeasure is comparable to zero under standard order iff it is a signed measure.

We first verify the claim made above: that if μ and ν are finite signed measures, the relation between them under standard order is the same as that given by the comparison of $\mu(A)$ and $\nu(A)$. Here μ and ν are identified, as usual, with the pseudomeasures (μ^+, μ^-), (ν^+, ν^-) respectively. (In particular, the ordering of finite definite integrals is the same as the standard ordering of the corresponding indefinite integrals.)

Theorem: Let μ and ν be finite signed measures on space (A, Σ). Then $(\mu^+, \mu^-) \geqslant (\nu^+, \nu^-)$ iff $\mu(A) \geq \nu(A)$.

Proof: Since $\geqslant$ is a vector partial order, $(\mu^+, \mu^-) \geqslant (\nu^+, \nu^-)$ iff

$$(\mu^+ + \nu^-, \nu^+ + \mu^-) = (\mu^+, \mu^-) - (\nu^+, \nu^-) \geqslant 0 \qquad (3.3.8)$$

Let (λ^+, λ^-) be the Jordan form of $(\mu^+ + \nu^-, \ \nu^+ + \mu^-)$. By the equivalence criterion,

$$\lambda^+ + \nu^+ + \mu^- = \lambda^- + \mu^+ + \nu^- \qquad (3.3.9)$$

Since all measures in (9) are finite, we obtain

$$\lambda^+(A) - \lambda^-(A) = [\mu^+(A) - \mu^-(A)] - [\nu^+(A) - \nu^-(A)]$$

$$= \mu(A) - \nu(A) \qquad (3.3.10)$$

Since (8) holds iff $\lambda^+(A) - \lambda^-(A) \geq 0$, (10) completes the proof. ∎

Strict inequality and indifference take a simple form for standard order. The following results are immediate from the definitions.

Theorem: Let ψ be a pseudomeasure on (A, Σ).

(i) $\psi > 0$ iff $\psi^+(A) > \psi^-(A)$;
(ii) $\psi \sim 0$ iff $\psi^+(A) = \psi^-(A)$ *and* these are finite;
(iii) $\psi, 0$ are not comparable iff $\psi^+(A) = \psi^-(A) = \infty$.

As with any vector partial order, we then have for pseudomeasures ψ_1 and ψ_2:

$$\psi_1 > \psi_2 \quad \text{iff} \quad (\psi_1 - \psi_2) > 0$$

$$\psi_1 \sim \psi_2 \quad \text{iff} \quad (\psi_1 - \psi_2) \sim 0$$

and ψ_1, ψ_2 not comparable iff $\psi_1 - \psi_2, 0$ not comparable.

Theorem: Standard order extends narrow order.

Proof: Let $\psi_1 > \psi_2$ (narrow order); then $(\psi_1 - \psi_2) > 0$; i.e., $(\psi_1 - \psi_2)^-(A) = 0$, while $(\psi_1 - \psi_2)^+(A) > 0$. Hence $(\psi_1 - \psi_2) > 0$, so that $\psi_1 > \psi_2$.

The corresponding result for indifference is trivial, since $\geq$ is antisymmetric. ∎

If Σ is not trivial, then this extension is proper; i.e., there are ψ_1, ψ_2 that are noncomparable under narrow order but comparable under standard order. An example is given in the proof that narrow order is incomplete. The same example shows that standard order is *not* antisymmetric except in the trivial case $\Sigma = \{\emptyset, A\}$.

Theorem: Standard order is incomplete except when Σ is a finite σ-field.

Proof: Let Σ be finite. Then it is generated by a finite partition $\{A_1, \ldots, A_n\}$. If μ is a σ-finite measure, then $\mu(A_i)$ is finite for all $i = 1 \ldots n$, so μ is in fact finite. Hence $\psi^+(A)$, $\psi^-(A)$ are both finite for all pseudomeasures ψ, and all pairs are comparable: $\geqslant$ is complete.

Conversely, let Σ be infinite. Let $\mathfrak{F} \subseteq \Sigma$ be infinite and *countable,* and let $\mathcal{G}$ consist of all $\mathfrak{F}$-sets, along with their complements. For each $a \in A$, let $E(a)$ be the intersection of all $\mathcal{G}$-sets to which a belongs. It may be shown that these sets $E(a)$ form an infinite measurable partition (details are omitted).

There is thus an infinite sequence $(E_1, E_2, \ldots)$ of nonempty, measurable, mutually disjoint sets. Choose a point $a_n \in E_n$ for all $n = 1, 2, \ldots$ and define the measures μ, ν by: $\mu(E) =$ number of points $a_n \in E$ for which n is *odd,* and $\nu(E) =$ number of points $a_n \in E$ for which n is *even,* for all measurable E. Thus $\mu(E_{2m-1}) = \nu(E_{2m}) = 1$, and $\mu(E_{2m}) = \nu(E_{2m-1}) = 0$, all $m = 1, 2, 3, \ldots$. Thus μ and ν are σ-finite, infinite, and mutually singular. Hence the pseudomeasure (μ, ν) is not comparable to 0 under standard order: $\geqslant$ is incomplete. ∎

The case where Σ is finite is not very interesting from our present point of view because in this case all pseudomeasures are finite signed measures, and Ψ is isomorphic to ordinary n-space for some $n = 0, 1, \ldots$, as we have noted above. Thus we may say: whenever pseudomeasures are interesting, standard order is incomplete.

APPLICATIONS

We consider here only the more important category of applications— the use of pseudomeasures to represent preferences. Let P be a set of conceivable alternative options in a certain situation, and let $\geqslant$ be a preference ordering on P ($\geqslant$ is a partial order, i.e., a transitive reflexive relation). We represent this ordering by a pseudomeasure-valued utility function $p \to \psi(p)$, mapping P into Ψ, the set of pseudomeasures over some space (A, Σ). (Set A might be completely unconnected with P, but usually there is some connection that makes the representation "natural.") That is, for any two options $p_1, p_2 \in P$, we have

$$p_1 \geqslant p_2 \quad \text{iff} \quad \psi(p_1) \geqslant \psi(p_2) \tag{3.3.11}$$

Here $\geqslant$ on the left is the preference ordering, while $\geqslant$ on the right is *standard ordering* on the space Ψ.[10]

---

10. This ambiguous usage should cause no confusion. We also consider below some "nonfaithful" representations, in which either the "if" or the "only if" of (11) is relaxed.

The size comparison between two given pseudomeasures is usually quite simple to make. For the great bulk of applications in this book (and this will probably be true in further applications as well) $\psi(p)$ takes the form of an integral in which either the integrand or the measure does not depend on p. Thus in the fixed-integrand case we need only worry about whether a statement of the following form is true:

$$\int f \, d\mu_1 \geq \int f \, d\mu_2 \qquad (3.3.12)$$

(All indefinite integrals are over space A.) In the fixed-measure case, the relevant statements are all of the form

$$\int f_1 \, d\mu \geq \int f_2 \, d\mu \qquad (3.3.13)$$

In turn, both these statements are logically equivalent to a statement of the form

$$\int f \, d\mu \geq 0 \qquad (3.3.14)$$

To go from (13) to (14), let $f = f_1 - f_2$. To go from (12) to (14), let $\mu = \mu_1 - \mu_2$. (If μ_1, μ_2 are both infinite measures, interpret $\mu_1 - \mu_2$ as the pseudomeasure (μ_1, μ_2).) If μ in (14) should turn out to be a signed measure, the following simple but important result shows that the size comparison of pseudomeasures reduces to the evaluation of an ordinary definite integral.

Theorem: Standard Integral Theorem. Let μ be a σ-finite signed measure on space (A, Σ), and let $f : A \to$ reals be measurable. Then (14) (in the sense of standard order) is true iff the *definite* integral

$$\int_A f \, d\mu \qquad (3.3.15)$$

is well defined and nonnegative.

Proof: The Jordan form of $\psi = \int f \, d\mu$ is

$$(\psi^+, \psi^-) = \left(\int f^+ \, d\mu^+ + \int f^- \, d\mu^-, \int f^- \, d\mu^+ + \int f^+ \, d\mu^- \right)$$

Relation (14) is true iff $\psi^+(A) \geq \psi^-(A) < \infty$, i.e., iff the double inequality

$$\int_A f^+ \, d\mu^+ + \int_A f^- \, d\mu^- \geq \int_A f^- \, d\mu^+ + \int_A f^+ \, d\mu < \infty$$

holds. But this is precisely the condition for (15) to be well defined and nonnegative. ∎

Consider the procedure of Ramsey in his now famous article.[11] Here each conceivable policy p is a time path of consumption and labor over the entire positive half-line. The total utility resulting from policy p has the form

$$\int_0^\infty g(p, t) \, dt \qquad (3.3.16)$$

where $g(p, t)$ is the momentary utility from consumption net of the disutility from labor at instant t under policy p (we need not be concerned with the exact form of this function). The trouble is that (16) will diverge for good policies. Accordingly, Ramsey uses, not (16), but

$$\int_0^\infty [b - g(p, t)] \, dt \qquad (3.3.17)$$

as the objective function. Here b is "bliss," the highest attainable momentary net utility level; (17) is the shortfall from constant bliss and is to be minimized. The integral is finite except for very poor policies, which may be ignored. Preference is now represented by an ordinary real-valued utility function, viz., the negative of (17).

We now treat the same problem by pseudomeasures. Let (A, Σ) be the positive half-line with Borel field. The utility assigned to policy p is the pseudomeasure $\psi(p) = \int g(p, t) \, dt$. ($dt$ of course refers to Lebesgue measure on the positive half-line.) For two policies, p_1 and p_2, we then have $p_1 \succeq p_2$ iff $\int g(p_1, t) \, dt \succeq \int g(p_2, t) \, dt$ (standard order). By the standard integral theorem, this is true iff the *definite* integral relation

$$\int_0^\infty [g(p_1, t) - g(p_2, t)] \, dt \geq 0 \qquad (3.3.18)$$

holds, the integral being well defined. It is easy to see that (18) determines exactly the same ordering among policies as (17), with the minor exception that (17) does not discriminate adequately among alternative very poor policies for which it equals $+\infty$.

Thus the Ramsey approach is in this case essentially the same as the pseudomeasure approach. What is gained by using the latter? First, we avoid the slightly ad hoc "bliss" procedure, which may not be available for other problems. But, more important, we have a unified procedure, which works for multidimensional and abstract spaces, which works when the measure rather than the integrand varies (as in Bernoullian utility discussed below), etc.

We also sharply distinguish the Ramsey approach from the "over-

11. F. P. Ramsey, A mathematical theory of saving, *Econ. J.* 38 (Dec. 1928):543–59. Reprinted in K. J. Arrow and T. Scitovsky, eds., *Readings in Welfare Economics* (Irwin, Homewood, Ill., 1969), pp. 619–33.

taking" approach that grew out of it. Overtaking depends essentially on the order or metric properties of the real line; the Ramsey approach does not. It turns out that, just as the Ramsey approach is a special case of *standard ordering* of pseudomeasures, the overtaking approach is a special case of *extended ordering*. This is discussed below.

The following simple problem offers further insights into the use of standard order. Let (A, Σ, μ) be a probability measure space; i.e., μ is a measure with $\mu(A) = 1$. Let $f:A \to$ reals be measurable, and let the definite integral $\int_A f \, d\mu$ exist and be finite, with value c.[12] Consider the problem of minimizing

$$v(x) = \int_A [f(a) - x]^2 \, \mu(da) \qquad (3.3.19)$$

over real numbers x. It is well known that the unique minimizer for this problem is $x = c$, *provided* the integral (19) is finite for all x. If (19) is infinite for some x, it is infinite for all x, so that every real number is a minimizer. We now show that the use of pseudomeasures allows this proviso to be dropped.

Thus we now rewrite $v(x)$ as $\psi(x)$ and interpret (19) as the *indefinite* integral. There are two preliminary minor points to note. First, we are minimizing, so "smaller" is "better"; but under standard ordering "larger" is "better" in an obvious sense. This difficulty may be remedied in either of two equivalent ways: insert a minus sign in front of (19) to convert it to a maximum problem, or use *reverse* standard ordering rather than (direct) standard ordering, defined by $\psi_1 \succcurlyeq \psi_2$ in the reverse sense iff $\psi_1 \preccurlyeq \psi_2$ in the direct sense. We shall use the latter approach. Second, remember that standard order is in general not complete, so that there are two possible senses in which a solution may be optimal: it may be *best,* or it may be merely *unsurpassed.* In the following theorem the stronger of these two senses may be asserted.

Theorem: Let μ, f, and c be as above. The problem of minimizing $\psi(x)$ over real numbers x has a *unique best* solution, viz., $x = c$. ("Minimization" is understood in the sense of reverse standard ordering.)

Proof: Let real $x \neq c$, and consider the *definite* integral

$$\int_A [(f(a) - x)^2 - (f(a) - c)^2] \, \mu(da)$$

$$= \int_A [2(c - x)f(a) + x^2 - c^2] \, \mu(da) = (c - x)^2 > 0$$

---

12. In this probabilistic context f is usually called a random variable, with expectation c.

Since this is well defined and positive, it follows that the *indefinite* integral satisfies

$$\int [(f(a) - x)^2 - (f(a) - c)^2]\, \mu(da) > 0$$

(standard order), by the standard integral theorem. Thus

$$\psi(x) > \psi(c) \tag{3.3.20}$$

(standard order) for all $x \neq c$. Thus c is best under reverse standard ordering. Because indifference is precluded in (20), c is the unique number with this property. ∎

In probability terms, then, we may say: the second moment of f about x is uniquely minimized when x = expectation f (even if the second moment is infinite under conventional calculation). For a similar result concerning the median see Section 9.4.

BERNOULLIAN UTILITY UNDER STANDARD ORDER

Let (A, Σ) be, as usual, a measurable space fixed throughout the discussion, and let **P** be the set of all probability measures on this space. We are concerned with preference orderings over **P**. The modern discussion of this subject arises from the observation[13] that under certain quite plausible assumptions concerning the preference ordering $\succcurlyeq$, a rather strong conclusion could be drawn (for Σ finite): a measurable function $u{:}A \rightarrow$ reals exists such that the mapping $p \rightarrow \int_A u\, dp$, which assigns to each $p \in$ **P** the expectation of u with respect to p, is a utility function that represents $\succcurlyeq$. This is the expected utility theorem.

This result has been generalized in various ways, but these generalizations always run up against the obstacle that the integral $\int_A u\, dp$ must be well defined and finite. In practice this means either that u is bounded or the p's must be restricted to a small subset of **P** (once we go to an infinite σ-field; if Σ is finite, u is automatically bounded).[14]

Both these restrictions are objectionable. The restriction of **P** to

---

13. In J. von Neumann and O. Morgenstern, *Theory of Games and Economic Behavior,* 2nd ed. (Princeton Univ. Press, Princeton, N.J., 1947), pp. 617–28. The axiomatic discussion was flawed, one of the basic axioms being concealed implicitly in an operation, as noted by E. Malinvaud (repr. on p. 271 of *Readings,* mentioned in fn. 14). The idea of maximizing expected utility originates with Daniel Bernoulli, 1738.

14. In so-called "mixture spaces" the utility function need not be bounded, but here it is not expressed in the form of an integral. See I. N. Herstein and J. Milnor, An axiomatic approach to measurable utility, *Econometrica* 21 (Apr. 1953):291–97, repr. in P. Newman, ed., *Readings in Mathematical Economics,* Vol. 1, (Johns Hopkins Press, Baltimore, 1968), pp. 264–70. In our lemma below we use a theorem of Fishburn that is very similar to the Herstein-Milnor result.

finitely concentrated probabilities (see below) simply does not allow enough scope. The objections to bounded u require more discussion.

Consider the following "Archimedean" postulate. Let $a_1, a_2, a_3 \in A$, with $a_1 > a_2 > a_3$; then a number $x, 0 < x < 1$, exists such that

$$[(1 - x)a_1 + x a_3] > a_2 \qquad (3.3.21)$$

Here a_i refers to the probability measure with all mass simply concentrated on point a_i, $i = 1, 2, 3$, and (21) states that some probability mixture of a_1 and a_3 is preferred to a_2. Suppose a_1 and a_2 are situations differing only in some trivial respect, e.g., having this morning's egg boiled for 3 minutes vs. 3 minutes and 1 second, while a_3 is a horrendous situation such as a world pandemic or thermonuclear war. One may argue that (21) is still satisfied by some x very very close to 0 in value. The point is controversial.

Now suppose the expected utility condition holds, with a function u that is *bounded below*. We claim this has a consequence that is *less* plausible than (21) by an order of magnitude. Without loss of generality let $u(a_1) = 1$, $u(a_2) = 0$, and let $-M$ be a lower bound for u, where M is a large positive real number. Then, for any choice a_3, (21) is satisfied by $x = 1/(M + 2)$, since the left side of (21) then has utility at least equal to $1/(M + 2) > 0$. That is, the mixture proportions x and $1 - x$ may be chosen *in advance* of knowing a_3, no matter how horrible. The Archimedean postulate allows x to depend on a_3, so that progressively more horrible situations may be counterbalanced by being given progressively less weight.

There is a similar implausibility argument for u bounded *above*. Start with the "dual" Archimedean postulate that replaces all $>$ signs by $<$, let a_1, a_2 differ trivially as above, and let a_3 be some highly desirable situation such as universal salvation or utopia.

In general, bounded utility appears to characterize orderings with a certain pettifogging quality, in which there is no Pascalian wager, no Faustian aspiration, no Promethean ambition. To exclude these preferences would be to exclude the values of many makers and shapers of history, not all of whom are irrational.

What these thoughts amount to is this: any axiom system implying that "rational" preference orderings satisfy the expected utility condition with bounded utility function is simply too restrictive. But if the boundedness restriction on u is removed, what sense is one to make of the integral $\int_A u \, dp$? Our recommendation should not be too surprising: reinterpret this as an *indefinite* integral, and let size ordering among these entities be given by *standard ordering* of pseudomeasures.

But why should one do this? Just as the ordinary expected utility condition needs justification, so too does this standard ordering condition.

We now give a set of axioms that implies it. These axioms have about the same general level of plausibility as those in customary use—even a bit more, since they are weakened to the point where they do not imply that u is bounded.

The main limitation imposed is that Σ be generated by a *countable partition* of A. This limitation is regrettable but still allows the main point to come through: an axiomatic basis exists for the use of unbounded utility functions and standard ordering in the treatment of uncertainty.[15]

Without real loss of generality, we may assume that each element in the partition generating Σ is singleton. Thus we get A countable, and Σ = all subsets of A. This is the space on which we work. Each probability measure p on (A, Σ) is completely determined by its values on the singleton sets $\{a\}$. In fact, it will sometimes be convenient in what follows to think of probabilities as point functions (with domain A), and we write $p(a)$ instead of the technically correct $p(\{a\})$.

Probability measure p is said to be *finitely concentrated* iff there is a finite set $E \subseteq A$ such that $p(A \setminus E) = 0$; i.e., $p(a) = 0$ for all but a finite number of points $a \in A$. The set of all probabilities will be denoted by **P** as above, while the subset of finitely concentrated probabilities will be denoted by **F**.

In axiom 4 the following concept is used. Real sequence $x_1, x_2, \ldots$ is *monotone* iff it is either nonincreasing or nondecreasing: either $x_1 \geq x_2 \geq \ldots$, or $x_1 \leq x_2 \leq \ldots$.

Axioms concerning partial order $\succcurlyeq$ on **P** follow:

Axiom 1: Any two finitely concentrated probability measures are comparable.

Axiom 2: Let $p_1, q_1 \in$ **P**, let $p_2, q_2 \in$ **F**, with $p_1 \sim q_1$ and $p_2 > q_2$; let x be a number, $0 < x < 1$; then

$$[(1 - x)p_1 + xp_2] > [(1 - x)q_1 + xq_2] \qquad (3.3.22)$$

Axiom 3: Let $p_1, q_1 \in$ **P** with $p_1 > q_1$; then $p_2, q_2 \in$ **F** and a number $x, 0 < x < 1$, exist such that $p_2 < q_2$ and (22) is true.

Axiom 4: Let $p, q \in$ **P**; let $p_k, q_k, k = 1, 2, \ldots$, be two sequences of finitely concentrated probability measures such that for all $a \in A$ the

---

15. Actually, the following theorems and proofs generalize easily to the class of all discrete probabilities (i.e., concentrated on a countable set) over a *general* measurable space (with all singletons measurable); but the resulting utility function u may not be measurable. To guarantee the existence of u measurable, add the following axiom to axioms 1–5 below:

For all $a_o \in A$, $\{a \mid p_a > p_{a_o}\} \in \Sigma$, where p_a is the probability measure simply concentrated at point a.

(In the present special case Σ = all subsets, this axiom is superfluous.)

three sequences $(p_k(a))$, $(q_k(a))$, and $(p_k(a) - q_k(a))$, $k = 1, 2, \ldots$, are monotone, and such that for all $a \in A$,

$$\lim_{k \to \infty} p_k(a) = p(a), \qquad \lim_{k \to \infty} q_k(a) = q(a)$$

and such that, for all $k = 1, 2, \ldots$, $p_k > q_k$; then it is *false* that $q > p$.

Axiom 5: Let $\succeq'$ be any partial order on **P** that satisfies axioms 1–4 and that extends $\succeq$ (i.e., $p > q$ implies $p \succ' q$, and $p \sim q$ implies $p \sim' q$); then $\succeq$ and $\succeq'$ are identical.

Axiom 1 is a weak completeness axiom. Axiom 2 is a form of the strong independence axiom and asserts roughly that mixing in a pair of indifferent probabilities does not disturb order of preference. Axiom 3 is very weak and asserts roughly that for any $p_1 > q_1$ we can find finitely concentrated $p_2 \prec q_2$ for which the preference intensity is not infinitely stronger than the original. Axiom 4 is an Archimedean axiom of sorts and asserts that under certain conditions, if sequences p_k, q_k converge to p, q respectively, it cannot happen that preferences between p_k and q_k all run in one direction and preference between p, q runs in the opposite direction. (The monotonicity clause in axiom 4 has no intuitive appeal in itself, but note that its insertion weakens the axiom and thereby makes axiom 4 more plausible in the logical sense.)[16] Finally, axiom 5, like axiom 1, is a weak completeness axiom and asserts that $\succeq$ has maximal comparability in the class of partial orders satisfying axioms 1–4. (The assumption that *any* two probabilities are comparable—which we do not make—would imply both axioms 1 and 5.)

The following lemma asserts that axioms 1–4 alone guarantee the existence of a function u that provides a "nonfaithful" representation of $\succeq$ in the sense of Aumann.[17]

Lemma: Let **P** be the set of all probability measures on (A, Σ), where A is countable and $\Sigma = $ all subsets of A; let $\succeq$ be a partial ordering on **P** satisfying axioms 1–4. Then a function $u : A \to$ reals exists such that for all $p, q \in$ **P**,

$$p > q \quad \text{implies} \quad \int u \, dp > \int u \, dq \tag{3.3.23}$$

$$p \sim q \quad \text{implies} \quad \int u \, dp \sim \int u \, dq \tag{3.3.24}$$

Here $>$ and $\sim$ on the left refer to the partial order on **P**, while $>$ and $\sim$

---

16. If the monotonicity clause is dropped from axiom 4, it may be shown that the resulting set of axioms is equivalent to the existence of a *bounded* utility function—which of course deprives this analysis of its essential point.

17. R. J. Aumann, Utility theory without the completeness axiom, *Econometrica* 30 (1962):445–62.

on the right refer to standard order on the space of pseudomeasures over (A, Σ).

Proof: Let us first prove another Archimedean condition: if q, p, $p' \in F$ (i.e., they are finitely concentrated), and $p' > q > p$, then a number x, $0 < x < 1$, exists such that

$$(1 - x)p + xp' < q \tag{3.3.25}$$

To show this, define the two sequences p_k, q_k, $k = 1, 2, \ldots$, by $p_k = [k/(k + 1)]p + [1/(k + 1)]p'$ and $q_k = q$. The p_k, q_k are all finitely concentrated; they converge pointwise to p, q respectively, and the monotonicity clause of axiom 4 is satisfied. But the conclusion of axiom 4 is false, since $q > p$; hence the remaining premise of axiom 4 must be false, so that a k_o exists for which $p_{k_o} > q_{k_o}$ is false. By axiom 1 it follows that $p_{k_o} \lesssim q_{k_o}$. There are two cases: If $p_{k_o} < q_{k_o}$, then (25) is verified with $x = 1/(k_o + 1)$; and if $p_{k_o} \sim q_{k_o}$, we apply axiom 2 to obtain

$$\left[\left(\frac{k_o + 1}{k_o + 2}\right)p_{k_o} + \left(\frac{1}{k_o + 2}\right)p\right] < \left[\left(\frac{k_o + 1}{k_o + 2}\right)q_{k_o} + \left(\frac{1}{k_o + 2}\right)q\right]$$

which verifies (25) with $x = 1/(k_o + 2)$.

By a similar argument (interchanging the roles of p and p') we can show the existence of a number x such that (25') is true, where (25') is obtained from (25) by substituting $>$ for $<$.

Axioms 1 and 2, along with conclusions (25) and (25') and Theorem 8.2 of Fishburn,[18] imply the existence of a function $u:A \rightarrow$ reals such that, for all *finitely concentrated* p, q we have

$$p \gtrsim q \quad \text{iff} \quad \int_A u \, dp \geq \int_A u \, dq \tag{3.3.26}$$

We will show that this u satisfies (23) and (24).

To prove (23), let $p, q \in P$ with $p > q$ and define the function $f:A \rightarrow$ reals by

$$f(a) = u(a)[q(a) - p(a)] \tag{3.3.27}$$

The hardest part of the proof will be to show that the sum of the *positive* terms of $f(a)$, summed over $a \in A$, is *finite*. Arguing by contradiction, suppose that the sum of $f^+(a)$ is $+\infty$. Then there exist a number $\delta > 0$ and an enumeration $a_0, a_1, \ldots$ of the points of A such that

---

18. P. C. Fishburn, *Utility Theory for Decision Making* (Wiley, New York, 1970), p. 107.

$$f(a_0) + f(a_1) + \cdots + f(a_n) \geq \delta \qquad (3.3.28)$$

for all $n = 0, 1, 2, \ldots$. (Let a_0 be any point with $f(a_0) > 0$, and let $\delta = f(a_0)$; then enumerate the remaining positive and negative terms and choose enough positive terms to overbalance the first negative term, enough positive terms after that to overbalance the next negative term, etc.)

Next define a sequence (p_k), $k = 1, 2, \ldots$, of finitely concentrated probabilities as follows:

$$p_k(a_0) = p(a_0) + [p(a_{k+1}) + p(a_{k+2}) + \cdots]$$

$$p_k(a_i) = p(a_i) \qquad \text{for } i = 1, \ldots, k$$

$$p_k(a_i) = 0 \qquad \text{for } i > k \qquad (3.3.29)$$

A sequence (q_k), $k = 1, 2, \ldots$, is defined similarly, with q taking the role of p in (29).

We claim that $p_k \succsim q_k$ for an infinite number of k-values. For suppose this were false; then by axiom 1, $q_k > p_k$ for all k past a certain value k_o. Furthermore, we have

$$\lim_{k \to \infty} p_k(a) = p(a), \qquad \lim_{k \to \infty} q_k(a) = q(a)$$

for all $a \in A$. Also, each of the three sequences $(p_k(a))$, $(q_k(a))$, and $(p_k(a) - q_k(a))$, $k = 1, 2, \ldots$, is monotone for all $a \in A$ except possibly for $(p_k(a_0) - q_k(a_0))$ (since, for $a = a_i$, $i \neq 0$, each of these sequences is 0 for $k < i$ and is constant for $k \geq i$). As for a_0, we note that any real sequence has a monotone subsequence (converging to its lim sup); hence there is a subsequence that satisfies the monotonicity clause for *all* $a \in A$, *including* a_0. Axiom 4 now implies that $p > q$ is false: contradiction. Hence indeed $p_k \succsim q_k$ infinitely often. Let $k_1 < k_2 < \cdots$ be the k-values for which this is true.

Now apply (26) to each such k_n. Evaluating the integrals in (26) (which are just finite sums) and substituting from (29) and then (27), we obtain

$$f(a_0) + f(a_1) + \cdots + f(a_{k_n}) + u(a_0)\left[\sum_{i=k_n+1}^{\infty} (q(a_i) - p(a_i))\right] \leq 0 \qquad (3.3.30)$$

for $n = 1, 2, \ldots$. This, however, contradicts (28), for as $n \to \infty$ the sum of the f-terms remains $\geq \delta > 0$, while the bracketed expression in (30) converges to 0; hence the left side of (30) is positive for sufficiently large n.

We have now achieved our contradiction and conclude that indeed the sum of the positive terms of $f(a)$ must be finite.

Next consider the definite integral

$$\int_A u \, d(q - p) = \left[\int_A u^+ \, d(q - p)^+ + \int_A u^- \, d(q - p)^- \right]$$
$$- \left[\int_A u^- \, d(q - p)^+ + \int_A u^+ \, d(q - p)^- \right]$$

The sum of the first two integrals on the right equals the sum of $f^+(a)$ over $a \in A$. Since this is finite, the integral is well defined; furthermore, the sum of $f(a)$ over $a \in A$ converges to the same number (possibly $-\infty$) regardless of the order of summation, and this number is the value of the integral. We will now show that this value is nonpositive.

Let $a_0, a_1, \ldots$ be any enumeration of the points of A, and define p_k, q_k as in (29). The argument above shows that $p_k \succcurlyeq q_k$ infinitely often; hence, by (26) again, we obtain (30). As $n \to \infty$, the left side of (30) converges to the sum of $f(a)$. Hence this is ≤ 0. We have proved that

$$\int_A u \, d(q - p) \leq 0 \qquad (3.3.31)$$

In turn, this implies

$$\int u \, dp \succcurlyeq \int u \, dq \qquad (3.3.32)$$

(standard order), by the standard integral theorem. To establish (23), we must strengthen this to *strict* preference. By axiom 3, there exist p', $q' \in F$ and a number x, $0 < x < 1$, with

$$p' \prec q' \qquad (3.3.33)$$

and

$$(1 - x)p + xp' \succ (1 - x)q + xq' \qquad (3.3.34)$$

Relations (33) and (26) yield

$$\int_A u \, dp' < \int_A u \, dq' \qquad (3.3.35)$$

while the same argument that led from $p > q$ to (31) now leads from (34) to

$$\int_A u \, d[(1 - x)q + xq' - (1 - x)p - xp'] \leq 0 \qquad (3.3.36)$$

Inequalities (36) and (35) in turn yield the respective inequalities:

$$\int_A u \, d(q - p) \leq \left(\frac{x}{1 - x} \right) \int_A u \, d(p' - q') < 0 \qquad (3.3.37)$$

Thus the integral in (31) is actually negative, which strengthens (32) to strict preference. This proves (23).

The proof of (24) now follows easily. Let $p \sim q$. If u is constant then (24) is trivial. If u is not constant, then p', $q' \in \mathbf{F}$ exist, with $p' > q'$. For any $x, 0 < x < 1$, it then follows by axiom 2 that (34) is true. The argument above then yields the *left* inequality in (37). Since x can be arbitrarily close to 0, (31) must be true, which yields (32). Interchanging the roles of p and q, the opposite inequality must also hold. Hence $\int u\ dp \sim \int u\ dq$ (standard order). ∎

We now come to the main result.

Theorem: Let $\mathbf{P}$ be the set of all probability measures on (A, Σ), where A is countable and Σ = all subsets of A; let $\succcurlyeq$ be a partial ordering on $\mathbf{P}$. Then each of the following conditions implies the other:

 (i) $\succcurlyeq$ satisfies axioms 1–5;
 (ii) A function $u{:}A \rightarrow$ reals exists such that for all $p, q \in \mathbf{P}$,

$$p \succcurlyeq q \quad \text{iff} \quad \int u\ dp \succcurlyeq \int u\ dq \qquad (3.3.38)$$

(Here $\succcurlyeq$ on the left refers to the partial order on $\mathbf{P}$; on the right it refers to standard order on the space of pseudomeasures over (A, Σ).)

Proof: Let function u satisfy (38); we must show that $\succcurlyeq$ on $\mathbf{P}$ satisfies axioms 1–5.

If $p, q \in \mathbf{F}$, the definite integral $\int_A u\ d(p - q)$ is well defined; hence $\int u\ dp$, $\int u\ dq$ are comparable under standard order, by the standard integral theorem; hence p, q are comparable by (38). This proves axiom 1.

Let p_1, q_1, p_2, q_2 satisfy the premises of axiom 2, so that, by (38), $\int u\ dp_1 \sim \int u\ dq_1$ and $\int u\ dp_2 > \int u\ dq_2$ (standard order).

By elementary pseudomeasure operations we find, for $0 < x < 1$, that

$$\int u\ d[(1 - x)p_1 + xp_2] > \int u\ d[(1 - x)q_1 + xq_2]$$

With (38) this yields (22) and proves axiom 2.

Let $p_1 > q_1$; it follows from (38) that u cannot be constant, for this would imply universal indifference; hence $p_2, q_2 \in \mathbf{F}$ exist, with $p_2 \prec q_2$. Choose any positive x less than

$$\bar{x} = \int_A u\ d(p_1 - q_1) \bigg/ \left[\int_A u\ d(p_1 - q_1) + \int_A u\ d(q_2 - p_2) \right] \qquad (3.3.39)$$

(Both definite integrals in (39) are > 0, and $\int_A u\ d(q_2 - p_2) < \infty$. If

$\int_A u \, d(p_1 - q_1) = \infty$, interpret $\bar{x}$ as 1; in any case $0 < \bar{x} \le 1$, so that x exists.) One easily verifies that

$$\int_A u \, d[(1 - x)p_1 + xp_2 - (1 - x)q_1 - xq_2] > 0$$

This, with (38) and the standard integral theorem, yields (22) and proves axiom 3.

Establishing axiom 4 is the difficult part of the proof. We argue by contradiction. Let p, q, and sequences p_k, q_k, $k = 1, 2, \ldots$, satisfy all the premises of axiom 4, but also let $q > p$. For each k define the function $f_k : A \to$ reals by $f_k(a) = u(a)[q_k(a) - p_k(a)]$, and similarly define f as in (27). Let μ be counting measure on (A, Σ), so that integration of f_k with respect to μ is the same as summation of $f_k(a)$ over $a \in A$, which in turn is the same as integration of u with respect to signed measure $q_k - p_k$. Thus

$$\int_A f_k \, d\mu = \int_A u \, d(q_k - p_k) \tag{3.3.40}$$

$k = 1, 2, \ldots$. A similar relation holds for f, p, and q (just drop the subscript k in (40)).

We shall deduce the contradiction

$$0 < \int_A f \, d\mu = \int_A (\liminf_{k \to \infty} f_k) \, d\mu \le \liminf_{k \to \infty} \int_A f_k \, d\mu \le 0 \tag{3.3.41}$$

The equality in (41) arises from the fact that

$$\lim_{k \to \infty} f_k(a) = f(a)$$

for all $a \in A$, so that the integrands are equal. The left inequality in (41) arises from the fact that $q > p$, along with (38), the standard integral theorem, and (40) without subscript k. The right inequality in (41) arises from the fact that for each k, $p_k > q_k$, so that by the same argument, $\int_A f_k \, d\mu < 0$ for all $k = 1, 2, \ldots$.

This leaves the middle inequality of (41) to be verified. This is the conclusion of Fatou's lemma and may be asserted *if* we can show the existence of a function $g : A \to$ reals such that $f_k \ge g$ for all k and such that

$$\int_A g \, d\mu > -\infty \tag{3.3.42}$$

We construct g as follows. Define the two subsets of A:

$$E = \{a \mid p_1(a) \ne q_1(a)\}, \qquad N = \{a \mid f(a) < 0\}$$

and let

$$g(a) = -|u(a)| \qquad \text{if } a \in E$$

$$g(a) = f(a) \qquad\quad \text{if } a \in N\backslash E$$

$$g(a) = 0 \qquad\qquad\; \text{if } a \in A\backslash(N \cup E) \qquad (3.3.43)$$

We first verify that for each $a \in A$, we have

$$f_k(a) \geq g(a) \qquad\qquad\qquad (3.3.44)$$

$k = 1, 2, \ldots$. This is obvious for $a \in E$, since p_k, q_k are probabilities. For $a \notin E$ we utilize the fact that $(p_k(a) - q_k(a))$, $k = 1, 2, \ldots$, is a monotone sequence. This implies that $f_k(a)$, $k = 1, 2, \ldots$, is a monotone sequence for each point a. Since $f_1(a) = 0$ for $a \notin E$, it follows that $f_k(a)$ lies between 0 and the limit $f(a)$ for each k. For $a \in N\backslash E$ we have $f(a) < 0$, so that (44) follows from (43). For $a \in A\backslash(N \cup E)$ we have $f(a) \geq 0$, so that $f_k(a) \geq 0$, again verifying (44).

Finally, we verify (42), which is true iff the sum of the *negative* terms of $g(a)$ over $a \in A$ is finite. The set E is finite, since p_1, q_1 are finitely concentrated. The set $A\backslash(N \cup E)$ makes no contribution to the sum. On the set $N\backslash E$ we have $g = f$, but the sum of the negative terms of $f(a)$ is finite by the left inequality in (41). Thus (42) is verified. We may now assert Fatou's lemma, yielding the contradiction (41). This proves axiom 4.

Let $\succeq'$ be another partial order on **P** satisfying the premises of axiom 5. Then $\succeq'$ satisfies axioms 1–4; hence, by the preceding lemma, a function $u':A \to$ reals exists such that for all $p, q \in$ **P**,

$$\text{if } p \succeq' q \qquad \text{then} \qquad \int u' \, dp \geq \int u' \, dq \qquad (3.3.45)$$

(standard order). Furthermore, if $p, q \in$ **F**, then the "if ... then" of (45) may be strengthened to "if and only if," by axiom 1.

Relation $\succeq'$ extends $\succeq$, and $\succeq$ is complete on **F** (proof of axiom 1 above). It follows that $\succeq'$ and $\succeq$ coincide when restricted to **F**. Then u and u' are two Bernoullian utilities representing the same ordering on **F**. It follows that u' is a positive affine transformation of u: real numbers x, y, with $x > 0$, exist such that $u'(a) = xu(a) + y$ for all $a \in A$.[19] But then, for *any* $p, q \in$ **P**, we have

$$\int_A u' \, d(p - q) = x \int_A u \, d(p - q) + \int_A y \, d(p - q)$$

$$= x \int_A u \, d(p - q)$$

(provided the left integral is well defined). It follows by the standard integral theorem that the conclusion in (45) implies

---

19. This result is well known. See, for example, Fishburn, *Utility Theory*, p. 107.

$$\int u \ dp \succcurlyeq \int u \ dq \qquad (3.3.46)$$

(standard order). In turn, (46) implies $p \succcurlyeq q$, by (38), which in turn implies $p \succcurlyeq' q$, since $\succcurlyeq'$ extends $\succcurlyeq$. This with (45) closes a circle of implications and shows that for all $p, q \in \mathbf{P}$, $p \succcurlyeq' q$ iff $p \succcurlyeq q$. This proves axiom 5.

Half the proof is now complete; the other half now follows rapidly. Let $\succcurlyeq$ satisfy axioms 1–5. By the preceding lemma, there exists a function $u{:}A \rightarrow$ reals "nonfaithfully" representing $\succcurlyeq$ by (23) and (24). Let $\succcurlyeq'$ be the partial order on $\mathbf{P}$ determined by u according to (38). We show that $\succcurlyeq'$ satisfies the premises of axiom 5. By the first half of this proof $\succcurlyeq'$ satisfies axioms 1–4. Furthermore, if $p > q$, then $\int u \ dp > \int u \ dq$ by (23), which in turn implies $p \succ' q$ by (38); and if $p \sim q$, then $\int u \ dp \sim \int u \ dq$, by (24), which in turn implies $p \sim' q$, by (38). Thus $\succcurlyeq'$ extends $\succcurlyeq$. The premises of axiom 5 being satisfied, it follows that $\succcurlyeq$ and $\succcurlyeq'$ are identical. This completes the proof! ∎

A few concluding comments follow. If universe set A is not merely countable but *finite,* any utility function u must be bounded, and standard order reduces to the ordinary size comparison of definite integrals: we are back to the conventional expected utility condition. There is a corresponding simplification on the axiomatic front in this case. Every probability measure is now finitely concentrated ($\mathbf{F} = \mathbf{P}$), so that axiom 1 now asserts the completeness of $\succcurlyeq$. As mentioned above, this implies axiom 5 (Proof: If $\succcurlyeq$ is complete and $\succcurlyeq'$ extends $\succcurlyeq$, then $\succcurlyeq' = \succcurlyeq$!); hence axiom 5 may be discarded. Also axiom 2 now implies axiom 3 (Hint: Let $p_2 = q_1, q_2 = p_1$, $x = 1/3$; mix $(p_1 + q_1)/2$ into both sides of $p_1 > q_1$); hence axiom 3 may also be discarded. We are left with the conventional three axioms of completeness (axiom 1), strong independence (axiom 2), and Archimedes (axiom 4). (Exercise: Show that the monotonicity clause in axiom 4 may now be deleted to yield an axiom logically equivalent to the original.)

Going back to the general case, let $\succcurlyeq$ have a representation (38) with u unbounded. What conventional axioms will $\succcurlyeq$ not satisfy? Strong independence still stands, but $\succcurlyeq$ is definitely not complete. For there will be a sequence of points $a_1, a_2, \ldots$ in A (not necessarily exhaustive) such that $u(a_n) > 2^n$, all n (if u is unbounded above), or such that $u(a_n) < -2^n$, all n (if u is unbounded below). Let $p(a_{2n}) = q(a_{2n-1}) = 2^{-n}$, all $n = 1, 2, \ldots$, with zero values elsewhere; then $\int u \, dp$, $\int u \, dq$ are not comparable under standard order, so p, q are not comparable under $\succcurlyeq$.

Furthermore, the Archimedean axioms cannot hold in full generality. Specifically, if u is unbounded above, we show that (25) cannot hold for every probability triple $p' > q > p$ (it does hold for $p', p, q \in \mathbf{F}$ as proved above).   Take a sequence a_1, a_2, ... with $u(a_n) > 2^n$, all n, and say, $u(a_2) > u(a_1)$. Let $p'(a_n) = 2^{-n}$, all n (the "Petersburg" distribution), and let $q = a_2, p = a_1$; then $p' > q > p$, but we can see that (25) is false for every x, $0 < x < 1$. Similarly, (25') is false for u unbounded below.

These considerations give a clue as to how one might set about modifying existing models that are too strong in that they imply bounded utility. For example, in Arrow's model it would be interesting to see the effect of relaxing his "monotone continuity" assumption, which has an Archimedean flavor.[20]

3.4. EXTENDED ORDERING OF PSEUDOMEASURES

The virtues of standard order may be summarized again as follows: (i) It resolves the blurring of preferability relations which arises when objective function integrals are infinite; (ii) it extends the scope of comparability by admitting policies whose integrals are not well defined; and (iii) when alternative policies have well-defined and finite integrals, the standard ordering criterion reduces to the ordinary size comparison of definite integrals.

The one disquieting property standard order has is that generally it is incomplete, and this raises the question: Is it worthwhile to extend standard order to make more pairs of pseudomeasures comparable? The affirmative is based on the feeling that one *should* be able to compare any two options; the negative is based on the feeling that any filling in of the gaps left by standard order involves arbitrary decisions lacking the intuitive appeal of standard order.

Let us examine these issues. First, the order to be concerned with is not standard order per se, but is that which it induces on the set of feasible alternatives. That is, although standard order is not complete, it is conceivable that in any "nonartificial" problem for any pair of feasible alternatives p_1 and p_2, the corresponding pseudomeasures ψ_1 and ψ_2 *are* comparable. This is probably the case for most problems using pseudomeasure evaluations. However, there are "natural" problems, even in classical location theory, for which noncomparability arises. The Löschian problem on the unbounded plane is an example.

Second, completeness is not crucial. From the point of view of the theory of choice, the ideal situation obtains when, for any of the range of problems under consideration, there is a unique best choice. For this to occur, it is in general neither necessary nor sufficient that the order be complete.

20. K. J. Arrow, *Essays in the Theory of Risk-Bearing* (Markham, Chicago, 1971), pp. 48–49, 63–65.

Consider a pseudomeasure ψ that is not comparable to 0 under standard order, so that $\psi^+(A) = \psi^-(A) = \infty$. Let (P, N) be a Hahn decomposition for ψ. We can think of ψ intuitively as an infinite positive mass placed on P, coupled with an infinite negative mass placed on N. One possible way of achieving comparability with 0 would be to "cancel" patches of negative mass in N against patches of positive mass in p and to come up with some kind of "net" mass, which may be positive, negative, or zero. The problem is to determine the method of "matching up" the patches to be canceled. This involves some more or less arbitrary rule; but if the space A has some structure in addition to its σ-field Σ (in particular, if it has a metric), there are some fairly "natural" ways this can be done.

Consider, e.g., a space A with measurable partition $\{A_1, A_2, \ldots\}$ and measures μ, ν with values on these sets as follows:

$$\begin{array}{ccccc} & A_1 & A_2 & A_3 & A_4 \ldots \\ \mu & 2 & 0 & 2 & 0 \ldots \\ \nu & 0 & 1 & 0 & 1 \ldots \end{array} \qquad (3.4.1)$$

Since μ and ν are both infinite and mutually singular, they are not comparable under standard order. Yet μ *seems* to be bigger. Actually, if the partition is arbitrary this feeling is an illusion. By clumping A_2, A_4, A_6 together, etc., we can make ν seem bigger. But if the partition is somehow naturally ordered as it stands, we may try canceling 1 against 2 with a net bigness rating to μ. An example might be where A is the nonnegative real numbers, and the A_n are a sequence of intervals in natural order.

We now consider some orderings based on the principle just outlined. Unlike narrow and standard order, which are uniquely determined by (A, Σ), there will in general be many of these "extended orderings," and it is a matter of ad hoc judgment as to which, if any, are to be considered "correct." The extended order is determined by an extension class, which we now define.

Definition: Let (A, Σ) be a measurable space. A collection of measurable sets $\mathfrak{F}$ is an *extension class* iff

(i) for all F_1, $F_2 \in \mathfrak{F}$, a set $F_3 \in \mathfrak{F}$ exists such that $F_1 \cup F_2 \subseteq F_3$;
(ii) there is a countable subcollection $\mathfrak{F}'$ that covers A (i.e., $A = \cup \mathfrak{F}'$).

We give some examples:

1. The class consisting of the universe set A alone is an extension class.
2. More generally, any collection of measurable sets including A is an extension class.

3. More interesting examples arise when A is a metric space (and all open and closed discs are measurable). The collection of all closed discs $\{a \mid d(a, a^o) \leq r\}$ is an extension class, as is the collection of all open discs.
4. Let A be the nonnegative real line. The class of all sets of the form $\{a \mid 0 \leq a \leq a^o\}$, for $a^o \in A$, is an extension class.

One or two more preliminary concepts are needed.

Definition: Let ψ be a pseudomeasure on (A, Σ). The *restricted domain* of ψ is the class of all measurable sets E such that $\psi^+(E)$ and $\psi^-(E)$ are not both infinite.

We denote this class by Σ_ψ. Clearly Σ_ψ coincides with Σ iff at least one of the two measures ψ^+, ψ^- is finite, i.e., iff ψ is a signed measure.

Definition: The *value function* of pseudomeasure ψ has domain Σ_ψ, and to all $E \in \Sigma_\psi$ assigns the value $\psi^+(E) - \psi^-(E)$.

Without risk of confusion, we denote the value of the value function of ψ at the set $E \in \Sigma_\psi$ by the symbol $\psi(E)$. The latter is therefore an extended real number for any such E. When pseudomeasure ψ is in fact a signed measure, its value function is precisely that signed measure in the ordinary sense of the term: a countably additive function on the σ-field Σ. In all other cases, however, the domain Σ_ψ of the value function is no longer a σ-field.

We now come to the vector partial order on Ψ determined by an extension class $\mathfrak{F}$. As with standard order, we first prove a preliminary result to guarantee the consistency of the forthcoming definition.

Lemma: Let $\mathfrak{F}$ be an extension class on measurable space (A, Σ). The following set of pseudomeasures is a convex cone: the set of all pseudomeasures ψ satisfying

(i) $$\mathfrak{F} \subseteq \Sigma_\psi \tag{3.4.2}$$

(ii) for all $\epsilon > 0$ there is an $\mathfrak{F}$-set F_ϵ such that for all $\mathfrak{F}$-sets F containing F_ϵ,

$$\psi(F) > -\epsilon \tag{3.4.3}$$

Proof: The zero pseudomeasure satisfies (2) and (3). Let ψ satisfy them, and let b be a positive real number. The restricted domains of ψ and $b\psi$ are the same, and for all $E \in \Sigma_\psi$, $(b\psi)(E) = b\psi(E)$. Given $\epsilon > 0$, choose $F_{(\epsilon/b)}$; then for $\mathfrak{F}$-set F containing this set we have $\psi(F) > -\epsilon/b$, so that $(b\psi)(F) > -\epsilon$. Hence $b\psi$ satisfies (2) and (3).

It remains to show that if ψ_1 and ψ_2 satisfy (2) and (3), so does $\psi_1 + \psi_2$. First, ψ_1^- and ψ_2^- are finite on all $\mathfrak{F}$-sets. To see this, let $F_1 \in \mathfrak{F}$ be such that $\psi_1(F) > -1$ for all $F \in \mathfrak{F}$ containing F_1. Choosing an arbitrary $F \in \mathfrak{F}$, there is an $\mathfrak{F}$-set $F' \supseteq (F \cup F_1)$, hence $\psi_1(F') > -1$, implying $\psi_1^-(F)$ finite. A similar argument works for ψ_2. Since $\psi_1^- + \psi_2^- \geq (\psi_1 + \psi_2)^-$, it follows that the latter is also finite on all $\mathfrak{F}$-sets. Hence $\psi_1 + \psi_2$ satisfies (2).

As for (3), we start with the equality

$$(\psi_1 + \psi_2)^+ + \psi_1^- + \psi_2^- = (\psi_1 + \psi_2)^- + \psi_1^+ + \psi_2^+ \qquad (3.4.4)$$

which follows from the equivalence criterion. For $F \in \mathfrak{F}$ all the lower variations in (4) are finite, hence we may transpose them and combine them with the upper variations to obtain

$$(\psi_1 + \psi_2)(F) = \psi_1(F) + \psi_2(F) \qquad (3.4.5)$$

Now for given $\epsilon > 0$, choose F_1, $F_2 \in \mathfrak{F}$, so that $\psi_1(F) > -\epsilon/2$ if $F \supseteq F_1$, and $\psi_2(F) > -\epsilon/2$ if $F \supseteq F_2$. There is an $\mathfrak{F}$-set $F_\epsilon \supseteq (F_1 \cup F_2)$, and for any $\mathfrak{F}$-set F containing F_ϵ we have, from (5),

$$(\psi_1 + \psi_2)(F) > -\epsilon/2 - \epsilon/2 = -\epsilon$$

since $F \supseteq F_1$ and $F \supseteq F_2$. Hence $\psi_1 + \psi_2$ satisfies (3). ∎

Definition: Let $\mathfrak{F}$ be an extension class on measurable space (A, Σ). The *extended order determined by* $\mathfrak{F}$, $\succcurlyeq_{\mathfrak{F}}$, on the space of pseudomeasures, is the vector partial order with positive cone $\{\psi \mid \psi$ satisfies (2) and (3)$\}$.

The intuitive notions underlying this construction are as follows. The extension class $\mathfrak{F}$ determines a generalized sort of convergence toward the universe set A via successively larger sets $F \in \mathfrak{F}$. Condition (3) then states roughly that ψ^+ catches up to ψ^- as $\mathfrak{F}$-sets get larger.

Our next result is the crucial property of extended order.

Theorem: For any extension class $\mathfrak{F}$ on measurable space (A, Σ), $\succcurlyeq_{\mathfrak{F}}$ is an extension of standard order $\succcurlyeq$.

Proof: First we note that for any extension class $\mathfrak{F}$, there is a sequence (F_n), $n = 1, 2, \ldots$, (finite or infinite) of $\mathfrak{F}$-sets such that $F_1 \subseteq F_2 \subseteq \cdots$ and whose union is A. To see this, let $\{F'_1, F'_2, \ldots\}$ be a countable collection of $\mathfrak{F}$-sets whose union is A. Then, successively, there are $\mathfrak{F}$-sets $F_1 \supseteq F'_1$, $F_2 \supseteq (F_1 \cup F'_2)$, $F_3 \supseteq (F_2 \cup F'_3)$, etc., and these unprimed F_n's satisfy the stated conditions.

Now let $\psi \succ 0$ (standard order), so that $\psi^+(A) > \psi^-(A)$; ψ is a signed measure, so its restricted domain is all of Σ. Thus ψ satisfies (2). If $\psi^+(A) = \infty$, the sequence $\psi^+(F_n)$ surpasses any finite number as $n \to \infty$.

Hence for N sufficiently large, $\psi^+(F_N) \geq \psi^-(A) + 1$, since $\psi^-(A)$ is finite. If $\psi^+(A)$ is finite, then

$$\psi^+(A) > \tfrac{1}{2}[\psi^+(A) + \psi^-(A)] > \psi^-(A) \qquad (3.4.6)$$

The sequence $\psi^+(F_n)$ approaches $\psi^+(A)$ as $n \to \infty$. Hence for N sufficiently large, $\psi^+(F_N)$ surpasses the middle term in (6).

Now let F be an $\mathfrak{F}$-set containing F_N. We get

$$\psi(F) = \psi^+(F) - \psi^-(F) \geq \psi^+(F_N) - \psi^-(A) \geq b \qquad (3.4.7)$$

where b is a fixed positive number not dependent on F. This implies that $\psi \succ_{\mathfrak{F}} 0$. Also $(-\psi) \succeq_{\mathfrak{F}} 0$ is false, since any $\mathfrak{F}$-set has a larger $\mathfrak{F}$-set F satisfying (7), so that $(-\psi)(F) \leq -b$. Thus $\psi >_{\mathfrak{F}} 0$.

For any pair of pseudomeasures, $\psi_1 > \psi_2$ implies $(\psi_1 - \psi_2) > 0$, so that $(\psi_1 - \psi_2) >_{\mathfrak{F}} 0$, which implies $\psi_1 >_{\mathfrak{F}} \psi_2$. This establishes the desired implication for strict inequality.

Next, let $\psi \sim 0$, so that $\psi^+(A) = \psi^-(A) < \infty$. Again (2) is satisfied. Taking an $\mathfrak{F}$-sequence $F_1 \subseteq F_2 \subseteq \cdots$ whose union is A, both $\psi^+(F_n)$ and $\psi^-(F_n)$ increase to their common limit $\psi^+(A)$. Hence for any $\epsilon > 0$ there is an F_n such that, for any $\mathfrak{F}$-set F containing F_n, both $\psi^+(F)$ and $\psi^-(F)$ lie in the interval $[\psi^+(A) - \epsilon/2, \psi^+(A)]$. Hence $|\psi(F)| = |\psi^+(F) - \psi^-(F)| < \epsilon$, which implies $\psi \sim_{\mathfrak{F}} 0$.

Finally, $\psi_1 \sim \psi_2$ implies $(\psi_1 - \psi_2) \sim 0$, hence $(\psi_1 - \psi_2) \sim_{\mathfrak{F}} 0$, so that $\psi_1 \sim_{\mathfrak{F}} \psi_2$. ∎

This is a very comforting theorem, but it does not guarantee that any particular extended order $\succeq_{\mathfrak{F}}$ is a *proper* extension of $\succeq_{\mathfrak{F}}$, i.e., that some pairs of pseudomeasures not comparable under $\succeq$ become comparable under $\succeq_{\mathfrak{F}}$. And in fact there is one case when $\succeq_{\mathfrak{F}}$ is definitely not a proper extension—when the universe set A belongs to $\mathfrak{F}$. For in this case, by (2) the only possible pseudomeasures comparable to 0 are signed measures, which are already all comparable under standard order; hence $\succeq_{\mathfrak{F}}$ and $\succeq$ coincide. We can only hope for a proper extension when A does not belong to $\mathfrak{F}$.

Exercise: Let A be countable, let Σ be all subsets, let $\mathfrak{F}$ be all finite subsets. Show that the extended ordering $\succeq_{\mathfrak{F}}$ coincides with $\succeq$.

We now show that the "overtaking" criterion developed in recent years[21] is just a special extended order. Let p_1 and p_2 be alternative development policies leading to "payoff streams" $\int_0^\infty f(p_i, t)\, dt, i = 1, 2$

---

21. See C. C. von Weizsäcker, Existence of optimal programmes of accumulation for an infinite time horizon, *Rev. Econ. Stud.* 32(Apr. 1965):85–104; *Rev. Econ. Stud.* 34, Jan. 1967 (entire issue is on Optimal Infinite Programmes). For comparison with the earlier and distinct approach of Ramsey see discussion at (3.16)–(3.18).

respectively. Then policy p_1 is said to *catch up to* policy p_2 iff

$$\lim_{t^0 \to \infty} \inf \int_0^{t^0} [f(p_1, t) - f(p_2, t)] \, dt \geq 0 \qquad (3.4.8)$$

That is, for all $\epsilon > 0$ there is a $t^o(\epsilon)$ such that for all $t \geq t^o(\epsilon)$ the integral in (8) exceeds $-\epsilon$.[22] But this is precisely the extended ordering $\succcurlyeq_{\mathfrak{F}}$ that arises from the extension class $\mathfrak{F}$ consisting of all closed intervals $[0, a], a \in A$ (where universe set A here is the nonnegative real half-line).

This account is oversimplified in one respect. There are a number of minor variants of this criterion, some found in the literature, others of which can be devised. However, most of the others turn out not to be extensions of standard order. Now standard ordering is intuitively much more compelling than any overtaking variant. Hence if some pairs of pseudomeasures are given one order by $\succcurlyeq$ and a different order by some other criterion, this constitutes grounds for dropping the other criterion as counterintuitive. We therefore do not bother with any variants of the overtaking criterion other than "catching up to."

This criterion *is* a proper extension of standard order. For example, the pair of measures μ, ν in (1) are not comparable under $\succcurlyeq$. But—taking the ordered partition $(A_1, A_2, \ldots)$ to represent successive intervals on the half-line $[0, \infty)$—they *are* comparable under "catching up," and in fact $\mu \succ_{\mathfrak{F}} \nu$.

The catching up criterion appears to be somewhat specialized, and it is not immediately clear how to generalize it to spaces other than the real half-line. We now show, however, that it can be reinterpreted in a way that generalizes to *any* metric space.

Theorem: Let A be the nonnegative real half-line; let $\mathfrak{F}$ be the extension class of all closed intervals $[0, a], 0 \leq a < \infty$; and let $\mathfrak{G}$ be the extension class of all closed intervals $[a_1, a_2], 0 \leq a_1 \leq a_2 < \infty$. Then the extended orderings determined by these, $\succcurlyeq_{\mathfrak{F}}$ and $\succcurlyeq_{\mathfrak{G}}$, are identical.

Proof: Special case of next theorem. ∎

Theorem: Let $\mathfrak{F}$ and $\mathfrak{G}$ be two extension classes on measurable space (A, Σ) satisfying:

(i) $\mathfrak{F} \subseteq \mathfrak{G}$,
(ii) for all $G \in \mathfrak{G}$ there is an $F \in \mathfrak{F}$ with $F \supseteq G$,
(iii) any $\mathfrak{G}$-set containing an $\mathfrak{F}$-set is itself an $\mathfrak{F}$-set.

Then $\succcurlyeq_{\mathfrak{F}}$ and $\succcurlyeq_{\mathfrak{G}}$ are identical.

---

22. D. Gale, On optimal development in a multi-sector economy, *Rev. Econ. Stud.*, 34(Jan. 1967): 1–18, esp. pp. 2–3.

Proof: We use an obvious notation for $\mathfrak{F}$-sets and $\mathfrak{G}$-sets. Let $\psi \succapprox_{\mathfrak{F}} 0$, so that $\mathfrak{F} \subseteq \Sigma_\psi$; and $F \supseteq F_\epsilon$ implies $\psi(F) > -\epsilon$, all $\epsilon > 0$. For any G, there is an $F \supseteq G$; since $F \in \Sigma_\psi$, $G \in \Sigma_\psi$; thus $\mathfrak{G} \subseteq \Sigma_\psi$. Also, F_ϵ is itself a $\mathfrak{G}$-set; and if $G \supseteq F_\epsilon$, then $G \in \mathfrak{F}$, so that $\psi(G) > -\epsilon$. This proves that $\psi \succapprox_{\mathfrak{G}} 0$.

Conversely, let $\psi \succapprox_{\mathfrak{G}} 0$. Since $\mathfrak{G} \subseteq \Sigma_\psi$, $\mathfrak{F} \subseteq \Sigma_\psi$. $G \supseteq G_\epsilon$ implies $\psi(G) > -\epsilon$. For F_ϵ we choose any $\mathfrak{F}$-set containing G_ϵ. Then if $F \supseteq F_\epsilon$, $F \supseteq G_\epsilon$; hence $\psi(F) > -\epsilon$, since F is also a $\mathfrak{G}$-set. This proves that $\psi \succapprox_{\mathfrak{F}} 0$. ∎

Thus "catching up" is also the order determined by the class of closed bounded intervals on the real half-line. This suggests that for any metric space (in which all closed discs are measurable) a natural generalization of the catching up criterion would be to use the extended order determined by the class of all closed discs. We shall actually use this procedure for the Löschian problem in which the universe set A is the plane. One could also use open discs and open intervals throughout instead of closed.[23]

We now return to the study of extended orders in general.

Theorem: Let (A, Σ) be a measurable space, and ψ_1, ψ_2 two pseudomeasures not comparable under standard order. Then an extension class $\mathfrak{F}$ exists such that $\psi_1 >_{\mathfrak{F}} \psi_2$.

Proof: Let $\psi = \psi_1 - \psi_2$. We must show that $\psi >_{\mathfrak{F}} 0$ for some extension class $\mathfrak{F}$. Let (P, N) be a Hahn decomposition for ψ. Since ψ is not comparable to 0 under standard order, ψ^+ and ψ^- are both infinite; in particular, $\psi^+(P) = \infty$.

Let $\{N_1, N_2, \ldots\}$ be a countable measurable partition of N such that $\psi^-(N_i)$ is finite, all i. For $\mathfrak{F}$ we take the class $\{P \cup N_1, P \cup N_1 \cup N_2, \ldots\}$. This is clearly an extension class. Also, for each set $F \in \mathfrak{F}$, $\psi^+(F) = \infty$, and $\psi^-(F) < \infty$, which implies $\psi >_{\mathfrak{F}} 0$. ∎

By symmetry there is another extension class $\mathfrak{G}$ giving the opposite inequality: $\psi_2 >_{\mathfrak{G}} \psi_1$. This underscores the great diversity among the possible extended orders and the need for some "rational" selection among them (in the occasional cases in which standard order does not suffice).

Any pseudomeasure not already comparable to 0 can be made com-

---

23. Concerning closed versus open discs, we can prove the following result: The pseudomeasure orderings determined by the following three extension classes—the class of open discs; the class of closed discs; and the class of all discs, open and closed—are *identical, provided* at least one of the following conditions is satisfied: *either* (i) every closed disc is *compact, or* (ii) any two points belong to a set *isometric* to the real line. (Both these conditions are satisfied by the Euclidean and city-block metrics, for example. For "compactness" see Section 7.4; for "isometry" see Section 2.7.)

parable under the appropriate extension class $\mathfrak{F}$. In general, $\mathfrak{F}$ depends on the pseudomeasure. Can one make the stronger claim that there is an $\mathfrak{F}$ that simultaneously makes all pseudomeasures comparable to 0 (hence to each other)? Our last theorem shows that the answer is no except in a trivial case.

Theorem: Let (A, Σ) be a measurable space. No extended order on the set of pseudomeasures is complete unless Σ is a finite σ-field.

Proof: If Σ is finite, standard order is already complete, and all extended orders coincide with $\succcurlyeq$.

Conversely, suppose there is an extension class $\mathfrak{F}$ such that $\succcurlyeq_{\mathfrak{F}}$ is complete. Then $\mathfrak{F} \subseteq \Sigma_{\psi}$ for all pseudomeasures ψ, by (2). Suppose there were a set $F \in \mathfrak{F}$ containing an infinite number of measurable sets. The proof that standard order is incomplete if Σ is infinite shows how to construct a pseudomeasure ψ such that $\psi^{+}(F) = \psi^{-}(F) = \infty$. Then F would not belong to the restricted domain of ψ. This contradiction shows that each $\mathfrak{F}$-set contains at most a *finite* number of measurable sets.

Let $F_1 \subseteq F_2 \subseteq \cdots$ be an increasing sequence of $\mathfrak{F}$-sets whose union is A. We may assume that $F_1 \neq \phi$, $F_{n+1} \backslash F_n \neq \phi$ for all $n = 1, 2, \ldots$. We assume this sequence is infinite and reach a contradiction. Let ψ be a pseudomeasure such that $\psi(F_1) = 1$, $\psi(F_{n+1} \backslash F_n) = 2$ if n is even, and $\psi(F_{n+1} \backslash F_n) = -2$ if n is odd. (Since the number of measurable sets in each of these sets is finite, it is trivial to construct such a ψ.) Then $\psi(F_n) = 1$ if n is odd, and $\psi(F_n) = -1$ if n is even.

Now let F be any $\mathfrak{F}$-set. The measurable sets contained in F are generated by a finite partition $\mathcal{G}$ of F. Each $G \in \mathcal{G}$ is contained in some F_n of the sequence $F_1 \subseteq F_2 \subseteq \cdots$; hence F is contained in the union of these, which is another F_n. Also, $\psi(F_n)$ and $\psi(F_{n+1})$ take on the values $+1$, -1 in some order. Hence for any $F \in \mathfrak{F}$, there are $\mathfrak{F}$-sets F', F'', each containing F, with $\psi(F') = 1$, $\psi(F'') = -1$. This shows that ψ is not comparable to 0 under $\succcurlyeq_{\mathfrak{F}}$, contradicting the assumption of completeness of $\succcurlyeq_{\mathfrak{F}}$.

Thus the sequence $F_1 \subseteq F_2 \subseteq \cdots$ is finite, so Σ is finite. ∎

Note that for the case in which some $\succcurlyeq_{\mathfrak{F}}$ is complete, standard order is itself already complete. Thus the situation is this: if standard order is not complete, while some of the gaps can be filled by using one or another extended order, it is impossible to fill all of the gaps. We close on this slightly pessimistic note.

CONCLUSION

Standard order on the vector space of pseudomeasures over measurable space (A, Σ) has great intuitive appeal as a representation of preferences. It appears to carry one satisfactorily through the great bulk of problems that arise. (Standard order suffices for 99% of this book; only in the last subsection of the last section of the last chapter do we go beyond it.)

When the incompleteness of standard order causes trouble, one can use an extended ordering to fill in some of the gaps. The problem here is to choose the appropriate extension. The most appealing suggested choice is the catching-up or overtaking criterion. This has been applied to some special cases, and the generalization suggested here is to the extended order determined by the closed discs (or the open discs) of a metric space. The intuitive idea here is that nearby positive and negative masses may be canceled. In all interesting cases even extended ordering remains incomplete.

4

Feasibility

4.1. INTRODUCTION

We have argued that the world can be described as a measure μ over the space (Ω, Σ), where Ω is the set of all possible histories (see Section 2.5). Here, for any measurable set of histories H, $\mu(H)$ is the total "mass" embodied in this set, where mass is to be interpreted in the broad sense discussed in Chapter 2.

This is the point of view of an omniscient observer who describes the world after the entire drama has unfolded itself (at time $+ \infty$, so to speak). From the point of view of someone living and acting *in* the world, μ is not given in all detail. Rather he has some power or freedom to *choose* how the world will develop. This may be represented formally by a set **M** of measures over (Ω, Σ). The interpretation is that for any $\mu \in$ **M,** there is some feasible plan of action by which he can guarantee that history will unfold according to the description μ, but that no feasible plan of action will attain any μ not belonging to **M.** The set **M** will vary from person to person and also will vary for the same person at different times. It will be called the *feasible set* of person p at time t.

For a beggar, the feasible set will be relatively "small." That is, he has so little power that the various alternative measures μ in **M** will be "very similar" to each other; his actions make "very little" difference. For an emperor the feasible set will be relatively "large."

There are various oversimplifications and conceptual difficulties involved in the notion of feasible set as just presented. First, it neglects uncertainty. One does not actually know the full effects of any actions one might try to undertake. This uncertainty may be represented as a probability over the universe set $\mathbf{M}^o$ of all measures over (Ω, Σ). This is a two-level measure (see Section 2.8) representing the effect of one attempted line of action. The feasible set itself will then be a set of such two-level probability measures. This is a fairly complicated construction, but something like it appears necessary to handle the problem of uncertainty adequately. In this chapter we shall, by and large, pass over the problem of uncertainty to avoid undue complexity. The assumption will be that one has perfect information concerning the consequences of any plan of action. The feasible set **M**, instead of being a set of probabilities over M^o, is then merely a subset of M^o.

A second difficulty concerns the interaction of several agents. Does not the power of person p_1 depend on the actions of other persons $p_2, p_3, \ldots$? What if they make incompatible choices? We examine in some detail how a plan of action is translated into a measure μ under conditions of certainty. Actions include, in the first instance, motions of the body that affect the environment—planting, harvesting, and eating; weaving, carrying, and building, etc. The plan gives the time schedule for these actions, starting from the time t_o at which the plan begins. The plan also includes actions affecting oneself, which enable one to carry out the other actions at the appropriate time, e.g., locomotion to be at the right location for an action, self-maintenance activities, self-training regimens to develop the skills needed for some future action. These actions set up causal chains that reverberate into the future. The certainty assumption means that one can predict these effects perfectly (including the unfolding of history that continues after one's death; it also means perfect information concerning the history of the world previous to t_o).

Now introduce other peope into the environment. Just as with the natural environment, other people are affected by one's actions, in particular by speech and by the writing and sending of messages. The assumption of certainty means that one can predict perfectly the responses of others to one's actions, which may be to ignore these actions, to engage in cooperative activities with oneself, to attack oneself, to take action affecting a third person, etc. We again have an unfolding of causal chains, except that these chains now involve the activities of other people.

Thus one's power does indeed depend on the actions of others; but since their responses are by assumption known with certainty, no problem of incompatible choices arises.

The assumption of certainty is a gross simplification, yet often a useful first approximation. It is not necessarily a worse approximation for someone living in society than it is for a Robinson Crusoe. A high degree of predictability is a *sine qua non* for social existence, and one of the prime functions of social institutions is to insure this predictability. We know that under normal conditions a storekeeper will sell us any item on the shelf at its stated price, a fire department will respond to an alarm, an oncoming motorist will yield our right of way. An employer knows his employee will follow orders within a certain broad zone of legitimate authority. In fact, the general course of civilization has probably been to increase the overall predictability of the future. The incursion of droughts, floods, and other natural fluctuations has been damped; epidemics are less threatening. The improvement of transportation and the rise of insurance, both private and social, pools the risks of individual misfortune over the entire society and establishes a subsistence floor that tends to rise over time. Violence tends to decline *within* a region with the territorial spread of the nation-state: Philadelphia and New York do not make war on each other as Athens and Sparta did.[1]

Assuming certainty then, our problem in this chapter will be to describe feasible sets **M** in various plausible and convenient ways; **M** is a very complicated object even under certainty (remember that each of its points is a complete description of the world), and one needs good schematic ways for characterizing it, at least approximately.

In general, we shall try to characterize **M** by "whittling down" from above; i.e., there will be a number of simpler sets, each one consisting of all measures satisfying some test criterion. Any feasible measure must satisfy all these tests, so that **M** is included in the intersection of all these sets. Hopefully, it is equal to their intersection; if not, further tests are needed. Thus a given measure μ may be infeasible because it violates a natural law, because of technical ignorance, because resources are not available, because it violates a legal statute, or because the person lacks the money or authority to induce needed actions by other people. Then **M** will consist of the measures μ satisfying all these criteria (and perhaps other criteria not listed).

Several of these broad criteria could themselves be expressed as the conjunction of simpler ones. Thus to satisfy the natural-law criteria, the measure may have to satisfy conservation laws, maximal density constraints, dynamical laws in the form of differential equation systems, etc. To comply with legal statutes, it must satisfy zoning laws, traffic laws, housing laws, antipollution laws, etc.

Our aim is to make a quick survey of these various exclusion criteria. Obviously, this must be superficial; anything more would require expertise in dozens of different specialties. We stress features that are analytically tractable and at the same time catch broad structural aspects of the various criteria. In particular, these will include most of the feasible sets used in the rest of the book. (Budget constraints will be discussed in Chapter 6.)

4.2. UNCONTROLLABLE REGIONS

Person p at time t_o has feasible set **M**. He cannot do anything about the world before time t_o; the past has already happened. How is this fact reflected in the set **M**?

---

1. The decline of international violence is more dubious. See L. F. Richardson, *Statistics of Deadly Quarrels,* Q. Wright and C. C. Lienau, eds. (Boxwood Press, Pittsburgh, 1960).

Let μ' and μ'' be feasible. Recall that the cross-sectional measure at time t gives the distribution of mass over $R \times S$ at that time. It must be true that for any $t < t_o$, the cross-sectional measures at t determined by μ' and μ'' are the same. Thus

$$\mu'\{h \mid h(t) \in E\} = \mu''\{h \mid h(t) \in E\} \tag{4.2.1}$$

for any measurable $E \subseteq R \times S$, any $t < t_o$, and any two μ', $\mu'' \in \mathbf{M}$. A similar equality holds for double cross-sectional measures, etc., provided all times are in the past of t_o. Next consider the production and consumption measures, which are on universe set $R \times S \times T$ (Section 2.5). If we consider the *past half-space* $R \times S \times \{t \mid t < t_o\}$, then the production measures λ_1' and λ_1'', derived from μ', $\mu'' \in \mathbf{M}$ respectively, must be identical when restricted to this half-space, since they both equal the actual pattern of production realized in the past; similarly for consumption measures λ_2', λ_2''.

All these remarks are implied by the following principle, which expresses with complete generality the notion of the uncontrollability of the past. It is convenient here to add the artificial point z_o to $R \times S$ and let the history h take on the value z_o at the times before it is "born" and after it "dies." With this convention, h has as its domain the entire time axis T, with range in $(R \times S) \cup \{z_o\}$.

Two histories h' and h'' are *identical before* t_o iff $h'(t) = h''(t)$ for all $t < t_o$. Let H be a set of histories. H is t_o-*past-determined* iff whenever $h' \in H$ and h' and h'' are identical before t_o, then $h'' \in H$.

We now state the *past uncontrollability principle:* any two feasible measures have identical values on all measurable t_o-past-determined sets.

The set appearing in (1) is t_o-past-determined; so is the set determining production or consumption on any measurable $G \subseteq [R \times S \times \{t \mid t < t_o\}]$. This shows that the remarks above are implied by the past uncontrollability principle.

Not only the past but a portion of the future will be uncontrollable. Consider a fire station located at s_o at time t_o. The locations that can be reached by time $t > t_o$ by a fire engine starting at t_o depend on the maximal speed of the engine, street layout, traffic congestion, etc. There will be an accessible region that in general expands with increasing t, the whole set of accessible points in Space-Time forming roughly a "cone" with its apex at (s_o, t_o) and opening into the future.[2]

Without speedier communications the entire subset of $S \times T$ outside this cone is uncontrollable by the engine dispatcher at time t_o. Maximal speed limitations imply that such uncontrollable regions exist in general.

---

2. This concept has become familiar through relativity theory, where the finite speed of light plays the role of the finite fire engine speed. See H. Minkowski, Space and Time, in *The Principle of Relativity* (Dover, New York, 1923), p. 84.

Even within this cone there will be aspects that are virtually uncontrollable. The great processes of nature, earthquakes, hurricanes, etc., are still in this category. For all but a handful of people, the tides in the affairs of men—war, revolution, depression, religious movements, fashion cycles—must be considered uncontrollable.

We conclude with an abstract definition of uncontrollability that includes as special cases everything discussed in this section.

Definition: Let **M** be a set of measures on the space (A, Σ). Set $E \in \Sigma$ is *uncontrollable* with respect to **M** iff $\mu(E)$ has the same value for all $\mu \in$ **M**.

4.3. CROSS-SECTIONAL CONSTRAINTS

Cross-sectional constraints are those that exclude measures whose cross-section as of time t fails to satisfy certain conditions. In more detail we start with a measure μ on the space of histories (Ω, Σ) and then examine its cross-section μ_t, which is given by $\mu_t(E) = \mu\{h \mid h(t) \in E\}$ for all measurable E. Of course μ_t is a measure on the space $(R \times S, \Sigma_r \times \Sigma_s)$. There will in general be many measures μ with the same cross-section at t; and if μ_t fails, all these measures are infeasible. Some of the conditions to be imposed may have to be met by μ_t for all time instants t, others perhaps for only some t. To simplify notation we drop the subscript t. The measures are still over universe set $R \times S$.

INTEGER VALUES AND FINITE CONCENTRATION

For a certain subset R' of Resources, in which objects come in "natural units" such as people, cattle, and cars, μ may take on only integer values (or be infinite) when restricted to $R' \times S$. The constraint here is one of purely semantic origin; any other measures would be meaningless.

Definition: Measure μ on (A, Σ) is *concentrated* on set $E \subseteq A$ iff $\mu(A) > 0$ and $\mu(F) = 0$ for all measurable F disjoint from E (E itself need not be measurable).

We have already met several related concepts. Measure μ is *simply concentrated* iff it is concentrated on some singleton set. Similarly, μ is *finitely concentrated* iff it is concentrated on some finite set.

Restrictions to finitely concentrated measures arise in two ways. First, certain resource types are just naturally not spread out over space and are well represented as being confined to a finite number of locations.

One thinks of the natural units mentioned above and perhaps others.[3] Second, even though it is technically feasible to spread a resource continuously over Space, budgets may not permit this; e.g., an overhead cost may be associated with each location at which the resource is situated. In many problems the multiplicity is also specified; e.g., given n policemen, deploy them to minimize total crimes. The case $n = 1$ is especially interesting; the classic plant-location problem of Weber is a special case.

There is a close connection between integer-valuedness, atomicity, and finite concentration. Recall that μ is atomic iff $\mu(A) \neq 0$, and for all $E \in \Sigma$, either $\mu(E) = 0$ or $\mu(A \setminus E) = 0$. Every simply concentrated measure is atomic.

Theorem: Let μ be a bounded measure, not identically zero, on (A, Σ). If μ takes on just a finite number of values, then there is a finite measurable partition $\{A_1, \ldots, A_n\}$ such that μ restricted to each A_i is atomic.

Proof: Since μ is bounded, a countable measurable partition $\{A_0, A_1, \ldots\}$ exists such that μ restricted to A_0 is nonatomic and μ restricted to A_i is atomic for $i = 1, 2, \ldots$. This partition must in fact be finite, since otherwise $\mu(A_1)$, $\mu(A_1 \cup A_2), \ldots$ would give an infinite number of different μ-values. On the nonatomic part, μ takes on all values between 0 and $\mu(A_0)$. Hence $\mu(A_0) = 0$. Since $\mu \neq 0$, there is at least one A_i, say $i = 1$, and μ restricted to $A_0 \cup A_1$ remains atomic. ∎

The converse of this theorem is also true but of less interest to us. Now if μ is integer valued and bounded (we do not consider the unbounded case), it takes on just a finite number of values, hence is atomic on all A_i for some partition $\{A_1, \ldots, A_n\}$. Atomic measures are not always simply concentrated; however, there is one very common condition under which the two concepts coincide.

Definition: Let (A, Σ) be a measurable space; Σ is *countably generated* iff there is a countable subclass $\mathcal{G} \subset \Sigma$ that generates Σ.

For example, the Borel field on the real line is countably generated, since it is generated by the collection $\{a \mid a < x\}$, x rational; similarly for the n-dimensional Borel field, $n = 2, 3, \ldots$. But this property will often not hold for more complex σ-fields, such as the one over the space of histories Ω or those involved in multilayer measures.

---

3. Cf. the discussion of "indivisibility," Section 4.8.

Theorem: If μ is an atomic measure on (A, Σ) and Σ is countably generated, then μ is simply concentrated.

Proof: Let $\{G_1, G_2, \ldots\}$ generate Σ. For each G_n exactly one of $\mu(G_n)$, $\mu(A \backslash G_n)$ is positive, the other being zero. Let F_n be either G_n or $A \backslash G_n$, chosen so that $\mu(F_n) > 0$, $\mu(A \backslash F_n) = 0$. Let $F = \bigcap_{n=1}^{\infty} F_n$. Then $\mu(A \backslash F) = \mu[\bigcup_{n=1}^{\infty}(A \backslash F_n)] \leq \mu(A \backslash F_1) + \mu(A \backslash F_2) + \cdots = 0$, so that $\mu(A \backslash F) = 0$. Hence $\mu(F) > 0$.

Therefore F is nonempty, so there exists $a_o \in F$. We show that μ is concentrated on the set $\{a_o\}$. It suffices to show that for any measurable E with $a_o \not\in E$, we have E, F disjoint; for then $\mu(E) = 0$ is immediate. Consider the class Σ' of all measurable sets that either contain F or are disjoint from F; Σ' is closed under complements and countable unions, so that it is a σ-field. Furthermore, $G_n \in \Sigma'$ for all n. It follows that $\Sigma' = \Sigma$. Since E above does not contain F, it is disjoint from F and the proof is complete. ∎

We conclude that, if μ on (A, Σ) is integer valued and bounded and Σ is countably generated, then μ is finitely concentrated. Thus in most cases of interest, integer-valuedness is a strengthening of the finite concentration condition.

SPACE CAPACITY

Let $r \in R$ be some particular resource type and let F be a region of Space. The amount of r that can be squeezed into F may well have a finite upper limit, where we "run out of space." Let $\nu_r(F)$ be this upper limit.

If F_1 and F_2 are two disjoint regions, we must have

$$\nu_r(F_1 \cup F_2) \leq \nu_r(F_1) + \nu_r(F_2) \tag{4.3.1}$$

For if (1) were false, there would be no way to approach capacity on $F_1 \cup F_2$ without exceeding capacity on one of the two subregions. Condition (1) is called *finite subadditivity*. It is possible that the inequality in (1) may sometimes be strict, which means that capacity cannot be reached in F_1 and F_2 simultaneously. The most interesting case, however, is where (1) becomes an equality for all disjoint regions F_1, F_2, so that ν_r is finitely additive. It is not unreasonable that ν_r should be countably additive, so that for any countable packing of regions, the capacity of the union is the sum of the capacities of the individual regions. We may also safely assume that $\nu_r(\phi) = 0$. With these assumptions, ν becomes a measure over S, the *capacity measure for resource r*.

Now let μ be a cross-sectional measure. We assume that all singleton sets are measurable in R: $\{r_o\} \in \Sigma_r$, for all $r_o \in R$. It must then be true that

$$\mu(\{r\} \times F) \leq \nu_r(F) \qquad (4.3.2)$$

for all resources r and regions F. Condition (2) states that the total mass of r in region F does not exceed the capacity of that region for r. But (2) is not stringent enough. Each resource type is taking up space on its own, and the region must have a global capacity sufficient to accommodate them all simultaneously. This suggests the following approach.

First, we postulate a general capacity measure α over Space (S, Σ_s). This measure will be called *ideal area*. It may or may not coincide with ordinary physical area (or volume). Next we postulate a function $f : R \times S \rightarrow$ reals, having the interpretation: $f(r, s)$ is the "amount of space" needed per unit of resource r at location s; f is nonnegative and assumed to be measurable with respect to $\Sigma_r \times \Sigma_s$. This description of f is vague; the precise role of f (and α) is given in the next inequality, which gives the total system of capacity constraints that must be satisfied by any feasible cross-section μ.

$$\int_{R \times F} f \, d\mu \leq \alpha(F) \qquad (4.3.3)$$

for all regions F. The left side of (3) is a plausible expression for the total "demand for space" by all resources together that occupy region F, and (3) requires that μ be small enough so that this total demand does not exceed the "space" available in any region.

The most interesting special case arises when f does not depend on its s-coordinate: $f(r, s) = f(r, s') \; (= f(r)$, say) for all $r \in R$, s, $s' \in S$. Then $f(r)$ may be thought of as the reciprocal of the maximal density to which resource r can be squeezed. If in addition just one single resource type r_o is distributed over S (i.e., $\mu[(R \backslash \{r_o\}) \times S] = 0$), then (3) reduces to (2):

$$\int_{R \times F} f \, d\mu = \int_{\{r_o\} \times F} f \, d\mu = f(r_o)\mu(\{r_o\} \times F) \leq \alpha(F)$$

which is the same as (2) if we define $\nu_r = \alpha/f(r)$.

Note that (3) allows the possibility of f being zero sometimes. A resource type r for which $f(r, s) = 0$, all $s \in S$, will be called *non-space-using*. Postulating that certain resources are non-space-using is a simplifying approximation that is often useful. In this case the constraints (2) simply disappear.

Consider some real-world examples of maximal capacity constraints. For most resource types there will be some physical density beyond which the resource will be destroyed. Here α will be ordinary physical volume (or perhaps area). For people this capacity is sometimes approached: (3)

must be near equality inside subway trains during the New York City rush hours.[4]

Legal statutes often have the effect of placing capacities well below the physical limit. These may be referred to as *anticongestion laws,* whether that is their primary purpose or not. Fire laws will limit occupancy of halls; no-standing laws will limit occupancy of buses and movies. Zoning and housing laws requiring open spaces, minimal lot sizes, maximal lot coverage, maximal building height and bulk, etc., all have the effect of spreading people out and reducing actual occupancy far below what would be physically possible.

The fact that we have here social rather than physical constraints raises no difficulties. The interpretation of (3) is different: The measures μ that violate it are illegal, not necessarily physically impossible. The function f used in these constraints will in general be larger than for the physical constraints; more space is legally required per unit resource than is physically necessary. Since laws vary from place to place, $f(r, s)$ will in general vary with its s-coordinate (rising in places where laws are "tougher"). For physical constraints one can probably make do with an f depending only on r.

There is one essential difference between natural laws and statute laws as far as feasibility is concerned. Natural laws cannot be violated (by definition), but one can often violate statutes (by committing a crime). Thus to treat statutes as constraints is to restrict the feasible set unduly; e.g., one may decide to park illegally and run the risk of getting a ticket. For the most part we ignore this point and treat statutory "constraints" as binding.

Exclusions are a limiting case of maximal capacity constraints. They specify that a measure must take the value zero on certain sets. Examples are laws against trespassing, zoning laws, segregation laws, and curfews. Consider a curfew, for instance. It specifies that $\mu(E \times F) = 0$, where E is the set of "unauthorized personnel" and F is, say, the streets of a certain town. Cross-sectional measures μ_t must satisfy this condition for those times t at which the curfew is in effect, say nighttimes over a certain time interval. Exclusions may be handled formally by letting f take on the value $+\infty$ in (3). If α is σ-finite (as we may assume from its interpretation), it is easy to show that (3) implies $\mu\{(r, s) \mid f(r, s) = \infty\} = 0$. Thus all that needs doing is to set f equal to infinity on the excluded sets.

Finally, the *limited variety* constraint is a special kind of exclusion. Some resource types may not be able to exist. Commodities may not be producible except in a limited number of qualities and styles—unicorns

---

4. For other examples see A. D. Biderman, M. Louria, and J. Bacchus, *Historical Incidents of Extreme Overcrowding* (Bur. Soc. Sci. Res., Washington, D.C., 1963).

and philosophers' stones are not found in nature, etc. Let E (assumed measurable) be the subset of R consisting of all these excluded resources. Then $\mu(E \times S)$ must be zero for any feasible cross-section μ.

RESOURCE CAPACITY

The quantities of various resource types available may be limited. These limitations may change over time as resources are created or destroyed, but for any given time t we may postulate a *resource-capacity measure* ν_t on the space (R, Σ_r). Any feasible cross-section μ_t must then satisfy

$$\mu_t(E \times S) \leq \nu_t(E) \tag{4.3.4}$$

for all $E \in \Sigma_r$, i.e., the left marginal of μ_t cannot exceed ν_t.

Limited variety constraints are a special case of (4) (as well as a special case of exclusions). They may be represented by setting ν_t equal to zero on the appropriate sets. There will in general also be constraints of the form (4) for subregions and not merely for S as a whole. This will occur whenever resources are tied up in particular regions and cannot be transported elsewhere with infinite speed.

DISALLOWED CONFIGURATIONS

Recall that a configuration is simply a measure on universe set $R \times F$, R being the set of resources and F a region of S. We have defined when two configurations are to be considered of the same type, and also have defined the notion of an abstract configuration type, which may be exemplified in various actual regions (see Section 2.7). (These concepts involve a metric on S.)

Suppose that certain abstract configuration types are set aside as "disallowed." These would, if exemplified, violate some natural or human law. For example, the following might be illegal configurations: two bars within distance x_1, a bar and a church within distance x_2, or a house without a fire hydrant within distance x_3. A cross-sectional measure μ is to be considered infeasible if, for any region F, μ restricted to $R \times F$ exemplifies a disallowed configuration type. Thus a separate test must be passed for each region.

A less stringent version of this test works with a set of "allowable" abstract configuration types. A cross-section μ passes the test if there is a countable collection of regions $\mathfrak{F}$ which together cover S, such that μ restricted to $R \times F$ exemplifies an allowable configuration type for each $F \in \mathfrak{F}$. That is, it must be possible to represent μ as a "patching" of allowable configurations.

These two versions will be called the *strict* and *loose constructionist* versions of the configuration test respectively. This approach is very general and flexible in either version. However, it does not allow for spatial variation in what is allowable or is not; thus it is probably most useful in connection with natural laws or within the domain of a single legal system.

4.4. INTERTEMPORAL CONSTRAINTS

We now go on to feasibility conditions involving several different time instants. The possible feasibility conditions are much richer than for cross-sections. One broad class of conditions requires that certain sets of histories have measure zero. This resembles the concept of *exclusion* discussed above. In fact, exclusions are just a special case requiring that the set of histories occupying $G \subseteq R \times S$ at time t have measure zero.

NONINTERACTIVE SYSTEMS

We start with a very simple case. Suppose for each pair of positive numbers t_0, t_1 ($t_0 < t_1$), there is a function $f_{t_0 t_1} : R \times S \rightarrow R \times S$ expressing a *dynamical law:* a history in state (r, s) at moment t_0 will move to state $f_{t_0 t_1}(r, s)$ at moment t_1.[5] This means that the entire future of a history is determined by the state it occupies at any one moment. We then specify that the set of all histories violating this dynamical law has measure zero (assuming this set is measurable).

A condition of this sort is implausible because it does not allow for interaction: the course of a history depends only on its past, not on the environment. More generally, we look for a rule by which the "rate of change" of a history in state (r, s) at time t depends also on the cross-sectional measure at time t. (This would require putting some structure on R to define the notion of a rate.) For example, the acceleration of a particle under gravity depends on the distribution of mass over S; the behavior of a person may depend on his observation of the distribution of behavior of other people.

BARRIERS

Of the innumerable forms such a rule might take, we mention just one type. A *barrier* for a certain state (r, s) is a configuration preventing

---

5. The family of functions f must satisfy the consistency condition: $f_{t_0 t_2}(r, s) = f_{t_1 t_2}(f_{t_0 t_1}(r, s))$. Death could be represented by letting f take on the "nonexistence" value z_o, with $f_{t_0 t_1}(z_o) = z_o$; birth could be represented by also introducing "backward causation": $t_0 > t_1$.

that state from changing in certain ways. Barriers are quite pervasive; some occur in nature, some are added by man, some are removed by man. A house serves as a barrier against the weather, preventing air in an unpleasant state from gaining access. Umbrellas are barriers against rain, thimbles against needles. The skin is a barrier against infection. Walls, fences, locks, guards, and watchdogs are barriers against trespassing.

Transportation construction in general may be thought of as barrier removal or barrier circumvention. Barrier removal occurs when the rough surface of the earth is smoothed, as in road and rail construction, tunneling, and bridging. Barrier circumvention occurs when an alternative medium is developed enabling one to bypass the former barrier—as in air and sea travel, pipelines and power lines, broadcasting.

By slight extension of the meaning, one can speak of barriers to entering certain occupations, certain industries, or certain social statuses (citizenship, marriage, political office, etc.) The formal analyses of these situations is similar to that of barriers in the strict sense.

A POLLUTION MODEL

The following model reverts back to the assumption of no interaction with the environment but allows several histories to spread out from single points. Models of this sort may be suitable as representations of the diffusion and transformation of substances, as in air pollution studies. Our aim, however, is mainly to illustrate how the concepts we have been working with might be applied, and we make no attempt to take realistic complications into account.[6]

We start with our basic sets of Resources, Space, and Time (R, S, T), with σ-fields Σ_r, Σ_s, Σ_t respectively. Suppose a unit mass of resource r_0 is released at location s_0 at instant t_0. What has happened to it by time t_1? We assume the answer is given by a function $f : R_0 \times S_0 \times T_0 \times T_1 \times (\Sigma_{r_1} \times \Sigma_{s_1}) \to$ reals.

Subscripts 0 and 1 will be used with R, S, and T, and points belonging to them, to distinguish "origins" from "destinations." Origins, denoted by 0, are points in $R \times S \times T$ at which mass is released into circulation; destinations, denoted by 1, are points where mass is found after circulating awhile. $\Sigma_{r_1} \times \Sigma_{s_1}$ is parenthesized because it is a product σ-field, while the other crosses stand for cartesian products.

6. For an interesting attempt to model air circulation and pollution in an actual region (Los Angeles) see F. N. Frenkiel, Atmospheric pollution and zoning in an urban area, *Scientific Monthly* 82(Apr. 1956):194–203. H. Reiquam, Sulfur: simulated long-range transport in the atmosphere, *Science* 170(Oct. 16, 1970):318–20, deals with northwestern Europe.

The function f is a conditional measure; i.e.,

(i) for fixed r_0, s_0, t_0, t_1, $f(r_0, s_0, t_0, t_1, \cdot)$ is a measure on the space $(R_1 \times S_1, \Sigma_{r_1} \times \Sigma_{s_1})$;

(ii) for fixed $G \in \Sigma_{r_1} \times \Sigma_{s_1}$, $f(\cdot, \cdot, \cdot, \cdot, G)$ is a measurable function with respect to its domain space $(R_0 \times S_0 \times T_0 \times T_1, \Sigma_{r_0} \times \Sigma_{s_0} \times \Sigma_{t_0} \times \Sigma_{t_1})$.

The interpretation of f is this: $f(r_0, s_0, t_0, t_1, G)$ is the total mass in resource-location states in G at moment t_1, which arises from the release of unit mass of resource r_0 at location s_0 at moment t_0. That is, the unit mass will become diffused, perhaps spreading over various locations and also perhaps becoming changed into different resource types. Thus we get a changing series of cross-sectional measures depending on t_1, the moment of observation.

We require that f satisfy the following consistency condition: if $t_0 < t_1 < t_2$ are three moments, then

$$f(r_0, s_0, t_0, t_2, G) = \int_{R_1 \times S_1} f(r_1, s_1, t_1, t_2, G)\, f(r_0, s_0, t_0, t_1, dr_1, ds_1)$$

(4.4.1)

for all $r_0 \in R_0$, $s_0 \in S_0$, $G \in \Sigma_{r_2} \times \Sigma_{s_2}$. The left side of (1) gives the mass on set G at time t_2. The right side gives the same thing indirectly: first, by finding the entire distribution on $R_1 \times S_1$ (at the intermediate time t_1); finding the contribution to G at t_2 by unit mass at (r_1, s_1) at t_1; and then integrating (in effect, taking the limit of the weighted sum of these contributions).[7] If $t_1 < t_0$, then f is identically zero: there is no mass before the date of release. If $t_1 = t_0$, then f is simply concentrated, with unit mass concentrated at the single point (r_0, s_0): This merely gives the initial condition at t_0.

The conservation of matter is expressed by the condition

$$f(r_0, s_0, t_0, t_1, R_1 \times S_1) = 1$$

(4.4.2)

for all r_0, s_0, t_0, t_1 for which $t_1 \geq t_0$. Condition (2) states that the same total mass is present in some form somewhere at any time instant after release. However, we may want to incorporate the gradual absorption or "death" of some of the histories starting at (r_0, s_0) at time t_0. In this

---

7. In the theory of Markov processes, which the present model resembles, relations of the form (1) are known as Chapman-Kolmogorov equations. If the family of measures represented by f are all simply concentrated, this entire construction reduces, in effect, to the dynamical law system discussed above, and (1) reduces to the consistency condition for dynamical laws.

case the value of f in (2) will be a nonincreasing function of t_1, holding r_0, s_0, t_0 fixed and equaling one at $t_1 = t_0$.[8]

The second half of this model is a measure ν on the space $(R_0 \times S_0 \times T_0, \Sigma_{r_0} \times \Sigma_{s_0} \times \Sigma_{t_0})$, representing the pattern of release of resources (pollutants). On rectangles the interpretation of ν is as follows: $\nu(E \times F \times G)$ is the total mass of resources of types E released in region F in period G. Note that ν can incorporate the possibility of positive quantities emanating at single locations (Con Edison plants?) as well as continuous releases over S, and also the possibility of positive "gobs" appearing at single time instants as well as continuous releases over T.

Given the measurable set $G \subseteq R_1 \times S_1$, what is the total mass embodied in the resource-location pairs in G, at moment t_1, as a result of the release pattern ν? The answer is

$$\lambda(t_1, G) = \int_{R_0 \times S_0 \times T_0} f(r_0, s_0, t_0, t_1, G) \, \nu(dr_0, ds_0, dt_0) \qquad (4.4.3)$$

Here again (3) may be thought of as the limit of a weighted sum of the contributions to G from the various triples (r_0, s_0, t_0), the weights being determined by ν. For fixed t_1, $\lambda(t_1, \cdot)$ is in fact a measure on $R_1 \times S_1$. If the conservation law (2) is in effect, it follows from (3) that

$$\lambda(t_1, R_1 \times S_1) = \nu[R_0 \times S_0 \times \{t \mid t \leq t_1\}]$$

That is, the total mass found at moment t_1 equals the total mass released not later than t_1, as one would expect. One can go a step further and define a total "exposure" measure ρ over $R_1 \times S_1 \times T_1$:

$$\rho(B) = \int_{T_1} \lambda[t_1, \{(r_1, s_1) \mid (r_1, s_1, t_1) \in B\}] \, dt_1 \qquad (4.4.4)$$

for all $B \in \Sigma_{r_1} \times \Sigma_{s_1} \times \Sigma_{t_1}$.[9] The integration in (4) is with respect to ordinary Lebesgue measure on the real line t_1, representing "quantity of time." The interpretation of ρ on rectangles is as follows: $\rho(E \times F \times G)$ equals total mass-time of exposure to resources (i.e., pollutants) of types E in region F during period G. Unlike most of our other measures, ρ is in mass-time units (e.g., ton-hours, man-days) rather than in mass units; ρ is needed to evaluate cumulative exposure effects.

Let us give a plausible concrete example for the "diffusion" function f. We simplify by assuming that just a single resource type is involved and

---

8. Remember also that "mass" as we are using the term need not coincide with physical mass. If it does not, (2) is not the same as the physicist's "conservation of mass."

9. The well-definedness of (3) and the fact that λ is a measure for each t_1 follow from our assumption that f is a finite conditional measure. For the validity of (4) we also need to assume that ν is abcont, as of course it would be in this model.

no transformations in R occur; i.e., no chemical transformations occur, but only spatial motions. This simplification is reflected formally by dropping all references to R in the preceding expressions: f becomes a function with domain $S_0 \times T_0 \times T_1 \times \Sigma_{s_1}$, ν becomes a measure on $S_0 \times T_0$, etc.

Space is taken to be the plane, with cartesian coordinates (x, y); Σ_s is the Borel field for the plane. With these preliminaries, we now give f in the form of an indefinite integral:

$$f(x_0, y_0, t_0, t_1, G) = \int_G \frac{1}{2\pi ab(t_1 - t_0)}$$

$$\cdot \exp\left[-\frac{(x_1 - x_0 - c(t_1 - t_0))^2}{2a^2(t_1 - t_0)} - \frac{(y_1 - y_0)^2}{2b^2(t_1 - t_0)} - k(t_1 - t_0)\right] dx_1 dy_1$$

$$(4.4.5)$$

for all $(x_0, y_0) \in S_0$, all $t_0 < t_1$, all $G \in \Sigma_{s_1}$. The integration in (5) is with respect to 2-dimensional Lebesgue measure. Here $\exp[z]$ stands for e^z; a, b, c, k are real constants satisfying $a > 0$, $b > 0$, $k \geq 0$. Formula (5) is valid only for $t_1 > t_0$. We have already mentioned that for $t_1 < t_0$, f is identically zero; and for $t_1 = t_0$, f is simply concentrated with unit mass concentrated at (x_0, y_0). For any $t_1 > t_0$, then, (5) is a bivariate normal distribution with mean at $[x_0 + c(t_1 - t_0), y_0]$, variance of $a^2(t_1 - t_0)$ in the x-direction, variance $b^2(t_1 - t_0)$ in the y-direction, and covariance zero. The mass over all S at moment t_1 is equal to $\exp[-k \cdot (t_1 - t_0)]$, so that the conservation law (2) is satisfied iff $k = 0$.

We might interpret (5) as follows. A unit mass released at location (x_0, y_0) at time t_0 is subjected to random and systematic forces. The latter consists of wind blowing in the x-direction at velocity c. The former causes the mass to spread out in a normal distribution pattern. The whole distribution moves with the wind and also keeps spreading, variances being proportional to elapsed time. The k term is thrown in to allow for possible disappearance of mass through absorption. It may be verified that f defined by (5) satisfies the Chapman-Kolmogorov equation (1) (with R being deleted).

The model we have outlined is one fragment of a larger system. The release measure ν, e.g., will in its turn be derived from the distribution of activities over $S \times T$, along with their associated production measures. Conversely, the resulting measures λ given by (3) condition the environment and thereby affect the feasibility of activities that might run in various places. This whole system of relations provides a test that must be passed by any feasible measure μ on the space of histories.

4.5. ACTIVITY DISTRIBUTIONS

The feasibility test to be presented now constitutes a generalization of the standard model of activity analysis.[10] We concentrate here on the formal development.

We have already discussed a "configuration" test for cross-sections μ_t, in which either μ_t restricted to any region must not be disallowed (strict constructionism) or μ_t must be "patched" together by allowable restrictions (loose constructionism). The same sort of test could be constructed for intertemporal feasibility, with activity types taking the place of configuration types.

Tests of this sort, however, are rather clumsy to work with because we must compare patterns spread over S (or $S \times T$). It would be much more convenient if the test involved only the comparison of single points, so to speak. The following is an attempt to carry out this construction.

We postulate a set of *allowable activity types* **Q**. In accordance with our aim, we consider only "simply located" activity types, i.e., activities having just a single location at any one moment. We also restrict our attention to *sedentary* activities, so that the single location is fixed over T. This second assumption is less crucial and could be relaxed, but it is convenient.

Let us spell out what these restrictions amount to. Taking an activity to be a measure over a set of histories, these will all have an identical, constant itinerary. Hence the histories are distinguished only by their transmutation paths. We may then simply identify an activity as a measure over the universe set Ω, of all transmutation paths (= functions whose domain is a closed time interval and that take values in R).

Another concept of activity identifies it with a production or consumption measure (or the pair of them). These are over a subset of $R \times S \times T$. But because of the very special kinds of activities we are considering, all reference to S may be suppressed. For simplicity we always take the universe set to be $R \times T$.

The distinction between activity and activity type has been slurred over in the preceding discussion. Recall that an *activity* is defined in terms of "real" S and T, while an *activity type* is defined with "abstract" sets S' and T' in place of these (S' is a metric space and T' has the structure of the real line). An activity type is then exemplified in an actual activity iff there is a measure-preserving spatial isometry and time translation between the abstract and real spaces.

In the present context the situation is much simpler. Since the

---

10. The classic reference is T. C. Koopmans, ed., *Activity Analysis of Production and Allocation* (Wiley, New York, 1951).

regions in which activities are located are single points, there is no isometry problem and we can ignore metrical considerations entirely. We
could still use an abstract T set, but it will be more convenient not to do
so. Finally, no confusion will arise in this discussion if we drop the distinction between "activity" and "activity type" and simply refer to them
both as "activity."

We summarize this discussion in the following definition, which combines the various special activity concepts.

Definition: An *activity* is a triple, consisting of a measure ρ on the
space of transmutation paths (Ω_r, Σ') and a pair of measures λ_1 and λ_2
on $(R \times T, \Sigma_r \times \Sigma_t)$.

Here ρ gives the amounts of "capital goods" and "materials" tied up in
the activity, while λ_1 and λ_2 are the production and consumption measures respectively. We need not be concerned with the nature of the
σ-field Σ' on Ω_r; for the present discussion, it suffices to know that it
exists. It is understood that this definition is only for the present discussion; in other cases we may wish to revert to the more general activity
concept discussed in Chapter 2.

We use the letter q to designate an activity and write, e.g., $\lambda_2(q, G)$
for the value of consumption in activity q on set $G \in \Sigma_r \times \Sigma_t$. As stated
above, **Q** will designate the set of *allowable activities*, which in general
will be a small subset of the set of *all* triples of measures $(\rho, \lambda_1, \lambda_2)$.

Q itself will now be made into a measurable space by placing a σ-field
Σ_q on it. The conditions to be placed on Σ_q will be indicated below.

Definition: An *assignment* is a measure ν on the space $(S \times \mathbf{Q}, \Sigma_s \times \Sigma_q)$.

Assignment ν describes how activities are distributed over the world. On
rectangles it may be interpreted as follows: $\nu(F \times G)$ is the total amount
of activities of types G situated in region F. This somewhat vague characterization will be pinned down below. Note that ν is a (generalized)
2-layer measure, since the elements of **Q** are themselves (triples of)
measures.

We now give the basic mathematical justification for the procedures
to be used in this section. We state it in abstract form to make it self-
contained.

Theorem: Let (Q, Σ_q), (S, Σ_s), and (B, Σ_b) be measurable spaces; let
ν be a measure on $(S \times Q, \Sigma_s \times \Sigma_q)$; let $\lambda: Q \times \Sigma_b \to$ extended reals be

an abcont conditional measure. Then for any measurable subset G of $(S \times B, \Sigma_s \times \Sigma_b)$ the integral

$$\int_{S \times Q} \lambda[q, \{b \mid (s, b) \in G\}] \, \nu(ds, dq) = \mu(G) \qquad (4.5.1)$$

is well defined, and the resulting set function μ is a measure on $(S \times B, \Sigma_s \times \Sigma_b)$.

Proof: We show that the conditions for the existence of a product measure on the space $(S \times Q \times B, \Sigma_s \times \Sigma_q \times \Sigma_b)$ are satisfied. We consider this as a product of *two* spaces, the first being $(S \times Q, \Sigma_s \times \Sigma_q)$ and the second being (B, Σ_b); ν is a measure over the first space.

Define $\lambda': S \times Q \times \Sigma_b \to$ extended reals by the rule $\lambda'(s, q, E) = \lambda(q, E)$, all $s \in S$, $q \in Q$, $E \in \Sigma_b$. One easily verifies that λ' is abcont conditional. Hence μ', defined by

$$\mu'(H) = \int_{S \times Q} \lambda'[s, q, \{b \mid (s, q, b) \in H\}] \, \nu(ds, dq) \qquad (4.5.2)$$

for any $H \in \Sigma_s \times \Sigma_q \times \Sigma_b$, is a measure on $(S \times Q \times B)$. However, μ in (1) is merely the marginal of this measure on the component space $(S \times B, \Sigma_s \times \Sigma_b)$; i.e., $\mu(G) = \mu'(G \times Q)$ for any measurable $G \subseteq (S \times B)$. This may be verified by substituting $G \times Q$ for H in (2); simplification yields (1). Hence μ itself is a measure. ∎

Let us now interpret this theorem concretely; S, Q, and ν have the meanings already discussed. Let $(B, \Sigma_b) = (R \times T, \Sigma_r \times \Sigma_t)$; then we may interpret λ to be the family of production measures λ_1, so that $\lambda(q, \cdot) = \lambda_1(q, \cdot)$ is the production measure on $R \times T$ associated with the activity q.

Now μ in (1) becomes a measure on $(R \times S \times T, \Sigma_r \times \Sigma_s \times \Sigma_t)$.[11] How is this to be interpreted? Contemplation of (1), and the nature of ν and λ, suggests that μ is the *total production measure resulting from the activity assignment ν*. Thus if we take any region F and consider for each activity the mass produced of resource types E in time period G, then $\mu(E \times F \times G)$ is the limit of the weighted sum of these masses, the weights being provided by the assignment ν restricted to $F \times Q$.

Precisely the same interpretation, but with λ_2 instead of λ_1, now yields a measure μ on $R \times S \times T$ that is to be interpreted as the total *consumption* measure resulting from the assignment ν. Finally, we interpret (B, Σ_b) to be the space of transmutation paths (Ω_r, Σ') and λ to be

---

11. Actually $S \times B = S \times R \times T$ is the order of the components of the cartesian product in (1). No confusion should arise if we permute them to the usual alphabetical order.

the family of measures ρ on this space; i.e., $\lambda(q, \cdot) = \rho(q, \cdot)$ is the "capital-goods" measure associated with the activity q.

With this interpretation, μ in (1) becomes a measure on the space $(S \times \Omega_r, \Sigma_s \times \Sigma')$. How is this to be interpreted? First, consider the set $S \times \Omega_r$. A moment's reflection shows this can be identified with the set of all histories that have constant itineraries. We call these the *sedentary* histories, for the point $(s_o, h_r) \in S \times \Omega_r$ corresponds naturally to the history h whose transmutation path is h_r and whose itinerary has the constant value s_o over the time interval in which it exists. The natural interpretation of μ here is as the *distribution of mass over the space of sedentary histories*. Thus letting F be a region and H a measurable set of transmutation paths, $\mu(F \times H)$ equals the total mass embodied in sedentary histories that are located in F and have transmutation paths in H. (If at the cost of further complications we had introduced activities involving nonsedentary histories, the μ here would come out to be the basic world-description measure over the space of *all* histories Ω.)

To summarize: Any activity assignment ν determines a triple of measures, all by (1). When we substitute the production and consumption measures, λ_1 and λ_2 respectively, of the various activities in $\mathbf{Q}$ for λ in (1), we come out with the production and consumption measures over $R \times S \times T$ that are yielded by this assignment ν. When we substitute the "capital-goods" measures ρ of the various activities for λ, we come out with the mass distribution over sedentary histories. We abbreviate these three measures as μ_1, μ_2, and μ_0 respectively. Thus μ_1 and μ_2 are on the space $(R \times S \times T, \Sigma_r \times \Sigma_s \times \Sigma_t)$, while μ_0 is on the space $(S \times \Omega_r, \Sigma_s \times \Sigma')$.

Before going on to discuss the feasibility tests arising from this analysis, let us see what it reduces to in a very simple case: where all four sets, $\mathbf{Q}$, R, S, T, are *finite* (and all subsets are measurable). This case is of interest for two reasons. First, it gives a heuristic guideline to the analysis just completed. Second, it shows how everything boils down to what is essentially ordinary activity analysis.

The value of assignment ν at the singleton set $\{(s, q)\}$ will be written simply as ν_{sq}, and we adopt a similar notation for all other measures. In fact, in this simple case the measures can be thought of as ordinary point functions, and the notation underlines this fact; ν_{sq} is the "level" at which activity q is running at location s.

For convenience we let λ stand for either λ_1 or λ_2. Then λ_{qrt} equals total production (or consumption) of resource r at time t in activity q. In (1) we choose the singleton set $\{(r, s, t)\}$ for G. Then the integral (1) reduces to a simple summation over activities $q \in \mathbf{Q}$: $\mu_{rst} = \nu_{sq_1}\lambda_{q_1rt} + \nu_{sq_2}\lambda_{q_2rt} + \cdots$. In this case the interpretation of μ is obvious: the total production (or consumption) of resource r at location s at time t, obtained by taking a weighted sum of production (consumption) of r at t for each allowable activity, the weights being the levels of the various activities at s as indicated by assignment ν.

Things are slightly more complicated for the capital-goods measure ρ. For simplicity we ignore "births" and "deaths," and assume that all transmutation paths exist at all times. If N is the number of time points, a transmutation path may be written as an N-tuple $(r_1, \ldots, r_N)$ in R, r_t being its resource state at time t. Then Ω_r may be identified with the product space R^N and $\rho_{q,r_1,\ldots,r_N}$ is the mass embodied in the transmutation path $(r_1, \ldots, r_N)$ in activity q. Now we choose the singleton $\{(s, r_1, \ldots, r_N)\}$ for G in (1). The integral (1) again reduces to a summation over activities $q \in \mathbf{Q}$:

$$\mu_{s,r_1,\ldots,r_N} = \nu_{sq_1}\rho_{q_1,r_1,\ldots,r_N} + \nu_{sq_2}\rho_{q_2,r_1,\ldots,r_N} + \cdots$$

This $\mu_{s,r_1,\ldots,r_N}$ is the total mass embodied in the sedentary history whose location is fixed at s and which runs through the sequence of resource states $(r_1, \ldots, r_N)$ over T. This again is a weighted sum of the mass embodied in transmutation path $(r_1, \ldots, r_N)$ for each allowable activity, the weights being the levels of the various activities at s.

We now return to the general case. So far we have said nothing about feasibility, but we now propose a test that has a vague resemblance to the configuration test (loose constructionist version). A measure passes that test if it is a countable patching of allowable configurations. Here we have a corresponding set of allowable activities $\mathbf{Q}$, and we consider only measures that can be built up from the activities of $\mathbf{Q}$. We interpret "built up" to mean that there is an assignment ν such that the measure μ to be tested is determined by ν according to (1). That is, considering (1) as a function that assigns a measure μ to every measure ν, the measures passing this test are those in the range of the function.

This might be considered too easy a test, since ν is not restricted in any way. One natural restriction that might be placed on ν is that it be σ-finite or even bounded.

Another constraint on ν that suggests itself is an areal capacity restriction. We have discussed this in connection with cross-sectional constraints, where we postulated a "demand-for-space" function $f: R \times S \to$ reals, which restricted the possible cross-sections μ_t at moment t. Now the measure μ_0 on the space of sedentary histories $S \times \Omega_r$ determines a cross-sectional measure μ_t for every moment t, hence must satisfy an areal capacity constraint for all t. We must have

$$\int_{F \times \Omega_r} f(h_r(t), s)\, \mu_0(ds, dh_r) \le \alpha(F)$$

for all regions F and all moments t, where α is ideal area. This is essentially a restatement of (3.3).

It could be argued, however, that this understates the demand for space. An activity needs not only space to store its "capital goods" at any moment but also "aisle space" or "elbow room," i.e., enough extra vacant area to carry out the manipulations and transformations in which it is involved. This suggests we should attach a "demand-for-space" function directly to activities per se. Thus we write $f:S \times \mathbf{Q} \rightarrow$ reals, f nonnegative, measurable with respect to $\Sigma_s \times \Sigma_q$. Intuitively, $f(s, q)$ is to be understood as the space demand by (unit level of) activity q at location s. We must satisfy

$$\int_{F \times \mathbf{Q}} f \, dv \le \alpha(F) \tag{4.5.3}$$

for all regions F. This places a direct restriction on the possible assignments v and therefore a further indirect restriction on the measures μ_0, μ_1, μ_2 that must satisfy (1) from some v.

We now turn to the individual measures μ_0, μ_1, μ_2. Our ultimate aim is to establish feasibility conditions for measures μ on the space of histories Ω. How does such a μ relate to this triple of measures? As for μ_0, its universe set is the set of sedentary histories $(S \times \Omega_r)$, which is a subset of Ω. The feasibility condition on μ, then, is that its restriction to $S \times \Omega_r$ be an allowable μ_0.[12]

As for the production and consumption measures μ_1 and μ_2, these represent "births" and "deaths" of histories respectively. Thus a finished product is produced in a manufacturing process at a certain place and time. This initiates a new history, which perhaps moves into transportation to be consumed elsewhere. The smoke emitted from a chimney initiates histories involved in atmospheric circulation. We must have, for measurable $(E \times F \times G) \subseteq (R \times S \times T)$, $\mu_1(E \times F \times G)$ equal to the total mass embodied in histories that originate in period G in region F at a resource type in E. A similar relation holds for μ_2 and the mass of histories terminating in $E \times F \times G$.[13]

For μ to pass the feasibility test, then, there must be an assignment

---

12. This is the case if all the mass embodied in the set $S \times \Omega_r$ comes from a system of activities in the activity analysis framework of this section. Another more complex possibility is that there are several superimposed systems in operation, e.g., one of the activity analysis form, one of the diffusion process form discussed above, and perhaps others. In this case the measure μ will be the sum of the mass distributions involved in the several systems.

13. Again, if the activity analysis system is not the only one in operation, μ will relate to the sum of the production measures from the various systems, not to μ_1 alone; similarly for μ_2.

v yielding measures μ_0, μ_1, μ_2, which are *simultaneously* compatible with μ.

4.6. NEIGHBORHOOD EFFECTS

We now consider some of the presuppositions implicit in the preceding construction. These are worth studying on their own and not merely in connection with the activity analysis model.

Let F and G be two disjoint regions. The possible processes that can go on in region G can be influenced by what goes on in region F. For example, the sound, light, heat, or substances emanating from F may condition the environment of G. These influences are sometimes called *neighborhood effects*. We may expect in a general way that neighborhood effects will become stronger the closer F and G are to each other, while they tend to disappear between distant regions. (Some influences, e.g., radioactive fallout, have worldwide effects. If we include in the concept of neighborhood effect the deliberate propagation of influence via the transport-communications system, in addition to the "natural" influences just mentioned, even distant regions will be palpably influenced by each other.)

Now consider the opposite case where there are no neighborhood effects. This may be assumed as a simplifying approximation when influences are sufficiently weak. But how exactly does one formulate the concept "no neighborhood effects occur?" The next few paragraphs attempt to pin down this notion.

The concept of direct summation of measure spaces has already been considered as a special case of "patching" (Section 2.4). Explicitly, we have the following definition.

Definition: Let (A_n, Σ_n, μ_n), $n = 1, 2, \ldots$, be a countable collection of measure spaces, where the A_n's form a packing: $A_m \cap A_n = \emptyset$ if $m \neq n$. The *direct sum* of these spaces is the triple (A, Σ, μ), where

(i) $A = A_1 \cup A_2 \cup \cdots$;
(ii) Σ consists of all sets of the form $E_1 \cup E_2 \cup \cdots$, where $E_n \in \Sigma_n$ for all $n = 1, 2, \ldots$;
(iii) μ has domain Σ, and for $E = E_1 \cup E_2 \cup \cdots, E_n \in \Sigma_n$,

$$\mu(E) = \mu_1(E_1) + \mu_2(E_2) + \cdots \qquad (4.6.1)$$

With this definition, (A, Σ, μ) is a measure space. It is not difficult to show that Σ is closed under countable unions and differences and that $A \in \Sigma$, so that Σ is a σ-field with universe set A. For each n, Σ_n is the restriction of Σ to subsets of A_n. Finally, the disjointness of the A_n's guarantees that for any $E \in \Sigma$, its representation in the form $E_1 \cup E_2 \cup \cdots$ is unique. Then μ as given by (1) is well defined and μ_n is the restriction of μ to A_n. The fact that μ is a measure is a simple consequence of the patching theorem.

We use the symbol $\oplus$ for direct sums. Thus

$$\Sigma = \Sigma_1 \oplus \Sigma_2 \oplus \cdots, \qquad \mu = \mu_1 \oplus \mu_2 \oplus \cdots$$

Next we want to extend this definition to the case where there is a whole *set* of measures $\mathbf{M}_n$ defined on each space (A_n, Σ_n), not merely the single measure μ_n. Write $(A_n, \Sigma_n, \mathbf{M}_n)$ for the measurable space together with the set of measures. Again we assume that the A_n's form a packing.

Definition: The *direct sum* $(A_1, \Sigma_1, \mathbf{M}_1) \oplus (A_2, \Sigma_2, \mathbf{M}_2) \oplus \cdots$ is the triple $(A, \Sigma, \mathbf{M})$, where A and Σ are formed as above and $\mathbf{M}$ is the set of all measures μ formed according to (1), where $\mu_1, \mu_2, \ldots$ are selected from $\mathbf{M}_1, \mathbf{M}_2, \ldots$ in all possible ways.

Now let $(S \times A, \Sigma_s \times \Sigma_a)$ be a product measurable space. (We later interpret S to be physical Space, but for the time being let us proceed abstractly.) Let $\mathbf{M}$ be a set of measures on this space.

Definition: $(S \times A, \Sigma_s \times \Sigma_a, \mathbf{M})$ is *countably rectangular* iff, for any countable measurable partition $\{S_1, S_2, \ldots\}$ of S, it is the direct sum

$$(S_1 \times A, \Sigma_{s_1} \times \Sigma_a, \mathbf{M}_1) \oplus (S_2 \times A, \Sigma_{s_2} \times \Sigma_a, \mathbf{M}_2) \oplus \cdots$$

Here Σ_{s_n} is the restriction of Σ_s to subsets of S_n, and $\mathbf{M}_n$ is the set of all restrictions to S_n of the measures $\mu \in \mathbf{M}$.

Here are some examples:

1. Let S consist of two points, A of one point: $S = \{s_1, s_2\}$, $A = \{a\}$. All subsets are measurable. Let $\mathbf{M}$ consist of the four measures μ_{ij}, $i = 1, 2$; $j = 1, 2$, whose values on the two points of $S \times A$ are given by $\mu_{ij}\{(s_k, a)\} = i$ if $k = 1$, and $= j$ if $k = 2$. Then $\mathbf{M}$ is countably rectangular. But if any one of these measures is deleted, the remaining trio is not countably rectangular.
2. The set of *all* measures, and the set of all σ-*finite* measures, on $S \times A$ are both countably rectangular.
3. The set of all bounded measures is not countably rectangular if the σ-field Σ_s is infinite. (The proofs of these statements are left as exercises.)

Under countable rectangularity the set $\mathbf{M}$ is built up from component sets in roughly the same way a rectangle set in a cartesian product space is built up from the "sides" of the rectangle.

Now let us interpret S as Space. The set A will be given a variety of interpretations, but in all cases the set $\mathbf{M}$ will be some "allowable" set of measures. This apparatus is designed to capture the intuitive notion that if there are no neighborhood effects, any region can be "autonomously" assigned its own allowable set of measures, this set not depending at all on what is chosen elsewhere. Collection $\mathbf{M}_n$ plays the role of this autonomous set for the region S_n, and the "countable rectangularity" property expresses precisely the fact that the choices from the respective sets $\mathbf{M}_n$ can be made freely and independently of each other.

We now take (A, Σ_a) to be the space of allowable activities (Q, Σ_q), and take $\mathbf{M}$ to be the allowable assignments ν on $S \times Q$. Then in general this will be countably rectangular in the activity analysis setup. We have already noted this if the assignment ν can be any measure or any σ-finite measure. Slightly less obvious is the fact that, even if a space-capacity constraint of the form (5.3) (repeated as (2) below) is imposed, the resulting set of allowable assignments retains this property.

Theorem: Let (S, Σ_s) and (Q, Σ_q) be measurable spaces, $f:S \times Q \to$ reals nonnegative and measurable, α a measure on S. Then the set of measures ν on $S \times Q$ that satisfy

$$\int_{F \times Q} f \, d\nu \leq \alpha(F) \qquad (4.6.2)$$

for all measurable $F \subseteq S$, is countably rectangular.

Proof: Let $\{S_1, S_2, \ldots\}$ be a countable measurable partition of S. For any ν satisfying (2), ν_n, its restriction to $S_n \times Q$, satisfies the same condition for all measurable $F \subseteq S_n$. Now let ν_n be such a measure on $S_n \times Q$, $n = 1, 2, \ldots$, and consider the direct sum $\nu' = \nu_1 \oplus \nu_2 \oplus \cdots$. For any measurable $F \subseteq S$ we have

$$\int_{F \times Q} f \, d\nu' = \int_{(F \cap S_1) \times Q} f \, d\nu' + \int_{(F \cap S_2) \times Q} f \, d\nu' + \cdots$$

$$= \int_{(F \cap S_1) \times Q} f \, d\nu_1 + \int_{(F \cap S_2) \times Q} f \, d\nu_2 + \cdots$$

$$\leq \alpha(F \cap S_1) + \alpha(F \cap S_2) + \cdots = \alpha(F)$$

Thus ν' satisfies (2). This proves countable rectangularity. ∎

Next we turn to the sets of allowable measures μ_0, μ_1, μ_2 determined by the allowable assignments of (5.1). The following result applies to all three measures.

Theorem: Let measure μ on $S \times B$ be determined by ν on $S \times Q$ by the rule

$$\mu(G) = \int_{S \times Q} \lambda[q, \{b \mid (s, b) \in G\}] \, \nu(ds, dq) \qquad (4.6.3)$$

where G is a measurable subset of $S \times B$, and $\lambda : Q \times \Sigma_b \to$ extended reals is an abcont conditional measure. Then if the set of allowable measures ν is countably rectangular, the same is true for the resulting set of allowable measures μ.

Proof: Let $\{S_1, S_2, \ldots\}$ be a countable measurable partition of S. If μ satisfies (3) for some allowable ν, its restriction to $S_n \times B$ satisfies (3) for all measurable $G \subseteq S_n \times B$, with S_n substituted for S, and ν_n, the restriction of ν to $S_n \times Q$, substituted for ν. Now let μ_n be such a measure on $S_n \times B$, determined by ν_n on $S_n \times Q$, $n = 1, 2, \ldots$. Consider the direct sums $\mu' = \mu_1 \oplus \mu_2 \oplus \cdots$, and $\nu' = \nu_1 \oplus \nu_2 \oplus \cdots$. By the countable rectangularity assumption, ν' is allowable.

For any measurable $G \subseteq S \times B$ we have

$$\mu'(G) = \mu'[G \cap (S_1 \times B)] + \mu'[G \cap (S_2 \times B)] + \cdots$$

$$= \mu_1[G \cap (S_1 \times B)] + \mu_2[G \cap (S_2 \times B)] + \cdots$$

$$= \int_{S_1 \times Q} \lambda[q, \{b \mid (s, b) \in G\}] \, \nu_1(ds, dq) + \cdots$$

$$= \int_{S \times Q} \lambda[q, \{b \mid (s, b) \in G\}] \, \nu'(ds, dq)$$

Hence μ' is allowable. Thus the set of allowable measures μ is countably rectangular. ■

The assumption of no neighborhood effects, therefore, pervades the activity analysis model we have constructed. How realistic is this? There will always be some neighborhood effects, so the real question is whether these effects are unimportant enough to be ignored. The answer appears to depend on the scale of observation. On a "person-sized" level, neighborhood effects are so vital that any model ignoring them would be useless. Chopping a person in half will rapidly affect his functioning: each half needs the "neighborhood effects" emanating from the other. Similarly, the technical possibilities in half a machine or half a house will be affected if the other half is sheared off.

At the level of the ordinary urban neighborhood, the neighborhood effects are still important but not nearly as momentous. "Urban problems" are in large measure the reflection of these interdependencies, resulting from the proximity of masses of people to each other. Going up the scale to the economywide and worldwide levels, neighborhood effects are much attenuated. We might expect, then, that the activity analysis model described here could be fairly applicable to economywide technical possibilities, would be less so at the urban neighborhood level, and would be poor as a model for individual household possibilities.

There is one consideration that vitiates conclusions of this sort. It is the system of constraints as a whole that is subject to criticism, not any particular subsystem in isolation. If neighborhood effects are not taken into account in the activity analysis subsystem, they may be taken account of elsewhere in such a way that the set of measures passing all feasibility tests reflects the existence of these effects.

Other "neighborhood effect" concepts are not captured by the countable rectangularity property. We sometimes want an asymmetrical concept in which region E has an effect on region F but not vice versa. Countable rectangularity is symmetric in the sense that no ordering distinctions are made among the components of a direct sum. We briefly indicate how these one-way effects might be represented. This is done by bringing in Time explicitly.

First, we need a slight weakening of the countable rectangularity concept. Let $\mathbf{M}$ be a set of measures on space (A, Σ), and let E, F be two disjoint measurable subsets of A. Then $\mathbf{M}$ is said to be *rectangular* with respect to the pair of sets E, F iff

$$\mathbf{M}_{E \cup F} = \mathbf{M}_E \oplus \mathbf{M}_F \qquad (4.6.4)$$

where $\mathbf{M}_E$ is the set of restrictions to E of the measures $\mu \in \mathbf{M}$, and similarly for $\mathbf{M}_F$ and $\mathbf{M}_{E \cup F}$. Countable rectangularity implies that (4) is true for any such E, F.

Now let $\mathbf{M}$ be the set of allowable measures on the space $(B \times S \times T, \Sigma_b \times \Sigma_s \times \Sigma_t)$. Here S and T are Space and Time; B might be the resource set R or some more complicated set, depending on the problem in hand. Let $F_1, F_2 \subseteq S$ be two disjoint regions.

Now we define: there are *no neighborhood effects from F_1 to F_2 across time instant t_o* iff $\mathbf{M}$ is rectangular with respect to the two sets

$$(B \times F_1 \times \{t \mid t < t_o\}) \quad \text{and} \quad (B \times F_2 \times \{t \mid t > t_o\})$$

This amounts to saying that what can happen on F_2 after time t_o is not affected by what happens on F_1 prior to t_o. This concept is not symmetrical, since there might still be neighborhood effects across t_o from F_2 to F_1.

4.7. SUPERPOSITION AND RETURNS TO SCALE

Definition: Let **M** be a nonempty set of measures on space (A, Σ). Set **M** is said to be *additive* iff, whenever μ_1 and μ_2 belong to **M**, then $\mu_1 + \mu_2$ belongs to **M**; and **M** is said to be a *cone* iff, whenever $\mu \in$ **M** and $c \geq 0$ is a real number, $c\mu \in$ **M**.

An additive set of measures is also said to obey the *superposition principle*, since two members of it may be "superimposed" to form a third member. A nonempty **M** that is both additive and conical is a convex cone.[14]

The set of *all* measures, of all σ-*finite* measures, and of all *bounded* measures on (A, Σ) are examples of sets having both these properties. Now consider the set of all allowable assignments ν on the space $S \times Q$ in the activity analysis model. This determines a set of measures μ on the space $S \times B$, where B and μ have various interpretations, by the integration formula (5.1).

It follows at once from elementary integration theorems that if ν' determines μ' and ν'' determines μ'' via (5.1), then $\nu' + \nu''$ determines $\mu' + \mu''$. Hence if the set of allowable assignments ν is additive, so is the resulting set of measures μ. Similarly, if the set of allowable ν's is a cone, so is the resulting set of allowable μ's.

Are these conditions realistic? As noted in the discussion of neighborhood effects, any conclusion is to be treated with caution: Even if we decide (as we shall) that these conditions are not very defensible, it still does not follow that we should reject feasibility tests that assume them. The system of feasibility tests as a whole must be confronted. As a simple example, consider the activity analysis model in which the set of allowable assignments is unrestricted. This allows, say, cross-sectional measures of arbitrarily high density, which is not realistic. But there are other feasibility tests that exclude excessive densities—in particular, space capacity constraints. It may be very convenient to keep the unrestricted activity analysis model as one subsystem of constraints, and no objections need arise to the system of constraints as a whole.

With this caution in mind, we pose the question in the following form. Given the set of allowable configuration types or activity types, is it reasonable to suppose that this set is additive and/or conical?

There is one minor difficulty involved in the concept of additivity here, which we illustrate with the set of allowable configuration types. A configuration type is a measure on a universe set of the form $R \times F$, where F is an "abstract" region that is a measurable and a metric space.

---

14. For further analysis of these and other properties of "production sets," in the context of n-space, see G. Debreu, *Theory of Value* (Wiley, New York, 1959), pp. 39–42.

Now consider μ_1 and μ_2, defined on $R \times F_1$, $R \times F_2$ respectively. Since in general F_1 and F_2 are not the same, the sum $\mu_1 + \mu_2$ is not well defined. We can, however, proceed as follows. Suppose there is a measurability-preserving isometry between F_1 and F_2, say $f:F_1 \rightarrow F_2$. This, with μ_1, induces a measure μ_1' on $R \times F_2$; and we now define the sum of μ_1 and μ_2 as $\mu_1' + \mu_2$, a measure on $R \times F_2$. If there are several different isometries between F_1 and F_2, in general each will lead to a different summation operation. If there is no isometry, the sum is not defined. These complications reduce the usefulness of the additivity concept in this context.

By contrast, the condition that the set of allowable configuration types or activity types is a cone is perfectly well defined. For configurations, this reads: if μ on $R \times F$ is an allowable configuration type, so is $k\mu$ on $R \times F$, k being a nonnegative real number. We devote most of our attention to the question of reasonableness of this condition and the corresponding condition for activity types.

Let us connect this with the concept of *scale*. Recall that in our discussion of scale (Section 2.7) we distinguished a number of different concepts, in particular the notions of a k-fold expansion in the *intensive,* the *extensive,* and the *duplicative* sense. Now suppose the set **M** of allowable configuration types or activity types has the following property: if $\mu \in$ **M** and μ' is the k-fold expansion of μ in the x-sense, $\mu' \in$ **M**. In this case we say that **M** has *constant returns to scale in the x-sense.* We discuss each of the various senses in turn.

A moment's reflection shows that the two conditions "**M** is a cone," and "**M** has constant returns to scale in the *intensive* sense," are the same. How reasonable is this property? That is, if μ is allowable, is it reasonable that $k\mu$ should be allowable for any real number $k \geq 0$?

There are two cases. (i) For $k > 1$, arbitrarily high densities would be allowable. But presumably at some point it would become physically impossible to squeeze that mass into the given space; and even before one reaches this density, the increasing concentration of mass will in general lead to interactions (neighborhood effects!) that prevent an exact proportional change of mass everywhere. (ii) For $k < 1$, this last objection still holds, in reverse. There may be "threshold effects" or "critical masses" that prevent one from halving mass everywhere and maintaining feasibility.

When $k \geq 0$ is an *integer* and $\mu \in$ **M**, then $k\mu \in$ **M** follows both from the condition that **M** is a cone and from the condition that **M** is additive. Thus the objections against arbitrarily high k values are objections against both these conditions.

Now consider constant returns to scale in the *extensive* sense. Here we introduce the areal measure α on Space, and a k-fold expansion involves a similarity mapping that multiplies *area,* as well as all masses, by k. All densities (with respect to α) remain the same, but now another difficulty arises. If volume expands by a factor of k, surface area expands by $k^{2/3}$ and length by $k^{1/3}$. These nonproportional changes in general make it impossible to maintain an extensive scale change. For example, if a house is doubled in linear dimension, its "capacity" (roughly measured by volume) octuples; but the rate of heat loss (roughly proportional to surface area) only quadruples, so the heating plant need not expand in proportion. Absolute scale changes *do* make a difference, and constant returns to scale in the extensive sense must also be rejected as a general rule.[15]

This brings us to constant returns to scale in the *duplicative* sense. Here an activity type or configuration type is placed "side by side" with itself. Specifically, if μ is an allowable configuration with universe set $R \times F$, a k-fold expansion of μ is a measure μ' on $R \times F'$ such that there is a partition $\{F_1', F_2', \ldots, F_k'\}$ of F' into k pieces, and μ' restricted to each piece $R \times F_i'$, $i = 1, \ldots, k$, is a *duplicate* of μ (i.e., there is a measurability-preserving isometry f from F to F_i', and μ' restricted is the measure induced by f from μ). A similar definition holds for activity types.

Note first that k must be an integer for this definition to be meaningful. A hen and a half does not lay an egg and a half if one hen lays one egg. This gives at best a weaker condition than constant returns per se and might be called constant *integer* returns to scale.

Also, unlike the other scale concepts, many distinct configuration types are k-fold expansions of the same μ because nothing is said about the metric relations of the pieces F_i' to each other, but only about their internal structure. The k pieces may be close to each other or scattered. Under constant duplicative returns, all these k-fold expansions would be allowable.

Constant duplicative returns is a corollary of one of the feasibility test systems we have discussed: the configuration criterion, in the loose constructionist version. A cross-sectional measure passes this test iff it is a (countable) patching of exemplifications of allowable configuration types. If μ is allowable, a k-fold expansion is such a patching (in fact it is a direct sum of the k-duplicates). A similar statement holds for activities in place of configurations.

The basic weakness of the constant duplicative returns assumption is that shared by this loose constructionist version: the ignoring of neighborhood effects. In general, what is feasible in a region depends on the environment of that region and cannot simply be drawn from a fixed list

---

15. There is much literature on the effects of extensive scale changes, both in engineering and biology. In the former it goes under the titles "dimensional analysis" or "theory of models." Both aspects are treated in D'Arcy Thompson's classic work, *On Growth and Form* (Cambridge Univ. Press, 1917).

of allowable possibilities. However, it may be a fair approximation in some situations, and as such it appears to be the least objectionable of the three senses of "constant returns to scale" we have discussed.

4.8. INDIVISIBILITY

A long tradition in economic theory connects departures from "constant returns to scale" with "indivisibility." Another somewhat more recent literature denies the connection. Our aim here is, not to resolve this issue once and for all, but to clarify it by distinguishing the many different concepts named by these terms. It is likely that much of the controversy arises from the confusion of meanings of the same term in the minds of different participants (or of the same participant). In Section 4.7 we have distinguished several different meanings of constant returns to scale. In this section we do the same for indivisibility.[16]

One may distinguish at least six different meanings of the term "indivisibility," many of which have already been discussed:

1. As a synonym for *integer-valuedness*.
2. As the requirement that a certain measure be finitely concentrated, in particular, that it be *simply concentrated*. The man who "flung himself upon his horse, and rode madly off in all directions" was violating an indivisibility constraint in this sense.[17]
3. As a *limited variety* constraint. This may be due to natural laws, as in biology, where only a limited range of organic forms are viable; it may be due to lack of knowledge of how to make certain resource types or configurations; or it may be due to a high overhead cost of starting new product lines or to lack of raw materials.
4. As the condition that certain configurations *cannot be split* into two spatially separated halves. This is of course the original meaning of the term "indivisibility."
5. As the condition that certain configurations cannot be split without destroying their functioning, as with "organic wholes."
6. As the condition that certain configurations cannot be spatially *segre-*

---

16. For the argument that "nonconstant returns to scale" result from "indivisibilities" see, e.g., F. H. Knight, *Risk, Uncertainty and Profit* (Houghton Mifflin, Boston, 1921), pp. 98, 177; A. P. Lerner, *The Economics of Control* (Macmillan, New York, 1946), pp. 68–69, 143; T. C. Koopmans, *Three Essays on the State of Economic Science* (McGraw-Hill, New York, 1957), pp. 150–54. For criticisms and further discussion see P. A. Samuelson, *Foundations of Economic Analysis* (Harvard Univ. Press, Cambridge, 1947), pp. 84–85; E. H. Chamberlin, *The Theory of Monopolistic Competition*, 7th ed. (Harvard Univ. Press, Cambridge 1956), App. B; H. Leibenstein, The proportionality controversy and the theory of production, *Quart. J. Econ.* 69(Nov. 1955):619–25; D. Schwartzman, The methodology of the theory of returns to scale, *Oxford Econ. Pap.* NS 10(Feb. 1958):98–105.

17. Quotation from Stephen Leacock.

gated by resource components; e.g., a metal that cannot be extracted from its ore.

Some of these spatial interpretations have temporal equivalents, e.g., "noninterruptibility" constraints.

Surveying these interpretations and referring back to our critical discussion of returns to scale, we note that none of these conditions was used in the argument. Constant intensive returns was rejected because of threshold and congestion effects, constant extensive returns because of length-area-volume nonproportionalities, and constant duplicative returns because of neighborhood effects. (Threshold, congestion, and length-area-volume phenomena can themselves probably be regarded as special manifestations of neighborhood effects.) Thus it would appear that nonconstant returns to scale (in any sense) can appear without indivisibilities (in any sense of the term).

In fact, one might be well advised to reverse the standard argument and derive certain kinds of indivisibility conditions from the nonconstancy of returns to scale. Suppose length-area-volume effects make an organic form viable only in a limited size range. This is an example of nonconstant extensive returns and leads to an indivisibility of type 3. The existence of neighborhood effects underlies indivisibilities of type 5 and perhaps also of types 4 and 6.

4.9. SPATIAL CONTROL

Most of our discussion has been of tests that any feasible measure μ on the space of histories Ω must satisfy. In this section we take a different point of view and consider some means by which the acting person carries out his choice among the feasible alternatives. The discussion will be entirely informal.

We are concerned with spatial control, i.e., the control of the movements of things. It appears that spatial control underlies control in general.[18] To explain: We act upon the world exclusively through motions of the body (including speech, which is a motion of the diaphragm, vocal cords, etc.). These acts influence objects or other people. We cause things to interact, as a rule, by placing them in proximity. A plan of action may be in large part described as a schedule, bringing people and/or objects together at various points of Space-Time to interact in desired ways, the outputs of some of these processes being stored or transported to arrive at other Space-Time points where they serve as inputs for other processes.

---

18. "Man whilst operating can only apply or withdraw natural bodies; nature internally performs the rest." *Novum Organum,* Book I, Aphorism 4, in Francis Bacon, *Advancement of Learning and Novum Organum* (Willey Book Co., New York, 1900).

(These processes include not only "production" in the ordinary sense of the term, but residential processes, education, dances, political meetings, etc.)

For such a plan to be feasible, the transportation and storage facilities must be available when and where needed, and it must be possible to carry out the processes with the scheduled factors. This will generally involve persuading other people to cooperate, e.g., by exhortation or offers of services or money.

This account omits one important aspect of spatial control. We must not only *move* things but also *prevent* motion. Everyone is aware of the fact that transportation incurs a cost, i.e., precludes some alternative opportunities by requiring the sacrifice of resources, time, and effort. The prevention of movement also incurs a cost. Perhaps this is noted so rarely because its manifestations are too obvious to comment upon. A few examples illustrate this point.

First, there is a need to maintain *altitude*. That is, if things are to be brought into spatial proximity to interact, they must be at about the same distance from sea level. In the presence of gravity, the means of accomplishing this in almost all cases is to provide a horizontal surface that gives common support to the various interacting entities. The crust of the earth is available for this purpose but has certain drawbacks. It may depart too far from the horizontal, so that we must incur costs, either to flatten it, or to prevent things from rolling downhill, or both. It may not provide adequate support, as in marsh or swamp, not to mention open water, and so must be either reinforced or unused. Finally, there may not be enough of it in the right places, so that more surface must be constructed at considerable cost. The prime example of this is multiple-story structures, but most furniture also serves the function of providing extra horizontal surface where needed: beds, chairs, tables, shelves.

Second, we not only must bring the right things together to interact but also must keep the wrong things away. We have already commented on the function of barriers in keeping out the weather, unauthorized personnel, etc. The entire institution of private property may be construed as a system of selective barriers, denying access to all except those authorized by the owner of the property or those having special access rights (easements, search warrants, etc.). Nor is this merely a capitalistic arrangement: the phenomenon of "too many cooks spoiling the broth" is a universal technological problem, requiring the limitation of access rights in any economic system.

It should be clear from these examples that the prevention of motion is as fundamental a task as the provision of transportation. There is a close analogy here between the Resources set and Space. We devote effort not only to transforming things from less to more desirable resource states but also to maintaining things in their present state: to the prevention or slowing of depreciation. Much of our total effort is of this treadmill variety, merely stopping things from getting worse: most eating, sleeping, exercise, medical care, haircuts, laundering.

Similarly, in Space we try not only to move things to better locations (transportation) but also to prevent or slow their moving to worse locations. This could be called *location maintenance.* We have already examined the special case of altitude maintenance and now discuss others. We continue to use the term *barrier* as a general name for any mechanism or institution that maintains location.

Consider the very general class of barriers we may call *walls.* These prevent various resource types from moving through the border of a certain region. They may be classified according to the kinds of resources they bar and whether they function to keep things in the region, or out of the region, or both. Thus a country may bar immigration, emigration, or both. Glass is a barrier to the passage of air but lets light through. A locked door is permeable to someone with the key; it is a barrier to others.

Storage facilities and packaging in general are all walls in this sense (cans, sacks, silos, barns, crates). These serve the double functions of protecting the contents by barring contaminants, the weather, pilferers, etc., from entering and of holding the contents in place by barring exit. A special class of walls (including clothing, window shades, and sound-proofing) serves the function of insuring privacy, i.e., prevents the dissemination of certain light or sound patterns that might be perceived by outsiders.

Brakes are barriers that prevent or limit the mobility of specific things. These include anchors, hobbles, paperweights, and ball and chain as well as ordinary vehicle brakes.

Bindings are mechanisms that prevent or limit the *relative* motion of different things. These need not be barriers as we have been using the term, since the entire configuration of things bound together can move as a group relative to the earth. In fact, brakes may be considered the special case of bindings in which the earth itself is one of the objects. Bindings include adhesives, bolts, nails, zippers, string; they may also include packaging and storage facilities insofar as they hold things within an integument. But typically we want not merely to bring things together but to hold them at proper relative distances. This is done by using a structural frame, which is a rigid body or one with a limited number of degrees of freedom for motion (e.g., buildings and the metallic or wooden parts of machines); the skeleton plays a similar role in organisms. These again are forms of bindings.

Much effort goes into the design of properly selective barriers (i.e., those preventing the passage of some things and not others) and of barriers that can be controlled to vary their selective power as desired. This involves both technological research and institutional arrangements (guards, customs inspectors, censors). The evolution of military technology is to an extent a race between ever more penetrating offensive weapons and the finding of barriers to stop them, from the sword and shield to the missile and antimissile.[19]

There are usually many ways of building barriers to accomplish a certain function. Roofs and umbrellas are substitute barriers against the rain. To stop a pollutant emitted at location s_1 from reaching a person at s_2, we can place a barrier at the source (smoke control), at the recipient (gas mask), or at an intermediate point (insulated house with filter).

We also may have options to erect a barrier or to take some other action that obviates the need for the barrier. Two mutually hostile groups can migrate away from each other, or they can stay put and erect barriers to reduce contacts ("separation" vs. "segregation").[20] Or instead of soundproofing to insure privacy, sounds can be masked by creating artificial noise. This has been used in connection with church confessionals and physicians' examining rooms (not to mention gangland "rubouts").[21] Multiple-purpose barriers are not uncommon. Thus the Great Wall of China was built to keep the nomads out but also, and perhaps primarily, to keep the Chinese in.[22]

We now turn briefly from the prevention to the promotion of motion, i.e., to transportation. Transportation is defined broadly to include any deliberate effort to change location. It therefore includes communication (the transportation of letters, electromagnetic waves, and other resource types designed especially to convey information) and, for the most part, public utilities (largely concerned with the movement of water, gas, electricity, and sewage).

A transportation system may be classified into the channel, transmitter, receiver, power source, vehicle, and cargo. Not all these components are present in all systems. In automotive transportation the channel is the road, the power source is internal combustion, the transmitter and receiver are parking facilities. In radio we have the transmitting station and the radio receiver, with transportation of electromagnetic waves; there is no vehicle or channel in this case. In the sewer system the channel is the network of sewer pipes, the transmitters are the various toilet fa-

---

19. Similarly, for the coevolution of the safe and safe-cracking techniques see E. H. Sutherland and D. R. Cressey, *Principles of Criminology*, 7th ed. (Lippincott, Philadelphia, 1966), p. 275.

20. Cf. *Genesis* 13:6–11.

21. *New York Times*, May 10, 1964, p. 40.

22. O. Lattimore, *Inner Asian Frontiers of China*, 2nd ed. (Am. Geogr. Soc., New York, 1951), p. 240.

cilities, etc., the receiver may be a treatment plant, the power source is usually gravity, the cargo is sewage, and there is no vehicle.

Transportation construction refers to the building of channels, transmitters, and receivers; it may be thought of as barrier removal or circumvention. Consider the road system. When completed, it establishes a more or less unobstructed surface connecting any two sites. The internal "road" system of buildings—the corridors, stairs, and elevators—may be thought of as a fine-structured extension of the road system proper; together they connect any two "rooms" in the economy.

While road-building reduces barriers to travel along its length, it tends to create new barriers transversely. For example, the building of a bridge creates a barrier to ships too tall to clear it (and thus establishes a lower head of navigation).[23] When roads intersect in a grid system, cross-traffic creates very considerable interference in the form of slowdowns and extra fuel consumption.[24] This sometimes makes it advisable to invest extra resources to reduce the interference, e.g., by overpasses, clover-leaf intersections, or traffic lights. The tradition that the poor live "on the other side of the tracks" indicates that a transport artery may function as a social barrier.[25]

---

23. A. E. Smailes, *The Geography of Towns,* 5th ed. (Hutchinson Univ. Libr., London, 1966), pp. 47, 54.

24. D. M. Winch, *The Economics of Highway Planning* (Univ. Toronto Press, Toronto, 1963), pp. 67–68.

25. In the reconstruction of Paris under Napoleon III and Baron Haussmann it is said that boulevards were planned to break up working-class neighborhoods to reduce their revolutionary potential (as well as to provide easy access for government troops). H. Malet, *Le Baron Haussmann et la Rénovation de Paris* (Les Editions Municipales, Paris, 1973), pp. 187–89. E. J. Hobsbawm, Cities and Insurrections, in his *Revolutionaries: Contemporary Essays* (Pantheon Books, New York, 1973), p. 226.

5

The Allocation of Effort

5.1. INTRODUCTION
Consider the following problems:

1. A person has a certain sum of money, and a number of projects in which to invest it. The return from each project depends on the amount invested, and the problem is to split the money among projects so as to maximize the total return.
2. The return from an activity depends on the amount of time devoted to it. Split the 24 hours of a day among activities so as to maximize the total return.
3. The expected return from oil exploration depends on the region explored and the intensity of search effort there. Allocate a given total searching effort so as to maximize the expected total return.
4. The crime rate in an urban district depends on the district and the number of policemen patrolling it. Distribute the police force over Space so as to minimize total crimes.
5. Again there is a range of possible activities. Some are productive, earning money but with a disutility attached to participating in them; some are consumptive, yielding utility for the spending of money. The problem is to maximize total net utility, subject to total spending being equal to total earning.

All these problems have the following formal structure: maximize

$$f_1(x_1) + \cdots + f_n(x_n) \tag{5.1.1}$$

subject to

$$x_1 + \cdots + x_n = X \tag{5.1.2}$$

Here $f_i(x)$ is the return from allocating an amount x to project i. Equation (2) is the fundamental "budget" constraint, stating the total amount available to allocate; X can be time, money, effort, resources. The "proj-

219

ects" $i = 1, \ldots, n$ can be regions of Space, periods of Time, activities. The individual x_i may be required to be nonnegative, but not necessarily. In problem 5, for example, we could measure the amount devoted to activity i by the money spent; in the case of productive activities this would be negative. Total spending equal to total earning is then represented as $X = 0$. The return functions f_i can also be negative. In problem 4, $f_i(x)$ would be *minus* the number of crimes in district i with x policemen assigned there.

Problems of the form (1)–(2) are among the simplest and most ubiquitous of all. One popular definition of economics takes it to be the study of the relationship "... between ends and scarce means which have alternative uses,"[1] which may be construed as the study of problems involving constraints of the form (2). Though this seems rather narrow, it indicates the pervasiveness of this condition.

Our aim in the first part of this chapter is to study problem (1)–(2) under conditions of extreme generality. More exactly, we assume that the set of possible projects forms a general measure space, not just a finite set. Such generalization is clearly in order for many problems. The distribution of resources over S or T is over a continuum. The number of possible alternative investment opportunities will often be infinite.[2]

Our results will generalize existing work in several directions: (i) very weak restrictions on the nature of the payoff functions f; (ii) very flexible feasibility conditions, including the possibility of negative investments; (iii) no restriction that the measure be over n-space or that it be nonatomic; no topological or metrical conditions imposed on it; and (iv) pseudomeasure-valued utilities.

All these generalizations are of interest for one application or another. However, we use just one constraint, while other formulations allow several.

---

1. L. Robbins, *An Essay on the Nature and Significance of Economic Science,* 2nd ed. (Macmillan, London, 1952), p. 16.

2. Some work has been done in generalizing (1)–(2) to a continuum. The calculus of variations may be applied in some special cases. Work beyond this point was begun by Bernard Koopman in 1956. See J. De Guenin, Optimum distribution of effort: An extension of the Koopman basic theory, *Oper. Res.* 9(Jan.–Feb. 1961):1–7. Another exposition may be found in S. Karlin, *Mathematical Methods and Theory in Games, Programming, and Economics* (Addison-Wesley, Reading, Mass., 1959), Vol. 2, Ch. 8, where the connection with the Neyman-Pearson lemma is stressed. Also see M. E. Yaari, On the existence of an optimal plan in a continuous-time allocation process, *Econometrica* 32 (Oct. 1964):576–90; R. J. Aumann and M. Perles, A variational problem arising in economics, *J. Math. Anal. Appl.* 11 (1965):488–503; and D. H. Wagner and L. D. Stone, Optimization of allocations under a coverability condition, *SIAM J. Control* 12 (Aug. 1974):373–79.

5.2. FORMULATING THE PROBLEM

We start with the following ingredients: a measure space (A, Σ, μ), where μ is σ-finite, and a measurable function $f{:}A \times$ reals $\rightarrow$ reals. (Where the real numbers are concerned, measurability refers, as always, to the Borel field.)

The problem is to find a measurable function $\delta{:}A \rightarrow$ reals that maximizes the utility function

$$U(\delta) = \int f(a, \delta(a)) \, \mu(da) \qquad (5.2.1)$$

over a certain feasible set of such functions δ.

The expression (1) is an indefinite integral yielding a signed measure over the space A, or, more generally, a pseudomeasure in case (1) is not well defined in the ordinary sense. Integral (1) is always well defined as a pseudomeasure because μ is σ-finite and the integrand $f(a, \delta(a))$ is finite and measurable. (Measurability follows from the fact that it is the composition of two measurable functions: $a \rightarrow (a, \delta(a))$ and f itself.)

We use standard ordering of pseudomeasures to rank alternative functions δ. In the present case, by the standard integral theorem this means that δ_1 is at least as preferred as δ_2 iff

$$\int_A [f(a, \delta_1(a)) - f(a, \delta_2(a))] \, \mu(da) \qquad (5.2.2)$$

is well defined as an ordinary definite integral and is ≥ 0. This is possible even if (1) is not well defined in the ordinary sense for either δ_1 or δ_2. If (2) is not well defined, δ_1 and δ_2 are not comparable under standard order. The possibility of noncomparability makes it important to recall the distinction between a given feasible δ^* being *best* ($\delta^* \succcurlyeq \delta$ for all feasible δ) and being merely *unsurpassed* (there is no feasible $\delta \succ \delta^*$).

If the integral (1) is finite for all feasible δ, standard order reduces to the ordinary comparison of definite integrals, and there is no need to bring in pseudomeasures. The reader who is troubled by pseudomeasures has the option of adding conditions insuring that (1) is always finite and can then obtain a special case of most of the following theorems. Here, as elsewhere, the use of pseudomeasures simplifies and generalizes by enabling us to drop superfluous conditions.

The utility function (1.1) is a special case of (1). Let A consist of just n points, $\Sigma =$ all subsets of A, and let μ have the value *one* on each point (counting measure). In this case δ is an n-tuple $(\delta_1, \ldots, \delta_n)$, $f(a, \delta(a))$ for the ith point a may be written $f_i(\delta_i)$, and the integral (1) reduces to the finite sum (1.1).

Symbols A, f, μ, and δ may be interpreted as follows: A is the set of alternative "projects" among which we are allocating, and may be a set of locations, times, activities, etc. For fixed $a \in A$, $f(a, \cdot)$ is a real-valued function of a real variable, which gives the "payoff density" yielded from the "investment density" $\delta(a)$ applied to point a; μ may be areal measure over S, or time measure over T, or some other measure such that (1) gives the utility. If it has a positive value at a single point, then in general a nonzero payoff may be obtained from that point. Finally, δ gives the density of the distribution of the resource, money, time, effort, etc., over the alternatives of A. That is, the distribution of the resource being allocated is given by the indefinite integral over A:

$$\int \delta \, d\mu \qquad (5.2.3)$$

We have said nothing about the feasibility conditions for δ, except that it must be real valued and measurable. We shall always require that (3) be a finite signed measure; i.e., any feasible δ satisfies the condition

$$\left| \int_A \delta \, d\mu \right| < \infty \qquad (5.2.4)$$

Now consider the apparently much more specialized condition:

$$\int_A \delta \, d\mu = 0 \qquad (5.2.5)$$

Condition (5) seems very narrow. Only problem 5 of Section 5.1 satisfies it (total earning = total spending, so total net resource endowment $X = 0$). The others have positive total resource endowments. However, we now show that *any* allocation problem satisfying feasibility condition (4) can be converted into another allocation problem satisfying (5).

Theorem: Given A, Σ, μ, f, **M**, where (A, Σ, μ) is a measure space, μ is σ-finite, $f: A \times$ reals $\rightarrow$ reals is measurable, and **M** is a set of measurable functions $\delta: A \rightarrow$ reals, all satisfying (4), the problem being to maximize (1) over $\delta \in$ **M**.

Then A', Σ', μ', f', **M'** exist satisfying the same conditions and for which (5) is satisfied for all $\delta' \in$ **M'**, such that there is a 1–1 correspondence between **M** and **M'** that preserves preferability relations. (That is, if δ_1, $\delta_2 \in$ **M** correspond to δ_1', $\delta_2' \in$ **M'** respectively, then $\delta_1 \succeq \delta_2$ iff $\delta_1' \succeq' \delta_2'$. Here the preference relation $\succeq$ comes from (2), while $\succeq'$ comes from (2) with f', μ' substituted for f, μ.)

Proof: Let z_o be an artificial point not belonging to A. Define

(i) $A' = A \cup \{z_o\}$; $\Sigma' = \{G \mid G \subseteq A' \text{ and } G \setminus \{z_o\} \in \Sigma\}$;

(ii) $\mu'(G) = \mu(G)$ if $z_o \notin G$, $\mu'(G) = \mu(G \setminus \{z_o\}) + 1$ if $z_o \in G$, all $G \in \Sigma'$;

(iii) $f'(a, x) = f(a, x)$ for all $a \in A$, all real x; $f'(z_o, x) = 0$, all real x).

Let $\mathbf{M}'$ be the set of all functions $\delta':A' \rightarrow$ reals whose restriction to A belongs to $\mathbf{M}$ and which satisfy

$$\delta'(z_o) = - \int_A \delta' \, d\mu \qquad (5.2.6)$$

Note that $\delta'(z_o)$ is finite by condition (4). Now μ' restricted to A coincides with μ, and $\mu'\{z_o\} = 1$. Hence for $\delta' \in \mathbf{M}'$,

$$\int_{A'} \delta' \, d\mu' = \int_A \delta' \, d\mu + \delta'(z_o)\mu'\{z_o\} = 0$$

from (6). Hence δ' satisfies (5).

The correspondence between δ' and its restriction to A is 1–1 between $\mathbf{M}'$ and $\mathbf{M}$. The definitions of f' and μ' imply that preferability is preserved. ■

Briefly, the new problem is obtained from the old by adding a point z_o to A, making $\{z_o\}$ measurable, giving it measure *one,* setting $f = 0$ on it, and giving any δ a value on z_o that just cancels the surplus or deficit of $\int \delta \, d\mu$ on A. (A procedure similar to this is very common in finite problems, where it takes the form of adding "disposal activities," "slack variables," etc.)

This result is very useful because condition (5) is mathematically convenient. Our standard procedure will be as follows. The heavy mathematical work will be on problems with the special condition (5). Having obtained a result, we then go to the general problem. This is translated into a problem satisfying (5) by means of the recipe in the proof just given. The result is applied to the translated problem and usually yields a more general theorem for the general problem.

We now turn to a more specific system of feasibility conditions. Let two measurable functions b, $c{:}A \rightarrow$ extended reals be given, as well as two numbers, L_o and L^o, which may also be infinite. In terms of these, the feasible set consists of those functions δ that are measurable, real valued, satisfy (4), and also satisfy the two conditions

$$b \leq \delta \leq c \qquad (5.2.7)$$

$$L_o \leq \int_A \delta \, d\mu \leq L^o \qquad (5.2.8)$$

That is, for all $a \in A$, $\delta(a)$ satisfies the double inequality $b(a) \leq \delta(a) \leq c(a)$, and in addition its integral satisfies the double inequality (8). Allowing b, c, L_o or L^o to be infinite is simply a device for removing some of these constraints; e.g., setting $L_o = -\infty$ is the same as simply removing the left inequality in (8).

This system of constraints is very flexible; e.g., if the density δ must by its nature be nonnegative, this may be indicated formally by setting b identically zero. Or, if there is no lower limit to δ at any point, set b identically equal to $-\infty$. A total resource constraint that must be satisfied with equality, as in (1.2), is indicated by setting $L_o = L^o$ and both equal to total available net resources. The function c is an *investment capacity* constraint, limiting the amount of resource that can be squeezed into the various subsets of A. The function b is an *investment requirement* constraint in that it places a lower bound (positive, negative, or zero) on investment over the various subsets of A.

This is the feasibility system that will occupy most of our attention. We now prove a result specializing the theorem above and showing that this problem can be transformed into one with a simpler system of constraints.

Theorem: Given measure space (A, Σ, μ), μ σ-finite, measurable functions $f{:}A \times$ reals $\rightarrow$ reals, b, $c{:}A \rightarrow$ extended reals, and extended real numbers L_o, L^o, consider the problem of maximizing (1) over those measurable real-valued functions δ that satisfy (4), (7), and (8).

Then A', Σ', μ', f', b', c' exist, with analogous properties, from which the following problem is formulated: Maximize (1′) over measurable real-valued functions δ' satisfying (5′) and (7′). (The primes indicate that f', b', c', etc., are to be substituted for f, b, c, etc.) There is a 1–1 correspondence between the feasible sets of these two problems which preserves preferability relations.

Before going on to the proof, note the effect of this theorem. The constraint (8) is eliminated and replaced by (5), so that instead of being confined to the interval $[L_o, L^o]$, $\int_A \delta \, d\mu$ must be zero. This is very convenient mathematically.

Proof: Take an artificial point z_o not belonging to A, and define A', Σ', μ', and f' exactly as in the proof of the theorem above. Let b' and c' be the functions identical to b and c respectively, when restricted to A, and for which

$$b'(z_o) = -L^o, \qquad c'(z_o) = -L_o \qquad (5.2.9)$$

We show that with these definitions, the feasible sets of the original and transformed problems are in 1–1 correspondence. With each $\delta{:}A \rightarrow$ reals feasible for the original problem, associate the function $\delta'{:}A' \rightarrow$ reals that coincides with δ on A and satisfies (6) for $\delta'(z_o)$. Any such δ' is feasible for the transformed problem, since (5') follows from (6) just as in the preceding proof. Also, (7') is satisfied for all $a \in A$, and as for z_o the condition

$$b'(z_o) \leq \delta'(z_o) \leq c'(z_o) \tag{5.2.10}$$

is an immediate consequence of (6), (8), and (9). This proves δ' is feasible for the transformed problem.

Conversely, let δ' be transformed-feasible. Its restriction to A is then feasible for the original problem, since (8) follows from (9), (10), and (5'), and the other feasibility conditions are obviously satisfied. Furthermore, the function $\delta'':A' \to$ reals associated with the restriction of δ' is δ' itself; this follows again from (5'). We have proved that the original feasible set is mapped *onto* the transformed feasible set.

Finally, if two functions δ_1 and δ_2 are unequal, their extensions are obviously unequal. This proves we indeed have a 1–1 correspondence. That preferability relations are preserved follows from the way f' and μ' are defined. ∎

We make one final preliminary point. Given a measure space (A, Σ, μ), recall that a condition is said to hold *almost everywhere* or for *almost all* $a \in A$ iff there is a set $E \in \Sigma$ of measure zero (a *null* set) such that the condition holds for all $a \notin E$. (The case $E = \emptyset$ is not excluded; here the condition would hold for *all* $a \in A$.) Now let $\delta_1, \delta_2:A \to$ reals both be feasible and be equal almost everywhere: $\mu\{a \mid \delta_1(a) \neq \delta_2(a)\} = 0$. We can see that

$$\int f(a, \delta_1(a))\, \mu(da) = \int f(a, \delta_2(a))\, \mu(da)$$

and

$$\int \delta_1\, d\mu = \int \delta_2\, d\mu \tag{5.2.11}$$

so that δ_1 and δ_2 yield the same utility function and resource distribution. In effect, δ_1 and δ_2 are two different representations of the same solution (11). We could systematically ignore exceptions to rules occurring within null sets only; e.g., the constraint (7) could be weakened to $b(a) \leq \delta(a) \leq c(a)$ for *almost* all $a \in A$ without altering the problem in any essential way. In any case, we should be prepared to find the following discussion well seasoned with the phrases "almost everywhere," "almost all."

5.3. SUFFICIENT CONDITIONS FOR OPTIMALITY

A feature that characterizes a very wide class of optimization problems is the role played by "multipliers" or "shadow prices." These are numbers associated with the constraints of the problem from which special conditions are formed, either necessary or sufficient for optimality, and which sometimes allow us to transform the original problem into a new and simpler one. These "prices" are especially useful in economic and social science problems because they not only expedite the solution but suggest institutional arrangements that will lead the economy to carry out the solution in practice.

In the allocation-of-effort problem we would expect that a price could be associated with the total resource constraint in such a way that someone taking account of the cost of the resources allocated to the various projects (as well as the payoff from these projects) would be led to the optimal solution. A number of qualifications must be added, but this idea is a red thread running through the following results.

We first give a very general condition guaranteeing that a feasible solution is best. We are dealing with a partially ordered utility function, so the conclusion that δ is *best* is much stronger than the conclusion that it be merely *unsurpassed.*

Theorem: Let (A, Σ, μ) be a measure space, μ σ-finite; let $f:A \times$ reals $\rightarrow$ reals be measurable; and let **M** be a collection of measurable functions $\delta:A \rightarrow$ reals such that

$$\int_A \delta \, d\mu = 0 \qquad (5.3.1)$$

for all $\delta \in$ **M**. Let $\delta^o \in$ **M** be a function and p^o be a real number such that, for all $\delta \in$ **M**,

$$f(a, \delta^o(a)) - p^o \delta^o(a) \geq f(a, \delta(a)) - p^o \delta(a) \qquad (5.3.2)$$

for almost all $a \in A$.

Then δ^o is *best* for the problem of maximizing

$$\int f(a, \delta(a)) \, \mu(da) \qquad (5.3.3)$$

over $\delta \in$ **M** (maximization refers to standard ordering of pseudomeasures, here and throughout this discussion).

Proof: We must show that for any feasible δ,

$$\int_A [f(a, \delta^o(a)) - f(a, \delta(a))] \, \mu(da) \qquad (5.3.4)$$

is well defined as an ordinary definite integral and is ≥ 0.

From (2) we have

$$f(a, \delta^o(a)) - f(a, \delta(a)) \geq p^o[\delta^o(a) - \delta(a)]$$

almost everywhere. By (1) the integral of $p^o[\delta^o(a) - \delta(a)]$ is zero, and this fact is all we need. ∎

(2) is implied by the assertion that for each $a \in A$ (except possibly on a null set), the investment density $\delta^o(a)$ is chosen to maximize $f(a, x) - p^o x$ over the feasible investment levels x. The first term gives the payoff density, and the second reflects the "cost" of using up the resource. (Note that either f or p^o, or both, can be negative.)

We can derive a generalization of this result.

Theorem: Let A, Σ, μ, f, and **M** be as above, except that the feasibility condition (1) is replaced by the weaker condition,

$$\left| \int_A \delta \, d\mu \right| < \infty \qquad (5.3.5)$$

for all $\delta \in$ **M**. Let $\delta^o \in$ **M** be a function and p^o be a real number such that (2) holds, and also

$$p^o \int_A (\delta^o - \delta) \, d\mu \geq 0 \qquad (5.3.6)$$

for all $\delta \in$ **M**. Then δ^o is best for the problem of maximizing (3) over $\delta \in$ **M**.

Proof: We take an artificial point z_0 and transform this problem into its equivalent on $A' = A \cup \{z_0\}$ (see Section 5.2). This translated problem is in proper form for the theorem just given, and we need merely verify that condition (2) holds on $A \cup \{z_0\}$. By assumption this is true on A. For the point z_0 we have that $f(z_0)$ is identically zero, $\delta'(z_0) = -\int_A \delta \, dv$, and similarly for $\delta^{o'}(z_0)$. Condition (2) for the point z_0 is then precisely condition (6). ∎

The extra imposed condition (6) is easy to interpret. If p^o is positive, there is no feasible δ for which $\int_A \delta \, d\mu > \int_A \delta^o \, d\mu$. In commonsense terms, this states that if the resource is valuable, as much as possible should be allocated. If p^o is negative, the "resource" is illth rather than wealth, and as little as possible should be used. Finally, if $\int_A \delta^o \, d\mu$ is neither the highest nor the lowest possible value attainable, then (6) implies $p^o = 0$. In this case the resource is a "free good," and for each $a \in A$ we simply choose $\delta_o(a)$ to maximize $f(a, x)$ over attainable investment levels x without worrying about resource cost.

We can also find a sufficient condition for δ^o to be the unique best solution. But we must be careful in interpreting the concept of "uniqueness" here. According to previous discussion we may identify two functions δ_1 and δ_2 that are equal almost everywhere. Let us say that δ_1 and δ_2 are *essentially distinct* iff $\mu\{a \mid \delta_1(a) \neq \delta_2(a)\} > 0$. Then, in line with our discussion, we say that δ^o is the *unique* best solution iff δ^o is best and there is no essentially distinct δ that is also best. There may obviously be more than one best solution. For example, if f is identically zero, any feasible solution is best.

We now give the uniqueness condition. Going back to the first sufficiency theorem, suppose that all the premises hold and in addition the following: for each $\delta \in M$ that is essentially distinct from δ^o, there is a set E_δ of positive measure such that (2) holds with *strict* inequality for all $a \in E_\delta$. Then δ^o is the *unique* best solution.

The proof is simple. Reasoning as before, we find the integral (4) is *positive,* not merely nonnegative. Hence $\delta^o > \delta$ for all $\delta \in M$ essentially distinct from δ^o.

This uniqueness condition immediately generalizes to the case where (1) is replaced by (5). Let all the premises of the second sufficiency theorem hold and, in addition, the following: for each $\delta \in M$ essentially distinct from δ^o, *either* there is a set E_δ as above *or* (6) holds with strict inequality (or both). Then δ^o is the unique best solution.

The proof consists in translating this problem into the one in which (1) holds, then applying the E_δ condition to this translated problem. In doing so, note that the singleton set $\{z_o\}$ has positive measure $\mu'\{z_o\} = 1$. This shows that strict inequality in (6) for all δ insures uniqueness.

Finally, we mention a much weaker sufficient condition for δ^o being best. In the theorem above p^o is chosen in advance and (2) must be satisfied for all δ. But it suffices that for any feasible δ there exists a p^o (which may be different for different δ's) such that (2) is satisfied for this p^o and δ. This variable p^o lacks the appeal of the shadow price interpretation, however.

5.4. NECESSARY CONDITIONS FOR OPTIMALITY

The sufficient conditions we have just stated are very convenient to use where they hold. Unfortunately, it is easy to find problems whose optimal solution does not satisfy any such condition; i.e., the stated conditions are not always *necessary*.

Here is a simple example. Let A consist of two points: $A = \{a_1, a_2\}$, Σ = all subsets, $\mu = 1$ on each point; let the payoff function be given by $f(a_1, x) = x^2$, $f(a_2, x) = -2x^2$; the investment function is given by (x_1, x_2), with the constraint $x_1 + x_2 = 0$. Thus the problem is of the finite form (1.1)–(1.2): maximize

$$x_1^2 - 2x_2^2 \qquad (5.4.1)$$

subject to

$$x_1 + x_2 = 0 \qquad (5.4.2)$$

The unique best solution is obviously $x_1 = x_2 = 0$. Now the sufficient condition (3.2) requires that there be a real number p^o such that (for point a_1) $0 \geq x^2 - p^o x$ for all real x. Obviously there is no such p^o. (However, this example does not violate the weaker condition mentioned at the end of Section 5.3.)

We now investigate necessary conditions for optimality. Broadly speaking, these have been found more useful than sufficient conditions because it is hard to find simple sufficient conditions that cover a very wide range of problems. Establishing necessary conditions is much more difficult than establishing sufficient conditions. Compare the length of proofs in this section with that preceding.

The classical example of a necessary condition is that a function maximized in the interior of a domain have a derivative of zero at the maximizing point if it is differentiable there. Necessary conditions in general are used just as this one is; viz., we narrow the search for an optimal solution to those (hopefully few) points that satisfy the necessary condition and then try by other means to test them directly for optimality.

We concentrate on the special class of allocation problems discussed above, where the feasible solutions are those lying between two functions $b, c:A \to$ extended reals and integrating to zero. Afterward, some of the key results will be generalized to the problem where the constraint $L_o \leq \int_A \delta \, d\mu \leq L^o$ replaces the condition that the integral be zero. A number of preliminary concepts and lemmas will be needed before we can get down to serious business.

Recall that the *supremum* of a set of extended real numbers is the smallest extended real number not less than any of the numbers in the set (it is the "least upper bound" of the set). The supremum of a *function* $g:A \to$ extended reals is defined to be the supremum of its range $\{g(a) \mid a \in A\}$.

We use the notation $g \mid E$ to represent the restriction of g to the subdomain $E \subseteq A$. The supremum of $g \mid E$ is less than or equal to the supremum of g itself. Suppose the domain A of g is the universe set of a measure space (A, Σ, μ). We consider all possible restrictions $g \mid E$ such that $\mu(A \backslash E) = 0$ and take the supremum of each one. The infimum of the resulting set of extended real numbers is called the *essential supremum* of g. To put it another way, the essential supremum of g is the largest extended real number x such that $x \leq \sup \{g(a) \mid a \in E\}$ for all $E \in \Sigma$ such that $\mu(A \backslash E) = 0$.

One special case may be noted. If μ is the identically zero measure, A itself is a null set, and we may take $E = \phi$. The range of $g \mid E$ is then the empty set ϕ. Applying the definition of supremum literally yields sup $\phi = -\infty$. Thus the essential supremum of any function g is $-\infty$ in this case.

The *essential infimum* of a function $g{:}A \rightarrow$ extended reals, with respect to (A, Σ, μ), is defined analogously. Just switch the words "supremum" and "infimum," and the words "greatest" and "least," in the preceding discussion. Or, equivalently, we could define it by the rule:

$$\text{essential infimum of } g = -\text{essential supremum } (-g)$$

We use the standard abbreviations "ess sup" and "ess inf" for these concepts.

The following result, whose proof is omitted, will be needed later.

Lemma: Given measure space (A, Σ, μ) and function $g{:}A \rightarrow$ extended reals, let $\{A_0, A_1, \ldots\}$ be a countable measurable partition (or even just a covering) of A. Then

$$\text{ess sup } g = \sup\{\text{ess sup}(g \mid A_n) \mid n = 0, 1, \ldots\} \qquad (5.4.3)$$

$$\text{ess inf } g = \inf\{\text{ess inf}(g \mid A_n) \mid n = 0, 1, \ldots\} \qquad (5.4.4)$$

Here ess sup$(g \mid A_n)$ refers to the function $g \mid A_n$ and the measure space (A_n, Σ_n, μ_n), which is the restriction of (A, Σ, μ) to A_n, and similarly for ess inf$(g \mid A_n)$.

Note that measurability of g is nowhere mentioned. Indeed, these concepts are perfectly well defined, and the lemma correct, for any function g, measurable or not. This is important because the functions g and h which we define below are not necessarily measurable.

Next let f be a real-valued function whose domain is an interval of real numbers $[b, c]$. The endpoints of the interval may or may not be included, and we may have $b = c$. Function f is *continuous* at the point $x_o \in [b, c]$ iff for any sequence $x_1, x_2, \ldots$ of points of $[b, c]$ whose limit is x_o and any number $\epsilon > 0$ there is an integer N such that for all $n > N$

$$-\epsilon < f(x_o) - f(x_n) < \epsilon \qquad (5.4.5)$$

Now (5) is a double inequality. If we drop the left inequality in (5), keeping everything else the same, we get a weaker concept; f is then said to be *lower semicontinuous* at the point x_o. Intuitively, lower semicontinuity at x_o allows f to take a "sudden" jump downward, but not upward, at x_o. (Continuity prohibits sudden jumps in either direction.) Function f is *continuous,* or *lower semicontinuous,* iff it is continuous, or lower semicontinuous, at every point of its domain respectively.

There is another way to characterize lower semicontinuity that is more useful (though less intuitive) then the definition just given. An *open interval* of real numbers is an interval not containing its endpoints (this is the same as an *open disc* on the real line). An *open set* on $[b, c]$ is the intersection of $[b, c]$ with any union of open intervals. Then $f:[b, c] \to$ reals is lower semicontinuous iff $\{x \mid f(x) > y\}$ is an open set on $[b, c]$ for all real numbers y. We omit the proof that the two lower semicontinuity concepts are the same.

We need the following result.

Lemma: Let $f:[b, c] \to$ reals be lower semicontinuous with $b < c$. Then

$$\sup f = \sup(f \mid E) \qquad (5.4.6)$$

where E is the set of *rational* numbers in $[b, c]$.

Proof: Obviously, $\sup f \geq \sup(f \mid E)$. Conversely, for any number $y < \sup f$, the set $\{x \mid f(x) > y\}$ is open and nonempty; hence there is a rational number x_o belonging to it: $f(x_o) > y$. ∎

Next we need several concepts related to but more general than the concept of derivative of a function. Let $f:[b, c] \to$ reals again have a real interval domain, which may or may not include its endpoints, and let x be a point of the domain $\neq c$.

Definition: The *lower right derivate* of f at the point x is the limit, as ϵ goes to zero from above, of

$$\inf\{[f(x + y) - f(x)]/y \mid 0 < y < \epsilon\} \qquad (5.4.7)$$

That is, given $\epsilon > 0$, we find for each point y in the open interval[3] $(0, \epsilon)$ the value of $[f(x + y) - f(x)]/y$ (which is the average slope of f from x to $x + y$) and take the infimum of this set of values. Having done this for each $\epsilon > 0$, we take the limit as $\epsilon \to 0$.

This concept is well defined for *any* real-valued function and *any* domain point except the right endpoint c, but it may take on an infinite value. For the infimum of any set of real numbers is some extended real number, so (7) is well defined for fixed $\epsilon > 0$. Furthermore, one sees that (7) is nondecreasing as $\epsilon \to 0$ and hence has a limit in the extended real numbers. This of course contrasts with differentiability, which is not a universal property.

---

3. To be precise, we never choose ϵ larger than $c - x$. This insures that $x + y$ remains in the domain of f, so $f(x + y)$ is well defined.

We use the notation $D_+ f(x)$ for the lower right derivate of f at x. If the right endpoint c is part of the domain, we make the convention that $D_+ f(c) = -\infty$.

Similarly for all domain points x, other than the left endpoint b, we have the following.

Definition: The *upper left derivate* of f at the point x is the limit, as ϵ goes to zero from above, of

$$\sup\{[f(x) - f(x - y)]/y \mid 0 < y < \epsilon\} \qquad (5.4.8)$$

An argument similar to that just given shows this is always well defined in the extended real numbers. We use the notation $D^- f(x)$ for the upper left derivate of f at x. If the left endpoint b is in the domain, we make the convention that $D^- f(b) = +\infty$.

One may also define the upper right derivate: replace inf by sup in (7); and the lower left derivate: replace sup by inf in (8). But we do not need these concepts. A function is *differentiable* at interior point x iff all four of these quantities are equal; their common value is then the *derivative* of f at x.[4]

Next we need the following result concerning atomic measures.

Lemma: Let (A, Σ, μ) be a measure space with μ atomic; let $(f_1, f_2, \ldots)$ be a sequence of measurable functions on A, taking values in the extended real numbers. Then there is *exactly one* sequence of extended real numbers $(x_1, x_2, \ldots)$ satisfying

$$\mu\{a \mid f_n(a) = x_n \quad \text{for all} \quad n = 1, 2, \ldots\} \neq 0 \qquad (5.4.9)$$

Proof: By definition of "atomic," exactly one of the two numbers $\mu(E), \mu(A \backslash E)$ is zero for any choice of $E \in \Sigma$. Clearly there can be at *most* one sequence of numbers satisfying (9), hence we must show there is at *least* one.

Take any measurable function f and consider the supremum x_o of the set of numbers x satisfying

$$\mu\{a \mid f(a) < x\} = 0 \qquad (5.4.10)$$

If $x_o > -\infty$, take a sequence (x_n) rising to x_o. Since (10) holds for each x_n, it holds for x_o itself. It also obviously holds if $x_o = -\infty$. Now consider the condition

$$\mu\{a \mid f(a) > x\} = 0 \qquad (5.4.11)$$

---

4. On derivates see E. J. McShane and T. A. Botts, *Real Analysis* (Van Nostrand, Princeton, 1959), pp. 110–11.

This obviously holds for x_o if $x_o = \infty$. If $x_o < \infty$, take a sequence (x_n) decreasing to x_o; then (11) holds for each such $x_n > x_o$ (since the complement of the set in (11) has positive measure). Hence (11) holds again for x_o. We have thus shown that

$$\mu\{a \mid f(a) \neq x_o\} = 0 \tag{5.4.12}$$

Now for each f_n let x_n be the corresponding number satisfying (12). The set $\{a \mid f_n(a) \neq x_n$ for at least one $n = 1, 2, \ldots\}$ is the union of the countable collection of null sets $\{a \mid f_n(a) \neq x_n\}$, $n = 1, 2, \ldots$, hence has measure zero itself. Therefore its complement is not a null set. But this is exactly the statement (9). ∎

Let us now get down to business. The allocation problem will be determined by the measure space (A, Σ, μ), the payoff function $f:A \times$ reals $\rightarrow$ reals, and the lower and upper capacity functions b, $c:A \rightarrow$ extended reals.

We make the following convention. If a specific point $a \in A$ is chosen, $f(a, \cdot)$ is a function of a real variable. In referring to it, we always take this function to be restricted to the interval $[b(a), c(a)]$. (The endpoint $b(a)$ is to be included iff it is finite, and similarly for $c(a)$.) Thus the statement that $f(a, \cdot)$ is lower semicontinuous refers to this function with the domain $[b(a), c(a)]$, and similarly for derivates of this function; in particular we have at the endpoints that $D^- f(a, b(a)) = +\infty$ and $D_+ f(a, c(a)) = -\infty$.

To explain the formulation of the following result, recall that since μ is σ-finite, there is a countable measurable partition $\{A_0, A_1, \ldots\}$ of A such that μ restricted to A_0 is nonatomic, while each A_n, $n = 1, 2, \ldots$, is an *atom* (i.e., μ restricted to A_n is atomic). Set A_0 is the *nonatomic* part of A, and $A \setminus A_0$ is the *atomic part*. This partition is unique up to null sets. (We do not assume that atoms are simply concentrated. Doing so would simplify some of the following proofs.)

We postulate that the given feasible density δ^o is *unsurpassed* in the allocation problem. This is a weaker assumption, and therefore it yields a stronger result, than if we postulated that δ^o were *best*.

The following is called a lemma rather than a theorem because it lacks immediate intuitive appeal. The result is quite powerful, however, and implies all the other results we obtain in this section.

Lemma: Let (A, Σ, μ) be a measure space with μ σ-finite; let $f:A \times$ reals $\rightarrow$ reals, and let b, $c:A \rightarrow$ extended reals be measurable functions. Let feasible δ^o be unsurpassed for the problem of maximizing

$$\int f(a, \delta(a)) \, \mu(da) \tag{5.4.13}$$

subject to $\delta:A \to$ reals being measurable and satisfying

$$b \le \delta \le c \qquad (5.4.14)$$

and

$$\int_A \delta \, d\mu = 0 \qquad (5.4.15)$$

Let A_0, $A\backslash A_0$ be the nonatomic and atomic parts of A respectively. Let $f(a, \cdot)$ be lower semicontinuous for all $a \in A_0$.

Define two functions g, $h:A \to$ extended reals as follows: For $a \in A_0$,

$$g(a) = \sup\{[\,f(a, \delta^o(a) + y) - f(a, \delta^o(a))]/y\} \qquad (5.4.16)$$

the supremum being taken over all y in the open interval $(0, \ c(a) - \delta^o(a))$;

$$h(a) = \inf\{[\,f(a, \delta^o(a)) - f(a, \delta^o(a) - y)]/y\} \qquad (5.4.17)$$

the infimum being taken over all y in the open interval $(0, \delta^o(a) - b(a))$. For $a \in A\backslash A_0$,

$$g(a) = D_+ f(a, \delta^o(a)) \qquad (5.4.18)$$

$$h(a) = D^- f(a, \delta^o(a)) \qquad (5.4.19)$$

Then for any pair of disjoint sets $G, H \in \Sigma$, we have

$$\operatorname{ess\,sup}(g \mid G) \le \operatorname{ess\,inf}(h \mid H) \qquad (5.4.20)$$

Proof: It suffices to prove (20) for the special case where $G \subseteq A_m$, $H \subseteq A_n$, for some $m, n = 0, 1, 2, \ldots$. For suppose this has been proved for each m, n, and let G, H be any disjoint measurable sets. By assumption we have

$$\operatorname{ess\,sup}[g \mid (G \cap A_m)] \le \operatorname{ess\,inf}[h \mid (H \cap A_n)]$$

for all $m, n = 0, 1, 2, \ldots$. Then by (3) and (4),

$$\operatorname{ess\,sup}(g \mid G) = \sup_m [\operatorname{ess\,sup}(g \mid (G \cap A_m))]$$

$$\le \inf_n [\operatorname{ess\,inf}(h \mid (H \cap A_n))] = \operatorname{ess\,inf}(h \mid H)$$

since $(G \cap A_m)$, $m = 0, 1, \ldots$, partitions G, and $(H \cap A_n)$, $n = 0, 1, \ldots$, partitions H. Thus (20) would be proved in general.

There are four cases to consider:

(i) $G \subseteq A_0, H \subseteq A_0$;
(ii) $G \subseteq A_m, H \subseteq A_n$ for $m, n \ne 0$;

(iii) $G \subseteq A_m, H \subseteq A_0$ for $m \neq 0$;
(iv) $G \subseteq A_0, H \subseteq A_n$ for $n \neq 0$.

In each case we assume that (20) is *false* and show there is a feasible δ that surpasses δ^o, giving a contradiction.

Case (i). $G \subseteq A_0, H \subseteq A_0$:
For each positive real number y define the functions $g_y, h_y : A \to$ extended reals by

$$g_y(a) = [f(a, \delta^o(a) + y) - f(a, \delta^o(a))]/y \qquad (5.4.21)$$

if $c(a) - \delta^o(a) > y$; $= -\infty$ otherwise;

$$h_y(a) = [f(a, \delta^o(a)) - f(a, \delta^o(a) - y)]/y \qquad (5.4.22)$$

if $\delta^o(a) - b(a) > y$; $= +\infty$ otherwise.

These are all measurable functions, since f, b, c, and δ^o are all measurable. Furthermore, when restricted to A_0,

$$g = \sup g_y, \qquad h = \inf h_y \qquad (5.4.23)$$

the sup and inf being taken over all positive *real* y.

We now show that (23) remains true even if y merely ranges over the positive *rational* numbers. Take a point $a \in A_0$ and consider $g_y(a)$ as a function of y with domain the open interval $(0, c(a) - \delta^o(a))$. Since f is lower semicontinuous, it follows easily that the y-function $g_y(a)$ is lower semicontinuous (use the definition given by the right half of (5)). If $c(a) > \delta^o(a)$, it follows from (6) that (23) is true for g when the supremum is taken over the positive rational values of y. If $c(a) = \delta^o(a)$, this is still true, the two suprema both being $-\infty$. This proves the contention for g.

As for $h(a)$, we consider $h_y(a)$ as a function of y, with domain $(0, \delta^o(a) - b(a))$. We then verify that *minus* $h_y(a)$ is lower semicontinuous. We then verify as above that $\sup(-h_y(a))$ is the same whether taken over positive real y or positive rational y. But $\sup(-h_y(a)) = -\inf(h_y(a)) = -h(a)$, which proves the contention for h.

Now assuming that (20) is false, choose a real number x satisfying

$$\text{ess sup}(g \mid G) > x > \text{ess inf}(h \mid H) \qquad (5.4.24)$$

and let

$$G' = G \cap \{a \mid g(a) > x\}, \qquad H' = H \cap \{a \mid h(a) < x\}$$

$$G_y = G \cap \{a \mid g_y(a) > x\}, \qquad H_y = H \cap \{a \mid h_y(a) < x\} \qquad (5.4.25)$$

for all positive real y.

Since we are operating within A_0, (23) is true for y ranging over the positive rationals. Hence $G' = \cup G_y$ and $H' = \cup H_y$, the union taken over the positive rationals. As unions of a countable number of measurable sets, G' and H' are themselves measurable. Also $\mu(G') > 0$, $\mu(H') > 0$; for if not, (24) would be false. It follows that there must be positive (rational) numbers y_1 and y_2 such that

$$\mu(G_{y_1}) > 0, \qquad \mu(H_{y_2}) > 0 \tag{5.4.26}$$

since a countable union of null sets is a null set.

Now μ, being σ-finite and nonatomic on G_{y_1}, takes on all values between 0 and $\mu(G_{y_1})$ on this set, and similarly for H_{y_2}. Hence, from (26) we can find measurable subsets $G'' \subseteq G_{y_1}$ and $H'' \subseteq H_{y_2}$ such that

$$y_1 \mu(G'') = y_2 \mu(H'') \tag{5.4.27}$$

with this common value being positive and finite.

We are now ready to construct another feasible density δ^{oo}, which will surpass δ^o. Let

$$\delta^{oo}(a) = \delta^o(a) + y_1, \qquad \text{if } a \in G''$$

$$\delta^{oo}(a) = \delta^o(a) - y_2, \qquad \text{if } a \in H'' \tag{5.4.28}$$

$$\delta^{oo}(a) = \delta^o(a), \qquad \text{if } a \notin G'' \cup H''$$

This is well defined, since G and H (hence G'' and H'') are disjoint. First, we verify that δ^{oo} is feasible; it is real valued and measurable. Also,

$$\int_A (\delta^{oo} - \delta^o) \, d\mu = y_1 \mu(G'') - y_2 \mu(H'') = 0$$

from (28) and (27). Thus it satisfies condition (15).

Next, if $a \in G''$, then $g_{y_1}(a) > x > -\infty$, by (25). Hence $y_1 < c(a) - \delta^o(a)$, by (21). This means that adding y_1 to $\delta^o(a)$ on G'' does not violate the feasibility condition $\delta(a) \le c(a)$. Similarly, if $a \in H''$, then $h_{y_2}(a) < x < \infty$, by (25). Hence $y_2 < \delta^o(a) - b(a)$, by (22), so that subtracting y_2 on H'' does not violate the feasibility condition $\delta(a) \ge b(a)$. Thus (14) remains satisfied, and δ^{oo} is feasible.

Comparing utility from δ^{oo} and δ^o, we have

$$\int_A [f(a, \delta^{oo}(a)) - f(a, \delta^o(a))] \, \mu(da) = \int_{G''} y_1 g_{y_1} \, d\mu - \int_{H''} y_2 h_{y_2} \, d\mu$$

from (28), (21), and (22),

$$> \int_{G''} y_1 x \, d\mu - \int_{H''} y_2 x \, d\mu$$

since $G'' \subseteq G_{y_1}$ and $H'' \subseteq H_{y_2}$ and $\mu(G'')$, $\mu(H'')$ are positive,

$$= xy_1\mu(G'') - xy_2\mu(H'') = 0$$

from (27). Thus δ^{oo} is preferred to δ^o under standard ordering of pseudo-measures. The denial of (20) thus contradicts the premise that δ^o is unsurpassed. This establishes (20) for $G \subseteq A_0, H \subseteq A_0$.

Case (ii). $G \subseteq A_m, H \subseteq A_n$, where $m, n \neq 0$:
As above, we assume (20) is false, so that

$$\text{ess sup}(g \mid G) > \text{ess inf}(h \mid H) \tag{5.4.29}$$

For this to hold, G and H must have positive measure, so that they are atoms. Applying (9) to the function $c - \delta^o$ restricted to G, there must be a unique constant ϵ_0 such that

$$\mu[G \cap \{a \mid c(a) - \delta^o(a) = \epsilon_0\}] > 0 \tag{5.4.30}$$

Since $c - \delta^o$ is nonnegative, $\epsilon_0 \geq 0$. In fact, $\epsilon_0 > 0$; for if $c(a) = \delta^o(a)$, then $g(a) = D_+f(a, \delta^o(a)) = -\infty$; hence $\epsilon_0 = 0$ would imply ess sup$\cdot$ $(g \mid G) = -\infty$, contrary to (29). A similar argument shows there is a unique positive constant η_0 such that

$$\mu[H \cap \{a \mid \delta^o(a) - b(a) = \eta_0\}] > 0 \tag{5.4.31}$$

Label the sets in (30) and (31) by G', H' respectively. These are subatoms of G and H. A measure that is atomic and σ-finite is bounded, so that $\infty > \mu(G') > 0, \infty > \mu(H') > 0$.

Now take two sequences of positive numbers, beginning with ϵ_0, η_0 respectively and decreasing to zero:

$$\epsilon_0 > \epsilon_1 > \epsilon_2 > \cdots, \lim \epsilon_n = 0, \qquad \eta_0 > \eta_1 > \eta_2 > \cdots, \lim \eta_n = 0$$

and chosen so that

$$\epsilon_n\mu(G') = \eta_n\mu(H') \tag{5.4.32}$$

for all $n > 0$.

Consider the sequence of functions g_{ϵ_n}, $n = 1, 2, \ldots$, given by (21), all restricted to the domain G'. The condition $\epsilon_n < \epsilon_0$ guarantees that these are all finite. Similarly, consider the sequence h_{η_n}, $n = 1, 2, \ldots$, given by (22), all restricted to the domain H'. These are also all finite, since $\eta_n < \eta_0$.

Applying (9) once again, we find there must be two unique sequences of constants, say $c_1, c_2, \ldots$ and $b_1, b_2, \ldots$ such that

$$\mu[G' \cap \{a \mid g_{\epsilon_n}(a) = c_n \quad \text{for all} \quad n \subseteq 1, 2, \ldots\}] > 0 \tag{5.4.33}$$

$$\mu[H' \cap \{a \mid h_{\eta_n}(a) = b_n \quad \text{for all} \quad n = 1, 2, \ldots\}] > 0 \tag{5.4.34}$$

These constants must be finite.

Label the sets in (33) and (34) by G'', H'' respectively. Also let

$$c_o = \lim \inf c_n = \lim \inf g_{\epsilon_n}(a) \tag{5.4.35}$$

$$b_o = \lim \sup b_n = \lim \sup h_{\eta_n}(a) \tag{5.4.36}$$

Equation (35) is valid for all $a \in G''$, (36) for all $a \in H''$. We also have

$$\lim \inf g_{\epsilon_n}(a) \geq D_+ f(a, \delta^o(a)) \tag{5.4.37}$$

$$\lim \sup h_{\eta_n}(a) \leq D^- f(a, \delta^o(a)) \tag{5.4.38}$$

for $a \in G''$, $a \in H''$ respectively. Inequality (37) results from the fact that D_+ is the limit of infima taken over entire intervals, while the left side of (37) is the limit of infima taken over subsets of these intervals (viz., the points ϵ_n). A similar argument establishes (38).

Now the right-hand sides of (37) and (38) are nothing but g and h respectively. Hence

$$c_o \geq \operatorname{ess\,sup}(g \mid G'') = \operatorname{ess\,sup}(g \mid G)$$
$$> \operatorname{ess\,inf}(h \mid H) = \operatorname{ess\,inf}(h \mid H'') \geq b_o \tag{5.4.39}$$

from (35), (37), (29), (36), and (38).

Since $c_o > b_o$, an n' must exist such that $c_{n'} > b_{n'}$. We are now ready to construct a feasible density δ^{oo} that surpasses δ^o. Let

$$\delta^{oo}(a) = \delta^o(a) + \epsilon_{n'}, \quad \text{if } a \in G''$$
$$\delta^{oo}(a) = \delta^o(a) - \eta_{n'}, \quad \text{if } a \in H'' \tag{5.4.40}$$
$$\delta^{oo}(a) = \delta^o(a), \quad \text{if } a \notin G'' \cup H''$$

This is feasible. First,

$$\int_A (\delta^{oo} - \delta^o) \, d\mu = \epsilon_{n'} \mu(G'') - \eta_{n'} \mu(H'')$$

$$= \epsilon_{n'} \mu(G') - \eta_{n'} \mu(H') = 0$$

from (32). Second, for $a \in G''$, $\delta^o(a) + \epsilon_{n'} < \delta^o(a) + \epsilon_0 = c(a)$, so the upper bound constraint $\delta \leq c$ remains satisfied. Similarly, $\delta^o(a) - \eta_{n'} > \delta^o(a) - \eta_0 = b(a)$ for $a \in H''$, so the lower bound constraint $\delta \geq b$ remains satisfied. This proves feasibility.

Comparing utilities, we obtain

$$\int_A [f(a, \delta^{oo}(a)) - f(a, \delta^o(a))] \, \mu(da) = \int_{G''} \epsilon_{n'} g_{\epsilon_{n'}} \, d\mu - \int_{H''} \eta_{n'} h_{\eta_{n'}} \, d\mu$$

from (40), (21), and (22),

$$= \epsilon_{n'} c_{n'} \mu(G'') - \eta_{n'} b_{n'} \mu(H'')$$

from (33) and (34),

$$> b_{n'}[\epsilon_{n'}\mu(G'') - \eta_{n'}\mu(H'')] = 0$$

from (32), since $c_{n'} > b_{n'}$, $\mu(G'') = \mu(G')$, $\mu(H'') = \mu(H')$, and all terms are positive. Thus δ^{oo} is preferred to δ^o under standard ordering of pseudomeasures. The denial of (20) again leads to contradiction.

Case (iii). $G \subseteq A_m, H \subseteq A_0, m \neq 0$:

The proof for this case combines the techniques of the two preceding cases, and we merely outline the procedure. As before we assume (20) is false. Let x be a real number satisfying

$$\text{ess sup}(g \mid G) > x > \text{ess inf}(h \mid H) \qquad (5.4.41)$$

Reasoning as in case (i), we can find a positive (rational) number y_2 such that $\mu(H_{y_2}) > 0$, where $H_{y_2} = H \cap \{a \mid h_{y_2}(a) < x\}$, and h_{y_2} is given by (22).

Next we find the number ϵ_0 and the set G' as in (30), and again take a positive sequence decreasing to zero: $\epsilon_0 > \epsilon_1 > \epsilon_2 > \ldots$; lim $\epsilon_n = 0$. As in case (ii), we find a unique sequence of real numbers c_1, $c_2, \ldots$ such that $\mu(G'') > 0$, where

$$G'' = G' \cap \{a \mid g_{\epsilon_n}(a) = c_n \quad \text{for all} \quad n = 1, 2, \ldots\}$$

(cf. (33)). Continuing, we find that (cf. (39))

$$\liminf c_n \geq \text{ess sup}(g \mid G) \qquad (5.4.42)$$

Now choose n' so large that

$$\epsilon_n < y_2\mu(H_{y_2})/\mu(G'') \qquad (5.4.43)$$

for all $n > n'$. This can always be done, since lim $\epsilon_n = 0$ and the right side of (43) is positive ($\mu(G'')$ is finite, since μ is σ-finite and atomic on G''). Inequalities (42) and (41) imply that there is an $n'' > n'$ for which

$$c_{n''} > x \qquad (5.4.44)$$

Since μ is σ-finite and nonatomic on H_{y_2}, it takes on all values between 0 and $\mu(H_{y_2})$. Hence there is a measurable subset $H'' \subseteq H_{y_2}$ such that

$$\epsilon_{n''}\mu(G'') = y_2\mu(H'') \qquad (5.4.45)$$

since $\epsilon_{n''}$ satisfies inequality (43). The common value in (45) is positive and finite.

We now construct a new feasible density δ^{oo} as follows:

$$\delta^{oo}(a) = \delta^o(a) + \epsilon_{n''}, \qquad \text{if } a \in G''$$

$$\delta^{oo}(a) = \delta^o(a) - y_2, \qquad \text{if } a \in H''$$

$$\delta^{oo}(a) = \delta^o(a), \qquad\qquad \text{if } a \notin G'' \cup H''$$

To show feasibility, we have, first,

$$\int_A (\delta^{oo} - \delta^o) \, d\mu = \epsilon_{n''} \mu(G'') - y_2 \mu(H'') = 0$$

from (45). Arguments already given in cases (i) and (ii) show that the bounding constraints $b \le \delta \le c$ remain valid for δ^{oo}; hence it is feasible. Comparing utilities,

$$\int_A [f(a, \delta^{oo}(a)) - f(a, \delta^o(a))] \, \mu(da) = \int_{G''} \epsilon_{n''} g_{\epsilon_{n''}} \, d\mu - \int_{H''} y_2 h_{y_2} \, d\mu$$

$$> \epsilon_{n''} c_{n''} \mu(G'') - y_2 x \mu(H'') > x[\epsilon_{n''} \mu(G'') - y_2 \mu(H'')] = 0$$

from (21), (22), (25), (44), and (45). Hence δ^{oo} is preferred to δ^o under standard ordering of pseudomeasures. This contradiction establishes (20) for case (iii).

Case (iv). $G \subseteq A_0, H \subseteq A_n, n \ne 0$:
This is completely symmetric with case (iii): One finds a set G_{y_1} as in case (i), a sequence $b_1, b_2, \dots$ as in (ii), etc. Details are left as an exercise.

Thus (20) has been verified in all four special cases. By the argument that begins this proof, (20) is now established in general. ∎

This has been a long and tedious proof, and the result itself does not look prepossessing. A few intuitive remarks may be in order, then, to indicate what the lemma says and why it "should" be true.

The function g represents, roughly, the return per unit increase in investment at various points of A. The function h represents, again roughly, the loss in return per unit disinvestment. If (20) is violated for a pair of disjoint sets G, H, this means there are subsets G', H', of positive measure such that g on G' is higher than h on H'. Then if we transfer some mass from H' to G', the net gain on the latter set outweighs the net loss on the former, resulting in a new feasible density that surpasses δ^o. The main burden of the proof just given is to find the appropriate subsets and the appropriate mass to transfer.

We stress the generality of this result. The only special conditions imposed on A, Σ, μ, b, c, or f, aside from measurability, are that μ be σ-finite, f real, and $f(a, \cdot)$ lower semicontinuous for $a \in A_0$. These are very weak conditions and will nearly always be satisfied in practice.

Our aim now is to use this result to derive the existence of a number that behaves somewhat like a shadow price.

Lemma: Let (A, Σ, μ), f, b, c, and δ^o satisfy the conditions of the preceding lemma, with δ^o unsurpassed for the problem of maximizing (13) subject to (14) and (15). Let the functions g, $h:A \rightarrow$ extended reals be defined as above by (16)–(19).

Then an extended real number p^o and a set $E \in \Sigma$ exist such that E is either an atom or a null set, and

$$g(a) \leq p^o \leq h(a) \qquad (5.4.46)$$

for all $a \in A \backslash E$. This may be expressed by the statement: Except for at most one atom, (46) is true almost everywhere.

Proof: Let $\{A_0, A_1, \ldots\}$ be the decomposition of A into nonatomic part A_0 and atoms $A_1, A_2, \ldots$. Let $g^n = \text{ess sup }(g \mid A_n)$, $h^n = \text{ess inf }(h \mid A_n)$ for $n = 0, 1, \ldots$. The preceding lemma then states that

$$g^m \leq h^n \qquad (5.4.47)$$

for all $m, n = 0, 1, \ldots$ *such that* $m \neq n$. It gives no information if $m = n$, since then there is no disjointness of A_m, A_n.

First we suppose that

$$\sup g^m \leq \inf h^n \qquad (5.4.48)$$

Choose any number p^o between these bounds. Then (46) is true for all $a \in A_n$ except possibly for a null set $E_n \subset A_n$, $n = 0, 1, \ldots$. Hence (46) is true everywhere except for the set $E = \cup E_n$. Since E is a null set, the lemma is established if (48) is true.

Now let (48) be false. There must be a pair of indices m' and n' for which $g^{m'} > h^{n'}$. By (47) these indices must be equal: $m' = n'$. But then for all $m \neq n'$, $n \neq n'$, we have

$$g^m \leq h^{n'} < g^{n'} \leq h^n \qquad (5.4.49)$$

by (47) again. Hence if we *exclude* the set $A_{n'}$, (48) will be reestablished for the rest of A. Thus (46) will be true almost everywhere on $A \backslash A_{n'}$.

It remains to prove only that the anomalous $A_{n'}$ must be an atom, i.e., that $n' \neq 0$. Suppose on the contrary that $g^0 > h^0$. Choose two real numbers x, y, so that $g^0 > x > y > h^0$, and let

$$G = A_0 \cap \{a \mid g(a) > x\}, \qquad H = A_0 \cap \{a \mid h(a) < y\}$$

As in (i) of the preceding proof we conclude that G and H are measurable sets. If $\mu(G \cap H) = 0$, then

$$\text{ess sup }[g \mid (G \backslash H)] = g^0 > h^0 = \text{ess inf}[h \mid (H \backslash G)] \qquad (5.4.50)$$

The first equality in (50) is obtained by noting that the essential supremum of g on A_0 is the same as on G, since $g^0 > x$. This in turn is the same as on $G \backslash H$, since this just removes a null set, and similarly for the last equality in (50). But (50) contradicts the preceding lemma, since $G \backslash H$ and $H \backslash G$ are disjoint.

If $\mu(G \cap H) > 0$, we can split $G \cap H$ into two pieces of positive measure F_1 and F_2, since μ is nonatomic on A_0. But then

$$\text{ess sup}\,(g \mid F_1) \geq x > y \geq \text{ess inf}\,(h \mid F_2)$$

This again contradicts the preceding lemma, since F_1 and F_2 are disjoint. Hence $n' \neq 0$, and the offending set $A_{n'}$ must be an atom. ∎

The exceptional case, where (46) fails for some $A_{n'}$, will be referred to as the case of the *anomalous atom*. Here is a trivial problem in which it arises. Let A consist of just one point: $A = \{a_o\}$, of measure *one;* let the payoff function for this point be $f(x) = |x|$; the bounds satisfy $b < 0 < c$. Because of the integral constraint (15) there is just one feasible, hence optimal, solution; viz., $\delta(a_o) = 0$. The space consists of one atom, and we verify that $g(a_o) = +1, h(a_o) = -1$, so (46) cannot be true for a_o.

The anomalous atom situation is related to but not identical with the well-known case in which there can be "increasing marginal returns" on (at most) one alternative project at the optimal allocation.[5] This occurs in the example beginning this section, (1)–(2), in which the second derivative is positive on point a_1. But this latter situation involves "second-order" conditions, while the anomalous atom involves "first-order" conditions.

The number p of this lemma will turn out to behave much like a shadow price. In this connection it is desirable that it be finite. This is not guaranteed, and it is not always possible to find a finite p satisfying (46), but there are several simple conditions that imply there is such a real number.

One such sufficient condition is that the anomalous atom case occur. For then the inequalities (49) let in daylight between sup g^n and inf h^n (both sup and inf over $n \neq n'$) and allow us to pick a *real* number between them.

A second condition that insures that the p in (46) will be finite is that

$$\mu[A_0 \cap \{a \mid \delta^o(a) < c(a)\}] > 0 \tag{5.4.51}$$

$$\mu[A_0 \cap \{a \mid \delta^o(a) > b(a)\}] > 0 \tag{5.4.52}$$

are both true; i.e., there is a nonatomic set of positive measure on which δ^o is strictly below its upper bound, *and* a similar set on which δ^o is strictly above its lower bound. For, from their definitions, $g > -\infty$ on the set in (51), and $h < \infty$ on the set in (52). Hence $-\infty < g^0 \leq p \leq h^0 < \infty$, and p is finite.

---

5. G. J. Stigler, *The Theory of Price,* rev. ed. (Macmillan, New York, 1952), pp 119–20.

We now introduce a differentiability condition. We make the following conventions. First, let $b(a) < \delta^o(a) < c(a)$ for a certain point $a \in A$; $f(a, \cdot)$ is said to be *differentiable* at the point $\delta^o(a)$ iff

$$\lim_{y \to 0} [f(a, \delta^o(a) + y) - f(a, \delta^o(a))]/y \qquad (5.4.53)$$

exists. This is the usual definition, except that the values $\pm\infty$ are possible. These occur with a vertical tangent at $\delta^o(a)$, $+\infty$ if $f(a, \cdot)$ is increasing at $\delta^o(a)$, $-\infty$ if it is decreasing. Next, if $\delta^o(a) = c(a)$, we say that $f(a, \cdot)$ is differentiable at $\delta^o(a)$ iff the limit (53) exists when $y \to 0$ through *negative* values *and* this limit $= -\infty$. Finally, if $\delta^o(a) = b(a)$, we say that $f(a, \cdot)$ is differentiable at $\delta^o(a)$ iff the limit exists when $y \to 0$ through *positive* values *and* this limit $= +\infty$. When f is differentiable, the value of the limit in (53) is called the *derivative* and denoted $Df(a, \delta^o(a))$.

The next result seems sufficiently interesting to be labeled a theorem.

Theorem: Given measure space (A, Σ, μ), μ σ-finite, and measurable functions $f:A \times \text{reals} \to \text{reals}$, b, $c:A \to$ extended reals, with $f(a, \cdot)$ lower semicontinuous on the nonatomic part of A, let δ^o be unsurpassed for the problem of maximizing $\int f(a, \delta(a)) \, \mu(da)$ over measurable functions $\delta:A \to$ reals that satisfy $b \leq \delta \leq c$ and $\int_A \delta \, d\mu = 0$. Let g, $h:A \to$ extended reals be defined as usual by (16)–(19). In addition, let there be a set E of positive measure such that $f(a, \cdot)$ is differentiable at $\delta^o(a)$ for all $a \in E$.

Then $Df(a, \delta^o(a))$ is equal to a constant p^o almost everywhere on E, and p^o is the unique number that satisfies

$$g(a) \leq p \leq h(a) \qquad (5.4.54)$$

almost everywhere.

Proof: Let $\{A_0, A_1, A_2, \ldots\}$ be a decomposition of A, with μ nonatomic on A_0 and atomic on A_n, $n = 1, 2, \ldots$. Since $\mu(E) > 0$, $\mu(E \cap A_n) > 0$ for at least one value of $n = 0, 1, \ldots$.

First, take the case when $\mu(E \cap A_{n'}) > 0$ for some $n' \neq 0$. Then for any $a \in A_{n'}$ for which $Df(a, \delta^o(a))$ exists, we have

$$g(a) = D_+ f(a, \delta^o(a)) = Df(a, \delta^o(a)) = D^- f(a, \delta^o(a)) = h(a) \qquad (5.4.55)$$

from the definitions of these functions. (If $\delta^o(a) = c(a)$, the common value in (55) is $-\infty$; if $\delta^o(a) = b(a)$, the common value is $+\infty$.)

Now $Df(a, \delta^o(a))$, as a function of a with domain E, is measurable, since it is the limit of a sequence of measurable functions g_{y_i} given by (21), or h_{y_i} given by (22), as $y_i \to 0$ through positive values. Also $A_{n'}$, hence $E \cap A_{n'}$, is an atom. Invoking the lemma (9), we find that

$Df(a, \delta^o(a))$ (hence g and h) is equal to a constant p^o almost everywhere on $E \cap A_{n'}$. It follows that

$$\text{ess sup}(g \mid A_{n'}) = \text{ess inf}(h \mid A_{n'}) = p^o$$

Using the abbreviations $g^n = \text{ess sup}(g \mid A_n)$, $h^n = \text{ess inf}(h \mid A_n)$, it follows from (20) that for all $n_1 \neq n', n_2 \neq n'$,

$$g^{n_1} \leq h^{n'} = p^o = g^{n'} \leq h^{n_2} \tag{5.4.56}$$

so that

$$\sup g^n = p^o = \inf h^n \tag{5.4.57}$$

and (54) is true almost everywhere.

Next take the case when $\mu(E \cap A_0) > 0$. For any $a \in A_0$ for which $Df(a, \delta^o(a))$ exists we have

$$g(a) \geq Df(a, \delta^o(a)) \geq h(a) \tag{5.4.58}$$

For on A_0, g is the supremum of the functions g_y, while Df is their limit as $y \to 0$; and h is the infimum of the functions h_y, while Df is again their limit as $y \to 0$. It follows from (58) that $g^0 \geq h^0$. But from the preceding lemma (46), the opposite inequality is also true, so $g^0 = h^0$ ($= p^o$, say). It then follows from (58) that $Df(a, \delta^o(a)) = p^o$ almost everywhere on A_0.

Also, the same argument (56) establishes (57) again. Hence (57) is established in all cases for a unique number p^o, and (54) is true for this number. The arguments just given show that $Df(a, \delta^o(a))$ is equal to this number p^o almost everywhere on $E \cap A_n$ for all $n = 0, 1, \ldots$, hence almost everywhere on E. ∎

Note that the anomalous atom case cannot arise here. It is precluded by the condition of differentiability on a set of positive measure. This theorem yields a third simple sufficient condition for p^o to be finite, viz., that there be a set of positive measure on which $Df(a, \delta^o(a))$ exists and is finite. For $Df(a, \delta^o(a)) = p^o$ almost everywhere, and the condition just given insures that p^o is finite.

Finally, we want to show that p^o acts as a shadow price, i.e., that $\delta^o(a)$ maximizes

$$f(a, x) - p^o x \tag{5.4.59}$$

over all real x satisfying $b(a) \leq x \leq c(a)$, for almost all $a \in A$. Here (59) may be interpreted as giving the "payoff density" $f(a, x)$ minus the "resource cost," $p^o x$.

We make a special convention as to the meaning of maximization in case p^o is infinite in (59). Namely, if $p^o = -\infty$, then "maximizing" means taking x as large as possible, so that $x = c(a)$ is the maximizer of (59) (if $c(a)$ is finite). And if $p^o = +\infty$, then x must be taken as small as possible, so that if $b(a)$ is finite, $x = b(a)$ is the maximizer.

For finite p^o the condition that $\delta^o(a)$ maximize (59) will be recognized as the *sufficient* condition for optimality given by (3.2), specialized to the particular feasibility conditions of the problem we are studying in this section.[6] This suggests the question: Does (3.2) still suffice for "bestness" if the p^o appearing in it is infinite, and we interpret maximization according to the convention just mentioned? The answer is yes for any arbitrary feasibility conditions that include

$$\int_A \delta \, d\mu = 0 \qquad (5.4.60)$$

We demonstrate this. The feasible set **M** consists of measurable functions $\delta : A \to$ reals, all of which satisfy (60). The sufficient condition is that for all δ,

$$f(a, \delta^o(a)) - p^o \delta^o(a) \geq f(a, \delta(a)) - p^o \delta(a) \qquad (5.4.61)$$

for almost all $a \in A$. Using our convention, if $p^o = +\infty$, (61) reduces to

$$\delta^o(a) \leq \delta(a) \qquad (5.4.62)$$

But (60) implies that

$$\int_A (\delta - \delta^o) \, d\mu = 0 \qquad (5.4.63)$$

and this together with (62) means that $\delta = \delta^o$ almost everywhere. Hence δ^o is best because the feasible set is trivial: δ^o is the only feasible solution. Similarly, if $p^o = -\infty$, (61) reduces to (62) reversed, and the same argument applies. This completes the demonstration.

We are thus on the verge of establishing a necessary *and* sufficient condition for optimality. One consequence of this will be that unsurpassed and best solutions coincide for this problem. For, starting from an unsurpassed δ^o, we derive a condition that suffices for δ^o to be best.

Now to establish the shadow price condition (59)! For this we need an extra condition on the atoms of μ. The trouble is that g and h are defined on the atoms in such a way that they depend only on the immediate neighborhood of $\delta^o(a)$, whereas (59) asserts something about the entire range $[b(a), c(a)]$. The assumption of concavity will bridge the gap.

---

6. Condition (59) is even a bit stronger than the sufficient condition (3.2). In (59) the null set for which $\delta^o(a)$ does not maximize is chosen once and for all. But the null set for which (3.2) fails depends on δ, and conceivably there is no null set E such that (3.2) holds for all δ and all $a \in A \backslash E$.

We define this abstractly. Let f be a real-valued function whose domain is a real interval $[b, c]$ (endpoints may or may not be included, and $b = c$ is possible); f is said to be *concave* iff for any numbers x, y in its domain and any number t in the interval $[0, 1]$, we have

$$f(tx + (1 - t)y) \geq tf(x) + (1 - t)f(y)$$

A concave function may be shown to be continuous, except possibly at the endpoints of its domain, where a "sudden" downward jump is possible. Thus a concave function (defined on an interval) is always *lower semi*continuous. We also state without proof the following well-known facts about concave functions:

$$\frac{f(x + y_1) - f(x)}{y_1} \leq D_+ f(x) \leq D^- f(x) \leq \frac{f(x) - f(x - y_2)}{y_2} \quad (5.4.64)$$

for any positive real y_1, y_2 and real x such that $x - y_2$, x, and $x + y_1$ are in the domain of f. (Our special conventions concerning D_+ and D^- insure that (64) holds even if x is an endpoint of the domain.)

Let us return to the problem in hand. In stating that $f(a, \cdot)$ is concave, we follow our standing convention of taking the domain of this function to be restricted to $[b(a), c(a)]$, ($b(a)$ to be included iff it is finite, and similarly for $c(a)$).

The following theorem may be taken to be the main result of this section.

Theorem: Let (A, Σ, μ) be a measure space, with μ σ-finite; let $f : A \times$ reals $\to$ reals, and b, $c : A \to$ extended reals be measurable; let $f(a, \cdot)$ be lower semicontinuous on the nonatomic part of A and concave on the rest of A; let δ^o be feasible for the problem of maximizing

$$\int f(a, \delta(a)) \, \mu(da) \quad (5.4.65)$$

over measurable functions $\delta : A \to$ reals which satisfy

$$b \leq \delta \leq c \quad (5.4.66)$$

$$\int_A \delta \, d\mu = 0 \quad (5.4.67)$$

Then the following conditions are logically equivalent:

(i) δ^o is unsurpassed for this problem;
(ii) δ^o is best for this problem;
(iii) there is an extended real number p^o and a null set E such that $\delta^o(a)$ maximizes

$$f(a, x) - p^o x \quad (5.4.68)$$

over all real x in the closed interval $[b(a), c(a)]$, for all $a \in A \backslash E$.

Proof: (iii) implies (ii) is already contained in the sufficiency theorem (3.1)–(3.3), plus the argument of (61)–(63). Condition (ii) implies (i) by definition. It remains to show that (i) implies (iii).

Let $\{A_0, A_1, \ldots\}$ be a decomposition, so that μ is nonatomic and f is lower semicontinuous on A_0, while μ is atomic and f concave on each A_n, $n \neq 0$.

We show that concavity precludes the occurrence of an anomalous atom in (46). Choose any atom A_n. First, $g \mid A_n$ is measurable (g is given by (16) and (18)). This follows from the fact that, by concavity, $g_y(a)$ given by (21) is nonincreasing in y for fixed $a \in A_n$. Hence the lower right derivate $D_+ f(a, \delta^o(a))$, which is $g(a)$, is the limit of any sequence $g_{y_i}(a)$, y_i going to zero through positive values. Thus $g \mid A_n$ is the limit of a sequence of measurable functions $g_{y_i} \mid A_n$, and so is measurable itself. A similar argument establishes the measurability of $h \mid A_n$, with h given by (17) and (19).

The lemma (9) then implies that g and h are constant almost everywhere on A_n. These constants must equal g^n, the essential supremum of $g \mid A_n$, and h^n, the essential infimum of $h \mid A_n$.

There is a point $a \in A_n$ such that $g(a) = g^n$ and $h(a) = h^n$, since these relations hold almost everywhere on A_n, and $\mu(A_n) > 0$. But $g(a)$ and $h(a)$ equal $D_+ f(a, \delta^o(a))$ and $D^- f(a, \delta^o(a))$ respectively. Then (64), middle, implies that $g^n \leq h^n$.

This is true for every $n = 1, 2, \ldots$. We also have $g^0 \leq h^0$ from (46), and $g^{n_1} \leq h^{n_2}$ for all $n_1 \neq n_2$, from (20). Hence $\sup g^m \leq \inf h^n$ (both taken over all $m, n = 0, 1 \ldots$), and (46) is true almost everywhere. There is no anomalous atom.

Thus there is an extended real number p^o such that

$$g(a) \leq p^o \leq h(a) \qquad (5.4.69)$$

for almost all $a \in A$. We now show that this p^o satisfies the condition (iii) almost everywhere.

First, suppose $p^o = -\infty$. Then $g = -\infty$ almost everywhere, from (69). Let $g(a) = -\infty$ for some $a \in A \backslash A_0$. Then $f(a, \cdot)$ is concave, and it follows that there cannot be any positive y_1 satisfying (64), left. That is, $\delta^o(a)$ is at the upper limit: $\delta^o(a) = c(a)$. If $g(a) = -\infty$ for some a in the nonatomic part A_0, it follows from the definition of g, (16), that $\delta^o(a)$ is again at the upper limit. Thus $\delta^o = c$ almost everywhere. But by our convention for p^o infinite, this is precisely the condition that δ^o maximize (68) for $p^o = -\infty$, almost everywhere. Hence (iii) is established in this case.

Next, suppose $p^0 = +\infty$. Then $h = +\infty$ almost everywhere. An argument similar to that just given shows that $\delta^o = b$ almost everywhere, which is the condition for (iii) to be satisfied when $p^o = +\infty$.

It remains to establish (iii) when p^o is finite. In this case, the condition that $\delta^o(a)$ maximize (68) is equivalent to the following double inequality:

$$\frac{f(a, \delta^o(a) + y_1) - f(a, \delta^o(a))}{y_1} \le p^o \le \frac{f(a, \delta^o(a)) - f(a, \delta^o(a) - y_2)}{y_2}$$

(5.4.70)

which must hold for all positive y_1 such that $\delta^o(a) + y_1 \le c(a)$, and for all positive y_2 such that $\delta^o(a) - y_2 \ge b(a)$. We now show that (69) implies (70). If $a \in A \backslash A_0$, then $f(a, \cdot)$ is concave, and the implication follows at once from (64).

Let $a \in A_0$; $g(a)$ is defined by (16) as the supremum of the left side of (70), as y_1 varies over the *open* interval $(0, c(a) - \delta^o(a))$. Similarly, $h(a)$ is, by (17), the infimum of the right side of (70), as y_2 ranges over $(0, \delta^o(a) - b(a))$. Inequality (69) then implies that (70) holds for all *interior* points of $[b(a), c(a)]$. But it must then hold for the endpoints as well because $f(a, \cdot)$, being lower semicontinuous, makes no sudden upward jump at $b(a)$ or $c(a)$. This establishes the implication in general.

Since (69) holds for almost all $a \in A$, so does (70). This completes the proof that (i) implies (iii). ■

Thus under the special assumptions made concerning f, a necessary and sufficient condition that feasible δ^o be unsurpassed or best, for the problem of maximizing (65) subject to (66) and (67), is that a "shadow price" p^o exists under which (except for a null set), for each "project" $a \in A$ separately, $\delta^o(a)$ is chosen to maximize the "payoff" $f(a, x)$, net of the "resource cost" $p^o x$.

This result is important. First, it suggests an efficient method for finding an optimal solution (if there is one). Namely, choose an arbitrary number p, and for each $a \in A$ choose $\delta^o(a)$ to maximize $f(a, x) - px$ over the feasible interval $[b(a), c(a)]$, *disregarding* the total resource constraint (67). If by chance (67) is satisfied by this process, we have found an optimal solution. If not, adjust p to a new value p' and try again.

How should p be adjusted? It is easily seen from (68) that the maximizing value of x is a nonincreasing function of p. (We are implicitly assuming, for simplicity, that there is a unique maximizer of (68) for each p and $a \in A$.) Hence, if total resource availability is exceeded by the trial solution, raise the tentative shadow price p, and lower it in the opposite case. This simple monotonic relation between p and total resource demand makes it easy to "zero in" on the proper p (again assuming that there *is* an optimal solution). The necessity of the shadow price condition also guarantees that we will not overlook the optimal solution by this procedure.

Furthermore, the shadow price condition suggests institutional arrangements for arriving at an optimal solution. For example, if one overall organization is responsible for this allocation, separate divisions might be responsible for separate subsets of projects $a \in A$. The head office might dictate the tentative shadow price to the divisions, note the consequent resource demand, adjust the price accordingly, etc.[7] Going a step further the free market itself is an institutional mechanism for carrying out the price-adjustment process discussed above.

Two special cases in which the results of the preceding theorem are valid may be noted. The first is when μ is nonatomic (as well as σ-finite), and $f(a, \cdot)$ is lower semicontinuous for all $a \in A$. The second is when $f(a, \cdot)$ is concave for all $a \in A$ (with no assumptions on μ other than σ-finiteness). The validity of the first case is obvious: since μ has no atomic part, no concavity assumption is needed. The validity of the second follows from the fact that a concave function is lower semicontinuous, so that if $f(a, \cdot)$ is concave everywhere, the premises of the theorem are certainly fulfilled.

Finally, we recall that the condition (67) is much less narrow than it' appears. We can formulate an apparently much more general theorem, which falls out as an immediate corollary.

Theorem: Let all the premises of the preceding theorem be fulfilled, except that the feasibility condition (67) is replaced by $L_o \leq \int_A \delta \, d\mu \leq L^o$. (Here L_o and L^o are two extended real numbers; $\int_A \delta \, d\mu$ is still required to be finite, however.) Then the following are equivalent:

(i) δ^o is unsurpassed for this problem;
(ii) δ^o is best for this problem;
(iii) there is an extended real number p^o and a null set E such that $\delta^o(a)$ maximizes

$$f(a, x) - p^o x \qquad (5.4.71)$$

over all real x in the closed interval $[b(a), c(a)]$ for all $a \in A \backslash E$. Furthermore, if $\int_A \delta^o \, d\mu > L_o$, then $p^o \geq 0$; and if $\int_A \delta^o \, d\mu < L^o$, then $p^o \leq 0$.

Proof: (ii) implies (i) by definition, and (iii) implies (ii) by a sufficiency theorem already given (cf. (3.5)–(3.6)). To show that (i) implies (iii), we add a point z_o to A and transform this problem into one of the preceding type by the recipe given in Section 5.2.

---

7. Cf. G. B. Dantzig and P. Wolfe, The decomposition algorithm for linear programs *Econometrica*, 29(Oct. 1961):767–78.

The singleton set $\{z_o\}$ is an atom, since $\mu'\{z_o\} = 1 > 0$; $f(z_o, \cdot)$ is identically zero, and this is a concave function, so the premises are fulfilled on $A' = A \cup \{z_o\}$. By the preceding theorem, a number p^o exists such that $\delta^{o'}(a')$ maximizes (71) for almost all $a' \in A'$. In particular, it must maximize (71) for $a' = z_o$, since $\mu\{z_o\} > 0$. Since $f(z_o, \cdot)$ is identically zero, (71) reduces to

$$-p^o x \tag{5.4.72}$$

If $p^o > 0$, then the maximizer of (72) is as small as possible: $\delta^{o'}(z_o) = b'(z_o)$. But recalling the translation recipe, (2.9), this is simply

$$-\int_A \delta^o \, d\mu = -L^o \tag{5.4.73}$$

Similarly, if $p^o < 0$, we must have $\delta_o(z_o) = c'(z_o)$, which is to say, by (2.9) again,

$$-\int_A \delta^o \, d\mu = -L_o \tag{5.4.74}$$

Equations (73) and (74) give the two extra conditions on p^o. ∎

Thus if δ^o is optimal for this more general problem, we get a shadow price condition of the same type as above, with an extra sign condition on p^o, depending on where $\int_A \delta^o \, d\mu$ is located in the interval $[L_o, L^o]$. The economic interpretation of these sign conditions is the same as in Section 5.3.

Note added in proof:
The approach taken by Wagner and Stone[8]—involving the Lyapunov-Blackwell convexity theorem, the separating hyperplane theorem, and the von Neumann–Aumann measurable choice theorem—seems to hold great promise for generalizing the foregoing results and simplifying the proofs, at least insofar as the "nonatomic" parts of the arguments are concerned.

To handle the pseudomeasure-valued objective function, one might subtract the constant $\int f(a, \delta^o(a)) \, \mu(da)$, where δ^o is unsurpassed, which leaves one with "ordinary" integrals, at least in a "neighborhood" of δ^o. Another technical problem is the coordination of this approach with a different argument for the σ-atomic complement in a general measure

---

8. D. H. Wagner and L. D. Stone, Necessity and existence results on constrained optimization of separable functionals by a multiplier rule, *SIAM J. Control* 12(Aug. 1974): 356–72.

space A (cf. the proof beginning on p. 234 and the final existence proof in Section 5.6).

If these technical problems can be surmounted, the following generalizations appear likely:

(i) integral constraints of the form

$$L_{io} \leq \int_A g_i(a, \delta(a))\, \mu(da) \leq L_i^o, i = 1, \ldots, n$$

with a corresponding "Lagrangean" function

$$f(a, x) - p_1 g_1(a, x) - \cdots - p_n g_n(a, x)$$

generalizing (71) (this is the "general" Neyman-Pearson problem);

(ii) δ having its range in a rather general space, with constraints $\delta(a) \in E(a)$, all $a \in A$, generalizing (66);

(iii) dropping of lower semicontinuity conditions on $f(a, \cdot)$.

In exchange some weak topological assumptions may have to be added, such as "condition (α)" of Wagner and Stone.[9]

5.5. EXISTENCE OF FEASIBLE SOLUTIONS

Until now we have been examining conditions that imply or are implied by the fact that a given feasible solution δ^o is optimal. We now take up a different task, that of proving an optimal solution exists for a given problem. First, we start with the simpler task (i.e., simpler for a given problem) of proving at least one feasible solution exists. Even this is by no means trivial for the problem with which we have been dealing, characterized by constraints (4.66)–(4.67).

Theorem: Let (A, Σ, μ) be a measure space, with μ σ-finite; let b, $c:A \rightarrow$ extended reals be measurable functions. The following conditions are logically equivalent:

(i) A measurable function $\delta:A \rightarrow$ reals exists such that

$$b \leq \delta \leq c \qquad (5.5.1)$$

$$\int_A \delta\, d\mu = 0 \qquad (5.5.2)$$

---

9. Wagner and Stone, Necessity, p. 363.

(ii) $b \le c$, $b < \infty$, $c > -\infty$, and

$$\int_A b \, d\mu \le 0 \le \int_A c \, d\mu \qquad (5.5.3)$$

Proof: That (i) implies (ii) is obvious. Conversely, assume (ii). First, because μ is σ-finite, a *positive* measurable function $k:A \to$ reals exists such that

$$\infty > \int_A k \, d\mu \ge 0 \qquad (5.5.4)$$

If $\mu = 0$, take $k = 1$ everywhere. Otherwise, let $\{G_1, G_2, \ldots\}$ be a countable measurable partition with $\infty > \mu(G_n) > 0$ for all n. On G_n let k be equal to the constant $1/[2^n \mu(G_n)]$. This function fulfills the stated conditions.

Next define the *median function m* as the one that picks the middle in size order of three extended real numbers; thus $m(3, -2, \infty) = 3$, $m(-\infty, \infty, 17) = 17$, etc.

Then for each real number x, define the function $m_x:A \to$ extended reals by $m_x(a) = m(b(a), c(a), xk(a))$. Because $b \le c$, we find that

$$m_x = \max[b, \min(c, xk)] = \min[\max(b, xk), c] \qquad (5.5.5)$$

Thus m_x is measurable and also finite, since $b(a)$ and $c(a)$ are never both infinite of the same sign. We now show that if

$$\int_A b \, d\mu < 0 < \int_A c \, d\mu \qquad (5.5.6)$$

then m_x will be a feasible solution for (1) and (2), for some real number x. (If (6) is false, then either $\int_A b \, d\mu = 0$ or $\int_A c \, d\mu = 0$. In either case we get an immediate feasible solution. In the first case, e.g., set $\delta(a) = b(a)$ whenever $b(a) > -\infty$; and on the null set where $b = -\infty$, choose $\delta = \min[c, 0]$. Hence finding a feasible solution in case (6) will prove the theorem.)

First of all, by (5), $b \le m_x \le c$. Next we show that $\int_A m_x \, d\mu$ exists and is finite for all real x. We have

$$-(c^- + |xk|) \le \min(c, xk) \le m_x \le \max(b, xk) \le b^+ + |xk|$$

Also, $-\int_A (c^- + |xk|) \, d\mu > -\infty$, and $\int_A (b^+ + |xk|) \, d\mu < \infty$ because of (3) and (4), which shows that $\int_A m_x \, d\mu$ is finite.

Next we show that $\int_A m_x \, d\mu$, as a function of x, is continuous. Let $x_1, x_2, \ldots$ be a sequence, either increasing or decreasing, whose limit is the real number x. By the monotone convergence theorem we have

$$\lim_n \int_A m_{x_n} \, d\mu = \int_A \lim_n m_{x_n} \, d\mu = \int_A m_x \, d\mu$$

proving continuity.

If $x_1, x_2, \ldots$ is a sequence increasing to $+\infty$, then m_x increases to c, so

$$\lim_n \int_A m_{x_n} \, d\mu = \int_A c \, d\mu > 0$$

If $x_1, x_2, \ldots$ is a sequence decreasing to $-\infty$, then m_x decreases to b, so

$$\lim_n \int_A m_{x_n} \, d\mu = \int_A b \, d\mu < 0$$

Both these results are again by monotone convergence. The inequalities are from (6).

Hence for sufficiently large real x, $\int_A m_x \, d\mu$ is positive; and for sufficiently small real x, it is negative. Since it is continuous, there must then be an x-value for which $\int_A m_x \, d\mu = 0$, and this m_x is feasible. Hence (ii) implies (i). ∎

As usual, this theorem has an immediate generalization.

Theorem: Let (A, Σ, μ) be a measure space with μ σ-finite, let $b, c\colon A \to$ extended reals be measurable, and let L_o, L^o be two extended real numbers. The following are equivalent:

(i) A measurable function $\delta\colon A \to$ reals exists such that $b \le \delta \le c$, $\int_A \delta \, d\mu$ is finite, and $L_o \le \int_A \delta \, d\mu \le L^o$;

(ii) $\infty > b \le c > -\infty$, $\infty > L_o \le L^o > -\infty$, $\int_A b \, d\mu \le L^o$, and $\int_A c \, d\mu \ge L_o$.

This is an easy corollary of the preceding theorem, by the now familiar procedure of transforming this problem into one for which (2) holds. We leave the details as an exercise.

5.6. EXISTENCE OF OPTIMAL SOLUTIONS

We now come to the much more difficult problem of proving the existence of optimal solutions. A number of assumptions will be made that are more restrictive than those made previously. In particular, the bounding functions b and c will have *finite* integrals, and the payoff functions $f(a, \cdot)$ will be *continuous;* as usual, this refers to the interval $[b(a), c(a)]$. Even so, much work is involved.

We first prove existence under the assumption that μ is nonatomic.

The basic procedure is to find functions satisfying the sufficient condition for optimality and then show that one is feasible. Next we go to the opposite case where μ is σ-atomic, using an entirely different procedure. Finally, we combine these results to prove existence under general (σ-finite) μ.

Standard ordering of pseudomeasures is still used for ordering utilities, and existence is proved for best solutions.

Theorem: Let (A, Σ, μ) be a measure space with μ σ-finite and *nonatomic;* let $b, c:A \rightarrow$ reals and $f:A \times$ reals $\rightarrow$ reals be measurable; let $f(a, \cdot)$ be continuous for all $a \in A$. Assume $b \leq c$, and let L be a real number such that

$$-\infty < \int_A b \, d\mu \leq L \leq \int_A c \, d\mu < \infty \tag{5.6.1}$$

Then the problem, maximize $\int f(a, \delta(a)) \, \mu(da)$ over measurable functions $\delta:A \rightarrow$ reals, subject to

$$b \leq \delta \leq c \tag{5.6.2}$$

$$\int_A \delta \, d\mu = L \tag{5.6.3}$$

has a *best* solution.

Proof: For each $a \in A$ and each extended real number p, let $E_{a,p}$ be the set of real numbers x that maximize the expression

$$f(a, x) - px \tag{5.6.4}$$

over the closed interval $[b(a), c(a)]$. By our convention concerning infinite p-values, we have $E_{a,\infty} = \{b(a)\}$ and $E_{a,-\infty} = \{c(a)\}$. Since (4) is continuous in x (for finite p) and the maximization is over a closed bounded interval, the sets $E_{a,p}$ are in all cases nonempty, closed, and bounded. Hence they themselves have a minimum value and a maximum value for all (a, p). Define the functions $\beta, \gamma:A \times$ extended reals $\rightarrow$ reals by

$$\beta(a, p) = \min E_{a,p}, \qquad \gamma(a, p) = \max E_{a,p} \tag{5.6.5}$$

Let $-\infty < p_1 < p_2 < \infty$. We have the following chain of relations:

$$c(a) = \gamma(a, -\infty) = \beta(a, -\infty) \geq \gamma(a, p_1) \geq \beta(a, p_1)$$

$$\geq \gamma(a, p_2) \geq \beta(a, p_2) \geq \gamma(a, \infty) = \beta(a, \infty) = b(a) \tag{5.6.6}$$

The equalities in (6) follow from $E_{a,\infty} = \{b(a)\}$, $E_{a,-\infty} = \{c(a)\}$. The middle inequality in (6) is the only one that needs proving. From the definition of $E_{a,p}$ we have

$$f(a, \beta(a, p_1)) - p_1\beta(a, p_1) \geq f(a, \gamma(a, p_2)) - p_1\gamma(a, p_2)$$

$$f(a, \gamma(a, p_2)) - p_2\gamma(a, p_2) \geq f(a, \beta(a, p_1)) - p_2\beta(a, p_1)$$

Adding these two inequalities and simplifying, we get $\beta(a, p_1) \geq \gamma(a, p_2)$. This establishes (6). Thus for fixed $a \in A$, $\beta(a, \cdot)$ and $\gamma(a, \cdot)$ are nonincreasing functions.

Next we show that $\gamma(a, \cdot)$ is *continuous from the left;* i.e., if p_1, $p_2, \ldots$ is an *increasing* sequence whose limit is p_o (possibly $+\infty$), then the limit of $\gamma(a, p_n)$ is $\gamma(a, p_o)$. First, let p_o be finite. The sequence $y_n = \gamma(a, p_n)$ is nonincreasing, hence it has a limit $y_o \geq \gamma(a, p_o)$. We must show this is an equality; to do this, it suffices to prove that y_o maximizes (4), since $\gamma(a, p_o)$ is the largest number that does so.

Now (4) is jointly continuous in x and p ($a \in A$ is fixed), hence

$$\lim_{n \to \infty} [f(a, y_n) - p_n y_n] = f(a, y_o) - p_o y_o \qquad (5.6.7)$$

For any $x \in [b(a), c(a)]$ we have $f(a, y_n) - p_n y_n \geq f(a, x) - p_n x$, since y_n maximizes (4). Hence by (7),

$$f(a, y_o) - p_o y_o \geq \lim_{n \to \infty} [f(a, x) - p_n x] = f(a, x) - p_o x$$

Thus y_o maximizes (4) for $p = p_o$, so $\gamma(a, p)$ is continuous from the left for any finite p.

Next let $p_o = +\infty$. Since $f(a, \cdot)$ is continuous on the closed bounded interval $[b(a), c(a)]$, it has a finite upper bound N. For any $\epsilon > 0$ choose p finite and greater than $[N - f(a, b(a))]/\epsilon$. Then for any $c(a) \geq x \geq b(a) + \epsilon$, we have

$$p[x - b(a)] \geq p\epsilon > N - f(a, b(a)) \geq f(a, x) - f(a, b(a))$$

so that

$$f(a, b(a)) - pb(a) > f(a, x) - px \qquad (5.6.8)$$

Inequality (8) shows that no such x can maximize (4), hence $\gamma(a, p) < b(a) + \epsilon$. Thus if $p_1, p_2, \ldots$ increases without bound, $\gamma(a, p_n)$ approaches $b(a) = \gamma(a, \infty)$. This proves that $\gamma(a, \cdot)$ is continuous from the left. A similar argument shows that $\beta(a, \cdot)$ is *continuous from the right.*

Next we show that for fixed p, $\gamma(a, p)$ is a *measurable* function of a. We split A into two measurable pieces and consider each separately. On the set $\{a \mid b(a) = c(a)\}$, $\gamma(\cdot, p) = b = c$ for any p and so is measurable.

Consider the complementary set $E = \{a \mid b(a) < c(a)\}$. To show that $\gamma(\cdot, p)$ is measurable, it suffices to show that the sets $E \cap \{a \mid \gamma(a, p) < y\}$ are all measurable as y ranges over the *rational* numbers. Now, for fixed a, p, (4) is a continuous, hence lower semicontinuous, function of x. Hence its supremum on any interval $[b, c]$ (with or without the endpoints, and

$b < c$) equals its supremum over the *rational* numbers on that interval; see (4.6). One then verifies that for any rational y,

$$E \cap \{a \mid \gamma(a,p) < y\} =$$

$$E \cap \left\{ a \left| \begin{array}{l} \sup\{f(a,x) - px \mid x \text{ rational}, x < y\} \\ > \sup\{f(a,x) - px \mid x \text{ rational}, x \geq y\} \end{array} \right. \right\} \qquad (5.6.9)$$

In (9) a and p are held fixed, and the two sups are taken over x as indicated. For this formula only, we define $f(a,x)$ to be $-\infty$ if x is not in the closed interval $[b(a), c(a)]$; and we note that for fixed x the function $f(\cdot, x)$ thus defined is measurable. In verifying (9) there are five cases to consider, depending on whether y is in the interval $[b(a), c(a)]$, at an endpoint, or on either side of it. This interval is nondegenerate for $a \in E$, hence it always contains a rational point. We omit details.

Since the rational numbers are countable, the two sups in (9) are over a countable number of measurable functions, hence they are themselves measurable functions of a on E (p, y fixed). Hence the right side of (9) is a measurable set. This proves $\gamma(\cdot, p)$ is measurable. A similar argument proves that $\beta(\cdot, p)$ is measurable.

Next consider the integral $\int_A \gamma(a,p) \, \mu(da)$ as a function of p. For any $-\infty \leq p \leq \infty$ this is well defined and in fact finite, by (6) and (1). It is also nonincreasing in p, since $\gamma(a,p)$ is nonincreasing for each $a \in A$. Using the monotone convergence theorem, it follows from the left-continuity of $\gamma(a, \cdot)$ that $\int_A \gamma(a, \cdot) \, \mu(da)$ is also left-continuous. (Take a sequence $p_1, p_2, \ldots$ increasing to p_o; then $\gamma(a, p_n) \to \gamma(a, p_o)$ by left-continuity; monotone convergence then yields $\int_A \gamma(a, p_n) \, \mu(da) \to \int_A \gamma(a, p_o) \, \mu(da)$.) A similar argument shows that $\int_A \beta(a, \cdot) \, \mu(da)$ is finite, nonincreasing, and *right*-continuous.

We are now ready to construct the optimal solution. Let

$$p^o = \sup\left\{ p \left| \int_A \gamma(a,p) \, \mu(da) \geq L \right. \right\} \qquad (5.6.10)$$

Since $\int_A \gamma(a, \cdot) \, \mu(da)$ is continuous from the left, it follows that

$$\int_A \gamma(a, p^o) \, \mu(da) \geq L \qquad (5.6.11)$$

if $p^o > -\infty$. Inequality (11) is also true if $p^o = -\infty$, by (6) and (1). Next we show that

$$\int_A \beta(a, p^o) \, \mu(da) \leq L \qquad (5.6.12)$$

If $p^o = +\infty$, then (12) follows from (6) and (1). If $p^o < \infty$, then for every $p > p^o$ we have

$$\int_A \beta(a,p) \, \mu(da) \le \int_A \gamma(a,p) \, \mu(da) < L$$

from (10). Then (12) follows from the right-continuity of $\int_A \beta(a, \cdot) \, \mu(da)$. This p^o turns out to be the shadow price of the optimal solution.

Now consider the indefinite integral

$$\int [\gamma(a,p^o) - \beta(a,p^o)] \, \mu(da) \tag{5.6.13}$$

This is finite. Also, since μ is nonatomic, (13) is a nonatomic measure. Hence it takes on every value between 0 and its value on A, inclusive of these bounds. Now

$$L - \int_A \beta(a,p^o) \, \mu(da) \tag{5.6.14}$$

lies between these bounds, from (11) and (12). Hence there is a measurable set F such that (14) equals the value of (13) at F. This yields

$$\int_F \gamma(a,p^o) \, \mu(da) + \int_{A\backslash F} \beta(a,p^o) \, \mu(da) = L \tag{5.6.15}$$

We now claim that the function $\delta^o{:}A \rightarrow$ reals, coinciding with $\gamma(\cdot, p^o)$ on F and with $\beta(\cdot, p^o)$ on $A\backslash F$, is best. For it is measurable and satisfies (2), since both $\gamma(\cdot, p^o)$ and $\beta(\cdot, p^o)$ satisfy these conditions. Also it satisfies (3), by (15). Hence δ^o is feasible. Also, it satisfies the sufficient condition for "bestness," since both $\gamma(a,p^o)$ and $\beta(a,p^o)$ maximize (4) for $p = p^o$. ∎

We now remove the condition that μ is nonatomic. In its place, however, we are obliged to add a further condition on f, viz., that $|f(a,x)| \le \theta(a)$, where θ is some function with a finite integral. One consequence of this new condition may be noted: it guarantees that the utility function is a finite signed measure for all feasible δ. Hence standard ordering of pseudomeasures reduces to the ordinary comparison of definite integrals, and the distinction between "best" and "unsurpassed" disappears. To emphasize this point, we write the utility functions in the following theorem and proof in the form of *definite* integrals.

Theorem: Let (A, Σ, μ) be a measure space with μ σ-finite; let $b, c{:}A \rightarrow$ reals and $f{:}A \times$ reals $\rightarrow$ reals be measurable; let $f(a, \cdot)$ be continuous for all $a \in A$. Assume $b \le c$, and

$$-\infty < \int_A b \, d\mu \le 0 \le \int_A c \, d\mu < \infty \tag{5.6.16}$$

Also assume a measurable function $\theta\colon A \to$ reals exists such that $\int_A \theta\, d\mu$ is finite, and

$$| f(a,x) | \leq \theta(a) \qquad (5.6.17)$$

for all $x \in [b(a), c(a)]$, $a \in A$.

Then the problem, maximize

$$\int_A f(a, \delta(a))\, \mu(da) \qquad (5.6.18)$$

over measurable functions $\delta\colon A \to$ reals, subject to

$$b \leq \delta \leq c \qquad (5.6.19)$$

$$\int_A \delta\, d\mu = 0 \qquad (5.6.20)$$

has a best solution.

Proof: This proof is divided into two parts. In the first we assume μ is *σ-atomic;* i.e., a countable measurable partition $\{A_1, A_2, \ldots\}$ exists such that μ restricted to each A_n is atomic. Since μ is also σ-finite, it must be bounded on each A_n.

First, we show that total utility depends only on how mass is distributed *among* the atoms and, given this, is independent of how mass is distributed *within* the atoms. That is, suppose δ_1 and δ_2 are two densities such that

$$\int_{A_n} \delta_1\, d\mu = \int_{A_n} \delta_2\, d\mu = \lambda_n \qquad (5.6.21)$$

say, for all $n = 1, 2, \ldots$. Then

$$\int_A f(a, \delta_1(a))\, \mu(da) = \int_A f(a, \delta_2(a))\, \mu(da) \qquad (5.6.22)$$

To show this, we invoke the lemma (4.9), stating that δ_1 and δ_2 must each be equal to constants almost everywhere on any atom: say $\delta_1 = d_{1n}$, $\delta_2 = d_{2n}$ on A_n almost everywhere. From (21) we obtain $d_{1n}\mu(A_n) = d_{2n}\mu(A_n) = \lambda_n$, which means that $d_{1n} = d_{2n}$ for all n. Hence $\delta_1 = \delta_2$ almost everywhere, so that (22) is correct. Thus utility depends only on the sequence $(\lambda_1, \lambda_2, \ldots)$ and is in fact given by

$$g_1(\lambda_1) + g_2(\lambda_2) + \cdots \qquad (5.6.23)$$

where

$$g_n(\lambda_n) = \int_{A_n} f[a, \lambda_n/\mu(A_n)]\, \mu(da) \qquad (5.6.24)$$

This reduces the problem to two simpler issues. First, for what sequences $(\lambda_1, \lambda_2, \ldots)$ are there feasible densities δ such that (21) is satisfied for all $n = 1, 2, \ldots$? Second, among these feasible sequences is there one that maximizes (23)?

The first question is easily answered. Integrating the constraint (19) over A_n, we obtain

$$b_n \leq \lambda_n \leq c_n \qquad (5.6.25)$$

for all n, where $b_n = \int_{A_n} b \, d\mu$, $c_n = \int_{A_n} c \, d\mu$. Furthermore, the constraint (20) implies that

$$\lambda_1 + \lambda_2 + \cdots = 0 \qquad (5.6.26)$$

Thus (25) and (26) give necessary conditions for any feasible sequence $(\lambda_1, \lambda_2, \ldots)$. Conversely, they are also sufficient for a feasible δ to exist yielding this sequence. For if (25) is satisfied, one easily sees that some weighted average $t_n b + (1 - t_n) c = \delta$ will satisfy (21) for n. The δ thus constructed automatically satisfies (19) and satisfies (20) because of (26).

We have thus reduced the problem to one with a countable number of unknowns: Maximize (23) over real sequences $(\lambda_1, \lambda_2, \ldots)$ satisfying (25) and (26).

Let $\lambda^k = (\lambda_1^k, \lambda_2^k, \ldots)$, $k = 1, 2, \ldots$, be a sequence of these feasible sequences such that the value of (23) approaches its supremum as $k \to \infty$. We first give a standard argument to show there is a subsequence λ^{k_1}, $\lambda^{k_2}, \ldots$ such that, for all $n = 1, 2, \ldots$, the sequence $\lambda_n^{k_1}, \lambda_n^{k_2}, \ldots$ has a limit λ_n^o.

First, the sequence $\lambda_1^1, \lambda_1^2, \ldots$ is all contained in the closed, bounded interval $[b_1, c_1]$. Hence there is a convergent subsequence. By the same argument there is a subsequence of this subsequence such that the λ_2 values converge. Continuing, we get a sequence of sequences, each a subsequence of its predecessor, the nth subsequence having convergent λ_n values. Finally, one takes the "diagonal", viz., the qth term of the qth subsequence. This yields a subsequence converging for *all* $n = 1, 2, \ldots$.

We now show that the resulting limiting sequence $(\lambda_1^o, \lambda_2^o, \ldots)$ is optimal. First we prove feasibility. Inequality (25) is satisfied by λ_n^o, since it is the limit of a sequence in that interval $[b_n, c_n]$. We must show that

$$\lambda_1^o + \lambda_2^o + \cdots = 0 \qquad (5.6.27)$$

Let $\lambda^{k_1}, \lambda^{k_2}, \ldots$ be the subsequence converging to $\lambda^o = \lambda_1^o, \lambda_2^o, \ldots$. Think of

$$\lambda_1^{k_i} + \lambda_2^{k_i} + \cdots \qquad (5.6.28)$$

as the integral of a function $\lambda^{k_i}:\{1, 2, \ldots\} \to$ reals, all subsets of the positive integers being measurable, and having the counting measure: $\nu\{n\} = 1$

for all $n = 1, 2, \ldots$. We also have for all i and all n, $|\lambda_n^{k_i}| \le |b_n| + |c_n|$, by (25); and the sum over all n of $|b_n| + |c_n|$ is finite, by (16). Hence we may invoke the dominated convergence theorem and assert that the limit of the sums in (28) as $k_i \to \infty$ is the sum of the limits. But all sums in (28) equal zero, by (26). Hence (27) is true. This proves that λ^o is feasible.

It remains to show that $(\lambda_1^o, \lambda_2^o, \ldots)$ maximizes (23) over the set of feasible sequences, (25)–(26). First, we show that the function g_n given by (24), over the domain (25), is *continuous*. Let $\lambda^1, \lambda^2, \ldots$ be a sequence of numbers with limit λ, all satisfying (25). Since $f(a, \cdot)$ is continuous,

$$\lim_{k \to \infty} f(a, \lambda^k/\mu(A_n)) = f(a, \lambda/\mu(A_n))$$

for all $a \in A_n$. Also, $|f(a, \lambda^k/\mu(A_n))| \le \theta(a)$, and

$$\int_{A_n} \theta \, d\mu = \theta_n < \infty \tag{5.6.29}$$

by (17). Hence we may invoke the dominated convergence theorem again and assert that

$$\lim_{k \to \infty} g_n(\lambda^k) = g_n(\lambda)$$

Hence g_n is continuous for all $n = 1, 2, \ldots$. Next, for each i think of the sum

$$g_1(\lambda_1^{k_i}) + g_2(\lambda_2^{k_i}) + \cdots \tag{5.6.30}$$

as the integral of a function with domain $\{1, 2, \ldots\}$, the measure on this space being counting measure, as in (28). We have for all i and all n, $|g_n(\lambda_n^{k_i})| \le \theta_n$, from (24) and (29). Also, $\theta_1 + \theta_2 + \cdots = \int_A \theta \, d\mu < \infty$. Hence we may invoke dominated convergence yet a third time and assert that the limit of the sums in (30) as $k_i \to \infty$ is the sum of the limits. Now $\lambda_n^{k_i} \to \lambda_n^o$ for all n, as $k_i \to \infty$. By the continuity of g_n it follows that

$$\lim_{k_i \to \infty} g_n(\lambda_n^{k_i}) = g_n(\lambda_n^o)$$

for all $n = 1, 2, \ldots$. Hence the limit of the sums in (30) is

$$g_1(\lambda_1^o) + g_2(\lambda_2^o) + \cdots \tag{5.6.31}$$

But the limit of the sums in (30) is also the supremum of (23) over all feasible $(\lambda_1, \lambda_2, \ldots)$, by the construction of the original sequence of sequences, $\lambda^1, \lambda^2, \ldots$. Hence (31) is the maximum of (23), and $(\lambda_1^o, \lambda_2^o, \ldots)$ is optimal. Any feasible δ yielding this sequence via (21) is then a best solution.

This completes the first half of the proof.

We now drop the restriction that μ be σ-atomic. Since μ is σ-finite, there is a countable measurable partition $\{A_0, A_1, \ldots\}$ such that μ is nonatomic on A_0 and atomic on each $A_n, n = 1, 2, \ldots$

Let $b_0 = \int_{A_0} b \, d\mu$ and $c_0 = \int_{A_0} c \, d\mu$, and let λ_0 be a number with

$$b_0 \leq \lambda_0 \leq c_0 \qquad (5.6.32)$$

Consider the problem of maximizing (18), *with A_0 in place of A,* over measurable functions $\delta : A_0 \to$ reals satisfying (19) and

$$\int_{A_0} \delta \, d\mu = \lambda_0 \qquad (5.6.33)$$

The preceding theorem states that a best solution δ^o exists to this problem, since μ restricted to A_0 is nonatomic. For this best solution the utility function has the value

$$\int_{A_0} f(a, \delta^o(a)) \, \mu(da) \qquad (5.6.34)$$

Now δ^o, hence (34), depends on λ_0. We write $g_0(\lambda_0)$ for the value (34) as a function of λ_0. The domain of g_0 is $[b_0, c_0]$. Because of the special assumption (17), we interpret (34) as a (finite) definite integral, hence g_0 is real valued rather than pseudomeasure valued as in the general case.

Consider now the problem of maximizing

$$g_0(\lambda_0) + g_1(\lambda_1) + \cdots \qquad (5.6.35)$$

over all sequences $(\lambda_0, \lambda_1, \ldots)$ satisfying

$$b_n \leq \lambda_n \leq c_n \qquad (5.6.36)$$

for all $n = 0, 1, 2, \ldots$, and

$$\lambda_0 + \lambda_1 + \cdots = 0 \qquad (5.6.37)$$

Here λ_n, b_n, c_n, and g_n, $n = 1, 2, \ldots$, have exactly the same meanings as before, while $\lambda_0, b_0, c_0,$ and g_0 have just been defined.

If $(\lambda_0^o, \lambda_1^o, \ldots)$ is an optimal solution to (35)–(37), we can construct an optimal solution δ^o to the original problem (18)–(20) as follows. On A_0, let δ^o coincide with the optimal solution to the nonatomic problem (32)–(33), with parameter $\lambda_0 = \lambda_0^o$. On A_n for $n > 0$, choose any feasible δ satisfying (21) for λ_n^o. The resulting function δ^o is clearly feasible. It is also optimal, since the utility function (18) for any feasible δ does not exceed (35), where the λ_n's are determined from δ by (21); for $\delta = \delta^o$ the utility function is equal to (35), which is the maximum of its possible values.

It suffices, then, to show that (35)–(37) has an optimal solution. This is of exactly the same form as the problem of maximizing (23) subject to (25) and (26), with one possible exception: we do not know whether the function g_0 is *continuous*. If this could be shown, the first half of this proof demonstrates the existence of an optimal $(\lambda_0^o, \lambda_1^o, \ldots)$, and we would be finished.

We now show that g_0 *is* continuous. As a first step we show it is *concave*.[10] Let L_1, L_2, L_3 satisfy

$$\int_{A_0} b \, d\mu \leq L_1 < L_2 < L_3 \leq \int_{A_0} c \, d\mu \qquad (5.6.38)$$

and let $\delta_i^o : A_0 \to$ reals be the optimal solution for the parameter L_i, $i = 1, 2, 3$. The proof of the preceding theorem shows that these optimal solutions have shadow prices. Hence for δ_2^o an extended real number p^o exists such that $\delta_2^o(a)$ maximizes $f(a, x) - p^o x$ over $x \in [b(a), c(a)]$ for almost all $a \in A_0$. This p^o must be finite; for if $p^o = +\infty$, then $\delta_2^o = b$ almost everywhere on A_0, which contradicts (38); similarly, $\delta_2^o = c$ almost everywhere on A_0 if $p^o = -\infty$, again contradicting (38). It follows that

$$f(a, \delta_2^o(a)) - p^o \delta_2^o(a) \geq f(a, \delta_i^o(a)) - p^o \delta_i^o(a)$$

almost everywhere on A_0, $i = 1, 3$. Integration over A_0 yields $g_0(L_2) - g_0(L_i) \geq p^o(L_2 - L_i)$, $i = 1, 3$, so that

$$\frac{g_0(L_2) - g_0(L_1)}{L_2 - L_1} \geq p^o \geq \frac{g_0(L_3) - g_0(L_2)}{L_3 - L_2}$$

for all $L_1 < L_2 < L_3$, a condition equivalent to concavity.

Since a concave function is continuous at all interior points, the only thing left to prove is that g_o is continuous at the endpoints b_0 and c_0. Because g_0 is concave, to establish continuity at b_0 it suffices to prove that

$$g_0(b_0) = \lim_{k \to \infty} g_0(L_k) \qquad (5.6.39)$$

for *any one* sequence $L_1, L_2, \ldots$ converging to b_0.

We reintroduce the function $\gamma(a, p)$ given by (5). In the preceding proof it was established that for fixed $a \in A_0$, $\gamma(a, p)$ has the limit $b(a)$ as $p \to \infty$. Also, for fixed p, $\gamma(\cdot, p)$ is measurable, and

$$\int_{A_0} \gamma(a, p) \, \mu(da)$$

as a function of p, approaches

---

10. The concavity and continuity of g_0 are of some independent interest, yielding a "comparative statics" result for the preceding theorem, as the parameter L varies in (3).

$$\int_{A_0} \gamma(a, \infty) \, \mu(da) = \int_{A_0} b \, d\mu = b_0$$

as $p \to \infty$.

Let $p_1, p_2, \ldots$ be a sequence increasing without bound, and define $L_1, L_2, \ldots$ by

$$L_k = \int_{A_0} \gamma(a, p_k) \, \mu(da) \tag{5.6.40}$$

The sequence $L_1, L_2, \ldots$ then converges to b_0. Also,

$$g_0(L_k) = \int_{A_0} f(a, \gamma(a, p_k)) \, \mu(da)$$

This follows from the fact that the function $\delta_k^o(a) = \gamma(a, p_k)$ has the shadow price p_k and satisfies the resource constraint (40), hence it is optimal for L_k. Now for each $a \in A_0$

$$\lim_{k \to \infty} f(a, \gamma(a, p_k)) = f(a, b(a))$$

by the continuity of $f(a, \cdot)$. Hence, by (17), we may apply the dominated convergence theorem and conclude that

$$\lim_{k \to \infty} g_0(L_k) = \int_{A_0} f(a, b(a)) \, \mu(da) = g_0(b_0)$$

This proves (39) for the sequence $L_1, L_2, \ldots$ and establishes the continuity of g_0 at b_0. Continuity of g_0 at c_0 is proved by a similar argument, with $\beta(a, p)$ in place of $\gamma(a, p)$ and $p \to -\infty$.

This supplies the missing link in the proof, and we conclude that a best solution δ^o exists in the general case. ∎

As usual, there is an immediate generalization. If (20) in the preceding theorem (or (3) in the one before that) is replaced by the condition

$$L_o \leq \int_A \delta \, d\mu \leq L^o \tag{5.6.41}$$

with $\int_A \delta \, d\mu$ required to be finite, and (16) in the preceding theorem (or (1) in the one before that) is replaced by the condition

$$-\infty < \int_A b \, d\mu - L^o \leq 0 \leq \int_A c \, d\mu - L_o < \infty$$

then a best solution still exists in these respective cases.

The proof—which consists as always in transforming this problem into an equivalent one in which (20) and (16), or (3) and (1), hold—is left as an exercise.

We now give an example of a problem not having an optimal solution. Let $A = \{1, 2, 3, \ldots\}$, Σ = all subsets, μ the counting measure; $b(n) = 0$ and $c(n) = 1$ for all $n = 1, 2, \ldots$. The payoff function is $f(n, x) = -x/n^2$. The density function, in addition to satisfying $0 \leq \delta(n) \leq 1$ for all n, must satisfy $\delta(1) + \delta(2) + \cdots = 1$.

We can verify that all the premises of the preceding theorem are satisfied (take $\theta(n) = 1/n^2$ in (17); $L_o = L^o = 1$ in (41)), with one exception: $\int_A c \, d\mu = c(1) + c(2) + \cdots = \infty$.

There is no optimal solution to this problem, since any given feasible solution can be improved. To see this, let δ be feasible and choose any n for which $\delta(n) > 0$. Alter δ by replacing $\delta(n)$ by 0, and $\delta(n + 1)$ by $\delta(n + 1) + \delta(n)$, everything else the same. This remains feasible, and the change in the utility function is $\delta(n)[n^{-2} - (n + 1)^{-2}] > 0$, so δ is non-optimal. This example gives a certain insight into the role of the finiteness condition on $\int_A b \, d\mu$ and $\int_A c \, d\mu$.

5.7. UNIQUENESS OF OPTIMAL SOLUTIONS

By uniqueness we mean the property that there is *at most one* optimal solution. (The ordinary word "uniqueness" sometimes carries the connotation of "exactly one"; but we are not concerned here with existence, only with nonduplication of solutions.)

As before, we identify any two densities that are unequal only on a null set. Thus to say there is at most one optimal solution is to say: if δ_1 and δ_2 are both optimal solutions, $\mu\{a \mid \delta_1(a) \neq \delta_2(a)\} = 0$.

We need two new concepts for the following result.

Definition: f:reals $\rightarrow$ reals is *strictly concave* iff for any two distinct real numbers x, y and any $0 < t < 1$,

$$f(tx + (1 - t)y) > tf(x) + (1 - t)f(y) \qquad (5.7.1)$$

This is a bit stronger than concavity per se, because of the strict inequality in (1). A linear function is concave but not strictly concave.

Definition: Let $\mathbf{M}$ be a set of real-valued functions, all with domain A. Set $\mathbf{M}$ is *convex* iff for any δ_1, $\delta_2 \in \mathbf{M}$ and any $0 < t < 1$, the function $t\delta_1 + (1 - t)\delta_2$ belongs to $\mathbf{M}$.

For example, the feasible sets we have been dealing with throughout Sections 5.4, 5.5, and 5.6 are all convex.

Utility ordering is still that of standard ordering of pseudomeasures. The distinction between "best" and "unsurpassed" must again be stressed, because it is critical in the following result.

Theorem: Let (A, Σ, μ) be a measure space with μ σ-finite; let $f{:}A \times$ reals $\rightarrow$ reals be measurable and such that for all $a \in A$, $f(a, \cdot)$ is strictly concave; let **M** be a convex set of real-valued measurable functions. Then the problem, maximize

$$\int f(a, \delta(a))\, \mu(da) \qquad (5.7.2)$$

over $\delta \in$ **M**, has at most one *best* solution.

Proof: Let δ_1 and δ_2 both be best. Then for any $\delta \in$ **M** we have

$$\int_A [f(a, \delta_i(a)) - f(a, \delta(a))]\, \mu(da) \geq 0$$

$i = 1, 2$. Adding these two inequalities, we get

$$\int_A [f(a, \delta_1(a)) + f(a, \delta_2(a)) - 2f(a, \delta(a))]\, \mu(da) \geq 0 \qquad (5.7.3)$$

Now consider the function $\delta = \frac{1}{2}\delta_1 + \frac{1}{2}\delta_2$. This $\delta \in$ **M**, by convexity. For this δ the integrand in (3) is never positive, and is in fact negative on the set $\{a \mid \delta_1(a) \neq \delta_2(a)\}$, by strict concavity. Hence this set has measure zero, which establishes uniqueness. ∎

It is *not* true that there must be at most one *unsurpassed* solution, as the following counterexample shows.

Let (A, Σ, μ) be Lebesgue measure on the real line. Let $f(a, x) = -x^2 + 2x$ if $a \geq 0$; let $f(a, x) = -x^2 - 2x$ if $a < 0$; and let **M** be the set of constant functions whose single value lies in the closed interval $[-1, 1]$. Set **M** is obviously convex, and one sees that $f(a, \cdot)$ is strictly concave for all $a \in A$.

Now let x_1, x_2 be two numbers in $[-1, 1]$ with $x_1 > x_2$, and let ψ_1, ψ_2 be the pseudomeasures obtained by substituting the corresponding functions in (2). Since $f(a, \cdot)$ is increasing for $a \geq 0$ and decreasing for $a < 0$, $(\psi_1 - \psi_2)^+$ is a multiple of Lebesgue measure truncated to the positive half-line, while $(\psi_1 - \psi_2)^-$ is a multiple of Lebesgue measure truncated to the negative half-line. It follows that

$$(\psi_1 - \psi_2)^+(A) = (\psi_1 - \psi_2)^-(A) = \infty \qquad (5.7.4)$$

Hence *all* feasible solutions are unsurpassed because, by (4), no two of them are comparable.

5.8. POLICE, CRIMINALS, AND VICTIMS

We now apply the preceding theory to the problem of the spatial distribution of crime. This section goes beyond the simple optimization framework of the rest of this chapter, in that several different populations with diverse motives are interacting. Thus we are in a "game" situation, and what is optimal for one agent may not be optimal for another.

There is a population of potential *victims,* a population of potential *criminals,* who commit crimes upon the victims when the opportunity presents itself, and a population of *policemen,* who try to prevent criminals from perpetrating their misdeeds. The three populations inhabit the measure space (S, Σ, α), α being ideal area over Space, S. If v, c, and p are the densities of the three respective populations (density with respect to α), the *density of crimes* at location s is given by a function

$$f(v(s), c(s), p(s)) \qquad (5.8.1)$$

and total crime is then given by $\int_S f \, d\alpha$. We would expect f to be an increasing function of v and c and a decreasing function of p.

Consider for illustrative purposes the crime function

$$f(v, c, p) = vce^{-p} \qquad (5.8.2)$$

A semiplausible rationalization for (2) might run as follows. For $p = 0$, a crime occurs if there is an "encounter" between a potential victim and criminal, and with "random" movements the frequency of encounters should be proportional to the product of the densities vc.[11] Next suppose the commission of a crime is inhibited if a policeman is present within a certain "surveillance radius." If policemen are randomly distributed, the probability of no policeman being present within the critical radius declines exponentially with police density, and this gives (2). (Units of measurement for v, c, and p may be chosen to avoid multiplicative constants, as in (2).)

Whatever one thinks of such arguments, it is still illuminating to discuss the consequences of (2), or more generally (1), under various behavioral assumptions. We assume that victims and police distribute themselves over S so as to reduce crimes, while criminals distribute themselves so as to increase crimes.

One further objection to this setup should be mentioned. Should not these population distributions be integer valued or finitely concentrated?

---

11. Two similar cases come to mind. The law of mass action in chemistry takes reaction rate to be proportional to the concentrations of the reagents. In the Lotka-Volterra theory of predator-prey interaction, encounter frequency is again proportional to the product of the species densities. A. J. Lotka, *Elements of Mathematical Biology* (Dover, New York, 1956), pp. 88–95.

If so, they are unlikely to have density functions. The answer is that c, v, and p are best thought of, not as densities for cross-sectional distributions, but for the time averages arising from the random perambulations of the populations. Thus a measure μ, where $\mu(E)$ is the *expected* number of people of a certain type in region E, can very well have an areal density.

Before launching into details, we briefly consider some specific interpretations of this general model. "Crime" is a rather heterogeneous category, and not all types of crimes would be well represented by the model. First, there are numerous "crimes without victims":[12] gambling, traffic in drugs, prostitution, etc. In some of these cases the frequency might be described by (1) and (2) (e.g., "random" encounters with streetwalkers) but we would not expect the "victims" to be motivated to reduce the incidence of such "crimes." Second, there are crimes that do not require a specific encounter for their commission, e.g., counterfeiting or antitrust law violation.

Burglary, larceny, robbery, and rape are examples of types of crime that do not have these disqualifying features, and their incidence might be approximately represented by a function of the form (1). One might want to reinterpret v, c, or p in some of these cases. For example, in burglary the spatial distribution of (movable) *wealth* would seem more relevant than the spatial distribution of people, so v should perhaps be taken as wealth density rather than population density.

Certain noncrime situations may also be represented by this model. Consider military attacks against targets (installations, opposing forces, civilians). Let v be the density of targets, c the density of, say, bombing, and p the density of "defense equipment"; the above model might then predict the volume of destruction in terms of these three distributions. The controllers of p and v are motivated to reduce destruction, the controllers of c to increase it. Hence we might expect to find the spatial distributions here similar to those resulting from crime incidence.

Again, consider the following "imitation-snob" situation. There is a "high-prestige" and a "low-prestige" population. The high-prestigers want to avoid contacts with the low-prestigers, while the latter want to increase contacts with the former. Interpreting criminals and victims to be the low- and high-prestige populations respectively, and crimes to be contacts between the two populations, we get something like the model above. The police might enter as harassers of the low-prestige population in its attempts to increase contacts.

Our aim here is to develop and explore theoretical models, not to tailor them closely to any particular real-world situation. (For crimes

---

12. See E. M. Schur, *Crimes without Victims* (Prentice-Hall, Englewood Cliffs, N.J., 1965).

such an attempt would in any case be difficult because of the spotty quality of most crime data.)[13]

We now return to the formal model, which has not yet been completely specified. For simplicity we assume that the three populations are mutually exclusive and that no transformation from one to the other is possible. Thus we ignore the possibility that victims themselves can inhibit crimes by surveillance, the possibility that some criminals can be victimized by other criminals, etc.

Two cases will be explored. In the first there are no police (the anarchistic or "Wild West" case), and the two populations, victims and criminals, are freely mobile over S. In the second case the distribution of victims is *fixed*, and the remaining two populations are freely mobile. This might occur, for example, if crime is of minor importance so that it exerts no locational pull on the population at large. Another interpretation is that the population distribution of victims adjusts very slowly compared to the other two populations, so that it may be considered fixed in the short run.

A given population tries to reduce, or increase, crime. What does this mean? There are (at least) two interpretations: the *individualistic* and the *collusive*. If criminals act collusively the entire body of criminals distributes itself so as to maximize total crime; if they act individualistically, they will move from places where the density of crimes *per criminal* is low to where it is high. Similarly, if victims collude, they will distribute themselves to minimize total crime; if they are individualists, they move from places where the density of crimes *per victim* is high to where it is low. For some crime functions f the resulting distributions are the same under either assumption, but in general they will differ.

For the police the most plausible assumption is collusion: they are distributed by central headquarters to minimize total crime. For victims the individualistic assumption is more plausible: each potential victim moves to reduce the incidence of crime on himself. For criminals both possibilities are plausible, depending on whether crime is "petty," or "organized" by some criminal mastermind.

We analyze just three of the many possible combinations: (i) no police, both victims and criminals are individualists; (ii) no police, both victims and criminals collude; (iii) victim distribution fixed, both police and criminals collude.

---

13. U.S. President's Commission on Law Enforcement and Administration of Justice, *The Challenge of Crime in a Free Society* (Dutton, New York, 1968); M. E. Wolfgang, Urban Crime, in J. Q. Wilson, ed., *The Metropolitan Enigma,* Ch. 8 (Harvard Univ. Press, Cambridge, 1968), esp. pp. 253–63, 276–81.

Case (i): With no police we have a crime density function $f:\text{reals}^2 \to$ reals, viz., $f(v, c)$ is the crime density at a location if victim density there is v and criminal density is c. All functions are assumed to be measurable, finite, and nonnegative. We also assume that $f(0, c) = f(v, 0) = 0$ and that the right-hand partial derivatives $D_1 f(0, c)$ and $D_2 f(v, 0)$ exist for all $c, v \geq 0$. We use the notation

$$D_1 f(0, c) = \lim_{v \to 0^+} [f(v, c) - f(0, c)]/v$$

$$D_2 f(v, 0) = \lim_{c \to 0^+} [f(v, c) - f(v, 0)]/c$$

To avoid trivialities, we assume that the total population of victims V and criminals C are fixed positive real numbers, as is the total available area $\alpha(S)$. The constraints on the density functions v and c are that they be nonnegative real measurable and satisfy

$$\int_S v \, d\alpha = V, \qquad \int_S c \, d\alpha = C \tag{5.8.3}$$

Definition: The pair of feasible densities v^o, $c^o:S \to$ reals is an *individualistic equilibrium pair* iff there is a null set $E \in \Sigma$, and two real numbers $k_v, k_c \geq 0$ such that for all $s \in S \backslash E$,

$$f(v^o(s), c^o(s))/v^o(s) = k_v \qquad \text{if } v^o(s) > 0 \tag{5.8.4}$$

$$f(v^o(s), c^o(s))/c^o(s) = k_c \qquad \text{if } c^o(s) > 0 \tag{5.8.5}$$

$$D_1 f(0, c^o(s)) \geq k_v \qquad \text{if } v^o(s) = 0 \tag{5.8.6}$$

$$D_2 f(v^o(s), 0) \leq k_c \qquad \text{if } c^o(s) = 0 \tag{5.8.7}$$

The intuitive meaning of (4)–(7) is as follows. We interpret the "incidence of crime" on any victim at location s to be the *crimes per victim* at that point, which is the left side of (4) if $v(s) > 0$. If $v(s) = 0$, the natural interpretation is $D_1 f(0, c(s))$, which is what crimes per victim would be for a low-density migration there. Then (4) and (6) are precisely the conditions under which no potential victim can, by moving, reduce the incidence of crime on himself. Similarly, we take the "gain from crime" for any criminal at location s to be the *crimes per criminal* at that point, which is the left side of (5) or (7) for $c(s) > 0$, $c(s) = 0$ respectively. Then (5) and (7) are the conditions that no criminal can gain from moving. As usual, we allow exceptions on a set of measure zero. Two density functions v_1 and v_2, which differ only on a null set, are taken to be identical, and similarly for c_1 and c_2.

Case (ii): Now we go to collusive criminal-victim interaction. The set of feasible densities is again given by (3).

Definition: The pair of feasible densities v^o, c^o is a *collusive equilibrium pair* iff

(i) $\int f(v^o, c^o)\, d\alpha$ is unsurpassed in the set of pseudomeasures $\int f(v^o, c)\, d\alpha$, c ranging over the feasible criminal densities;

(ii) $-\int f(v^o, c^o)\, d\alpha$ is unsurpassed in the set of pseudomeasures $-\int f(v, c^o)\, d\alpha$, v ranging over the feasible victim densities.

That is, *given* the distribution v^o, criminals arrange themselves over S so that no other arrangement of criminals leads to a distribution of total crimes surpassing the one resulting from c^o; conversely, *given* c^o, v^o is chosen so that *minus* the distribution of total crimes is not surpassed by that resulting from any other victim distribution. The reason for the "minus" is of course that victims are trying to reduce total crime, which is equivalent to trying to increase minus total crime.

We must invoke pseudomeasures only if the total crime integral can be unbounded. For the present application we may safely assume that for any f arising in practice, all integrals are finite. Nonetheless, we give the more general definition above because the results obtained are valid for it and no extra work is involved.

If all integrals are finite, the above definition may be restated in simpler form: the feasible pair (v^o, c^o) is a *collusive equilibrium pair* iff

$$\int_S f(v^o, c)\, d\alpha \le \int_S f(v^o, c^o)\, d\alpha \le \int_S f(v, c^o)\, d\alpha \qquad (5.8.8)$$

for all feasible v, c. The left inequality in (8) states that given v^o, c^o is chosen to maximize total crime; the right inequality states that given c^o, v^o is chosen to minimize total crime.

Inequality (8) is precisely the saddlepoint condition that constitutes an equilibrium in 2-person zero-sum games. Since both sides are colluding, we have in effect just two decision makers; and the whole problem may be thought of as a game between a maximizing player, Crim, and a minimizing player, Vic, the payoff to Crim under the strategy pair (v, c) being $\int_S f(v, c)\, d\alpha$.

The difference between (4)–(7) on the one hand, and (8) on the other, is that in the former the individual victim or criminal does not take account of the effects of his moves on the gains or losses of his "colleagues." In this respect the difference is vaguely similar to that of a competitive vs. monopolistic industry respectively. To put the matter in a slightly different and slightly inaccurate way, under individualism *average* gains or losses are equated over S, while under collusion *marginal* gains or losses are equated.

Consider the very simple crime function

$$f(v, c) = vc \qquad (5.8.9)$$

which is (2) with p set equal to zero. One easily verifies that the pair of *uniform distributions*,

$$v^o(s) = V/\alpha(S), \qquad c^o(s) = C/\alpha(S) \qquad (5.8.10)$$

(almost everywhere) is an equilibrium pair under both definitions ($k_v = C/\alpha(S)$, $k_c = V/\alpha(S)$, (6) and (7) are satisfied with equality, and (8) is satisfied with equality for *any* feasible v, c, total crime being $VC/\alpha(S)$).

One also suspects that (10) is the *only* equilibrium pair under either definition. A nonrigorous argument would go as follows. Suppose that v were not constant almost everywhere. Then the criminals would all crowd into the region where v was densest. But then victims would not be in equilibrium, since they could move into the region vacated by criminals. A similar argument applies if c were not constant almost everywhere.[14]

Our aim now is to generalize this argument and make it rigorous. We prove this separately for the individualistic and collusive cases. Each of these cases has its own appropriate class of functions for which the statement is proved, and (9) belongs to both classes.

Theorem: Given measure space (S, Σ, α) with $\infty > \alpha(S) > 0$, positive real numbers V and C, and a measurable function f:nonnegative reals$^2 \rightarrow$ nonnegative reals that satisfies:

(i) $f(v, c) = 0$ iff either $v = 0$ or $c = 0$ (or both);

(ii) the right-hand partial derivatives $D_1 f(0, c)$ and $D_2 f(v, 0)$ exist for all real $v, c \geq 0$, and if $v > 0$, then $D_2 f(v, 0) > 0$;

(iii) $f(Vx, Cx)/x$ is a strictly increasing function of x $(x > 0)$.

Then (10) is the unique individualistic equilibrium pair.

Proof: We verify at once that (10) satisfies (4)–(7) almost everywhere: (4)–(5) follows from the constancy of v^o and c^o, while (6)–(7) are trivial because v^o and c^o are positive. Hence it remains only to show the uniqueness of this solution.

Let (v^o, c^o) be an individualistic equilibrium pair, and let $E = \{s \mid v^o(s) > 0\}$, $F = \{s \mid c^o(s) > 0\}$. Suppose first that $k_c = 0$. It follows

---

14. Using this argument as a guide we can develop a *dynamic* model of redistribution of the two populations from a nonequilibrium position. We do not go into this.

that $\alpha(E \cap F) = 0$, for otherwise (5) would be violated, since $f(v, c) > 0$ if $v > 0$ and $c > 0$. Also, we must have $\alpha(E \backslash F) = 0$; for otherwise (7) would be violated, since $D_2 f(v, 0) > 0$ on this set. But this means that $\alpha(E) = 0$, which implies the feasibility condition (3) is violated for v^o, since $V > 0$. We have a contradiction, and it follows that $k_c > 0$.

This implies that $\alpha(F \backslash E) = 0$; for otherwise (5) would be violated, since $f(0, c) = 0$. Hence $\alpha(E \cap F) > 0$; for otherwise $\alpha(F) = 0$, violating (3) for c^o. Now for almost all points s of $E \cap F$ we have

$$k_v v^o(s) = f(v^o(s), c^o(s)) = k_c c^o(s) \qquad (5.8.11)$$

from (4) and (5). Since $\alpha(E \cap F) > 0$, there exists s satisfying (11). Hence $k_v > 0$. It follows that $\alpha(E \backslash F) = 0$; for otherwise (4) would be violated, since $f(v, 0) = 0$. Also $\alpha(A \backslash (E \cup F)) = 0$; for otherwise (6) would be violated, since $D_1 f(0, 0) = 0$.

We have now shown that $v^o > 0$ and $c^o > 0$ almost everywhere, hence (11) is valid almost everywhere. Integrating (11) over S, we find, from (3), that $V k_v = C k_c$. Hence

$$v^o / V = c^o / C \qquad (5.8.12)$$

almost everywhere. Letting $x(s)$ be the common ratio in (12) at the point s, we find from (11) that

$$V k_v = f(V x(s), C x(s)) / x(s) = C k_c \qquad (5.8.13)$$

almost everywhere. The middle term in (13) is strictly increasing in x, hence there is just one solution x^o: v^o and c^o must be constant almost everywhere, which yields (10). ∎

We now give the corresponding result for collusive equilibrium. The uniqueness, rather than the existence, of equilibrium is the more interesting condition, and that is established in the following theorem.

Theorem: Let (S, Σ, α) be a measure space with $\infty > \alpha(S) > 0$ and α nonatomic; let V and C be positive real numbers; let measurable function f:nonnegative reals^2 → reals satisfy:

   (i) for any fixed real $c \geq 0$, $f(v, c)$ is continuous in v and differentiable with respect to v for $v > 0$ (notation: $D_1 f(v, c)$);

  (ii) for any fixed real $v \geq 0$, $f(v, c)$ is continuous in c and differentiable with respect to c for $c > 0$ (notation: $D_2 f(v, c)$);

 (iii) for any fixed real $c > 0$, $D_2 f(v, c)$ is strictly increasing in v;

 (iv) for any fixed real $v > 0$, $D_1 f(v, c)$ is strictly increasing in c.

Then (10) is the only possible collusive equilibrium pair.

Proof: Let v^o, c^o be a collusive equilibrium pair. Then c^o is unsurpassed for the problem of maximizing

$$\int f(v^o(s), c(s)) \, \alpha(ds) \tag{5.8.14}$$

over the nonnegative densities $c:S \to$ reals satisfying $\int_S c \, d\alpha = C$. Since α is nonatomic and $f(v, \cdot)$ is continuous, all $v \geq 0$, we have as a necessary condition for this that an extended real number k and a null set $E \in \Sigma$ exist such that, for all $s \in S \backslash E$, $c^o(s)$ maximizes

$$f(v^o(s), x) - kx \tag{5.8.15}$$

over nonnegative real x (Section 5.4. Recall the convention concerning infinite k values, pp. 244–45.)

The "shadow price" k must in fact be finite. For if $k = -\infty$, (15) has no maximizer; while if $k = +\infty$, c^o would be zero almost everywhere, which violates $\int_S c^o \, d\alpha = C > 0$.

We now show that if $s_1, s_2 \in S \backslash E$ and $v^o(s_1) > v^o(s_2)$, then

$$\text{either } c^o(s_1) < c^o(s_2), \quad \text{or } c^o(s_1) = c^o(s_2) = 0 \tag{5.8.16}$$

Condition (16) will be demonstrated by eliminating two possibilities.

Possibility A: $c^o(s_1) = c^o(s_2) > 0$. Since (15) is differentiable in x, the derivative must be zero at these respective points, i.e.,

$$D_2 f(v^o(s_i), c^o(s_i)) = k \tag{5.8.17}$$

$i = 1, 2$. But $D_2 f(\cdot, c')$ is strictly increasing, c' being the common positive value of $c^o(s_1) = c^o(s_2)$. Hence at least one of the two equations in (17) must be invalid.

Possibility B: $c^o(s_2) > c^o(s_1)$. Consider

$$f(v^o(s_1), x) - f(v^o(s_2), x) \tag{5.8.18}$$

as a function of the nonnegative real variable x. Function (18) is continuous and has a positive derivative for all $x > 0$. Hence, using the mean value theorem, (18) is strictly increasing in x. It follows that

$$f[v^o(s_1), c^o(s_2)] - f[v^o(s_2), c^o(s_2)] > f[v^o(s_1), c^o(s_1)] - f[v^o(s_2), c^o(s_1)] \tag{5.8.19}$$

Also, since $c^o(s_i)$ maximizes (15) for $s = s_i$, $i = 1, 2$, we get

$$f[v^o(s_1), c^o(s_1)] - kc^o(s_1) \geq f[v^o(s_1), c^o(s_2)] - kc^o(s_2) \tag{5.8.20}$$

and also (20) with subscripts 1 and 2 interchanged. Adding these three inequalities (19)–(20) and simplifying, we get the contradiction $0 > 0$. This eliminates possibility (B) and establishes (16).

This entire argument may now be repeated with the roles of v^o and

c^o interchanged, the only difference being that v^o is maximizing the *negative* of (14). Since $D_1[-f(v, c)]$ is strictly *decreasing* in c, we obtain the analog of (16) again, but with one inequality sign reversed: If s_1, $s_2 \in S \backslash E'$ (E' being a certain null set) and $c^o(s_1) > c^o(s_2)$, then

$$\text{either } v^o(s_1) < v^o(s_2), \text{ or } v^o(s_1) = v^o(s_2) = 0 \qquad (5.8.21)$$

Finally, suppose two points s_1, $s_2 \in S \backslash (E \cup E')$ exist such that $v^o(s_1) > v^o(s_2)$. We cannot also have $c^o(s_1) > c^o(s_2)$, for then (21) would lead to a contradiction. Hence $c^o(s_1) = c^o(s_2) = 0$, by (16). For any other point $s \in S \backslash (E \cup E')$, choose s_1 or s_2, depending on which s_i satisfies $v^o(s_i) \neq v^o(s)$. The argument just given then shows that $c^o(s) = 0$. Hence if v^o is not constant on $S \backslash (E \cup E')$, c^o is identically zero on this set. Since $E \cup E'$ is a null set, this gives the contradiction $0 = \int_S c^o \, d\alpha = C > 0$. Hence v^o is constant almost everywhere. A similar argument shows that c^o is constant almost everywhere. ∎

This does not prove that the pair of uniform densities (10) *is* a collusive equilibrium pair: There is the possibility that no such pair exists. To test whether (10) is a collusive equilibrium pair is not difficult. Under the premises of this theorem, the shadow price conditions are both necessary and sufficient for equilibrium (p. 249). Hence (10) is such a pair iff there are numbers k_1 and k_2 such that $C/\alpha(S)$ maximizes

$$f(V/\alpha(S), x) - k_1 x \qquad (5.8.22)$$

over $x \geq 0$, and $V/\alpha(S)$ maximizes

$$-f(x, C/\alpha(S)) - k_2 x \qquad (5.8.23)$$

over $x \geq 0$. If conditions (22) and (23) are added to the premises, we may assert that the pair of uniform distributions is the unique collusive equilibrium pair.

Some real-world situations illustrate these results, at least approximately. The policy of dispersing population to reduce losses from air attack is an example. If carried out to the limit, both targets and attackers would be uniformly distributed. In a uniform environment, a predator and a prey species would tend to become uniformly distributed. The distribution of Christians and lions in the Roman arena must have been roughly uniform.

Case (iii): Now we introduce police. The distribution of victims is given and fixed, while police and criminals are freely mobile. The former try to reduce, the latter to increase, total crimes. As the crime density function we take

$$f(v, c, p) = g(v)ce^{-p} \qquad (5.8.24)$$

where g is strictly increasing, nonnegative, and finite. This is just a slight generalization of (2). One can obtain results for more general functions, but (24) leads to a very simple and elegant equilibrium with some provocative implications.

In the following result, premise (26) is introduced to avoid an uninteresting complication; it is quite weak. Also premise (25) could have been deduced from rather weak assumptions, but we take the simpler course of assuming it outright. All logarithms are to base e.

Theorem: Let (S, Σ, α) be a measure space, with $\infty > \alpha(S) > 0$; let C and P be positive real numbers; let $v:S \rightarrow$ nonnegative reals be measurable, with $\int_S v \, d\alpha > 0$ (i.e., there are some victims); let g:nonnegative reals $\rightarrow$ nonnegative reals be strictly increasing. Assume:

(i) there is exactly one number L satisfying

$$\int_{\{s \mid g(v(s)) > L\}} \log\left[g(v(s))/L\right] \alpha(ds) = P \qquad (5.8.25)$$

(ii) for this L, $\alpha\{s \mid g(v(s)) = L\} = 0$. $\qquad (5.8.26)$

Then there is *exactly one* collusive equilibrium pair (c^o, p^o) for the problem, maximize over c, minimize over p: $\int g(v(s))c(s)e^{-p(s)} \alpha(ds)$. Here $c:S \rightarrow$ reals and $p:S \rightarrow$ reals must be nonnegative measurable and must satisfy

$$\int_S c \, d\alpha = C, \qquad \int_S p \, d\alpha = P \qquad (5.8.27)$$

Apart from a null set, c^o and p^o have the following form:

if $g(v(s)) \leq L$, then $c^o(s) = p^o(s) = 0$ $\qquad (5.8.28)$

If $g(v(s)) > L$, then

$$c^o(s) = C/\alpha\{s \mid g(v(s)) > L\} \qquad (5.8.29)$$

(a constant), and

$$p^o(s) = \log\left[g(v(s))/L\right] \qquad (5.8.30)$$

Proof: Let $g(v(s)) = h(s)$. First we show that (28)–(30) is the only possible collusive equilibrium pair. Let c^o, p^o be such a pair. The conditions for the existence of shadow prices are satisfied, since $h(s)xe^{-p^o(s)}$ is concave in x, and $-h(s)c^o(s)e^{-x}$ is concave in x. Hence there are extended real numbers k_c and k_p such that (except on a null set) $c^o(s)$ maximizes

$$h(s)e^{-p^o(s)}x - k_c x \qquad (5.8.31)$$

over $x \geq 0$, and $p^o(s)$ maximizes

$$-h(s)c^o(s)e^{-x} - k_p x \tag{5.8.32}$$

over $x \geq 0$.

We once and for all exclude the null set on which (31) or (32) is not maximized. Thus "all s" means "all s in the complement of this set"; "there is a point" refers to the complement, etc.

First, $k_c < \infty$; for if not, $c^o(s) = 0$, all s, violating (27). Similarly, $k_p < \infty$.

Next, $k_c > 0$. To see this, note that the assumptions on v and g imply that $\{s \mid h(s) > 0\}$ has positive measure. If $k_c \leq 0$, then on this set (31) would be strictly increasing in x, hence have no maximizer.

Next, $k_p > 0$. For if $h(s) = 0$, then $c^o(s) = 0$, since $k_c > 0$ in (31). From (27) there is a point s_1 for which $c^o(s_1) > 0$; hence $h(s_1) > 0$ also. Now if $k_p \leq 0$, then for point s_1 (32) would be strictly increasing, hence have no maximizer.

If $c^o(s) = 0$, then $p^o(s) = 0$. This follows from $k_p > 0$ in (32). If $c^o(s) > 0$, then (32) is strictly concave in x and hence has (at most) one maximizer. This maximizer is zero iff the slope at $x = 0$ is nonpositive. Thus we have

$$p^o(s) = 0 \qquad \text{iff} \qquad h(s)c^o(s) \leq k_p \tag{5.8.33}$$

If the slope at $x = 0$ is positive, the maximizer of (32) is obtained by setting the derivative equal to zero. Thus if $h(s)c^o(s) > k_p$, then

$$p^o(s) = \log[h(s)c^o(s)/k_p] \tag{5.8.34}$$

Now the set $\{s \mid p^o(s) > 0\}$ has positive measure from (27). Hence $c^o(s)$ maximizes (31) on this set. Substituting from (34) into (31), we find that $c^o(s)$ maximizes

$$[(k_p/c^o(s)) - k_c]x \tag{5.8.35}$$

over $x \geq 0$, for any s such that $p^o(s) > 0$. Since $c^o(s) > 0$, the bracketed expression in (35) must be zero:

$$c^o(s) = k_p/k_c \tag{5.8.36}$$

so that c^o is *constant* on the set $\{s \mid p^o(s) > 0\}$.

Next, the two sets $\{s \mid p^o(s) > 0\}$ and $\{s \mid h(s) > k_p/c^o(s)\}$ are the same. Hence from (34) and (27),

$$\int_{\{s \mid h(s) > k_p/c^o(s)\}} \log[h(s)c^o(s)/k_p]\,\alpha(ds) = P \tag{5.8.37}$$

From (37) and (25) we obtain $k_p/c^o(s) = L$ if $p^o(s) > 0$, so that, from (36),

$$k_c = L \tag{5.8.38}$$

Substituting (38) and (36) into (34), we obtain

$$p^o(s) = \log[h(s)/L] \qquad (5.8.39)$$

wherever $p^o(s) > 0$ (so that $h(s) > L$ on this set).

Now consider the maximization of (31). If $c^o(s) = 0$, then $p^o(s) = 0$, and the fact that $x = 0$ maximizes (31) implies $h(s) \leq k_c$. Hence, from (38), if $c^o(s) = 0$, then $h(s) \leq L$.

Next let $c^o(s) > 0$. The case where $p^o(s) > 0$ also holds has already been discussed, yielding (35) and (36). Suppose $p^o(s) = 0$. The fact that $x > 0$ maximizes (31) yields $h(s) = k_c = L$. But from (26) this occurs only on a set of measure zero.

We are now finished, as the two sets $E = \{s \mid p^o(s) > 0, \ c^o(s) > 0\}$ and $F = \{s \mid p^o(s) = 0, \ c^o(s) = 0\}$ together exhaust S, except for a null set. On set F, $g(v(s)) = h(s) \leq L$, and on set E, $g(v(s)) > L$. Condition (30) is the same as (39). Also, c^o is constant on the set $\{s \mid g(v(s)) > L\}$, and zero off it, so (29) follows from (27).

To show that (28)–(30) actually gives an equilibrium pair, we need only verify the shadow price conditions (31) and (32) for some k_c, k_p, since these are sufficient for unsurpassedness (in fact, for bestness). The pair (c^o, p^o) given by (28)–(30) do indeed maximize (31) and (32) respectively for $k_c = L$, $k_p = LC/\alpha\{s \mid g(v(s)) > L\}$. Verification is left as an exercise. ∎

The equilibrium solution (28)–(30) may be characterized as follows. There are two radically different regimes, a *high-density* regime I (characterized by victim densities satisfying $g(v) > L$) and a *low-density* regime II. In regime II there are no police, no criminals, and no crimes. In regime I, while density of police rises with that of victims, the density of criminals is uniform; so is the density of crimes, as one verifies by substituting (29) and (30) into (24).

This leads to the surprising conclusion that the most crime-ridden victims are those living at *intermediate* densities. For since crime is uniformly distributed in regime I, crimes *per victim* must be inversely proportional to the density of victims. Starting at the highest victim densities, crimes per victim rise as victim density falls, reaching a peak and then suddenly falling to zero as the critical density is passed and regime II is entered.

Here is another slightly paradoxical implication. Suppose there is an anticrime drive, and the total police force P is expanded. Since the integrand in (25) is nonincreasing in L, the new equilibrium L must be lower. Total crime, which equals CL, does indeed fall. But in the process the critical victim density falls, and regime I—which is $\{s \mid g(v(s)) > L\}$—expands at the expense of regime II. People who were living at densities just below the old critical density will suddenly find themselves engulfed in a crime wave, crimes per victim jumping from zero to the highest level in the system. All this is the result of increased law enforcement!

The explanation, of course, is that the increased "heat" on criminals in the old regime I induces them to disperse into the "greener pastures" of regime II. The "spillover" effect of law enforcement in one community on the crime rate in neighboring communities has been recognized. It is sometimes claimed that better law enforcement decreases crime in neighboring communities.[15] This may be true with respect to the *apprehension* of wanted criminals. But insofar as the police serve the function of *deterring* potential criminals from committing crimes, the argument just given indicates that better law enforcement may well *increase* crime in the environs.

This sort of two regime equilibrium is by no means unusual in game theory.[16] But we would expect any such effect to be blurred when applied to the real world. In general we do not find the intermediate density peaking of crimes per victim as predicted by this model (bank robbery may conform to this pattern, if we define "victim" density properly). Instead, the usual pattern is for crimes per victim to rise with size of place and to be inversely related to distance from the central city. However, there are many exceptions, and different types of crime have distinct patterns.[17]

The preceding model could have gone wrong in any number of ways. The three populations are not fixed in size and are not homogeneous. The crime function may be misspecified. Movement costs have been ignored. Finally, the motivations of the three populations may have been misspecified. In particular, it is not at all clear that police are allocated to minimize (an index of) total crime.[18,19]

---

15. For example, C. M. Tiebout, A pure theory of local expenditures, *J. Polit. Econ.* 64 (Oct. 1956):423.

16. See, e.g., M. Dresher, *Games of Strategy* (Prentice-Hall, Englewood Cliffs, N.J., 1961), pp. 124–27.

17. E. H. Sutherland and D. R. Cressey, *Principles of Criminology,* 7th ed. (Lippincott, Philadelphia, 1966), pp. 187–91.

18. On alternative criteria for the distribution of police, see C. S. Shoup, Standards for distributing a free governmental service: Crime prevention, *Public Finance* 19, no. 4 (1964): 383–92; and his *Public Finance* (Aldine, Chicago, 1969), pp. 115–18.

19. An early version of Section 5.8 may be found in the deleted Chapter 5 of the original manuscript of A. M. Faden *Essays in Spatial Economics,* Ph.D. Diss., Dept. Econ., Columbia Univ., 1967.

6

Markets

This chapter will take up two major topics: the set of feasible trading possibilities open to an individual, and the equilibrium of markets—in particular, the real-estate market. The first topic continues the feasibility discussion of Chapter 4, budget constraints being special kinds of institutional constraints. Budget constraints merit special treatment because of their peculiar form and great importance generally.

Our aim in this chapter is to bring these concepts within our measure-histories-activities framework. We touch on only a few high points and special cases. A comprehensive treatment is out of the question, since most of economics could be encompassed in this chapter.

6.1. BUDGET CONSTRAINTS

Consider the balance sheet of a given person, firm, or other agent at time t. The major items may be classified into physical assets, financial assets, liabilities, and net worth. *Physical assets* include all things owned by the agent in question, such as land, buildings, equipment, inventories, household goods. *Financial assets* include all monetary claims on other agents, such as cash (a claim on the banking system), accounts receivable, bonds, promissory notes. (Corporate stock is a borderline asset. In closely held firms we can think of the stock as representing the net assets of the corporation itself; in large publicly owned companies it functions more like a debt with uncertain face value and should perhaps be classified as a financial asset.) *Liabilities* are the claims against the agent in question by other agents. Thus every liability on one balance sheet is matched by a financial asset on some other balance sheet, and vice versa. The totality of all financial assets on all balance sheets combined (including government bodies, churches, universities, etc., and all foreigners as well) should equal the totality of all liabilities if no slipups in accounting have been made. Finally, *net worth* equals the value of physical assets plus financial assets minus liabilities. It follows that the totality of all net worths on all balance sheets combined should equal the totality of the values of physical assets on all balance sheets, since total financial assets cancel against total liabilities.

We comment briefly on this scheme. The neat dichotomization of assets becomes a little ragged upon examination. Actually, nearly all assets represent claims of one sort or another—in particular, claims against "trespass" in a generalized sense: to own a commodity means that no one else has the right to use it. It is a simplifying idealization to substitute the commodity itself for the bundle of rights and claims entailed by its ownership. A number of "intangible" assets do not fit easily into either the physical or financial category (e.g., patents, franchises, easements) but may be expressed in terms of claims and rights. "Goodwill" is not even a claim but is a reflection of the habits of trading partners.

Also, a number of items not customarily included among assets perhaps should be. These include government-owned resources placed at public disposal free or for a nominal fee, e.g., parks, roads, police and fire protection, the judicial system. The person himself (and perhaps some dependents) might be included among his physical assets: he owns his own body.

Finally, there is the important category of *control* of assets as opposed to ownership of assets. This includes *rentals* (of land, labor, etc.) and the holding of office and will be discussed further below.

Having discussed these complications briefly, we go to the opposite extreme and simplify the balance sheet for purposes of analysis. We assume there is just one homogeneous kind of financial asset (call it "bonds") that accumulates interest at rate $k(t)$ at time t. Interest is compounded continuously; $k(t)$ is assumed to be continuous and nonnegative.

Focusing on one economic agent, let $b(t)$ be his *net bondholding* at time t. This is defined as financial assets minus liabilities. If $b(t) > 0$, the agent in question is a net creditor at time t; if $b(t) < 0$, the agent is a net debtor. The sum of $b(t)$ over all agents must be identically zero for any time t.

Three influences are assumed to change b over time: the accumulation of interest, the sale of physical assets (which raises b), and the purchase of physical assets (which lowers b). Let $f_1(t)$, $f_2(t)$ be the rate at which physical assets are being sold and purchased respectively at time t, in dollar terms. These are assumed to be continuous, nonnegative functions for the time being.

We then have the differential equation:

$$Db(t) = k(t)b(t) + f_1(t) - f_2(t) \qquad (6.1.1)$$

The only term that needs comment in (1) is the interest term $k(t)b(t)$. This has the sign of $b(t)$, indicating that interest payments are positive for creditors and negative for debtors, so that (1) is correct for both these cases. Realistically, there should be several other terms on the right side: wages and rentals, taxes and transfers, etc. These are all being ignored for simplicity's sake.

A simple transformation allows us to get rid of the interest term in (1). Define *discounted net bondholding* to be the function $b':T \to$ reals given by

$$b'(t) = b(t)\exp\left[-\int_0^t k(t)\, dt\right] \qquad (6.1.2)$$

The integral in (2) is the ordinary Riemann integral. If $t < 0$, the standard convention of elementary calculus that $\int_0^t = -\int_t^0$ is followed. Similarly, we define *discounted sales* and *purchases* f_1' and f_2' by (2), with f_1, f_2 in place of b respectively. Equations (1) and (2) then imply that

$$Db'(t) = f_1'(t) - f_2'(t) \qquad (6.1.3)$$

so that the interest term drops out. This simplification is useful, and we use discounted values wherever possible. (If there are multiple interest rates or if the rate varies with b, this simplification is not available.)

We have taken sales and purchases to be *rates* that are continuous functions of Time. Let us now generalize this. Transactions occur not only (if at all) continuously but also in lumps. To incorporate this possibility, we take sales and purchases to be (bounded) *measures* λ_1 and λ_2 on universe set T. Thus $\lambda_1(G) =$ value of sales in period G for any Borel set G on the real line. We can use either current dollars or discounted dollars as our measurement units; for simplicity we use discounted dollars.

The differential equation (3) (or rather, its integral) then generalizes as follows. For any two moments t' and t'', with $t' < t''$, we have

$$b(t'') - b(t') = \lambda_1[t',t'') - \lambda_2[t',t'') \qquad (6.1.4)$$

Here $[t',t'')$ is the time interval between t' and t'', including the past endpoint t' but excluding the future endpoint t''; $b(t)$ is discounted net bondholding (the prime has been dropped).

A conventional element is involved in (4). If a "lumpy" transaction of positive measure occurs at the point t, it is arbitrary whether $b(t)$ is defined to include or exclude the transaction at t itself. According to (4) it *excludes* the transaction, and this makes b continuous from the past but not necessarily from the future.[1]

---

1. If $\lambda_2 = 0$, (4) states that b is a distribution function for λ_1 in the wide sense. This is so because we have defined distribution functions to be continuous from below, which in this case means from the past.

We now resolve the sale and purchase measure (given in dollar terms) into *prices* and *quantities*. That is, total sales will be a composite of sales in different markets at different prices. But what is a market? At the least, to identify a market one must know *what* is being sold, *where* it is being sold, and *when* it is being sold. This gives a triple (r, s, t) and suggests that markets be identified with points of $R \times S \times T$. The set of all markets will then be a subset $E_o \subseteq (R \times S \times T)$. Point (r, s, t) will belong to E_o iff the resource r is being sold at location s at time t. We assume that $E_o \in (\Sigma_r \times \Sigma_s \times \Sigma_t)$.

We first make the competitive assumption that a single ruling price prevails in each market. That is, a function $p{:}E_o \to$ reals exists, where $p(r, s, t)$ is the price at which resource r is sold at location s at time t; p is assumed to be measurable. In what follows we also assume p to be nonnegative, although negative prices (for "illth", or noxious "resources") can easily be handled. Here, as above, we distinguish between *current prices p* (those at which sales actually occur) and *discounted prices* p': $p'(r, s, t) = p(r, s, t)\exp[- \int_0^t k(t)\, dt]$.

The *sales* (or *exports*) of an agent over the set of markets will be given by a measure μ_1 on universe set E_o; μ_1 is in terms of physical quantities, not dollar values, and $\mu_1[E_o \cap (F \vee G \times H)]$ is the total mass of resources of types F sold in region G in period H. The *value* sold over various resource types, regions, and periods may now be expressed as an indefinite integral over universe set E_o:

$$\int p\, d\mu_1 \tag{6.1.5}$$

The value measure (5) is in either current or discounted dollars, depending on whether p is current or discounted prices. For simplicity, take (5) to be in discounted terms. The relation between (5) and the value of sales measure λ_1 is then given by

$$\lambda_1(G) = \int_{[E_o \cap (R \times S \times G)]} p\, d\mu_1 \tag{6.1.6}$$

for any $G \in \Sigma_t$. Equation (6) simply states that the (discounted) value of total sales in time period G is that over all Resources and all Space in that period. (If (5) is extended to universe set $R \times S \times T$ by defining it to be zero on $(R \times S \times T)\backslash E_o$, then λ_1 is the marginal on component space T of this extended measure.)

Similarly, there will be a (physical quantity) *purchase* (or *import*) measure μ_2 over E_o, leading to a value-of-purchase measure (5) over E_o, whose relation to λ_2 is given by (6) (μ_1 being replaced by μ_2 in (5) and (6)).

The analysis to this point consists essentially in having set out a number of accounting identities, and no *constraints* on the actions of the agent have yet been mentioned. The problem may be expressed as follows. What combinations of sale and purchase measures (μ_1, μ_2) are financially available to the agent? Or, in short, what are the exchange possibilities? There are many possible answers to this question, depending on institutional arrangements—in particular, on the structure of the capital market.

One simple and popular (though not very realistic) approach is to assume perfect information, including a knowledge of the time t^o at which the agent in question will die. The constraint then takes the simple form:

$$b(t^o) \geq 0 \qquad (6.1.7)$$

That is, the agent must have repaid all debts by the time he expires; or, more exactly, the amount owed to him must be at least as large as the amount he owes to others at this time.

We express (7) in terms of the sales and purchase measures μ_1 and μ_2. Consider the options open to the agent at time $t_o < t^o$, when he starts with an initial net bondholding of $b(t_o)$, which may be negative (starting in debt), positive, or zero. Substituting from (6) and (4) into (7), we obtain

$$b(t_o) + \int_{[E_o \cap (R \times S \times [t_o, t^o])]} p \, d\mu_1 \geq \int_{[E_o \cap (R \times S \times [t_o, t^o])]} p \, d\mu_2 \qquad (6.1.8)$$

which has a very simple interpretation. The right-hand term is the discounted value of all purchases made in the interval $t_o \leq t < t^o$; the middle term is the discounted value of all sales in that interval. Then (8) states that the present value of purchases cannot exceed the present value of sales plus (the present value of) the initial credit balance. This is a direct generalization of the familiar linear consumer budget constraint $y \geq p_1 x_1 + \cdots + p_n x_n$, where the "income" term y may be interpreted as net credit, and x_i is net purchases of commodity i; if $x_i < 0$, this indicates a sale rather than a purchase.

The form of (8) allows a lumpy transaction to occur at the initial point t_o, but not at the point of expiration t^o. This is an artifact of our definitions and could be altered if desired by a minor modification of (8).

The constraint (7) or (8) is attractive because of its linearity but rather tenuous from the viewpoint of realism. There is, first, the problem of what to do with entities that do not have a natural lifespan, such as corporations or government bodies. But even apart from this, this constraint allows arbitrarily high indebtedness at any time before t^o, which is clearly untrue of any existing credit system.

A better approximation to reality is obtained by introducing *collateral* requirements. These allow indebtedness (negative bondholding) up to a point determined by other assets. The other balance sheet categories must now be taken into account.

Suppose the agent's physical assets have been appraised and that $v(t)$ is the *value of physical assets* at time t. Also let $w(t)$ be his *net worth* at that time. The basic balance sheet identity is

$$w(t) = v(t) + b(t) \qquad (6.1.9)$$

This can be measured in either current or discounted dollars.

A simple form of collateral constraint is then

$$-b(t) \geq cw(t) \qquad (6.1.10)$$

for all $t \in T$, where c is some positive real constant. That is, one may go into debt up to some multiple of one's net worth. From (9), (10) can also be written in the equivalent form

$$-b(t) \geq v(t)c/(1 + c) \qquad (6.1.11)$$

all t, so that one can borrow a fraction of every dollar's worth of physical assets.

Still more realistic would be a condition taking account of the fact that physical assets vary considerably in their ability to serve as collateral. Best of all is real estate, which is easily appraised, durable, and cannot be absconded with. On the other hand, "human capital" (the value of a person's own body as measured, say, by the discounted value of net future earnings) is poor collateral because it is hard to appraise and its mobility and long payback period make repayment difficult to enforce. For this reason students find it difficult to obtain unsecured long-term educational loans. (Under other institutional conditions human capital could function well as collateral; e.g., consider indentured servitude.) This realistic complication could be represented by replacing the right side of (11) with a weighted sum, each class of assets multiplied by the appropriate fraction corresponding to its collateral-serving ability.

Special kinds of economic agents have special kinds of budget conditions constraining them. Government bodies are limited by legislative appropriations, banks by reserve requirements. (To express the latter, one must distinguish the various categories of financial assets and liabilities; it will no longer do to lump them together as "bonds" as we have been doing.)

6.2. RENTALS

We now add the possibility of rental transactions to those of sale and purchase. Rentals are very important, much more so than might be inferred from their modest share of national income.

Abstractly, a rental transaction occurs when one relinquishes *control* but not *ownership* of an object. The most important type of rental by far is the *employment relation,* in which the worker places himself (within limits) at the disposal of the employer without relinquishing ownership of his own body; i.e., without becoming the employer's slave. Then we have real-estate rentals, leading to the ordinary *landlord-tenant relation*. And there are a large number of miscellaneous rental markets; e.g., for cars, furniture, machinery, costumes.

A number of points need clarification. First, what is ownership and what is control? We are not concerned here with any strict legal definitions, but with the functional concepts as they relate to the set of feasible options open to an agent.

To *control* an object for a certain time interval, as the term is used above, means to secure the acquiescence of other people not to interfere with one's use of the object or try to use it themselves. In the employment relation the "object" is another person, and the relation entails a willingness to obey orders within a certain "legitimate" range.

Ownership may now be defined in terms of control. To *own* an object means either to have permanent control of it or (in case it is rented out) to reacquire permanent control at some stipulated future date. In brief, the owner of an object is the agent to whom control ultimately reverts.

Realistic complications cloud these neat concepts. Control is a matter of degree: much of the legal system consists of restrictions on the uses to which an agent can put the objects he owns or rents. Restrictions are especially important in the case of rentals, for the owner will rarely relinquish control without stipulating limits on the uses to which the rented object is to be put. If nothing else, the owner has an interest in the maintenance of the rented object, since it will eventually revert to his own use.

The essence of the rental relationship, then, is *serial control*. The owner contracts with someone to give up control of the object for a limited time in exchange for a rental payment. (In the employment relation this of course is the wage.) There may be a whole sequence of such renters with whom the owner contracts in turn, as when a worker moves from job to job, or a landlord rents to tenant after tenant. Another possibility (if this is permitted by the rental contract) is for the renter in turn to relinquish control to a third party, giving a *subleasing* arrangement; he in turn could rent to a fourth party, etc.[2]

The same object may concurrently be changing ownership through outright sales. The times ownership changes hands need not coincide with the times control changes hands. Finally, ownership or control at any stage can be exercised jointly, with power centered in some committee of separate interests. The pattern of control can become rather tangled.

We briefly consider the relation between rentals and *services*. Ac-

---

2. The real-estate market is especially rich in having diverse agents with interests in the same parcel: owners, developers, builders, tenants, holders of easements, government agencies. See R. Turvey, *The Economics of Real Property* (Allen and Unwin, London, 1957), pp. 4–5; and W. L. C. Wheaton, Public and Private Agents of Change in Urban Expansion, in M. M. Webber et al., *Explorations into Urban Structure* (Univ. Pa. Press, Philadelphia, 1964), pp. 171–75.

cording to our previous discussion (Section 2.7) a service activity is one in which the histories that enter the activity are owned by different agents. Now if *B* rents an object he owns to *A* (the "object" may be *B* himself as an employee), and *A* uses this object (together with others that *A* owns) to run an activity, we may speak of *B* providing a service for *A*. Thus a worker provides labor services, a landlord provides housing services, etc. The "rental" and the "service" are just two aspects of the same relation— the former concentrating on the transaction that brings the factors together, the latter on the activity in which the factors jointly participate.

However, rental is not the only way diversely owned factors are brought together. The issue revolves around which agent is in control of the process, the recipient of service *A* or the provider *B*. Consider the service of watch repairing. Here the owner *A* surrenders control of his watch to the repairman *B*. But far from receiving a rental payment for this surrender of control, *A* actually pays *B* for the service rendered.

The relation in which *A* and *B* stand in this case is not that of employer and worker or tenant and landlord, as in the case of rentals, but that of *bailor* and *bailee* respectively. Without worrying about the legal niceties involved,[3] we refer to the general relation that obtains here as a *bailment* relation. In bailments, *A* relinquishes control of an object he owns to *B*, *B* performs a service that benefits the object, *B* returns the object to its owner *A*, and *A* pays *B* for the service. In rentals it is *B* who relinquishes control to *A*, who uses the object for his own benefit and again pays *B* for this service.

Bailment relations are very common, perhaps almost as common as rentals. Most repair services are bailments, including "repairs" to *A* himself, as by surgeons and barbers. Storage services provided by warehouses and transportation services provided by postmen or by common carriers for a person or his goods are other examples. The large publicly owned corporation may be thought of as involving a bailment relation in which the physical assets of the corporation are turned over to the control of management by the stockholders collectively.[4]

We have so far divided services into rental and bailment types, depending on who is in control. But control is a matter of degree, and there are borderline cases where one type blends into the other. An unskilled

---

3. Cf. Bailment, in H. C. Black, *Black's Law Dictionary*, rev. 4th ed. (West Publ. Co., St. Paul, 1968), pp. 179–80. None of the meanings described in this article is quite identical with the one we are using.

4. The "separation of ownership and control" as a social problem was first broached in connection with corporations. See A. A. Berle, Jr., and G. C. Means, *The Modern Corporation and Private Property* (Commerce Clearing House, New York, 1932). Our analysis indicates this separation is a universal phenomenon. It is true that in the corporate case there are special institutional obstacles to having the assets revert to the control of their legal owners. But compare this situation with the case of self-perpetuating boards of trustees or church hierarchies, where there are no legal owners at all!

worker and a surgeon both provide services; the first seems clearly involved in a rental relation, the second in a bailment. At skill levels intermediate between these two the control pattern will shift gradually from
one of these forms to the other. We thus get situations of shared control.

Finally, consider social activities such as parties, picnics and beach
outings, where the various participants provide each other with "com-
panionship" services. These do not fit either of the categories above, and
control itself (in the sense of a single agent coordinating the factors entering the activity, without interference from any other agent) may not exist.

Having examined some institutional features of rental and related
markets, we turn to the problem of the determination of rental *prices*.
Assuming competitive markets, one's first impulse is to imitate the structure of the sales market and postulate a rental price function π, whose
domain is a subset of $R \times S \times T$: $\pi(r, s, t)$ is the rental price for resource
type r at location s at time t. Function π would have the dimension of
money per unit mass *per unit time* (e.g., wage in dollars per man-hour,
land rent in dollars per acre-year).

This may be a fair approximation, and many rental markets appear
to have a structure resembling this. But it has one basic shortcoming, viz.,
that the rent does not depend on how the resource is going to be used by
the renter. Since the owner will eventually regain control of his property,
he will not be indifferent between uses that leave his property dilapidated,
leave it unaffected, or enhance its value. On the contrary, a premium
would be required for him to rent to someone who will dilapidate his
property, while he will be willing to accept a lower rent from someone
who will return his property in improved condition. Indeed, if the improvement is big enough, he will be willing to accept a *negative* rent, i.e.,
to pay to have his property used by the other person. In this case the
direction of service is reversed, and the rental relation has in fact become
a bailment relation.

Thus a worker would be willing to accept a lower wage on a job that
affords training opportunities than on one that does not.[5] If the training
opportunities were sufficiently rich, he might even accept a negative wage
(we would probably call this "going to school" rather than "working"; the
borderline between these cases is not sharp).

"Dilapidation" and "enhancement" refer to the position of the
rented object in Resource space R upon reversion to the owner. But the
same analysis applies to physical Space S. A car-rental agency will demand a premium payment from someone who wants to return the car at
some out-of-the-way place. The potential renter, furthermore, may be
concerned not only with the end state of the rented object but with the

---

5. See G. S. Becker, *Human Capital* (Nat. Bur. Econ. Res., New York, 1964), Ch. 2, for
an extended analysis.

entire time path over which it moves and the activity in which it participates. In the employment relation the worker will be concerned with the
pleasantness or unpleasantness of working conditions and require a
premium for working under poor conditions. The landlord may require
that his premises not be used for certain disapproved activities, or at least
that he be paid a premium if they are.

Discrimination may be considered a special case of preferences concerning alternative activities into which one's rented property enters. It
refers to preferences among alternative individuals or types of people
participating in these activities. Discrimination per se refers to a preference for nonassociation with someone, while "nepotism" refers to a preference for association.[6] We speak of this as a special case of preferences
over activities because the definition of "activity" gives the distribution of
mass over the entities participating in it and so will specify the types of
people involved. One may discriminate, in the first instance, with reference to a trading partner, and second, with reference to that person's associates.

Discrimination exists also on the sales market, but it is probably
more important in relation to rentals because the association between
trading partners is closer and lasts longer when rentals are involved, and
attraction or aversion is therefore likely to be more salient. For this reason we omitted any discussion of discrimination in connection with outright sales.

In summary, it would appear that any model of the rental market that
postulates a rental function $\pi(r, s, t)$ is inadequate. (The cases where such
a rental function does obtain seem to be those where the uses of the rented
object are so circumscribed, by custom or by explicit agreement, that one
need not be concerned by their variation.)

What, then, does an adequate representation of the rental market
look like? The following model incorporates some of the considerations
discussed above. It makes rentals depend, not on points of $R \times S \times T$,
but of $(R \times S \times T)^2$. For $t_1 < t_2$, $\pi(r_1, s_1, t_1, r_2, s_2, t_2)$ is the price to be
paid for attaining control of a unit of resource r_1 at location s_1 at time t_1,
and for relinquishing control of a unit of r_2 at s_2 at t_2. Typically, the mass
will flow along a history whose graph connects (r_1, s_1, t_1) and (r_2, s_2, t_2),
so that the "same" object is returned; but this formulation is somewhat
more general (e.g., one may borrow a cup of sugar one day and return a—
presumably different—cup of sugar the next day). By allowing π to take
on negative values, bailments as well as rentals may be encompassed.

Function π has the dimensions "dollars per unit mass," just as prices
in general do. We avoid complications by taking these to be *discounted*

---

6. G. S. Becker, *The Economics of Discrimination,* 2nd ed. (Univ. Chicago Press,
Chicago, 1971), p. 15.

dollars. Otherwise, e.g., we must worry about whether the rental is to be paid at the beginning or the end of the period or in periodic payments.

If one abstracts from legal complexities, tax liabilities, credit considerations, control restrictions, market frictions and imperfections, etc., a rental transaction may be thought of as a combination of two sale transactions. The agent acquiring temporary control in effect buys the object at the beginning of the rental interval and sells it back at the end of the period. In fact, if all the appropriate markets exist and the just-mentioned complications do not occur, one can give an informal argument for the following equality. For $t_1 < t_2$,

$$\pi(r_1, s_1, t_1, r_2, s_2, t_2) = p(r_1, s_1, t_1) - p(r_2, s_2, t_2) \tag{6.2.1}$$

all prices being measured in discounted dollars. For if the left side were larger than the right, one could buy a unit of r_1 at (s_1, t_1), immediately rent it out, receive back a unit of r_2 at (s_2, t_2), immediately resell it, and emerge with a positive profit. If the left side were smaller than the right, one could acquire control over a unit of r_1 at (s_1, t_1), immediately sell it, then buy a unit of r_2 at (s_2, t_2) and hand it over to complete the rental transaction, again making a positive profit. With perfect information, arbitrage assures that neither of these inequalities obtains.

6.3. IMPERFECT MARKETS

Until now we have been making the following competitive assumptions. (i) The agent is faced with a price system $p:E_o \to$ reals that does not depend on how much he buys or sells, and (ii) there is a unique discount rate $k:T \to$ reals that does not depend on the creditor-debtor position of the agent. We now briefly discuss weakening one or both of these conditions.

Let us first abandon condition (i), while keeping the perfect capital market assumption (ii). Let μ_1 and μ_2 be sales and purchases (in physical, not value, terms). It will be convenient to consider *net sales* $\mu = \mu_1 - \mu_2$, where μ is a *signed* measure, assumed bounded. Under competitive assumption (i) the *net revenue* obtained from μ will be

$$f(\mu) = \int_{E_o} p \, d\mu \tag{6.3.1}$$

Here p is assumed to be bounded measurable. Both p and net revenue are measured in discounted dollars.

The function f defined by (1) is *linear;* i.e., $f(\mu' + \mu'') = f(\mu') + f(\mu'')$, and $f(c\mu) = cf(\mu)$ for any two bounded signed measures μ', μ'' and any real number c. But for imperfect markets, net revenue will in general be a *nonlinear* function of net sales, and the problem arises of how to represent such functions in a convenient and plausible way.

Of the many possibilities, we consider here only representation by *densities*. This is a natural generalization of (1) that appears to correspond quite well with the generalization from perfect to imperfect markets. We define things abstractly. Let (A, Σ) be a measurable space and **M** be the set of all σ-finite signed measures on it. Let $g:A \times$ reals $\rightarrow$ reals be bounded measurable and α be a fixed bounded measure on (A, Σ). In terms of g and α we define the function $f:\mathbf{M}' \rightarrow$ reals by

$$f(\mu) = \int_A g(a, \delta(a))\ \alpha(da) \qquad (6.3.2)$$

Here **M**' is the subset of **M** consisting of those signed measures that are *absolutely continuous* with respect to α, and δ is the density of μ with respect to α:

$$\mu = \int \delta\ d\alpha \qquad (6.3.3)$$

Density δ exists by the Radon-Nikodym theorem. It is not unique, but any two densities for the same μ differ at most on a set of α-measure zero; hence they give the same value in (2), so that f is well defined.

To see the connection between (1) and (2), consider the function $g(a, \delta) = p(a) \cdot \delta$. By (3), (2) then reduces to (1) (with A in place of E_o).[7] But g in (2) can also be nonlinear in δ, which leads to a nonlinear f.

The interpretation of (2) is as follows: $g(a, \cdot)$ is the function giving net revenue density in terms of net sales density at the point $a \in A$. Thus for $\delta(a) > 0$, $g(a, \delta(a))/\delta(a)$ is the demand curve at a, and for $\delta(a) < 0$ it is the supply curve. In both cases, quantity is the independent variable and price is the dependent variable. All densities are with respect to α.

The interpretation of α depends on the space A. Suppose first that we are dealing with just a single commodity and that A is a bounded region of physical Space. (Net sales and net revenue may be thought of as steady flows per unit time). Then the natural interpretation for α is *areal measure*, and $g(a, \cdot)$ gives revenue in dollars per acre (per year) in terms of sales in, say, tons per acre (per year). For some problems this may be too restrictive. Suppose, for example, there are cities (represented as geometric points) at which one can garner positive revenue. Ordinary areal measure assigns measure zero to single points, which precludes representation in the form (2). This is easily remedied. To areal measure per se, we add a measure assigning unit mass to any point at which a city exists and let this sum be α. Then representation by (2) is again possible; if there is a city at point a_o, $g(a_o, \cdot)$ gives net revenue per year accruing at a_o in terms of sales per year at that point.

Next let A be a bounded subset of Space-Time, $S \times T$; we are again

---

7. Strictly speaking, (1) is not a special case of (2) because the function g just given is not bounded and (2) is valid only for the subset **M**'.

dealing with a single commodity. There is again a "natural" interpretation for α, viz., as the product measure formed from areal measure on S and Lebesgue measure on T. Just as above, if markets are concentrated so that positive revenue accrues on a set of product measure zero, α may be modified to make representation in the form (2) possible.

Finally, we come to the case we started with, where $A = E_o \subseteq (R \times S \times T)$. In this case there does not seem to be any natural measure α because there is nothing for R that corresponds to area for S and "quantity of time" (Lebesgue measure) for T. This creates no difficulty if there are just a finite number of resource types or even a countable number, say $\{r_1, r_2, \ldots\}$; for in this case we choose a measure assigning a positive mass to each $\{r_n\}$, say 2^{-n}, and then take the product of this with the S and T measures. And even in the general case there may be an α for which representation (2) is plausible.

Note that (2) has the same form as the utility function of the allocation-of-effort problem, Chapter 5. Thus the problem of maximizing total net revenue in an imperfect market system, with a given endowment of goods, is encompassed in the results of that chapter.

However, (2) does have one rather important shortcoming as a representation of imperfect markets. While it allows variable prices, each price depends only on the quantity forthcoming in its own market (all "cross-elasticities" equal zero). In reality, we would expect that a greater sale in a market would depress price not only there but in markets for similar resource types at nearby space-time points. We do not go into the problem of representing this phenomenon.

We now turn to imperfections in the capital market. It will still be assumed that there is just one type of financial asset, "bonds," so that the model remains highly simplified. But the discount rate k will now be "personalized," and will depend on the creditor position of the agent in question (as well as on the time). A plausible method is to let k depend on b/v, the ratio of net bondholding to the value of physical assets of the agent. Thus the personal discount rate at time t will be

$$k(t, b(t)/v(t)) \tag{6.3.4}$$

where k is a decreasing (or at least nonincreasing) function of its second argument. This reflects the following real-world situation. First, take the case of a creditor ($b > 0$, hence $b/v > 0$). As he extends more and more credit, the investment opportunities become progressively less attractive, so that the average rate of return k declines. Next consider a debtor (b and $b/v < 0$). For small debts one can rely on relatives and friends. Then one might try commercial banks. After the line of credit is exhausted here, one might try the "friendly finance" companies. And if even this does not suffice, there are always loan sharks and racketeers (who have a comparative advantage in enforcing collection of debt). At each stage a person's credit rating becomes shakier, and lenders compensate by charging higher interest rates. Thus k rises as b/v becomes more negative, and it is therefore again a decreasing function of its second argument.

The collateral constraint (1.11) may be interpreted in terms of (4). Suppose there is some negative value of b/v at which k goes to $+\infty$, which is to say that no more credit is forthcoming from any source beyond this point. We then get a lower bound constraint on b/v, which in (1.11) is equal to $-c/(1 + c)$.

With (4), one can no longer speak unambiguously of "discounted dollars" because the size of discount itself depends on the agent's actions. The basic differential equation (1.1) connecting bondholding and sales becomes

$$Db(t) = k[t, b(t)/v(t)]b(t) + f(t) \tag{6.3.5}$$

where $f(t)$ is the net rate (in current dollars) at which physical assets are being sold at time t (net rate means sales minus purchases). In general, (5) will no longer be integrable in elementary form.

Then (5), combined with given initial condition $b(t_o)$, $v(t_o)$ and either constraints (1.7) or (1.10)–(1.11), yields a system of conditions that indicate what triples of time paths $(f(t), b(t), v(t))$ are feasible.

Finally, let us combine imperfections in the commodity and capital markets. One additional problem now arises. Prices and revenues in commodity markets can no longer be expressed in discounted dollars; instead, we express them in current dollars.[8]

We must complete our system of conditions by expressing the current dollar net rate of sales, $f(t)$ in (5), in terms of the signed measure of net sales μ. (As above, the universe set of μ is $E_o \subseteq (R \times S \times T)$, the set of triples (r, s, t) for which markets exist; μ is measured in mass units.) A measure α on E_o is needed to express net revenue in the form (2). For this, we postulate that a measure β on $(R \times S, \Sigma_r \times \Sigma_s)$ has already been arrived at by some such process as discussed above. The product measure of β on $R \times S$ and Lebesgue measure on T, restricted to E_o, will be taken for α.

Signed measures μ that are not absolutely continuous with respect to α are dismissed at once as infeasible. For any other μ we form the density function $\delta = d\mu/d\alpha$ and substitute in (2), which can be expressed as an iterated integral, first with respect to β over $R \times S$, then with respect to Lebesgue measure over T. The first iteration is the one yielding the

---

8. One could insert an intermediate step here, deflating by a price index $P(t)$ to get measurements in "constant dollars." For consistency, one then must subtract the "rate of inflation" $DP(t)/P(t)$ from (4) to get the "real" rate of discount. One then proceeds as above, with "real" or "constant" values in place of "current" values.

net rate of sales function:

$$f(t) = \int_{\{(r,s) \mid (r,s,t) \in E_0\}} g[r, s, t, \delta(r, s, t)] \, \beta(dr, ds) \qquad (6.3.6)$$

Here g is net revenue density, per unit time per "unit β." Performing the integration, we get the net revenue per unit time, which is $f(t)$. Then (6) substituted in (5) yields the basic relation between net bondholding $b(t)$ and net sales μ. This, with the other conditions mentioned after (5), then gives the set of feasible triples $(\mu, b(t), v(t))$.

6.4. THE REAL-ESTATE MARKET

The real-estate market distributes the control and ownership of Space-Time among economic agents. That is, while a typical commodity market is characterized by a triple (r, s, t), we are now dealing with pairs (s, t). *The* real-estate market is then the ensemble of all these separate point markets. The "homogeneity" of S and T, as opposed to the "heterogeneity" of R, makes the structure of the market here a good deal simpler than in the commodity case.

There are a number of different interpretations as to what exactly is being sold or rented in these markets. Real estate—or "land"—refers ambiguously to the earth itself, with its soil, forests, waters, minerals, air, etc.; to the products of human construction more or less permanently affixed to it, such as buildings, roads, bridges; or to the Space-Time continuum that they occupy.

One can distinguish conceptually between control of these various components. Control over a portion of Space-Time is the right to exclude trespass by other agents, with their properties and associated activities. This is not quite the same as control over the use to which a building in the region may be put. In practice, of course, it would be inconvenient to have these rights in the hands of agents with opposing wills. Hence *control* is generally vested in one agent, although there may be separate markets for the components of a real-estate package, and *ownership* may be scattered among several agents.

For most of our analysis it does not matter which of the various possible interpretations is used: whether the real-estate transaction is just for the Space-Time "shell," or whether the constructed or natural contents of the region are part of the package. We ignore the ambiguity whenever the analysis applies to any of the interpretations.

The real-estate market has two further simplifying peculiarities as compared with commodity markets. First, the amount of Space-Time anywhere is fixed. Second, the control or ownership of each point (s, t) is usually in the hands of *one* agent. This suggests that we represent the control or ownership of any one agent as a (measurable) *subset* of $S \times T$, rather than as a measure over $S \times T$.

In more detail, we suppose that each agent i chooses a subset $E_i \in (\Sigma_s \times \Sigma_t)$. The chosen subsets for different agents are disjoint, so that the collection of all the E_i's is a packing. Thus E_i is that portion of Space-Time under the exclusive control (or ownership) of agent i. (Later we generalize to allow for joint ownership or control, but the model just given will serve for most of this section.)

As an example, E_i might be of the form $F \times G$, where F is a region of S and G is an interval of T, say (t_1, t_2). This means that agent i acquires region F at moment t_1, keeps it until moment t_2, at which time he divests himself of it, and has no other portion of $S \times T$. More generally, E_i might be a union of such "rectangles," different regions being held for different time intervals. Thus E_i might consist of pieces scattered all over the world, as might be the case if agent i is an international corporation. The size, shape, number, and duration of holdings are all determined by the set E_i.

We next consider feasibility restrictions on the possible sets E that can be chosen. Some restrictions hold for all agents, while others hold for selected types of agents. We first consider universal restrictions. A number of these arise from the need for informational economy. This leads to a restriction in the variety of possible transactions. Thus real-estate transactions are almost universally of the rectangular form mentioned above: a parcel F is acquired for a time interval G. (This includes the case of outright sale of a parcel. If the parcel is never resold, the time interval G extends to the infinite future.) There are usually further restrictions on F and G. For F in particular, it is typical to partition a portion of the earth's surface into *lots,* with the stipulation that the lot must change hands as a unit. If F is a lot and F_i is the region held by agent i, then either $F \subseteq F_i$, or $F \cap F_i = \emptyset$.

As for the vertical dimension, one can imagine the lot F as actually representing a three-dimensional cone, with its apex at the center of the earth and projecting through F at the earth's surface to infinity. But subsoil rights are sometimes transacted for separately, and it is not at all clear how high a person's air rights extend. Would someone have the right to build a structure on his property so tall that it interfered with airline flight paths?

The available sets E would be restricted to unions of allowable rectangles. A good portion of the earth's surface is not available for transactions at any given time, e.g., unclaimed territory, the high seas, and the public domain, including the road and street system. There may be maximal limits to the holdings of any one agent in certain regions, a result of land reform movements. As for particularistic restrictions on landholdings, these have been applied historically to aliens and certain minority groups such as the Jews in Russia or the Japanese in California.

Private restrictive covenants will limit the market still further for certain groups.

We now go a step beyond our analysis of commodity markets (which stopped with a discussion of feasibility conditions) and investigate the full conditions of equilibrium in the real-estate market. This involves a discussion of the *preferences* of the market participants and the conditions under which a pricing system clears the market. We end with a proof of the existence of equilibrium under certain simplifying assumptions.

Ideally, we want a model that simultaneously determines the pattern of ownership over Space-Time, the pattern of control, and the pattern of land uses because these three systems are interrelated. An agent wants control of a certain region to operate certain activities there. Which activities are feasible depends on a number of factors. Among these are technical knowledge, budgetary limits, legal constraints such as housing and zoning laws, activities in adjacent regions (in the case of neighborhood effects), and, in particular, the capital endowment resulting from previous land uses on the same site. The desirability of various regions to an agent is a reflection of the desirability of these possible uses to him.

As for ownership, there is (apart from any "psychic income" received from having a stake in the land), a comparative advantage from the agent's owning land that he controls because the inevitable frictions and inefficiencies that accompany the rental relationship are avoided. In the real world a very close association exists between the pattern of ownership and the pattern of control, which results from this phenomenon.

The structure of equilibrium in the real-estate market needs further specification. We suppose there is a perfect capital market, so that all prices and rentals may be measured in *discounted* dollars. Then we postulate that at equilibrium there will be a *rental measure* μ over a measurable subset E_o of $(S \times T, \Sigma_s \times \Sigma_t)$, with the interpretation $\mu(E) =$ rental (in discounted dollars) for the control of Space-Time "region" $E \subseteq E_o$.

A number of comments follow. First, we should distinguish carefully between regions per se and "regions" of Space-Time, which are measurable subsets of S and $S \times T$ respectively. The context will make clear which we are talking about, and quotation marks will not be used.

Second, it may not be possible to assign a rental value to all measurable subsets of E_o in an empirically meaningful way. If parcels are always transacted for as units, there is no rental value for a fraction of a parcel. This difficulty is easily remedied: we simply aggregate μ to the appropriate sub-σ-field of $\Sigma_s \times \Sigma_t$. If E_o is partitioned into rectangles $F \times G$, F ranging over the minimal subdivision units and G over the minimal time intervals for which transactions occur, the appropriate σ-field is the one generated by this partition. Since no difficulties arise, we suppose this aggregation has been done but retain our original notation.

The last comment is more substantive. By postulating μ, we have willfully fallen into the trap we warned against in the discussion of rental markets. The rental for region E should in principle depend on the activities the tenant operates in E. If he returns it with a dilapidated capital endowment, a higher rental would presumably be charged in compensation. If he returns it in improved condition, the rent would be lower, possibly negative (as when an owner turns his land over to a developer). The rental measure μ, however, implies that rent does not depend on land use.

This is done mainly for simplicity's sake, but we might justify it as an approximation under certain special conditions. One condition is that the activities contemplated have no "construction" or "mining" components, so that alternative activities would have little differential effect on capital endowment. Once the structures are in place, alternative office activities, manufacturing activities, or residential living activities probably do not make much difference in depreciation rates. The situation is different with farming, fishing, and forestry as well as mining and construction per se.

A second condition (this is more dubious) under which the effects of activities on rentals may conceivably be ignored is when the market is for Space-Time per se as separate from its contents. The rental is then "ground rent" only and is presumably affected largely by the overall "location" of the region in $S \times T$ and relatively little by the particular site characteristics such as the capital endowment. We are also implicitly assuming away neighborhood effects, since otherwise rentals would be affected by the activities operating in adjacent regions.

In any case, we postulate μ. In this section μ and all other measures are assumed to be finite. (A generalization will be discussed later.) Let us now go on to land values. For each time t, we suppose there is an equilibrium *land-value measure* μ_t, whose universe set F_t is a region of S, with the interpretation: $\mu_t(F)$ is the sale value (in discounted dollars) of region $F \subseteq F_t$.

Note that the land-value measures μ_t are over subsets of *Space*, while the rental measure μ is over a subset of *Space-Time*. An argument similar to that for (2.1) suggests the relation, for $t_1 < t_2$,

$$\mu[F \times \{t \mid t_1 \leq t < t_2\}] = \mu_{t_1}(F) - \mu_{t_2}(F) \qquad (6.4.1)$$

for any region $F \subseteq S$ for which all these markets exist. Equation (1) states that the value of a parcel at t_1 equals what you can get by selling it later at t_2, plus the rental obtained for the interval, when everything is measured in discounted dollars. This is not unreasonable, but it does assume away market frictions, ignorance, psychic income from ownership, etc.

An immediate consequence of (1) is the following. Let Space-Time region E be a disjoint union of n rectangles: $F_j \times \{t \mid t_{j1} \leq t < t_{j2}\}$,

$j = 1, \ldots, n$. One can attain ownership of just E by $2n$ transactions, buying at t_{j1} and selling at t_{j2}. Then (1) implies that the *net* expenditure for the ownership of E is exactly the same as the rental charge for E.

We now discuss the concrete organization of the market. Everything starts at time t_o, and there is an initial distribution of land ownership. Agent i owns region F_i, the collection of all the F_i's being a packing in S. The real-estate market then operates to create two measurable partitions of E_o: a partition by ownership, and a partition by control—say E_i' is the portion of Space-Time that comes to be owned by agent i, and E_i'' is the portion that comes to be controlled.

Here E_o is the region of $S \times T$ on which transactions can occur. None of E_o is assumed to occur prior to time t_o. We also suppose that each ownership region E_i' is a finite union of disjoint rectangles, as just discussed. The number of participants in the market will be assumed finite, except in one or two discussions below.

6.5. REAL-ESTATE PREFERENCES

Now we come to *preferences*. It is assumed that each agent has a preference ordering over triples (E', E'', x), where E' and E'' are measurable subsets of E_o (E' in the union of rectangles form) and x is a real number. This triple represents the situation in which the agent owns region E', controls region E'', and has a net expenditure of x on real-estate transactions (in discounted dollars).

This is a somewhat unusual set of objects over which to express preferences, but it is just the right thing for capturing the options available to the participant in the real-estate market. Note that budgetary stringencies may be reflected in this preference order: a poverty-stricken agent, or one sailing close to the wind, will give relatively heavy weight to variations in x.

Given the land-value and rental measures, what will be the net expenditures of agent i if he chooses E' to own and E'' to control? The answer is

$$x = \mu(E'') - \mu_{t_o}(F_i) \qquad (6.5.1)$$

i.e., the rental for his control set E'', net of the value of his initial holding F_i. Relation (1) seems rather surprising, since it is independent of E'. To demonstrate (1), we think of the agent as making four transactions: (i) selling his initial holding, (ii) buying E', (iii) renting out $E' \setminus E''$ (the portion of E' he does not choose to control), and (iv) acquiring control over $E'' \setminus E'$ (which, with $E'' \cap E'$, gives him the set E'' for control). The net expenditures for these four transactions are the four respective terms in

$$-\mu_{t_o}(F_i) + \mu(E') - \mu(E' \setminus E'') + \mu(E'' \setminus E') \qquad (6.5.2)$$

which is the same as (1). (If the "bases" of some of the rectangles constituting E' overlap F_i at time t_o, a part of (i) and (ii) is a fictitious transaction in which the agent sells to himself. This of course will have no effect on the sum (2).)

Intuitively, E' does not enter (1) because it is bought and then rented (partly to the agent himself perhaps), hence it incurs a net expenditure of zero, by (4.1). This underlines the assumptions behind that equation, which can be put roughly as follows. The pattern of ownership really does not matter because markets operate without friction and nobody gets any psychic income from owning land per se.

We now specialize our assumptions concerning preferences to bring them into line with this approach; viz., we assume that each agent is indifferent to variations in the first term of the triple (E', E'', x). This is a considerable simplification because it means that each agent has a well-defined preference order over pairs (E'', x). Assuming (for simplicity of notation if nothing else) that these orderings may be represented by utility functions, we have for each agent i a function

$$U_i(E, x) \qquad (6.5.3)$$

giving his preferences over combinations of E, the region of Space-Time that he controls, and x, the net discounted expenditures on real estate. Region $E = E''$ and x are connected by relation (1).

The conditions for equilibrium in the market for *control* of Space-Time may now be stated. Given initial holdings F_i at time t_o and utility functions (3) for all agents, the equilibrium consists of (i) a bounded rental measure μ on E_o, (ii) a bounded land-value measure μ_{t_o} for the initial time t_o, and (iii) a measurable partition (E_i) of E_o among the agents i; and

$$U_i[E, \mu(E) - \mu_{t_o}(F_i)] \qquad (6.5.4)$$

must be maximized at $E = E_i$ over all possible measurable subsets E of E_o for each agent i. That is, given the relevant prices for regions, no agent can choose a more preferred region than the one he actually has chosen, and these chosen regions partition the market.

Note that the market for *ownership* of real estate has dropped out of sight, except for the initial holdings. We shall in fact concentrate all our attention on the control market.

Even this simplified model seems still too general to give interesting results. We therefore consider a further specialization of (3), with the utility function U_i in the form

$$U_i(E, x) = V_i(E) - x \qquad (6.5.5)$$

for all agents i. Function (5) represents the assumption of "constant marginal utility of money" and may be taken as a reasonable approximation when real-estate expenditures are just a small fraction of one's total expenditures.

Function (5) leads to a great simplification in the conditions of equilibrium. Specializing (4), we find that the conditions for μ and the partition (E_i) are *independent* of the initial holdings F_i. The conditions are that E_i must maximize

$$V_i(E) - \mu(E) \tag{6.5.6}$$

over all measurable $E \subseteq E_o$ for each agent i.

In the special utility function (5), V_i is a *set* function, whose domain is the σ-field on E_o. If, in particular, V_i is a *bounded signed measure* for each i, we can and will prove the existence of an equilibrium for the real-estate market. But first let us contemplate the stated condition on V_i and discuss its plausibility.

First, V_i is allowed to be a signed measure, so that it might conceivably take on negative values for certain sets. This means that agent i would prefer *not* to have control over certain regions. Is this realistic? Control, in fact, is typically attended with some obligations or other disabilities, and these might occasionally outweigh the benefits of control. Examples are legal liability for accidents, maintaining nuisances, and paying property taxes (in the cases where these liabilities devolve on the tenant rather than the landlord), the onus of being a "slumlord," etc.

Thus, since it might have some applications and creates no mathematical complications, we keep the generality of using signed measures rather than measures per se. Note, however, that the equilibrium rental distribution might in this case also turn out to be a (proper) *signed* measure.

More interesting is the *additivity* condition on V_i. Letting E and F be two disjoint regions, is it plausible that

$$V_i(E \cup F) = V_i(E) + V_i(F)? \tag{6.5.7}$$

This says, roughly, that the desirability of controlling a region does not depend on what other regions agent i controls.

In general, (7) will not hold in the real world. One can think of several situations where the left side of (7) should be smaller than the right, and several others where it should be larger. The "smaller" case arises because regions can be *substitutes* for each other. Let E and F be alternative plots suitable for residential use by person i. He might have little need for both and therefore be hardly willing to pay more for E and F combined than for either alone. In general, after a certain amount of land at the right places and times has been acquired, an agent will not be able to make much use of additional land.

Conversely, the left side of (7) will exceed the right when the regions E and F are *complements*. They might be too small separately to accommodate a certain projected land use, but adequate together. In this case, agent *i* might be unwilling to pay much for one of these regions without the other, but a good deal for both together. "Too small" can refer either to *S* or *T* or both. Thus suppose *E* and *F* are both rectangles in *S* × *T*. They may be adjacent parcels over the same time interval, together adequate to accommodate a plant (or a complex of "linked" plants) of efficient size, but too small separately. They may be successive time intervals of the same parcel, each too short alone for a certain developmental project but adequate together. In this case agent *i* might be willing to pay a premium to have a long-term lease encompassing both *E* and *F*.

There are several real-world manifestations of this effect. Plants buy up excess land in hopes of inducing "linked" plants to settle there.[9] Large parcels tend to be worth more per square foot than small ones.[10] These phenomena would be hard to explain if (7) held.

In general, the *shape* of the region of control is an important factor in its value to an agent. Connected, "chunky" parcels, if small, tend to be preferred to fragmented or elongated ones because the uses planned by one agent will generally have heavy transport-communication flows among themselves: the movements of the agent himself, the flow of goods in an integrated plant system, the flow of messages between headquarters and field offices. A "tight" site pattern tends to save on these "connection" costs.

Certain exceptions, however, prove the rule that certain shapes facilitate interaction better than others. A multistage assembly line process (as in automobile manufacture) might be best housed in a long narrow plant site. A similar argument has been applied to agriculture in terms of ease of plowing and communication with the world at large.[11] These examples are spatial, but similar arguments apply to *T*. For the same total duration, one connected stretch is generally preferred to a number of small interrupted intervals.

These arguments apply mainly to "relatively small" regions. For "large" regions, the substitutability among neighboring points begins to outweigh the complementarity, and one prefers to have sites scattered, e.g., chain stores spread out rather than agglomerate. None of these "shape" effects would arise if the additivity condition (7) were valid. Nonetheless, we persist in assuming it as a mathematically tractable first approximation, which compromises between the two opposite tendencies just discussed.

---

9. J. R. P. Friedmann, *The Spatial Structure of Economic Development in the Tennessee Valley* (Univ. Chicago Press, Chicago, 1955), pp. 35, 42–43.

10. The premium is known as "plottage value." A. M. Weimer and H. Hoyt, *Principles of Real Estate,* 4th ed. (Ronald, New York, 1960), pp. 285–86.

11. C. P. Barnes, Economics of the long-lot farm, *Geogr. Rev.* 25(1935):298–301; M. Chisholm, *Rural Settlement and Land Use* (Hutchinson, London, 1962), p. 156.

6.6. EQUILIBRIUM IN THE REAL-ESTATE MARKET

We state the problem abstractly. Given measurable space (A, Σ) and n bounded signed measures $\mu_1, \ldots, \mu_n$, do n measurable subsets $E_1^o, \ldots, E_n^o$ and a bounded signed measure μ^o exist such that (i) the collection $\{E_1^o, \ldots, E_n^o\}$ is a partition of A, and (ii) for each $i = 1, \ldots, n$, E_i^o maximizes

$$\mu_i(E) - \mu^o(E) \qquad (6.6.1)$$

over all measurable subsets E?

The interpretation we have in mind is that A is the subset of $S \times T$ for which the real-estate market exists; μ^o is the rental (signed) measure; there are n participants in the market, and E_i^o is the region chosen by agent i for his control, the net cost to him being $\mu^o(E_i^o)$; (1) is the same as (5.6) and is the utility level of agent i if he acquires region E. (We have changed the notation V_i to μ_i, since we are assuming this function is a (signed) measure.) Conditions (i) and (ii) are then precisely the equilibrium conditions for the real-estate market.

We shall prove the existence of equilibrium. Note that the number of participants in the market is finite, and all the μ_i's are also finite. The first condition is essential; the second can be removed, but (1) then requires some reinterpretation. This will be done in Section 6.9. Our method of procedure will be indirect, and we shall prove several other properties of the equilibrium. These have independent economic meanings and are of interest in themselves. This procedure also paves the way for the generalization to pseudomeasures that comes later.

Recall that a *Hahn decomposition* of a signed measure μ is a pair of measurable sets (P, N) that partition universe set A and for which $\mu(E) \geq 0$ on all measurable sets $E \subseteq P$, and $\mu(F) \leq 0$, all measurable $F \subseteq N$. We now need a generalization of this concept to several signed measures.

Definition: Given measurable space (A, Σ) and signed measures $(\mu_1, \ldots, \mu_n)$, an *extended Hahn decomposition* is an n-tuple of measurable sets $(E_1, \ldots, E_n)$ which partitions A and for which

$$\mu_i(F) \geq \mu_j(F) \qquad (6.6.2)$$

on all measurable $F \subseteq E_i$, $i, j = 1, \ldots, n$.

That is, on E_i, μ_i is at least as large as any of the other signed measures μ_j, $j = 1, \ldots, n$.

The economic interpretation of the condition that $(E_1^o, \ldots, E_n^o)$ is an extended Hahn decomposition is that, on any measurable subset of E_i^o, agent i will at least match the bid of any other agent. This appears to be a reasonable alternative definition for a partition being an equilibrium for the real-estate market.

This definition of equilibrium looks quite different from the definition given by (1). For one thing, it involves no mention of any rental distribution μ^o. But our next result shows that in fact these two definitions are equivalent.

Theorem: Let μ^o and $\mu_1, \ldots, \mu_n$ be bounded signed measures on measurable space (A, Σ), and let $(E_1^o, \ldots, E_n^o)$ be measurable sets that partition A. Then μ^o satisfies one of the following two conditions iff it satisfies the other:

    (i) for each $i = 1, \ldots, n$, E_i^o maximizes

$$\mu_i(E) - \mu^o(E) \tag{6.6.3}$$

    over all $E \in \Sigma$;
    (ii) for all $i, j = 1, \ldots, n$ $(i \neq j)$,

$$\mu_i(F) \geq \mu^o(F) \geq \mu_j(F) \tag{6.6.4}$$

for any measurable $F \subseteq E_i^o$.

Furthermore, a bounded signed μ^o exists satisfying one (hence both) of these conditions iff $(E_1^o, \ldots, E_n^o)$ is an extended Hahn decomposition for $(\mu_1, \ldots, \mu_n)$.

Proof: Let E_i^o maximize (3) for all i, and take measurable $F \subseteq E_i^o$. Then

$$\mu_i(E_i^o) - \mu^o(E_i^o) \geq \mu_i(E_i^o \backslash F) - \mu^o(E_i^o \backslash F)$$

Simplification yields the left inequality in (4). Also,

$$\mu_j(E_j^o) - \mu^o(E_j^o) \geq \mu_j(E_j^o \cup F) - \mu^o(E_j^o \cup F)$$

Simplification yields the right inequality in (4).

Conversely, let (4) hold, and take any $E \in \Sigma$. Then (3) is the sum of n terms: $\mu_i(E \cap E_j^o) - \mu^o(E \cap E_j^o)$, $j = 1, \ldots, n$. For all $j \neq i$, these terms are ≤ 0, by the right inequality in (4). Hence

$$\mu_i(E) - \mu^o(E) \leq \mu_i(E \cap E_i^o) - \mu^o(E \cap E_i^o) \tag{6.6.5}$$

Also,

$$0 \leq \mu_i(E_i^o \backslash E) - \mu^o(E_i^o \backslash E) \tag{6.6.6}$$

by the left inequality of (4). Adding (5) and (6), we obtain $\mu_i(E) - \mu^o(E) \leq \mu_i(E_i^o) - \mu^o(E_i^o)$, so that E_i^o does indeed maximize (3). This proves the equivalence of conditions (i) and (ii).

Next suppose there is a μ^o satisfying these conditions. The extended Hahn decomposition condition (2) follows at once from (4).

Finally, suppose $(E_1^o, \ldots, E_n^o)$ is an extended Hahn decomposition for $(\mu_1, \ldots \mu_n)$. For μ^o choose the signed measure given by

$$\mu^o(E) = \mu_1(E \cap E_1^o) + \cdots + \mu_n(E \cap E_n^o) \qquad (6.6.7)$$

all $E \in \Sigma$; i.e., μ^o is the direct sum or patching of the μ_i, respectively restricted to E_i^o, $i = 1, \ldots, n$. Then for measurable $F \subseteq E_i$, we obtain $\mu^o(F) = \mu_i(F) \geq \mu_j(F)$, from (7) and (2). This implies (4). ∎

Condition (3) is the same as (1), so that (i) is precisely our original condition that μ^o and $(E_1^o, \ldots, E_n^o)$ should be an equilibrium for the real-estate market. We have just shown that there is a μ^o such that this is the case iff $(E_1^o, \ldots, E_n^o)$ is an extended Hahn decomposition. Thus in this sense the two equilibrium concepts coincide. In addition, the result just obtained states a necessary and sufficient condition for a given signed measure μ^o to be the rental distribution in the real-estate equilibrium. This is (4), and it too has a simple economic interpretation: the price at which something sells must lie between the highest and second-highest offers of the bidders in the market.

We now prove the existence of an equilibrium, by proving that an extended Hahn decomposition $(E_1^o, \ldots, E_n^o)$ exists. The preceding theorem then implies the existence of a rental signed measure μ^o, and together these satisfy the equilibrium conditions (1) or (3).

In fact, we prove something stronger: given signed measures $\mu_1, \ldots, \mu_n$ (not necessarily finite or even σ-finite), if each *pair* of these has an extended Hahn decomposition, then so does the whole n-tuple $(\mu_1, \ldots, \mu_n)$. To see that this implies the existence of a decomposition in the case we are considering, let μ_i and μ_j be bounded signed measures. Then $\mu_i - \mu_j$ is a signed measure. Hence by the ordinary Hahn decomposition theorem, there is a pair of measurable sets (P, N) that partitions A and for which $\mu_i(F) - \mu_j(F)(\geq, \leq)0$ for measurable $F \subseteq P$, $F \subseteq N$ respectively. But this is precisely the condition (2) for (P, N) to be an extended Hahn decomposition for the pair (μ_i, μ_j). Hence such a decomposition exists for any pair of bounded signed measures. The premise of the following theorem is therefore fulfilled, and we conclude that a decomposition exists for any n-tuple of *bounded* signed measures.

Theorem: Let $(\mu_1, \ldots, \mu_n)$ be an n-tuple of signed measures on space (A, Σ) such that each pair of these has an extended Hahn decomposition. Then the whole n-tuple has an extended Hahn decomposition.

Proof: Proof is by induction on n. The statement is true for $n = 2$ by assumption. For $n > 2$, we suppose it holds for $n - 1$, and prove it for n. Thus for $(\mu_1, \ldots, \mu_{n-1})$ there is a measurable $(n - 1)$-tuple $(E_1, \ldots, E_{n-1})$ that partitions A and for which

$$\mu_i(F) \geq \mu_j(F) \tag{6.6.8}$$

for all measurable $F \subseteq E_i$, $i, j = 1, \ldots, n - 1$.

For each $i = 1, \ldots, n - 1$ there is by assumption an extended Hahn decomposition (P_i, N_i) for the pair (μ_i, μ_n). Thus

$$\mu_i(F) \geq \mu_n(F) \tag{6.6.9}$$

for measurable $F \subseteq P_i$, and for $F \subseteq N_i$ the opposite inequality holds. Now define $(E_1^o, \ldots, E_n^o)$ by

$$E_i^o = E_i \cap P_i \tag{6.6.10}$$

for $i = 1, \ldots, n - 1$, and

$$E_n^o = (E_1 \cap N_1) \cup (E_2 \cap N_2) \cup \cdots \cup (E_{n-1} \cap N_{n-1}) \tag{6.6.11}$$

We now show that $(E_1^o, \ldots, E_n^o)$ is an extended Hahn decomposition for $(\mu_1, \ldots, \mu_n)$. First, one easily verifies that it is a measurable partition of A. Next let i and j not equal n, and take measurable $F \subseteq E_i^o$. Then $F \subseteq E_i$, so that (8) holds. Also, $F \subseteq P_i$, so that (9) holds. We have thus shown that (8) holds for all $i = 1, \ldots, n - 1$ and all j, *including $j = n$*, when $F \subseteq E_i^o$.

It only remains to show that

$$\mu_n(F) \geq \mu_j(F) \tag{6.6.12}$$

for all measurable $F \subseteq E_n^o$ and all $j = 1, \ldots, n - 1$. For any such F and j we find that

$$
\begin{aligned}
\mu_n(F) &= \mu_n(F \cap E_1 \cap N_1) + \cdots + \mu_n(F \cap E_{n-1} \cap N_{n-1}) \\
&\geq \mu_1(F \cap E_1 \cap N_1) + \cdots + \mu_{n-1}(F \cap E_{n-1} \cap N_{n-1}) \\
&\geq \mu_j(F \cap E_1 \cap N_1) + \cdots + \mu_j(F \cap E_{n-1} \cap N_{n-1}) = \mu_j(F)
\end{aligned}
\tag{6.6.13}
$$

which yields (12). The equalities in (13) come from (11); the first inequality in (13) arises from the fact that $(F \cap E_i \cap N_i) \subseteq N_i$, $i = 1, \ldots, n - 1$, so that (9) reversed holds for each such set; the second inequality arises from the fact that $(F \cap E_i \cap N_i) \subseteq E_i$ for $i = 1, \ldots, n - 1$, so that (8) holds for each such set.

We have shown that $(E_1^o, \ldots, E_n^o)$ given by (10) and (11) is an extended Hahn decomposition for $(\mu_1, \ldots, \mu_n)$. This completes the induction and the proof.[12] ∎

---

12. If $\mu_1, \ldots, \mu_n$ are all σ-finite, there is an alternative, but less straightforward, proof based on the Radon-Nikodym theorem. Cf. the proof of Theorem 2 of L. E. Dubins and E. H. Spanier, How to cut a cake fairly, *Am. Math. Monthly* 68(1961):1–17, repr. in P. Newman, ed., *Readings in Mathematical Economics,* Vol. 1 (Johns Hopkins Press, Baltimore, 1968).

We have now proved the existence of equilibrium in the real-estate market with a *finite* number of participants. What if the number of participants is infinite? This might occur with an unbounded S or T horizon. There is no trouble extending the definition of equilibrium and of extended Hahn decomposition to the case of an infinite number of participants with corresponding bounded signed measures μ_i. This will involve an infinite measurable partition of A, the pieces in 1–1 correspondence with the μ_i's, and the relations (1) or (2) holding for each piece. In fact, we make this very extension later.

But is it true that this equilibrium or decomposition always exists? The answer is no, as the following trivial counterexample demonstrates. Let the space A consist of just one point. Let $\mu_1, \mu_2, \ldots$ be an infinite sequence of measures, viz., $\mu_n(A) = 1 - 1/n, n = 1, 2, \ldots$. No equilibrium and no extended Hahn decomposition exist, since no matter to what participant n we assign the single point of A, he is "outbid" by participant $n + 1$.

Let us turn briefly to the question of *uniqueness* of equilibrium.

Theorem: Let $(\mu_1, \ldots, \mu_n)$ be an n-tuple of signed measures on space (A, Σ) with extended Hahn decomposition $(E_1^o, \ldots, E_n^o)$; let $(E_1, \ldots, E_n)$ be another n-tuple of measurable sets that partitions A. Then $(E_1, \ldots, E_n)$ is also an extended Hahn decomposition iff

$$\mu_i(F) = \mu_j(F) \qquad (6.6.14)$$

for all measurable $F \subseteq (E_i^o \cap E_j), i, j = 1, \ldots, n$.

Proof: Let $(E_1, \ldots, E_n)$ be another decomposition and take measurable $F \subseteq (E_i^o \cap E_j)$. Then $\mu_i(F) \geq \mu_j(F) \geq \mu_i(F)$, from (2), which yields (14).

Conversely, let (14) hold, and take measurable $F \subseteq E_i$.

$$\mu_i(F) = \mu_i(F \cap E_1^o) + \cdots + \mu_i(F \cap E_n^o)$$

$$= \mu_1(F \cap E_1^o) + \cdots + \mu_n(F \cap E_n^o)$$

$$\geq \mu_j(F \cap E_1^o) + \cdots + \mu_j(F \cap E_n^o) = \mu_j(F) \qquad (6.6.15)$$

for any $j = 1, \ldots, n$. The first and last equalities in (15) arise from the fact that $\{E_1^o, \ldots, E_n^o\}$ is a partition; the middle equality arises from (14) and the fact that $(F \cap E_k^o) \subseteq (E_k^o \cap E_i)$; the inequality in (15) arises from the fact that $(E_1^o, \ldots, E_n^o)$ is an extended decomposition.

Thus (15) implies that $\mu_i(F) \geq \mu_j(F)$ when $F \subseteq E_i$, so that $(E_1, \ldots, E_n)$ is indeed another decomposition. ∎

This theorem again has a simple economic interpretation. A given region F can be under different controllers in two equilibria iff their bids over this region are identical. Condition (4) for the rental distribution μ^o then implies that μ^o is identical over F to this common signed measure. The two agents are then indifferent about controlling F, the rental just canceling out the benefits they receive from this control.

Thus while multiple equilibria are possible, they are of a somewhat trivial character from a practical point of view. The extreme case occurs when all the μ_i's are identical (all agents have identical preferences). One easily verifies that *any* measurable partition into n pieces is an extended Hahn decomposition in this case and that it is an equilibrium; the rental distribution μ^o is this common signed measure, and all agents are indifferent among all possible regions.

It is very common for equilibria to satisfy one or another extremal condition that can be given a welfare interpretation of sorts. Our final result of this section is of this character.

Theorem: Let $(\mu_1, \ldots, \mu_n)$ be an n-tuple of bounded signed measures on space (A, Σ). The n-tuple of sets $(E_1^o, \ldots, E_n^o)$ is an extended Hahn decomposition iff it maximizes

$$\mu_1(E_1) + \cdots + \mu_n(E_n) \tag{6.6.16}$$

over all measurable n-tuples $(E_1, \ldots, E_n)$ that partition A.

Proof: Let $(E_1^o, \ldots, E_n^o)$ be an extended Hahn decomposition, and let $(E_1, \ldots, E_n)$ be another measurable n-tuple that partitions A. Then

$$\mu_i(E_i^o \cap E_j) \geq \mu_j(E_i^o \cap E_j) \tag{6.6.17}$$

for all $i, j = 1, \ldots, n$, by (2). Adding the inequalities (17) over the n^2 possible (i, j) pairs yields

$$\mu_1(E_1^o) + \cdots + \mu_n(E_n^o) \geq \mu_1(E_1) + \cdots + \mu_n(E_n) \tag{6.6.18}$$

so that $(E_1^o, \ldots, E_n^o)$ does indeed maximize (16).

Conversely, let $(E_1^o, \ldots, E_n^o)$ maximize (16), and take measurable $F \subseteq E_i^o$. Define $E_i = E_i^o \backslash F$, $E_j = E_j^o \cup F$, and $E_k = E_k^o$ for all $k \neq i$, $k \neq j$, $k = 1, \ldots, n$. Then the n-tuple $(E_1, \ldots, E_n)$ partitions A. We then have (18), which simplifies to

$$\mu_i(E_i^o) + \mu_j(E_j^o) \geq \mu_i(E_i^o \backslash F) + \mu_j(E_j^o \cup F)$$

and this in turn simplifies to $\mu_i(F) \geq \mu_j(F)$. Thus $(E_1^o, \ldots, E_n^o)$ is an extended Hahn decomposition. ∎

This again has a simple economic interpretation. If $(E_1, \ldots, E_n)$ represents the way Space-Time is partitioned among the n agents, (16) may be thought of as a kind of *social valuation* of this partition. It is the sum of the values each participant personally places on the region allotted to himself. (These values are all measured in the same units, viz., discounted dollars.)

Thus a given partition is an equilibrium for the real-estate market iff it maximizes the social valuation of land in discounted dollars, as given by (16). This is a useful observation, though one cannot draw welfare or policy conclusions from it without further assumptions.

In all the preceding analysis it was required that the n-tuples $(E_1, \ldots, E_n)$ form a *partition* of the universe set A, rather than just a *packing*. In other words, the entire region had to be allocated; the possibility of leaving part of it vacant was not allowed. At first glance this seems rather restrictive, especially since we allow signed measures, so that some agents would prefer not to control some regions. In the real world, would not a universally repugnant region be left uncontrolled?

But this apparent generalization is already contained in the preceding model. We simply add a "dummy" participant $n + 1$, with μ_{n+1} *identically zero*. Letting $(E_1^o, \ldots, E_{n+1}^o)$ be an extended Hahn decomposition for $(\mu_1, \ldots, \mu_{n+1})$, we find from (2) that all the signed measures $\mu_1, \ldots, \mu_{n+1}$ are *nonpositive* over E_{n+1}^o, while μ_i is *nonnegative* over E_i^o, $i = 1, \ldots, n$. The obvious interpretation of this phenomenon is that E_i^o is the region chosen by agent i, $i = 1, \ldots, n$, while E_{n+1}^o is the residue left vacant. All the preceding theorems remain valid for this situation.

6.7. JOINT CONTROL AND AGENT MEASURE SPACES

We now consider a few real generalizations. In the above analysis each agent obtains exclusive control of a Space-Time region. But in the real world there are numerous examples of joint control: partnerships, joint ventures, committee management, corporate stockholders. How is one to represent this?

It is not immediately obvious how to "split" a set E among several agents in proportion to their share of control. However, recall that with any set $E \subseteq A$ is associated its *indicator* function $I_E : A \to \{0, 1\}$, viz., $I_E(a) = 1$ if $a \in E$, $= 0$ if $a \in A \setminus E$. There is a 1–1 correspondence between the subsets of A and the set of all 0-1 functions on A by this association. Thus instead of representing a real-estate allocation by $(E_1, \ldots, E_n)$ (these forming a partition of A), we could just as well have represented it by the corresponding n-tuple of indicators $(I_{E_1}, \ldots, I_{E_n})$. For any $a \in A$, we have

$$I_{E_1}(a) + \cdots + I_{E_n}(a) = 1 \qquad (6.7.1)$$

Indeed, (1) is just another way of saying that $\{E_1, \ldots, E_n\}$ is a partition.

The advantage of using indicators is that it suggests the proper generalization to joint control. Namely, a measurable function $h_i:A \to [0, 1]$ is associated with agent i, taking values in the closed interval of real numbers between 0 and 1. The intended interpretation is that $h_i(a)$ is the *fraction* of point a controlled by agent i. These n functions must satisfy

$$h_1(a) + \cdots + h_n(a) = 1 \qquad (6.7.2)$$

for all $a \in A$, since the total of all fractional controls must add to 1.[13] Exclusive control is precisely the special case in which all functions h_i take on only the values 0 or 1. They are then all indicator functions, and (2) reduces to (1).

The entire structure of the real-estate market model generalizes in a corresponding way. Instead of preference orderings over pairs (E, x), as in (5.3), the various agents have preferences over pairs (h, x). If μ^o is the rental signed measure, the rental for the control pattern given by h will be

$$\int_A h \, d\mu^o \qquad (6.7.3)$$

The special assumption we made, that $U_i(E, x)$ is of the form $\mu_i(E) - x$, where μ_i is a bounded signed measure, now becomes

$$U_i(h, x) = \int_A h \, d\mu_i - x$$

again for a bounded signed μ_i. Market *equilibrium* is given by rental distribution μ^o and a measurable n-tuple $(h_1^o, \ldots, h_n^o)$ satisfying (2) such that no agent i prefers any control function to h_i^o when confronted with μ^o.

When the only allowable functions h are indicators, everything reduces to the original model with exclusive control. There is no difficulty extending this model to a countably infinite, or even uncountable, number of participants. In the former case we get an infinite series on the left of (2), which must converge to 1 for all $a \in A$. In the latter case we use the concept of summation of an arbitrary collection of numbers (Section 2.2): for each $a \in A$, all but a countable number of the values $h(a)$ equal 0, and the remainder form an infinite series, again converging to 1.

A slightly different approach uses the concept of a *measure space of agents* (B, Σ', ν).[14] Here B is the set of agents. It comes supplied with a

---

13. These functions are formally the same as the "fuzzy sets" of L. Zadeh. The idea of using functions to generalize sets in this way is routine in measure theory, where it forms the basis of the "Daniell" approach. See L. H. Loomis, *An Introduction to Abstract Harmonic Analysis* (Van Nostrand, Princeton, N.J., 1953).

14. The first use of this concept is by R. J. Aumann, Markets with a continuum of traders, *Econometrica* 32(Jan.–Apr. 1964):39–50. repr. in Newman, ed., *Readings.* An extensive literature has grown up since then. For recent work see W. Hildenbrand, *Core and Equilibria of a Large Economy* (Princeton Univ. Press, Princeton, N.J., 1974).

σ-field Σ' on which is defined measure ν. Intuitively, $\nu(E)$ gives the "influence" of the set of agents E. To pin down this notion, we need a corresponding generalization of the concept of allocation in the real-estate market A. In the finite case this was an n-tuple of functions $(h_1, \ldots, h_n)$ satisfying (2). This may also be written as a single function

$$h:\{1, \ldots, n\} \times A \rightarrow [0, 1] \qquad (6.7.4)$$

For the set of traders B, an allocation will be a measurable function

$$h:B \times A \rightarrow [0, 1] \qquad (6.7.5)$$

with the rough intuitive meaning: $h(b, a)$ is the fraction of land at $a \in A$ controlled "per unit influence" of agent $b \in B$. This rather vague notion and the one above are explicated formally by the requirement

$$\int_B h(b, a) \, \nu(db) = 1 \qquad (6.7.6)$$

for all $a \in A$. Thus (6) and (5) generalize (2) and (4) respectively and indeed reduce to them when B consists of just n points ("agents"), $\Sigma' =$ all subsets, and ν is counting measure.

The generality of the "measure space of agents" approach may be illustrated by the case where ν is nonatomic (all singleton sets $\{b\}$ being measurable). Here no single agent, or even any countable number of agents, has positive influence. As Aumann points out,[15] this is exemplified in the concept of *perfect competition,* where each agent has negligible influence. The obscurities that remain in the present formalism are matched by the obscurities present in that popular concept.

For each agent $b \in B$ there is a preference order over pairs $(h(b, \cdot), x)$. Here $h(b, \cdot)$ is a function with domain A, and gives the control pattern for agent b. The rental "per unit influence" for this control is given by (3) just as above. An *equilibrium* in the real-estate market consists of a rental distribution μ^o, which is a bounded signed measure over (A, Σ), and a measurable control function h^o of the form (5) such that (6) is satisfied and such that, for all agents b (except possibly for a set of ν-measure zero), $h^o(b, \cdot)$ is at least as preferred as any other control $h(b, \cdot):A \rightarrow [0, 1]$ by agent b.

This entire approach, using functions h to represent joint control, applies just as well to joint *ownership.* Furthermore, there is nothing that restricts its use to the real-estate market. Consider, for example, the distribution of *physical assets* among economic agents at some time instant t ("distribution" may refer either to ownership or to control). In this case the set A is a subset of $R \times S$, rather than of $S \times T$ as in the

---

15. *Econometrica* 32(1964):39.

real-estate market. For each agent i we again have a function $h_i:A \rightarrow$ [0, 1], with the interpretation: $h_i(r, s)$ is the fraction of resource type r at location s controlled (or owned, as the case may be) by agent i at time t.

Suppose an agent's control pattern is given by $h:A \rightarrow [0, 1]$. If there is some natural "quantity" measure on the space (A, Σ), the control pattern may be expressed in terms of a *measure* rather than a point function. For example, let $A \subseteq R \times S$, and let h_i describe the control of physical assets by agent i at time t. Let μ_t be the cross-sectional measure at time t (so that μ_t is over $(R \times S, \Sigma_r \times \Sigma_s)$, and $\mu_t(E \times F)$ is the total mass of resources of types E in region F at time t), and let μ_t' be μ_t restricted to A. Then $\mu_{it} = \int h_i \, d\mu_t'$, an indefinite integral over A, expresses the control pattern of i at t as a measure: $\mu_{it}(G)$ is the total mass of resource type location pairs in G controlled by agent i at time t for all measurable $G \subseteq A \subseteq (R \times S)$.

Again, let α be "areal" measure on $(S \times T, \Sigma_s \times \Sigma_t)$. (If S is taken as 3-space, α may be 4-dimensional Lebesgue measure, say in units of "cubic-feet-days.") Let α' be α restricted to E_o, the portion of Space-Time on the market. Then $\alpha_i = \int h_i \, d\alpha'$, an indefinite integral over E_o, expresses i's real-estate control in measure form: $\alpha_i(G)$ is the amount of cubic-feet-days in region G controlled by agent i for all measurable $G \subseteq E_o \subseteq (S \times T)$.

This is all very well and sometimes useful, but note that in our entire discussion of the real-estate market it was never necessary to mention or use the concept of areal measure. All that was needed to define an equilibrium were the preference orderings over pairs (E, x), or more generally, (h, x). And to prove existence and other properties of equilibrium, what was needed was a specialization of the form of these preference orders. None of this involves areal measure. (Preferences among regions will of course be influenced by the areal capacities of these regions, among other factors. But this does not gainsay the fact that once preferences are given, areal measure per se plays no role. It is a fifth wheel.) This point is important in connection with Alonso's theory, where areal measure plays a key role.

6.8. COMPARISON WITH ALONSO'S THEORY

One of the leading theories of the real-estate market, and deservedly so, is that of William Alonso. We now compare his theory with the one presented above. Our conclusion will be that when certain kinks are straightened out, it becomes a special case of our own.

In his book of 1964 and preceding publications, Alonso develops his theory in the context of a featureless plain with a single point of attraction, which may be thought of as the central business district of a city.[16] But his underlying real-estate model does not really depend on this context, and indeed, in a later article he briefly indicates a generalization.[17] We concentrate on the book, while stressing the general features of the theory.

Two specializations may be noted at once. First, it is a theory of *Space* rather than Space-Time. This creates no difficulties of comparison when viewed formally. All the "regions" in our model may be thought of as subsets of S if desired, rather than subsets of $S \times T$.

Second, Alonso's is a *pure control* theory. Only the behavior of potential tenants is analyzed in detail, while landlords are assumed to auction off their land passively to the highest bidder. Our own theory has, in effect, made the same simplification, which loses a number of important real-world phenomena, e.g., discrimination, market frictions, ownership preferences. Our "intermediate level" model, in which the preferences of agent i are summarized in a utility function (5.3) of the form

$$U_i(E, x) \tag{6.8.1}$$

(where E is the region controlled and x is total *net* expenditure on real estate), does allow for one aspect of owner-controller interaction: the formation of real-estate prices affects the real wealth of existing landlords, and this "wealth effect" influences their preferences over regions (which in turn affects prices—we have a simultaneous equations situation). However, we proved no theorems at this level of generality but passed on to the "constant marginal utility of money" formulation (5.5), $V_i(E) - x$, in which wealth effects disappear.

Alonso comes up with a preference order for agent i of the form

$$U_i(s, p) \tag{6.8.2}$$

where s is the point where he locates and p is the rent per acre he must pay. (In the Thünen context of the model, U_i depends on s only through its distance from the point of attraction, so that (2) is a context-free generalization of what appears in his text. The same is true of all the other functions written below that involve location.) The indifference surfaces of (2) are called *bid-price curves*.

Function (2) bears comparison with (1). It is a sort of "single point" version of (1), the single location s contrasting with the region E and the rent density p with total expenditure x.

---

16. This is a special case of what we shall call Thünen systems, which are examined in great detail in Chapter 8.

17. W. Alonso, *Location and Land Use* (Harvard Univ. Press, Cambridge, 1964). The heart of the theory is in Chapters 3, 4, 5 and Appendices A and B. W. Alonso, A reformulation of classical location theory and its relation to rent theory, *Reg. Sci. Assoc. Pap.* 19(1967):41.

Function (2) is derived by Alonso from an underlying preference ordering of agent i. This takes two forms, depending on whether i is a consumer or a firm. We just consider the consumer case. Agent i is then assumed to have a preference ordering represented by the utility function

$$U_i'(s, q, z) \qquad (6.8.3)$$

where s again is location, q is acres of land controlled at location s, and z is all other goods consumed except land. Let π be the prices of the goods z, $k(s)$ the transportation expenditure incurred by the agent as a function of location, and y_i the given income of agent i. The following budget condition must be satisfied:

$$y_i = k(s) + \pi z + pq \qquad (6.8.4)$$

Now suppose s and p are given. The maximum of (3) over pairs (q, z) that satisfy (4) will then depend on (s, p). This is the function (2).

Before continuing the analysis, note that our utility function (1) can be derived from an underlying preference ordering in a similar manner. We did not do so before to avoid distracting attention from the essential features of the real-estate market. To facilitate comparison with (3) and (4), we postulate a single location s through which all commerce between the controlled region E and the rest of the world occurs. (This plays the same role as Alonso's "front door"; see below.) We then postulate a utility function

$$U_i'(s, E, z) \qquad (6.8.5)$$

where E is the region controlled by agent i and z is all other goods, as above. (Actually, z should be written as a signed measure over $R \times S \times T$; we do not do so in order to keep things as similar as possible to (3) and (4).) Also, let y_i be the agent's wealth in nonreal-estate assets.

Suppose we are given E, the region agent i chooses to control, and x, his total net expenditure on real estate. The following budget condition is analogous to (4):

$$y_i = k(s) + \pi z + x \qquad (6.8.6)$$

All terms in (6) are measured in discounted dollars. The maximum of (5) over all (s, z) satisfying (6) will then depend on (E, x). This is the utility function (1).

Let us now return to the Alonso utility function (3) and compare it with our (5). The region E in (5) can of course be any measurable subset of the universe set. Thus it can be the union of any number of disconnected pieces, each of these being of more or less arbitrary size and shape. By contrast, (3) refers to location "at" a single point s (or, in the Thünen context, "at" a single distance from the point of attraction). Thus there is no way of representing the preferences of an agent contemplating *multiple* locations—say a family wanting both a town house and a country house or wanting both a house and a business site. There is also no way of representing preferences regarding *shape* of lot, as opposed to *size*, which is represented by q.[18]

A second point concerning (3) is logically more serious because it involves an actual inconsistency. If the acreage of land controlled is positive ($q > 0$), it must be spread over more than a single point (and even over more than a single distance from the point of attraction). Hence there is no clear meaning to the concept of being located "at" the single point s (or "at" a single distance from the point of attraction).

Alonso is well aware of this problem of the "extended point." Before looking at his methods for dealing with it, we contemplate an alternative method. This involves postulating a *measure space of agents* (B, Σ', ν), with ν nonatomic in the manner of Aumann. We might argue intuitively as follows. Since any one agent now has zero influence, he may be thought of as located literally "at" the single point s, even though his "acreage per unit influence" q is positive.

Whatever one thinks of this argument, it is not difficult to write formally the equilibrium conditions resulting from it. Let $p^o:S \to$ reals be the *equilibrium rent-density* function. Each agent $b \in B$ maximizes (3) over s, q, z, subject to condition (4), with $p(s) = p^o(s)$. (Subscript i is replaced by b in (3) and (4).) Let $s(b)$, $q(b)$ be the locational and acreage choices of agent b. Then we must have for any region E,

$$\int_{\{b \mid s(b) \in E\}} q \, d\nu = \alpha(E) \tag{6.8.7}$$

which states that $\alpha(E)$, the total area of E, must equal the total demand for land by those agents who locate in E.

Alonso does not take this tack. Indeed, the Aumann approach requires an uncountable number of agents, while Alonso always works with a finite number. Instead, he makes an ad hoc assumption depending essentially on the Thünen context, viz., that (s, q) refers to a ring-shaped region of area q, concentric with the point of attraction, whose inner radius is the distance of s from the point of attraction (this assumption is dropped in his Appendix B).

The "extended point" problem causes one other bit of trouble. Since there is no single location s "at" which the agent chooses, there will in general not be any single rent density $p(s)$ either. This means that the term pq in (4) must be replaced by an integral, giving total expenditure for control of the ring (s, q) chosen by the agent (this is done in his Appen-

---

18. Shape is considered by Alonso in Appendix B of his book. We discuss this below.

dix A). It also means that the derived utility function $U_i(s, p)$—which is (2)—with its "bid-price" level surfaces, has no clear meaning.

Finally, in his Appendix B Alonso postulates a more general utility function than (3) to take account of shape and avoid making the ad hoc assumption just mentioned. The agent i chooses a point location s, called his "front door," and a region E to control. His preferences are expressed by the utility function

$$U_i'\left[s, \int_E f(d'(s, y))\alpha(dy), z\right] \tag{6.8.8}$$

The q-argument in (3) has been replaced by an integral, the other two arguments s and z remaining the same. As usual, α is areal measure. Now q can be expressed in terms of region E: $q = \alpha(E) = \int_E 1\,d\alpha$, so that (8) can be thought of as a generalization of (3) in which 1 has been replaced by a nonconstant integrand. In this integrand, d' is the (Euclidean) metric on the plane, and f is some positive, strictly decreasing, function. In effect, (8) makes the value of a point y inversely related to its distance from the agent's "front door."

With this change, Alonso's model becomes a special case of our own, for (8) is a specialization of the utility function (5). The budget condition in both cases is given by (6). We can thus obtain the derived utility function (1): $U_i(E, x)$, etc.

We now give an informal, nonrigorous argument to indicate how (8) gives information concerning the shape of the region E. Let rent density function $p^o:S \to$ positive reals be given. Suppose s, z, and x are chosen in advance. Subject to these values and to the budget constraint (6), we are to choose E to maximize (8). Any conditions derived from this problem will be necessary for optimality in the original problem. Assuming (8) to be increasing in its middle argument, one easily verifies that this special problem simplifies to: maximize $\int_E f(d'(s, y))\,\alpha(dy)$ over E, subject to $\int_E p^o\,d\alpha = x$. This can be turned into an allocation-of-effort problem as in Chapter 5. Omitting details, the optimal set E^o has the following form:

$$E^o = \{y \mid f(d'(s, y))/p^o(y) \geq c\} \tag{6.8.9}$$

for some constant c. Along the borderline of E^o the inequality of (9) becomes an equality.

Alonso derives this condition—that the ratio of $f(d'(s, y))$ to $p^o(y)$ is constant along the borderline. But at this point his analysis falters. Assuming rent density p^o to be a decreasing function of distance from the point of attraction, he concludes that the optimal region is egg shaped. (Recall that we are on the Euclidean plane.) But this is impossible in market equilibrium, since the plane—or any circular disc on it—cannot be partitioned into egg-shaped regions. The correct conclusion is that utility functions of the form (8) *preclude* the possibility that rent density has the property just mentioned. On the contrary, the real-estate market equilibrium (if it exists) must be such that, with the given p^o, the agents choose regions that partition the market among themselves.

In conclusion, in its most developed form Alonso's model becomes a special case of our own. Earlier versions involve either inconsistencies or ad hoc assumptions.

The following program remains to be carried out. For plausible utility functions $U_i(E, x)$ find the conditions under which a real-estate market equilibrium exists, describe its form, determine the equilibrium rental measure μ^o, and find any interesting extremal or other properties that the equilibrium possesses.

Among the plausible utility functions are those derived from an underlying utility of the form (8), with budget constraint (6). Since Alonso's exposition is flawed, it is not clear under what conditions an equilibrium even exists. More general functions than (8) should also be considered. For one thing, it seems unlikely that an agent with preferences represented by (8) would ever choose a region of multiple scattered locations, a very common real-world phenomenon. Rather, he would choose a region tightly clustered around his "front door." To generalize, we need the possibility of numerous "front doors," or perhaps an entirely different form of preference ordering.

This program *has* been carried out in two cases. The first is where $U_i(E, x) = \mu_i(E) - x$ for some bounded signed measure μ_i (agents $i = 1, \ldots, n$), which yields the theory of Section 6.6 (and which we generalize in Section 6.9). The second case is the Thünen equilibrium of Section 8.7.

6.9. PSEUDOMEASURE TREATMENT OF THE REAL-ESTATE MARKET

We return to the real-estate equilibrium model of Section 6.6 and for convenience repeat it here. We are given a measurable space (A, Σ) and n bounded signed measures $\mu_1, \ldots, \mu_n$, one for each agent in the market. An equilibrium consists of a *rental distribution* μ^o, which is a bounded signed measure on (A, Σ), and an n-tuple of measurable sets $(E_1^o, \ldots, E_n^o)$ that partition A and are such that E_i^o maximizes

$$\mu_i(E) - \mu^o(E) \qquad (6.9.1)$$

over all $E \in \Sigma$ for each $i = 1, \ldots, n$. The interpretation, of course, is that E_i^o is the region chosen by agent i, while (1) is the utility he attaches to the control of region E, $\mu^o(E)$ being the net expenditure he incurs for this control.

We want to generalize this setup in two directions. First, to consider the case of a (countably) *infinite* number of agents. Second, to consider what happens if μ^o or some or all of the μ_i are *infinite* (σ-finite) signed measures.

Before launching into details, we briefly discuss the question of whether such generalizations have any possible applications (we use the term "applications" as usual in a rather liberal sense). Consider models with unbounded Space or Time horizons. On the "endless plain" of location theory, for example, where the same patterns are repeated indefinitely (as in the Löschian system), we may reasonably presume that the number of agents and total rentals will be infinite. With an unbounded T horizon we may have an infinite sequence of generations. Whether rentals become infinite with unbounded T is more dubious (recall that everything is measured in discounted dollars, so that the present value of total rentals may well be finite even for unbounded T).

It is a little harder to find a rationale for the μ_i in the preferences of the individual agents to be infinite. Nonetheless, such a preference order might be reasonable for organizations such as corporations or governments, which are potentially immortal or can extend their control indefinitely over infinite S. Furthermore, there are other interpretations. If the i's are interpreted as activities or land uses rather than as agents, then the real-estate equilibrium may be thought of as the result of a global competition among alternative uses for the allocation of Space-Time. In this case, under the appropriate world system the μ_i may well be infinite.

We now proceed to the generalizations. An immediate difficulty arises. In (1) the meaningless expression $\infty - \infty$ may arise for certain values of E; also, there may be several values of E yielding $+\infty$ (or $-\infty$). Are these to be considered indifferent, or is it possible to discriminate among them?

The reader familiar with Chapter 3 will notice we have a situation tailor-made for the application of pseudomeasures. Namely, we interpret the difference in (1) in the sense of pseudomeasures and interpret "maximization" of (1) as referring to one or another of the orderings on the space of pseudomeasures discussed in Chapter 3.

When this is done, the entire theory of Section 6.6 generalizes directly from bounded signed measures to pseudomeasures. With the exception of one theorem, it also generalizes from a finite to a countable number of agents. Things generalize not only theorem by theorem but even proof by proof, so that we might say that the natural realm in which the theory of Section 6.6 is valid is the realm of pseudomeasures.[19]

---

19. There is one minor exception to the statement that this section generalizes Section 6.6. The existence theorem there is valid for *arbitrary* signed measures, not merely σ-finite ones. Since pseudomeasures generalize only σ-finite signed measures, this theorem is not completely encompassed in the present results. The existence theorem is also the exceptional theorem mentioned above.

In what follows we make a still further generalization and take not merely the differences $\mu_i - \mu^o$ but also μ_i and μ^o themselves to be pseudomeasures. This may have direct application in case there are "infinitely dispreferred" as well as "infinitely preferred" regions. However, our main object in doing this is that it facilitates the following development to plunge completely into the realm of pseudomeasures: proofs are smoother and theorems are simpler to state (as well as more general).

Before proceeding, we need a convention and one or two definitions. The convention arises from our dealing with a countable rather than a finite number of agents. Formerly, we had an n-tuple of signed measures $(\mu_1, \ldots, \mu_n)$ and a corresponding n-tuple of sets $(E_1, \ldots, E_n)$. Now we have a sequence of pseudomeasures $(\psi_1, \psi_2, \ldots)$ and a corresponding sequence of sets $(E_1, E_2, \ldots)$. These sequences are countable, possibly finite. We now insist that these two sequences be of the same length: either both infinite or both finite of length n, so that there is in any case a 1–1 correspondence between ψ_i and E_i. Both these cases will be encompassed by the single notation $(\psi_1, \psi_2, \ldots)$, $(E_1, E_2, \ldots)$. These sequences are of finite or infinite, but in any case equal, length.

We give the definitions. We refer to the triple (A, Σ, ψ) as a *pseudomeasure space* iff (A, Σ) is a measurable space and ψ is a pseudomeasure on this space. Sometimes we use the notation (A, Σ, μ, ν) for the same thing, where (μ, ν) is any representation of the pseudomeasure ψ as a pair of σ-finite measures.

Definition: Let (A, Σ, ψ) be a pseudomeasure space, and let E_1 be a measurable subset of A. The *restriction* of ψ to E_1 is the pseudomeasure space (E_1, Σ_1, ψ_1), where Σ_1 is Σ restricted to subsets of E_1, and $\psi_1 = (\mu_1, \nu_1)$; here μ_1 and ν_1 are the restrictions of μ and ν to E_1, respectively, where (μ, ν) is any representative of ψ.

That is, given ψ, choose any one of its forms (μ, ν); restrict the two measures to E_1, getting μ_1 and ν_1; and then ψ_1 is the pseudomeasure to which the pair (μ_1, ν_1) belongs.

It must be shown that this is a bona fide definition; i.e., the resulting pseudomeasure ψ_1 must be the same no matter what pair representing ψ is chosen. This is easily verified. The basic equivalence theorem for pseudomeasures states that $(\mu, \nu) = (\mu', \nu')$ (i.e., they represent the same pseudomeasure) iff

$$\mu + \nu' = \nu + \mu' \tag{6.9.2}$$

Now suppose $\psi = (\mu, \nu) = (\mu', \nu')$, so that (2) holds. Restricting all measures to E_1, we have $\mu_1 + \nu_1' = \nu_1 + \mu_1'$, so that $\psi_1 = (\mu_1, \nu_1) = (\mu_1', \nu_1')$. Thus the definition is sound. Sometimes we refer to ψ_1 itself, instead of (E_1, Σ_1, ψ_1), as the restriction of ψ to E_1.

Next, we need the concept of a *direct sum of pseudomeasures*. Recall what this concept means for measures. Let (E_i, Σ_i, μ_i), $i = 1, 2, \ldots$, be a (finite or infinite) sequence of measure spaces, the universe sets $E_1, E_2, \ldots$ being mutually disjoint. The *direct sum* of these spaces, written

$$(E_1, \Sigma_1, \mu_1) \oplus (E_2, \Sigma_2, \mu_2) \oplus \cdots \qquad (6.9.3)$$

or $\oplus_i(E_i, \Sigma_i, \mu_i)$, is the measure space whose universe set $\oplus_i E_i = E_1 \cup E_2 \cup \cdots$, whose σ-field $\oplus_i \Sigma_i$ consists of all sets of the form $F_1 \cup F_2 \cup \cdots$, where $F_i \in \Sigma_i$, $i = 1, 2, \ldots$, and whose measure is given by

$$(\oplus_i \mu_i)(F_1 \cup F_2 \cup \cdots) = \mu_1(F_1) + \mu_2(F_2) + \cdots \qquad (6.9.4)$$

Definition: Let (E_i, Σ_i, ψ_i), $i = 1, 2, \ldots$, be a (finite or infinite) sequence of pseudomeasure spaces, the universe sets $E_1, E_2, \ldots$ being mutually disjoint. The *direct sum* of these is the pseudomeasure space $(\oplus_i E_i, \oplus_i \Sigma_i, \oplus_i \psi_i)$, where $\oplus_i E_i$ and $\oplus_i \Sigma_i$ are defined just as above, while $\oplus_i \psi_i = (\oplus_i \mu_i, \oplus_i \nu_i)$, (μ_i, ν_i) being any representative of ψ_i, $i = 1, 2, \ldots$.

That is, for each ψ_i choose any one of its forms (μ_i, ν_i), take the direct sum of the sequence of measures $\mu_1, \mu_2, \ldots$, and do the same for $\nu_1, \nu_2, \ldots$; this yields a pair of measures, and $\oplus_i \psi_i$ is the pseudomeasure to which this pair belongs.

Again, it must be shown that this is a bona fide definition. Note first of all that the direct sum of σ-finite measures is a σ-finite measure. To see this, take the direct sum in (3), with each μ_i σ-finite. Each E_i then has a countable measurable partition $\{E_{i1}, E_{i2}, \ldots\}$ such that $\mu_i(E_{ij})$ is finite, all $i, j = 1, 2, \ldots$. The collection of all the sets E_{ij} is then a countable measurable partition of $\oplus_i E_i$; furthermore, by (4), μ_i coincides with $\oplus_i \mu_i$ on E_{ij}. Hence the latter measure is σ-finite.

This proves the pair $(\oplus_i \mu_i, \oplus_i \nu_i)$ does indeed represent a pseudomeasure. We must now show that the resulting pseudomeasure is the same no matter what pair representing each component ψ_i is chosen. To prove this, for each i let (μ_i', ν_i') be another pair representing ψ_i. By the equivalence theorem,

$$\mu_i + \nu_i' = \nu_i + \mu_i' \qquad (6.9.5)$$

for each $i = 1, 2, \ldots$. Taking direct sums, we obtain

$$(\oplus_i \mu_i) + (\oplus_i \nu_i') = \oplus_i(\mu_i + \nu_i') = \oplus_i(\nu_i + \mu_i') = (\oplus_i \nu_i) + (\oplus_i \mu_i') \qquad (6.9.6)$$

The middle equality in (6) comes from (5); the first and last equalities are easy consequences of (4). Hence, again using the equivalence theorem, we obtain $(\oplus_i \mu_i, \oplus_i \nu_i) = (\oplus_i \mu_i', \oplus_i \nu_i')$. Hence the direct sum $\oplus_i \psi_i$ is indeed well defined.

Now let a fixed measurable space (A, Σ) be given, and let $(\psi_1, \psi_2, \ldots)$ be a sequence of pseudomeasures on (A, Σ). For any corresponding sequence of measurable sets $(E_1, E_2, \ldots)$ that partition A, we define a new pseudomeasure on (A, Σ) as follows. First, restrict each ψ_i to its corresponding set E_i. (Recall our convention concerning equality of length of these sequences.) Second, take the direct sum of these restrictions. The result is a pseudomeasure that is intuitively obtained by patching together pieces of the original ψ_i's. We denote this pseudomeasure by ψ_E, so that

$$\psi_E = (\psi_1 \mid E_1) \oplus (\psi_2 \mid E_2) \oplus \cdots \tag{6.9.7}$$

$\psi_i \mid E_i$ being the restriction of ψ_i to E_i.

This construction bears a strong resemblance to (6.16). Recall that (6.16), which is $\mu_1(E_1) + \cdots + \mu_n(E_n)$, was the *social valuation* of the allocation $(E_1, \ldots, E_n)$, each agent placing his own evaluation (in discounted dollars) on the region he controls. Formula (7) is the natural generalization of (6.16), with everything now being in terms of pseudomeasures rather than numbers.

The last theorem of Section 6.6 states that the n-tuple $(E_1^o, \ldots, E_n^o)$ partitioning A maximized social valuation iff it was an extended Hahn decomposition for the n-tuple of signed measures $(\mu_1, \ldots, \mu_n)$. Is there a corresponding generalization involving (7)?

The answer is yes if things are defined in the right way. First, we must generalize the decomposition concept to pseudomeasures. Second, we must specify in what sense(s) the term "maximization" is to be understood: What ordering is being referred to—narrow ordering of pseudomeasures? standard ordering? etc., and is the "maximizer" greatest or merely unsurpassed? We get a richer as well as more general theorem by distinguishing these alternative meanings.

Definition: Let (A, Σ) be a measurable space, and $(\psi_1, \psi_2, \ldots)$ a sequence of pseudomeasures on (A, Σ). The corresponding sequence of measurable sets $(E_1^o, E_2^o, \ldots)$ is an *extended Hahn decomposition* for $(\psi_1, \psi_2, \ldots)$ iff $\{E_1^o, E_2^o, \ldots\}$ partitions A, and for all $i, j = 1, 2, \ldots,$

$$(\psi_i - \psi_j)^-(E_i^o) = 0 \tag{6.9.8}$$

Here ψ^- denotes the lower variation of ψ, just as ψ^+ denotes the upper variation; (8) states that ψ_i is at least as large as ψ_j (in the sense of narrow ordering of pseudomeasures) when both are restricted to E_i^o. In the case where ψ_i and ψ_j are both bounded signed measures, (8) reduces to (6.2); hence this is indeed a generalization of the same concept defined in Section 6.6.

Now let $(\psi_1, \psi_2, \ldots)$ be a sequence of pseudomeasures on measurable space (A, Σ). We consider all possible corresponding sequences of measurable sets $(E_1, E_2, \ldots)$ such that $\{E_1, E_2, \ldots\}$ partitions A. With each such sequence is associated a pseudomeasure ψ_E by the rule (7), giving us a set of pseudomeasures Ψ_E. Let $(E_1^o, E_2^o, \ldots)$ be one particular such sequence, with its associated pseudomeasure $\psi_{E^o} \in \Psi_E$.

Theorem: Each of the following five statements implies the other four:

(i) $(E_1^o, E_2^o, \ldots)$ is an extended Hahn decomposition for $(\psi_1, \psi_2, \ldots)$;

(ii) ψ_{E^o} is *greatest* in the set Ψ_E, in the sense of narrow ordering of pseudomeasures;

(iii) ψ_{E^o} is *greatest* in Ψ_E, in the sense of standard ordering;

(iv) ψ_{E^o} is *unsurpassed* in Ψ_E, in the sense of standard ordering;

(v) ψ_{E^o} is *unsurpassed* in Ψ_E, in the sense of narrow ordering.

Proof: (ii) *implies* (iii) *implies* (iv) *implies* (v): These follow at once from the fact that standard ordering extends narrow ordering and from the definitions of "greatest" and "unsurpassed." To complete the proof, we show that (i) implies (ii), and (v) implies (i).

Let (i) be true, and let $(E_1, E_2, \ldots)$ be another feasible sequence. For any $i, j = 1, 2, \ldots$ it follows from (7) that ψ_{E^o} restricted to E_i^o is the same as ψ_i restricted to E_i^o, and ψ_E restricted to E_j is the same as ψ_j restricted to E_j. Hence

$$(\psi_{E^o} - \psi_E)^-(E_i^o \cap E_j) = (\psi_i - \psi_j)^-(E_i^o \cap E_j) = 0 \qquad (6.9.9)$$

The last equality in (9) follows from (8).

Summing (9) over all pairs (i, j), where i, j range independently over $1, 2, \ldots$, we obtain $(\psi_{E^o} - \psi_E)^-(A) = 0$, so that, by definition, $\psi_{E^o} \geq \psi_E$ (narrow order). Since ψ_E was an arbitrary member of Ψ_E, ψ_{E^o} is greatest under narrow ordering. This proves that (i) implies (ii).

The last part is a little more difficult. Assume that (i) is *false*, so that indices m, n exist for which

$$(\psi_m - \psi_n)^-(E_m^o) > 0 \qquad (6.9.10)$$

We now construct a feasible sequence $(E_1, E_2, \ldots)$ whose associated pseudomeasure ψ_E surpasses ψ_{E^o} (narrow order), proving (v) false.

Let (P, N) be a Hahn decomposition for the pseudomeasure $\psi_m - \psi_n$. Define $(E_1, E_2, \ldots)$ by

$$E_m = E_m^o \cap P, \qquad E_n = E_n^o \cup (E_m^o \cap N), \qquad E_i = E_i^o \qquad (6.9.11)$$

for all $i = 1, 2, \ldots$ other than $i = m$ or $i = n$.

Note that $m \neq n$, from (10). Hence (11) is well defined and the sequence $(E_1, E_2, \ldots)$ so defined is measurable and partitions A.

Pseudomeasure ψ_E derived from this sequence coincides with ψ_n when both are restricted to E_n. Also, ψ_{E^o} coincides with ψ_m when both are restricted to E_m^o. Now $E_m^o \cap N$ is a subset of both E_n and E_m^o. Hence

$$(\psi_{E^o} - \psi_E)^-(E_m^o \cap N) = (\psi_m - \psi_n)^-(E_m^o \cap N) = (\psi_m - \psi_n)^-(E_m^o) > 0 \tag{6.9.12}$$

The first equality in (12) arises from substitution; the second from the fact that (P, N) is a Hahn decomposition of $\psi_m - \psi_n$; the inequality is (10).

Now (12) implies that $\psi_{E^o} \not\geq \psi_E$ (narrow order). It remains to show that $\psi_E \geq \psi_{E^o}$, which will prove that ψ_{E^o} is surpassed. We do this by proving that

$$(\psi_E - \psi_{E^o})^-(E_i) = 0 \tag{6.9.13}$$

for all $i = 1, 2, \ldots$. Equality (13) is in fact immediate for all i other than $i = n$, since when restricted to E_i, both ψ_E and ψ_{E^o} coincide with ψ_i, hence with each other.

This leaves E_n; we consider separately its two pieces E_n^o and $(E_m^o \cap N)$. On E_n^o, ψ_E and ψ_{E^o} again both coincide with ψ_n. On $(E_m^o \cap N)$, ψ_{E^o} coincides with ψ_m, and ψ_E with ψ_n. Hence

$$(\psi_E - \psi_{E^o})^-(E_m^o \cap N) = (\psi_n - \psi_m)^-(E_m^o \cap N) \leq (\psi_n - \psi_m)^-(N) = 0$$

Hence (13) is true for all i. Adding over i, we obtain $(\psi_E - \psi_{E^o})^-(A) = 0$, which is $\psi_E \geq \psi_{E^o}$. Combined with $\psi_{E^o} \not\geq \psi_E$, we find that ψ_{E^o} is surpassed, so (v) is false; hence (v) implies (i). ∎

The corresponding theorem in Section 6.6 asserts the equivalence of $(E_1^o, E_2^o, \ldots)$ being an extended Hahn decomposition and maximizing $\mu_1(E_1) + \cdots + \mu_n(E_n)$. These statements are specializations of (i), and (iii) or (iv), respectively: the maximization refers only to "standard ordering" in the realm of bounded signed measures, i.e., based on the value assigned to the universe set A. By adding the specializations of statements (ii) and (v), we get a stronger theorem in the realm of bounded signed measures. Thus we find that an extended Hahn decomposition maximizes social valuation not only on A but on *every* measurable set simultaneously; this follows from (i) implying (ii).

To the five logically equivalent statements just mentioned two more can be added; any of the five statements implies and is implied by the condition that "ψ_{E^o} is greatest in the set Ψ_E under any *extended* ordering of pseudomeasures." This is the first statement; the second is obtained by replacing "greatest" by "unsurpassed." These may be inserted in the chain of implications between statements (iii) and (iv); this follows from the fact that any extended ordering is an extension of standard ordering.

The economic interpretation of this theorem is the same as in Section 6.6. It expresses the extended Hahn decomposition property as an extremal property. Since, as will next be shown, a sequence $(E_1, E_2, \ldots)$ is such a decomposition for $(\psi_1, \psi_2, \ldots)$ iff it is a market equilibrium, it also will have demonstrated an equivalence between market equilibrium and the maximization of social valuation, just as in Section 6.6.

Let us return to the real-estate market. We now have a (finite or infinite) sequence of agents. A market equilibrium will consist of a corresponding sequence of measurable sets $(E_1^o, E_2^o, \ldots)$ and a rental pseudomeasure ψ^o on (A, Σ). Here E_i^o will be the region falling to control of agent i, and these sets must partition A. Agent i will prefer region E_i^o at least as well as any other region, given ψ^o. But here a slight problem arises. How should the preferences of agent i over regions be represented? In the bounded signed measure case, agent i had the utility function

$$U_i(E) = \mu_i(E) - \mu^o(E) \qquad (6.9.14)$$

In the general case one naturally expects utility to be pseudomeasure valued (and to reduce to (14) when all pseudomeasures are in fact bounded signed measures). One's first impulse is to assign to set E a pseudomeasure restricted to E, but this will not do: pseudomeasures are comparable only if they are defined on the same measurable space. Thus if $U_i(E)$ were a pseudomeasure with universe set E, no two regions would be comparable.

This difficulty is easily resolved: with region E we associate the pseudomeasure

$$[(\psi_i - \psi^o) \mid E] \oplus [0 \mid (A \setminus E)] \qquad (6.9.15)$$

This is the direct sum of $(\psi_i - \psi^o)$ restricted to E, and the *zero* pseudomeasure restricted to $A \setminus E$. All pseudomeasures (15) are over the space (A, Σ). Pseudomeasure ψ_i is the generalization of μ_i and may be thought of intuitively as giving the value to agent i of the various regions, gross of any rental outlay. (More exactly, the pseudomeasure $[\psi_i \mid E] \oplus [0 \mid (A \setminus E)]$ gives i's gross evaluation of region E.)

One easily verifies that if ψ_i and ψ^o are both bounded signed measures, (15) under standard ordering in effect reduces to (14), with $\mu_i = \psi_i^+ - \psi_i^-, \mu^o = (\psi^o)^+ - (\psi^o)^-$; i.e., both (15) and (14) determine the same preference ordering over regions in this case. Thus (15) seems to be the natural generalization of (14).

Consider the result of adding ψ^o to (15). It is

$$[\psi_i \mid E] \oplus [\psi^o \mid (A \setminus E)] \qquad (6.9.16)$$

The verification of (16) rests on two observations. First, that

$$\psi = [\psi \mid E] \oplus [\psi \mid (A \backslash E)] \qquad (6.9.17)$$

is an identity for any pseudomeasure, in particular for ψ^o. Second, that

$$[\psi_i \mid E] \oplus [\psi^o \mid (A \backslash E)] = [[(\psi_i - \psi^o) \mid E] \oplus [0 \mid (A \backslash E)]]$$
$$+ [[(\psi^o \mid E] \oplus [\psi^o \mid (A \backslash E)]] \qquad (6.9.18)$$

which again is a special case of a theorem concerning sums of direct sums. The verification of (17) and (18) is left as an exercise. The corresponding transformation of (14) is

$$\mu_i(E) + \mu^o(A \backslash E) \qquad (6.9.19)$$

Since (15) and (16) differ only by a constant, they determine the same ordering over regions E, whether we use narrow order or standard order (or any other partial order on the vector space of pseudomeasures that is determined by a convex cone). This follows from the fact that, for any three pseudomeasures ψ^o, ψ^1, ψ^2 over (A, Σ), $\psi^1 \succeq \psi^2$ iff $\psi^1 + \psi^o \succeq \psi^2 + \psi^o$, where $\succeq$ stands for any such partial ordering. Hence the utility of region E for agent i could just as well be given by (16) as by (15). (Similarly, in the special case of bounded signed measures, $U_i(E)$ given by (19) yields the same preference ordering over regions as $U_i(E)$ given by (14).)

Now let $(E_1^o, E_2^o, \ldots)$, with ψ^o, be a real-estate market equilibrium. This means that for each i, E_i^o maximizes (15) or, equivalently, (16). But what does "maximize" mean? Does it refer to narrow or standard ordering, and does E_i^o maximize in the sense of being greatest, or merely unsurpassed?

Our preceding theorem furnishes a complete and satisfying answer to these questions. For the utility functions (15) or (16), all these senses of the term "maximize" are *equivalent:* E_i^o maximizes in one of these senses iff it maximizes in any other sense. Furthermore, by the same theorem a necessary and sufficient condition that E_i^o maximize in any (hence all) of these senses is that the pair $(E_i^o, A \backslash E_i^o)$ be an *extended Hahn decomposition* for the pair of pseudomeasures (ψ_i, ψ^o).

The preceding theorem refers to sequences of pseudomeasures $(\psi_1, \psi_2, \ldots)$ and corresponding sequences of measurable sets $(E_1, E_2, \ldots)$ that partition A. The statements above are nothing but the special case in which these sequences are merely pairs. The statement that $(E_i^o, A \backslash E_i^o)$ is a decomposition for the pair (ψ_i, ψ^o) comes from the utility function (16). If instead we use (15), we find that, equivalently, it is an extended Hahn decomposition for the pair $(\psi_i - \psi^o, 0)$. Furthermore, it follows at once from the definitions that these statements are true iff $(E_i^o, A \backslash E_i^o)$ is a Hahn decomposition for the pseudomeasure $\psi_i - \psi^o$. Thus we have seven or eight logically equivalent conditions on the set E_i^o.

With these preliminary comments out of the way, we now state a result that directly generalizes the first theorem of Section 6.6.

Theorem: Let $(\psi_1, \psi_2, \ldots)$ be a sequence of pseudomeasures on space (A, Σ), and let $(E_1^o, E_2^o, \ldots)$ be a corresponding sequence of measurable sets that partition A. Then pseudomeasure ψ^o satisfies one of the following two conditions iff it satisfies the other:

(i) For each $i = 1, 2, \ldots$ $(E_i^o, A \backslash E_i^o)$ is an extended Hahn decomposition for the pair (ψ_i, ψ^o);

(ii) for all $i, j = 1, 2, \ldots$ (with $i \neq j$) we have

$$(\psi_i \mid E_i^o) \geq (\psi^o \mid E_i^o) \geq (\psi_j \mid E_i^o) \qquad (6.9.20)$$

(i.e., when all are restricted to E_i^o, the three pseudomeasures are narrowly ordered as indicated).

Furthermore, a ψ^o exists satisfying one (hence both) of these conditions iff $(E_1^o, E_2^o, \ldots)$ is an extended Hahn decomposition for $(\psi_1, \psi_2, \ldots)$.

Proof: Condition (ii) may be rewritten as

$$(\psi_i - \psi^o)^-(E_i^o) = 0, \qquad (\psi^o - \psi_j)^-(E_i^o) = 0 \qquad (6.9.21)$$

while condition (i) is

$$(\psi_i - \psi^o)^-(E_i^o) = 0, \qquad (\psi^o - \psi_i)^-(E_j^o) = 0 \qquad (6.9.22)$$

both holding for all $i, j, i \neq j$. The left conditions in (21) and (22) are identical. The right conditions are also identical except for the interchange of indices i and j. This proves conditions (i) and (ii) imply each other.

Next suppose ψ^o exists satisfying these conditions. From (20) we obtain

$$(\psi_i - \psi_j)^-(E_i^o) = 0 \qquad (6.9.23)$$

for all i, j (with $i \neq j$). But (23) is exactly the condition (8) that $(E_1^o, E_2^o, \ldots)$ be an extended Hahn decomposition for $(\psi_1, \psi_2, \ldots)$.

Finally, suppose that $(E_1^o, E_2^o, \ldots)$ is an extended Hahn decomposition for $(\psi_1, \psi_2, \ldots)$. Let ψ^o be the pseudomeasure

$$(\psi_1 \mid E_1^o) \oplus (\psi_2 \mid E_2^o) \oplus \cdots \qquad (6.9.24)$$

We then obtain

$$(\psi^o \mid E_i^o) = (\psi_i \mid E_i^o) \geq (\psi_j \mid E_i^o) \qquad (6.9.25)$$

for all $i, j = 1, 2, \ldots$. The equality in (25) comes from (24); the inequality is the same as the decomposition condition (23). Since (25) implies (20), the proof is complete. ∎

According to the discussion preceding this theorem, condition (i) holds iff E_i^o maximizes the utility of agent i, (15) or (16), for all i, i.e., iff $(E_1^o, E_2^o, \ldots)$ combined with ψ^o is a real-estate equilibrium. Hence this theorem states that the condition of being a real-estate equilibrium is logically equivalent to being an extended Hahn decomposition for a given sequence $(\psi_1, \psi_2, \ldots)$.

Furthermore, (20) gives a necessary and sufficient condition for ψ^o to serve as the rental pseudomeasure for the given equilibrium $(E_1^o, E_2^o, \ldots)$. The economic interpretation of (20) is the same as in Section 6.6: for each E_i^o, ψ^o lies between the highest bid and all the others. This generalizes (6.4).

Our next result generalizes the "uniqueness" theorem of Section 6.6 concerning decompositions of $(\psi_1, \psi_2, \ldots)$. In view of the theorems above, this also fixes the extent to which market equilibria and maximizers of "social valuation" are unique.

Theorem: Let $(\psi_1, \psi_2, \ldots)$ be a sequence of pseudomeasures on space (A, Σ), with extended Hahn decomposition $(E_1^o, E_2^o, \ldots)$; let $(E_1, E_2, \ldots)$ be another sequence of measurable sets that partitions A (and has the same length as $(E_1^o, E_2^o, \ldots)$).

Then $(E_1, E_2, \ldots)$ is also an extended Hahn decomposition iff

$$| \psi_i - \psi_j | (E_i^o \cap E_j) = 0 \qquad (6.9.26)$$

for all $i, j = 1, 2, \ldots$.[20]

Proof: Let $(E_1, E_2, \ldots)$ also be an extended Hahn decomposition, so that $(\psi_j - \psi_i)^-(E_j) = 0$, all $i, j = 1, 2, \ldots$. This can also be written as $(\psi_i - \psi_j)^+(E_j) = 0$. Also, $(\psi_i - \psi_j)^-(E_i^o) = 0$, since $(E_1^o, E_2^o, \ldots)$ is a decomposition. Hence

$$| \psi_i - \psi_j | (E_i^o \cap E_j) = (\psi_i - \psi_j)^+(E_i^o \cap E_j) + (\psi_i - \psi_j)^-(E_i^o \cap E_j)$$

$$\leq (\psi_i - \psi_j)^+(E_j) + (\psi_i - \psi_j)^-(E_i^o) = 0$$

for all $i, j = 1, 2, \ldots$. This yields (26).

Conversely, let (26) hold. For all $i, k = 1, 2, \ldots$ we obtain

$$(\psi_i - \psi_k)^-(E_k^o \cap E_i) \leq | \psi_i - \psi_k | (E_k^o \cap E_i) = 0 \qquad (6.9.27)$$

from (26). Also, for all $i, j, k = 1, 2, \ldots$ we obtain

$$(\psi_k - \psi_j)^-(E_k^o \cap E_i) \leq (\psi_k - \psi_j)^-(E_k^o) = 0 \qquad (6.9.28)$$

---

20. Recall that for any pseudomeasure ψ, $| \psi |$ is a *measure* called the *total variation* of ψ and is equal to $\psi^+ + \psi^-$.

since $(E_1^o, E_2^o, \ldots)$ is a decomposition. But also

$$(\psi_i - \psi_j)^- \le (\psi_i - \psi_k)^- + (\psi_k - \psi_j)^- \qquad (6.9.29)$$

is true for any three pseudomeasures ψ_i, ψ_j, ψ_k. (This follows from the minimizing property of the Jordan form of a pseudomeasure.) From (27)–(29) we obtain

$$(\psi_i - \psi_j)^-(E_k^o \cap E_i) = 0 \qquad (6.9.30)$$

for all i, j, $k = 1, 2, \ldots$. Finally, adding (30) over all k, we obtain $(\psi_i - \psi_j)^-(E_i) = 0$ for all $i, j = 1, 2, \ldots$, so that $(E_1, E_2, \ldots)$ is indeed another extended Hahn decomposition. ∎

Condition (26) is a direct generalization of (6.14), the economic interpretation of which carries over to (26).

Our final result generalizes the existence theorem of Section 6.6. We have saved this for last because it is the only result that demands just a *finite* number of agents. Every other result generalizes to the *countable* case. Indeed, the counterexample already given in Section 6.6 shows that a countably infinite number of pseudomeasures may not have an extended Hahn decomposition.

Theorem: Any n-tuple of pseudomeasures $(\psi_1, \ldots, \psi_n)$ on space (A, Σ) has an extended Hahn decomposition.

Proof: Proof is by induction on n. First, for $n = 2$ let (P, N) be a Hahn decomposition for the pseudomeasure $\psi_1 - \psi_2$. Then $(\psi_1 - \psi_2)^-(P) = 0$, and also

$$(\psi_2 - \psi_1)^-(N) = (\psi_1 - \psi_2)^+(N) = 0$$

so that (P, N) is also an *extended* Hahn decomposition for the pair (ψ_1, ψ_2).

Next, assuming the statement holds for $n - 1$, we prove it for n. We have, then, a measurable $(n - 1)$-tuple $(E_1, \ldots, E_{n-1})$ that partitions A and for which

$$(\psi_i - \psi_j)^-(E_i) = 0 \qquad (6.9.31)$$

for all $i, j = 1, \ldots, n - 1$.

For each $i = 1, \ldots, n - 1$ let (P_i, N_i) be a Hahn decomposition for $\psi_i - \psi_n$, and define

$$E_i^o = E_i \cap P_i \qquad (6.9.32)$$

for $i = 1, \ldots, n - 1$, and

$$E_n^o = (E_1 \cap N_1) \cup (E_2 \cap N_2) \cup \cdots \cup (E_{n-1} \cap N_{n-1}) \qquad (6.9.33)$$

We prove that $(E_1^o, \ldots, E_n^o)$ is an extended Hahn decomposition for $(\psi_1, \ldots, \psi_n)$. It clearly partitions A.

For $i \neq n$, $j \neq n$ we have

$$(\psi_i - \psi_j)^-(E_i^o) \leq (\psi_i - \psi_j)^-(E_i) = 0$$

$$(\psi_i - \psi_n)^-(E_i^o) \leq (\psi_i - \psi_n)^-(P_i) = 0$$

both from (32). It only remains to show that

$$(\psi_n - \psi_i)^-(E_n^o) = 0 \tag{6.9.34}$$

for all $i \neq n$.

For any $i, j = 1, \ldots, n - 1$ we have

$$(\psi_j - \psi_i)^-(E_j \cap N_j) = 0 \tag{6.9.35}$$

from (31). Also

$$(\psi_n - \psi_j)^-(E_j \cap N_j) \leq (\psi_n - \psi_j)^-(N_j) = 0 \tag{6.9.36}$$

(35), (36), and (29) imply that

$$(\psi_n - \psi_i)^-(E_j \cap N_j) = 0 \tag{6.9.37}$$

Adding (37) over all $j = 1, \ldots, n - 1$ and noting (33), we obtain (34). This shows that $(E_1^o, \ldots, E_n^o)$ is indeed a decomposition for $(\psi_1, \ldots, \psi_n)$ and completes the induction and the proof. ∎

APPENDIX: THE VECTOR LATTICE OF PSEUDOMEASURES

We indicate some algebraic consequences of the results of the last section in an appendix because they are not applied in this book, but they are of interest for the further mathematical development of pseudomeasure theory.

Consider the vector space of all pseudomeasures Ψ on measurable space (A, Σ). Let $\{\psi_1, \ldots, \psi_n\}$ be a finite nonempty subset of Ψ, and suppose that ψ_o has the following properties: (i) $\psi_o \geq \psi_i$ for all $i = 1, \ldots, n$ ($\geq$ refers to *narrow order* throughout this section); and (ii) for any ψ, if $\psi \geq \psi_i$ for all $i = 1, \ldots, n$, then $\psi \geq \psi_o$.

Definition: Such a pseudomeasure ψ_o (if it exists) will be called the *supremum* of $\{\psi_1, \ldots, \psi_n\}$. Similarly, a ψ_o satisfying (i) and (ii), but with $\geq$ replaced by $\leq$, will be called the *infimum* of $\{\psi_1, \ldots, \psi_n\}$. The abbreviations sup and inf are used for these operations.

Set $\{\psi_1, \ldots, \psi_n\}$ has *at most one* supremum; for if ψ_o', ψ_o'' both satisfy (i) and (ii), then $\psi_o' \geq \psi_o'' \geq \psi_o'$. Hence $\psi_o' = \psi_o''$, since narrow order is antisymmetric. Similarly, it has at most one infimum.

Theorem: Any nonempty finite set of pseudomeasures $\{\psi_1, \ldots, \psi_n\}$ on space (A, Σ) has a supremum and an infimum. The supremum is given by

$$\psi_o = (\psi_1 \mid E_1^o) \oplus \cdots \oplus (\psi_n \mid E_n^o) \tag{6.9.38}$$

where $(E_1^o, \ldots, E_n^o)$ is any extended Hahn decomposition of $(\psi_1, \ldots, \psi_n)$. The infimum is given by

$$- \sup\{-\psi_1, \ldots, -\psi_n\} \tag{6.9.39}$$

Proof: We know that an extended Hahn decomposition exists, so there is at least one pseudomeasure of the form (38). We also know that (38) is greatest under narrow ordering in the set Ψ_E of all pseudomeasures of the same form, with $(E_1, \ldots, E_n)$ in place of $(E_1^o, \ldots, E_n^o)$, where $(E_1, \ldots, E_n)$ is a measurable n-tuple that partitions A.

In particular, take the n-tuple $(\phi, \ldots, A, \ldots, \phi)$, with the universe set in place i, the empty set everywhere else. The pseudomeasure of form (38) corresponding to this is simply ψ_i. Hence $\psi_o \geq \psi_i$, $i = 1, \ldots, n$.

Next suppose $\psi \geq \psi_i$, $i = 1, \ldots, n$, for some ψ. This implies $(\psi - \psi_i)^-(E_i^o) = 0$, $i = 1, \ldots, n$. Now ψ_o coincides with ψ_i when both are restricted to E_i^o. Hence $(\psi - \psi_o)^-(E_i^o) = 0$ for all i. Adding over i, we obtain $(\psi - \psi_o)^-(A) = 0$; i.e., $\psi \geq \psi_o$. This proves that ψ_o is indeed the supremum.

Now let ψ_{oo} abbreviate (39). We then have $-\psi_{oo} \geq -\psi_i$, all i, and if $\psi \geq -\psi_i$, all i, then $\psi \geq -\psi_{oo}$. These are the same as $\psi_{oo} \leq \psi_i$, all i; and if $-\psi \leq \psi_i$, all i, then $-\psi \leq \psi_{oo}$, which in turn is the same as (i) and (ii) with signs reversed. Hence ψ_{oo} is the infimum. ∎

A number of corollaries implicit in this theorem illuminate the concepts of the preceding discussion. First, the supremum is defined for a *set*, while the expression (38) is in terms of a particular ordering $(\psi_1, \ldots, \psi_n)$ of the elements of this set. A moment's reflection shows, however, that the ordering is irrelevant. Indeed, a permutation of $(\psi_1, \ldots, \psi_n)$ leads to a corresponding permutation of the decomposition $(E_1^o, \ldots, E_n^o)$. This leads merely to a change in the order of the summands in (38) and yields the same pseudomeasure.

Second, even though there may be many extended Hahn decompositions for $(\psi_1, \ldots, \psi_n)$, these all must yield the same pseudomeasure (38), since the supremum is unique.

$\mathrm{Sup}\{\psi_1, \ldots, \psi_n\}$ is exactly what was referred to above as the *social valuation* of the real-estate equilibrium. This provides a concrete interpretation.

The fact that the sup and inf always exist means that Ψ is not only a vector space but a *lattice* (with respect to narrow ordering).[21]

We conclude with some (fairly difficult) exercises:

1. Show that for any two pseudomeasures $\psi_1, \psi_2 \in \Psi$,

$$\sup\{\psi_1, \psi_2\} = \tfrac{1}{2}(\psi_1 + \psi_2 + |\psi_1 - \psi_2|)$$

$$\inf\{\psi_1, \psi_2\} = \tfrac{1}{2}(\psi_1 + \psi_2 - |\psi_1 - \psi_2|)$$

Here the total variation $|\psi_1 - \psi_2|$ is of course a measure, which may be identified as usual with the pseudomeasure $(|\psi_1 - \psi_2|, 0)$. (Hint: The Hahn decomposition of $\psi_1 - \psi_2$ is the same as the extended Hahn decomposition of (ψ_1, ψ_2).)

2. If ψ_1, ψ_2 are measures, show that these operations coincide with the ordinary sup and inf of measures (defined in Section 3.1).

3. Show that Ψ under narrow order is in fact a *distributive* lattice; i.e., for any three pseudomeasures ψ_1, ψ_2, ψ_3,

$$\inf\{\psi_1, \sup\{\psi_2, \psi_3\}\} = \sup\{\inf\{\psi_1, \psi_2\}, \inf\{\psi_1, \psi_3\}\}$$

and a similar equality holds for inf and sup interchanged. (Hint: First do the special case $\psi_1 = 0$; take a Hahn decomposition of $\psi_2 - \psi_3$ and do each half separately.)

4. Under *standard* ordering show that Ψ (partitioned into indifference classes) is *not* a lattice unless Σ is a finite σ-field. In fact, show that ψ_1, ψ_2 have a least upper bound under standard order iff they are comparable under standard order, i.e., iff $\psi_1 - \psi_2$ is a signed measure.

---

21. G. Birkhoff, *Lattice Theory,* Vol. 25, 3rd ed. (Am. Math. Soc. Colloq. Publ., Providence, R.I., 1967).

7

The Transportation and
Transhipment Problems

7.1. THE TRANSPORTATION PROBLEM: INTRODUCTION

The transportation problem[1] with m sources and n sinks is: find mn nonnegative numbers x_{ij}, $i = 1, \ldots, m$, $j = 1, \ldots, n$, satisfying

$$x_{i1} + x_{i2} + \cdots + x_{in} \leq \alpha_i \qquad (7.1.1)$$

$i = 1, \ldots, m$,

$$x_{1j} + x_{2j} + \cdots + x_{mj} \geq \beta_j \qquad (7.1.2)$$

$j = 1, \ldots, n$, and minimizing the sum of

$$f_{ij} x_{ij} \qquad (7.1.3)$$

over all mn terms of this form, $i = 1, \ldots, m$, $j = 1, \ldots, n$. Here the numbers α_i, β_j, and f_{ij} are given parameters (α_i, $\beta_j \geq 0$).

The most straightforward interpretation of this problem is the following: x_{ij} is the quantity of a certain commodity moving from a source i (say a manufacturing plant) to a sink j (say a market where the good is sold), and f_{ij} is *unit* transport cost incurred by this movement, so that (3) is *total* transport cost for the source-sink pair (i, j). The problem, then, is to minimize the grand total of costs over all such pairs. Source i has a *capacity* α_i, and the constraints (1) state that the total shipments from a source cannot exceed its capacity. There are m such constraints, one for each source. Sink j has a *requirement* β_j, and the n constraints (2) state that total shipments into sink j must not fall below its requirement. Besides this interpretation (from which the transportation problem gets its name) there are a remarkable number of others concerning resource assignments, scheduling, etc.[2]

---

1. A. M. Faden, The Abstract Transportation Problem, in A. M. Zarley, ed., *Papers in Quantitative Economics*, Vol. 2 (Univ. Press of Kansas, Lawrence, 1971), pp. 147–75, is a less advanced version of Sections 7.1–7.5.

2. S. Vajda, *Readings in Mathematical Programming*, 2nd ed. (Wiley, New York, 1962); G. B. Dantzig, *Linear Programming and Extensions* (Princeton Univ. Press, Princeton, 1963).

Now consider the following problem involving measures. We are given two measure spaces (A, Σ', μ') and (B, Σ'', μ''); (A, Σ') will be called the *source space,* and (B, Σ'') the *sink space.* Measure μ' will be called the *capacity measure,* and μ'' the *requirement measure.* We assume throughout this chapter that μ' and μ'' are σ-finite. We are also given a *cost function* $f: A \times B \to$ reals, assumed measurable with respect to the product σ-field, $\Sigma' \times \Sigma''$ on $A \times B$.

The problem is to find a measure λ on the space $(A \times B, \Sigma' \times \Sigma'')$ satisfying

$$\lambda(E \times B) \le \mu'(E) \qquad (7.1.4)$$

for all $E \in \Sigma'$,

$$\lambda(A \times F) \ge \mu''(F) \qquad (7.1.5)$$

for all $F \in \Sigma''$, and minimizing

$$\int f \, d\lambda \qquad (7.1.6)$$

Here (6) is an indefinite integral over space $A \times B$, and "minimization" is to be understood in the sense of (reverse) standard ordering of pseudo-measures. Of course, if the *definite* integral

$$\int_{A \times B} f \, d\lambda \qquad (7.1.7)$$

is well defined and finite for all feasible λ, this reduces to the ordinary minimization of (7). But there is no a priori guarantee that (7) *will* be finite or even well defined, without special conditions on f, μ', and μ''.

Formulas (4), (5), and (6) reduce to (1), (2), and (3) respectively iff both σ-fields Σ' and Σ'' are finite. To be precise, let Σ' be generated by a partition $\{A_1, \ldots, A_m\}$ of A, and Σ'' by a partition $\{B_1, \ldots, B_n\}$ of B. Then it is a simple exercise to verify the preceding statement: $\mu'(A_i) = \alpha_i$, $\mu''(B_j) = \beta_j$, etc. This shows that we *are* dealing with a bona fide generalization of the ordinary transportation problem.

The interpretations that can be given to (4)–(6) include all those for the ordinary "discrete" problem, and the greater flexibility that one attains with measures enables one to fit the real situation that much more closely. For example, in the transportation interpretation, one may now treat the case where sources are spread more or less continuously over the surface of the earth (as in agricultural production), where sinks are (as in the sale of consumer goods to a diffused population), or both. The best the ordinary problem (1)–(3) can do is to aggregate these distributions, e.g., to treat countries as if located at single points in international trade models.

Again, in some interpretations the sources and sinks correspond to time instants rather than locations; here the measure-theoretic formulation allows us to work with continuous time, rather than having to lump things into discrete periods. This has clear theoretical advantages; it may even be advantageous in practical applications, since continuous models are often simpler than discrete models.

One of the most important nontransport interpretations of the transportation problem refers to the assignment of resources to activities. Here the "sources" are the various kinds of resources available, and the "sinks" are the various activities. The measure-theoretic generalization is especially welcome in this interpretation to allow for the infinite variety of resources and activities. In Chapter 8 we show that Thünen systems can be represented by just such a model.

We now examine (4)–(6) more closely: $\mu'(E)$ gives the total capacity of the set of sources E, while $\lambda(E \times B)$ gives the total outflow from these sources. Then (4) is the condition that outflow not exceed capacity for any measurable set of sources. A similar relation (5) holds between $\mu''(F)$, the requirement for the set of sinks F, and $\lambda(A \times F)$, the total inflow into these sinks.

It will be convenient to formulate the constraints in terms of marginals. Recall that the *left marginal* of $(A \times B, \Sigma' \times \Sigma'', \lambda)$ is the measure λ' on (A, Σ'), given by $\lambda'(E) = \lambda(E \times B)$ for all $E \in \Sigma'$. Similarly, the *right marginal* is the measure λ'' on (B, Σ''), given by $\lambda''(F) = \lambda(A \times F)$ for all $F \in \Sigma''$. It follows at once that the constraints (4)–(5) can be written in the very simple form:

$$\lambda' \leq \mu' \tag{7.1.8}$$

$$\lambda'' \geq \mu'' \tag{7.1.9}$$

respectively. We refer to λ' and λ'' as the *outflow* and *inflow* measures respectively, and to λ itself as the *flow* measure.

Now we discuss a point that was glossed over. For (6) to be well defined as a pseudomeasure, λ must be σ-finite. Is this guaranteed? We examine the situation in some generality because it recurs several times in this chapter.

Let (A_i, Σ_i, μ_i), $i = 1, 2$, be two measure spaces, and $g: A_1 \to A_2$ measurable such that the following relation obtains:

$$\mu_2(E) = \mu_1\{a_1 \mid g(a_1) \in E\} \tag{7.1.10}$$

for all $E \in \Sigma_2$; i.e., μ_2 is the measure induced from μ_1 by g.

Now if μ_2 is σ-finite, then μ_1 is σ-finite. To show this, let $\mathcal{G}$ be a countable measurable partition of A_2 such that $\mu_2(G)$ is finite for all $G \in \mathcal{G}$. The collection of sets $\{a_1 \mid g(a_1) \in G\}$, $G \in \mathcal{G}$, is then a countable measurable partition of A_1 and, by (10), μ_1 is finite on each of these sets. Hence μ_1 is σ-finite if μ_2 is. This completes the proof.

The converse of this statement is not necessarily true. As an example, let μ_1 be any infinite σ-finite measure, and let A_2 consist of a single point. Then $\mu_2(A_2) = \infty$, and this is clearly not σ-finite.[3]

Now consider the transportation problem. The left marginal λ' is the measure induced from λ by the projection $g(a, b) = a$. Hence if λ' is σ-finite, so is λ. But we are given that μ' is σ-finite, and it follows from (8) that λ' is σ-finite. We conclude that any feasible flow λ is σ-finite and (6) is well defined. Note, however, that the *right* marginal λ'' is not necessarily σ-finite.

It is common practice in analyzing the transportation problem to replace some of the inequality signs in (1) and (2) by equalities. We also consider the consequences of replacing one or both of the inequalities (8)–(9) by equalities. This gives altogether four variants of the transportation problem. We label these types I, II, III, and IV, defined as follows:

	$\lambda'' = \mu''$	$\lambda'' \geq \mu''$
$\lambda' = \mu'$	I	III
$\lambda' \leq \mu'$	II	IV

Thus in types I and II, requirements must be met exactly, without oversupply. In types I and III capacity must be fully utilized. Type IV is our original problem, given by (8) and (9). In all four variants the objective remains the same: to minimize (6). The parallel analysis of these four types is quite instructive, and they exhibit a surprising degree of individuality.

7.2. THE TRANSPORTATION PROBLEM:
EXISTENCE OF FEASIBLE SOLUTIONS

The first problem we tackle is this: under what conditions does a flow measure λ exist satisfying the feasibility conditions (1.8) and (1.9) or their equality-constrained counterparts? The objective function (1.6) plays no part in this discussion.

For the *ordinary* transportation problem we have the following well-known results: (1.1) and (1.2) have a feasible solution iff total capacity is at least as large as total requirements; i.e., iff

$$\alpha_1 + \cdots + \alpha_m \geq \beta_1 + \cdots + \beta_n \qquad (7.2.1)$$

Furthermore, this remains true in the case where all inequality signs in (1.1) or in (1.2) (but not both) are replaced by equalities. Finally, if all

---

3. The measure induced by a σ-finite measure is always *abcont*, even if not σ-finite.

constraints are equalities, a solution exists iff (1) is satisfied with equality. Thus (1) is necessary and sufficient for the existence of feasible solutions in variants II, III, and IV; and (1) with equality in variant I.[4]

Our main feasibility result is that these conditions carry over completely to the measure-theoretic transportation problem. The demonstration of this is by no means trivial, especially when the capacity and requirement measures are infinite. Our first result establishes feasibility for the ordinary transportation problem (variant I) extended to the case where the number of sources and sinks is *countable*.[5]

Lemma: Let $\alpha_1, \alpha_2, \ldots$ and $\beta_1, \beta_2, \ldots$ be two sequences of nonnegative real numbers such that

$$\alpha_1 + \alpha_2 + \cdots = \beta_1 + \beta_2 + \cdots \qquad (7.2.2)$$

Then nonnegative numbers x_{ij} exist satisfying

$$x_{i1} + x_{i2} + \cdots = \alpha_i \qquad (7.2.3)$$

for all α_i, and

$$x_{1j} + x_{2j} + \cdots = \beta_j \qquad (7.2.4)$$

for all β_j. (The two sequences may be finite or infinite and need not be equal in length; the common sum in (2) may be finite or infinite; i indexes the sequence (α_i), j the sequence (β_j).)

Proof: Define $m_i = \alpha_1 + \cdots + \alpha_i$, $n_j = \beta_1 + \cdots + \beta_j$, $m_0 = n_0 = 0$; then let

$$x_{ij} = \min(m_i, n_j) - \min(m_i, n_{j-1}) - \min(m_{i-1}, n_j) + \min(m_{i-1}, n_{j-1}) \qquad (7.2.5)$$

$i = 1, 2, \ldots$, $j = 1, 2, \ldots$. We show that (5) gives the desired feasible solution. First, we show that all the numbers x_{ij} are nonnegative.

Suppose that $m_{i-1} \leq n_{j-1}$. Then the last two terms on the right of (5) cancel, and the difference of the first two is clearly nonnegative. (Remember that $n_j \geq n_{j-1}$.) A similar argument obtains if $m_{i-1} \geq n_{j-1}$. Hence x_{ij} is nonnegative.

Next we verify that for all indices j of the sequence (β_j) and all indices i of the sequence (α_i), we have

$$x_{i1} + x_{i2} + \cdots + x_{ij} = \min(m_i, n_j) - \min(m_{i-1}, n_j) \qquad (7.2.6)$$

Proceed by induction on j. For $j = 1$, (6) follows immediately from (5),

---

4. There are certain complications if mixtures of equality and inequality constraints appear *within* the capacity block, (1.1). We do not discuss these.

5. I seem to recall seeing this result in the literature, but I have lost the reference.

since $n_0 = 0$. Supposing (6) true for $j - 1$ in place of j, we add (5) to it and obtain (6) per se. Hence (6) is true in general.

Now in (6) let j increase indefinitely (if (β_j) is an infinite sequence) or to its maximum value (if (β_j) is a finite sequence). In either case we find that $\lim n_j \geq m_i$ because of (2). But this means that in the limit the right side of (6) simplifies to $m_i - m_{i-1} = \alpha_i$. Thus (3) is verified. The same argument with i and j interchanged verifies (4). ∎

As given by (5), x_{ij} has a very simple interpretation: it is precisely the "northwest corner" solution for the ordinary transportation problem.[6] Specifically, one starts by making x_{11} as large as possible without automatically violating (3) or (4); i.e., take $x_{11} = \min(\alpha_1, \beta_1)$. If $x_{11} = \alpha_1$, then all the other x_{1j} must be set equal to zero to satisfy (3) for $i = 1$; similarly, if $x_{11} = \beta_1$, all the other $x_{i1} = 0$. Now go to the as yet undetermined x_{ij} for which $i + j$ is as small as possible, and make *it* as large as possible, subject to not automatically violating (3) or (4). This recursive procedure yields (5). We have shown that it still yields a feasible solution even for a countable number of sources and sinks, provided (2) holds.

The northwest corner solution will play an important role in Chapter 8. In fact, for Thünen systems a certain (generalized) northwest corner solution is not only feasible but *optimal* and encapsulates in a striking way the main structural features of such systems.

We now come to the main result. Essential use is made of the *product measure* theorem. Recall that if (A, Σ', μ') and (B, Σ'', μ'') are measure spaces with μ, ν σ-finite or even arbitrary, a measure λ exists on the product space $(A \times B, \Sigma' \times \Sigma'')$ with the property that $\lambda(E \times F) = \mu'(E)\mu''(F)$ for all $E \in \Sigma'$, $F \in \Sigma''$. This is called a product measure and is denoted $\mu' \times \mu''$.

Theorem: Let (A, Σ', μ') and (B, Σ'', μ'') be σ-finite measure spaces that are the source and sink spaces respectively of a transportation problem; let $B \neq \phi$. A feasible flow measure λ exists for this problem iff

$$\mu'(A) = \mu''(B) \qquad (7.2.7)$$

in variant I, and iff

$$\mu'(A) \geq \mu''(B) \qquad (7.2.8)$$

in variants II, III, and IV.

Proof: The "only if" part is simple to demonstrate. Letting λ be

6. First proposed by G. B. Dantzig, Application of the Simplex Method to a Transportation Problem, Ch. 23 in T. C. Koopmans, ed., *Activity Analysis of Production and Allocation* (Wiley, New York, 1951), pp. 361–62.

feasible, we find that $\mu'(A) \geq \lambda(A \times B) \geq \mu''(B)$ in variants II, III, IV, while the same holds with equalities in variant I.

Now we demonstrate the "if" part. Any λ feasible for II or III is also feasible for IV. Hence it suffices to prove the existence of feasible λ for types II and III only, under assumption (8) (as well as feasibility for type I under (7)). We consider three cases, depending on the magnitude of $\mu''(B)$.

Case (i): $\mu''(B) = 0$. For variant I, (7) implies that $\mu'(A) = 0$ also. Hence the identically zero measure on $(A \times B, \Sigma' \times \Sigma'')$ is feasible. $\lambda = 0$ is also feasible for II, since only requirements must be satisfied with equality. As for III, choose an arbitrary point $b_o \in B$, and define λ as follows: $\lambda(G) = \mu'\{a \mid (a, b_o) \in G\}$ for all $G \in \Sigma' \times \Sigma''$. This λ is well defined, because "cross-sections" of measurable sets are measurable. It is easily verified to be a measure, and $\lambda(E \times B) = \mu'(E)$ for all $E \in \Sigma'$. Hence the conditions for transportation variant III are satisfied.

Case (ii): $\infty > \mu''(B) > 0$. For variants I and III, take λ proportional to the product measure: $\lambda = (\mu' \times \mu'')/\mu''(B)$. We then have $\lambda(E \times B) = \mu'(E)$, while $\lambda(A \times F) (=, \geq) \mu''(F)$ for variants (I, III) because of (7) and (8) respectively. Hence λ is feasible.

Variant II is slightly more complicated. If $\mu'(A) = \infty$, then because μ' is σ-finite hence abcont, a measure $\tilde{\mu}$ exists on A such that $\tilde{\mu} \leq \mu'$ and

$$\infty > \tilde{\mu}(A) \geq \mu''(B) \tag{7.2.9}$$

If $\mu'(A)$ is finite, we simply take $\tilde{\mu} = \mu'$; (9) then follows from (8). Now define λ by $\lambda = (\tilde{\mu} \times \mu'')/\tilde{\mu}(A)$. Then $\lambda(A \times F) = \mu''(F)$, all $F \in \Sigma''$. Also,

$$\lambda(E \times B) = \tilde{\mu}(E)\mu''(B)/\tilde{\mu}(A) \leq \tilde{\mu}(E) \leq \mu'(E)$$

all $E \in \Sigma'$. Hence the conditions for variant II are satisfied.

Case (iii): $\mu''(B) = \infty$. This is the hard part. From (7) or (8) we obtain $\mu'(A) = \infty$ also. Hence it suffices to find a feasible λ for variant I because this λ will also be feasible for II, III, and IV.

Since μ' is σ-finite hence abcont, it may be written as a countable sum of finite nonzero measures: $\mu' = \mu'_1 + \mu'_2 + \cdots$. Similarly, for μ'' we may write $\mu'' = \mu''_1 + \mu''_2 + \cdots$, all μ''_j finite, nonzero. Define the sequences (α_i), $i = 1, 2, \ldots$, (β_j), $j = 1, 2, \ldots$, by $\alpha_i = \mu'_i(A)$, $\beta_j = \mu''_j(B)$. These are positive real numbers.

Now invoke the preceding lemma; (2) is satisfied, both sides summing to $+\infty$. Hence nonnegative real x_{ij}, $i, j = 1, 2, \ldots$, exist satisfying (3) and (4).

Define λ as the sum of the product measures $(\mu'_i \times \mu''_j)x_{ij}/(\alpha_i\beta_j)$, the summation extending over all pairs (i, j), $i, j = 1, 2, \ldots$. We show that

λ is feasible for variant I. For $E \in \Sigma'$, $\lambda(E \times B)$ is the sum of all terms of the form

$$\mu_i'(E)\, \mu_j''(B)\, x_{ij}/(\alpha_i \beta_j) = \mu_i'(E)\, x_{ij}/\alpha_i$$

Summing first over j and using (3), we obtain $\mu_i'(E)$. Summing this over i, we obtain $\mu'(E)$. Hence the capacity constraint is satisfied: $\lambda(E \times B) = \mu'(E)$. A similar argument with i and j interchanged shows that $\lambda(A \times F) = \mu''(F)$, all $F \in \Sigma''$. Hence λ is feasible for variant I, hence feasible for II, III, and IV. This completes case (iii) and the proof. ∎

Actually, we see that this proof is valid even if μ' and μ'' are not σ-finite but merely abcont. In this case the λ constructed in the proof is also abcont. As an exercise, show there is *no* feasible solution in any variant if μ' is σ-finite and μ'' is *non*abcont, even if (7) and (8) obtain. (An example of a nonabcont measure is counting measure on the real line.)

7.3. THE TRANSPORTATION PROBLEM: DUALITY

Linear programs come in pairs, each being the *dual* of the other. The dual of the ordinary transportation problem (1.1)–(1.3) is: find $m + n$ nonnegative numbers $p_1, \ldots, p_m, q_1, \ldots, q_n$ satisfying

$$q_j - p_i \leq f_{ij} \tag{7.3.1}$$

$i = 1, \ldots, m; j = 1, \ldots, n$, and maximizing

$$\beta_1 q_1 + \cdots + \beta_n q_n - \alpha_1 p_1 - \cdots - \alpha_m p_m \tag{7.3.2}$$

This pair of programs has the following properties, shared by any pair of dual programs. If any feasible solutions are substituted in their respective objective functions, (1.3) and (2), then the value of the minimizing objective (1.3) never falls below the value of the maximizing objective (2). A pair of feasible solutions are jointly optimal for their respective problems iff the values they impart to the objective functions are *equal*.

This equality obtains iff these feasible solutions satisfy the following "complementary slackness" conditions. There is a natural 1–1 correspondence between the constraints of one problem and the variables of its dual. For the pair above, the constraints (1.1) correspond to the variables $p_i, i = 1, \ldots, m$; (1.2) corresponds to $q_j, j = 1, \ldots, n$; and (1) corresponds to $x_{ij}, i = 1, \ldots, m; j = 1, \ldots, n$. The complementary slackness condition then states that if a variable is *positive,* its corresponding constraint is satisfied with *equality*.

The question now arises, Does the duality construction carry over to the *measure-theoretic* transportation problem, and does the resulting pair have properties analogous to those just mentioned? The answer is yes, up to a point.

Consider the transportation problem determined by the pair of σ-finite measure spaces (A, Σ', μ'), (B, Σ'', μ''), with the constraints $\lambda' \leq \mu'$ and $\lambda'' \geq \mu''$, [see (1.8) and (1.9)], and the objective of minimizing $\int f \, d\lambda$ [see (1.6)] over feasible flow measures λ ($f : A \times B \rightarrow$ reals is measurable).

We define the *dual* of this problem as follows. Find nonnegative measurable functions $p : A \rightarrow$ reals and $q : B \rightarrow$ reals that satisfy

$$q(b) - p(a) \leq f(a, b) \qquad (7.3.3)$$

for all $a \in A$, $b \in B$, for which the definite integrals

$$\int_A p \, d\mu', \qquad \int_B q \, d\mu'' \qquad (7.3.4)$$

are both well defined and finite, and which maximize

$$\int_B q \, d\mu'' - \int_A p \, d\mu' \qquad (7.3.5)$$

over all pairs (p, q) satisfying (3) and (4).

Let us compare this with the ordinary dual, (1) and (2). We have already noted that the measure-theoretic reduces to the ordinary transportation problem exactly when the σ-fields Σ' and Σ'' are both finite. The same is true for the duals. More precisely, the situation is as follows. Let Σ', Σ'' be generated by the partitions $\{A_1, \ldots, A_m\}$, $\{B_1, \ldots, B_n\}$ respectively. Since p is measurable, it must be constant on each set A_i; let p_i be this value on A_i. Similarly, q has a constant value q_j on B_j, and f has a constant value f_{ij} on $A_i \times B_j$. Then (3) and (5) reduce to (1) and (2) respectively.

Condition (4) has no explicit counterpart in the ordinary dual; but since the integrals reduce to finite sums, it is automatically satisfied in the ordinary dual. One might ask, however, why (4) should be included. Would not (3) and (5) alone be an adequate generalization of the ordinary dual?

One difficulty that arises if (4) is dropped is that the objective function (5) might no longer be well defined for all feasible pairs (p, q). This difficulty is easily surmounted as follows. First, assume that A and B are disjoint (this involves no real loss of generality, since common points can be formally distinguished). Next interpret the expressions in (4) as *indefinite* integrals in the sense of pseudomeasures. Finally, interpret (5) as the *direct sum:*

$$\int q \, d\mu'' \oplus \int (-p) \, d\mu' \qquad (7.3.6)$$

Then (6) is a pseudomeasure over $A \cup B$, and "maximization" is to be understood in the sense of standard order. If now (4) also holds, then everything is finite and the standard ordering of (6) reduces to the ordinary size ordering of the definite integrals (5). Thus a perfectly reasonable problem results even if (4) is dropped.

Our main reason for inserting (4) is that with its aid we can prove that many ordinary duality properties generalize to the measure-theoretic case, whereas without it we can prove less. We refer to (3) and (5) alone, without (4), as the dual in the *wide sense*.

The entire discussion of duality to this point has been framed in terms of the *in*equality-constrained transportation problem, i.e., variant IV. For the other three variants we define the dual exactly as above, (3)–(5), but relax the nonnegativity constraints on p and/or q. Specifically, if requirements must be met exactly (variants I and II), then q is allowed to take on negative values; and if capacity must be utilized fully (variants I and III), then p is allowed to take on negative values. This is completely analogous to what happens for the corresponding variants of the ordinary transportation problem, and indeed for dual linear programs in general: equality constraints correspond to dual variables without sign restrictions.

With these definitions, the following theorems hold for all four variants of the transportation problem, each with its particular dual. The measure-theoretic transportation problem (in any variant) will be called the *primal*. This and its dual are determined by the σ-finite source and sink spaces, (A, Σ', μ') and (B, Σ'', μ''), and by the measurable cost function $f : A \times B \to$ reals.

Theorem: Let flow measure λ be feasible for the primal, and functions (p, q) be feasible for the dual. Then $\int_{A \times B} f \, d\lambda$ is well defined, and

$$\int_{A \times B} f \, d\lambda \geq \int_B q \, d\mu'' - \int_A p \, d\mu' \qquad (7.3.7)$$

Proof: We have

$$\int_A p \, d\mu' \geq \int_A p \, d\lambda' < \infty \qquad (7.3.8)$$

To show this, we consider two cases. In problem variants I and III, $\mu' = \lambda'$ and (8) is trivial; in variants II and IV, $\mu' \geq \lambda'$ and $p \geq 0$, so (8) again is valid. (The right inequality in (8) follows from the left, by (4).) Similarly,

$$\int_B q \, d\mu'' \leq \int_B q \, d\lambda'' > -\infty \qquad (7.3.9)$$

For $\mu'' = \lambda''$ in variants I and II, while $\mu'' \leq \lambda''$ and $q \geq 0$ in variants III and IV. The right inequality in (9) again follows from the left. From (8) and (9) we obtain

$$\int_B q \, d\mu'' - \int_A p \, d\mu' \leq \int_B q \, d\lambda'' - \int_A p \, d\lambda' > -\infty \qquad (7.3.10)$$

Now define the functions p_1, $q_1 : A \times B \to$ reals by $p_1(a, b) = p(a)$, and $q_1(a, b) = q(b)$, all $a \in A$, $b \in B$. We find that

$$\int_A p \, d\lambda' = \int_{A \times B} p_1 \, d\lambda, \qquad \int_B q \, d\lambda'' = \int_{A \times B} q_1 \, d\lambda \qquad (7.3.11)$$

by the induced integrals theorem. (Verify this separately for p^+ and p_1^+, and for p^- and p_1^-, and subtract; similarly for q, q_1.) From (10) and (11) we obtain

$$\int_B q \, d\lambda'' - \int_A p \, d\lambda' = \int_{A \times B} (q_1 - p_1) \, d\lambda > -\infty \qquad (7.3.12)$$

Since $q_1 - p_1 \leq f$ from (3), it follows that $\int_{A \times B} f \, d\lambda$ is well defined, for

$$\int_{A \times B} f^- \, d\lambda \leq \int_{A \times B} (q_1 - p_1)^- \, d\lambda < \infty$$

from (12). And, in fact, from (3) we obtain

$$\int_{A \times B} (q_1 - p_1) \, d\lambda \leq \int_{A \times B} f \, d\lambda \qquad (7.3.13)$$

Finally, (10), (12), and (13) together yield (7). ■

This theorem generalizes the ordinary duality property that a feasible value for the maximizing problem never exceeds a feasible value for the minimizing problem. Note that the mere existence of a dual-feasible pair (p, q) implies a condition on every feasible λ: that $\int_{A \times B} f \, d\lambda$ be well defined, or that the indefinite integral $\int f \, d\lambda$ be a signed measure, not a proper pseudomeasure.

We now introduce a concept that generalizes the notion of "complementary slackness."

Definition: Let flow measure λ on $(A \times B, \Sigma' \times \Sigma'')$ and the pair of functions $p : A \to$ reals, $q : B \to$ reals be feasible for the transportation problem and its dual respectively. Pair (p, q) is a *measure potential* for λ iff the following conditions are satisfied:

(i) $\lambda\{(a, b) \mid q(b) - p(a) < f(a, b)\} = 0;$ \qquad (7.3.14)

(ii) λ' and μ' coincide on subsets of $\{a \mid p(a) > 0\}$;   (7.3.15)

(iii) λ'' and μ'' coincide on subsets of $\{b \mid q(b) > 0\}$.   (7.3.16)

Condition (14) states that no flow occurs on the set where (3) is satisfied with strict inequality. Condition (15) states that capacity is fully utilized on the set of sources where p is positive. Condition (16) states that requirements are met exactly on the set of sinks where q is positive.

This definition is meant to apply to all four variants of the transportation problem. Note, however, that in variants I and III we have $\lambda' = \mu'$, so that (15) is automatically satisfied and may be dropped without changing the definition. Similarly, in variants I and II we have $\lambda'' = \mu''$, so that (16) must be true and may be dropped. As an exercise, the reader might verify that this reduces to the ordinary "complementary slackness" conditions when both σ-fields Σ' and Σ'' are finite.

Note that dual feasibility of (p, q) (i.e., satisfaction of conditions (3) and (4)) is a requirement for measure potentiality. If as suggested above, (4) is dropped as a dual constraint, this gives rise to a correspondingly weaker concept here, which may be called *measure potentiality in the wide sense*. For the time being, however, we stick with the original concept.

Theorem: Let λ^o be feasible for the transportation problem, and (p^o, q^o) feasible for its dual (given by (3) *and* (4)). Then (p^o, q^o) is a measure potential for λ^o iff

$$\int_{A \times B} f \, d\lambda^o = \int_B q^o \, d\mu'' - \int_A p^o \, d\mu' \qquad (7.3.17)$$

Proof: Let (p^o, q^o) be a measure potential for λ^o. Reviewing the preceding proof, we find that at (8), (9), and (13) the weak inequalities become equalities because of (15), (16), and (14) respectively. Hence (7) is satisfied with equality; this is (17).

Conversely, let (17) be true. Then all the integrals appearing in the preceding proof must be finite, and the weak inequalities of (8), (9), and (13) must all be equalities. But finite equality in (8) implies condition (15). This is trivial in variants I and III; in variants II and IV it follows from the facts $p^o \geq 0$, $\lambda^{o'} \leq \mu'$. Similarly, finite equality in (9) and (13) implies conditions (16) and (14) respectively. Hence (p^o, q^o) is a measure potential for λ^o. ∎

Theorem: If the pair (p^o, q^o) is a measure potential for λ^o, then λ^o is a *best* solution for the transportation problem, and (p^o, q^o) is a *best* solution for its dual ("best" in the sense of standard order).

Proof: Let λ and (p, q) be any other solutions feasible for the transportation problem and its dual respectively. From the two preceding theorems we obtain

$$\int_{A \times B} f \, d\lambda \geq \int_B q^o \, d\mu'' - \int_A p^o \, d\mu'$$

$$= \int_{A \times B} f \, d\lambda^o \geq \int_B q \, d\mu'' - \int_A p \, d\mu' \qquad (7.3.18)$$

All integrals in (18), except possibly the leftmost, are finite, by (4). Now, when the objective functions are well defined as definite integrals and finite for at least one of the two solutions being compared, standard ordering reduces to ordinary size ordering of definite integrals. Recalling that we are maximizing in the dual and minimizing in the primal, it follows at once from (18) that (p^o, q^o) and λ^o are best for their respective problems. ■

All these results are direct generalizations of duality relations that hold for the ordinary transportation problem. The finiteness condition (4) is essential to the preceding demonstrations. What happens if it is relaxed? It turns out that we can still deduce a weakened form of the conclusion of the preceding theorem, with "unsurpassed" in place of "best." Specifically we have the following result, which has an application in the theory of market regions (Section 9.5).

Theorem: Let (p, q) be a measure potential for λ^o in the wide sense, i.e., without (4), but with *at least one* of the two definite integrals in (4) well defined and finite. Then λ^o is *unsurpassed* for the transportation problem (in any variant; "unsurpassed" refers to reverse standard ordering of pseudomeasures).

Proof: We argue by contradiction. Suppose the conclusion were false; then there is another feasible measure λ surpassing λ^o. That is,

$$\int\!\!\int f d\lambda < \int\!\!\int f \, d\lambda^o \qquad (7.3.19)$$

(standard order; remember that we are minimizing).

Define the functions p_1, q_1 on $A \times B$ by the rules $p_1(a, b) = p(a)$, $q_1(a, b) = q(b)$. From (3) we obtain

$$\int (q_1 - p_1) \, d\lambda \leq \int\!\!\int f \, d\lambda \qquad (7.3.20)$$

($\leq$ refers to narrow order). Condition (14) holds for λ^o, so that inequality (3) is actually an equality λ^o-almost everywhere. From this we get the pseudomeasure equality

$$\int (q_1 - p_1) \, d\lambda^o = \iint f \, d\lambda^o \qquad (7.3.21)$$

Let ψ be the pseudomeasure (λ, λ^o). Then

$$\int (q_1 - p_1) \, d\psi \le \iint f \, d\psi < 0 \qquad (7.3.22)$$

(The left inequality in (22) arises on subtracting (21) from (20); the right inequality in (22) is the same as (19).) Since standard order is an extension of narrow order, (22) implies that

$$\int q_1 \, d\psi < \int p_1 \, d\psi \qquad (7.3.23)$$

The rest of this proof is devoted to showing that (23) is *false*. This contradiction will complete the proof.

One should keep in mind during what follows that the upper variation ψ^+ is just an ordinary measure on $A \times B$; hence it has left and right marginals, which we denote by $\psi^{+\prime}$, $\psi^{+\prime\prime}$ respectively. Similarly, the lower variation ψ^- has marginals $\psi^{-\prime}$, $\psi^{-\prime\prime}$.

Consider the following indefinite integral with its Jordan form:

$$\int p_1 \, d\psi = \left[\int p_1^+ \, d\psi^+ + \int p_1^- \, d\psi^-, \quad \int p_1^- \, d\psi^+ + \int p_1^+ \, d\psi^- \right]$$

We show that, for the following *definite* integrals,

$$\int_{A \times B} p_1^+ \, d\psi^+ = \int_A p^+ \, d(\psi^{+\prime}) \le \int_A p^+ \, d(\psi^{-\prime}) = \int_{A \times B} p_1^+ \, d\psi^-$$
$$\qquad (7.3.24)$$

$$\int_{A \times B} p_1^- \, d\psi^- = \int_A p^- \, d(\psi^{-\prime}) = \int_A p^- \, d(\psi^{+\prime}) = \int_{A \times B} p_1^- \, d\psi^+$$
$$\qquad (7.3.25)$$

The outer equalities in (24) and (25) all follow from the induced integrals theorem. For the middle relations in (24) and (25) consider two cases.

Case (i). Variant I or III:
Here

$$\lambda' = \lambda^{o\prime} = \mu' \qquad (7.3.26)$$

By the equivalence theorem for pseudomeasures we have

$$\psi^+ + \lambda^o = \psi^- + \lambda \qquad (7.3.27)$$

Since μ' is σ-finite, there is a countable measurable partition G of A such that $\mu'(G) < \infty$, all $G \in \mathsf{G}$. For any such G and any $E \in \Sigma'$ we have

$$\psi^{+\prime}(E \cap G) + \lambda^{o\prime}(E \cap G) = \psi^{-\prime}(E \cap G) + \lambda'(E \cap G)$$

on taking left marginals in (27). By (26) the λ', $\lambda^{o\prime}$ terms drop out. Adding over $G \in \mathsf{G}$, we obtain

$$\psi^{+\prime}(E) = \psi^{-\prime}(E) \qquad (7.3.28)$$

Thus the left marginals of ψ^+, ψ^- are equal, and the middle relations of (24) and (25) are established with equality.

Case (ii). Variant II or IV:

Here $p \geq 0$, so $p^- = 0$ and (25) is trivial. Also, since (15) holds for λ^o, we have $\lambda^{o\prime}(E) = \mu'(E) \geq \lambda'(E)$ for any measurable $E \subseteq \{a \mid p(a) > 0\}$. An argument similar to (26)–(28) now yields $\psi^{+\prime}(E) \leq \psi^{-\prime}(E)$ for all such E. This establishes the inequality in (24).

Now (24) and (25) imply that the following statement is *false:*

$$\int p_1 \, d\psi > 0 \tag{7.3.29}$$

For (29) is true iff the sum of the two left integrals exceeds the sum of the two right integrals in (24) and (25).

Furthermore, we have:

$$\text{If } \int_A p \, d\mu' \text{ is finite,} \quad \text{then } \int p_1 \, d\psi \leq 0 \tag{7.3.30}$$

To see this, note that $\psi^- \leq \lambda^o$ (minimizing property of Jordan form); hence $\psi^{-\prime} \leq \lambda^{o\prime}$, implying

$$\int_A p^{\pm} \, d(\psi^{-\prime}) \leq \int_A p^{\pm} \, d(\lambda^{o\prime}) \leq \int_A p^{\pm} \, d\mu' < \infty$$

This result shows that all the integrals in (24) and (25) are finite; (24) and (25) then yield the conclusion of (30).

We now run through a similar argument with q in place of p. Only the high points will be mentioned. We show that

$$\int_{A \times B} q_1^+ \, d\psi^+ = \int_B q^+ \, d(\psi^{+\prime\prime}) \geq \int_B q^+ \, d(\psi^{-\prime\prime}) = \int_{A \times B} q_1^+ \, d\psi^- \tag{7.3.31}$$

$$\int_{A \times B} q_1^- \, d\psi^- = \int_B q^- \, d(\psi^{-\prime\prime}) = \int_B q^- \, d(\psi^{+\prime\prime}) = \int_{A \times B} q_1^- \, d\psi^+ \tag{7.3.32}$$

The outer equalities in (31) and (32) again follow from the induced integrals theorem. For the middle relations there are again two cases to consider.

Case (i). Variant I or II:

Here we have

$$\lambda'' = \lambda^{o\prime\prime} = \mu'' \tag{7.3.33}$$

An argument similar to (26)–(28) shows that $\psi^{+\prime\prime} = \psi^{-\prime\prime}$, which establishes the middle relations of (31) and (32) with equality.

Case (ii). Variant III or IV:

Here $q \geq 0$, so $q^- = 0$ and (32) is trivial. Also, since (16) holds for λ^o, we have

$$\lambda^{o\prime\prime}(F) = \mu''(F) \leq \lambda''(F)$$

for any measurable $F \subseteq \{b \mid q(b) > 0\}$. There is one subtlety at this point, since we cannot assume that λ'' or $\lambda^{o\prime\prime}$ is σ-finite, but the argument still goes through as follows. Let $\mathcal{G}$ be a countable measurable partition of B such that $\mu''(G) < \infty$, all $G \in \mathcal{G}$. Then for any such F and G we have (noting that $\lambda^{o\prime\prime}(F \cap G)$ is finite),

$$\psi^{+\prime\prime}(F \cap G) = \psi^{-\prime\prime}(F \cap G) + \lambda''(F \cap G) - \lambda^{o\prime\prime}(F \cap G)$$

$$\geq \psi^{-\prime\prime}(F \cap G)$$

Adding over $G \in \mathcal{G}$, we obtain $\psi^{+\prime\prime}(F) \geq \psi^{-\prime\prime}(F)$ for all measurable $F \subseteq \{b \mid q(b) > 0\}$, which establishes the inequality in (31).

Parallel to the argument above, (31) and (32) now imply that the following statement is *false*:

$$\int q_1 \, d\psi < 0 \tag{7.3.34}$$

and furthermore,

$$\text{If } \int_B q \, d\mu'' \text{ is finite, then } \int q_1 \, d\psi \geqslant 0 \tag{7.3.35}$$

Condition (35) follows from the observations

$$\int_B q^{\pm} \, d(\psi^{-\prime\prime}) \leq \int_B q^{\pm} \, d(\lambda^{o\prime\prime}) = \int_B q^{\pm} \, d\mu'' < \infty \tag{7.3.36}$$

(The equality in (36) follows from (33) in variant I or II; it follows from $q \geq 0$ and (16) in variant III or IV.) Then (36), (31), and (32) imply that

$$\int_{A \times B} q_1^+ \, d\psi^- + \int_{A \times B} q_1^- \, d\psi^+ < \infty$$

and this fact, together with (31) and (32), yields the conclusion in (35).

Finally, at least one of the premises in (30) or (35) is true by assumption; hence at least one of the conclusions in (30) or (35) is true. Thus combining things with (23), either $\int q_1 \, d\psi < \int p_1 \, d\psi \leqslant 0$, which contradicts the falsity of (34), or $0 \leqslant \int q_1 \, d\psi < \int p_1 \, d\psi$, which contradicts the falsity of (29). This contradiction completes the proof. ∎

7.4. THE TRANSPORTATION PROBLEM: EXISTENCE OF OPTIMAL SOLUTIONS

Topological concepts have scarcely been used so far in this book. This has been a matter of deliberate policy to underline the fact that, contrary to popular belief, measure theory is far more significant than topology as a groundwork for social science. But for the rest of this section topology is essential. Indeed, we know of no general method for proving existence of optimal solutions to the measure-theoretic transportation problem without using topological concepts. Nor do we know of any topology-free method for constructing measure potentials from optimal solutions.

We do not go deeply into the subject but simply list those concepts and theorems that are actually used in the sequel.[7]

Given a fixed set A, let $\mathfrak{I}$ be a collection of subsets of A. Then $\mathfrak{I}$ is a *topology* over A iff (i) $A \in \mathfrak{I}$ and the empty set $\phi \in \mathfrak{I}$; (ii) if G_1, $G_2 \in \mathfrak{I}$, then $G_1 \cap G_2 \in \mathfrak{I}$; (iii) if $\mathcal{G} \subseteq \mathfrak{I}$, then $\cup \, \mathcal{G} \in \mathfrak{I}$. (In words: $\mathfrak{I}$ is closed under arbitrary unions and finite intersections and owns A and ϕ.) The pair $(A,\mathfrak{I})$ is a *topological space* and the members of $\mathfrak{I}$ are called *open sets*. Set F is said to be *closed* iff $A \backslash F$ is open.

Let $\mathcal{G}$ be any class of subsets of A. From this, construct $\mathcal{G}'$, the class of intersections of *finite* subclasses of $\mathcal{G}$; and then construct $\mathcal{G}''$, the class of unions of arbitrary subclasses of $\mathcal{G}'$, together with A and ϕ. One can show that $\mathcal{G}''$ is a topology, the topology *generated by* $\mathcal{G}$; $\mathcal{G}$ is called a *subbase*, and $\mathcal{G}'$ a *base* for this topology.[8]

Let $(A,\mathfrak{I})$ be a topological space, and let $E \subseteq A$. The *relative topology* on E is the class of all sets of the form $E \cap G$, G ranging over the open sets of A. This collection of sets makes E a topological space in its own right. This is the construction implicitly referred to below when we speak of the "usual topology of the rationals," or "topological completeness of a closed subset of A."

Here is an example. Let A be the real line, and let $\mathcal{G}$ be the class of all open intervals. The topology generated by $\mathcal{G}$ is the *usual topology* for the real line, the one implicitly used in ordinary discussions of continuity and convergence of real numbers.

This example has a far-reaching generalization. Recall that (A, d) is a *metric space* iff the function $d : A \times A \to$ reals satisfies $d(a, a) = 0$, $d(a_1, a_2) > 0$ if $a_1 \neq a_2$, $d(a_1, a_2) = d(a_2, a_1)$, $d(a_1, a_2) + d(a_2, a_3) \geq$

---

7. For further information see, e.g., J. L. Kelley, *General Topology* (Van Nostrand, Princeton, 1955).

8. For topologies the process of generation can be written in two steps, as just indicated. For σ-fields it cannot be written in even a countable number of steps. In both cases, however, the basic concept is the same: the intersection of all topologies (respectively σ-fields) containing the given class $\mathcal{G}$.

$d(a_1, a_3)$. Now let $\mathcal{G}$ be the class of all *open discs,* i.e., all sets of the form $\{a \mid d(a_o, a) < x\}$, for $a_o \in A$, x real and positive. The *topology generated by the metric d* is the one generated by this $\mathcal{G}$. The topology generated by the Euclidean metric in n-space, $n = 1, 2, \ldots$, is the usual topology for this space and will be assumed if no contrary assumptions are made.

Given a topological space $(A, \mathcal{J})$ one can ask: Does a metric d on A exist that generates $\mathcal{J}$ as described above? Iff this is the case $\mathcal{J}$ is said to be *metrizable.*

We need the concept of *completeness* of a metric space. A sequence $a_1, a_2, \ldots$ from (A, d) has the *Cauchy property* iff for all $\epsilon > 0$ there is an integer N such that, for all integers, $m, n > N$, $d(a_m, a_n) < \epsilon$: roughly speaking, the points get indefinitely close to each other. Space (A, d) is *complete* iff any such Cauchy sequence converges to a point $a_o \in A$; i.e., for all $\epsilon > 0$ an N exists such that, for all $n > N$, $d(a_o, a_n) < \epsilon$. For example, the real numbers are complete and the rationals are not, under the usual topology. A space $(A, \mathcal{J})$ that is not only metrizable but generated by a complete metric is said to be *topologically complete.* The real line and n-space in general are topologically complete under the usual topologies, and the same is true for any closed or open subsets of these spaces.

Space $(A, \mathcal{J})$ is *separable* iff there is a countable subset A' such that any nonempty open set meets A': $G \in \mathcal{J}$ and $G \neq \phi$ implies $G \cap A' \neq \phi$. For any n, n-space is separable (e.g., A' may be chosen to be the set of n-tuples with rational coordinates). Indeed, *any* subset of n-space is separable.

A subset K of a metrizable topological space is said to be *compact* iff every sequence from it contains a subsequence converging to a point of K. Any closed bounded subset of n-space, for example, is compact.

We need a few continuity concepts, confining our attention to real-valued functions. Let $(A, \mathcal{J})$ be a topological space, and f a real-valued function with domain A; f is *lower semicontinuous* iff every set of the form $\{a \mid f(a) > x\}$, x real, is open; f is *upper semicontinuous* iff $-f$ is lower semicontinuous or, equivalently, iff all sets of the form $\{a \mid f(a) < x\}$ are open. Finally, f is *continuous* iff it is both lower and upper semicontinuous.

Let $(A, \mathcal{J}')$ and $(B, \mathcal{J}'')$ be two topological spaces. On the cartesian product $A \times B$ one defines a *product topology,* written $\mathcal{J}' \times \mathcal{J}''$, essentially in the same way one defines product σ-fields. Specifically, rectangles $E \times F$ are called *open* iff $E \in \mathcal{J}'$ and $F \in \mathcal{J}''$; $\mathcal{J}' \times \mathcal{J}''$ is then defined to be the topology generated by the open rectangles. One can show, in fact, that $G \in \mathcal{J}' \times \mathcal{J}''$ iff G is the union of some subclass of open rectangles.

The discussion to this point has been purely topological. We also need some concepts that depend on both topological and measure-theoretic ideas. The *Borel field* of topological space $(A, \Im)$ is the σ-field generated by $\Im$.[9] Its members are called *Borel sets*.

Let $(A, \Im)$ be a metrizable topological space, let Σ be its Borel field, and let μ be a bounded measure on Σ. μ is said to be *tight* iff for every positive number ϵ there is a compact set K that $\mu(A \backslash K) < \epsilon$. Let **M** be a collection of measures on Σ; **M** is *uniformly tight* iff (i) there is a real number N such that $\mu(A) \leq N$ for all $\mu \in$ **M**, and (ii) for every positive number ϵ there is a compact K such that $\mu(A \backslash K) < \epsilon$ for all $\mu \in$ **M**. (The *same K* must serve for all μ.)

With $(A, \Im)$ and Σ as above, let μ_o be a bounded measure on Σ, and $\mu_1, \mu_2, \ldots$ a sequence of such measures. This sequence is said to *converge weakly* to μ_o iff for every $g:A \rightarrow$ reals that is bounded and continuous we have

$$\lim_{n \rightarrow \infty} \int_A g \, d\mu_n = \int_A g \, d\mu_o \qquad (7.4.1)$$

in the ordinary sense of limit of a sequence of real numbers.

With these definitions out of the way, we are ready to proceed. The following is an omnibus theorem, covering the alternative variants of the transportation problem under alternative assumptions. We first briefly contemplate the practical import of some of these assumptions. Any subset of *n*-space is separable (in the relative topology), and any subset that can be expressed as a countable intersection of open sets (a so-called G_δ-set) is topologically complete (Aleksandrov's theorem; in particular, closed or open subsets of *n*-space are G_δ). Hence these conditions constitute no real restriction when imposed on (physical) Space or Time or other spaces built up from these in a simple manner. It is unclear what restrictions they imply when imposed on topologies on more complex spaces, such as Resources, Histories, or Activities. The boundedness of f, μ', and μ'' means that these results will often not apply to problems involving infinite S or T horizons, but they constitute no real restriction when applied to "practical" problems in the narrow sense of the term.

These boundedness assumptions also imply that the objective function is well defined and finite as a definite integral for any feasible flow λ. Hence standard order reduces to the ordinary comparison of definite integrals. We may therefore drop the distinction between "best" and "unsurpassed" solutions and simply speak of optimal solutions.

---

9. If A is the real line with the usual topology, one can show that its Borel field as here defined coincides with its Borel field as defined in Chapter 2. The same is true for *n*-space.

Theorem: Omnibus Existence Theorem.   Let (A, Σ', μ'), (B, Σ'', μ'') be the source and sink spaces for the transportation problem, $B \neq \emptyset$, with cost function $f : A \times B \to$ reals, where

(i) μ', μ'', and f are all bounded and $\mu'(A) \geq \mu''(B)$;

(ii) topologies $\mathfrak{J}'$, $\mathfrak{J}''$ exist on A, B, with Σ', Σ'' as their Borel fields respectively; $\mathfrak{J}'$, $\mathfrak{J}''$ are both topologically complete and separable; and f is lower semicontinuous with respect to $\mathfrak{J}' \times \mathfrak{J}''$;

(iii) one of the following extra assumptions is made:
   (a) no extra assumption, or
   (b) $f \geq 0$, or
   (c) $(B, \mathfrak{J}'')$ is compact, or
   (d) $\mu'(A) = \mu''(B)$.

Then we have

	II	IV	III	I
(a)	E	×	×	×
(b)	E	E	×	×
(c)	E	E	E	×
(d)	E	E	E	E

where E indicates that an *optimal* solution always exists to the transportation problem for the given combination of variant I, II, III, or IV and assumption (a), (b), (c), or (d), and × indicates that sometimes such a solution does not exist.

Proof: There are sixteen things to prove and, of these, twelve can be disposed of rapidly, either directly or as a consequence of the other four. First, we discuss the counterexamples for the × entries.

Cases (Ia), (Ib), (Ic): Let A and B be singleton sets, with $\mu'(A) > \mu''(B)$. Then variant I does not have even a feasible, let alone an optimal, solution.

Cases (IIIa), (IIIb): Let A be singleton, with $\mu'(A) = 1$; let B be the integers $\{1, 2, \ldots\}$, with $\mathfrak{J}'' = \Sigma'' =$ all subsets of B, and μ'' identically zero; finally, let unit transport cost to point $n \in B$ be $1/n$, $n = 1, 2, \ldots$.

Since B is countable, $\mathfrak{J}''$ is separable; it is also topologically complete, since it is generated by the complete metric d given by $d(m, n) = 1$ if $m \neq n$. Thus the stated problem satisfies all premises of the theorem, including (a) and (b) of (iii).

There is no optimal solution, for cost can be reduced below any $\epsilon > 0$ by shipping one unit to a sufficiently large $n \in B$; but zero cost cannot be attained, since $f > 0$ and $\lambda(A \times B) = \mu'(A) > 0$.

Case (IVa): Let everything be as in the counterexample just given, except that unit transport cost to point $n \in B$ is $(1 - n)/n$.

Cost can be reduced below any real number > -1 by shipping one unit to sufficiently large $n \in B$, but cost of -1 cannot be attained. This is clear if $\lambda = 0$; while if $\lambda(A \times B) > 0$, then since $f > -1$, we have

$$\int_{A \times B} f \, d\lambda > \int_{A \times B} -1 \, d\lambda = -\lambda(A \times B) \geq -\mu'(A) = -1$$

Hence again there is no optimal solution.

Cases (IIb), (IIc), (IId): This finishes the $\times$'s, and we start on the E's. First, existence in cases (IIb), (IIc), and (IId) obviously follows from existence in case (IIa).

Cases (Id), (IIId), (IVd): Existence in these cases follows from existence in case (IId). To show this, we prove that any feasible flow measure λ must satisfy the transportation problem constraints with *equality* in all four variants. Suppose there were an $E \in \Sigma'$ such that $\mu'(E) > \lambda'(E)$. Adding the inequality $\mu'(A \setminus E) \geq \lambda'(A \setminus E)$, we obtain $\mu'(A) > \lambda'(A)$ (the *strict* inequality carries over because all measures involved are finite). But $\lambda'(A) = \lambda(A \times B) \geq \mu''(B)$, so $\mu'(A) > \mu''(B)$, contradicting premise (d). Thus we must have $\lambda' = \mu'$; a similar argument establishes $\lambda'' = \mu''$. Hence all four variants have the same set of feasible solutions. If an optimal solution exists for any of them, therefore, it must exist for all, under premise (d).

This leaves the four cases (IIa), (IIIc), (IVc), and (IVb).

Case (IVb): We show that the existence of an optimal solution in case IIb implies its existence in IVb. Let λ be feasible for variant IV, so that $\lambda'' \geq \mu''$. Thus μ'' is absolutely continuous with respect to λ''. Since all measures are also finite, we may invoke the Radon-Nikodym theorem and infer the existence of a function $g : B \to$ reals satisfying

$$\mu'' = \int g \, d\lambda'' \qquad (7.4.2)$$

This g must take on values in the closed interval $[0, 1]$, except possibly on a Σ''-set of λ''-measure zero. Altering it to zero on this set, which does not invalidate (2), we thus have $0 \leq g(b) \leq 1$ for all $b \in B$. Now define $h : A \times B \to [0, 1]$ by $h(a, b) = g(b)$, then define λ_1 by

$$\lambda_1 = \int h \, d\lambda \qquad (7.4.3)$$

This is an indefinite integral over $A \times B$, and λ_1 is therefore another flow measure on $(A \times B, \Sigma' \times \Sigma'')$. We now show it is feasible for variant II. First, $\lambda_1 \leq \lambda$, since $h \leq 1$; hence

$$\lambda_1' \leq \lambda' \leq \mu' \tag{7.4.4}$$

Also

$$\lambda_1'' = \int g \, d\lambda'' \tag{7.4.5}$$

from (3) by the induced integrals theorem. Relations (4), (5), and (2) show that λ_1 is feasible for variant II. Also, since $\lambda_1 \leq \lambda$ and $f \geq 0$, we obtain

$$\int_{A \times B} f \, d\lambda_1 \leq \int_{A \times B} f \, d\lambda \tag{7.4.6}$$

Thus assuming $f \geq 0$, we have shown that for any flow λ feasible in variant IV, a flow λ_1 exists that meets the more stringent conditions of variant II and whose transportation cost is no higher than that of λ, by (6). It follows that any solution optimal for variant II will also be optimal for variant IV, under premise (b).

Cases (IIa), (IIIc), (IVc): This is the last and most difficult part.[10] The proof goes through several stages; the first is to show that the set of feasible solutions in these cases is uniformly tight.

From the fact that $\mathfrak{I}'$ and $\mathfrak{I}''$ are both separable and metrizable, it may be shown that $\Sigma' \times \Sigma''$ is the Borel field of $\mathfrak{I}' \times \mathfrak{I}''$. Also, it is known that any bounded measure on the Borel field of a topologically complete and separable topological space is *tight*.[11] Hence for any $\epsilon > 0$ there are compact sets K', K'' (contained in A, B respectively) such that

$$\mu'(A \backslash K') < \epsilon/2, \qquad \mu''(B \backslash K'') < \epsilon/2 \tag{7.4.7}$$

Now let λ be a feasible flow measure. We always have $\lambda'(A \backslash K') \leq \mu'(A \backslash K')$. In variant II we also have

$$\lambda''(B \backslash K'') = \mu''(B \backslash K'') \tag{7.4.8}$$

Furthermore, if premise (c) holds, B itself is compact and we may choose $K'' = B$ to satisfy (7) and (8). Hence in all the three cases (IIa), (IIIc), and (IVc), K' and K'' may be chosen so that

$$\lambda'(A \backslash K') + \lambda''(B \backslash K'') < \epsilon \tag{7.4.9}$$

for all feasible λ. Next, consider the number $\lambda[(A \times B) \backslash (K' \times K'')]$. This does not exceed the left side of (9), since

---

10. Results from the theory of weak convergence of measures are used in this part of the proof. On this theory see P. Billingsley, *Convergence of Probability Measures* (Wiley, New York, 1968), Ch. 1; and K. R. Parthasarathy, *Probability Measures on Metric Spaces* (Academic Press, New York, 1967).

11. Ulam's theorem; see Billingsley, *Convergence*, pp. 5–6.

$$(A \times B)\backslash(K' \times K'') \subseteq [(A\backslash K') \times B] \cup [A \times (B\backslash K'')]$$

Also, $K' \times K''$ is itself compact in the product topology $\mathfrak{J}' \times \mathfrak{J}''$ (Tihonov's theorem). Finally, $\lambda(A \times B) \leq \mu'(A) < \infty$ for all feasible λ. This shows that the set of feasible solutions is indeed uniformly tight in all three cases.

We know there is at least one feasible solution (Section 7.2). Hence there is a sequence of them, $\lambda_1, \lambda_2, \ldots$, with the property

$$\lim_n \int_{A \times B} f \, d\lambda_n = V^o \tag{7.4.10}$$

where V^o is the infimum of the attainable values of the objective function.

Since $\mathfrak{J}'$ and $\mathfrak{J}''$ are both metrizable, the same may be shown to be true for the product topology. We now invoke the basic theorem of Prohorov-Varadarajan:[12] if $\lambda_1, \lambda_2, \ldots$ is a sequence from a uniformly tight set of measures on the Borel field of a metrizable topological space, then there is a subsequence that converges weakly to some measure λ^o (not necessarily a member of the set).

Let λ^o be this weak limit of a subsequence of the $\lambda_1, \lambda_2, \ldots$ that satisfies (10). We prove this λ^o is in fact optimal for the transportation problem by showing that it is feasible and that

$$\int_{A \times B} f \, d\lambda^o \leq V^o \tag{7.4.11}$$

We first prove (11). For convenience we retain the notation λ_1, $\lambda_2, \ldots$ for the convergent subsequence of the original sequence. Invoking a theorem of A. D. Aleksandrov, it follows that

$$\lambda^o(G) \leq \liminf_n \lambda_n(G) \tag{7.4.12}$$

for all open sets $G \subseteq A \times B$. Let us now temporarily make the additional assumption that $f \geq 0$. We then have

$$\int_{A \times B} f \, d\lambda^o = \int_0^\infty \lambda^o\{(a,b) \mid f(a,b) > t\} \, dt$$

$$\leq \int_0^\infty \liminf_n \lambda_n\{(a,b) \mid f(a,b) > t\} \, dt$$

$$\leq \liminf_n \int_0^\infty \lambda_n\{(a,b) \mid f(a,b) > t\} \, dt$$

$$= \liminf_n \int_{A \times B} f \, d\lambda_n = V^o \tag{7.4.13}$$

---

12. Billingsley, *Convergence*, p. 32.

The last equality in (13) comes from (10). The other two equalities invoke the Young integral, which is an ordinary Riemann integral. The first inequality results from Aleksandrov's theorem (12) and the fact that since f is lower semicontinuous, the set $\{(a,b) \mid f(a,b) > t\}$ is open. The second inequality is from Fatou's lemma.

Now drop the assumption that $f \geq 0$. Since f is bounded below, a real number x exists such that $f + x \geq 0$. The argument of (13) now yields

$$\int_{A \times B} (f + x) \, d\lambda^o \leq \liminf_n \int_{A \times B} (f + x) \, d\lambda_n \qquad (7.4.14)$$

But weak convergence implies that $\lim \lambda_n(A \times B) = \lambda^o(A \times B)$. (Substitute the constant function $g = 1$ in (1).) It follows that the x's drop out of (14). Thus we obtain (11).

The final step is to prove that λ^o is feasible. Since $\lambda_1, \lambda_2, \ldots$ converges weakly to λ^o, it follows that the marginals converge weakly to the respective marginals of λ^o (by the Mann-Wald theorem):[13] $\lim \lambda_n' = \lambda^{o\prime}$, $\lim \lambda_n'' = \lambda^{o\prime\prime}$, where lim stands for weak convergence. For all n the marginals satisfy the feasibility constraints for the transportation problem:

$$\lambda_n' \, (=, \leq) \, \mu', \qquad \lambda_n'' \, (=, \geq) \, \mu'' \qquad (7.4.15)$$

the particular signs depending on the variant in question. We must show that the marginals of λ^o satisfy the same constraints.

The following result will be used: let μ, ν be two bounded measures on the Borel field of a metrizable topological space C. If

$$\int_C g \, d\mu \geq \int_C g \, d\nu \qquad (7.4.16)$$

for all bounded nonnegative continuous functions g, then $\mu \geq \nu$.[14] It follows that if $\mu_1, \mu_2, \ldots$ and $\nu_1, \nu_2, \ldots$ are two sequences of bounded measures on this Borel field, converging weakly to bounded measures μ, ν respectively, and $\mu_n \geq \nu_n$ for all n, then $\mu \geq \nu$. For,

$$\int_C g \, d\mu = \lim_n \int_C g \, d\mu_n \geq \lim_n \int_C g \, d\nu_n = \int_C g \, d\nu$$

if g is bounded nonnegative continuous; thus (16) holds, yielding $\mu \geq \nu$.

Now consider the two sequences $\mu', \mu', \ldots$ and $\lambda_1', \lambda_2', \ldots$. The first is a constant sequence, converging weakly to μ'; the second converges weakly to $\lambda^{o\prime}$. Since $\mu' \geq \lambda_n'$, all n, we obtain $\mu' \geq \lambda^{o\prime}$ by the above result. Similar arguments show that the relations (15) get reproduced with $\lambda^{o\prime}$,

---

13. Billingsley, *Convergence*, pp. 30–31.
14. Billingsley, *Convergence*, p. 9, has a similar theorem, but with equalities in place of inequalities; a simple twist of his proof yields the statement just made.

$\lambda^{o\prime\prime}$ in place of λ'_n, $\lambda_n{}''$ respectively. Hence λ^o is indeed feasible. Since it also attains the infimum of the objective function, by (11), it is optimal. ∎

These results can be generalized in a number of ways. First, the condition that the unit transport cost function f be bounded can be relaxed. The fact that f is bounded *above* was used only to guarantee that $\int_{A \times B} f \, d\lambda^o$ is finite. Hence we need assume merely that f is bounded *below* and that $\int_{A \times B} f \, d\lambda$ is finite for at least one feasible λ.

The condition that f be bounded *below* can in turn be weakened to the following: a measurable function $h : A \to$ reals exists such that $h(a) \leq f(a, b)$, all $a \in A$, $b \in B$, and $\int_A h \, d\mu'$ is finite. We omit the proof of this statement.

Second, the condition that $(A, \mathfrak{I}')$ and $(B, \mathfrak{I}'')$ be topologically complete and separable can be weakened to their being merely *Borel subsets* of such spaces. For such subsets still remain separable metrizable, and any bounded measure on such a space still remains tight.[15] The proof then proceeds exactly as above. Just about any subsets of n-space, with the usual topology, that arise in practice would fulfill this condition.

Finally, the simple constraints of the transportation problem can be complicated considerably without invalidating the preceding proof. Suppose, for example, that shipments are not allowed between certain source-sink pairs (say because a road does not exist or the resources at the sources are unsuitable for the activities at the sinks). More generally, there may be upper limits on the flows between certain pairs (e.g., due to limited road capacity). Or, there may be lower limits. The following result gives the generalization.

Theorem: Let a transportation problem satisfy all the premises of the preceding theorem for one of the "E" cases (IIa, etc.). In addition to the usual constraints, any feasible flow λ is required to satisfy the following:

$$\lambda(G_i) \leq x_i \qquad (7.4.17)$$

all $i \in I$, and

$$\lambda(F_j) \geq y_j \qquad (7.4.18)$$

all $j \in J$. Here I and J are arbitrary sets indexing the constraints, the x_i and y_j are given real numbers, the G_i and F_j are given *open* and *closed* subsets respectively of $A \times B$.

Then, *provided* at least one flow λ exists satisfying the transportation

---

15. Parthasarathy, *Probability Measures*, pp. 29–30.

problem constraints augmented by (17) and (18), an optimal solution exists for this system of constraints.

Proof: The proof proceeds exactly as in the omnibus theorem, with these additional comments. The set of feasible solutions here is a subset of the original feasible set, hence it inherits the uniform tightness of the original. As above, there is a sequence of feasible flows $\lambda_1, \lambda_2, \ldots$ converging weakly to a bounded flow λ^o, and the objective function $\int_{A \times B} f \, d\lambda$ attains its infimum at λ^o.

It remains to prove that λ^o is feasible. It satisfies the original transportation problem constraints, by the proof above; and we must show it satisfies (17) and (18). But

$$\liminf_n \lambda_n(G_i) \leq x_i$$

all $i \in I$, since each of $\lambda_1, \lambda_2, \ldots$ satisfies (17). This, with Aleksandrov's theorem (12), proves that λ^o satisfies (17). Similarly, we have

$$\lambda^o(F_j) \geq \limsup_n \lambda_n(F_j) \geq y_j \tag{7.4.19}$$

all $j \in J$. The left inequality in (19) is a corollary of (12) which holds for any *closed* set F_j; the right inequality arises from the fact that all $\lambda_1, \lambda_2, \ldots$ satisfy (18). Relation (19) implies that λ^o itself satisfies (18). Hence it is feasible. ∎

7.5. THE TRANSPORTATION PROBLEM: EXISTENCE OF POTENTIALS

We have seen that if a pair of functions $p:A \to$ reals, $q:B \to$ reals is a measure potential for flow measure λ, the latter is optimal for the transportation problem. In this section we tackle the converse question, Given an optimal solution to the transportation problem, does a pair of functions exist that is a measure potential for λ?

Experience indicates that questions of this sort are hard to answer, and this one is no exception. Our procedure will be to establish the existence of functions with a slightly different property, that of being a *topological* potential for λ. We begin by setting out the various concepts needed, and investigating the relation between the two "potential" concepts.

Let $(C, \mathfrak{J})$ be a topological space. Set $E \subseteq C$ is a *neighborhood* of point $c \in C$ iff there is an open set G for which $c \in G$ and $G \subseteq E$. Now let C also be a measure space with σ-field Σ and measure μ (we make no assumptions about the relation between $\mathfrak{J}$ and Σ); $c \in C$ is a *point of support* of the measure μ (with respect to $\mathfrak{J}$) iff every measurable neighborhood of c has positive μ-measure. The set of all points of support is called simply the *support* of μ.

Intuitively, the support of a measure is "where it's at." As examples, we take some familiar probability measures on the real line with $\mathfrak{J}$ and Σ the usual topology and Borel field. For a discrete distribution, taking positive mass on at most a finite number of points, the support is simply those points. For the Poisson distribution, it is the nonnegative integers. For the normal distribution, it is the entire line. It may be shown that the support of a measure is always a *closed* set. (Hint: Show that the complement is open.)

Now suppose we are given a transportation problem defined by the source and sink spaces (A, Σ', μ'), (B, Σ'', μ'') respectively and by unit cost function $f : A \times B \to$ reals. As always, μ', μ'' are σ-finite and f is measurable. Also assume that A and B are furnished with topologies $\mathfrak{J}'$, $\mathfrak{J}''$ respectively. The following definition applies to any of the variants I, II, III, or IV of the transportation problem. Recall that all variants have the same dual, (3.3)–(3.5), except for sign restriction: $p \geq 0$ in variants II and IV, $q \geq 0$ in variants III and IV.

Definition: Let flow measure λ on $(A \times B, \Sigma' \times \Sigma'')$ and the pair of functions $p : A \to$ reals, $q : B \to$ reals be feasible for the transportation problem and its dual respectively. Pair (p, q) is a *topological potential* for λ iff the following conditions are satisfied:

(i) if (a, b) is a point of support for λ, then

$$q(b) - p(a) = f(a, b) \tag{7.5.1}$$

(ii) if $a \in A$ is a point of support for $(\mu' - \lambda')$, then

$$p(a) = 0 \tag{7.5.2}$$

(iii) if $b \in B$ is a point of support for $(\lambda'' - \mu'')$, then

$$q(b) = 0 \tag{7.5.3}$$

A few clarifying comments are in order. "Point of support" in (1) refers to the product space $A \times B$, and it is relative to the product topology $\mathfrak{J}' \times \mathfrak{J}''$. In (2) and (3) we are dealing with measures on (A, Σ') and (B, Σ''), relative to the topologies $\mathfrak{J}'$ and $\mathfrak{J}''$ respectively. Measure $\mu' - \lambda'$ is capacity minus outflow, and so is the *unused capacity* measure on the source space A. Similarly, $\lambda'' - \mu''$ is inflow minus requirement and so is the *oversupply* measure on the sink space B. (Subtraction of measures is defined as in Section 3.1.)

This definition is to apply to all four variants. Note, however, that for variants I and III, $\mu' = \lambda'$; hence $\mu' - \lambda' = 0$ has no points of support. Then (2) is vacuously true and may be dropped from the conditions. Similarly, in variants I and II condition (3) is vacuously true and may be dropped. This is exactly as in the definition of measure potential.

There is indeed a striking parallelism between the two "potential" concepts. Conditions (1)–(3) have as much claim to generalize the complementary slackness conditions as do the corresponding (3.14)–(3.16). For the special case in which the σ-fields Σ' and Σ'' are finite (and coincide with the respective topologies $\mathfrak{I}'$ and $\mathfrak{I}''$), both potential concepts reduce to the ordinary complementary slackness conditions. That (a, b) is a point of support for λ generalizes the notion of a *positive* flow from source a to sink b. Complementary slackness requires that the dual relation for the pair (a, b) be fulfilled with equality, and this is just what (1) requires. Again, if there is unused capacity at a source, the dual variable must be zero; this is (2). Condition (3) generalizes the analogous condition for oversupplied sinks.

Topological potentials *in the wide sense* are defined in the same way as the corresponding wide-sense concept for measure potentials, viz., by dropping the requirement that (p, q) must satisfy the finiteness condition (3.4). In the following discussion of the relation between these two "potential" concepts, we understand them to be either both ordinary or both wide sense.

We are mainly interested in determining when a topological potential will also be a measure potential; for a topological potential is what we get from the theorems to come, while a measure potential is what we want. The following concept is needed.

Definition: A topological space has the *strong Lindelöf property* iff for every collection of open sets $\mathcal{G}$ there is a countable subcollection $\mathcal{G}' \subseteq \mathcal{G}$ such that $\cup \mathcal{G}' = \cup \mathcal{G}$.

Any subset of n-space with the topology generated by the Euclidean metric (indeed, any separable metrizable space) has this property, so that it includes many cases of practical interest.[16] The following theorems apply to all four variants of the transportation problem.

---

16. For readers familiar with general topology the following remarks will serve to place this concept. The strong (or "hereditary") Lindelöf property is implied by the possession of a *countable base* and in turn implies the *(weak) Lindelöf property* that every covering of the space by open sets contains a countable subcovering. One shows by counterexamples that neither of these implications can be reversed. But in a *metrizable* space these three properties are logically equivalent and also equivalent to *separability*. Note also that $\mathfrak{I}'$, $\mathfrak{I}''$ strong Lindelöf does not guarantee that $\mathfrak{I}' \times \mathfrak{I}''$ is even weak Lindelöf (e.g., the RHO or Sorgenfrey topology on the real line). See A. Wilansky, *Topology for Analysis* (Ginn, Waltham, Mass., 1970). This book ends with a remarkable table of implications among topological properties.

Theorem: Given a transportation problem, and given topologies $\mathfrak{J}'$, $\mathfrak{J}''$ on the source and sink spaces A, B respectively; if (p,q) is a *topological* potential for flow λ and the product space $(A \times B, \mathfrak{J}' \times \mathfrak{J}'')$ has the strong Lindelöf property, then (p,q) is a *measure* potential for λ.

Proof: We show that (1)–(3) imply the corresponding conditions for measure potentiality, (3.14)–(3.16) respectively. If

$$q(b) - p(a) < f(a,b) \tag{7.5.4}$$

for a point (a,b), then (a,b) is not a point of support for λ, by (1). Hence (a,b) has a measurable neighborhood $N_{(a,b)}$ of λ-measure zero. There is an open set $G_{(a,b)} \subseteq N_{(a,b)}$ with $(a,b) \in G_{(a,b)}$. Consider the collection $\mathcal{G}$ of all these open sets, one for each point (a,b) satisfying (4). By the strong Lindelöf property, there is a countable subcollection $\mathcal{G}'$ with $\cup \mathcal{G}' = \cup \mathcal{G}$. Let $\mathfrak{N}'$ be the corresponding subcollection of neighborhoods of measure zero; $\mathfrak{N}'$ is also countable. Then

$$\{(a,b) \mid q(b) - p(a) < f(a,b)\} \subseteq (\cup \mathcal{G}) = (\cup \mathcal{G}') \subseteq (\cup \mathfrak{N}')$$

Hence $\lambda\{(a,b) \mid q(b) - p(a) < f(a,b)\}$ does not exceed the sum of $\lambda(N_{(a,b)})$ over all members of $\mathfrak{N}'$. This sum being zero, we obtain (3.14).

Next we prove (3.15). It is easily verified that the component spaces $(A, \mathfrak{J}')$ and $(B, \mathfrak{J}'')$ inherit the strong Lindelöf property from $(A \times B, \mathfrak{J}' \times \mathfrak{J}'')$. If $p(a) > 0$ for point $a \in A$, then a is not a point of support for $\mu' - \lambda'$, by (2). Arguing as above, we find that $\{a \mid p(a) > 0\}$ is covered by a countable number of sets of $(\mu' - \lambda')$-measure zero. Hence $(\mu' - \lambda')\{a \mid p(a) > 0\} = 0$, which implies that $\mu' = \lambda'$ on subsets of $\{a \mid p(a) > 0\}$. This is (3.15).

In the same way we find that $(\lambda'' - \mu'')\{b \mid q(b) > 0\} = 0$, which implies that $\lambda'' = \mu''$ on subsets of $\{b \mid q(b) > 0\}$, yielding (3.16). ∎

A condition for the opposite implication is easier to find and to prove.

Theorem: Let (p,q) be a *measure* potential for flow λ. Let $\mathfrak{J}'$, $\mathfrak{J}''$ be topologies on the source and sink spaces A, B respectively such that

$$\{(a,b) \mid q(b) - p(a) < f(a,b)\} \in \mathfrak{J}' \times \mathfrak{J}'' \tag{7.5.5}$$

and (in variants II and IV)

$$\{a \mid p(a) > 0\} \in \mathfrak{J}' \tag{7.5.6}$$

and (in variants III and IV)

$$\{b \mid a(b) > 0\} \in \mathfrak{J}'' \tag{7.5.7}$$

Then (p,q) is a *topological* potential for λ.

Proof: We show that (3.14)–(3.16) imply the corresponding conditions (1)–(3) for topological potentiality respectively.

Let (1) be *false,* so that λ has a point of support (a_o, b_o) for which $q(b_o) - p(a_o) < f(a_o, b_o)$. The set $\{(a, b) \mid q(b) - p(a) < f(a, b)\}$ is then a measurable neighborhood of (a_o, b_o), by (5), hence has positive λ-measure. Thus (3.14) is false. This proves that (3.14) implies (1).

In variants I and III, (2) is vacuously true. In variants II and IV let (2) be false, so that $\mu' - \lambda'$ has a point of support a_o for which $p(a_o) > 0$. By (6), $\{a \mid p(a) > 0\}$ is a measurable neighborhood of a_o, hence has positive $\mu' - \lambda'$ measure. Thus (3.15) is false. It follows that (3.15) implies (2) in all cases.

Finally, (3) is vacuously true in variants I and II. In variants III and IV let (3) be false and conclude by an argument similar to that just given that (3.16) is false. Thus (3.16) implies (3). ∎

Note that in variant I, (5) alone suffices to insure that a measure potential is a topological potential, and in variants II and III only one extra condition is needed.

Potentials have been defined in terms of a *pair* of functions (p, q). It often happens, however (as in Chapter 8), that one of these functions arises naturally from the problem situation and has a natural interpretation, while the other does not. For this reason it is useful to have a concept involving just one function. Suppose, then, one is given the ingredients of a transportation problem: measure spaces (A, Σ', μ'), (B, Σ'', μ''), measurable cost function $f: A \times B \to$ reals, with topologies $\mathfrak{I}', \mathfrak{I}''$ on A, B respectively. Let λ be a feasible flow. Measurable function $p: A \to$ reals is a *left half-potential* for λ iff

$$p(a_o) + f(a_o, b_o) \le p(a) + f(a, b_o) \qquad (7.5.8)$$

for all $a, a_o \in A$, $b_o \in B$ such that (a_o, b_o) is a point of support for λ (relative to the topology $\mathfrak{I}' \times \mathfrak{I}''$). That is, for fixed b_o, the function $p(\cdot) + f(\cdot, b_o)$ attains its infimum at any point $a_o \in A$ that, paired with b_o, supports λ. Similarly, measurable function $q: B \to$ reals is a *right half-potential* for λ iff

$$q(b_o) - f(a_o, b_o) \ge q(b) - f(a_o, b) \qquad (7.5.9)$$

for all $a_o \in A$, $b, b_o \in B$ such that (a_o, b_o) supports λ.

One easily verifies that if (p, q) is a topological potential for λ, then p and q separately are left and right half-potentials for λ respectively. Indeed, from (3.3) and (1) we obtain $p(a_o) + f(a_o, b_o) = q(b_o) \le p(a) + f(a, b_o)$ whenever (a_o, b_o) supports λ. This yields (8), and a similar argument yields (9). Conversely, given a half-potential, we can lay down certain conditions under which the opposite half-potential exists, the two together being a potential (p, q). (Thus if p is given, q might be defined by $q(b) = \inf\{p(a) + f(a, b)\}$, the infimum taken over all $a \in A$.) The following theorem, with the rest of this section, accomplishes this task indirectly and at the same time relates these to another concept of interest.

Theorem: Let flow λ have a half-potential, and let (a_i, b_i), $i = 1, \ldots, n$, be points of support for λ $(n \geq 2)$. Then

$$f(a_1, b_1) + \cdots + f(a_n, b_n) \leq f(a_1, b_2) + \cdots + f(a_{n-1}, b_n) + f(a_n, b_1)$$
$$(7.5.10)$$

Proof: Suppose λ has a left half-potential p. Then

$$p(a_i) + f(a_i, b_i) \leq p(a_{i-1}) + f(a_{i-1}, b_i)$$

$i = 2, \ldots, n$, and also

$$p(a_1) + f(a_1, b_1) \leq p(a_n) + f(a_n, b_1)$$

all from (8). Adding these n inequalities, the p's drop out, and we are left with (10). The proof for a right half-potential is similar. ∎

Inequality (10) is sufficiently interesting to merit a name; call it the *circulation condition*. Intuitively, it says the following. Suppose we cyclically reassign sources to sinks, shifting some outflow from a_1 away from b_1 and to b_2, etc., and completing the circle by shifting a_n-outflow from b_n to b_1. This reassignment, which leaves all total source outflows and sink inflows unaltered, does not reduce total cost, according to (10).

For $n = 2$ in particular, the circulation condition bears a striking resemblance to the concept of *comparative advantage*. (Think of a_1 and a_2 as two countries or workers and of b_1 and b_2 as alternative activities in which they can engage.) Comparative advantage is usually expressed as an inequality among products or ratios, however, while the circulation condition is an inequality among sums or differences.

We now come to the demonstration of the existence of topological potentials. This goes through two main stages, each rather long. We start with λ^o, an optimal solution to the transportation problem; more precisely, λ^o is *unsurpassed* under (reverse) standard ordering of pseudo-measures. From this we deduce five inequalities involving the cost function f. The crucial one is exactly the circulation condition (11) below. This is used to establish the existence and major properties of the topological potential. The other inequalities are needed for (2), (3), and nonnegativity conditions on p and q.

From now on we use the abbreviation ab for $f(a, b)$.

Lemma: Let (A, Σ', μ') and (B, Σ'', μ'') be σ-finite measure spaces, and $f:A \times B \to$ reals measurable; let $\mathfrak{I}'$, $\mathfrak{I}''$ be topologies on A, B respectively such that $\mathfrak{I}' \subseteq \Sigma'$, $\mathfrak{I}'' \subseteq \Sigma''$,[17] and f is continuous with respect to $\mathfrak{I}' \times \mathfrak{I}''$. Let λ^o be unsurpassed (reverse standard order) for the transportation problem formed from the above, and let (a_i, b_i), $i = 1, \ldots, n$, be points of support for λ^o. Then

(i) $$a_1b_1 + \cdots + a_nb_n \le a_1b_2 + \cdots + a_{n-1}b_n + a_nb_1 \quad (7.5.11)$$

holds for all problem variants, I, II, III, IV ($n = 2, 3, \ldots$).

(ii) If $a_0 \in A$ is a point of support for $(\mu' - \lambda^{o'})$, then in variants II and IV we have

$$a_1b_1 + \cdots + a_nb_n \le a_0b_1 + \cdots + a_{n-1}b_n \quad (7.5.12)$$

$n = 1, 2, \ldots$, and in variant IV we also have

$$a_1b_1 + \cdots + a_nb_n \le a_0b_1 + \cdots + a_nb_{n+1} \quad (7.5.13)$$

$n = 0, 1, \ldots$ (For $n = 0$ the left side of (13) is zero.)

(iii) If $b_1 \in B$ is a point of support for $(\lambda^{o''} - \mu'')$ and *all* subsets of B belong to $\mathfrak{I}''$ (hence to Σ''), then

$$a_1b_1 + \cdots + a_nb_n \le a_1b_2 + \cdots + a_nb_{n+1} \quad (7.5.14)$$

$n = 1, 2, \ldots$, in variants III and IV, and

$$a_1b_1 + \cdots + a_nb_n \le a_1b_2 + \cdots + a_{n-1}b_n \quad (7.5.15)$$

$n = 1, 2, \ldots$, in variant IV. (For $n = 1$ the right side of (15) is zero.) (Neither the a_i's nor the b_i's need to be distinct in any of these cases.)

Proof: For each case (11)–(15), we construct a new feasible flow; the corresponding inequality is then deduced from the fact that this new flow cannot surpass λ^o. Only (11) will be proved in detail.

Given $\epsilon > 0$, there are open sets $E_1, \ldots, E_n \subseteq A$ and $F_1, \ldots, F_n \subseteq B$ satisfying $a_i \in E_i$, $b_i \in F_i$, $i = 1, \ldots, n$; and for all $a \in E_i$, $b \in F_i$,

$$|ab - a_ib_i| \le \epsilon \quad (7.5.16)$$

$i = 1, \ldots, n$; and for all $a \in E_i$, $b \in F_{i+1}$,

$$|ab - a_ib_{i+1}| \le \epsilon \quad (7.5.17)$$

$i = 1, \ldots, n$. (In (17) for $i = n$, F_{n+1} and b_{n+1} are to be understood as F_1, b_1 respectively.) For the continuity of f implies there are open sets owning a_i and b_i respectively such that (16) is satisfied for all (a, b) in their cartesian product, and there are open sets owning a_i and b_{i+1} such that

---

17. $\mathfrak{I}' \subseteq \Sigma'$ and $\mathfrak{I}'' \subseteq \Sigma''$ do not guarantee that $(\mathfrak{I}' \times \mathfrak{I}'') \subseteq (\Sigma' \times \Sigma'')$, nor need we make this stronger assumption.

(17) is similarly satisfied. This gives two open sets for each of the points $(a_1, \ldots, a_n, b_1, \ldots, b_n)$. Let E_i be the intersection of the two open sets for a_i, and construct F_i in the same way for b_i, $i = 1, \ldots, n$. With these, all the relations above are satisfied simultaneously.

Since $E_i \times F_i$ is a measurable neighborhood of (a_i, b_i), a point of support for λ^o, then $\lambda^o(E_i \times F_i) > 0$. In fact $\lambda^o(E_i \times F_i)$ may be infinite; but if so, there is a measurable set $G_i \subseteq E_i \times F_i$ such that $\infty > \lambda^o(G_i) > 0$, since λ^o is σ-finite. Choose a set G_i satisfying these conditions for each $i = 1, \ldots, n$.

Now define measures ν_i on $(A \times B, \Sigma' \times \Sigma'')$ as follows:

$$\nu_i(H) = \lambda^o(H \cap G_i)/\lambda^o(G_i) \qquad (7.5.18)$$

$H \in \Sigma' \times \Sigma''$, $i = 1, \ldots, n$. Consider the signed measure ν given by

$$(\nu_1' \times \nu_2'') + \cdots + (\nu_{n-1}' \times \nu_n'') + (\nu_n' \times \nu_1'') - \nu_1 - \cdots - \nu_n \qquad (7.5.19)$$

Here ν_1' is the left marginal of ν_1, and ν_2'' is the right marginal of ν_2; we are to form the product measure of these, add up n similar product measures, and then subtract the ν_i's of (18). All the summands in (19) have universe set $A \times B$. For each i, $\nu_i(A \times B) = 1$, so that (19) is well defined, and in fact bounded. Finally, consider the (signed) measure λ given by

$$\lambda = \lambda^o + y\nu \qquad (7.5.20)$$

where y is the positive real number $\min\{\lambda^o(G_i) \mid i = 1, \ldots, n\}/n$.

We prove this λ is a feasible flow. First, λ is nonnegative. For $y\nu_i(H) \le \lambda^o(H)/n$, all i, from the definition of y, so that

$$\lambda(H) \ge \lambda^o(H) - y\nu_1(H) - \cdots - y\nu_n(H) \ge 0$$

Next, for any $E \in \Sigma'$ we obtain $\nu(E \times B) = 0$ by direct substitution in (19). Similarly, $\nu(A \times F) = 0$ for any $F \in \Sigma''$. This means that λ^o and λ have the same marginals. Hence in any variant of the transportation problem the feasibility of λ^o implies the feasibility of λ.

It follows that λ cannot (downwardly) surpass λ^o. Now

$$\int f \, d\lambda - \int f \, d\lambda^o = y \int f \, d\nu$$

But the indefinite integral $\int f \, d\nu$ is actually well defined and finite as a *definite* integral over $A \times B$. For f is bounded on the set

$$(E_1 \times F_1) \cup \cdots \cup (E_n \times F_n) \cup (E_1 \times F_2) \cup \cdots \cup (E_n \times F_1)$$

by (16) and (17), while ν is zero off this set. Hence unsurpassedness under (reverse) standard ordering reduces to the condition

$$\int_{A \times B} f \, d\nu \ge 0 \qquad (7.5.21)$$

Expanding this by (19), we find that

$$\int_{A \times B} f \, dv_i \geq (a_i b_i - \epsilon) v_i (A \times B) = (a_i b_i - \epsilon) \qquad (7.5.22)$$

$i = 1, \ldots, n$, since v_i is zero off $E_i \times F_i$, and f is bounded below by $a_i b_i - \epsilon$ on $E_i \times F_i$, according to (16). Similarly,

$$\int_{A \times B} f \, d(v_i' \times v_{i+1}'') \leq (a_i b_{i+1} + \epsilon) v_i'(A) v_{i+1}''(B) = (a_i b_{i+1} + \epsilon)$$
$$(7.5.23)$$

$i = 1, \ldots, n$, since $f \leq a_i b_{i+1} + \epsilon$ on $E_i \times F_{i+1}$, by (17), while $v_i' \times v_{i+1}''$ is zero off this set (for $i = n, i + 1$ should be read as 1).

From (21)–(23) we obtain

$$a_1 b_2 + \cdots + a_n b_1 - a_1 b_1 - \cdots - a_n b_n \geq -2n\epsilon \qquad (7.5.24)$$

But ϵ is an arbitrary positive number. Hence (24) implies the circulation condition (11). The first inequality has been obtained.

We now sketch the proof of (12) and (13). It proceeds as above with the following differences. In addition to the open sets $E_i, F_i, i = 1, \ldots, n$, we find open sets $E_0 \subseteq A$, and—for (13)—$F_{n+1} \subseteq B$ such that $a_0 \in E_0$, $b_{n+1} \in F_{n+1}$; and (17) holds for $i = 0, 1, \ldots, n$.

The sets $G_i \subseteq (E_i \times F_i)$, $i = 1, \ldots, n$, are constructed as above. In addition we find that $(\mu' - \lambda^{o\prime})(E_0) > 0$, since E_0 is a measurable neighborhood of a_0, a point of support for $\mu' - \lambda^{o\prime}$. In fact $(\mu' - \lambda^{o\prime})(E_0)$ may be infinite, but there is always a measurable set $A' \subseteq E_0$ for which $\infty > (\mu' - \lambda^{o\prime})(A') > 0$, since these measures are σ-finite.

The measures v_i, $i = 1, \ldots, n$, are defined exactly as in (18). In addition, define the measure μ_0' on (A, Σ') by

$$\mu_0'(E) = [(\mu' - \lambda^{o\prime})(E \cap A')]/(\mu' - \lambda^{o\prime})(A')$$

all $E \in \Sigma'$. Also, for (13), define measure μ_{n+1}'' on (B, Σ'') by $\mu_{n+1}''(F) = 1$ if $b_{n+1} \in F$; $= 0$ if $b_{n+1} \notin F$, all $F \in \Sigma''$. Consider the signed measure v on $(A \times B, \Sigma' \times \Sigma'')$ given by

$$(\mu_0' \times v_1'') + (v_1' \times v_2'') + \cdots$$
$$+ (v_{n-1}' \times v_n'') - v_1 - \cdots - v_n \, [+(v_n' \times \mu_{n+1}'')] \qquad (7.5.25)$$

where the bracketed measure is to be included when considering (13), and omitted when considering (12). (This formula applies for $n \geq 1$. For $n = 0$, which arises only for (13), define v as $\mu_0' \times \mu_1''$.) Each of the measures in (25) has the value *one* at $A \times B$, so (25) is well defined and bounded. Finally, consider the (signed) measure λ given by

$$\lambda^o + zv \qquad (7.5.26)$$

where z is the positive real number $\min(y, (\mu' - \lambda^{o\prime})(A'))$, y being defined as in (20).

Nonnegativity of λ is proved as above. Next, substitute $E \times B$ into (25) (*inclusive* of the bracketed term), where $E \in \Sigma'$. The result is $\mu_0'(E)$. Hence,

$$\lambda'(E) = \lambda^{o\prime}(E) + z\mu_0'(E) \leq \lambda^{o\prime}(E) + (\mu' - \lambda^{o\prime})(E)$$

the inequality arising from the definition of z. Hence $\lambda' \leq \mu'$, i.e., λ satisfies the capacity constraints in variants II and IV. The same result holds *a fortiori* if the bracketed term in (25) is omitted, since this leads to a smaller λ.

Next, substitute $A \times F$ into (25) (*omitting* the bracketed term), where $F \in \Sigma''$. Everything cancels, hence $\lambda'' = \lambda^{o\prime\prime}$ and feasibility is preserved for the requirement constraints in any variant. Adding in the bracketed term, however, yields merely $\lambda'' \geq \lambda^{o\prime\prime}$, so that feasibility is preserved in variant IV. To summarize: λ given by (26) is feasible for variants II and IV if ν is defined by (25) *omitting* the bracketed term and is feasible for IV if ν *includes* the bracketed term.

Just as above, the feasibility of λ leads to $\int_{A \times B} f \, d\nu \geq 0$, and an argument similar to the one above gives the inequalities (12) or (13), depending on whether (25) omits or includes the bracketed term respectively. These inequalities are then valid for the variants in which λ is feasible.

Finally, the proof for (14)–(15) is very similar to that for (12)–(13). There is no point a_0, and consequently no E_0, A', or μ_0'. We choose F_1 to be the singleton set $\{b_1\}$, which can be done because every subset of B is open by assumption. Otherwise proceeding exactly as above, we define ν by (25), modified by the omission of the measure $\mu_0' \times \nu_1''$. The bracketed measure $\nu_n' \times \mu_{n+1}''$ is to be included when considering (14), and omitted when considering (15). Again λ is defined by (26), where z, however, is now the positive real number

$$\min(y, \lambda^{o\prime\prime}\{b_1\} - \mu''\{b_1\}) \qquad (7.5.27)$$

y being defined as in (20). To show that z is indeed positive, note that $\{b_1\}$ is itself a measurable neighborhood of b_1, hence has positive ($\lambda^{o\prime\prime} - \mu''$)-measure since b_1 is a point of support for that measure.

Nonnegativity of λ is proved as above. Next substitute $E \times B$ into the modified (25) (*inclusive* of the bracketed term), where $E \in \Sigma'$. Everything cancels, hence $\lambda' = \lambda^{o\prime}$ and feasibility is preserved for the capacity constraints in any variant. Omitting the bracketed term yields $\lambda' \leq \lambda^{o\prime}$, which preserves feasibility in variant IV.

Next substitute $A \times F$ into the modified (25) (*omitting* the bracketed term), where $F \in \Sigma''$. The result is $-\nu_1(A \times F)$. Now since $F_1 = \{b_1\}$, we have $G_1 \subseteq E_1 \times \{b_1\}$. Hence from (18), with $i = 1$, we find that

$\nu_1(A \times F) = 1$ if $b_1 \in F$, and $\nu_1(A \times F) = 0$ if $b_1 \notin F$. In this second case, $\lambda''(F) = \lambda^{o\prime\prime}(F)$, while on $\{b_1\}$ we have $\lambda''\{b_1\} = \lambda^{o\prime\prime}\{b_1\} - z \geq \mu''\{b_1\}$, from (27). Hence $\lambda'' \geq \mu''$, and λ is feasible for the requirement constraints in variants III and IV. Adding in the bracketed term of (25) can only increase λ and hence preserves this feasibility. We summarize: if the bracketed measure in modified (25) is *included*, the λ thus defined is feasible for variants III and IV; if it is *omitted*, λ is still feasible for IV.

The argument from unsurpassedness to (14) or (15) then follows the pattern laid down above. ∎

We now come to the main result. We prove the existence of a topological potential *in the wide sense*; i.e., the integrals $\int_A p \, d\mu'$, $\int_B q \, d\mu''$ need not be finite or even well defined. As mentioned below, a simple extra premise removes this qualification. Note that the premises for variants III and IV are somewhat stronger than for variants I and II.

Theorem: Let (A, Σ', μ') and (B, Σ'', μ'') be nonempty σ-finite measure spaces, and let $f\colon A \times B \to$ reals be measurable and bounded. Let λ^o be a measure on $(A \times B, \Sigma' \times \Sigma'')$ that is unsurpassed (reverse standard order) for the transportation problem determined by these.

Also let $\mathfrak{I}'$, $\mathfrak{I}''$ be topologies on A, B respectively such that $\mathfrak{I}' \subseteq \Sigma'$, $\mathfrak{I}'' \subseteq \Sigma''$, and f is continuous with respect to $\mathfrak{I}' \times \mathfrak{I}''$. (In variants III and IV make the additional assumption that all subsets of B belong to $\mathfrak{I}''$.)

Then there exist bounded functions $p\colon A \to$ reals, $q\colon B \to$ reals, p lower and q upper semicontinuous, such that (p, q) is a topological potential for λ^o in the wide sense.

Proof: We use different definitions of p for the different variants I, II, III, IV. An α-*sequence* is a sequence of the form $a_0, b_1, a_1, \ldots, b_n, a_n$ ($n = 0$ or 1 or $\ldots$), where (a_i, b_i) is a point of support of λ^o for $i = 1, \ldots, n$. That is, it consists of $2n + 1$ points, alternating from A and B, beginning and ending with an A-point. (For $n = 0$ the sequence consists of a single A-point a_0.) A β-*sequence* is an α-sequence with an extra B-point b_{n+1} tacked on at the end. (Thus the shortest β-sequence is of the form a_0, b_1.)

The *value* of an α- (β-) sequence $a_0, b_1, \ldots, a_n, (b_{n+1})$ is defined as

$$-a_0 b_1 + a_1 b_1 - a_1 b_2 + \cdots + a_n b_n (-a_n b_{n+1}) \qquad (7.5.28)$$

where the parenthetical term is included for β-sequences only. (Here ab abbreviates $f(a, b)$ as usual. The singleton sequence a_0 is taken to have the value zero.)

Now define the function p with domain A in three ways. The α-*definition* sets $p(a)$ equal to the supremum of the values of all α-sequences beginning with a, all $a \in A$. The β-*definition* is the same with all β-sequences beginning with a. The γ-*definition* is the maximum of these two, i.e., the supremum of the values of all α- *and* β-sequences beginning with a.

We now prove that under any of these definitions p is *bounded, lower semicontinuous,* and *measurable*. Function f is bounded, so $|f| \leq N$ for some real number N. Hence $p(a) \geq 0$ or $-ab$ on the α- or β-definitions respectively for any $a \in A$, $b \in B$, so that $p \geq -N$: p is bounded below.

Next we prove boundedness above. Sequences of length at most four have values not greater than $3N$. For sequences of length five or more we make use of the *circulation condition* (11), which is valid in all variants under our premises. Formula (28) may be rewritten as

$$[a_1b_1 - a_1b_2 + \cdots - a_{n-1}b_n + a_nb_n - a_nb_1] + a_nb_1 - a_0b_1(-a_nb_{n+1})$$
$$(7.5.29)$$

$n = 2, 3, \ldots$. But (11) states that the bracketed expression is nonpositive. Hence (29) never exceeds $3N$. Thus p is bounded.

Lower semicontinuity is proved as follows. Think of all points in (28) except a_0 as being fixed. Then (28) defines a family of real-valued functions with common domain A, each sequence b_1, $a_1, \ldots, b_n$, a_n, (b_{n+1}) indexing one such function. Each of these functions is continuous, since f is continuous. One easily shows that the supremum of a set of continuous functions is lower semicontinuous. But p on any definition is such a supremum.

The sets $\{a \mid p(a) > x\}$, x real, are open, by lower semicontinuity. But $\mathcal{J}' \subseteq \Sigma'$, hence these sets are also measurable. Thus p is measurable. (Similarly, an upper semicontinuous function on B is measurable; this fact is needed later.)

With these general results in hand, we proceed to each variant in turn.

Variant I: Use any of the three definitions of p, then define $q{:}B \rightarrow$ reals by

$$q(b) = \inf[p(a) + ab] \qquad (7.5.30)$$

the infimum taken over all points $a \in A$. Since p and f are bounded, q is bounded. Think of $p(a) + ab$ as a family of functions of b, indexed by a. Each of these functions is continuous since f is continuous. Then q, as the infimum of a set of continuous functions, is upper semicontinuous, hence also measurable.

It remains to show that (p, q) is a topological potential for λ^o in the wide sense. For variant I this reduces to

$$q(b) - p(a) \leq ab \qquad (7.5.31)$$

for all $a \in A$, $b \in B$, with equality if (a, b) supports λ^o. Inequality (31) is an immediate consequence of (30). Let (a_0, b_0) support λ^o. Then

$$p(a) \geq -ab_0 + a_0 b_0 + p(a_0) \qquad (7.5.32)$$

for any $a \in A$; for the right side of (32) is the supremum of the values of sequences (α-, β-, or both) beginning $(a, b_0, a_0, \dots)$, while the left is the supremum over a wider class of such sequences. It follows that the infimum of $p(a) + ab_0$ is attained at $a = a_0$. Hence

$$q(b_0) = p(a_0) + a_0 b_0 \qquad (7.5.33)$$

so that (31) is satisfied with equality.

Variant II: Use the α-definition of p, and define q again by (30). All the argument for variant I applies here also. To complete the proof for variant II, two additional facts must be established: $p \geq 0$, and $p(a_0) = 0$ for any point a_0 supporting $\mu' - \lambda^{o'}$.

For any $a \in A$, the singleton consisting of a alone is an α-sequence of value zero; hence $p \geq 0$. Let a_0 support $\mu' - \lambda^{o'}$. Then inequality (12) is valid and states that (28) (omitting the parenthetical term) is never positive; hence $p(a_0) = 0$.

Variant III: Use the β-definition of p, and define $q(b)$ by (30) if b is *not* a point of support for $\lambda^{o''} - \mu''$, while $q(b) = 0$ if b *is* a point of support for $\lambda^{o''} - \mu''$.

Function q is clearly bounded. Since $\mathfrak{I}''$, hence Σ'', own all subsets of B, q is automatically continuous and measurable. To show that (p, q) is a topological potential for λ^o in the wide sense we must demonstrate (31) (with equality if (a, b) supports λ^o), and also demonstrate that $q \geq 0$, with $q(b_0) = 0$ for all points of support of $\lambda^{o''} - \mu''$. This last fact is true by the definition of q.

For any $a \in A$, $b \in B$ we have $p(a) \geq -ab$, since the latter is the value of the β-sequence a, b; hence $q \geq 0$. The same inequality yields (31) *on* the support of $\lambda^{o''} - \mu''$, since $q = 0$ there; (31) follows from (30) *off* the support.

Finally, let (a_0, b_0) support λ^o. If $q(b_0)$ satisfies (30), the argument of (32) leads to the equality (33). This leaves the case where b_0 supports $\lambda^{o''} - \mu''$. Then inequality (14) is valid in the form

$$-a_0 b_0 \geq -a_0 b_1 + a_1 b_1 - a_1 b_2 + \cdots - a_n b_{n+1} \qquad (7.5.34)$$

$n = 0, 1, \dots$. Here $b_1, a_1, \dots, a_n, b_{n+1}$ are any points such that (a_i, b_i) supports $\lambda^o, i = 1, \dots, n$. But by (28), $p(a_0)$ is the supremum of the right expression in (34) over all such sequences. Hence $p(a_0) \leq -a_0 b_0$. Since the opposite inequality always holds, $p(a_0) = -a_0 b_0$. But this is (33), since $q(b_0) = 0$. Hence (31) holds with equality if (a, b) supports λ^o.

Variant IV: Use the γ-definition of p, and define q as in variant III. The same arguments as in variant III then show that q is bounded, continuous, measurable, nonnegative, and zero on the support of $\lambda^{o\prime\prime} - \mu\prime\prime$, and that (31) holds for all $a \in A$, $b \in B$. Three more facts need to be demonstrated: $p \geq 0$, $p = 0$ on the support of $\mu\prime - \lambda^{o\prime}$, and (33) holds for (a_0, b_0) supporting λ^o.

For any $a \in A$, $p(a)$ is not less than the value of the singleton sequence a, hence $p \geq 0$. Next, let a_0 support $\mu\prime - \lambda^{o\prime}$. Then inequalities (12) and (13) are valid and imply that (28) never exceeds zero, with or without the parenthetical term; hence $p(a_0) = 0$.

Finally, let (a_0, b_0) support λ^o. As above, if b_0 does *not* support $\lambda^{o\prime\prime} - \mu\prime\prime$, then $q(b_0)$ satisfies (30) and the argument of (32) yields (33). Suppose b_0 *does* support $\lambda^{o\prime\prime} - \mu\prime\prime$. Then inequalities (14) and (15) are valid in the form

$$-a_0 b_0 \geq -a_0 b_1 + a_1 b_1 - \cdots + a_n b_n (-a_n b_{n+1}) \qquad (7.5.35)$$

$n = 0, 1, \ldots$, with or without the parenthetical term. (Note that for $n = 0$, (15) in the form (35) is simply $-a_0 b_0 \geq 0$.) By (28), $p(a_0)$ on definition γ is the supremum of all the right-hand expressions in (35), for $b_1, a_1, \ldots, b_n, a_n, (b_{n+1})$, where (a_i, b_i) supports λ^o, $i = 1, \ldots, n$. Hence again $p(a_0) \leq -a_0 b_0$, and the same argument as in variant III shows that (33) is satisfied. This finishes variant IV and the proof. ∎

As noted above, (p, q) is a topological potential in the wide sense. But if we impose the additional premise that $\mu\prime$ and $\mu\prime\prime$ are *bounded*, (p, q) is a potential in the strict sense. For the functions p and q as constructed above are also bounded, hence the definite integrals $\int_A p \, d\mu\prime$ and $\int_B q \, d\mu\prime\prime$ are both well defined and finite.

The extra premise that all subsets of B belong to $\mathfrak{J}\prime\prime$ (imposed for variants III and IV) is somewhat restrictive (although natural in some cases, e.g., when B is a finite set). In variant IV it may be replaced by the extra premises that f is positive and $\lambda^{o\prime\prime}$ is σ-finite.[18] We show this as follows. Using the Radon-Nikodym theorem as in case (IVb) of the optimality existence theorem, we show the existence of a measure λ_1 on $\Sigma\prime \times \Sigma\prime\prime$ satisfying $\lambda_1 \leq \lambda^o$ and being feasible for variant II, hence for the original variant IV. Since λ_1 does not surpass λ^o, and $f > 0$, we must in fact have $\lambda_1 = \lambda^o$. Thus λ^o is feasible for variant II and, in fact, unsurpassed for it. Let (p, q) be a wide-sense topological potential for λ^o constructed as for variant II. To verify that this (p, q) is also a variant IV wide-sense potential, we must demonstrate two additional facts: $q \geq 0$,

---

18. In fact, the single extra premise $f > 0$ implies that $\lambda^{o\prime\prime} = \mu\prime\prime$, hence (see below) that a wide-sense topological potential exists. But the proof in this case is more complicated.

and $q = 0$ on the support of $\lambda^{o''} - \mu''$. This last property is trivial because the support is empty, since $\lambda^{o''} = \mu''$. Also q is defined by (30), so the nonnegativity of p and f imply the nonnegativity of q. This concludes the proof.

The functions p and q constructed in our main theorem are semicontinuous. We can strengthen this result by adding some extra premises concerning f. First, we need a few continuity concepts.

Let $(C, \mathfrak{J})$ be a topological space, and $\mathbf{G}$ a set of real-valued functions with common domain C. Set $\mathbf{G}$ is *equicontinuous* at the point $c_o \in C$ iff for every positive number ϵ there is a neighborhood N of c_o such that

$$| g(c) - g(c_o) | < \epsilon \qquad (7.5.36)$$

for all $c \in N$ and all $g \in \mathbf{G}$. (If $\mathbf{G}$ consists of just one function g, this is simply the definition of *continuity* of g at the point c_o. One may then show that g is continuous as defined previously iff it is continuous at every point of its domain as defined by (36).) Next suppose that $\mathfrak{J}$ is generated by a metric d. Set $\mathbf{G}$ is *uniformly equicontinuous* iff for all $\epsilon > 0$ there is a $\delta > 0$ such that

$$\text{if } d(c_1, c_2) < \delta, \quad \text{then} \quad | g(c_1) - g(c_2) | < \epsilon \qquad (7.5.37)$$

for all $c_1, c_2 \in C$ and all $g \in \mathbf{G}$. (If $\mathbf{G}$ consists of just one function g, this is simply the definition of *uniform* continuity of g.)

Consider the transport cost function $f : A \times B \to$ reals. Now f may be thought of as a family of functions $f(\cdot, b) : A \to$ reals, indexed by $b \in B$. Suppose this family is *equicontinuous* at the point $a' \in A$; i.e., for all $\epsilon > 0$ there is a neighborhood N of a' such that $| f(a, b) - f(a', b) | < \epsilon$ for all $a \in N$, $b \in B$. Then we claim that p, constructed under any of the definitions α, β, γ, is *continuous* at a'.

The proof of this statement rests on the easily verified fact that if the collection of functions $\mathbf{G}$ satisfies (36) and the supremum of $\mathbf{G}$ is finite, then $\sup \mathbf{G}$ is continuous at c_o. Now consider (28) as a function of a_0, the other points $b_1, a_1, \ldots, b_n, a_n, (b_{n+1})$ being parameters. This function differs only by a constant from the function $-f(\cdot, b_1)$. Hence the functions (28) are equicontinuous at a', so that p, which is the supremum of an appropriate subset of them, is continuous at a'. This concludes the proof.

Thus if the family $f(\cdot, b)$, $b \in B$ is equicontinuous at each point $a \in A$, it follows that p is *continuous,* not merely lower semicontinuous.

What about q? Function f may also be thought of as a family of functions $f(a, \cdot)$ with common domain B, indexed by $a \in A$. Suppose this family is equicontinuous at $b'' \in B$. Then we claim that q, given by (30), is continuous at b''.

The proof is virtually the same as that for p, on noting that, if $\mathbf{G}$ satisfies (36), then the infimum of $\mathbf{G}$ is also continuous at c_o, provided inf

G is finite. (This takes care of variants I and II; in variants III and IV all subsets of *B* belong to ℑ″, which makes any function *q*:*B* → reals continuous.)

Next suppose ℑ′, the topology of *A*, is generated by a metric *d*′ under which the family *f*(·, *b*), *b* ∈ *B* is *uniformly equicontinuous.* Then *p* is *uniformly continuous,* under any of the definitions *α*, *β*, *γ*.

The proof is similar to those above and is based on the fact (again easily verified) that if the family **G** satisfies (37) and sup **G** is finite, then sup **G** is uniformly continuous. Reversing the roles of *A* and *B*, we get a similar condition implying that *q*, given by (30), is uniformly continuous.

Our final theorem summarizes many of the preceding results. The boundedness assumptions guarantee that the objective function for the primal is always a finite definite integral. Hence we speak merely of an optimal solution, the distinction between "best" and "unsurpassed" disappearing in the present instance.

Theorem: Let (A, Σ', μ') and (B, Σ'', μ'') be bounded measure spaces, and let $f{:}A \times B \to$ reals be bounded measurable. Let λ^o be an optimal flow for the transportation problem determined by these.

Also let ℑ′, ℑ″ be topologies on *A*, *B* respectively such that $\Im' \subseteq \Sigma'$, $\Im'' \subseteq \Sigma''$, *f* is continuous with respect to the product topology $\Im' \times \Im''$, and the latter has the strong Lindelöf property. (In variants III and IV add the condition that all subsets of *B* belong to ℑ″.)

Functions $p^o{:}A \to$ reals, $q^o{:}B \to$ reals exist such that the pair (p^o, q^o) is best for the dual of the transportation problem, and the primal and dual values are equal:

$$\int_{A \times B} f \, d\lambda^o = \int_B q^o \, d\mu'' - \int_A p^o \, d\mu'$$

Proof: By the main result of this section a bounded pair (p^o, q^o) exists that is a topological potential in the wide sense for λ^o—in fact, a potential in the ordinary sense, since μ' and μ'' are also bounded. The strong Lindelöf property then implies that (p^o, q^o) is a measure potential for λ^o, from which the stated result follows (see (3.18)). ∎

We conclude our discussion of the transportation problem with a brief glance at the pioneering work of Kantorovitch.[19] (We use our notation and terminology to facilitate exposition.) He formulates a special

19. L. Kantorovitch, On the translocation of masses, *Comptes Rendus (Doklady) de l'Académie des Sciences de l'URSS* 37, no. 7–8 (1942):199–201. (Repr. in *Manage. Sci.* 5(Oct. 1958):1–4).

case of the measure-theoretic transportation problem (variant I)—special in that the source and sink spaces are the same: $(A, \Sigma') = (B, \Sigma'')$, the cost function f is nonnegative, μ' and μ'' are bounded, and a certain topological structure is imposed. Next he defines a feasible flow λ as being *potential* iff a (measurable) function $p{:}A \rightarrow$ reals exists such that

$$| p(b) - p(a) | \leq f(a, b)$$

for all $a, b \in A$ (remember that $A = B$), and $p(b) - p(a) = f(a, b)$ if (a, b) is a point of support for λ. His main assertion is that a flow λ is optimal iff it is potential.

This is quite instructive, both in its accomplishments and its errors. The definition of p is close to our concept of "topological potential" for λ, differing from it in the minor point that the absolute value appears on the left of the inequality, and differing in the major point that just *one* function p appears instead of a pair (p, q). That is, while the problem formulated is of the *transportation* form, the "potential" concept used is more appropriate for the *transshipment* problem considered below. This could not have happened if Kantorovitch did not identify the source and sink spaces. To place things into the transportation problem framework, think of these relations as defining conditions on the *pair* (p, p) rather than the single function p. Pair (p, p) is indeed a topological potential for λ if it satisfies these conditions.

As for Kantorovitch's assertion, the "if" part is correct. We indicate how this can be demonstrated within the framework of our theory. The topological assumptions he makes imply the strong Lindelöf property, so that (p, p) is also a measure potential. The boundedness assumptions then guarantee optimality of λ.

The "only if" part is erroneous. If λ is an optimal flow one can indeed demonstrate the existence of a topological potential (p, q) under his assumptions (by our proof above). (Here both p and q will have domain A, since source and sink spaces are the same.) But one cannot make the further assertion that $p = q$, as the following counterexample demonstrates. Let $A = \{a, b, c\}$, with all subsets open and measurable; $\mu'\{a\} = \mu''\{c\} = 1$; $\mu'\{b, c\} = \mu''\{a, b\} = 0$; $f(a, c) = 1$, $f = 0$ elsewhere. This satisfies all Kantorovitch's premises. The optimal flow is $\lambda^o\{(a, c)\} = 1$, and $\lambda^o = 0$ on all other singletons; in fact, this is the only feasible flow. Now suppose this flow were potential. We would have $p(c) - p(a) = 1$, since (a, c) is a point of support for λ^o. But also $p(c) - p(b) \leq 0$ and $p(b) - p(a) \leq 0$—contradiction!

This three-page paper of Kantorovitch is the true *locus classicus* for the measure-theoretic transportation problem. The problem itself, the key role of "potentials," and certain basic solution methods are all adumbrated here, even if the exposition is flawed. The energies of researchers have in the meantime been directed into other channels (mainly to the development of "ordinary" programming methods), so that the work of Kantorovitch appears to be the direct predecessor of this chapter, with a lag of some thirty-five years.

7.6. TRANSHIPMENT: INTRODUCTION

The transhipment problem with n locations is to find n^2 nonnegative numbers $x_{ij}, i,j = 1, \ldots, n$, satisfying

$$(x_{i1} + \cdots + x_{in}) - (x_{1i} + \cdots + x_{ni}) \leq \alpha_i \qquad (7.6.1)$$

$i = 1, \ldots, n$, and minimizing the sum of

$$f_{ij}x_{ij} \qquad (7.6.2)$$

over all n^2 terms of this form. Here the numbers α_i and $f_{ij}, i,j = 1, \ldots, n$, are given parameters.

The simplest interpretation is the following: x_{ij} is the quantity of a certain commodity moving from location i to location j. On the left of (1) the first parenthetical sum, $x_{i1} + \cdots + x_{in}$ (excluding x_{ii}), is the total quantity *exported from* location i to other locations; the second parenthetical sum (again excluding x_{ii}) is the total quantity *imported to* location i from other locations. Hence the left side of (1) may be thought of as *net exports* from location i. (The meaning of x_{ii} is problematical, but this term cancels out from the left side of (1) and therefore creates no interpretive difficulty.) α_i may be thought of as the *net capacity* of location i. If positive, it gives the amount by which exports from i may exceed imports to i; if negative, it gives the amount by which imports to i must exceed exports from i. Constraint (1) states that net exports cannot exceed net capacity; we could just as well have stipulated that net *imports* cannot fall below net *requirements* β_i, where β_i is simply $-\alpha_i$.

Compare the constraint system (1) with the constraints of the transportation problem (1.1)–(1.2). We can think of the locations i for which α_i is *positive* as being *sources*, those for which α_i is *negative* as being *sinks*. (If $\alpha_i = 0$, i may be placed in either category.) The transportation problem allows flows only from sources to sinks, while the transhipment problem allows flows between any two sites, including source to source, sink to sink, and sink to source. Thus it becomes possible to "tranship" a flow from a source to a sink through a series of intermediate locations.

The objective function (2) has the same form as that for the transportation problem (1.3), though unit transport cost f_{ij} must now be defined for all pairs of locations, not just for source-sink pairs. As with the transportation problem, we may distinguish variants of the transhipment problem. Thus (1)–(2) will be called the *inequality*-constrained variant. The *equality*-constrained variant simply replaces $\leq$ in (1) by $=$.

The transhipment problem was first formulated by Orden, who also showed that there is a *transportation* problem that is equivalent to the transhipment problem in a certain sense.[20] Indeed, several ways have been suggested for "reducing" transhipment to the transportation problem (or something resembling it). We explore one of these below.

First, it is worthwhile to compare the transportation and transhipment problems from the point of view of possible applications. We have mentioned the interpretation of transhipment points as locations of physical Space. Specifically, imagine a system of cities, thought of as points, linked by a system of roads. A road *directly links* cities i and j iff it starts at i, ends at j, and passes through no other city. We then let f_{ij} be transport cost incurred in moving a unit mass of the commodity from i to j over this direct link, or over the cheapest direct link if there are several. (If there is no direct link, set $f_{ij} = \infty$. Alternatively, we may resort to the artifice of making f_{ij} finite but "very large," so that traffic avoids this "link" if at all possible.)

Note the heroic linearity assumptions involved in the objective functions (2) or (1.3), or their generalizations to integrals. Congestion phenomena and scale economies (both very important in transportation) are ignored, and a doubling of traffic is assumed to double cost on any link. The concept of transport cost itself covers a motley collection of categories: fuel consumption, vehicle and road wear, travel time, risk of accident, discomfort, deterioration of cargo, traffic control costs, and perhaps vehicle and road construction costs as well as pollution, noise, and other disamenities if all social costs are to be included.[21]

Brushing aside all these conceptual difficulties, we postulate a unit cost f_{ij} associated with the cheapest direct link from location i to location j. Suppose f_{ij} is always finite, and let A be the set of all locations in the transhipment problem. The domain of f is then $A \times A$, which is the same as that of a *metric* on A. Is it reasonable to assume that f *is* a metric? Examine the conditions one by one. Recall that $d{:}A \times A \rightarrow$ reals is a metric iff $d(a, a) = 0$, $d(a_1, a_2) > 0$ if $a_1 \neq a_2$, $d(a_1, a_2) = d(a_2, a_1)$, and $d(a_1, a_2) + d(a_2, a_3) \geq d(a_1, a_3)$.

As for the first property, $f_{ii} = 0$ has no clear empirical meaning; it often is mathematically convenient to make this assumption. As for the second, while f_{ij} will usually be positive for $i \neq j$, it might be negative for some pairs (e.g., pleasure driving). Again, while f_{ij} and f_{ji} might be approximately equal, one can think of several reasons for inequality: going uphill, upstream, or upwind vs. downhill, downstream, or downwind;

20. A. Orden, The transhipment problem, *Manage. Sci.* 2(Apr. 1956):276–85. We have adopted Orden's spelling of the word "transhipment."

21. We do not deal with nonlinear objective functions in this chapter, although some of our results do generalize to this case. Note that a nonlinear objective can still be pseudomeasure valued, as in Chapter 5.

one-way streets; asymmetric bus routes; the ease of getting from little-known i to well-known j because of direction signs and road convergence.

This leaves the triangle inequality: Is it true that $f_{ij} + f_{jk} \geq f_{ik}$? Not necessarily—it may be less costly in going from i to k to tranship through j rather than take the direct link. It is almost obvious, in fact, that if the triangle inequality holds, there is no rationale for transhipment: One does at least as well to ship directly from sources to sinks. (See Section 7.10.)

In summary, the unit transport cost function f need not be a metric. On the other hand, f's that satisfy some or all of the metric postulates constitute interesting special cases. Even if f is a metric, however, it need not have any close resemblance to "real" geographic distance. The irregularities of nature, the construction of roads between some but not all places, irregular tariffs, and institutional barriers all conspire to weaken the relation between geographic distance and transport cost.

The points of the transhipment problem can also be interpreted as points of Time, or Space-Time, rather than points of Space.[22] Then f_{ij} becomes *storage* cost or combined transport-storage cost. (If j precedes i, we set $f_{ij} = \infty$, or to a number high enough to discourage traffic flow into the past.)

The relative advantages of the transhipment and transportation problem formulations may be summarized as follows. Because it allows connections between all pairs of points, the transhipment formulation allows the study of routing patterns and intermediate flows that escape the transportation formulation. However, the very fact that all points are treated symmetrically, rather than being dichotomized into sources and sinks, means that a number of important interpretations of the transportation problem do not carry over to transhipment. In particular, this applies to the assignment of resources or land (as sources) to alternative activities (as sinks). The major application we make of the transportation problem has this interpretation (see Chapter 8); hence the latter is much more important to us than transhipment.

Even on its own ground the transhipment formulation is not necessarily more useful than the transportation formulation. We have mentioned that the former can be "reduced" to the latter in a number of ways. The following considerations, while not necessarily constituting a "reduction," indicate another reason why the transportation formulation is often perfectly satisfactory (f is finite in the following discussion). Dichotomize the points of a transhipment problem into "sources" and "sinks" according to the sign of the net capacity α_i. Suppose that for any pair consisting

22. Cf. P. A. Samuelson, Intertemporal price equilibrium: A prologue to the theory of speculation, *Weltwirtsch. Archiv* 79(1975):181–221; C. H. Kriebel, Warehousing with transhipment under seasonal demand, *J. Reg. Sci.* 3 (1961): 57–69.

of a source i and a sink j there is a *shortest route* $(k_1, \ldots, k_m)$ in the following sense: $k_1 = i$, $k_m = j$, and the sum

$$g_{ij} = f_{k_1 k_2} + f_{k_2 k_3} + \cdots + f_{k_{m-1} k_m} \tag{7.6.3}$$

is a minimum over all possible finite sequences of points, $(k_1, \ldots, k_m)$ satisfying $k_1 = i$, $k_m = j$. Choose one such route for each source-sink pair (i, j). It may then be verified that the transhipment problem (1)–(2) "reduces" to the transportation problem (1.1)–(1.3) of all source-sink pairs, with capacities and requirements given by the absolute values $|\alpha_i|$, and unit transport costs g_{ij} given by (3). "Reduction" here means that if y_{ij}^o is an optimal flow for this transportation problem, an optimal flow for transhipment is obtained by shipping y_{ij}^o along each link of the shortest route from i to j, adding over all source-sink pairs (i, j).

If the shortest routes are easily found or otherwise uninteresting, one might just as well go to the transportation problem derived above, which gives the optimal origin-destination flow pattern and is easier to solve than the original.

To round out the discussion, we mention the problem of finding a shortest route from i to j, one that minimizes (3). As pointed out by Orden, this can be formulated as a special case of the transhipment problem,[23] viz., let $\alpha_i = +1$, $\alpha_j = -1$, $\alpha_k = 0$ for all other points. An optimal solution to this will yield one or more sequences $k_1, \ldots, k_m$, with $k_1 = i$, $k_m = j$ and positive flows between each successive pair. A little thought shows that each such sequence is a shortest route, and the minimal total cost for this problem is precisely g_{ij} of (3).

Does a shortest route always exist for any origin i, destination j? It does iff the following *cyclic positivity* condition is satisfied:

$$f_{k_1 k_2} + \cdots + f_{k_{m-1} k_m} + f_{k_m k_1} \geq 0 \tag{7.6.4}$$

for all finite sequences $(k_1, \ldots, k_m)$, $m = 2, 3, \ldots$. Condition (4) states that the sum of unit costs around a closed circle of links is never negative. For suppose (4) were false for some sequence. By going around the circle sufficiently often, one can drive cost below any negative number; hence there cannot be a minimum sum (3). Conversely, if cyclic positivity holds, there is no cost advantage to routes that include the same point more than once. But there are just a finite number of routes without repeated points from i to j, hence a shortest route exists. (For the generalized transhipment problems discussed below, where the number of points may be infinite, the situation is much more complicated. But (4) remains a *neces-*

---

23. Orden, Transhipment problem, pp. 283–85. There are several algorithms for finding a shortest route, e.g., G. B. Dantzig, On the shortest route through a network, *Manage. Sci.* 6 (Jan. 1960): 187–90.

sary condition for the existence of shortest routes.) Cyclic positivity is implied by, but weaker than, the triangle inequality. Indeed, with the triangle inequality, the pair (i, j) itself is a shortest route, and $g_{ij} = f_{ij}$ in (3).

Finally, the measure-theoretic treatment of the transportation problem seems to be easier than that for transhipment.[24]

In the following sections we study the transhipment problem in measure-theoretic form. We emphasize aspects in which transhipment is distinctive, where new and sometimes paradoxical phenomena appear. We also show how the use of pseudomeasures to formulate *constraints* arises naturally for transhipment. (Until now, pseudomeasures have been used only to represent *preferences,* with one small exception in Section 6.9.)

7.7. TRANSHIPMENT: MEASURE-THEORETIC FORMULATIONS

We give two different measure-theoretic formulations of the transhipment problem. For the first, the raw materials are a measurable space, (A, Σ); a σ-finite *signed* measure μ on this space; and a function $f{:}A \times A \to$ reals, measurable with respect to the product σ-field $\Sigma \times \Sigma$ on $A \times A$.

The problem is to find a *bounded* measure λ on $(A \times A, \Sigma \times \Sigma)$ satisfying

$$\lambda' - \lambda'' \le \mu \qquad\qquad (7.7.1)$$

and minimizing

$$\int f \, d\lambda \qquad\qquad (7.7.2)$$

Here λ', λ'' are the left and right marginals of λ respectively, so that (1) could also be written in the following less abbreviated form:

$$\lambda(E \times A) - \lambda(A \times E) \le \mu(E) \qquad\qquad (7.7.3)$$

all $E \in \Sigma$. Now (2) is an indefinite integral over $A \times A$, and "minimize" is to be understood in the sense of (reverse) standard ordering of pseudomeasures. This is the *inequality*-constrained variant; we obtain the *equality*-constrained variant by substituting $=$ for $\le$ in (1) and (3).

The signed measure μ is to be interpreted as *net capacity,* so that $\mu(E)$ is the amount by which the *gross outflow* from the points in set E

---

24. The most fruitful network problems have not been of the transhipment type (1)–(2), but of the following form: Given a flow capacity on each link of a network, maximize the flow from a given source to a given sink. See L. R. Ford, Jr., and D. R. Fulkerson, *Flows in Networks* (Princeton Univ. Press, Princeton, 1962). A measure-theoretic treatment of these problems can be given, but we do not do so in this book.

may exceed the *gross inflow* to those points. Of course $\mu(E)$ may be negative. λ is the flow measure, so that $\lambda(E \times F)$ gives the total mass that moves (directly) from origins in E to destinations in F. In particular, $\lambda(E \times A)$ gives the gross outflow from origins in E to all destinations (*including* destinations in the set E); similarly, $\lambda(A \times E)$ gives the gross inflow to destinations in E from all origins. Thus (3) is precisely the relation between inflow, outflow, and capacity mentioned above; f is unit costs, and (2) gives the total cost of flow λ.

Care should be taken to distinguish set functions (such as λ) defined on the product space $(A \times A, \Sigma \times \Sigma)$ from those (such as λ', λ'', and μ) defined on (A, Σ).

Problem (1)–(2) reduces to (6.1)–(6.2) precisely in the case when Σ is a finite σ-field, so that we indeed have a generalization of the original transhipment problem. In ordinary transhipment we distinguish "source" points from "sink" points by the sign of the net capacity. In the generalization (1) this role is played by the *Hahn decomposition* of net capacity μ. If (P, N) is a Hahn decomposition, then $\mu(E) \geq 0$ for all measurable $E \subseteq P, \mu(F) \leq 0$ for all measurable $F \subseteq N$, so that the points of P may be thought of as sources, the points of N as sinks.

Boundedness is a feasibility condition for λ. Indeed, if λ were unbounded, (1) would not be well defined, since we would have $\lambda'(A) = \lambda''(A) = \infty$. However, boundedness excludes some interesting theoretical situations. On the endless plane of location theory there will usually be a flow of infinite mass. The same is true with an unbounded time horizon in those cases where A is a subset of Time or Space-Time. In these cases a similar question arises concerning the adequacy of a signed measure to represent the concept of "net capacity." Suppose A is split as above into two pieces, a source space P and a sink space N. Since μ is a signed measure, at least one of the two numbers $\mu(P)$, $\mu(N)$ must be finite. But there are reasonable problems involving both infinite capacity on P and infinite requirements on N.

Our second measure-theoretic formulation enables us to deal with the situations just discussed. As might have been expected, the key lies in the introduction of pseudomeasures. (A, Σ) and f remain as above. The objective is still to minimize (2), but (1) and the boundedness condition are replaced. Instead of the signed measure μ we have a pseudomeasure ψ on (A, Σ) and the constraint is to find a measure λ on $(A \times A, \Sigma \times \Sigma)$ whose marginals λ' and λ'' are σ-finite and satisfy

$$(\lambda', \lambda'') \leq \psi \qquad (7.7.4)$$

That is, we form the pseudomeasure (λ', λ'') from the marginals of λ and constrain it to be less than or equal to ψ under *narrow* ordering. Letting (ψ^+, ψ^-) be the Jordan form of ψ, (4) may be written in less abbreviated form as

$$\lambda(E \times A) + \psi^-(E) \leq \lambda(A \times E) + \psi^+(E) \qquad (7.7.5)$$

for all $E \in \Sigma$. This is the *unbounded* formulation of the transhipment problem, (1) giving the *bounded* formulation.

When λ is bounded and ψ is a signed measure μ, one can see that (5) is the same as (3). Thus the constraint (1) is a special case of (4). But it cannot be said that the bounded formulation is merely a special case of the unbounded one, since the boundedness condition is present in one and absent in the other. As above there is also an equality-constrained variant: just substitute = for $\leq$ in (4) and (5).

The pseudomeasure (λ', λ'') may be thought of as net outflow or net exports, and the fact that it is a pseudomeasure allows the possibility that *gross* inflow and outflow for a region may both be infinite. Again ψ is net capacity. If (P, N) is a Hahn decomposition for ψ, we may think of it in the following way: ψ^+ gives the net outflow capacity on source space P, while ψ^- gives the net inflow requirement on sink space N.

The conditions that λ' and λ'' are σ-finite are needed to make (λ', λ'') well defined as a pseudomeasure. Recall that either of these implies that λ itself is σ-finite, so that (2) remains well defined as a pseudomeasure. (Preference among different λ's is expressed via (2) by *standard* ordering, while the constraint (4) involves *narrow* ordering. This disparity is essential to achieve a bona fide generalization of the ordinary transhipment problem.)

We have now set up the two measure-theoretic transhipment problems and will investigate feasibility and duality conditions for them. But first we derive some results concerning measures λ on a product space of the form $(A \times A, \Sigma \times \Sigma)$. The aim is to achieve a certain insight into the structural differences between the transportation and transhipment constraints. The following remarks are abstracted from any particular problem context, however. They also apply to *arbitrary* measures λ, not merely to σ-finite measures.

Definition: Measure λ on $(A \times A, \Sigma \times \Sigma)$ is a *translocation*[25] iff there is a measurable partition $\{P, N\}$ of A into two pieces such that $\lambda[(A \times A)\backslash(P \times N)] = 0$. (That is, $\lambda(P \times P) = \lambda(N \times N) = \lambda(N \times P) = 0$. The only possible flow is from P to N.)

Theorem: λ is a translocation iff its marginals λ' and λ'' are mutually singular.

---

25. The terminology (but not the meaning) is from Kantorovitch.

Proof: Let λ be a translocation with P, N as in the definition above. Then $\lambda'(N) = \lambda''(P) = 0$, so λ', λ'' are mutually singular. Conversely, let λ', λ'' be mutually singular, so that $\lambda'(N) = \lambda''(P) = 0$ for some partition $\{P, N\}$ of A. But this yields $\lambda(P \times P) = \lambda(N \times N) = \lambda(N \times P) = 0$, so λ is a translocation. ∎

We want a formula giving the *transhipment* associated with any flow measure λ. Intuitively, the transhipment in a region is given by the "overlap" between inflow and outflow. To be precise, let transhipment be represented by a measure θ on (A, Σ). We require that $\theta \le \lambda'$ and $\theta \le \lambda''$; i.e., transhipment in any region does not exceed gross outflow from that region and does not exceed gross inflow into that region respectively. The "overlap" is the largest measure meeting these conditions. But this is precisely the infimum of λ' and λ'', as defined in Section 3.1. Thus we have the following definition.

Definition: Given measure λ on $(A \times A, \Sigma \times \Sigma)$, the *transhipment* is the measure θ on (A, Σ) given by

$$\theta(E) = \inf(\lambda', \lambda'')(E) = \inf\{\lambda'(F) + \lambda''(E \backslash F) \mid F \subseteq E, F \in \Sigma\}$$
(7.7.6)

for all $E \in \Sigma$. Here λ', λ'' are the left and right marginals of λ respectively. We have repeated the explicit formula for $\inf(\lambda', \lambda'')$ for convenience.

This definition seems to capture quite well the intuitive notion of "transhipment." In particular, consider the case when λ is a translocation. Here the marginals are mutually singular; there is no overlap, and transhipment should be zero. Furthermore, the converse should be true. If transhipment is zero, inflow and outflow should be mutually singular, so that λ is a translocation. The following result confirms this expectation. Note that we are actually proving an abstract theorem: $\inf(\mu, \nu) = 0$ *iff* (μ, ν) *is a mutually singular pair.*

Theorem: λ has a zero transhipment iff λ is a translocation.

Proof: Let λ be a translocation, so that $\lambda'(N) = \lambda''(P) = 0$ for some partition $\{P, N\}$ of A, by the theorem above. Then $\theta(A) \le \lambda'(N) + \lambda''(P) = 0$, from (6) with $F = N$. Hence $\theta = 0$.

Conversely, let $\theta(A) = 0$. Then for each $n = 1, 2, \ldots$ a set $F_n \in \Sigma$ exists such that $\lambda'(F_n) + \lambda''(A \backslash F_n) \le 2^{-n}$, from (6). Let $F = \limsup F_n$. For each $n = 1, 2, \ldots$ we have

$$\lambda'(F) \le \lambda'(F_n \cup F_{n+1} \cup \cdots) \le 2^{-n} + 2^{-(n+1)} + \cdots = 2 \cdot 2^{-n}$$

Hence $\lambda'(F) = 0$. Also $A \setminus F = \lim \inf(A \setminus F_n)$, so that $\lambda''(A \setminus F)$ does not exceed the sum of

$$\lambda''[(A \setminus F_n) \cap (A \setminus F_{n+1}) \ldots] \tag{7.7.7}$$

over $n = 1, 2, \ldots$. But each term (7) equals zero, since it does not exceed $\lambda''(A \setminus F_k)$ for arbitrarily high k. Hence $\lambda''(A \setminus F) = 0$. This with $\lambda'(F) = 0$ shows that λ', λ'' are mutually singular, hence λ is a translocation, by the preceding theorem. ■

Measures λ', λ'' have been interpreted as the gross outflow and inflow respectively, associated with λ. There are a number of intuitive net flow concepts—particularly, one that nets out the "overlap" of λ' and λ'' from each of them, i.e., subtracts the transhipment. However, these measures may all be infinite, so subtraction is not a well-defined operation. But recall that in Section 3.1 we did define a subtraction operation that is valid for infinite measures. The concepts in that section, in fact, turn out to be admirably well suited to explicate the intuitive notions we are struggling with here.

Definition: Let λ be a measure on $(A \times A, \Sigma \times \Sigma)$. The *net outflow* and *net inflow* of λ are the respective measures λ_1 and λ_2 on (A, Σ) given by $(\lambda_1, \lambda_2) = J(\lambda', \lambda'')$. That is, λ_1 and λ_2 are the *upper* and *lower variations* respectively of the *Jordan decomposition* of the pair (λ', λ'').

That this is a reasonable definition follows from the basic relation between pairs of measures, their Jordan decompositions, and their infima, which in this case is

$$\lambda_1 = \lambda' - \inf(\lambda', \lambda'') = \lambda' - \theta$$

$$\lambda_2 = \lambda'' - \inf(\lambda', \lambda'') = \lambda'' - \theta$$

(see p. 130.) That is, the net outflows and inflows are indeed obtained by subtracting transhipment from gross outflows and inflows respectively. (If θ is finite, this reduces to ordinary subtraction.)

Another intuitively appealing property one would wish the net flow measures λ_1 and λ_2 to possess is that they be *mutually singular*. For in this case one can split A into two pieces P and N, which can be unambiguously labeled as the outflow and inflow sets respectively. (Here $\lambda_1(N) = \lambda_2(P) = 0$.) An obvious sufficient condition for this is that λ be a translocation; for then even the gross flows λ' and λ'' are mutually singular, hence *a fortiori* the net flows λ_1 and λ_2. In this case $\lambda_1 = \lambda'$ and $\lambda_2 = \lambda''$, since $\theta = 0$.) The following result shows that even if λ is not a translocation, mutual singularity is guaranteed under quite general conditions.

Theorem: Let measure λ on $(A \times A, \Sigma \times \Sigma)$ be *abcont*. Then its net flows λ_1 and λ_2 are mutually singular.

Proof: Marginal λ' is induced from λ by the projection $(a_1, a_2) \rightarrow a_1$. Since λ is abcont, so is λ'. Hence the pair (λ', λ'') is Hahn decomposable, implying that its Jordan decomposition (λ_1, λ_2) is a mutually singular pair. (See Section 3.1.) ∎

Finally, let us tie these concepts to the transportation-transhipment problem dichotomy. We show that the transportation problem (variant I) is essentially the transhipment problem (equality-constrained unbounded formulation) with an extra constraint thrown in, that λ be a translocation.

Start with the transhipment problem whose feasible set is determined by the pseudomeasure space (A, Σ, ψ). Measure λ is feasible iff the marginals λ', λ'' are σ-finite and

$$(\lambda', \lambda'') = \psi \qquad (7.7.8)$$

(8) is an equality between pseudomeasures (cf. (4)). Now add the additional constraint that λ must be a translocation. It follows that λ', λ'' are mutually singular, so that (λ', λ'') is the Jordan form of $\psi: \lambda' = \psi^+$ and $\lambda'' = \psi^-$. Let $\{P, N\}$ be a partition of A such that $\psi^+(N) = \psi^-(P) = 0$. Then λ is zero when restricted to $(A \times A)\backslash(P \times N)$. Let λ_o be λ restricted to $P \times N$, let μ' be ψ^+ restricted to P, and let μ'' be ψ^- restricted to N; also let Σ', Σ'' be Σ restricted to P, N respectively. Then we can see that λ_o is feasible for the (variant I) *transportation* problem, with source and sink spaces (P, Σ', μ'), (N, Σ'', μ'') respectively.

Conversely, given this transportation problem with feasible flow λ_o, this entire procedure may be reversed to yield a translocation λ satisfying (8). Furthermore, if $f: A \times A \rightarrow$ reals determines the transhipment objective function and f_o is f restricted to $P \times N$, then $\int f_o \, d\lambda_o$ yields the same ordering among feasible transport flows λ_o as $\int f \, d\lambda$ does among the corresponding translocations λ. This shows the essential equivalence between these two problems.

7.8. TRANSHIPMENT: FEASIBILITY

We now investigate the conditions under which feasible solutions exist for the bounded and unbounded formulations of the transhipment problem. The bounded case is well behaved, and the results are analogous to those obtained for the transportation problem. But the results for the unbounded case are "wild."

First we discuss the bounded formulation (7.1). Actually, we prove results for a somewhat more general problem: Let net capacity be a pseudo-measure ψ and not merely a σ-finite signed net capacity measure μ. This yields a problem somewhere in between the bounded and unbounded formulations:

Find a *bounded* measure λ on $(A \times A, \Sigma \times \Sigma)$ satisfying

$$(\lambda', \lambda'') \leq \psi \qquad (7.8.1)$$

Here the marginals λ', λ'', as well as the pseudomeasure ψ, are defined on the space (A, Σ) as usual. The left side of (1) could also be written as $\lambda'-\lambda''$, as in (7.1), but we prefer the pseudomeasure notation.

The reader may wonder why we used constraint (7.1) instead of the more general (1). The answer is contained in the following theorem. If ψ is a *proper* pseudomeasure (i.e., if ψ^+ and ψ^- are both infinite measures) there is *no* bounded λ satisfying (1). Hence pseudomeasures are actually useless here, and we might just as well use the signed measure formulation of (7.1), which is after all much closer to intuition than (1) is. But we need to formulate the problem (1) to prove this very fact.

Theorem: Given pseudomeasure space (A, Σ, ψ), a bounded measure λ on $(A \times A, \Sigma \times \Sigma)$ exists satisfying (1) iff

$$\psi^+(A) \geq \psi^-(A) < \infty \qquad (7.8.2)$$

Also, a bounded λ exists satisfying (1) *with equality* iff

$$\psi^+(A) = \psi^-(A) < \infty \qquad (7.8.3)$$

Proof: Let bounded λ satisfy (1), which may also be written

$$\lambda' + \psi^- \leq \lambda'' + \psi^+ \qquad (7.8.4)$$

We have $\lambda'(A) = \lambda(A \times A) = \lambda''(A) < \infty$. Hence, substituting A into (4), the λ terms drop out and we have $\psi^-(A) \leq \psi^+(A)$. Next, let (P, N) be a Hahn decomposition for ψ; then $\psi^-(P) = 0$, and $\psi^-(N) \leq \lambda''(N)$ results from substituting N into (4). Hence $\psi^-(A)$ is finite. This yields (2). If bounded λ satisfies (1), hence (4), *with equality,* the same argument yields (3).

Conversely, let (2) obtain, and consider the *transportation* problem with source space (A, Σ, ψ^+), sink space (A, Σ, ψ^-), and constraints

$$\lambda' \leq \psi^+, \qquad \lambda'' = \psi^- \qquad (7.8.5)$$

This is variant II, hence a feasible solution λ exists, by (2). Constraints (5) imply (4), which is (1). Also $\lambda''(A) = \psi^-(A)$, hence λ is bounded, again by (2). Thus λ is feasible for the bounded transhipment problem.

Finally, let (3) obtain, and consider the same transportation problem except that (5) has all equalities. This is variant I, hence a solution exists by (3). Now (5) yields (1) with equality. Again λ is bounded, since $\psi^-(A)$ is finite. ∎

When ψ is the signed measure μ as in (7.1), this theorem takes on a very simple form.

Theorem: A bounded measure exists satisfying (7.1) iff $\mu(A) \geq 0$. A bounded measure exists satisfying (7.1) *with equality* iff $\mu(A) = 0$.

Proof: Proof is immediate from (2)–(3), on noting that $\mu(A) = \mu^+(A) - \mu^-(A)$. ∎

Thus a solution exists iff total net capacity is nonnegative or zero in the inequality- or equality-constrained problems respectively. This bears comparison with the transportation problem result. Here a feasible solution exists iff total requirement does not exceed total capacity (in variants II, III, IV, which involve inequality constraints) or iff total requirement equals total capacity (in the all-equality-constraint variant I).

This brings us to the unbounded formulation of transhipment (7.4). The basic result is that a feasible solution *always* exists (unless Σ is finite). This is highly paradoxical, since a solution exists even when $\psi^+(A)$ is less than $\psi^-(A)$ (even when the former is zero and the latter infinite, in fact). We first prove the result and then give a rough explanation of "why" it is true. In the following we prove feasibility for the *equality*-constrained problem. The solution constructed automatically remains feasible for the weaker inequality constraint, so that feasibility holds in general.

Theorem: Let (A, Σ, ψ) be a pseudomeasure space, with Σ an infinite σ-field. Then a measure λ on $(A \times A, \Sigma \times \Sigma)$ exists such that the marginals λ', λ'' are σ-finite and

$$(\lambda', \lambda'') = \psi \qquad (7.8.6)$$

(equality in the sense of pseudomeasures).

Proof: Choose a representative (μ, ν) of ψ. Since Σ is infinite and μ, ν are σ-finite, an infinite countable measurable partition $\{A_1, A_2, \ldots\}$ of A into nonempty sets exists such that $\mu(A_n)$ and $\nu(A_n)$ are finite for all $n = 1, 2, \ldots$.

Choose a point $a_n \in A_n$ for each n, and define the set function λ, with domain $\Sigma \times \Sigma$, as follows. For each $G \in \Sigma \times \Sigma$ and each $n = 1, 2, \ldots$, form the quantity

$\mu\{a \mid a \in A_n \text{ and } (a, a_n) \in G\} + \nu\{a \mid a \in A_n \text{ and } (a_n, a) \in G\}$

$$+ \mu_n \cdot x_n(G) + \nu_n \cdot y_n(G) \quad (7.8.7)$$

Here μ_n and ν_n are abbreviations for $\mu(A_n)$, $\nu(A_n)$ respectively; $x_n(G)$ is the number of integers $k \geq n$ for which $(a_k, a_{k+1}) \in G$; $y_n(G)$ is the number of integers $k \geq n$ for which $(a_{k+1}, a_k) \in G$. (If the number of such integers is infinite, take $x_n(G)$ or $y_n(G)$ to be $+\infty$ and form (7) by the rules of extended real-valued arithmetic.) Finally, $\lambda(G)$ is defined as the sum of the quantities (7) over all $n = 1, 2, \ldots$.

We claim that λ constructed in this way is the desired measure. First of all, for fixed n each of the terms in (7) is routinely verified to be a measure on $\Sigma \times \Sigma$; hence λ, as a sum of measures, is itself a measure.

Now consider various sets $E \in \Sigma$ in relation to the points $a_1, a_2, \ldots$. First, if none of these points belongs to E, we calculate from (7) that

$$\lambda(E \times A) = \mu(E), \quad \lambda(A \times E) = \nu(E) \quad (7.8.8)$$

Second, if none of these points belongs to E, with the single exception of a_m, a more complicated calculation from (7) yields

$$\lambda(E \times A) = \mu(E) + \mu_1 + \cdots + \mu_m + \nu_1 + \cdots + \nu_m \quad (7.8.9)$$

$$\lambda(A \times E) = \nu(E) + \mu_1 + \cdots + \mu_m + \nu_1 + \cdots + \nu_m \quad (7.8.10)$$

In particular, $E = A_m$ has the property just mentioned. Hence $\lambda'(A_m) = \lambda(A_m \times A)$ is finite for all $m = 1, 2, \ldots$, from (9). Similarly, $\lambda''(A_m)$ is finite for all m, from (10). Hence the marginals λ', λ'' are σ-finite.

Furthermore, from (8)–(10) we find that

$$\lambda'(E) + \nu(E) = \lambda''(E) + \mu(E) \quad (7.8.11)$$

for any set $E \in \Sigma$ to which *at most one* of the points $a_1, a_2, \ldots$ belongs. However, *any* set $E \in \Sigma$ can be countably partitioned into sets of this type: $E = (E \cap A_1) \cup (E \cap A_2) \cup \cdots$. Hence, by summation (11) is true for all $E \in \Sigma$. But this implies (6), by the equivalence theorem for pseudomeasures. Hence λ is feasible. ∎

Measure λ "works" in the foregoing proof for the following reasons. The point a_n functions as a "depot" or "entrepôt" between A_n and the rest of A; a_1 absorbs any surplus or deficit arising in A_1; a_2 does the same for A_2, and also absorbs the net surplus or deficit at a_1; a_3 does the same for A_3, and also absorbs the *cumulative* net surplus or deficit at a_2, etc. In this way, each successive set A_n is brought into balance, while the overall surplus or deficit "escapes to, or from, infinity." The paradox arises precisely because there is no point at which the buck stops and accounts must be settled. Similar phenomena arise in other contexts, e.g., in the theory of Markov chains with an infinite number of states, or in the theory of economic growth with intergenerational transfers and an infinite succession of generations.

One might be tempted to regard this paradox as a *reductio ad absurdum* of the unbounded formulation of the transhipment problem; but this would be an error, or at least a premature judgment. The formulation itself arises in a natural way. And even though a paradoxical flow pattern is *feasible,* it involves a great deal of cross-hauling. We may presume, then, that no such flow would be *optimal,* unless the problem is formulated in a way that allows no avoidance of such flows (by making requirements exceed capacities). It is quite common for useful models to introduce artifacts of this sort. Finally, many former "paradoxes" are now accepted as valid, so one should be wary of making summary judgments about what cannot occur in the real world.

The premise that Σ is an infinite σ-field is essential in this theorem. Indeed, if Σ is finite, we are in effect back in the ordinary transhipment problem, and the "tame" feasibility results of the bounded formulation apply.

7.9. TRANSHIPMENT: DUALITY

We give a brief introduction to the duality theory of the transhipment problem, one that parallels the treatment of the transportation problem. Return to the ordinary transhipment problem (6.1)–(6.2); its linear programming dual is the following. Find nonnegative numbers $p_1, \ldots, p_n$ satisfying $p_j - p_i \leq f_{ij}, i, j = 1, \ldots, n$, and maximizing

$$-\alpha_1 p_1 - \cdots - \alpha_n p_n \qquad (7.9.1)$$

(Minus signs appear in the objective function (1) because we expressed the primal in terms of net *capacities* α_i, $i = 1, \ldots, n$. If we had used net *requirements* instead, we would get plus signs.)

The *dual* of the corresponding measure-theoretic problem (unbounded formulation) (7.4) and (7.2) is defined as follows. Find a measurable nonnegative function $p:A \rightarrow$ reals satisfying

$$p(a'') - p(a') \leq f(a', a'') \qquad (7.9.2)$$

for all $a', a'' \in A$, and maximizing

$$-\int p \, d\psi \qquad (7.9.3)$$

Here (3) is an indefinite integral over space (A, Σ), and "maximization" is to be understood in the sense of standard ordering.

This is for the *inequality*-constrained version. For the *equality*-constrained version the dual is the same, except that p need not be nonnegative. Finally, for the *bounded* formulation, everything is as above except for notation: The signed measure μ replaces the more general pseudomeasure ψ in (3).

The dual for the transportation problem introduced some other constraints making certain definite integrals well defined and finite. Conditions of this sort play a role here too, but it is convenient to introduce them separately.

The following theorem yields the basic duality inequality. It applies to both equality- and inequality-constrained versions and to both bounded and unbounded formulations. The notation for the latter will be used (for the former, replace ψ by μ). The expression $\int_A p \, d\psi$ means the following. It is defined iff the two definite integrals $\int_A p \, d\psi^+$ and $\int_A p \, d\psi^-$ are both well defined and not infinite of the same sign. In this case we set

$$\int_A p \, d\psi = \int_A p \, d\psi^+ - \int_A p \, d\psi^- \tag{7.9.4}$$

Equivalently, the expression is defined iff $\int p \, d\psi$ is a signed measure, and in this case $\int_A p \, d\psi$ is its value at A.

Theorem: Let measure λ on space $(A \times A, \Sigma \times \Sigma)$ be feasible for the transhipment problem, and let $p:A \to$ reals be feasible for the corresponding dual problem. Also assume that

$$\int_A p \, d\lambda', \qquad \int_A p \, d\lambda'' \tag{7.9.5}$$

are both well defined and finite definite integrals. Then the following two definite integrals are well defined, and the stated inequality holds between them.

$$\int_{A \times A} f \, d\lambda \geq - \int_A p \, d\psi \tag{7.9.6}$$

Proof: Let the functions $p', p'':A \times A \to$ reals be given by

$$p'(a', a'') = p(a'), \qquad p''(a', a'') = p(a'')$$

for all $a', a'' \in A$. Condition (2) then takes the form $p'' - p' \leq f$, and we obtain

$$\int_A p \, d\lambda'' - \int_A p \, d\lambda' = \int_{A \times A} p'' \, d\lambda - \int_{A \times A} p' \, d\lambda$$

$$= \int_{A \times A} (p'' - p') \, d\lambda \leq \int_{A \times A} f \, d\lambda \tag{7.9.7}$$

The first equality in (7) arises from the induced integrals theorem. Note that

$$\int_{A \times A} (p'' - p')^- \, d\lambda \geq \int_{A \times A} f^- \, d\lambda \qquad (7.9.8)$$

from (2). The left side of (8) is finite, from (5), hence so is the right, hence the last integral in (7) is well defined. The inequality in (7) then follows from (2).

Next we prove that

$$\int_A p \, d\lambda'' - \int_A p \, d\lambda' \geq \int_A p \, d\psi^- - \int_A p \, d\psi^+ \qquad (7.9.9)$$

the differences being well defined. There are two cases.

For the *inequality*-constrained variant, we have $p \geq 0$ and $(\lambda', \lambda'') \leq \psi$; i.e.,

$$\lambda' + \psi^- \leq \lambda'' + \psi^+ \qquad (7.9.10)$$

so that

$$\int_A p \, d\lambda' + \int_A p \, d\psi^- \leq \int_A p \, d\lambda'' + \int_A p \, d\psi^+ \qquad (7.9.11)$$

Letting (P, N) be a Hahn decomposition for ψ, we have $\psi^-(P) = 0$, while $\psi^-(E) \leq \lambda'(E) + \psi^-(E) \leq \lambda''(E) + \psi^+(E) = \lambda''(E)$ for any measurable $E \subseteq N$. Hence $\psi^- \leq \lambda''$. It follows from (5) that at least three of the integrals in (11) are finite. Hence it is permissible to rearrange terms to obtain (9).

For the *equality*-constrained variant, (10) holds with equality. Also, by the minimizing property of the Jordan form, we have $\psi^+ \leq \lambda'$ as well as $\psi^- \leq \lambda''$. It then follows from (5) that all the integrals appearing in (11) are finite. Hence (11) holds (with equality) and may be rearranged to yield (9) (with equality). Together, (7), (9), and (4) yield (6). ∎

Next we look for a condition under which the inequality (6) of this theorem becomes an equality.

Definition: Let measure λ on $(A \times A, \Sigma \times \Sigma)$ and the function $p : A \to$ reals be feasible for the transhipment problem and its dual respectively. Function p is a *(transhipment) measure potential* for λ iff the following two conditions are satisfied.

$$\lambda\{(a_1, a_2) \mid p(a_2) - p(a_1) < f(a_1, a_2)\} = 0 \qquad (7.9.12)$$

and, when restricted to the subset $\{a \mid p(a) > 0\}$ of A, the two pseudomeasures (λ', λ'') and ψ coincide.

This definition is meant to apply to both bounded and unbounded formulations and to both equality- and inequality-constrained variants of the transhipment problem. Note, however, that for the *equality*-constrained variant the second condition is satisfied trivially and may be dropped; measure-potentiality reduces to (12) alone. Condition (12) states that there is no flow on the set of origin-destination pairs for which the dual inequality (2) is strict. This and the other measure-potentiality condition generalize the complementary slackness conditions for transhipment.

Theorem: Let measure λ^o on $(A \times A, \Sigma \times \Sigma)$ be feasible for the transhipment problem, and let $p^o{:}A \to$ reals be feasible for its dual. Also let

$$\int_A p^o \, d\lambda^{o\prime}, \qquad \int_A p^o \, d\lambda^{o\prime\prime}$$

both be well defined and finite. Then p^o is a (transhipment) measure potential for λ^o iff

$$\int_{A \times A} f \, d\lambda^o = -\int_A p^o \, d\psi \tag{7.9.13}$$

Proof: Let p^o be a measure potential for λ^o. Reviewing the preceding proof, we find that the weak inequality in (7) is satisfied with equality because of (12). For the *equality*-constrained variant this already yields (13), since (9) is also satisfied with equality. For the *inequality*-constrained variant the fact that $(\lambda', \lambda'') = \psi$ on the set $\{a \mid p(a) > 0\}$ yields (10) with equality on this set. Hence (11) and (9) are satisfied with equality, since $p = 0$ off this set; this again yields (13).

Conversely, assume (13). All integrals in the preceding proof are then finite, and the weak inequalities in (7) and (9) are satisfied with equality. But equality in (7) implies (12), while equality in (9) implies that $(\lambda', \lambda'') = \psi$ when these are restricted to $\{a \mid p(a) > 0\}$ (a trivial implication in the equality-constrained variant). ∎

Theorem: Let λ^o be a *bounded* measure on space $(A \times A, \Sigma \times \Sigma)$, and let $p^o{:} A \to$ reals be a *bounded* function such that p^o is a (transhipment) measure potential for λ^o. Then λ^o is best for the *bounded* formulation of the transhipment problem.

Proof: Let λ be any other feasible solution for the transhipment problem (bounded formulation). We show that

$$\int_{A \times A} f \, d\lambda^o = -\int_A p^o \, d\mu \le \int_{A \times A} f \, d\lambda \tag{7.9.14}$$

all these definite integrals being well defined. (Here f is, as usual, the unit cost function, and μ is the net capacity signed measure.) First,

$$\int_A p^o \, d\lambda', \qquad \int_A p^o \, d\lambda'' \qquad (7.9.15)$$

are both well defined and finite, since p^o and λ are both bounded. This yields the inequality in (14), by (6). If λ is replaced by λ^o, (15) remains finite; and this, with the measure-potentiality premise, yields the equality in (14), by (13). Furthermore, this common value is finite. Then (14) implies that λ^o is best under (reverse) standard ordering of pseudomeasures. ∎

These results apply to both the equality- and inequality-constrained variants. Note that the last theorem applies only to the bounded formulation of the transhipment problem, however. In the unbounded formulation, the integrals (15) will not necessarily be well defined and finite for all feasible λ, which means that the inequality of (14) cannot be derived.

In connection with transhipment potentials one should mention the work of Beckmann.[26] This deals with commodity flow on the plane and makes essential use of vector analysis (gradients, curls, etc.). Here "flow" refers to "continuous" physical movement (as a fluid) and is not immediately reducible to the origin-destination form of the transhipment problem. Yet Beckmann arrives at a potential function similar to the transhipment potential. One hopes that future work will produce some kind of synthesis of these approaches.

7.10. TRANSHIPMENT UNDER THE TRIANGLE INEQUALITY

We would like to obtain results for transhipment analogous to those for the transportation problem (such as the existence of optimal solutions and of potentials associated with unsurpassed solutions). These results, however, seem quite hard to come by without making special assumptions.

In this section we assume that the unit cost function f obeys the *triangle inequality:* $f(a_1, a_2) + f(a_2, a_3) \geq f(a_1, a_3)$ for all a_1, a_2, $a_3 \in A$. Intuitively this states that there is no advantage to indirect shipment. To move from a_1 to a_3, one does not gain by going through a_2, and the same is true for circuitous routes involving several intermediate points. Now if one compares the transhipment and transportation formulations, one sees that transhipment differs essentially in that it allows such circuitous ship-

26. M. Beckmann, A continuous model of transportation, *Econometrica* 20(Oct. 1952):643–60; The partial equilibrium of a continuous space market, *Weltwirtsch. Archiv* 71(1953):73–87.

ments, while the transportation formulation forbids them. Since, under the triangle inequality, this extra freedom seems to do no good, one would expect optimal solutions to the transportation problem to be optimal for transhipment as well.

This expectation turns out to be correct, at least under certain limited circumstances. The key to the following proofs is the consideration of the dual function, the potential. We do not know if there is a more direct way of proving them.

We start with a relatively simple case. Given a measure space (A, Σ, ν), ν bounded, and a point $a_o \in A$, consider the problem of distributing a mass $\nu(A)$ concentrated at the point a_o over space A according to distribution ν. As a transportation problem on the product space $A \times A$ the problem is trivial; in fact, there is exactly one feasible solution,[27] given by (1) below, and it must be optimal. As a transhipment problem there are many feasible solutions, but under the triangle inequality one feels intuitively that (1) should still be optimal. And so it is.

Theorem: Given bounded measure space (A, Σ, ν) and point $a_o \in A$; let $f : A \times A \to$ reals be bounded, measurable, and obey the triangle inequality; and let $f(a_o, a_o) = 0$. Let measure λ^o on $(A \times A, \Sigma \times \Sigma)$ be given by

$$\lambda^o(G) = \nu\{a \mid (a_o, a) \in G\} \qquad (7.10.1)$$

all $G \in \Sigma \times \Sigma$. Then λ^o is best for the transhipment problem of minimizing

$$\int_{A \times A} f \, d\lambda \qquad (7.10.2)$$

over bounded measures λ satisfying

$$\lambda' - \lambda'' = \nu_{a_o} - \nu \qquad (7.10.3)$$

(Here ν_{a_o} is the measure of mass $\nu(A)$ simply concentrated on the point a_o.)

Proof: One easily verifies that λ^o is feasible for (3).

Define the function $p : A \to$ reals by $p(a) = f(a_o, a)$. We show that p is bounded and a measure potential for λ^o. From the preceding duality theory this implies that λ^o is best.

Bounded measurability of p follows from the corresponding properties of f. The triangle inequality implies that

---

27. Exercise: Prove this uniqueness assertion without making the assumption that $\{a_o\} \in \Sigma$.

$$p(a_2) - p(a_1) \leq f(a_1, a_2) \tag{7.10.4}$$

for any $a_1, a_2 \in A$. Finally,

$$\lambda^o\{(a_1, a_2) \mid p(a_2) - p(a_1) < f(a_1, a_2)\} = \nu\{a \mid p(a) - p(a_o) < f(a_o, a)\} \tag{7.10.5}$$

by (1). But $p(a_o) = 0$, and it is then clear that the set on the right side of (5) is empty: The common value in (5) is zero. This proves that p is indeed a bounded measure potential for λ^o. ∎

To make further progress we must introduce topology.

Definition: Let measure λ on $(A \times A, \Sigma \times \Sigma)$ and the function $p:A \to$ reals be feasible for the (equality-constrained) transhipment problem and its dual respectively. Let $\mathfrak{I}$ be a topology on A.

Function p is a *(transhipment) topological potential* for λ iff the following condition is satisfied. If (a_1, a_2) is a point of support for λ, then $p(a_2) - p(a_1) = f(a_1, a_2)$. ("Point of support" refers to the topology $\mathfrak{I} \times \mathfrak{I}$ and σ-field $\Sigma \times \Sigma$ on $A \times A$.) That is, (4) is satisfied for all (a_1, a_2) and is satisfied with equality for points of support.

This definition is appropriate for the equality-constrained variant of the transhipment problem. (For the inequality-constrained variant an extra condition is needed; viz., if a_o supports $[\psi - (\lambda', \lambda'')]$, then $p(a_o) = 0$. We do not discuss this, since it is not needed in what follows.)

Theorem: Let p be a (transhipment) *topological* potential for λ for the equality-constrained transhipment problem, and let $\mathfrak{I} \times \mathfrak{I}$ have the strong Lindelöf property. Then p is a (transhipment) *measure* potential for λ.

The proof of this theorem is virtually identical with that of the corresponding theorem in the transportation problem (p. 358) and will not be repeated here.

We are now ready for the next result, which generalizes a preceding theorem at the cost of attaching some topological strings.

Theorem: Let μ, ν be bounded measures on (A, Σ). Let $\mathfrak{I}$ be a topology on A such that $\mathfrak{I} \subseteq \Sigma$ and $\mathfrak{I} \times \mathfrak{I}$ has the strong Lindelöf property. Let $f:A \times A \to$ reals be bounded, continuous, measurable,[28] and obey the triangle inequality; and let $f(a, a) = 0$, all $a \in A$. Let measure λ^o on

---

28. Actually, the properties of $\mathfrak{I}$ imply that $(\mathfrak{I} \times \mathfrak{I}) \subseteq (\Sigma \times \Sigma)$, hence the measurability of f follows from its continuity. The same is true in the following theorem.

($A \times A, \Sigma \times \Sigma$) be best for the *transportation* problem of minimizing (2) subject to $\lambda' = \mu$, $\lambda'' = \nu$.

Then λ^o is best for the *transhipment* problem of minimizing (2) over bounded measures λ satisfying $\lambda' - \lambda'' = \mu - \nu$.

Proof: This is trivial for $A = \phi$, so we may assume A is not empty. First, it is clear that λ^o is feasible for the transhipment problem.

Since λ^o is transport optimal, $\Im \subseteq \Sigma$ and f is bounded continuous, it follows that λ^o has a bounded (transportation) topological potential: a pair of functions p^o, q^o:$A \to$ reals that are bounded measurable and for which

$$q^o(b) - p^o(a) \leq f(a, b) \tag{7.10.6}$$

all $a, b \in A$, with equality in (6) if (a, b) supports λ^o. Now define the function p:$A \to$ reals by

$$p(a) = \inf\{p^o(x) + f(x, a)\} \tag{7.10.7}$$

the infimum being taken over all $x \in A$. (Note the distinction between p and p^o.) We show that p is a bounded *transhipment* topological potential for λ^o.

Boundedness of p follows from boundedness of p^o and f (remember that $A \neq \phi$). For fixed x, the right side of (7) is a continuous function of $a \in A$, since f is continuous. Then p, as the infimum of a collection of continuous functions, is upper semicontinuous. Since $\Im \subseteq \Sigma$, it follows that p is measurable.

Next, we prove that p is dual feasible, i.e.,

$$p(a_2) - p(a_1) \leq f(a_1, a_2) \tag{7.10.8}$$

all $a_1, a_2 \in A$. For any $x \in A$ we have

$$p(a_2) \leq p^o(x) + f(x, a_2) \tag{7.10.9}$$

by (7). Also, by the triangle inequality,

$$f(x, a_2) \leq f(x, a_1) + f(a_1, a_2) \tag{7.10.10}$$

Adding (9) and (10) and simplifying, we obtain $p(a_2) - f(a_1, a_2) \leq p^o(x) + f(x, a_1)$. Taking the infimum over $x \in A$ on the right side, we obtain (8).

Finally, let (a_1, a_2) be a point of support for λ^o; we show that (8) is satisfied with equality. First, we show that for any $a \in A$, we have

$$p^o(a) \geq p(a) \geq q^o(a) \tag{7.10.11}$$

The left inequality in (11) follows from $p^o(a) = p^o(a) + f(a, a) \geq p(a)$. The right inequality in (11) follows from taking the infimum over $x \in A$ in $p^o(x) + f(x, a) \geq q^o(a)$, which in turn derives from (6).

From (11) we obtain

$$p(a_2) - p(a_1) \geq q^o(a_2) - p^o(a_1) = f(a_1, a_2) \qquad (7.10.12)$$

The equality in (12) arises from the fact that (a_1, a_2) supports λ^o. Relation (12) shows that (8) must be satisfied with equality.

This completes the proof that p is a transhipment topological potential for λ^o. Since $\mathfrak{I} \times \mathfrak{I}$ has the strong Lindelöf property, p is also a transhipment measure potential for λ^o. Since p is also bounded, λ^o is best for the transhipment problem. ∎

We conclude by using this result to establish a theorem on the existence of optimal solutions to the transhipment problem.

Theorem: Let μ be a signed measure on (A, Σ), with $\mu(A) = 0$. Let $\mathfrak{I}$ be a topology on A such that $\mathfrak{I}$ is separable and topologically complete and Σ is the Borel field of $\mathfrak{I}$. Let $f:A \times A \to$ reals be bounded, continuous, and obey the triangle inequality; and let $f(a, a) = 0$, all $a \in A$.

Then a best solution λ^o exists to the transhipment problem of minimizing (2) over bounded measures λ satisfying $\lambda' - \lambda'' = \mu$.

Proof: Consider the *transportation* problem with origin and destination spaces (A, Σ, μ^+) and (A, Σ, μ^-) respectively. By the results of Section 7.4 a best solution λ^o exists to this problem. (Note that $\mu^+(A) = \mu^-(A) < \infty$.)

The premises of the preceding theorem are also fulfilled. (Since $\mathfrak{I}$ is separable metrizable, it has a countable base, hence so does $\mathfrak{I} \times \mathfrak{I}$, hence $\mathfrak{I} \times \mathfrak{I}$ has the strong Lindelöf property.) Hence this λ^o is also best for the transhipment problem of minimizing (2) over bounded measures λ satisfying $\lambda' - \lambda'' = \mu^+ - \mu^- = \mu$. ∎

This theorem is *false* without the triangle inequality premise; e.g., let $A = [0, 1]$, $f(x, y) = (x - y)^2$, and ship unit mass from point 0 to point 1. The feasible flow shipping unit mass from $(k - 1)/n$ to k/n, $k = 1, \ldots, n$, attains a cost of $1/n$. Yet cost zero is unattainable, so there is *no* optimal flow.

7.11. THE SKEW TRANSHIPMENT PROBLEM

We return for a moment to the ordinary transhipment problem (6.1)–(6.2), with a finite number of locations; x_{ij} and x_{ji} are the flows from location i to location j and vice versa, respectively. Define y_{ij}, the *net flow* from i to j, by

$$y_{ij} = x_{ij} - x_{ji} \qquad (7.11.1)$$

$i, j = 1, \ldots, n$. We note at once that y_{ij} need not be nonnegative. In fact $y_{ij} = -y_{ji}$ for all $i, j = 1, \ldots, n$. That is, if the numbers y_{ij} were arrayed in matrix form, they would form a *skew-symmetric* matrix.

(The term "net flow" has been used above in an entirely different sense in Section 7.7, viz., as the net amount entering or leaving a given set of locations. It was represented by a measure λ_1 or λ_2 on (A, Σ). Here it refers to a net movement on a given set of *pairs* of locations. It will be represented below by a *signed* measure (or, more generally, a pseudomeasure) on the *product* space $(A \times A, \Sigma \times \Sigma)$. Hence no confusion should arise between these concepts.)

The basic transhipment constraint (6.1) takes on a simple form when written in terms of net flows, viz.,

$$y_{i1} + \cdots + y_{in} \leq \alpha_i \qquad (7.11.2)$$

$i = 1, \ldots, n$. What about the objective function (6.2)? Can it be written in terms of net flows? We distinguish two cases, depending on the nature of the "gross" flow pattern (x_{ij}). This pattern is said to have the *no-cross-hauling* property iff $\min(x_{ij}, x_{ji}) = 0$ for all pairs (i, j), $i, j = 1, \ldots, n$. That is, a positive flow never occurs in both directions between any pair of locations (in particular, $x_{ii} = 0$, all i). In the no-cross-hauling case we see that (1) can be solved for x; viz., $x_{ij} = \max(y_{ij}, 0)$. Hence, if we restrict ourselves to such flow patterns, the objective function (6.2) can be written: minimize the sum of

$$f_{ij} \max(y_{ij}, 0) \qquad (7.11.3)$$

over all n^2 pairs (i, j), $i, j = 1, \ldots, n$.

If cross-hauling occurs, the objective function cannot be written in terms of (y_{ij}) alone. However, little is gained by allowing cross-hauling. For consider the following two possible situations.

1. $f_{ij} + f_{ji} < 0$ for some pair (i, j). Then there is no optimal solution to the transhipment problem, because the cyclic positivity condition (6.4) is violated.

2. $f_{ij} + f_{ji} \geq 0$ for all pairs (i, j). Then there is no point to cross-hauling. If x_{ij} and x_{ji} are both positive for some pair (i, j), an equal reduction of both these numbers by $\min(x_{ij}, x_{ji})$ preserves feasibility and reduces transport cost, or at worst leaves it unchanged.

The problem of finding a skew-symmetric flow pattern $(y_{ij} = -y_{ji})$ that satisfies (2) and minimizes (3) will be called the *skew formulation* of the (ordinary) transhipment problem. The intuitive advantage of this over the ordinary formulation is that it automatically focuses attention on the flows without cross-hauling, which are the only interesting ones. Also, it can be argued that the *net* flow (y_{ij}) is really what one is looking for in transhipment problems in any case. Condition (2) gives the *inequality*-constrained variant; the *equality*-constrained variant is obtained by substituting = for ≤ in (2).

All the arguments above carry over very neatly to the *measure-theoretic* transhipment problem. The remainder of this section will be devoted to showing this in detail. First, we need a few new concepts.

Transposition on a product space $A \times A$ refers to the interchange of left and right. It will be denoted by *. Thus if G is a subset of $A \times A$, its *transpose* is the set $G^* = \{(a_1, a_2) \mid (a_2, a_1) \in G\}$. (Just "reflect" G through the "diagonal.") Similarly, the *transpose* of a function $f : A \times A \to$ reals is given by $f^*(a_1, a_2) = f(a_2, a_1)$. We now want to extend this concept to *set* functions. For this, the following simple result is needed.

Lemma: Let (A, Σ) be a measurable space. If $G \in \Sigma \times \Sigma$, then $G^* \in \Sigma \times \Sigma$.

Proof: Consider the class, $\mathcal{G}$, of all $G \in \Sigma \times \Sigma$ for which $G^* \in \Sigma \times \Sigma$. We show that $\mathcal{G}$ owns all measurable rectangles, and that it is closed under complementation and countable unions. This implies that $\mathcal{G} = \Sigma \times \Sigma$ and concludes the proof.

Let $E, F \in \Sigma$. $(E \times F)^* = F \times E \in \Sigma \times \Sigma$, hence $E \times F \in \mathcal{G}$.

Let $G \in \mathcal{G}$. $((A \times A) \backslash G)^* = (A \times A) \backslash G^*$. This last set belongs to $\Sigma \times \Sigma$, since G^* does, hence $(A \times A) \backslash G \in \mathcal{G}$.

Let $G_n \in \mathcal{G}$ for $n = 1, 2, \ldots$. $(G_1 \cup G_2 \cup \cdots)^* = G_1^* \cup G_2^* \cup \cdots$. This last set belongs to $\Sigma \times \Sigma$, since each G_n^* does, hence $G_1 \cup G_2 \cup \cdots \in \mathcal{G}$. ∎

Definition: Let $\sigma : \Sigma \times \Sigma \to$ extended reals be a set function whose domain is the product σ-field $\Sigma \times \Sigma$. The *transpose* of σ is the set function σ^* given by

$$\sigma^*(G) = \sigma(G^*) \tag{7.11.4}$$

for all $G \in \Sigma \times \Sigma$.

This is well defined, by the lemma just proved. It is easily established that σ^* is a measure or signed measure iff σ is a measure or signed measure respectively. Also, σ-finiteness of σ implies the same for σ^*.

Definition: Let σ be a pseudomeasure on $(A \times A, \Sigma \times \Sigma)$. The *transpose* of σ is the pseudomeasure $\sigma^* = (\mu^*, \nu^*)$, where (μ, ν) is any representative of σ.

For this to be a sound definition, σ^* must not depend on the particular representative of σ that is chosen. Let (μ_1, ν_1) be another representative, so that $\mu + \nu_1 = \nu + \mu_1$ (equivalence theorem). This implies $\mu^* + \nu_1^* = \nu^* + \mu_1^*$, by (4), so that $(\mu^*, \nu^*) = (\mu_1^*, \nu_1^*)$, and the same σ^* results. Hence the definition is bona fide.

Note that λ^* and σ^* remain defined on the product space $(A \times A, \Sigma \times \Sigma)$, in contrast to λ', λ'', λ_1, λ_2 discussed previously, which are defined on (A, Σ). Note also that double transposition restores the original: $(G^*)^* = G$, $(\sigma^*)^* = \sigma$, etc.

In terms of transposes we now define the "skew" concepts needed for the skew transhipment problem.

Definition: Let σ be a signed measure or pseudomeasure on $(A \times A, \Sigma \times \Sigma)$. Then σ is *skew* iff

$$\sigma^* = -\sigma \qquad (7.11.5)$$

For signed measure σ, (5) states that σ takes on values of opposite sign on sets that are transposes of each other. It follows that if $G \in \Sigma \times \Sigma$ is a *symmetric* set (i.e., $G = G^*$), then $\sigma(G) = 0$. In particular, the universe set $A \times A$ is symmetric, so that $\sigma(A \times A) = 0$. Thus a skew signed measure must be *bounded*.

For skew pseudomeasures we have the following result.

Theorem: Let σ be a pseudomeasure on $(A \times A, \Sigma \times \Sigma)$. Each of the following conditions implies the other four:

  (i) σ is skew;
 (ii) σ^+ and σ^- are transposes of each other;
(iii) σ has a representative (μ, ν) for which $\mu = \nu^*$;
 (iv) σ has a representative (μ, ν) for which $\mu + \mu^* = \nu + \nu^*$;
  (v) $\mu + \mu^* = \nu + \nu^*$ for every representative (μ, ν) of σ.

Proof: Obviously, (ii) *implies* (iii) and (v) *implies* (iv).

(iii) *implies* (i). Letting (μ, ν) be the representative of σ with property (iii), we obtain $\sigma^* = (\mu^*, \nu^*) = (\nu, \mu) = -\sigma$, which shows that σ is skew.

(i) *implies* (v). Letting $\sigma = (\mu, \nu)$, we obtain $(\mu^*, \nu^*) = \sigma^* = -\sigma = (\nu, \mu)$, and (v) follows from the equivalence theorem for pseudomeasures.

(iv) *implies* (ii). Letting (μ, ν) be the representative of σ with property (iv), we obtain

$$((\sigma^+)^*, (\sigma^-)^*) = (\mu^*, \nu^*) = (\nu, \mu) = (\sigma^-, \sigma^+) \qquad (7.11.6)$$

The middle equality of (6) arises from (iv) via the equivalence theorem; the left and right equalities arise from two different ways of writing σ^* and $-\sigma$ respectively.

The left and right pairs in (6) are both mutually singular; for if (P, N) is a Hahn decomposition for σ, so that $\sigma^+(N) = \sigma^-(P) = 0$, then $(\sigma^+)^*(N^*) = (\sigma^-)^*(P^*) = 0$; but $\{P^*, N^*\}$ is a measurable partition of $A \times A$, so the left pair is mutually singular. By the uniqueness of the Jordan form, it follows that $(\sigma^+)^* = \sigma^-$ and $(\sigma^-)^* = \sigma^+$, which is condition (ii).

We now have a closed circle of implications. ∎

In connection with condition (iii) of this theorem, note that (if Σ is nontrivial) not all representatives of a skew pseudomeasure satisfy $\mu = \nu^*$. For example, the zero pseudomeasure is skew, and its representatives are the pairs (μ, μ) for all σ-finite measures μ. But $\mu = \mu^*$ is not true for all such measures.

We are now ready for the skew formulation of the transhipment problem. The latter comes in bounded and unbounded formulations, and each of these can be "skewed." The bounded problem becomes one of finding the best of a feasible set of skew signed measures (these must be bounded, as noted above); the unbounded problem becomes one of finding the best of a feasible set of skew pseudomeasures.

We formulate the skew bounded problem first. Measurable space (A, Σ) is given with a σ-finite signed measure μ on it (net capacity), and a measurable function $f:A \times A \to$ reals (unit transport cost).

The problem is to find a *skew signed* measure σ on $(A \times A, \Sigma \times \Sigma)$ satisfying

$$\sigma' \leq \mu \qquad (7.11.7)$$

and minimizing

$$\int f \, d\sigma^+ \qquad (7.11.8)$$

Here σ' is, as usual, the left marginal of σ. Integral (8) is an indefinite integral over $A \times A$, and "minimization" is taken in its usual meaning of (reverse) standard ordering of pseudomeasures. Note that the upper variation σ^+ occurs in the objective function rather than σ itself.

Problem (7)–(8) may be compared with the skew formulation of the *ordinary* transhipment problem (2)–(3). It is not difficult to show that (7)–(8) reduce to (2)–(3) respectively, in the special case when Σ is a finite σ-field. The discussion there also provides a rationale for the particular form that (7)–(8) take.

We now take up the skew unbounded problem. For this we need a new concept, that of the *marginal* of a pseudomeasure. This in turn is a special case of the following.

Definition: Let (B, Σ', σ) be a pseudomeasure space, (C, Σ'') another measurable space, and $g{:}B \rightarrow C$ a measurable function. The *pseudomeasure induced* on (C, Σ'') by g from σ is defined iff the *measures* μ and ν, induced from σ^+ and σ^- respectively, are both σ-finite. In this case (μ, ν) is the induced pseudomeasure.

Starting with the pseudomeasure space $(A \times A, \Sigma \times \Sigma, \sigma)$, the *left marginal,* if it exists, is the pseudomeasure induced on the space (A, Σ) by the projection $g'(a', a'') = a'$, according to this definition. Similarly, the *right marginal* is that induced by the projection $g''(a', a'') = a''$. We use the notation σ', σ'' for these respective marginals, so that

$$\sigma' = [(\sigma^+)', (\sigma^-)'], \qquad \sigma'' = [(\sigma^+)'', (\sigma^-)''] \qquad (7.11.9)$$

Again, σ' is defined iff $(\sigma^+)'$ and $(\sigma^-)'$ are both σ-finite, and similarly for σ''. In the case where σ is a *bounded signed measure,* these marginals are all bounded; hence σ' and σ'' are always well defined. In fact, one easily verifies that in this case σ' and σ'' are bounded signed measures coinciding with the usual marginal concepts $\sigma'(E) = \sigma(E \times A)$, $\sigma''(E) = \sigma(A \times E)$, all $E \in \Sigma$. Hence the σ' in (7) may be looked upon as a special case of the definition just given.

The *skew unbounded* problem may now be stated. It is precisely the same as the skew-bounded problem (7)–(8), except that σ ranges over the set of skew pseudomeasures for which σ' exists and satisfies (7). Also, the given "net capacity" signed measure μ in (7) is replaced by the given pseudomeasure ψ.

Let us contemplate these skew formulations. One possibly disquieting feature is that left and right appear to be treated asymmetrically: σ' must exist and satisfy a certain condition, but not σ''. But this is an illusion, as the following result indicates.

Theorem: Let σ be a skew pseudomeasure on the product space $(A \times A, \Sigma \times \Sigma)$. Then σ' exists iff σ'' exists, and in this case,

$$\sigma' = -\sigma'' \qquad (7.11.10)$$

Proof: Let $E \in \Sigma$. Then

$$(\sigma^+)'(E) = \sigma^+(E \times A) = \sigma^-(A \times E) = (\sigma^-)''(E) \qquad (7.11.11)$$

The middle equality arises from σ^+ and σ^- being transposes. Thus we obtain $(\sigma^+)' = (\sigma^-)''$. Similarly, $(\sigma^-)' = (\sigma^+)''$. Hence the two pairs in (9) are interchanges of each other, and σ', σ'' (if they exist) are negatives of each other. Also σ' exists iff σ'' exists. ∎

Equality (10) holds in the special case where σ is a skew signed measure. Thus in both bounded and unbounded skew formulations one could just as easily have written things in terms of σ'' as of σ'.

We now want to relate the skew to the nonskew formulations. In discussing the ordinary transhipment problem, we noted the connection between skew flows and ordinary flows having the "no-cross-hauling" property. To carry this connection over to the measure-theoretic problem, we need a generalization of this property.

Definition: Measure λ on $(A \times A, \Sigma \times \Sigma)$ has the *no-cross-hauling property* iff λ and its transpose λ^* are mutually singular.

One easily verifies the following. If Σ is a finite σ-field, this property in effect reduces to the one mentioned above for ordinary transhipment: $\min(x_{ij}, x_{ji}) = 0$. Any translocation has the no-cross-hauling property, since λ has all its mass on a set $P \times N$, and λ^* on the disjoint set $N \times P$, $\{P, N\}$ being a partition of A. The converse of this is false, and even in a three-point space one can find a nontranslocation with this property (exercise).

Theorem: Given product measurable space $(A \times A, \Sigma \times \Sigma)$, let L be the set of all σ-finite measures λ on it with the no-cross-hauling property; let L_1 be the set of those L-measures whose marginals λ', λ'' are σ-finite; let L_2 be the set of those L_1-measures that are bounded. Also, let Ψ be the set of skew pseudomeasures σ on $(A \times A, \Sigma \times \Sigma)$; let Ψ_1 be the set of those Ψ-pseudomeasures for which the left marginal σ' exists; let Ψ_2 be the set of skew signed measures on $(A \times A, \Sigma \times \Sigma)$.

Let g assign to each $\sigma \in \Psi$ its upper variation:

$$g(\sigma) = \sigma^+ \tag{7.11.12}$$

Let h assign to each $\lambda \in L$ the pseudomeasure (λ, λ^*):

$$h(\lambda) = (\lambda, \lambda^*) \tag{7.11.13}$$

Then g and h both establish 1–1 correspondences between the three pairs (L and Ψ, L_1 and Ψ_1, and L_2 and Ψ_2) and are inverses of each other:

$$g(h(\lambda)) = \lambda, \qquad h(g(\sigma)) = \sigma \tag{7.11.14}$$

Finally, if σ and λ are corresponding members of Ψ_1 and L_1, then

$$\sigma' = (\lambda', \lambda'') \tag{7.11.15}$$

Proof: First, we show that the ranges of g and h are contained in the proper sets. If $\sigma \in \Psi$, the transpose of σ^+ is σ^-, and of course σ^+, σ^- are mutually singular; hence σ^+ has the no-cross-hauling property: $g(\sigma) \in L$. If, in addition, $\sigma \in \Psi_1$, then $(\sigma^+)'$ and $(\sigma^-)'$ are σ-finite. The latter equals $(\sigma^+)''$ (cf. (11)); hence σ^+ has σ-finite marginals: $g(\sigma) \in \mathbf{L}_1$. If, in addition, $\sigma \in \Psi_2$, it is bounded, so σ^+ is bounded: $g(\sigma) \in \mathbf{L}_2$. This proves that g maps things into the right sets.

If $\lambda \in \mathbf{L}$, then (λ, λ^*) is a skew pseudomeasure: $h(\lambda) \in \Psi$. Suppose, in addition, that $\lambda \in \mathbf{L}_1$, so that λ', λ'' are σ-finite. For any $E \in \Sigma$, we have $\lambda''(E) = \lambda(A \times E) = \lambda^*(E \times A) = (\lambda^*)'(E)$. Hence $\lambda'' = (\lambda^*)'$, and the latter is σ-finite; (λ, λ^*) is the Jordan form of $h(\lambda)$, hence $(h(\lambda))'$ exists: $h(\lambda) \in \Psi_1$. If, in addition, $\lambda \in \mathbf{L}_2$, λ is bounded, so (λ, λ^*) is a signed measure: $h(\lambda) \in \Psi_2$. This proves that h maps things into the right sets.

It remains only to establish (14) and (15). For any $\lambda \in \mathbf{L}$, $g(h(\lambda)) = (\lambda, \lambda^*)^+ = \lambda$, since (λ, λ^*) is the Jordan form of $h(\lambda)$. For any $\sigma \in \Psi$, $h(g(\sigma)) = (\sigma^+, (\sigma^+)^*) = (\sigma^+, \sigma^-) = \sigma$, since σ^- is the transpose of σ^+. Finally, $\sigma' = ((\sigma^+)', (\sigma^-)') = ((\sigma^+)', (\sigma^+)'') = (\lambda', \lambda'')$, if σ, λ are corresponding members of Ψ_1, $\mathbf{L}_1$. This yields (15). ∎

This long theorem has a very simple interpretation. Compare the skew bounded transhipment problem (7)–(8), for instance, with the problem that follows.

Find a bounded measure λ satisfying $(\lambda', \lambda'') \leq \mu$ and minimizing $\int f \, d\lambda$. If we *add the additional constraint* that λ has the no-cross-hauling property, we find that the mappings g or h, (12) or (13), establish a 1–1 correspondence between the set of measures λ feasible for this problem and the set of signed measures σ feasible for the preceding problem. Furthermore, the objective functions assign the same utility to corresponding λ and σ, since $\lambda = \sigma^+$. Thus these problems are equivalent to each other in a rather strong sense.

Similarly, the *unbounded* skew and nonskew problems are equivalent to each other in this sense if we add the no-cross-hauling constraint to the nonskew problem. The feasible sets in the unbounded problems are subsets of Ψ_1 and $\mathbf{L}_1$ respectively, just as they are subsets of Ψ_2 and $\mathbf{L}_2$ respectively in the bounded problems.

For the ordinary transhipment problem the mappings g and h take forms we have already encountered: g becomes $x_{ij} = \max(y_{ij}, 0)$, and h becomes $y_{ij} = x_{ij} - x_{ji}$.

Finally, we want to investigate the effects of restricting attention to flows with the no-cross-hauling property. For ordinary transhipment we pointed out that if an optimal flow exists at all, then some flow without cross-hauling is optimal. This property carries over to measure-theoretic transhipment.

First, consider the process of "reducing" the flow pattern (x_{ij}) by subtracting $\min(x_{ij}, x_{ji})$ from x_{ij} and x_{ji} if these are both positive. This leads to the no-cross-hauling flow whose value at (i, j) is $\max(x_{ij} - x_{ji}, 0)$. The following concept generalizes this operation.

Definition: Let λ be a σ-finite measure on $(A \times A, \Sigma \times \Sigma)$. The *no-cross-hauling reduction* of λ is the measure $(\lambda, \lambda^*)^+$. (That is, form the pseudomeasure (λ, λ^*), then take its upper variation.)

To show that $(\lambda, \lambda^*)^+$ does indeed have no cross-hauling, note first that (λ, λ^*) is skew. The upper variation of this is obtained by applying the mapping g, (12), whose range was proved to lie in the set of no-cross-hauling measures. We also have $(\lambda, \lambda^*)^+ \leq \lambda$, by the minimizing property of the Jordan form.

We can obtain an exact expression for the size of this reduction, which may be called the "cross-hauling."

Definition: Let λ be a σ-finite measure on $(A \times A, \Sigma \times \Sigma)$. The *cross-hauling* associated with λ is the measure κ given by $\kappa = \inf(\lambda, \lambda^*)$.

The following results show that these definitions capture quite well the intuitive meaning of these concepts. (Subtraction of measures is defined in Section 3.1; if κ is finite, this reduces to ordinary subtraction.) This theorem holds for any measure λ, not merely for σ-finite measures, if we interpret $(\lambda, \lambda^*)^+$ to refer to the upper variation of the Jordan decomposition of (λ, λ^*), which is well defined for any λ on $A \times A$.

Theorem: Let λ be a measure on $(A \times A, \Sigma \times \Sigma)$. The no-cross-hauling reduction of λ equals $\lambda - \kappa$. Also, λ has the no-cross-hauling property iff $\kappa = 0$.

Proof: The no-cross-hauling reduction of λ is the upper variation of the Jordan decomposition of the pair (λ, λ^*), and this is known to equal $\lambda - \inf(\lambda, \lambda^*)$ (Section 3.1). The second statement is simply a special case of the theorem that a pair (μ, ν) is mutually singular iff $\inf(\mu, \nu) = 0$, which was proved in Section 7.7. Here $\mu = \lambda, \nu = \lambda^*$. ∎

The cross-hauling measure κ also has the property of being *symmetric;* i.e., $\kappa(G) = \kappa(G^*)$ for any $G \in \Sigma \times \Sigma$ (this is easily established from the fact that λ and λ^* enter symmetrically into its definition). This implies that the left and right marginals of κ are equal: $\kappa' = \kappa''$.

With these preliminaries established, we are ready for our final result. This generalizes the argument given for ordinary transhipment and says in effect: in looking for an optimal flow, we might as well confine attention to flows without cross-hauling. The theorem applies to both bounded and unbounded formulations, and to both equality- and inequality-constrained variants.[29]

---

29. We have been stating constraints in the inequality form in this section. But if $\leq$ is replaced by $=$ in these formulas, the discussion is still valid word for word.

Theorem: If measure λ on space $(A \times A, \Sigma \times \Sigma)$ is best (or unsurpassed) for the transhipment problem, then its no-cross-hauling reduction is also best (or unsurpassed) respectively.

Proof: For convenience we use $-f$ in place of f in the objective function and treat the problem as one of *maximization*. We also find it convenient to treat the measures discussed (which are all σ-finite) as pseudomeasures, so that they may be subtracted freely even though they may be infinite.

Let λ be best for the transhipment problem. First,

$$[(\lambda, \lambda^*)^{+\prime}, (\lambda, \lambda^*)^{+\prime\prime}] = (\lambda', \lambda'') = ((\lambda + \kappa)', (\lambda + \kappa)'') \qquad (7.11.16)$$

the equalities being understood in the pseudomeasure sense. To prove (16), we first verify that all six measures appearing there are σ-finite. λ' and λ'' are σ-finite since λ is feasible; κ and $(\lambda, \lambda^*)^+$ are both $\leq \lambda$, hence their marginals are σ-finite, too. The right equality arises from $\kappa' = \kappa''$. Similarly, $(\lambda, \lambda^*)^+$ and λ differ by κ (by the preceding theorem), and the same argument establishes the left equality in (16). Since feasibility depends only on the value of the pseudomeasure formed from the marginals in this way, it follows that $\lambda + \kappa$ and $(\lambda, \lambda^*)^+$ are also feasible flows.

Since λ is best, we must have

$$\int (-f) \, d\lambda \gtrsim \int (-f) \, d(\lambda + \kappa) \qquad (7.11.17)$$

(Here $\gtrsim$ is the preferred-or-indifferent relation for standard order.) The same pseudomeasure may be added to both sides of (17) without disturbing the order relation. Let us add $\int f \, d\kappa$ to obtain

$$\int (-f) \, d(\lambda - \kappa) \gtrsim \int (-f) \, d\lambda \qquad (7.11.18)$$

But $\lambda - \kappa = (\lambda, \lambda^*)^+$, by the preceding theorem, and (18) states that this measure is at least as preferred as λ. Since λ is best, so is $(\lambda, \lambda^*)^+$.

Next, let λ be unsurpassed for the transhipment problem. Since λ is feasible, so is $(\lambda, \lambda^*)^+$, by (16). Suppose that $(\lambda, \lambda^*)^+$ is surpassed by some feasible measure ν:

$$\int (-f) \, d\nu > \int (-f) \, d(\lambda, \lambda^*)^+ \qquad (7.11.19)$$

Adding $\int (-f) \, d\kappa$ to both sides of (19), we obtain

$$\int (-f) \, d(\nu + \kappa) > \int (-f) \, d\lambda \qquad (7.11.20)$$

Since ν is feasible, the same argument leading to (16) establishes that $\nu + \kappa$ is feasible. Then (20) states that λ is surpassed by $\nu + \kappa$. This contradiction proves that $(\lambda, \lambda^*)^+$ is unsurpassed. ∎

8

The Theory of Thünen Systems

8.1. INTRODUCTION

In both nature and society one often finds the following kind of spatial pattern. There is a certain special location surrounded by a series of concentric "rings" or "shells." At the locations in any one ring the same activity is occurring. From ring to ring there is a tendency for activities to become less "intense" in some sense as one moves outward.

An example is provided by a sphere in gravitational equilibrium. Here the densest substances lie toward the center, and density declines as one moves outward, ending in an ever thinner atmosphere. The environs of a volcano provide a less clear-cut example, as does an organic cell with its nucleus and cytoplasm.

In the social world a pattern of agricultural land uses surrounds a city. These tend to decline in intensity with increasing distance from the city. Within the city itself are the highly intensive land uses of the central business district (CBD) and a gradual diminution of intensity as one moves outward. On the "micro" level, the fields of individual farms tend to be cultivated with diminishing intensity as one moves farther from the farmhouse. We might also refer to the distribution of onlookers at sports events and other spectacles.

We discuss these and other examples in greater detail later. They are adduced here merely to introduce the class of phenomena to be considered. In all cases there is a greater or lesser degree of distortion from the "ideal" pattern of concentric rings of homogeneous activities. The same general phenomenon can occur at very different scales of magnitude: from the spatial ordering of an individual household to the pattern of large geographic regions, and even (as we shall see) to the entire world economy.

These patterns will be called *Thünen systems,* after the man who first investigated one of them in society, viz., the pattern of agricultural land uses around a city in an isolated region.[1] We make no use of Thünen's specific formulations, however, because modern developments have corrected and generalized them considerably, and this chapter will generalize them even further.

In developing a theoretical model for Thünen systems, the first problem that arises is specifying precisely what is meant by such systems. The concepts "activity" and "intensity," even the concept of "concentric rings," were used above in a vague commonsense way and need explication. After doing this, we present a model that is both explanatory and optimizing; i.e., it shows how the Thünen pattern will arise from the behavior of individual agents and also demonstrates that this pattern solves an optimization problem.

We place greatest stress on this optimality feature of Thünen systems because it has been largely neglected in the past. It is shown that Thünen systems are optimal for a special case of the measure-theoretic transportation problem of Chapter 7, the *allotment-assignment problem*. (Certain special Thünen systems also optimize the *allocation-of-effort problem* discussed in Chapter 5.) The potentials for this transportation problem may be interpreted (in part) as *land values*, and this establishes the connection between the optimizing and explanatory aspects of the model.

The special location that is the center of symmetry of the Thünen rings will be called the *nucleus* (corresponding in the examples above to the city, the CBD, the farmhouse, etc.). This will play a basic role for most of our exposition. In the end, however, even the nucleus can be dispensed with. The essential point is that the desirability of a location can be summarized in a single real number. This is usually the distance from the nucleus but may be well-defined even if there is no nucleus. All these points will be elaborated on below.

8.2. IDEAL DISTANCES AND IDEAL WEIGHTS

We now reintroduce our three basic sets: Resources, Space, and Time (R, S, T). The formal model to follow makes no concrete assumptions about the nature of these sets, and the generality resulting from this fact is useful. For example, T can be interpreted as having a bounded horizon or as being discrete, rather than as being the whole real Time axis; S may be thought of as a limited region of the earth's surface rather than as all of Space. Similarly, R may be thought of as restricted to resource types that make sense in the Thünen context, e.g., those that are "transportable."

We suppose a real-valued function θ with domain $R \times T \times S \times S$

---

1. P. Hall, ed., *Von Thünen's Isolated State*, trans. C. M. Wartenberg (Pergamon Press, New York, 1966). The German original was first published in Hamburg, 1826.

exists, the *unit transport cost function*. Specifically, $\theta(r, t, s_1, s_2)$ is the cost of shipping unit mass of resource type r at time t from origin s_1 to destination s_2.

Definition: $\theta:R \times T \times S \times S \to$ reals is *factorable* iff two functions $g:R \times T \to$ reals and $h:S \times S \to$ reals exist such that

$$\theta(r, t, s_1, s_2) = g(r, t)h(s_1, s_2) \qquad (8.2.1)$$

for all $r \in R$, $t \in T$, s_1, $s_2 \in S$. Function $g(r, t)$ is called the *ideal* (or *economic*) *weight* of resource r at time t, and $h(s_1, s_2)$ is called the *ideal* (or *economic*) *distance* from s_1 to s_2.

Excluding the trivial case when θ is identically zero, we can establish that g and h are unique up to scalar multiplication. To be precise, if g and h satisfy (1), so does the pair gx, h/x (x being any nonzero real number) and these pairs are the only solutions. Also, if θ is nonnegative, and g, h exist satisfying (1), then nonnegative g, h exist satisfying (1). For the following discussion we assume g and h are nonnegative.

Consider the economic meaning of θ and the condition (1). First, there is the problem of what instant t refers to in the case of time-consuming trips. A simple convention takes t to be the average of time of departure from s_1 and time of arrival at s_2. (A more elaborate analysis would insert an extra time component, resource r departing from s_1 at t_1 and arriving at s_2 at t_2. But this elaboration is not needed for the problems of this chapter.)

The mass flowing through the transportation system will be represented by a measure μ on universe set $R \times T \times S \times S$: $\mu(E \times F \times G \times H)$ = total mass of resources of types E flowing at times F from sources in region G to sinks in region H. Given μ and θ, total transport cost incurred is assumed to be

$$\int_{R \times T \times S \times S} \theta \, d\mu \qquad (8.2.2)$$

This is a severe assumption, ignoring as it does large-lot economies in transportation, congestion, and other interaction effects. A few devices mentioned below help to overcome these limitations but are only partially successful.

The factorability condition (1) states in effect that no source-sink pair (s_1, s_2) has a comparative advantage over any other such pair for the shipment of any resource at any time vis-à-vis another resource at another time. This is again a strong condition, and it is easy to find situations where it breaks down. For example, let s_1 and s_2 have good pipeline connections and poor road connections, and vice versa for s_3 and s_4. Then s_1 and s_2 might be "closer" for oil transportation and "further apart" for passenger transportation than s_3 and s_4; θ is clearly not factorable in this case. Nonetheless, we assume factorability as a very useful first approximation.

The great simplification that arises from factorability is that the same spatial transport-cost pattern applies to all resource types and times and may be summarized in a single function having only spatial arguments, viz., the ideal distance function h.

We now examine the two ideal functions g and h, which arise from a plausible factorable transport-cost function θ. As noted, g and h are unique up to a scalar multiple, so that the ratios of nonzero values $g(r_1, t_1)/g(r_2, t_2)$ are uniquely determined by θ, and similarly for h. The resulting patterns need not have any close relation to physical weights or distances respectively, though there will presumably be some overall positive correlation between ideal and physical values.

Consider the weight function g. Resource types, which for given physical weight are bulky, valuable, heavily taxed, or need special handling, will tend to have relatively high ideal weights. Small-lot shipments of the same resource tend to cost more per unit weight than large-lot shipments. It seems at first that the linearity of (2) precludes taking account of this last phenomenon, but one possible device for doing so is to distinguish different-size packages of the same resource formally as distinct resource types, the larger packages having smaller ideal/physical weight ratios.

How will $g(r, t)$ vary with time for fixed r? The secular trend will usually be downward for two reasons. First, there are technological improvements in transport and communications, extensions of the various grids, more vehicles in existence, etc., all of which reduce real transportation costs.[2] The second reason is the need to *discount*. To make the cost contributions of different times comparable in the integral (2), they must all be discounted to the same moment. The easiest solution is to build the discount factor directly into the ideal weight function g. The same real cost in the far future is less weighty than in the near future, and discounting introduces an additional "levitational" force over time.

Congestion may sometimes be allowed for by adjusting ideal weight. For example, suppose one is studying a metropolitan area and congestion appears periodically at weekday rush hours. One can represent this by letting g rise at these times; things get "heavier" during rush hours. (This is another device for circumventing in part the restricted form of (2). An adequate theory of congestion would require total cost to be a nonlinear function of μ, however.)

Factorability implies that for any particular resource type r, unit transport costs rise or fall proportionally for all source-sink pairs over time. Thus transport innovation must reduce costs pro rata; a reduction

2. These phenomena are sometimes referred to as the "shrinking globe." We find it more convenient to leave the globe unshrunk and to represent them by a reduction in the other cost component, ideal weight. Thus one might speak of "levitating resources," as resources in general get "lighter" over time.

in one region but not in another would violate factorability. Similarly, congestion must raise costs proportionally on all routes. These unlikely circumstances underline the strength of the factorability assumption.

Turning to ideal distances, we note that the term "distance" is a misnomer, because h need not obey the metric postulates. In particular, the symmetry postulate may be violated, due to uphill vs. downhill movements, wind and water currents, one-way streets, tariffs on imports but not exports, etc. Ideal distances will be distorted from physical distances because of geographic irregularities, because some pairs of locations have "good connections" relative to others, because fares are not faithful reflections of distances, because of heavy taxation at border crossings, etc. Just as temporal variations of congestion can be allowed for by adjusting ideal weights, spatial variations can be allowed for by adjusting ideal distances. That is, if certain regions (such as the central portions of cities) are generally congested, ideal distances between points in these regions will be large relative to physical distances. Speaking broadly, ideal distance tends to increase less than in proportion to physical distance (except perhaps for very long trips) because "overhead costs" such as loading, packing, billing, getting up steam, etc. (which may be a substantial fraction of total transport costs) are spread over a larger physical distance. (For very long trips the factors of cumulative fatigue and spoilage and the need to carry large amounts of food and fuel work in the opposite direction. For rocket flights the fuel-carrying factor is crucial.)[3]

8.3. IDEAL DISTANCES IN THÜNEN SYSTEMS

We develop several variant models for Thünen systems. The one to which most attention will be devoted is the *entrepôt* model. In this section we concentrate on some of its formal characteristics and do not worry about its realism.

The distinguishing feature of entrepôt models is the existence of a special location, called the *nucleus*, having the property that all transportation flows have the nucleus either as origin or as destination. That is, the exports of any land use located anywhere in the system all go to the nucleus; the imports of that land use all come from the nucleus. The nucleus functions as an *entrepôt* in the sense that a shipment from location

---

3. The notions of "ideal distance" and "ideal weight" (the latter not depending on Time) stem from Alfred Weber (1909). While the ideal weight concept has been widely accepted, location theorists have severely criticized ideal distances. See E. M. Hoover, *Location Theory and the Shoe and Leather Industries* (Harvard Univ. Press, Cambridge, 1937), p. 40, n. 10; W. Isard, *Location and Space-Economy* (MIT Press, Cambridge, 1956), p. 109; and especially T. Palander, *Beiträge zur Standortstheorie*. (Almqvist och Wiksell, Uppsala, 1935), pp. 195–99. This is curious. It is true that "ideal distances" will generally be non-Euclidean, so that special constructions based on the Euclidean metric cannot be used. But it should be clear that the "ideal" concepts of weight and distance are entirely on a par and in fact only defined conjointly.

s_1 to s_2 can be accomplished in two steps: from s_1 to the nucleus and from the nucleus to s_2. "Foreign trade" (i.e., flows between locations in the Thünen system and locations outside it) is not excluded; but any such trade must be channeled through the nucleus, so that the nucleus also functions as a *gateway* between the system and the rest of the world.

For entrepôt models we postulate a transport cost function with a slightly modified factorability condition. Transport cost between two nonnuclear sites is irrelevant, since by assumption no such flows ever occur. Hence we need to postulate the factorability condition (2.1) only in the case where s_1 or s_2 is the nucleus. Formally, the unit transport cost function θ satisfies the following condition.

Two functions g_{in}, $g_{out}:R \times T \rightarrow$ reals and two functions h_{in}, $h_{out}:S \rightarrow$ reals exist such that

$$\theta(r, t, s, s_N) = g_{in}(r, t)h_{in}(s) \qquad (8.3.1)$$

$$\theta(r, t, s_N, s) = g_{out}(r, t)h_{out}(s) \qquad (8.3.2)$$

Here s_N denotes the nucleus. Comparing (1) with (2.1), we see that $h_{in}(s) = h(s, s_N)$. The second argument of h is fixed at s_N and is dropped for simplicity. Thus $h_{in}(s)$ is simply the ideal distance *from* location s *to* the nucleus. Similarly, $h_{out}(s)$ is the ideal distance *to* location s *from* the nucleus.

The two g functions have a different significance. We allow the same resource r at the same time t to have two different ideal weights, depending on whether it is flowing into or out of the nucleus. This is a further relaxation of the factorability condition (2.1).

Conditions (1) and (2) together are clearly weaker than (2.1). In fact they are a bit too weak for our purposes, and we now add the *symmetry* condition that

$$h_{out} = h_{in} \qquad (8.3.3)$$

That is, the ideal distance between the nucleus and any other location in the system is the same in both directions. We denote this common function by h. (The domain of h is now simply S, not $S \times S$ as it was in Section 8.2.)

Function $h(s)$ will be referred to simply as "the distance of s." It provides a general index of inaccessibility of locations in the entrepôt model. The fact that the relative locational advantages of different places can be summarized in a single number in this way is one essential precondition for the striking simplicity of the results obtained for Thünen systems.

An example or two will illustrate that the symmetry assumption is less restrictive than might appear at first glance. Suppose the nucleus is perched on top of a hill, so that it costs, say, twice as much to transport resource r at time t from location s to the nucleus as to go in the opposite direction. Then if (1) and (2) are satisfied, so is the symmetry condition. We merely take $g_{in} = 2g_{out}$, $h_{in} = h_{out}$, so that a resource is twice as "heavy" traveling to the nucleus as when traveling away from it. We have merely thrown the burden of representing cost differentials onto the weight function, leaving the distance function invariant, just as we did in the case of the "shrinking globe."

The situation opposite to the one just mentioned is probably more common in practice: all roads lead to Rome more readily than they lead away, because of asymmetries of information. In this case g_{in} is smaller than g_{out}, and the same argument applies.

Again, consider the rush hour phenomenon in big cities. In the morning it is easier to travel away from the central business district than toward it, and the reverse is true in the evening. Then g_{in} would be larger than g_{out} at morning times and smaller at evening times, and the symmetry condition (3) would not necessarily be violated.

We now suppose that conditions (1)–(3) obtain, so that the distance function $h:S \rightarrow$ reals is well defined. The region

$$\{s \mid h(s) < z\} \tag{8.3.4}$$

is then called the *open disc of radius z* (about the nucleus). (A similar concept has already been defined for metrics, but we may not be dealing with a genuine metric in this case. Still, the concept is well defined if h is.)

The shape of the region (4) will depend on the nature of the function h. Suppose S is the plane, and for convenience let the nucleus s_N be at the origin $(0,0)$. If h is derived from a Euclidean metric, the regions (4) will be circular discs centered on the origin. If $h(s) = |x| + |y|$ (where (x,y) are the cartesian coordinates of s), then the regions (4) will be *diamonds*, i.e., squares with sides at 45° to the X and Y axes. This arises from a city-block metric, which in turn may be thought of as arising from a road system permitting only motions parallel to the X or Y axis.

Another common case arises from a limited number of traffic arteries converging radially on the nucleus (road, rail, river). Travel is relatively easy along such radials and difficult off them. In this case the regions (4) will tend to be amoeboid shaped, with "pseudopods" projecting out along each artery. It is even possible for these regions to fall into several disconnected pieces. This occurs with limited-access transportation systems (highways with infrequent exits, railways, airports). Here the immediate neighborhood of a point of access to the transportation network may be an isolated outpost that is economically "close" to the nucleus though physically distant.

The significance of this discussion is that in Thünen systems land uses are arranged in "rings," which are set-theoretic differences of open discs (4) with different radii. Only in the Euclidean case will these literally be rings, i.e., annuli centered on the nucleus. In other cases these rings will be more or less irregular and even disconnected. We would expect, for example, that typically "urban" land uses would "sprawl" deep into the countryside along major radial arteries and would tend to appear in the vicinity of commuter railway stations and airports.

These diverse phenomena are all covered by the entrepôt model, which predicts the pattern of land uses in terms of *ideal* distances. The geographical implications will then depend on the shape of the regions (4). The model itself, however, does not need nor make any such assumptions but is formulated throughout in terms of ideal, not physical, distances.

8.4. LAND USES

The *spatial field*, with the particular structure just discussed, is one of the two basic ingredients constituting Thünen systems. The other is the set of *land uses* to be distributed over this field. We now discuss these —first, more or less formally and then with concrete interpretations and illustrations of the concepts involved.

FORMAL STRUCTURE

We use a version of the activity analysis model of Section 4.5 and review the salient concepts, using the notation of that section. An *activity* q is a triple of measures, ρ on (Ω_r, Σ') and λ_1 and λ_2 on $(R \times T, \Sigma_r \times \Sigma_t)$. Here Ω_r is the space of transmutation paths, and ρ represents the "capital-goods" structure of the activity; λ_1 is the "production" measure, describing the resource-time distribution of outputs; similarly, λ_2 is the "consumption" measure.

This describes one activity. Next, $\mathbf{Q}$ is the set of all feasible activities; and ν, a measure on $(S \times \mathbf{Q}, \Sigma_s \times \Sigma_q)$, describes the *assignment* of activities to locations. On measurable rectangles, $\nu(E \times F)$ is the "amount" of activities of types F located in region E. This determines the *total production measure* μ_1 over the space $(R \times S \times T, \Sigma_r \times \Sigma_s \times \Sigma_t)$ as follows:

$$\mu_1(G) = \int_{S \times \mathbf{Q}} \lambda_1[q, \{(r,t) \mid (r,s,t) \in G\}] \, \nu(ds, dq) \qquad (8.4.1)$$

for all $G \in \Sigma_r \times \Sigma_s \times \Sigma_t$. Here λ_1 is the function with domain $\mathbf{Q} \times (\Sigma_r \times \Sigma_t)$ for which $\lambda_1(q, \cdot)$ is the production measure associated with activity $q \in \mathbf{Q}$. The λ_1 is assumed to be an abcont conditional measure, which insures that the integral (1) is well defined and that μ_1 is a measure. Similarly, conditional measure λ_2 is constructed from the consumption measures of the various activities. Replacing λ_1 by λ_2 in (1), we obtain μ_2, the *total consumption measure* over $R \times S \times T$ determined by ν.

We now place these concepts in the Thünen context. The essential point is that all production (no matter when, where, or what is produced) must get shipped *to* the nucleus. Similarly, all consumption (over all T, S, and R) must come *from* the nucleus. Combined with our factorability assumptions, this yields an expression for the total transport cost incurred by an assignment ν. Furthermore, we are able to apply the concept of *ideal weight* to the activities themselves, not just to resource-time pairs; this simplifies things considerably.

We now spell out these statements. An ideal distance function $h:S \to$ reals exists giving the "inaccessibility" of any location from the nucleus, and two ideal weight functions $g_{in}, g_{out}:R \times T \to$ reals exist. We assume these functions are nonnegative and measurable. Define the *in-weight* of activity q as follows:

$$w_{in}(q) = \int_{R \times T} g_{in}(r, t)\, \lambda_1(q, dr, dt) \qquad (8.4.2)$$

The *out-weight* of activity q, written $w_{out}(q)$, is defined by (2) by substituting g_{out} for g_{in} and λ_2 for λ_1. Finally, the *weight* of activity q is the sum of these two:

$$w(q) = w_{in}(q) + w_{out}(q) \qquad (8.4.3)$$

The w_{in}, w_{out} and w are all extended real-valued functions with domain $\mathbf{Q}$. The conditions on g_{in}, g_{out} and λ_1, λ_2 insure that they are nonnegative and measurable.

We now show that the total transport cost incurred in a Thünen system under assignment measure ν is simply

$$\int_{S \times Q} h(s)w(q)\, \nu(ds, dq) \qquad (8.4.4)$$

Total transport cost is the sum of cost incurred on shipments into the nucleus plus cost incurred on shipments out of the nucleus. The in-shipment cost is given by

$$\int_{R \times S \times T} g_{in}(r, t)h(s)\, \mu_1(dr, ds, dt) \qquad (8.4.5)$$

since this is what (2.2) reduces to for the special case in hand. Here μ_1 is the total production measure as given by (1). Remember that all the mass of the distribution μ_1 must be shipped to the nucleus. A unit mass of resource type r located at s and shipped at moment t incurs a cost of $g_{in}(r, t)h(s)$.

We claim that (5) is equal to

$$\int_{S \times Q} h(s)w_{in}(q)\, \nu(ds, dq) \qquad (8.4.6)$$

To show this, we introduce the measure μ_1^* on the product space $S \times Q \times R \times T$ by the following iterated integral:

$$\mu_1^*(H) = \int_{S \times Q} \nu(ds, dq) \int_{R \times T} \lambda_1(q, dr, dt) I_H \qquad (8.4.7)$$

for all $H \in \Sigma_s \times \Sigma_q \times \Sigma_r \times \Sigma_t$ (here I_H is the indicator function). One verifies that μ_1 given by (1) is the marginal of μ_1^* on the component space $R \times S \times T$.[4]

It follows from the induced integrals theorem that (5) is equal to

$$\int_{S \times Q \times R \times T} g_{in}(r, t) h(s) \, \mu_1^*(ds, dq, dr, dt)$$

By (7) and Fubini's theorem, this in turn equals the iterated integral

$$\int_{S \times Q} \nu(ds, dq) \int_{R \times T} \lambda_1(q, dr, dt) \, g_{in}(r, t) h(s) \qquad (8.4.8)$$

Evaluating (8) from right to left, the integration over $R \times T$ yields the simple expression $w_{in}(q) h(s)$, by (2), so that (8) equals (6). We have proved that (5) and (6) are indeed equal.

The out-shipment cost is given by (5), with g_{out} and μ_2 replacing g_{in} and μ_1 respectively. The argument just given, with λ_2, μ_2^* replacing λ_1, μ_1^*, proves that the out-shipment cost is equal to

$$\int_{S \times Q} h(s) w_{out}(q) \, \nu(ds, dq) \qquad (8.4.9)$$

Finally, adding (6) and (9) and using (3), we see that *total* transport cost is indeed given by (4). This completes the proof.

In commonsense terms the argument just given amounts to the following. An activity determines a certain production and consumption pattern of resources over time. These incur transport costs per unit ideal distance as determined by their ideal weights, and this implicitly determines a weight for the activity itself, viz., the cost incurred by its inputs and outputs in moving unit distance. This activity weight is given by (2) and (3). It is then intuitively plausible that the total transport cost incurred by the spatial activity distribution ν should be given by (4), the integral of the activity weights multiplied by the ideal distances its inputs and outputs must travel.

Important advantages are obtained by this transformation. The expression (4) in terms of activities is much simpler than (5) plus the cor-

---

4. We take certain notational liberties in rearranging the order of the component spaces. No confusion should result from this, since there are no repetitions among these components.

responding expression for out-shipments. Using (4) and the other constructions discussed below, it is possible to dispense with explicit consideration of R and T and to work entirely with activities (and S). This is the natural approach for concrete applications and it again leads to great formal simplicity.

Note that the "capital-goods" structure of activities, given by the measure space $(\Omega_r, \Sigma', \rho)$ (where ρ depends on activity q), does not influence transportation cost. This is as it should be; ρ refers to the internal operation of these activities, and no spatial movement is involved. (Shipments of equipment, construction materials, etc., are already incorporated in λ_1 and λ_2.) In fact, ρ plays a very subordinate role in what follows and will be ignored except for occasional comments.

We now come to the question of constraints on the possible activity distributions ν. Just one kind of constraint will be imposed: a limit on areal capacity.[5] That is, activities demand "room" in which to operate; regions have a limited amount of room, and this limits the total amount of activities that can be squeezed into them.

In Section 4.5, the areal capacity constraint was written in the following form:

$$\int_{F \times Q} k \, d\nu \leq \alpha(F) \qquad (8.4.10)$$

for all regions F. Here α is a measure on physical Space (S, Σ_s), the *ideal areal* measure. The nonnegative measurable function $k{:}S \times Q \to$ reals gives the "demand for room" by activity q at location s. Then (10) states that the total demand for room in region F cannot exceed the capacity of that region.

We make the special assumption that $k = 1$ identically. Then (10) becomes

$$\nu(F \times Q) \leq \alpha(F) \qquad (8.4.11)$$

for all regions F.

The step from (10) to (11) is less restrictive than it appears to be. It amounts, essentially, to the assumption that function k is *factorable:* $k(s, q) = k_1(s)k_2(q)$ for some pair of positive functions $k_1{:}S \to$ reals and $k_2{:}Q \to$ reals. To see this, define two new measures, α' on S and ν' on $S \times Q$, as follows:

$$\alpha'(F) = \int_F [1/k_1(s)] \, \alpha(ds)$$

for all $F \in \Sigma_s$, and

---

5. An "allotment" constraint will be added later, but this constraint is "artificial" in a sense to be discussed below.

$$\nu'(G) = \int_G k_2(q)\,\nu(ds,dq) \tag{8.4.12}$$

for all $G \in \Sigma_s \times \Sigma_q$. Then from (10) we obtain for all $F \in \Sigma_s$,

$$\int_{F\times Q} k_1(s)\,\nu'(ds,dq) = \int_{F\times Q} k_1(s)k_2(q)\,\nu(ds,dq)$$

$$\leq \alpha(F) = \int_F k_1(s)\,\alpha'(ds)$$

Treating the left and right integrals as measures over S, we integrate the positive function $1/k_1$ with respect to them to obtain

$$\nu'(F \times Q) \leq \alpha'(F) \tag{8.4.13}$$

for all $F \in \Sigma_s$. Condition (13) has the same form as (11).

Now the units in which activities are measured are arbitrary, and the "amount" of activity has no intrinsic meaning. Suppose, then, we change measurement units as follows. Activity q (or, more precisely, unit level of activity q) is now redefined to be the triple $[\rho/k_2(q), \lambda_1/k_2(q), \lambda_2/k_2(q)]$, where $(\rho, \lambda_1, \lambda_2)$ is the original activity q. Then assignment ν in the original units is the same as ν' in the new units, ν and ν' being related by (12). Similarly, there is no intrinsic significance to the ideal areal measure α, and it might just as well be replaced by α', with corresponding changes in k to keep the constraint conditions invariant.

With these changes of units (10) becomes (13). We may, in fact, simply forget about the original measures ν and α and drop the primes in (13), obtaining (11). (Corresponding changes must also be made in the weight function $w(q)$; we suppose this has been done without changing notation.)

One can now give an intuitively appealing interpretation to the nondescript concept of "amount" of activities ν. Constraint (11) implies that ν and α are dimensionally comparable, so that ν may be thought of as given in "ideal" areal units—"acres" if you wish. Specifically, $\nu(F \times G)$ is the acreage required by the activities of types G which are operated in region F. Similarly, the measures ρ, λ_1, λ_2 have the dimensions "mass per unit area." For example, $\lambda_1(E \times H)$ would be the production of resources of types E in period H, in "tons per acre," say.

We have been discussing activities in general to this point. Let us refer to activities that have a *positive* demand for "room" as *land uses*. All the activities discussed in connection with Thünen systems will be land uses, as is clear from constraint (11). (Non-land-using activities could not be measured in acres, since positive amounts could be operating in regions of zero ideal area.)

The areal constraint (11) is expressed as an inequality. In what follows we find it convenient to express this as an *equality*. No real loss of generality is involved here, since we can add a special land use called "vacancy," which takes up the slack, if any.

This concludes our formal discussion of land uses. The two basic formulas we have arrived at are (4), the expression for total transportation cost in terms of distance, weight, and activity distribution, and (11), the areal constraint on activity distribution.

Note that (11) has the form of the *capacity constraint* in a measure-theoretic transportation problem, where the source space is (S, Σ_s, α) and the sink space is $(Q, \Sigma_q, ?)$, the question mark referring to an as yet unspecified requirement measure. Also, (4) has the form of the *objective function* for this problem, ν being the unknown "flow" measure. The only missing ingredient is the *requirement constraint,* and this role will be filled by the "allotment" mentioned in footnote 5.

INTERPRETATIONS AND ILLUSTRATIONS

Illustrations of theoretical concepts are always useful for making connections with the real world. For the land-use concept just developed they are especially important for two reasons. First, a great diversity of phenomena are encompassed by it, and this fact can be driven home only by examples. Second, the concept is unusual in several respects, and some of the associated terms (such as "production" and "consumption") are used in strange ways; all this needs elucidation.

A land use is longitudinal, stretching over the entire time horizon. Suppose that a site is successively vacant; used for farming; then for residing, manufacturing, and office activities; and ends up as a parking lot. This whole succession (with the construction and demolition that occurs between phases) must be considered to be *one* land use, not a series of land uses. The production and consumption measures on $R \times T$, λ_1 and λ_2, will concentrate mass on different resource types in different epochs, of course, and the history of the "goings-on" could be reconstructed in part from a knowledge of these two measures.

We examine these measures in more detail. In the entrepôt model all production is to be shipped to the nucleus. This means that we must include in "production" all resources that leave the site and travel to the nucleus. Consider a residential land use in the context of an urban Thünen system, with the CBD as the nucleus. Any household member who makes a trip downtown for work, shopping, recreation, or whatever must be considered to be "produced" at that time and "exported" to the nucleus. The same is true for other household "exports" (e.g., outgoing mail and telephone calls, garbage and sewage) insofar as they move to a centralized processing point. Similarly, people traveling from the CBD to the household must be considered to be "consumed" at the time of the trip and will be recorded in λ_2. The same is true for other resources coming in from downtown (consumer goods, water, gas, electricity, incoming mail, telephone calls). A round trip counts both as an export and an import. Every trip must be counted separately.

What about local trips to neighborhood facilities (routine grocery shopping, children's school trips, local movies)? These should *not* be counted.[6] The basic principle for distinguishing these trips from those mentioned above is this: λ_1 and λ_2 are to be constructed so that the land-use weight, as determined by (2) and (3), is an accurate reflection of the "pull" of the nucleus on this land use. Extra trips to the CBD increase this pull; i.e., a land use with more such trips would save more in transport costs by moving one unit of (ideal) distance closer to the CBD than would a land use with fewer such trips, all other imports and exports being the same. But a change in local trips would be irrelevant in this respect.

A land use is defined by the triple $(\rho, \lambda_1, \lambda_2)$, and any variation in any of these measures, however slight, yields a different land use. Consider the general category of *intensity* variations, for example. Corn can be grown in a continuum of different ways, with variations of fertilizer input per acre leading to variations of corn yield per acre. Each of these different input-output level combinations is to be considered a different land use.

Intensity variations manifest themselves in the levels of inflowing and outflowing traffic per acre and in the general degree of crowding of resources upon the site. One particular form that intensification takes is the phenomenon of *multiple-story* land uses, and this is important enough to deserve separate discussion.

An N-story structure provides a stack of N horizontal surfaces of support, on which N different processes can run simultaneously, one above the other.[7] This can be represented in at least two ways in terms of our categories. One approach identifies S with supporting surfaces in general, including the (land?) surface on the earth *and* floors above (and possibly below) it. From this point of view, land uses are inherently "single-story." Most are placed at ground level, some on floors above or below ground. Multiple-story construction (including bridges, tunnels, and pit mine construction) is then a way of *creating* new Space.

The second approach restricts S to the earth's surface. A site may be

6. If we are investigating the neighborhood Thünen system centered on the cluster of local facilities rather than the structure of the city as a whole, these local movements *should* be counted.

7. If the roof is utilized, there are $N + 1$ surfaces, the topmost being unsheltered. "Open space" may be thought of as a zero-story structure whose "roof" is the surface of the earth.

utilized for any of several land uses, some of which will be multiple story. The latter involve several processes stacked vertically, usually preceded by construction of the multiple-story structure that supports them. For example, a ten-story office building, with detailed specification of what goes on at each floor, would be a typical multiple-story land use. On the preceding approach, it would decompose into ten separate land uses. For the most part, we use the second approach.

Next consider *time displacement* as a form of variation among land uses; e.g., a trip is made sooner or later, a crop is harvested (and shipped) sooner or later. A special case is where the entire land use is shifted "rigidly" in time. To be precise, we must specify the structure of T as used in the model. Suppose T is the nonnegative real numbers, so that the Thünen system is taken to begin at some moment, time zero, but unfolds indefinitely into the future. Land use q' is then said to be a t_o-*forward displacement of* q ($t_o > 0$) iff for all $E \in \Sigma_r \times \Sigma_t$,

$$\lambda_i(q', E) = \lambda_i[q, \{(r, t) \mid t \geq 0, \text{and } (r, t + t_o) \in E\}]$$

$i = 1, 2$. That is, the production and consumption assigned to any measurable subset of $R \times T$ by q is the same as that assigned by q' to that set displaced forward t_o time units. Also, the $t \geq 0$ insures that q' neither produces nor consumes before moment t_o.

This rigid displacement might arise in a land speculation situation, in which the controller of a site knows what he wants to do but is waiting for the right moment to initiate operations. Just as with intensity variations, displacements are to be considered as different land uses.

What can be said about the weights of these various land uses? To find $w(q)$, one needs the measures λ_1, λ_2 associated with q, as well as the ideal weight functions g_{in}, g_{out}, and then uses formulas (2) and (3). Certain general observations concerning procedures and "tendencies" are in order.

First, λ_1 and λ_2 are the production and consumption on one "ideal acre" of S. Hence in measuring shipments to and from some actual land use, one must always adjust for this by dividing by the number of "ideal acres" on the site. As a first approximation one may identify ideal area with physical area, adjusting the former downward for sites with rough topography or poor drainage. In computing areas occupied by land uses, the accoutrements such as landscaped grounds and parking facilities should be included. This will diminish the computed land-use weight by increasing the denominator.

In general, the more "intensive" land uses tend to have greater weights, since λ_1 and λ_2 are larger. In particular, the weight of multiple-story land uses tends to rise with the number of stories. The imports and exports of such a land use are the sums of the imports and exports origi-

Chapter 8

nating on the various floors; e.g., the trips to and from an office building are the sums of those terminating on the first, second, third, etc., floors. The ideal area of the site occupied by such a land use, however, is that of the "ground floor" only, not the sum of the floor areas of the successive stories. Extremely high weights can thus be obtained via skyscraper construction.

As for time displacements, some tendency exists for forward displacement to make land uses lighter. This is a reflection of the tendency for ideal weights of resources to become lighter over time, owing to transportation improvements and discounting. Forward displacement shifts the masses distributed by λ_1 and λ_2 toward smaller values of the integrands g_{in} and g_{out}, reducing the integrals (2).

Turning attention to the ideal weight functions, we note the general tendency for the ideal weight/physical weight ratio to be higher for people than for nonhuman resources, since people demand more in the way of roominess, comfort, etc., for their own travel than for the shipment of their chattels. Thus trips by people are an important contributor to the weight of most land uses, and probably dominate in those involving "facilities" such as residences, churches, schools, hospitals, office buildings, and retail trade.

Ideal weight varies considerably from person to person. To assess what is involved, remember that "transportation cost" is a composite money valuation of many diverse components, viz., not only fares and fuel consumption but the value of time spent in traveling, risk of accident, discomfort and fatigue, etc. The potential traveler evaluates these dimensions in dollar terms, and it is this personal assessment that constitutes his transport cost and determines his ideal weight.[8,9] Thus we may expect idiosyncratic elements to enter into ideal weight: someone with a pathological fear of travel accidents will be very "heavy" on that account.

At the same time we may expect some regularities. Valuation of elapsed time will rise with foregone earnings, so that people with high wages (actual or imputed) will tend to be "heavy." Rich people will, on the average, be willing to pay more to avoid the same degree of accident risk and uncomfortable travel conditions than poor people. Thus we may expect that ideal weight will rise with both earned and unearned income, and more so per dollar of earned than of unearned income.

However, one factor goes counter to this tendency; viz., the rich tend to use speedier and more comfortable modes of transportation (e.g., airplanes, taxis, private automobiles). The automobile functions as a gen-

8. This refers to "voluntary" travel. For "involuntary" travel—as by children, prisoners, and military personnel—the relevant valuation is not done by the traveler himself.
9. This self-assessment feature of transportation cost raises certain questions concerning the welfare implications of the entrepôt model. We comment on this later.

eral "map shrinker" or, better, as a "levitator," reducing the ideal weights of those who customarily travel with it.

Ideal weight also varies over time for the same person. We have already discussed some general features of this time dependency. Over and above these are variations induced by changes in the opportunity value of time. Thus, weight probably rises when a person enters the labor force and it falls at retirement. In the shorter run, weight is lower during evenings and weekends when there are fewer earning opportunities.[10]

These factors all enter into the computation of the ideal weight functions g_{in} and g_{out}, which in turn enter into the computation of land-use weight via (2) and (3). If one deals with a land use that is roughly steady state over a long period, in which trips by people are the dominant weight influence, the following schematic may be helpful for measurement purposes:

$$\begin{bmatrix} \text{weight of} \\ \text{land use} \end{bmatrix} = \frac{1}{i} \begin{bmatrix} \text{mean ideal} \\ \text{weight of} \\ \text{triptakers} \end{bmatrix} \begin{bmatrix} \text{nuclear} \\ \text{trips per} \\ \text{person} \\ \text{per year} \end{bmatrix} \begin{bmatrix} \text{population} \\ \text{density on} \\ \text{site} \end{bmatrix} \begin{bmatrix} \text{real acres} \\ \text{ideal acres} \end{bmatrix}$$

Each of the right-hand factors should be roughly estimable. Here i is the discount rate, inserted to convert the flow to a present value; mean ideal weight is based on income, car ownership, etc., and is an average weighted by trip-taking propensities; a round trip counts as two trips; the real/ideal areal ratio is based on topography, drainage, etc.

Finally, let us note the realism, or lack thereof, of the land-use concept we are using. The main departure from realism appears to lie in the absence of restrictions on the possible assignments v (other than the areal capacity limitation (11)). Thus land uses can be mixed freely, and the presence of a distribution of uses in one region has no effect on what is feasible in an adjoining disjoint region. In short, neighborhood effects are excluded, as are the associated effects of "scale" and "indivisibility." As discussed in Section 4.6, the resulting departure from realism tends to be more severe the smaller the scale of the system under discussion.

Another unrealistic simplification arises in the form of the areal constraint (11) itself. This says in effect that the demand-for-room function $k(s,q)$ is factorable. In more picturesque language, no location has a *comparative advantage* over any other in relative suitability for any pair of land uses. It is easy to find exceptions; e.g., soil fertility is relevant for agricultural land uses but irrelevant for most urban land uses; hence *infertile* land has a comparative advantage for the latter. Marshy land has a

---

10. G. S. Becker, A theory of the allocation of time, *Econ. J.* 75 (Sept. 1965): 493–517.

comparative advantage for certain kinds of recreational uses, hilly land for residences, etc. On the institutional side, *zoning* is in effect the artificial introduction of comparative advantages by differential exclusions of certain land uses from certain regions. Some (but not all) forms of *real-estate taxation* have the same effect. All these phenomena are excluded by assumption. (Later we discuss the modifications induced by introducing some of them.)

A similar difficulty arises if the Thünen system starts up from some designated "time zero." If time zero precedes the settlement of the region one has only to contend with the geographic nonuniformities of nature. But if one places time zero *in medias res,* with a preceding period of settlement, further complications occur. Man himself creates differential advantages by building different structures in different places, by leaving other places vacant, and by distributing himself nonuniformly over the landscape. This point is important because the model has variables that refer to time zero (e.g., land values at that time) and not to other times.

8.5. THE ALLOTMENT-ASSIGNMENT PROBLEM

We have set up an apparatus of concepts for Thünen systems; it is now time to produce some models. Two kinds will be considered. One is behavioral, the interactions of many agents in the real-estate market leading to the Thünen configuration of land uses. The other involves optimization—specifically, the minimization of total transport cost over a certain set of possible assignments. This again leads to the same land-use pattern, so that the free market interaction of numerous agents leads to the minimization of transport cost.

In the theory of urban structure, a long controversy has raged on exactly this point: Is the metropolis laid out to minimize the "friction of space" (and *should* it be so laid out)? The literature has been ably reviewed by Alonso, who concludes that friction is not (and should not be) minimized because other desiderata such as roominess are also important.[11] While this is perfectly correct, the issue is not settled because the meaning of "minimization" is left unclear. Specifically, one must name the *set of alternatives* under consideration before one can say that the alternative actually chosen does or does not minimize a certain objective. In the following development, the set of alternatives is such that the free market *does* minimize total transport cost over that set. (Whether it should do so is something we discuss later.)

We first present the optimization model. The problem is to choose an

11. W. Alonso, *Location and Land Use* (Harvard Univ. Press, Cambridge, 1964), Ch. 1 and pp. 101–5. The term "friction of space" dates from R. M. Haig, 1926; it is sometimes taken to include total land value as well as total transportation cost.

assignment ν, which is formally a measure over the product space $(S \times Q,$ $\Sigma_s \times \Sigma_q)$, out of the feasible set of such assignments. The objective is to minimize total transport costs on shipments to and from the nucleus. According to our previous analysis, this is given by (4.4):

$$\int_{S \times Q} h(s)w(q)\, \nu(ds, dq) \qquad (8.5.1)$$

where $h:S \rightarrow$ reals and $w:Q \rightarrow$ reals are the ideal distance and land-use weight functions respectively. These are assumed to be measurable. Actually, all our basic results are still obtained with a much more general objective function than (1). We need the following concepts.

Definition: Let f be a real-valued function whose domain is 2-space (i.e., the plane, reals^2). Function f has *positive cross-differences* iff for all real numbers x_1, x_2, y_1, y_2 such that $x_1 < x_2$ and $y_1 < y_2$, we have

$$f(x_1, y_1) + f(x_2, y_2) > f(x_1, y_2) + f(x_2, y_1) \qquad (8.5.2)$$

Function f has *nonnegative cross-differences* iff the same condition holds with $\geq$ replacing $>$ in (2).

These definitions easily extend to the case where the domain of f is a rectangle

$$X \times Y \qquad (8.5.3)$$

X and Y being real intervals. Simply restrict x_1, x_2 to lie in X, and y_1, y_2 in Y. Now consider the integral

$$\int_{S \times Q} f(h(s), w(q))\, \nu(ds, dq) \qquad (8.5.4)$$

where f is a measurable function having positive (or perhaps nonnegative) cross-differences. (From this point we no longer write Q in boldface, since we are dealing with an abstract problem in which S and Q enter symmetrically.) Integral (1) is the special case of (4) in which f is simply the product $f(x, y) = xy$. (This function clearly satisfies (2).) Hence any general results obtained using (4) as the objective function will apply to (1) in particular. The domain of f in (4) will usually be the plane, but if the ranges of h and w are both bounded, it is possible (and sometimes advantageous) to let it be a rectangle with interval sides.[12]

Integrals (1) and (4) are written as definite integrals. In case they are infinite or not well defined, however, we interpret them as indefinite

---

12. Note that the f appearing in (4), whose domain is 2-space, does not have the same form as the unit transport cost function f of Chapter 7, whose domain is $A \times B$. The analog of the latter is the composite function $f(h(\cdot), w(\cdot))$, with domain $S \times Q$. However, below we convert this problem into another for which the f's *do* correspond.

integrals in the sense of pseudomeasures (ν being σ-finite), and "minimization" of (1) or (4) is taken in the sense of (reverse) standard ordering of pseudomeasures. For well-defined finite integrals this of course reduces to the ordinary size comparison of definite integrals.

Next we come to feasibility conditions on ν. First, there is the areal capacity constraint (4.11) (in equality form):

$$\nu(F \times Q) = \alpha(F) \qquad (8.5.5)$$

for all $F \in \Sigma_s$. Here ideal area α is formally a measure on (S, Σ_s). α is given and assumed to be σ-finite; (5) then guarantees that any feasible assignment ν will also be σ-finite.

We now have two-thirds of a transportation problem, with objective function (4) and capacity constraint (5); the requirements constraint is missing. It is possible to stop at this point and consider the "one-sided" transportation problem: minimize (4) over measures ν, subject to (5). Formally, a model of this sort has been constructed by Stevens, with an inequality constraint and in non-measure-theoretic terms.[13] This one-sided problem is in fact the special case of the transportation problem (variant III) in which the requirement measure is zero. Hence the theory of the problem is more or less encompassed in the results of Chapter 7.

In any case, this one-sided approach does not appear to get very far, and for deep results one must go on to the full "two-sided" transportation problem. We therefore add the following constraint:

$$\nu(S \times G) = \beta(G) \qquad (8.5.6)$$

for all $G \in \Sigma_q$. Here β is a given σ-finite measure on the space of land uses; β will be called the *allotment measure* and (6) the *allotment constraint*. The entire problem of minimizing (4) over assignments ν, subject to constraints (5) and (6), will be called the *allotment-assignment problem*.

Constraint (6) may be interpreted as follows. For any measurable set of activities G, a total acreage allotment $\beta(G)$ is specified, which must be met by any assignment; two acres must be devoted to turnip growing, five acres to education, etc. There is still freedom to shuffle these land uses around over S, but the totals are fixed. In contrast to the areal-capacity constraint (5), which represents a "real" restriction on possible assignments grounded in natural law or human institutions, (6) is best regarded as an "artificial" restriction added to attain certain results. (There is one exception: (6) is a "natural" restriction in *layout* problems, for which technology dictates the allotment, as in the separate processes of a manu-

---

13. B. H. Stevens, Location theory and programming models: The Von Thünen case, *Reg. Sci. Assoc. Pap.* 21 (1968):19–34. The model is on p. 26. The objective here is to maximize total bid rent or profit rather than to minimize total transport cost, but the formal structure is the same. M. Beckmann and T. Marschak, An activity analysis approach to location theory, *Kyklos* 8(1955):125–141, seem to have envisaged this model in a remark on p. 128.

facturing complex. But the Thünen framework is not well suited for layout problems.)[14]

Though artificial in the sense of not representing an actual constraint on behavior, (6) serves a function that arises from the "inner logic" of Thünen systems. Consider the matter in the following light. Thünen systems arise in a diversity of contexts, on all different scales. They have in common precisely the pattern of land uses: the ring structure and the ordering of uses. They do not have in common the particular land uses present in each and their relative proportions: in short, the *allotments* of land uses. For someone looking for a universal theory of Thünen systems, the allotment measures are the contingent features. It is then reasonable to treat allotments as exogenous and to set up a model that yields the Thünen pattern of land uses regardless of what the allotment is. And (6) does just this. The allotment β is given a priori, and we are to find the optimal assignment *within* that given allotment. The resulting pattern is (within very wide limits) independent of β.

This approach is not used by any other contemporary model builder in the Thünen tradition.[15] Rather, these authors try to predict the assignment of land uses without assuming the allotment in advance. In this sense our aim is narrower and more modest than theirs. But by the same token we cut through the aspects of these models, which (from the point of view of predicting the Thünen pattern) are irrelevant and distracting, and thus we attain a deeper understanding of that pattern. From this point of view our assumptions are much weaker than any of theirs.

The allotment-assignment problem, then, is given by objective function (4) to be minimized subject to constraints (5) and (6) on assignments v. This is formally a measure-theoretic transportation problem of variant I, i.e., with equality constraints (other variants could be used, but I is the simplest). The special feature of the allotment-assignment problem lies in the form of the integrand in (4), particularly in the fact that f has positive (or nonnegative) cross-differences.

This special feature enables us to make very strong statements concerning the nature of the solution. We need a few concepts for this. First, on the plane it will be convenient to say that point (x_1, y_1) is *south-*

14. Agricultural programs often contain acreage allotment restrictions but these typically apply farm by farm, not on a global basis for all Space as in (6).

15. Such as: E. S. Dunn, Jr., *The Location of Agricultural Production* (Univ. Florida Press, Gainesville, 1954); R. F. Muth, The spatial structure of the housing market, *Reg. Sci. Assoc. Pap.* 7:(1961):207–20; L. Wingo, Jr., *Transportation and Urban Land* (Resources for the Future, Inc., Washington, D.C., 1961); Alonso, *Location and Land Use;* E. S. Mills, *Studies in the Structure of the Urban Economy* (Johns Hopkins Press, Baltimore, 1972), Ch. 5–8. For recent work see A. Anas and D. S. Dendrinos, The New Urban Economics: A Brief Survey, pp. 23–51, and Coded Bibliography, pp. 293–307, in G. J. Papageorgiou, ed., *Mathematical Land Use Theory* (Heath, Lexington, Mass., 1976). Recent models transcend the Thünen framework, e.g., by incorporating multiple nuclei or externalities. Much of our conceptual critique of Alonso's model (Sect. 6.8) applies to the new urban economics generally.

west of (x_2, y_2) iff $x_1 < x_2$ and $y_1 < y_2$, *northwest* iff $x_1 < x_2$ and $y_1 > y_2$, etc. Next, given two subsets of the plane E_1 and E_2, E_1 is said to be *southwest* of E_2 iff every point of E_1 is southwest of every point of E_2 in the sense just defined. Next let functions $h:S \rightarrow$ reals and $w:Q \rightarrow$ reals be given, and let (s_1, q_1), (s_2, q_2) be two points of the cartesian product $S \times Q$; (s_1, q_1) is *southwest* of (s_2, q_2) iff $h(s_1) < h(s_2)$ and $w(q_1) < w(q_2)$. Finally, given two subsets of $S \times Q$, E_1 and E_2, E_1 is *southwest* of E_2 iff every point of E_1 is southwest of every point of E_2 in this sense. Using this last concept we have the following.

Definition: Let ν be a measure on the product space $(S \times Q,$ $\Sigma_s \times \Sigma_q)$, and let $h:S \rightarrow$ reals and $w:Q \rightarrow$ reals be functions. Measure ν satisfies the *measurable weight-falloff condition* iff there do not exist two sets E_1, $E_2 \in \Sigma_s \times \Sigma_q$, both of positive ν-measure, with E_1 southwest of E_2.

As with potentials, there is a corresponding topological concept. We suppose topologies $\mathfrak{J}_s$ and $\mathfrak{J}_q$ have been placed on S and Q respectively, making them topological spaces as well as measurable spaces. These determine a product topology, $\mathfrak{J}_s \times \mathfrak{J}_q$ on $S \times Q$, and this with $\Sigma_s \times \Sigma_q$ determines the support of measure ν.

Definition: Let ν be a measure on $(S \times Q, \Sigma_s \times \Sigma_q)$, and let $h:S \rightarrow$ reals and $w:Q \rightarrow$ reals be functions. Measure ν satisfies the *topological weight-falloff condition* iff there do not exist two points of support for ν, one of which is southwest of the other.

Roughly speaking both these concepts state intuitively that h and w are negatively correlated: ν tends to concentrate mass where h is high and w low, and vice versa. Note that h and w enter symmetrically into these definitions, so that, instead of speaking of weight w falling off as distance h rises, one could speak of distance falling off as weight rises.

The two functions h and w determine a single function mapping $S \times Q$ into the plane, viz., the one that assigns the value $(h(s), w(q))$ to the point (s, q). Assume that h and w are measurable; then this function is measurable. Hence for any measure ν over universe set $S \times Q$, it induces a measure λ over the plane: $\lambda(E) = \nu\{(s, q) \mid (h(s), w(q)) \in E\}$ for any Borel subset E of the plane.

Now the two weight-falloff definitions above apply just as well to λ as to ν (the plane being furnished with its usual topology and Borel field, and with h and w each replaced by the identity map $x \rightarrow x$ on the real line). Hence we have four apparently different concepts. But our next result shows that three of these conditions are logically equivalent.

Theorem: Let ν be a measure on $(S \times Q, \Sigma_s \times \Sigma_q)$, let $h{:}S \rightarrow$ reals and $w{:}Q \rightarrow$ reals be measurable, and let λ be the measure on the plane induced from ν by h and w. Then each of the following conditions implies the other two:

    (i) ν satisfies the measurable weight-falloff condition;
    (ii) λ satisfies the measurable weight-falloff condition;
    (iii) λ satisfies the topological weight-falloff condition.

Proof: (i) *implies* (ii): Let E_1, E_2 be two measurable subsets of the plane, with E_1 southwest of E_2. Their inverse images,

$$\{(s,q) \mid (h(s), w(q)) \in E_i\} \tag{8.5.7}$$

$i = 1, 2$, retain this southwest-northeast relation. Hence at least one of them has ν-measure zero, which implies $\lambda(E_i) = 0$ for at least one E_i. Thus λ satisfies measurable weight-falloff.

    (ii) *implies* (iii): Let z_1, z_2 be two points of the plane, with z_1 southwest of z_2. Open discs E_1, E_2 about z_1, z_2 respectively exist such that E_1 is southwest of E_2. Sets E_1 and E_2 cannot both have positive λ-measure, hence z_1 and z_2 cannot both support λ. Thus λ satisfies topological weight-falloff.

    (iii) *implies* (i): Let E_1, E_2 be two measurable subsets of $S \times Q$, with E_1 southwest of E_2, and consider their images in the plane:

$$F_i = \{(h(s), w(q)) \mid (s, q) \in E_i\}$$

$i = 1, 2$. F_1 is southwest of F_2. Hence at least one of these two sets, say F_j, cannot own any points of support for λ. Thus each point $(x, y) \in F_j$ has a measurable neighborhood of λ-measure zero. The usual topology of the plane has the strong Lindelöf property, so that F_j is contained in the union of a countable number of these neighborhoods. Call this union G; G is measurable, and $\lambda(G) = 0$. It follows that

$$0 = \nu\{(s, q) \mid (h(s), w(q)) \in G\} \geq \nu(E_j) \tag{8.5.8}$$

The equality in (8) arises from the fact that λ is induced from ν; the inequality arises from the fact that E_j is contained in the inverse image of F_j, which in turn is contained in the inverse image of G. Thus at least one of E_1, E_2 has ν-measure zero: ν satisfies measurable weight-falloff.

We now have a closed circle of implications. ∎

We speak simply of ν (or λ) satisfying *the* weight-falloff condition in the event that any (hence all) of the above three conditions obtains. What about the fourth condition, which is topological weight-falloff for ν? This depends on the topologies $\mathfrak{I}_s$ and $\mathfrak{I}_q$, which do not enter the definition of the other three concepts.

Theorem: Let ν be a measure on $(S \times Q, \Sigma_s \times \Sigma_q)$, let $h:S \to$ reals and $w:Q \to$ reals be measurable, and let λ be the measure on the plane induced from ν by h and w; also let $\mathfrak{I}_s$ and $\mathfrak{I}_q$ be topologies on S and Q respectively; then

(i) if ν (or λ) satisfies the weight-falloff condition and h and w are *continuous* functions, then ν satisfies the *topological* weight-falloff condition;

(ii) if ν satisfies the *topological* weight-falloff condition and $\mathfrak{I}_s \times \mathfrak{I}_q$ has the *strong Lindelöf property*, then ν (or λ) satisfies the weight-falloff condition.

Proof: (i) If (s, q) is a point of support for ν, then $(h(s), w(q))$ is a point of support for λ. To show this, let $(h(s), w(q)) \in E_1 \subseteq E_2$, where E_1 is open and E_2 measurable. The inverse images (7) are open for E_1 and measurable for E_2, since h, w are continuous and measurable. Hence the inverse image of E_2 has positive ν measure, implying $\lambda(E_2) > 0$. Thus $(h(s), w(q))$ supports λ.

Now let (s_i, q_i), $i = 1, 2$, be two points of support for ν. Since $(h(s_i), w(q_i))$, $i = 1, 2$, are both points of support for λ, they cannot stand in a southwest-northeast relation. Hence neither can (s_i, q_i), $i = 1, 2$, so that ν satisfies topological weight-falloff.

(ii) Let E_1, E_2 be two measurable subsets of $S \times Q$, with E_1 southwest of E_2. At least one of these two sets, say E_j, cannot own any points of support for ν. Utilizing the strong Lindelöf property as in the preceding proof ((iii) *implies* (i)), it follows that E_j is contained in a ν-null set. Thus at least one of E_1, E_2 has ν-measure zero: ν satisfies weight-falloff. ■

These results imply that if h and w are both continuous and $\mathfrak{I}_s \times \mathfrak{I}_q$ has the strong Lindelöf property, then any of these weight-falloff conditions implies the other three. The next result establishes a connection between weight-falloff and allotment-assignment.

Theorem: Let (S, Σ_s, α) and (Q, Σ_q, β) be σ-finite measure spaces. Let $\mathfrak{I}_s \subseteq \Sigma_s$ and $\mathfrak{I}_q \subseteq \Sigma_q$ be topologies on S and Q respectively. Let $h:S \to$ reals, $w:Q \to$ reals, and $f:$reals$^2 \to$ reals be functions such that the composite function $f(h(\cdot), w(\cdot)):S \times Q \to$ reals is *measurable* and *continous* with respect to $\mathfrak{I}_s \times \mathfrak{I}_q$. Finally, let f have *positive cross-differences*.

Then if measure ν^o is *unsurpassed* for the allotment-assignment problem of minimizing (4) (reverse standard order) subject to the constraints

(5) and (6), it follows that ν^o satisfies the *topological weight-falloff condition* (with respect to h and w).

Proof: The premises imply that ν^o satisfies the *circulation condition* (7.5.10) or (7.5.11). Thus if (s_1, q_1) and (s_2, q_2) are two points of support for ν^o, we have

$$f_{12} + f_{21} - f_{11} - f_{22} \geq 0 \qquad (8.5.9)$$

where f_{ij} abbreviates $f(h(s_i), w(q_j))$, $i, j = 1, 2$. We cannot have both $h(s_1) < h(s_2)$ and $w(q_1) < w(q_2)$ because in this case (9) would contradict the positive cross-differences condition (2). That is, (s_1, q_1) cannot be southwest of (s_2, q_2). Thus ν^o satisfies topological weight-falloff. ∎

Our next result is similar to this one. Though its proof is more complicated, it is also more interesting because it makes no continuity assumptions; indeed, it uses no topological concepts whatever. A function is *half-bounded* iff it is bounded below or bounded above (or both, i.e., bounded); a *set* is bounded iff the distance function between its points is bounded; the half-boundedness condition below is very weak.

Theorem: Let (S, Σ_s, α) and (Q, Σ_q, β) be σ-finite measure spaces. Let $h:S \to$ reals, $w:Q \to$ reals, $f:\text{reals}^2 \to$ reals be functions such that f has *positive cross-differences*, and the composite function $f(h(\cdot), w(\cdot))$: $S \times Q \to$ reals is measurable. Let measure ν^o be *best* for the allotment-assignment problem of minimizing (4) (reverse standard order) subject to constraints (5) and (6).

Then ν^o satisfies the *(measurable) weight-falloff condition* (with respect to h, w).

If ν^o is merely *unsurpassed*, the same conclusion follows provided one adds the premise that f is *half-bounded on any bounded subset of the plane* and that h, w are measurable.

Proof: Assume ν^o violates measurable weight-falloff, so that sets $F_1, F_2 \in \Sigma_s \times \Sigma_q$ of positive ν^o-measure exist, with F_1 southwest of F_2. Either of these may in fact have infinite measure, but in any case they will contain subsets $G_i \subseteq F_i$, $i = 1, 2$, of positive finite measure, since ν^o is σ-finite. Define the measures ν_1, ν_2 on $(S \times Q, \Sigma_s \times \Sigma_q)$ by

$$\nu_i(H) = \nu^o(H \cap G_i)/\nu^o(G_i) \qquad (8.5.10)$$

$i = 1, 2, H \in \Sigma_s \times \Sigma_q$, and define the signed measure ν by

$$\nu = (\nu_1' \times \nu_2'') + (\nu_2' \times \nu_1'') - \nu_1 - \nu_2 \qquad (8.5.11)$$

(Here ν_i', ν_i'' are the left and right marginals, respectively, of ν_i.) ν is well defined, since both ν_1 and ν_2 are bounded measures. Finally, consider the (signed) measure

$$\nu^o + z\nu \tag{8.5.12}$$

where $z = \min[\nu^o(G_1), \nu^o(G_2)]/2 > 0$. One easily verifies that (12) remains feasible for the allotment-assignment problem (cf. (7.5.20)).

Taking the case where ν^o is best, we reach a contradiction. Since (12) is feasible, we must have

$$\int f(h(s), w(q)) \nu(ds, dq) \geqslant 0 \tag{8.5.13}$$

(The integral in (13) is a pseudomeasure over $S \times Q$, and $\geqslant$ refers to standard ordering (cf. (7.5.21)).) The integral in (13) can be written as the sum of four indefinite integrals, corresponding to the splitting of ν into its four components (11). We now show that

$$\int_{S \times Q} f \, d(\nu'_1 \times \nu''_2) + \int_{S \times Q} f \, d(\nu'_2 \times \nu''_1) - \int_{S \times Q} f \, d\nu_1 - \int_{S \times Q} f \, d\nu_2 \tag{8.5.14}$$

is a well-defined expression; here "f" abbreviates $f(h(\cdot), w(\cdot))$. That is, each of the four definite integrals in (14) is well defined, and their sum is not of the form $\infty - \infty$. To see this, note that there are numbers x, y such that

$$h(s_1) \leq x \leq h(s_2) \tag{8.5.15}$$

$$w(q_1) \leq y \leq w(q_2) \tag{8.5.16}$$

for all $(s_1, q_1) \in G_1$, $(s_2, q_2) \in G_2$, where one of the inequality signs in (15) and one in (16) can be replaced by $<$. This follows from G_1 being southwest of G_2. Condition (15) determines a partition of S into two pieces, one set satisfying the left inequality, the other the right. Similarly, (16) splits Q into two pieces. Together these split $S \times Q$ into four pieces. Set G_1 is contained in the "southwest quadrant" of low h, w values; G_2 is contained in the "northeast quadrant" of high h, w values. It follows that the complement of the (southwest, northeast) quadrant has (ν_1-measure, ν_2-measure) zero respectively. And from this it follows that the complement of the (southeast, northwest) quadrant has ($\nu'_2 \times \nu''_1$-measure, $\nu'_1 \times \nu''_2$-measure) zero respectively. Thus the four components of ν are mutually singular in pairs. The indefinite integral (13) can therefore be expressed as a direct sum of four integrals over these quadrants. Being comparable to 0, the integral (13) must be a signed measure, so that (14) is indeed well defined. In fact, the relation (13) implies that the expression (14) is nonnegative, by the standard integral theorem.

Now consider the "four-dimensional" product-measure space

$$(S_1 \times Q_1, \Sigma_s \times \Sigma_q, \nu_1) \times (S_2 \times Q_2, \Sigma_s \times \Sigma_q, \nu_2)$$

Here S_1 and S_2 are replicas of S, and Q_1 and Q_2 are replicas of Q; the subscripts are added for clarity. We have

$$\int_{S \times Q} f \, dv_i = \int_{S_1 \times Q_1 \times S_2 \times Q_2} f_{ii} \, d(v_1 \times v_2) \qquad (8.5.17)$$

$i = 1, 2$, and

$$\int_{S \times Q} f \, d(v'_i \times v''_j) = \int_{S_1 \times Q_1 \times S_2 \times Q_2} f_{ij} \, d(v_1 \times v_2) \qquad (8.5.18)$$

where $(i, j) = (1, 2)$ and $(i, j) = (2, 1)$.

In (17) and (18) f on the left again abbreviates $f(h(\cdot), w(\cdot))$, while f_{ij} stands for $f(h(s_i), w(q_j))$, i and j ranging over 1, 2 (four cases). The four equations in (17) and (18) all arise from the induced integrals theorem, resulting from four different projections from the space $S_1 \times Q_1 \times S_2 \times Q_2$ to $S \times Q$. Thus (17) for $i = 1$ arises from the projection $(s_1, q_1, s_2, q_2) \rightarrow (s_1, q_1)$; for $i = 2$ it arises from $(s_1, q_1, s_2, q_2) \rightarrow (s_2, q_2)$. Equation (18) for $(i, j) = (1, 2)$ arises from $(s_1, q_1, s_2, q_2) \rightarrow (s_1, q_2)$; for $(i, j) = (2, 1)$ it arises from $(s_1, q_1, s_2, q_2) \rightarrow (s_2, q_1)$. The only difficulty in demonstrating all this arises in (18), where it must be shown, e.g., that $v'_1 \times v''_2$ is the measure induced from $v_1 \times v_2$ by the projection $(s_1, q_1, s_2, q_2) \rightarrow (s_1, q_2)$. This follows from direct substitution in the definition of product measure. The well-definedness of the left integrals in (17) and (18) implies the well-definedness of the right integrals and the stated equalities.

Equations (17), (18), and the nonnegativity of (14) then yield

$$\int_{S_1 \times Q_1 \times S_2 \times Q_2} (f_{12} + f_{21} - f_{11} - f_{22}) \, d(v_1 \times v_2) \geq 0 \qquad (8.5.19)$$

But this is a contradiction. First, $v_i[(S_i \times Q_i)\backslash G_i] = 0$, $i = 1, 2$, so that the complement of $G_1 \times G_2$ has $v_1 \times v_2$-measure zero. Second,

$$(v_1 \times v_2)(G_1 \times G_2) = v_1(G_1) \cdot v_2(G_2) = 1 > 0$$

Finally, the integrand in (19) is negative on $G_1 \times G_2$, since f has positive cross-differences. Hence the integral in (19) must be negative, a contradiction. This proves the first half of the theorem.

Now take the case where f is *half-bounded on bounded sets* and v^o is merely *unsurpassed,* and again assume that v^o violates measurable weight-falloff. Proceeding as above, we find a set G_1 southwest of a set G_2, both with positive finite v^o-measure. Each of these contains a subset of positive measure on which $f(h(\cdot), w(\cdot))$, as well as h and w themselves, are *bounded*, for the measurable sets

$$G_i \cap \{(s, q) \mid m \leq f(h(s), w(q)) < m + 1,$$

$$n \leq h(s) < n + 1, p \leq w(q) < p + 1\}$$

$m, n, p = 0, \pm 1, \pm 2, \ldots$, countably partition G_i, hence one of these has positive measure. For simplicity we designate these subsets by the same symbols, G_1 and G_2.

Define ν_1 and ν_2 as in (10) and consider the expression (14). The four measures appearing in these integrals are bounded. The complement of G_i has ν_i-measure zero, $i = 1, 2$, and $f(h(\cdot), w(\cdot))$ is bounded on $G_1 \cup G_2$; hence the last two integrals in (14) are well defined and finite. As for the first two, $\nu'_i \times \nu''_j$ are both zero off some set on which h and w are bounded, and the integrands are half-bounded on this set. It follows that they too are well defined and not infinite of opposite sign. Hence the whole expression (14) is well defined.

It follows that the indefinite integral (13) is a signed measure, not a proper pseudomeasure, and is therefore comparable to 0 under standard ordering. Just as above we construct the new feasible solution (12), and the unsurpassedness of ν^o together with comparability then yields relation (13). Hence (14) is nonnegative. The argument above then again yields a contradiction, and the last part of the theorem is proved. ∎

We make one immediate generalization: the premises on f need hold only for f restricted to the range of (h, w). The proof above still holds verbatim.

These theorems have a very simple intuitive meaning. Suppose ν fails to satisfy, say, topological weight-falloff, so that there are points of support (s_1, q_1), (s_2, q_2), the first southwest of the second. Then shift a mass of activities in the neighborhood of q_1 from the neighborhood of location s_1 to the neighborhood of location s_2, and shift a mass of activities in the neighborhood of q_2 in the opposite direction. This reshuffling does not affect the feasibility conditions (5) and (6), and the positive cross-difference condition on f implies that the total transport cost (4) has been reduced. Hence the original assignment ν has been surpassed. The proofs above are merely a rigorization of this informal argument.

If we are looking for optimal solutions to the allotment-assignment problem under the mild conditions stated above, these theorems say we might as well confine our attention to assignments ν satisfying some weight-falloff condition. But there is as yet no guarantee that such solutions will be best or even unsurpassed. We do not know at this stage whether such feasible assignments even exist. And even if they do, they may not be optimal, since the possibility remains that there are no optimal solutions.

We attack these difficulties by transforming the original allotment-assignment problem into a simpler one. Specifically, we "induce" the original problem, which is set in the product space $S \times Q$, into the plane by means of the functions h and w. Measure α on (S, Σ_s) is induced by h into a measure on the real line. Similarly, β on (Q, Σ_q) is induced by w into a measure on the real line. Finally, ν induces a measure λ on the plane via the combined function $(s, q) \rightarrow (h(s), w(q))$.

We retain the notation α, β for the measures on the real line induced by these respective original measures, and rely on context to distinguish them. The transformed allotment-assignment problem now reads as follows:

Find a measure λ on the plane that satisfies the constraints

$$\lambda'(E) = \alpha(E) \qquad (8.5.20)$$

$$\lambda''(E) = \beta(E) \qquad (8.5.21)$$

for all Borel sets E on the real line, and that minimizes

$$\int_{\text{reals}^2} f \, d\lambda \qquad (8.5.22)$$

Here λ', λ'' are the left and right marginals of λ respectively and are measures on the real line. Constraints (20) and (21) are the analogs of the areal-capacity and allotment constraints (5) and (6) respectively. The integrand f in the objective function (22) is the same as the f appearing in (4), and so has positive (or nonnegative) cross-differences.

The transformed measures α and β in (20) and (21) must be σ-finite. This is not implied by the σ-finiteness of the original α, β measures in (5) and (6) and must be explicitly postulated. In fact we make an even stronger assumption below.

This transformed problem has been placed on the plane. More generally, it could be placed on a rectangle $X \times Y$ (with interval sides) provided only that the ranges of h and w are contained in X, Y respectively. Then $X \times Y$ is the domain of f and the universe set of λ; X and Y are the universe sets of transformed α and β respectively.

The objective function (22) is written as a definite integral. But just as with (4), if it is not well defined or finite for certain feasible measures λ, it is to be interpreted as an indefinite integral pseudomeasure; and "minimization" is to be understood in the sense of (reverse) standard ordering.

The first thing to notice about this transformed problem is that, formally, it is just a special case of the allotment-assignment problem: both (S, Σ_s) and (Q, Σ_q) are the real line with its Borel field, and both h and w are the identity function. The preceding theorems then apply and take a very simple form:

Theorem: Let measurable f:reals^2 → reals have *positive cross-differences*, and let λ^o be *best* for the problem of minimizing (22) subject to (20) and (21); then λ^o satisfies the *weight-falloff condition*. The same conclusion holds if λ^o is merely *unsurpassed*, provided f is *half-bounded on bounded sets*.

We now investigate the feasibility and optimality relations between the original and the transformed allotment-assignment problems. The following property of induced pseudomeasures is needed.

Lemma: Let (B, Σ') and (C, Σ'') be measurable spaces, and $g:B \to C$ a measurable function. Let ν, ν^o be two measures on (B, Σ'), and λ, λ^o the measures on (C, Σ'') induced by g from ν, ν^o respectively, all four of these measures being σ-finite.

Then (λ^o, λ) is the pseudomeasure induced by g from pseudomeasure (ν^o, ν).

Proof: Recall (p. 398) that the pseudomeasure *induced* by g from ψ on (B, Σ') is (μ_1, μ_2), where μ_1, μ_2 are the measures induced by ψ^+, ψ^- respectively; this is defined iff μ_1, μ_2 are both σ-finite. For $\psi = (\nu^o, \nu)$ we have

$$\psi^+ + \nu = \psi^- + \nu^o \tag{8.5.23}$$

(equivalence theorem). This implies

$$\mu_1 + \lambda = \mu_2 + \lambda^o \tag{8.5.24}$$

To show this, let $E \in \Sigma''$, and apply (23) to $\{b \mid g(b) \in E\}$. The four terms in (23) are respectively equal to the four terms in (24) applied to E. Hence (24) is true.

Now $\psi^+ \le \nu^o$ and $\psi^- \le \nu$ (minimizing property of the Jordan form). Hence $\mu_1 \le \lambda^o$ and $\mu_2 \le \lambda$, so that μ_1 and μ_2 are σ-finite, and the induced pseudomeasure exists. From (24) we obtain $(\mu_1, \mu_2) = (\lambda^o, \lambda)$ by the equivalence theorem again. ∎

Theorem: Let (S, Σ_s, α) and (Q, Σ_q, β) be measure spaces, and let $h:S \to$ reals, $w:Q \to$ reals, and f:reals$^2 \to$ reals be measurable functions. Let α and β, as well as their namesakes induced on the real line by h and w respectively, be σ-finite. Let ν^o be a measure on $(S \times Q, \Sigma_s \times \Sigma_q)$ that is feasible for the original allotment-assignment problem (4)–(6), and let λ^o be the measure on the plane induced from ν^o by the mapping $(s,q) \to (h(s), w(q))$. Then

(i) λ^o is feasible for the transformed problem (20)–(22);

(ii) if λ^o is unsurpassed for the transformed problem, ν^o is unsursurpassed for the original problem.

Proof: (i) The feasibility condition on ν^o is that its left and right marginals coincide with α and β respectively. For any Borel set E on the real line we have

$$\lambda^{o\prime}(E) = \lambda^o(E \times \text{reals})$$

$$= \nu^o[\{s \mid h(s) \in E\} \times Q] = \alpha\{s \mid h(s) \in E\} \quad (8.5.25)$$

The right-hand term in (25), however, is simply $\alpha(E)$ for the induced measure α; this proves (20) for λ^o. A similar argument establishes (21). Thus λ^o is feasible for the transformed problem.

(ii) Assuming that ν^o is surpassed, we prove that λ^o is surpassed. Abbreviate the composite function $f(h(\cdot), w(\cdot)):S \times Q \to$ reals by k. Then by hypothesis a feasible ν exists such that

$$\int (-k) \, d\nu > \int (-k) \, d\nu^o \quad (8.5.26)$$

(These are indefinite integrals over $S \times Q$, and $>$ is the "greater than" relation for standard ordering of pseudomeasures. The minus sign is introduced to convert the objective from minimization to maximization.)

Relation (26) is equivalent to

$$\int k \, d\psi > 0 \quad (8.5.27)$$

where ψ is the pseudomeasure (ν^o, ν). From the definition of standard order, (27) is the same as

$$\int_{S \times Q} k^+ \, d\psi^+ + \int_{S \times Q} k^- \, d\psi^- > \int_{S \times Q} k^+ \, d\psi^- + \int_{S \times Q} k^- \, d\psi^+$$

$$(8.5.28)$$

where these are four ordinary definite integrals.

Let μ_1, μ_2 be the measures induced on the plane from ψ^+, ψ^- respectively, by the mapping $(s, q) \to (h(s), w(q))$. Then (28) implies

$$\int_{\text{reals}^2} f^+ \, d\mu_1 + \int_{\text{reals}^2} f^- \, d\mu_2 > \int_{\text{reals}^2} f^+ \, d\mu_2 + \int_{\text{reals}^2} f^- \, d\mu_1$$

$$(8.5.29)$$

since by the ordinary induced integrals theorem, the four integrals in (29) are equal to the integrals in (28) respectively from left to right.

We have $\psi^+ \leq \nu^o$ by the minimizing property of the Jordan form. Hence their induced measures stand in the same relation: $\mu_1 \leq \lambda^o$. A similar argument yields $\mu_2 \leq \lambda$, where λ is the measure on the plane induced from ν. Part (i) established that λ^o and λ were feasible for the transformed problem. Hence they, and therefore μ_1 and μ_2, are σ-finite. Then (29) implies

$$\iint d(\mu_1, \mu_2) > 0 \qquad (8.5.30)$$

in terms of pseudomeasure (μ_1, μ_2). This latter is the pseudomeasure induced from $\psi = (\nu^o, \nu)$, and we now invoke the preceding lemma to establish the pseudomeasure equality:

$$(\mu_1, \mu_2) = (\lambda^o, \lambda) \qquad (8.5.31)$$

Finally, (30) and (31) yield $\int(-f)\,d\lambda > \int(-f)\,d\lambda^o$. But $\int(-f)\,d\lambda$ is just the (negated) objective function (22) for the transformed problem, so λ^o is surpassed by λ. ∎

(If (4) is a well-defined, finite, definite integral for all feasible ν, part (ii) of this theorem can be proved in a few lines, since (22), for the λ induced from ν, is equal to (4) by the induced integrals theorem. In this simple case the distinction between "unsurpassed" and "best" disappears.)

In general, these results cannot be strengthened. One cannot infer the feasibility of ν^o from that of λ^o, nor the optimality (in any sense) of λ^o from that of ν^o. This latter inference, for example, is blocked by the following difficulty. To establish the optimality of λ^o one must consider all other feasible measures λ. But it is not necessarily the case that every such λ is the induced measure from some feasible ν. (In fact there may not be any such ν, feasible or not.) The optimality of ν^o tells nothing about such "uninduced" feasible measures λ, so that the optimality of λ^o cannot be inferred.

We now make a fairly detailed study of the transformed problem and then use the preceding theorems to draw conclusions about the original problem. Consider the following conditions on the original measures α and β.

Definition: Given measure α on (S, Σ_s), and measurable $h{:}S \to$ reals, α is *finite from below* iff

$$\alpha\{s \mid h(s) < x\} \quad \text{is finite} \qquad (8.5.32)$$

for all real numbers x. Similarly, given β on (Q, Σ_q) and measurable $w{:}Q \to$ reals, β is *finite from above* iff

$$\beta\{q \mid w(q) > x\} \quad \text{is finite} \qquad (8.5.33)$$

for all real x.

The interpretation of (32) is that the ideal area of the region *within* ideal distance x of the nucleus is finite for any real x. This is not implausible and does not preclude the possibility that the ideal area of Space as a whole is infinite. Similarly, (33) states that the allotment to the set of land uses of weight exceeding x is finite for any real x.

These properties can be stated in logically equivalent form in terms of the transformed measures α and β. Namely, (32) and (33) are the same as

$$\alpha\{y \mid y < x\} \quad \text{and} \quad \beta\{y \mid y > x\} \quad \text{are finite} \qquad (8.5.34)$$

for all real x respectively. That is, the α-measure of any left half-line and the β-measure of any right half-line are finite. Note that (34) implies the σ-finiteness of α and β.

These conditions are imposed because they insure the existence of a unique measure λ on the plane satisfying the weight-falloff condition with marginals α and β. As a preliminary, we establish a connection between the weight-falloff and *northwest corner* conditions. We have already encountered the latter condition for a measure λ on a product space $A \times B$, where A and B are both countable (Section 7.2). The general idea is that, given its marginals, λ have as much mass as possible concentrated into "corner" sets, these being defined in terms of certain complete orderings on A and B. In the present case, both A and B are the real line, which has a natural order. Then λ is to concentrate its mass in the "corner" with low A-values and high B-values. The following definition makes this precise.

Definition: Let λ be a measure on the plane, with left and right marginals λ', λ'' respectively; λ satisfies the *northwest corner condition* iff

$$\lambda\{(x, y) \mid x < x_1, y > y_1\} = \min[\lambda'\{x \mid x < x_1\}, \lambda''\{y \mid y > y_1\}]$$
$$(8.5.35)$$

for all pairs of real numbers (x_1, y_1).

The set $\{(x, y) \mid x < x_1, y > y_1\}$ is the quadrant of the plane "northwest" of the point (x_1, y_1). It can be seen that for any measure the left side of (35) never exceeds the right. Hence (35) is indeed the condition that the mass on these northwest sets be as large as possible.

Theorem: Let λ be a measure on the plane whose left marginal $\lambda' = \alpha$ is finite from below and whose right marginal $\lambda'' = \beta$ is finite from above; i.e., (34) holds. Then λ satisfies the weight-falloff condition iff it satisfies the northwest corner condition.

Proof: For any point (x_1, y_1) define the three sets

$$E = \{(x, y) \mid x < x_1, y > y_1\}$$
$$F = \{(x, y) \mid x < x_1, y \leq y_1\}$$
$$G = \{(x, y) \mid x \geq x_1, y > y_1\} \qquad (8.5.36)$$

Then

$$\lambda(E) + \lambda(F) = \lambda'\{x \mid x < x_1\} \tag{8.5.37}$$

$$\lambda(E) + \lambda(G) = \lambda''\{y \mid y > y_1\} \tag{8.5.38}$$

so that (35) takes the form

$$\lambda(E) = \min[\lambda(E) + \lambda(F), \lambda(E) + \lambda(G)] \tag{8.5.39}$$

Now let λ satisfy weight-falloff. Then either $\lambda(F) = 0$ or $\lambda(G) = 0$, since F is southwest of G. Either of these cases yields (39), so that λ satisfies northwest corner.

Conversely, let (39) be true for all (x_1, y_1). By (34), all terms in (37) and (38) are finite. Then (39) implies that either $\lambda(F) = 0$ or $\lambda(G) = 0$. Now let (x_2, y_2) and (x_3, y_3) be any two points, the first southwest of the second, and choose x_1 and y_1 such that $x_2 < x_1 < x_3$, $y_2 < y_1 < y_3$. Then F and G of (36) are measurable neighborhoods of (x_2, y_2) and (x_3, y_3) respectively, so these points cannot both support λ. This proves that λ satisfies weight-falloff. ■

Our next result is based on the theory of distribution functions. Owing to the fact that we are dealing with "northwest" rather than "southwest" sets, the standard theorem (p. 92) must be rephrased in a slightly different form.

Lemma: Let g:reals$^2 \to$ reals satisfy the following three conditions:

(i) For all real numbers x_1, x_2, y_1, y_2, with $x_1 < x_2$ and $y_1 < y_2$ we have

$$g(x_1, y_1) + g(x_2, y_2) \leq g(x_1, y_2) + g(x_2, y_1) \quad (8.5.40)$$

(ii) g is continuous from the "northwest": for any (x_1, y_1) and any $\epsilon > 0$, there is a $\delta > 0$ such that

$$|g(x, y) - g(x_1, y_1)| < \epsilon \tag{8.5.41}$$

for any (x, y) satisfying $x_1 - \delta \leq x \leq x_1$ and $y_1 + \delta \geq y \geq y_1$;

(iii) for fixed y, $g(x, y) \to 0$ as $x \to -\infty$, and for fixed x, $g(x, y) \to 0$ as $y \to +\infty$.

Then there is exactly one measure λ on the plane satisfying

$$\lambda\{(x, y) \mid x < x_1, y > y_1\} = g(x_1, y_1) \tag{8.5.42}$$

for all points (x_1, y_1).

Note the opposite sign orientations of x and y in parts (ii) and (iii). Condition (40) states that g has *nonpositive* cross-differences (cf. (2)), whereas the usual distribution functions have the opposite property. This lemma follows from the usual statement by "reflecting" the plane through the X-axis: $(x, y) \rightarrow (x, -y)$. We are now ready for the main result.

Theorem: Let α and β be two measures on the real line L satisfying (34): α is finite from below and β finite from above. Also let $\alpha(L) = \beta(L)$. Then there is *exactly one* measure λ on the plane having α and β as its left and right marginals and satisfying the *weight-falloff condition*.

Proof: Define g:reals$^2 \rightarrow$ reals by

$$g(x_1, y_1) = \min[\alpha\{x \mid x < x_1\}, \beta\{y \mid y > y_1\}] \qquad (8.5.43)$$

This is indeed finite, by (34). We now show that g satisfies (i), (ii), and (iii) of the preceding lemma.

(i) Choose real numbers $x_1 < x_2$ and $y_1 < y_2$. We clearly have $g(x_1, y_1) \leq g(x_2, y_1)$ and $g(x_2, y_2) \leq g(x_2, y_1)$. If $\alpha\{x \mid x < x_1\} \leq \beta\{y \mid y > y_2\}$, then also $g(x_1, y_1) = g(x_1, y_2)$. If $\alpha\{x \mid x < x_1\} \geq \beta\{y \mid y > y_2\}$, then also $g(x_2, y_2) = g(x_1, y_2)$. In either case the equality, combined with one of the inequalities, yields (40).

(ii) Note that $\alpha\{x \mid x < x_1\}$, as a function of x_1, is continuous from below; and $\beta\{y \mid y > y_1\}$, as a function of y_1, is continuous from above. (This follows from the basic continuity property of measures, p. 31.) That is, given (x_1, y_1), for any $\epsilon > 0$ there is a $\delta > 0$ such that

$$\left| \alpha\{x \mid x < x_2\} - \alpha\{x \mid x < x_1\} \right| < \epsilon \qquad (8.5.44)$$

$$\left| \beta\{y \mid y > y_2\} - \beta\{y \mid y > y_1\} \right| < \epsilon \qquad (8.5.45)$$

for all x_2, y_2 satisfying: $x_1 - \delta \leq x_2 \leq x_1$ and $y_1 + \delta \geq y_2 \geq y_1$. Then (44) and (45) together yield (41).

(iii) The limit of $\alpha\{x \mid x < x_1\}$ is zero as $x_1 \rightarrow -\infty$, and the limit of $\beta\{y \mid y > y_1\}$ is zero as $y_1 \rightarrow +\infty$. Hence the limit of $g(x_1, y_1)$ is zero in both cases, which is (iii).

Applying the lemma, we conclude that a measure λ exists satisfying (42). We now show that this λ has the required properties.

Let y_1 go to $-\infty$ in (42). The left side has the value $\lambda'\{x \mid x < x_1\}$ as limit, where λ' is the left marginal of λ. $\beta\{y \mid y > y_1\}$ has the value $\beta(L) = \alpha(L)$ as limit, which is at least as large as $\alpha\{x \mid x < x_1\}$ for any x_1. Hence $g(x_1, y_1)$ approaches $\alpha\{x \mid x < x_1\}$ as limit. This proves that $\lambda'\{x \mid x < x_1\} = \alpha\{x \mid x < x_1\}$ for all real x_1: λ' and α have the same distribution function and must therefore coincide.

Letting x_1 go to $+\infty$ in (42), a similar argument shows that $\lambda'' = \beta$. Thus α and β are indeed the left and right marginals of λ respectively. This being the case, (42) is the same as (35), so that λ satisfies the northwest corner condition. By the preceding theorem, it therefore satisfies weight-falloff.

The existence of λ has now been established. To prove uniqueness, let λ satisfy weight-falloff and have marginals α and β. By the preceding theorem, λ satisfies northwest corner. Hence λ satisfies the relation (42), where g is given by (43). But there is just one measure satisfying this relation, so λ is unique.[16] ∎

We now show that this result, establishing the existence and uniqueness of a weight-falloff measure on the plane, extends in part to the original problem on $S \times Q$.

Theorem: Uniqueness Theorem.   Let (S, Σ_s, α) and (Q, Σ_q, β) be measure spaces, and $h:S \to$ reals, $w:Q \to$ reals measurable functions such that α is finite from below and β finite from above (with respect to h, w respectively). Let $\alpha\{s \mid h(s) = x\} = 0$ for all real numbers x, and let Σ_q be the class of all sets of the form $\{q \mid w(q) \in E\}$, E ranging over the Borel field on the real line.

Then there is *at most one* measure ν on $(S \times Q, \Sigma_s \times \Sigma_q)$, with marginals α, β, satisfying the (measurable) weight-falloff condition.

Proof: Consider the class $\mathfrak{R}$ of sets of the form

$$F \times \{q \mid x < w(q) \leq y\} \tag{8.5.46}$$

where $F \in \Sigma_s$ and x, y are real numbers.

Class $\mathfrak{R}$ generates the σ-field $\Sigma_s \times \Sigma_q$. To show this, it suffices to prove that all measurable rectangles belong to Σ, the σ-field with universe set $S \times Q$ generated by $\mathfrak{R}$. Consider the class of Borel sets E on the real line having the property that $\{q \mid w(q) \in E\}$ belongs to the σ-field on Q generated by the sets $\{q \mid x < w(q) \leq y\}$, x, y real. This class is closed under countable unions and complements; it also includes all half-open intervals $\{z \mid x < z \leq y\}$, x, y real. But the latter generate the Borel field; hence the sets $\{q \mid x < w(q) \leq y\}$ generate the σ-field of sets $\{q \mid w(q) \in E\}$, E ranging over *all* real Borel sets. By assumption, this σ-field is Σ_q. Hence Σ owns all measurable rectangles, so that $\Sigma = \Sigma_s \times \Sigma_q$.

Next, $\mathfrak{R}$ is a semi-σ-ring (p. 80).   Indeed, $\emptyset \in \mathfrak{R}$, the intersection

---

16. Measures on the plane satisfying the northwest (or rather, southwest) corner condition have been studied from another point of view by M. Fréchet. See W. Feller, *An Introduction to Probability Theory and Its Applications*, Vol. 2 (Wiley, New York, 1966), pp. 162–63, problem 6.

of two $\mathfrak{R}$-sets is an $\mathfrak{R}$-set, and the difference of two $\mathfrak{R}$-sets can be expressed as the union of three disjoint $\mathfrak{R}$-sets (in fact, $\mathfrak{R}$ is a semiring).

Let ν_1, ν_2 be two measures with marginals α, β satisfying (measurable) weight-falloff. We prove that ν_1 and ν_2 must coincide on $\mathfrak{R}$ and that a countable number of $\mathfrak{R}$-sets exist whose union is $S \times Q$, such that ν_1 and ν_2 are finite on each. The basic extension theorem (p. 80) then guarantees that ν_1 and ν_2 are equal, and we are through.

First, the $\mathfrak{R}$-sets $S \times \{q \mid n < w(q) \leq n + 1\}$, $n = 0, \pm1, \pm2, \ldots$, cover $S \times Q$; and ν_1 and ν_2 must be finite on each, since their common right marginal β is finite from above. This proves half the statement in the paragraph above.

It remains to show that ν_1, ν_2 coincide on $\mathfrak{R}$. Since they satisfy measurable weight-falloff, the measures, λ_1, λ_2, induced from them on the plane by the mapping $(s, q) \to (h(s), w(q))$ must satisfy weight-falloff and must have as marginals the measures on the real line induced from α and β. These marginals are finite from below and above respectively, so that exactly one weight-falloff measure λ exists having them as marginals. Thus $\lambda = \lambda_1 = \lambda_2$. In fact, λ is the *northwest corner* measure with these marginals. It follows that

$$\nu_i\{(s, q) \mid h(s) < x', w(q) > x\} = \min[\alpha\{s \mid h(s) < x'\}, \beta\{q \mid w(q) > x\}] \tag{8.5.47}$$

is true for all real numbers, x, x', for $i = 1, 2$. For this is merely the northwest corner condition (35) expressed in terms of ν_i, α and β. Also, (47) holds for infinite x or x', as may be verified by direct substitution.

Next, for any number k between 0 and $\alpha(S)$ inclusive, an extended real number y' exists such that

$$\alpha\{s \mid h(s) < y'\} = k \tag{8.5.48}$$

To see this, take the supremum of the numbers y' for which the left side of (48) does not exceed k. For this value we have

$$\alpha\{s \mid h(s) < y'\} \leq k \leq \alpha\{s \mid h(s) \leq y'\} \tag{8.5.49}$$

from the "continuity" of measures. But by assumption, the left and right terms in (49) are equal, hence (48) follows.

Now choose any member of $\mathfrak{R}$. Given the numbers x and y in (46) choose x' and y' to satisfy

$$\alpha\{s \mid h(s) < x'\} = \beta\{q \mid w(q) > x\} \tag{8.5.50}$$

$$\alpha\{s \mid h(s) < y'\} = \beta\{q \mid w(q) > y\} \tag{8.5.51}$$

Such x' and y' exist, by (48), since the right-hand terms in (50) and (51) lie between 0 and $\beta(Q) = \nu_i(S \times Q) = \alpha(S)$.

From (47) and (50) we obtain

$$\nu_i\{(s,q) \mid h(s) < x', w(q) > x\} = \alpha\{s \mid h(s) < x'\} = \beta\{q \mid w(q) > x\}$$
(8.5.52)

for $i = 1, 2$. The common value in (52) is finite because x is finite, and β is finite from above. It follows that

$$\nu_i\{(s,q) \mid h(s) \geq x', \ w(q) > x\} = 0 = \nu_i\{(s,q) \mid h(s) < x', \ w(q) \leq x\}$$
(8.5.53)

$i = 1, 2$. To see this, note that the sum of the left-hand terms in (52) and (53) is $\nu_i\{(s,q) \mid w(q) > x\} = \beta\{q \mid w(q) > x\}$, and the first equality in (53) follows by subtraction. The second equality is proved similarly. The same argument applies with y, y' substituted for x, x' respectively, and we conclude that (53) remains true with these substitutions. Thus we have four equalities (53) for each $i = 1, 2$.

Now let $G = F \cap \{s \mid y' \leq h(s) < x'\}$. We then have

$$\nu_i[F \times \{q \mid x < w(q) \leq y\}] = \nu_i[G \times \{q \mid x < w(q) \leq y\}]$$
(8.5.54)

$i = 1, 2$. For the set of points (s, q) belonging to the left but not the right set in (54) is contained in the union of two of the four sets of measure zero of (53), as one verifies. Finally, we have

$$\nu_i[G \times \{q \mid x < w(q) \leq y\}] = \nu_i(G \times Q) = \alpha(G)$$
(8.5.55)

$i = 1, 2$. For the set of points (s, q) belonging to the middle but not to the left set in (55) is contained in the union of the other two sets of measure zero of (53), as one verifies.

Equations (54) and (55) show that ν_1 and ν_2 coincide on all $\Re$-sets. Hence they are identical. ∎

Note that the assumptions imposed on the two component spaces are quite different, unlike all the other theorems of this section. The condition that $\alpha\{s \mid h(s) = x\} = 0$ states that the measure on the real line induced by h from α is *nonatomic*, while the condition that Σ_q is all sets of the form $\{q \mid w(q) \in E\}$ states that Σ_q is the σ-field *inversely induced* by w from the real Borel field. From the symmetry of the allotment-assignment problem in S and Q, it is clear that these conditions could have been interchanged (making induced β nonatomic and Σ_s inversely induced) without invalidating the conclusion. But the form in which the theorem is stated is the one that applies neatly to realistic Thünen systems. Neither of these assumptions can be dropped without invalidating the conclusion. This will be illustrated later with counterexamples from simple Thünen systems.

Next we have an existence theorem for the original allotment-assignment problem similar to the one proved above for the transformed problem. A function is said to be *semicontinuous* iff it is either upper or lower semicontinuous (or both, i.e., continuous).

Theorem: Let (S, Σ_s, α) and (Q, Σ_q, β) be measure spaces, with $\alpha(S) = \beta(Q) < \infty$. Let Σ_s and Σ_q be the Borel fields of topologies $\mathfrak{I}_s$ and $\mathfrak{I}_q$ respectively, these making S and Q Borel subsets of topologically complete and separable spaces. Let the functions $h{:}S \to$ reals and $w{:}Q \to$ reals be semicontinuous.

Then a measure ν^o exists on $(S \times Q, \Sigma_s \times \Sigma_q)$ with marginals α and β, which satisfies the (measurable) weight-falloff condition (with respect to h, w).

Proof: First, let h and w be lower semicontinuous. Let $f{:}\text{reals}^2 \to$ reals be a function that has positive cross-differences and is bounded, continuous, and increasing in each argument. An example is

$$f(x, y) = (1 + e^{-x})^{-1}(1 + e^{-y})^{-1} \qquad (8.5.56)$$

We now show that the composite function $f(h(\cdot), w(\cdot)){:}S \times Q \to$ reals is lower semicontinuous. Let (s_o, q_o) belong to the set

$$\{(s, q) \mid f(h(s), w(q)) > z\} \qquad (8.5.57)$$

z being a real number. Since f is continuous, the set

$$\{(x, y) \mid f(x, y) > z\} \qquad (8.5.58)$$

is open in the plane. Point $(h(s_o), w(q_o))$ belongs to (58), hence there is a point (x_o, y_o) southwest of $(h(s_o), w(q_o))$ that belongs to (58). Consider the following subset of $S \times Q$:

$$\{s \mid h(s) > x_o\} \times \{q \mid w(q) > y_o\} \qquad (8.5.59)$$

By the lower semicontinuity of h and w, (59) is open. Point (s_o, q_o) belongs to (59) by construction. Finally, (59) is contained in (57), since f is increasing in its arguments. Hence (57) is an open set for any z, so that $f(h(\cdot), w(\cdot))$ is indeed lower semicontinuous.

It is also a bounded function, and these properties, with the other premises, imply that the allotment-assignment problem (4)–(6) has a best solution ν^o (Section 7.4). Since f has positive cross-differences, this ν^o satisfies the measurable weight-falloff condition (p. 427).

For the remaining three cases, replace f by f', where

$f'(x, y) = f(-x, -y)$ \quad if h, w are both upper semicontinuous

$f'(x, y) = -f(-x, y)$ \quad if h is lower and w upper semicontinuous

$f'(x, y) = -f(x, -y)$ \quad if h is upper and w lower semicontinuous

f being given by (56).

In all cases f' remains bounded continuous, with positive cross-differences (because the number of sign changes is even). And in all cases $f'(h(\cdot), w(\cdot))$ remains lower semicontinuous. This is clear for $f(-h(\cdot), -w(\cdot))$, since negation converts upper to lower semicontinuity. For $-f(-h(\cdot), w(\cdot))$, and $-f(h(\cdot), -w(\cdot))$, the two argument functions of f are now *upper* semicontinuous. In this case, reasoning similar to that above shows that the composite function $f(\cdots)$ is *upper* semicontinuous. (Reverse the inequality signs in (57)–(59), and take (x_o, y_o) northeast rather than southwest.) Hence $f'(\cdots) = -f(\cdots)$ is indeed lower semicontinuous.

As above, a measurable weight-falloff measure v^o then exists in all cases. ∎

Unlike the situation in the transformed problem, the v^o of this theorem need not be unique. A trivial example of this is where h or w is a constant and there are at least two feasible measures. For here *every* feasible measure v satisfies both weight-falloff conditions vacuously.

We have obtained conditions under which an optimal solution must satisfy a weight-falloff condition. Our next result is a converse, indicating conditions under which a weight-falloff measure is optimal.

Theorem: Let (S, Σ_s, α) and (Q, Σ_q, β) be *bounded* measure spaces. Let the functions $h:S \rightarrow$ reals and $w:Q \rightarrow$ reals be measurable, and let $f:\text{reals}^2 \rightarrow$ reals be *bounded lower semicontinuous,* with *nonnegative cross-differences.* Let measure v^o on $(S \times Q, \Sigma_s \times \Sigma_q)$ have marginals α and β and satisfy the *(measurable) weight-falloff condition.*

Then v^o is *best* for the allotment-assignment problem of minimizing

$$\int_{S \times Q} f(h(s), w(q)) \, v(ds, dq) \qquad (8.5.60)$$

over measures v with marginals α and β.

Proof: First, assume that f has *positive* cross-differences. Consider the allotment-assignment problem on the plane induced from the given problem. Since f is bounded lower semicontinuous, α and β are bounded, and $\alpha(S) = v^o(S \times Q) = \beta(Q)$, a best solution λ^{oo} exists to this transformed problem (cf. Section 7.4). And λ^{oo} must satisfy the weight-falloff condition, since f has positive cross-differences.

Since v^o satisfies weight-falloff, the measure λ^o on the plane induced from it by the mapping $(s, q) \rightarrow (h(s), w(q))$ must satisfy weight-falloff. Also, λ^o is feasible for the transformed problem. But there is only one weight-falloff measure feasible for the transformed problem, since α and β are bounded. Hence $\lambda^o = \lambda^{oo}$. Therefore λ^o is unsurpassed for the transformed problem, implying that ν^o is unsurpassed for the original. But (60) is well defined and finite for all feasible ν, so "unsurpassed" coincides with "best." This proves the theorem for the special case of positive cross-differences.

Now let f have merely *nonnegative* cross-differences. Choose a function g:reals$^2 \to$ reals that is bounded, lower semicontinuous, and has *positive* cross-differences (such as (56)). Then, for any positive real number ϵ, the function $f + \epsilon g$ has the same properties as g. Consider the perturbed allotment-assignment problem in which f in (60) is replaced by $f + \epsilon g$. By the results just proved, ν^o is best for this problem. Hence

$$\int_{S \times Q} (f + \epsilon g)(h(s), w(q)) \; \nu^o(ds, dq) \leq \int_{S \times Q} (f + \epsilon g)(h(s), w(q)) \; \nu(ds, dq)$$

for any other feasible measure ν and any real $\epsilon > 0$. Now let ϵ go to zero. By the dominated convergence theorem, the limit of the integral on each side is the integral of the limit. Hence ν^o remains best when $\epsilon = 0$. ∎

We show later that this theorem can be strengthened to some extent; viz., the premise that f is lower semicontinuous can be dropped. But the method of proof just used is quite instructive and completely different from the method to be used below, which involves the construction of a potential.

The resulting theory is fairly satisfactory, and the conditions under which it holds are not too onerous. The boundedness of f, however, is a nuisance; e.g., the product function, $f(x, y) = xy$ (which is the original form in which transport cost presented itself) is not bounded on the plane. This limitation is easily remedied if the ideal distance and weight functions h and w have bounded ranges. For then we can take the domain of the transformed problem to be, not the entire plane, but a rectangle with bounded intervals as sides. On such a set the product function (and most other functions of interest) will be bounded, and the preceding theorem can be applied.

Insight into the distinction between positive and nonnegative cross-differences can be gained by contemplating the case where f has *zero* cross-differences, i.e., where

$$f(x_1, y_1) + f(x_2, y_2) = f(x_1, y_2) + f(x_2, y_1) \qquad (8.5.61)$$

for all numbers x_1, x_2, y_1, y_2. Equation (61) holds iff f can be written as the sum of separate x- and y-functions:

$$f(x, y) = f_1(x) + f_2(y) \qquad (8.5.62)$$

(Proof: If (62) holds, then (61) is verified by substitution. Conversely, choose an arbitrary y_o, and define f_1, f_2 by $f_1(x) = f(x, y_o)$, $f_2(y) =$

$f(x, y) - f(x, y_o)$. For the definition of f_2 to be sound, the expression $f(x, y) - f(x, y_o)$ must not depend on x. But this is guaranteed by (61), and (62) follows at once.) But if (62) holds (and f, α, β are all bounded), then the objective function (60) is equal to $\int_S (f_1 \circ h) \, d\alpha + \int_Q (f_2 \circ w) \, d\beta$, by the induced integrals theorem ($\circ$ signifies the composition of functions). Thus transport cost depends only on the marginals α and β of ν. Since all feasible ν have the same marginals, they are *all* best solutions. Thus while positive cross-differences restrict best solutions to the weight-falloff measures, nonnegative cross-differences may allow others.

POTENTIALS

We now turn our attention to the construction of potentials. This is of interest not only for the further insights it furnishes concerning the optimality properties of weight-falloff measures but also because potentials have direct intuitive interpretations as land values and as "gross profits" on land uses.

Let measure ν^o on $(S \times Q, \Sigma_s \times \Sigma_q)$ have left and right marginals α and β, so that it is feasible for the allotment-assignment problem. Recall that a pair of measurable functions[17] $p{:}S \to$ reals and $k{:}Q \to$ reals is a *measure potential* for ν^o (in the wide sense) iff

$$k(q) - p(s) \leq f(h(s), w(q)) \qquad (8.5.63)$$

for all $s \in S, q \in Q$, and

$$\nu^o\{(s, q) \mid k(q) - p(s) < f(h(s), w(q))\} = 0 \qquad (8.5.64)$$

Now furnish S and Q with topologies $\mathfrak{I}_s$ and $\mathfrak{I}_q$ respectively. The pair of measurable functions (p, k) is said to be a *topological potential* for ν^o (in the wide sense) iff (63) holds for all $s \in S$, $q \in Q$; and if (s, q) is a point of support for ν^o, then (63) holds with *equality*.

We shall make essential use of the transformation of the allotment-assignment problem into the plane. The transformed problem is itself a special case of the allotment-assignment problem, and we may therefore contemplate potentials for *its* feasible solutions λ. Our first task will be the following. Let ν be feasible for the original problem, and λ the measure induced from ν; λ is feasible for the transformed problem. What

---

17. We write the potential as the pair (p, k) rather than (p, q) as in Chapter 7, to avoid confusion with points $q \in Q$. The stated definition is for variant I of the transportation problem, which we are using. Other variants add nonnegativity conditions on p or k. Also this is a "wide-sense" definition, so that there is no requirement that the integrals $\int_S p \, d\alpha, \int_Q k \, d\beta$ be well defined and finite. The same remarks apply to the next definition.

relations then hold between the properties of there being potentials (measure or topological) for λ and for ν?

First, consider one preliminary. We take the transformed problem to be defined, not necessarily on the whole plane, but on a rectangular subset of the plane $X \times Y$. The exact definition will be given later; for the present we need merely assume that X and Y contain the ranges of h and w respectively. A potential for the transformed problem is then a pair of measurable functions $p:X \to$ reals, $k:Y \to$ reals satisfying one or the other of the definitions above. Thus $X \times Y$ is the domain of f and the universe set of feasible measures λ.

Theorem: Let ν^o be a measure on $(S \times Q, \Sigma_s \times \Sigma_q)$ with left and right marginals α and β. Let X and Y be measurable subsets of the real line; and let $h:S \to X$, $w:Q \to Y$, $f:X \times Y \to$ reals, $p:X \to$ reals, and $k:Y \to$ reals be measurable functions. Let λ^o be the measure on $X \times Y$ induced from ν^o by the mapping $(s, q) \to (h(s), w(q))$, and consider the following three conditions:

(i) (p, k) is a topological potential for λ^o;
(ii) (p, k) is a measure potential for λ^o;
(iii) $(p \circ h, k \circ w)$ is a measure potential for ν^o.[18]

Then condition (i) implies condition (ii), which in turn implies condition (iii).

Proof: The usual topology on the plane, or any subset of the plane, has the strong Lindelöf property. This insures that any topological potential is a measure potential (Section 7.5). Thus (i) implies (ii).

Let condition (ii) be valid. Then $k(y) - p(x) \le f(x, y)$ for all $x \in X$, $y \in Y$. Letting $x = h(s)$, $y = w(q)$, we verify (63) for the pair of functions $(p \circ h, k \circ w)$. Also we have

$$\nu^o\{(s, q) \mid k(w(q)) - p(h(s)) < f(h(s), w(q))\}$$
$$= \lambda^o\{(x, y) \mid k(y) - p(x) < f(x, y)\} = 0 \qquad (8.5.65)$$

The right equality is from condition (ii), the left from the fact that the argument of ν^o is the inverse image of the argument of λ^o under the mapping $(s, q) \to (h(s), w(q))$. Condition (65) yields (64) for the pair of functions $(p \circ h, k \circ w)$. Thus (iii) is valid. ∎

This theorem is silent about topological potentials for ν^o. Indeed, this concept is not even defined, since nothing is said about any topologies on S or Q.

---

18. Here $p \circ h$ is the composite function whose value at $s \in S$ is $p(h(s))$; similarly for $k \circ w$.

Our plan of action is to construct a topological potential for the weight-falloff measure λ^o. If λ^o is induced from a measure ν^o feasible for the original allotment-assignment problem, the preceding theorem yields a measure potential for ν^o. From this we can make inferences concerning the optimality of ν^o.

In constructing a potential for λ^o we could make use of the theory developed for the transportation problem. Instead, we use a special procedure that utilizes the distinctive properties of weight-falloff and non-negative cross-differences. This not only allows us to weaken the assumptions needed, but the procedure is of interest in itself and has intuitive appeal.

We begin with an observation. A measure λ on the plane has all its mass concentrated on its support. That is, if E is the support of λ, then the complement of E has measure zero (E and its complement are Borel sets, since E is closed). The proof of this rests on the strong Lindelöf property of the usual topology of the plane. For every point of the complement of E has a measurable neighborhood of measure zero. A countable subcollection of these neighborhoods covers the complement of E, which therefore has measure zero.

Given measure λ on the plane, we restrict it to a rectangle $X \times Y$ as follows. X consists of all numbers x having the property that either there is a number y such that (x, y) supports λ or x is between two such numbers x_1 and x_2. Y is defined analogously, the roles of x and y being interchanged. X is an interval, which may be the entire real line or may be bounded below, or above, or both. If bounded, the endpoints may or may not be included in X. The same remarks apply to Y. The support E of λ is contained in $X \times Y$. Hence by the preceding argument, we are throwing away a set of measure zero. Finally, note that X and Y are empty iff λ has empty support. By the preceding argument this occurs iff $\lambda = 0$. We exclude this trivial case by assumption. Call $X \times Y$ the *support rectangle*.[19]

Now let λ^o be a measure on $X \times Y$ satisfying the weight-falloff condition. With the point of support (x, y) associate the value $x - y$. No two points of support have the same value, for if this were true of (x_i, y_i), $i = 1, 2$, we would have $x_1 - x_2 = y_1 - y_2 \neq 0$; the points would thus stand in a southwest-northeast relation, contradicting weight-falloff. The valuation thus determines a complete antisymmetric ordering of the points of support. We have, in fact, a "Maginot line" of points of support strung across the plane, running from northwest to southeast (possibly including vertical north-to-south stretches and/or horizontal west-to-east stretches).

---

19. When X and Y are constructed in this manner, the ranges of h and w may not be respectively contained in them. We shall deal with this minor complication in due course. For the time being we concentrate attention exclusively on the plane and forget about the original allotment-assignment problem.

This line may have gaps in it. A *gap* is defined as a pair of distinct points of support (x_1, y_1), (x_2, y_2) with no other points of support "between" them in the ordering. Wherever such a gap exists, connect the two points constituting it by a *straight line segment*. The union of the original support and all these line segments is called the *line of support L* of λ^o. It is easy to prove the following facts about the line of support. No two points of it stand in a southwest-northeast relation. For every $x \in X$, a $y \in Y$ exists such that $(x, y) \in L$. For given $x \in X$, there is either a unique $y \in Y$ such that $(x, y) \in L$, or a closed interval (possibly unbounded) of such y's. Similar statements apply with x and y interchanged. The line of support is contained in the rectangle of support.

Having furnished the line of support, the measure λ^o has completed its role in the construction of a topological potential for itself, and attention now passes to the cost function $f:X \times Y \to$ reals. From f we construct a function $p:X \to$ reals, which turns out to be (under certain conditions) the left half of a topological potential for λ^o.

Function $p(x)$ is defined as follows. Choose a fixed point $x_0 \in X$, and define $p(x_0) = 0$. For $x \neq x_0$ take a sequence (x_i, y_i), $i = 0, \ldots, n$, of points on the line of support, with $x_n = x$. This sequence is to be *monotone*, i.e., either strictly increasing in the value $x_i - y_i$ (for $x > x_0$) or strictly decreasing in $x_i - y_i$ (for $x < x_0$); n is any positive integer. With this sequence associate the real number

$$(x_0 y_1 - x_1 y_1) + (x_1 y_2 - x_2 y_2) + \cdots + (x_{n-1} y_n - x_n y_n) \qquad (8.5.66)$$

where we use the abbreviation xy for $f(x, y)$ here and below. Then $p(x)$ is defined as the *infimum* of (66) over all such monotone sequences from x_0 to x.

Lemma: Let measurable $f:X \times Y \to$ reals have nonnegative cross-differences. Then p is finite, measurable, and

$$p(x') + f(x', y') \leq p(x) + f(x, y') \qquad (8.5.67)$$

for any $x, x' \in X, y' \in Y$ such that $(x', y') \in L$, the line of support.

Proof: Let (x_i, y_i), $i = 0, \ldots, n$, be a monotone sequence of L-points for $x = x_n$. Then

$$(x_{i-1} y_0 - x_i y_0) \leq (x_{i-1} y_i - x_i y_i) \leq (x_{i-1} y_n - x_i y_n)$$

$i = 1, \ldots, n$. This follows at once from nonnegative cross-differences and the monotonicity of the sequence. Adding these inequalities over i, we obtain

$$(x_0 y_0 - x_n y_0) \leq z \leq (x_0 y_n - x_n y_n) \qquad (8.5.68)$$

where z is of the form (66). From this we obtain

$$(x_0y_0 - xy_0) \le p(x) \le (x_0y - xy) \tag{8.5.69}$$

where y_0 and y are any two numbers such that $(x_0, y_0) \in L$ and $(x, y) \in L$. The right inequality in (69) follows from the right of (68) by taking the infimum of z over all permissible sequences with fixed $(x, y) = (x_n, y_n) \in L$ (n any positive integer). The left inequality in (69) follows from the left of (68) by taking the infimum of z over all permissible sequences with fixed y_0, on noting that this does not restrict the range of values (66). Inequality (69) shows that p is indeed finite.

Inequality (67) follows at once for two special cases besides the obvious $x = x'$: for $x' = x_0$ from the left of (69), and for $x = x_0$ from the right of (69). This leaves the case where x_0, x, and x' are all distinct. There are three subcases, depending on which of these numbers is between the other two.

(i) If x_0 is between x and x', we have

$$p(x') \le x_0y' - x'y' \tag{8.5.70}$$

$$p(x) \ge (x_0y_0 - xy_0) \ge (x_0y' - xy') \tag{8.5.71}$$

Inequality (70) and the left inequality in (71) arise from (69); the right of (71) arises from nonnegative cross-differences. Together (70) and (71) yield (67).

(ii) If x is between x_0 and x', we consider monotone sequences (x_i, y_i), $i = 0, \ldots, n$, of L-points, with $(x_n, y_n) = (x', y')$, which include x among the x_i's, say $x = x_j$. Separate the corresponding expression (66) into two sums: the first $2j$ terms (ending with $-x_jy_j$), and the last $2n-2j$ terms (beginning with x_jy_{j+1}). This latter sum does not exceed $x_jy_n - x_ny_n$. (The argument for this is the same as that leading to the right-hand inequality of (68), except that we begin with x_j, not x_0.) Hence we have

$$p(x') \le (x_0y_1 - x_1y_1) + \cdots + (x_{j-1}y_j - x_jy_j) + x_jy_n - x_ny_n \tag{8.5.72}$$

Now (72) holds for all monotone sequences $(x_0, y_0), \ldots, (x_j, y_j)$ of L-points for which $x_j = x$. Taking the infimum over the corresponding sums (66), we obtain $p(x') \le p(x) + x_jy_n - x_ny_n$, which is the same as (67).

(iii) Finally, if x' is between x_0 and x, we note first that removing a point (x_j, y_j) (where $0 < j < n$) does not decrease the corresponding sum (66). For the change in (66) is

$$(x_{j-1}y_{j+1} - x_{j+1}y_{j+1}) - (x_{j-1}y_j - x_jy_j) - (x_jy_{j+1} - x_{j+1}y_{j+1})$$

which is ≥ 0, by nonnegative cross-differences and the monotonicity of the sequence $(x_i, y_i), i = 0, \ldots, n$, of L-points.

Now for any $\epsilon > 0$ we can find a sequence of L-points such that the sum (66) does not exceed $p(x) + \epsilon$. If (x', y') is not among these points, slip it into the sequence so as to preserve monotonicity: say $(x', y') = (x_j, y_j)$ after relabeling. By the observation just made, this insertion cannot increase (66), so it remains $\leq p(x) + \epsilon$. Now separate (66) into two sums as above. The first (which ends with the term $-x_j y_j$) is at least as large as $p(x')$. The second (which begins with $x_j y_{j+1}$) is at least as large as $x_j y_j - x_n y_j$. (Use the same argument that leads to the left inequality of (68), with j in place of 0.) Hence $p(x) + \epsilon \geq p(x') + x_j y_j - x_n y_j$. That is, $p(x) + \epsilon \geq p(x') + x'y' - xy'$. Since $\epsilon > 0$ is arbitrary, we again obtain (67).

It remains only to prove that p is measurable. We show that p restricted to the bounded interval from x_0 to x^o is measurable.[20] Since x^o is arbitrary, this implies that p itself is measurable.

For each $m = 1, 2, \ldots$, let (x_{mi}, y_{mi}), $i = 0, \ldots, n_m$, be a monotone sequence of L-points, with $x_{m0} = x_0$ and $x_{mn_m} = x^o$, such that

$$p(x^o) + (1/m) \geq (x_{m0} y_{m1} - x_{m1} y_{m1}) + \cdots + (x_{m,n_m-1} y_{mn_m} - x_{mn_m} y_{mn_m})$$
$$(8.5.73)$$

Now let (x_i, y_i), $i = 0, \ldots, n$, be a monotone sequence of L-points, with $x_n = x^o$. In terms of this sequence, we define the function p' on the (x_0, x^o) interval as follows. First, for each number x in the interval, choose a $y(x)$ for which $(x, y(x)) \in L$. Then

$$p'(x) = (x_0 y_1 - x_1 y_1) + \cdots + (x_{j-1} y_j - x_j y_j) + (x_j y(x) - xy(x))$$

Here j is such that $(x, y(x))$ lies between (x_j, y_j) and (x_{j+1}, y_{j+1}) (possibly coinciding with the latter) in the natural ordering of L-points.

Function p' is measurable. For $x_j y(x) = f(x_j, y(x))$ and $xy(x) = f(x, y(x))$ are both measurable, since f itself is measurable and $y(x)$, being monotone, is measurable. Thus p' is measurable on the interval from x_j to x_{j+1}, hence on the whole interval (x_0, x^o).

Now let p_m be the p'-function defined in terms of the sequence (x_{mi}, y_{mi}), $i = 0, \ldots, n_m$. We claim that for all x in the interval (x_0, x^o),

$$p_m(x) \geq p(x) \geq p_m(x) - (1/m) \qquad (8.5.74)$$

The left inequality in (74) is immediate. To prove the right inequality, insert the point $(x, y(x))$ in the sequence (x_{mi}, y_{mi}), $i = 0, \ldots, n_m$, and make the corresponding change on the right side of (73). Since this side does

---

20. We need not distinguish between open, closed, etc., intervals here, since the value of a function at isolated points does not affect its measurability.

not increase, (73) remains valid. On the right side the sum of terms up to $-xy(x)$ is $p_m(x)$, so we have

$$p(x^o) + (1/m) \geq p_m(x) + z \qquad (8.5.75)$$

Here z is the sum of the remaining terms; it begins with xy_{mj} for some j. Now let (x_i, y_i), $i = 0, \ldots, n$, be a monotone sequence of L-points, with $x_n = x$. We have

$$p(x^o) \leq (x_0 y_1 - x_1 y_1) + \cdots + (x_{n-1} y_n - x_n y_n) + z$$

Taking the infimum over all such sequences, we obtain

$$p(x^o) \leq p(x) + z \qquad (8.5.76)$$

Together (75) and (76) yield the right inequality of (74). But (74) implies that $p(x)$ is the limit of $p_m(x)$ as $m \to \infty$, for all x in the interval (x_0, x^o). As the limit of a sequence of measurable functions, p itself (restricted to (x_0, x^o)) is measurable. ∎

This lemma implies that p as defined by (66) is a *left half-potential* for λ^o (see (7.5.8)). For, if (x', y') supports λ^o, then $(x', y') \in L$, and (67) is then the half-potential condition for p.

Theorem: Let α, β be σ-finite measures on the real intervals X, Y respectively, and let measurable $f: X \times Y \to$ reals have nonnegative cross-differences. Let λ^o satisfy the weight-falloff condition and be feasible for the allotment-assignment problem of minimizing $\int f \, d\lambda$, subject to the constraints $\lambda' = \alpha$, $\lambda'' = \beta$. Let $X \times Y$ be the rectangle of support for λ^o.

Then a (p, k) exists that is both a topological and measure potential for λ^o (wide sense).

Proof: Any topological potential is a measure potential here, hence we need only construct the former. Construct L, the line of support for λ^o, and then construct $p: X \to$ reals according to (66). Now define the function $k: Y \to$ reals by

$$k(y) = \inf\{p(x) + f(x, y)\} \qquad (8.5.77)$$

the infimum taken over all $x \in X$. Function k is finite; for, by (67), the infimum is attained at any x such that $(x, y) \in L$, and such an x exists for each y.

It follows at once from (77) that

$$k(y) - p(x) \leq f(x, y) \qquad (8.5.78)$$

for all $x \in X$, $y \in Y$. Furthermore, (78) is satisfied with equality if

(x, y) supports λ^o. For in this case $(x, y) \in L$, and the infimum of (77) is attained at this x.

Next we show that k is measurable. For each $y \in Y$ choose an $x(y)$ such that $(x(y), y) \in L$. Then

$$k(y) = p(x(y)) + f(x(y), y) \qquad (8.5.79)$$

Function $x(y)$ is monotone, hence measurable. Also, f and p are measurable, and (79) then shows that k is measurable.

Thus the pair (p, k) is a topological potential for λ^o. ∎

The properties of p and k, other than measurability, are unspecified. However, if one makes further assumptions about f one can say more. (In (iv) function p is convex iff $-p$ is concave.)

Theorem: Assume the premises of the preceding theorem and construct the potential (p, k) according to its proof. Then

   (i) If f is bounded, p and k are *bounded*.

  (ii) Consider f as a family of functions $f(\cdot, y){:}X \rightarrow$ reals indexed by $y \in Y$. If these functions are all strictly (increasing, decreasing), p is *strictly (decreasing, increasing)* respectively.

 (iii) Consider f as a family of functions $f(x, \cdot){:}Y \rightarrow$ reals indexed by $x \in X$. If these functions are all strictly (increasing, decreasing), k is *strictly (increasing, decreasing)* respectively.

 (iv) If the functions $f(\cdot, y)$, $y \in Y$, are all concave, then p is *convex;* if the functions $f(x, \cdot)$, $x \in X$, are all concave, then k is *concave.*

  (v) If the functions $f(\cdot, y)$, $y \in Y$, are all upper semicontinuous, then p is *lower semicontinuous;* if the functions $f(x, \cdot)$, $x \in X$, are all upper semicontinuous, then k is *upper semicontinuous.*

 (vi) If the family in (ii) is equicontinuous at $x \in X$, then p is *continuous* at x; if the family in (iii) is equicontinuous at $y \in Y$, then k is *continuous* at y.

(vii) If the family in ((ii), (iii)) is uniformly equicontinuous, then (p, k) is *uniformly continuous* respectively.

Proof: (i) Boundedness of p follows from (69), and boundedness of k from that of p and f.

(ii) Let $f(\cdot, y)$ be strictly increasing for all $y \in Y$, and let $x, x' \in X$ satisfy $x < x'$. Choosing $y' \in Y$ so that $(x', y') \in L$, we obtain from (67), $p(x) - p(x') \geq f(x', y') - f(x, y') > 0$, so p is strictly decreasing. If $f(\cdot, y)$ is strictly decreasing, all $y \in Y$, let $x > x'$. The same argument yields $p(x) > p(x')$ again, so p is strictly increasing.

(iii) Since the infimum in (77) is attained at any x such that $(x, y) \in L$, we have

$$k(y') - k(y) \geq f(x', y') - f(x', y) \qquad (8.5.80)$$

valid for all $x' \in X$, y, $y' \in Y$ such that $(x', y') \in L$. If now $f(x, \cdot)$ is strictly increasing, all $x \in X$, choose $y' > y$. Then (80) implies that $k(y') > k(y)$, so k is strictly increasing. If $f(x, \cdot)$ is strictly decreasing, choose $y' < y$. Again (80) yields $k(y') > k(y)$, so k is strictly decreasing.

(iv) For each $x \in X$ there is a $y \in Y$ such that $(x, y) \in L$. Since (77) is satisfied with equality at such a point (x, y), we obtain

$$p(x) = \sup\{k(y) - f(x, y)\} \qquad (8.5.81)$$

the supremum taken over all $y \in Y$. For fixed y, $k(y) - f(\cdot, y)$ is convex. Hence p, as the supremum of a family of convex functions, is convex. Similarly, from (77), k is the infimum of a family of concave functions, hence is itself concave.

(v) For fixed $y \in Y$, $k(y) - f(\cdot, y)$ is lower semicontinuous. From (81), p is the supremum of a family of such functions, hence is itself lower semicontinuous. Similarly, from (77), k is the infimum of a family of upper semicontinuous functions, hence is itself upper semicontinuous.

(vi) and (vii) Treating p, k again as the supremum, infimum of a certain family of functions, repeat the arguments given in Section 7.5, pp. 369–70. ∎

Under certain conditions p and k can be expressed as *line integrals* along the line of support. These are defined as follows. Let g be a bounded measurable function whose domain is a subset of the plane containing the line of support, and let $x_1 \leq x_2$, where $x_1, x_2 \in X$. Then we define

$$\oint_{x_1}^{x_2} g \, dx = \int_{(x_1, x_2)} g(x, y(x)) \, dx \qquad (8.5.82)$$

On the right of (82), $y(x)$ is any function such that $(x, y(x)) \in L$ for all $x \in X$. The integral is then over the open interval (x_1, x_2) with respect to Lebesgue measure. For this to be a bona fide definition, the value of the integral must not depend on the particular function $y(x)$ chosen. That this is so may be seen as follows. For given $x \in X$, the set of numbers y such that $(x, y) \in L$ is either a singleton or an entire interval, the interiors of two such intervals being disjoint. Since each such interval has a rational number in its interior, there are at most a countable number of them. Hence, except for a countable number of x-values, $y(x)$ is uniquely determined. But the Lebesgue measure of a countable set is zero, so changes in $y(x)$ on this set do not affect (82). Hence the definition is sound. This is called the *integral of g along the line of support with respect to x, from x_1 to x_2.*

It is also convenient to define this in the case when $x_1 > x_2$ by the rule

$$\oint_{x_1}^{x_2} g \, dx = - \oint_{x_2}^{x_1} g \, dx.$$

The *line integral of g with respect to y, from y_1 to y_2 ($y_1 < y_2$)* is defined analogously, with $g(x(y), y)$ replacing $g(x, y(x))$ on the right of (82).

In the following theorem recall that $D_i f(x, y)$ represents the partial derivative of f at the point (x, y) with respect to its ith argument, $i = 1, 2$.

Theorem: Let all the premises of the theorem above (p. 450) be satisfied, and let the potential (p, k) be defined by (66) and (77). Let G be an open subset of the plane containing $X \times Y$, and let $f':G \to$ reals coincide with f on $X \times Y$. Then

(i) if $D_1 f'(x, y)$ exists and is continuous on G, we have

$$p(x_2) - p(x_1) = \oint_{x_2}^{x_1} D_1 f(x, y) \, dx \qquad (8.5.83)$$

for all $x_1, x_2 \in X$;

(ii) if $D_2 f'(x, y)$ exists and is continuous on G, we have

$$k(y_2) - k(y_1) = \oint_{y_1}^{y_2} D_2 f(x, y) \, dy \qquad (8.5.84)$$

for all $y_1, y_2 \in Y$.

Proof: (i) Note first that the integrands $D_i f$ in (83) and (84) exist and coincide with $D_i f'$, $i = 1, 2$ respectively. To prove (83), we need to demonstrate only the special case where $x_2 = x_0$, x_0 being the special number used in the definition of p; for then (83) follows in general by subtraction. Thus we must show that

$$-p(x^o) = \oint_{x_0}^{x^o} D_1 f(x, y) \, dx \qquad (8.5.85)$$

for all $x^o \in X$.

Take numbers y_0, y^o such that $(x_0, y_0) \in L$ and $(x^o, y^o) \in L$ and consider the closed, bounded rectangle F having the points (x_0, y_0), (x^o, y^o), (x^o, y_0), and (x_0, y^o) as corners. By assumption, $D_1 f'$ is continuous on F; hence it is bounded and *uniformly* continuous on F. Boundedness insures that the integral (85) is well defined. By uniform continuity for all $m = 1, 2, \ldots$ there is a $\delta_m > 0$ such that if $|y' - y''| < \delta_m$, then

$$|D_1 f(x, y') - D_1 f(x, y'')| \leq 1/m \qquad (8.5.86)$$

for all x between x_0 and x^o, all y', y'' between y_0 and y^o.

Now take a sequence of points along the line of support from (x_0, y_0) to $(x_n, y_n) = (x^o, y^o)$ such that $|y_{i-1} - y_i| < \delta_m$ for $i = 1, \ldots, n$. Then for each $i = 1, \ldots, n$,

$$\left| \int_{x_i}^{x_{i-1}} D_1 f(x, y_i) \, dx - \oint_{x_i}^{x_{i-1}} D_1 f(x, y) \, dx \right| \leq \frac{1}{m} |x_i - x_{i-1}|$$

from (86), since for each x between x_{i-1} and x_i, the $y(x)$ such that $(x, y(x)) \in L$ differs from y_i by less than δ_m. Now

$$\int_{x_i}^{x_{i-1}} D_1 f(x, y_i) \, dx = f(x_{i-1}, y_i) - f(x_i, y_i) \qquad (8.5.87)$$

The right of (87) is a typical pair of terms in the summation (66) for the sequence (x_i, y_i), $i = 0, \ldots, n$. It follows that the sum (66) differs from the line integral

$$\oint_{x^o}^{x_0} D_1 f(x, y) \, dx \qquad (8.5.88)$$

by at most $|x^o - x_0|/m$ if the differences between successive y_i's are all less than δ_m.

Now take a sequence, the mth member of which is itself a sequence of points along L whose corresponding sum (66) is within $1/m$ of $p(x^o)$. Add to this mth sequence, if necessary, sufficient extra points so that successive y_i's differ by less than δ_m. These additions do not increase (66), so it remains within $1/m$ of $p(x^o)$; it is also within $|x^o - x_0|/m$ of (88). Letting $m \to \infty$, we conclude that (88) equals $p(x^o)$. This is the same as (85), which implies (83).

(ii) To prove (84), we need yet another expression for k, one similar to (66). Specifically, we now show that

$$k(y^o) = \inf[(x_{n-1}y_n - x_{n-1}y_{n-1}) + \cdots + (x_1 y_2 - x_1 y_1) + x_0 y_1] \qquad (8.5.89)$$

for all $y^o \in Y$ (xy abbreviates $f(x, y)$). Here the infimum is to be taken over all monotone sequences of points (x_i, y_i), $i = 0, \ldots, n$, along the line of support such that $y_n = y^o$ (n can be any positive integer).

To prove (89), note that the bracketed expression is simply $z + x_n y_n$, where z is the sum (66). Since $k(y^o) = p(x_n) + x_n y_n \leq z + x_n y_n$, this proves (89) with the sign $\leq$ substituted for $=$. To prove the opposite inequality, choose a number x^o for which $(x^o, y^o) \in L$, then a sequence of L-points such that (66) comes within ϵ of $p(x^o)$:

$$p(x^o) + \epsilon \geq (x_0 y_1 - x_1 y_1) + \cdots + (x_{n-1} y_n - x_n y_n) \qquad (8.5.90)$$

Here $x_n = x^o$. We may also assume that $y_n = y^o$; for if not, insert the point (x^o, y^o) in the sequence. This does not disturb the validity of (90), but does make $(x_{n-1} y_n - x_n y_n) = 0$, since $x_{n-1} = x_n = x^o$. Hence we can keep or delete the last parenthetical term in (90), allowing us to make $-x^o y^o$ the last term. Adding $x^o y^o$ to both sides and letting $\epsilon \to 0$, we obtain (89) with the opposite inequality. This proves (89).

Now choose a number y_0 such that $(x_0, y_0) \in L$. Note that $k(y_0) = x_0 y_0$, since $p(x_0) = 0$. Hence

$$k(y^o) - k(y_0) = \inf[(x_{n-1} y_n - x_{n-1} y_{n-1}) + \cdots + (x_0 y_1 - x_0 y_0)]$$

At this point we repeat the argument of the first half of this proof, interchanging the roles of x and y, to conclude that

$$k(y^o) - k(y_0) = \oint_{y_0}^{y^o} D_2 f(x, y) \, dy$$

Letting $y^o = y_2$ and then $y^o = y_1$ and subtracting, we obtain (84). ∎

Note that in (83) x_1 is on top, while in (84) y_2 is on top. For the product function $f(x, y) = xy$, we have $D_1 f(x, y) = y$, $D_2 f(x, y) = x$, so that the line integrals (83), (84) take on an especially simple form in this case.

We now return to the problem of showing that weight-falloff measures are optimal for allotment-assignment.

Theorem: Let (S, Σ_s, α) and (Q, Σ_q, β) be bounded measure spaces, let $h:S \to$ reals and $w:Q \to$ reals be measurable, and let ν^o be a measure on $(S \times Q, \Sigma_s \times \Sigma_q)$, with marginals α and β, satisfying the (measurable) weight-falloff condition (with respect to h, w). Let bounded measurable f:reals$^2 \to$ reals have nonnegative cross-differences.

Then ν^o is best for the allotment-assignment problem of minimizing

$$\int_{S \times Q} f(h(s), w(q)) \, \nu(ds, dq) \qquad (8.5.91)$$

over measures ν with marginals α and β.

Proof: Let λ^o be the measure on the plane induced from ν^o by the mapping $(s, q) \to (h(s), w(q))$. Since ν^o satisfies measurable weight-falloff, so does λ^o. Let $X \times Y$ be the rectangle of support for λ^o. If $X \times Y = \phi$, then $\nu^o = 0$, so $\alpha = \beta = 0$ and the theorem is trivial. We may, therefore, assume that $X \times Y$ is not empty. Let $S_1 = \{s \mid h(s) \in X\}$ and $Q_1 = \{q \mid w(q) \in Y\}$. The complement of $X \times Y$ on the plane has λ^o-measure zero; hence

$$\nu^o((S \times Q)\backslash(S_1 \times Q_1)) = 0, \qquad \alpha(S\backslash S_1) = 0, \qquad \beta(Q\backslash Q_1) = 0$$

Now consider the modified allotment-assignment problem in which α and h are restricted to S_1, β and w are restricted to Q_1, feasible measures ν have universe set $S_1 \times Q_1$, and the integral (91) is over $S_1 \times Q_1$. Any measure feasible for the original problem has the form $\nu \oplus 0$, where ν is feasible for the modified problem, and 0 is the zero measure on

$$(S \times Q)\backslash(S_1 \times Q_1)$$

This establishes a 1–1 correspondence between the feasible solutions to these two problems, and the values of the objective function (91) for corresponding solutions are equal. Hence we need prove only that the restriction of ν^o to $S_1 \times Q_1$ is best for the modified problem.

Let $\alpha_1, \beta_1, h_1, w_1$, and ν_1^o denote the appropriate restrictions of these functions for the modified problem. Also let f_1, λ_1^o be the restrictions of f and λ^o to $X \times Y$. Since λ_1^o satisfies weight-falloff and f_1 has nonnegative cross-differences, a measure potential (p, k) exists for λ_1^o. Hence $(p \circ h_1, k \circ w_1)$ is a measure potential for ν_1^o, since ν_1^o induces λ_1^o.

Also f_1 is bounded, so that p and k and therefore $p \circ h_1$ and $k \circ w_1$ are bounded. It follows that the integrals $\int_{S_1} p \circ h_1 \, d\alpha_1$ and $\int_{Q_1} k \circ w_1 \, d\beta_1$ are well defined and finite. Hence $(p \circ h_1, k \circ w_1)$ is a measure potential for ν_1^o in the *narrow* sense. This implies that ν_1^o is best for the modified allotment-assignment problem, so ν^o is best for the original. ∎

This strengthens a previous result (p. 442). Finally, we conclude with a theorem on the existence of optimal solutions.

Theorem: Let (S, Σ_s, α) and (Q, Σ_q, β) be measure spaces with $\alpha(S) = \beta(Q) < \infty$. Let Σ_s and Σ_q be the Borel fields of topologies $\mathfrak{J}_s$ and $\mathfrak{J}_q$ respectively, these being topologically complete and separable. Let $h{:}S \to$ reals and $w{:}Q \to$ reals be semicontinuous, and let $f{:}$reals$^2 \to$ reals be bounded measurable with nonnegative cross-differences.

Then a *best* solution exists to the allotment-assignment problem of minimizing (91) over measures ν with marginals α and β.

Proof: By a preceding theorem (p. 441), a measure ν^o with marginals α and β exists which satisfies measurable weight-falloff. By the theorem above, ν^o is best. ∎

8.6. APPLICATIONS OF ALLOTMENT-ASSIGNMENT

The preceding section has been written at a rather abstract level, and the hurried reader may have trouble interpreting the results for the allotment-assignment problem in terms of a concrete Thünen system. We now give several illustrations, and show that the allotment-assignment problem has applications well beyond the range of models contemplated in Sections 8.1–8.4.

We begin with the original Thünen model. Here Space is a circular disc of finite radius r, its center being the nucleus. (S, Σ_s, α) is ordinary two-dimensional Lebesgue measure on the plane, restricted to S, so that the area of the entire system, $\alpha(S)$, is πr^2. The ideal distance of location s, $h(s)$, is simply the ordinary Euclidean distance of s from the nucleus. The set of land uses Q is finite, say $\{q_1, \ldots, q_n\}$. It is natural in this case to let Σ_q be the class of all subsets of Q. The ideal weight function w is now simply an n-tuple, $(w_1, \ldots, w_n)$, w_i being the weight of land use q_i, $i = 1, \ldots, n$. We assume to begin with that all these w_i's are positive and distinct, and we may suppose that land uses are numbered in order of decreasing weight: $w_1 > w_2 > \cdots > w_n > 0$. The allotment measure β over (Q, Σ_s) is also characterized as an n-tuple $(\beta_1, \ldots, \beta_n)$, β_i being the allotment of the singleton set $\{q_i\}$. We assume that

$$\beta(Q) = \beta_1 + \cdots + \beta_n = \pi r^2 = \alpha(S) \tag{8.6.1}$$

That is, the acreage allotted to all activities together just uses up the area of the system. Equation (1) is a necessary condition for the existence of a feasible solution to the allotment-assignment problem with marginals α, β.

Now consider assignments ν on $(S \times Q, \Sigma_s \times \Sigma_q)$. Geometrically, $S \times Q$ consists of n replicas of the circular disc S, one for each q_i. Label these replicas $S_1, \ldots, S_n$. (Thus $S_i = S \times \{q_i\}$, which we may identify with S.) Then ν is the direct sum of n measures $\nu_1, \ldots, \nu_n$, ν_i being over S_i. In fact, ν_i is the areal distribution of land use q_i over Space. For feasibility one must have

$$\nu_i(S_i) = \beta_i \tag{8.6.2}$$

$i = 1, \ldots, n$, and

$$\nu_1(F) + \cdots + \nu_n(F) = \alpha(F) \tag{8.6.3}$$

for all $F \in \Sigma_s$. Equation (2) states that the acreage occupied by each land use meets its allotment, while (3) states that the area of any region is just exhausted by its assignment to the various land uses. (The sets F on the left of (3) are replicas of the region $F \subseteq S$ and should, strictly speaking, be distinguished by subscripts i.)

Among these feasible ν's, consider one very special assignment: the *Thünen assignment* ν^o. This is characterized by an n-tuple $r_1, \ldots, r_n$, where

$$\pi r_i^2 = \beta_1 + \cdots + \beta_i \qquad (8.6.4)$$

$i = 1, \ldots, n$ (r_i is the radius of the borderline between the zones occupied by land uses q_i and q_{i+1}; note $r_n = r$). Then ν_i^o is defined to be 2-dimensional Lebesgue measure *truncated* to the circular ring

$$\{s \mid r_{i-1} \leq h(s) \leq r_i\} \qquad (8.6.5)$$

$i = 1, \ldots, n$ (for $i = 1$, $r_0 = 0$). The feasibility of ν^o is easily checked; e.g., the area of region (5) is $\pi r_i^2 - \pi r_{i-1}^2 = \beta_i$, which establishes (2).

It is customary to represent ν^o by collapsing the replicas $S_1, \ldots, S_n$ onto the disc S. Measures $\nu_1^o, \ldots, \nu_n^o$ are mutually singular, and one can say, roughly, that land use q_i occupies the ring-shaped region (5). This gives the familiar picture of Thünen systems with concentric rings of land uses.

The Thünen assignment ν^o satisfies the *measurable weight-falloff condition*. This is almost obvious, since weights fall as one moves outward from ring to ring. To prove it, let E_1, E_2 be two sets of positive ν^o-measure, with E_1 southwest of E_2. This implies that $\nu^o(E_1 \cap S_i) > 0$, $\nu^o(E_2 \cap S_j) > 0$ for some i, j, with $i > j$ (since $w_i < w_j$). Hence there must be points $(s_1, q_i) \in E_1, (s_2, q_j) \in E_2$, such that

$$h(s_1) \geq r_{i-1} \geq r_j \geq h(s_2) \qquad (8.6.6)$$

Relation (6) contradicts the assumption that E_1 is southwest of E_2, and the proof is complete.

Furthermore, ν^o is the only feasible assignment satisfying the measurable weight-falloff condition. This follows from the uniqueness theorem (p. 438), whose premises are satisfied: Σ_q is the σ-field inversely induced by w from the real Borel field, and

$$\alpha\{s \mid h(s) = x\} = 0 \qquad (8.6.7)$$

for all real x. Equation (7) states that the area of the set of locations *exactly at* distance x from the nucleus is zero.

It follows (p. 455) that the Thünen assignment is the unique feasible assignment that minimizes total transportation costs, given by

$$\int_{S \times Q} h(s)w(q) \, \nu(ds, dq) \qquad (8.6.8)$$

(Uniqueness of the minimizer follows from p. 427. Exercise: Evaluate (8) explicitly in terms of $\beta_1, \ldots, \beta_n$ for $\nu = \nu^o$.)

Furthermore, we can calculate the potential (p, k) explicitly for the Thünen assignment. Integral (8) is the special case of the allotment-assignment problem for which $f(x, y)$ is the product function xy. This is continuously differentiable, hence we may use the line integral formulas (5.83) and (5.84) to calculate p and k.

The line of support L, along which the integrals are to be taken, is a "staircase" polygon in the rectangle of support $X \times Y$, going horizontally from $(0, w_1)$ to (r_1, w_1), then down vertically to (r_1, w_2), then over to (r_2, w_2), etc., ending at (r_n, w_n).[21] Applying (5.83), we find that p (whose domain is $X = \{x \mid 0 \leq x \leq r\}$) is a decreasing, convex, polygonal function, whose slope is $-w_i$ on the segment from r_{i-1} to r_i, $i = 1, \ldots, n$; p is unique up to an additive constant.[22]

The composition $p \circ h$ determines a real-valued function on Space which is a left half-potential for the allotment-assignment problem. This may be interpreted as land value, or more precisely as *land-value density*, taken per unit of ideal area ("dollars per acre"). Differences in land values at different sites reflect differences in locational advantages. In particular, the decline in land-value density at a rate of w_i per unit distance as one moves outward from the nucleus matches exactly the *increase* in transport cost incurred per unit ideal distance per unit ideal area.

Applying (5.84), function k (whose domain is $Y = \{y \mid w_n \leq y \leq w_1\}$) is an increasing, concave, polygonal function, whose slope is r_i on the segment from w_{i+1} to w_i, $i = 1, \ldots, n - 1$. Actually, only the values of k at the points $w_1, \ldots, w_n$ are relevant, since the composite function $k \circ w$, which determines a right half-potential on $\mathbf{Q}$, takes on only these values. The interpretation of k is also less clear than that of p, since k corresponds to the "artificial" allotment constraint, while p corresponds to the "natural" areal capacity constraint. $k(w_i)$ may be thought of as the *gross cost density* associated with land use q_i, the sum of transport cost and land value per unit area.

We have spelled out the application of the allotment-assignment analysis to this simple Thünen model in considerable detail. For the models that follow the discussions are briefer, leaving the task of filling in the details of the argument to the reader.

Suppose the preceding Thünen system is modified by having a total allotment that is *less* than the total areal capacity (change the middle equality sign of (1) to $<$). This is no longer in allotment-assignment form, but can be brought into this form by the device of adding an artificial "vacancy" land use q_{n+1} to the set $\mathbf{Q}$. This q_{n+1} has an ideal weight w_{n+1} of zero and is given an allotment β_{n+1}, which just exhausts the surplus areal capacity.

The above analysis now applies verbatim to this modified system,

---

21. The plane in which the rectangle of support lies should not be confused with the plane in which S lies. The former is an intrinsic feature of any allotment-assignment problem; the latter is an accidental feature of the special allotment-assignment problem now under discussion.

22. This ambiguity of the additive constant is characteristic for variant I transportation problems, of which allotment-assignment is a special case. It has an economic interpretation, as we shall see.

and we find that q_{n+1} occupies the outermost ring, since it is the lightest land use. In other words, the occupied land is crowded in as closely as possible to the nucleus, the surplus vacant land being on the outskirts. The only other difference is that land-value density $p \circ h$ is constant in the vacant ring beyond radius r_n (since the slope of p is $-w_{n+1} = 0$ there).[23]

Concerning the meaning of "vacancy," it should be stressed that it refers to *permanent* vacancy, i.e., over the entire time horizon of the system. Suppose, for example, that Time is a certain 100-year interval, and that a given site is unoccupied for 99 years but does trade with the nucleus in the final year. The land use assigned to this site is not the vacancy land use because ideal weight involves an integration over all Time and will therefore be positive in this case, whereas it must be zero for "vacancy."

Going a step further, one can allow some (or all) the weights w_i to be *negative*. Formally, this creates no difficulties, and the unique minimizer of (8) is again the Thünen assignment: concentric rings are occupied by land uses in order of decreasing weight, the outermost zone being occupied by the land use whose weight is most negative. The half-potential p is still convex but no longer decreasing; rather, it decreases at distances occupied by land uses of positive weight and increases at distances occupied by land uses of negative weight.

How does one interpret negative weights? One possibility is to think of these land uses as being *repelled* from the nucleus (say by pollution, crime, or other aspects of the urban syndrome) rather than as being *attracted* by transportation linkage. The land use of highest negative weight is the one most repelled, and it will naturally settle in the outermost ring.

There is also a purely formal use for negative weights. Suppose the problem given is to *maximize* rather than minimize (8). This is equivalent to negating the weights and minimizing, which is again in allotment-assignment form. It follows that the solution to maximizing (8) is the *anti-Thünen* assignment, obtained by reversing the distance ordering of land uses: the lightest one in the closest ring, the heaviest in the outermost ring, etc.

We now drop the assumption that Space is a circular disc, and instead let it be *any* bounded measurable subset of the plane. Now α is still 2-dimensional Lebesgue measure restricted to S, we still assume that $\beta(\mathbf{Q}) = \alpha(S)$, and $h(s)$ is still the Euclidean distance from the nucleus to location s.

With these weaker assumptions, there is still a unique feasible assignment satisfying the measurable weight-falloff condition. This assignment has virtually the same form as the Thünen assignment ν^o; viz., land use q_i

---

23. If the allotment-assignment problem were formulated with an areal capacity *inequality* constraint (variant II or IV rather than I), this constant would be zero: Unused land is valueless.

will occupy part of a ring-shaped zone (centered on the nucleus) of the form (5), the q_i ranging outward in order of decreasing weight. The novelty is that not all the points of the plane satisfying (5) are available, but only those in S. In general, (4) will also be false. Instead, we have

$$\alpha[S \cap \{s \mid h(s) \le r_i\}] = \beta_1 + \cdots + \beta_i \qquad (8.6.9)$$

$i = 1, \ldots, n$. Here r_i, the distance of the borderline between the land-use zones for q_i and q_{i+1}, will generally be larger than the r_i of (4). One must go further out to get the same area, since pieces of the plane are missing.

The existence of a weight-falloff measure follows from the existence of $r_1, \ldots, r_n$ satisfying (9), and this in turn can be proved using the argument of (5.48)–(5.49). The uniqueness theorem also still applies. This measure still minimizes (8). The potential (p, k) is the same as before. In short, removing an arbitrary piece of land from the system makes no qualitative difference in the solution.

The foregoing generalization applies to cases where a portion of the region is unavailable for occupancy. For example, this may be due to bodies of water, poor drainage, irregular terrain, or other natural adversities.[24] Or, some land may be preempted for public use or lie outside the legal jurisdiction of the system. For another application, suppose the system is subject to a zoning ordinance. This invalidates the allotment-assignment formulation because some land uses can be assigned to some zones but not to others. However, if we confine attention to any *one* zone, the formulation becomes revalidated, the set of land uses **Q** being replaced by **Q**′, which is the subset of uses allowed in that particular zone. We may therefore expect that the Thünen ring 'pattern will be present *within* any particular zone but that land uses, and the radii of the borderline between them, will vary from zone to zone.

We now drop the assumption that α is 2-dimensional Lebesgue measure. Instead, we let it be *any* measure such that $\alpha(S) = \beta(\mathbf{Q})$ is finite, where S is still a bounded subset of the plane (this could represent differential "fertility" of land or the fact that land in some places has been augmented by the building of multiple-story structures). A feasible assignment still exists, satisfying the measurable weight-falloff condition, and it has virtually the same Thünen form as above. The only novelty that can arise would be if

$$\alpha\{s \mid h(s) = r_i\} > 0 \qquad (8.6.10)$$

for some $i = 1, \ldots, n - 1$, r_i being the radius of the borderline between

---

24. The New York metropolitan region is fragmented by bodies of water. Yet population density maps indicate fairly close conformity to a Thünen ring pattern (centered on midtown Manhattan) that is unaffected by these interruptions. This is just what one would expect from the above analysis.

land uses q_i and q_{i+1}. For suppose x acres of this borderline are to be assigned to activity q_i, where x is positive but less than (10). Then, provided α on $\{s \mid h(s) = r_i\}$ is not simply concentrated, there will be more than one way to apportion the land of (10) between q_i and q_{i+1} without violating weight-falloff. Thus uniqueness may break down with this more general areal capacity measure α. Any of these weight-falloff measures is best for the allotment-assignment problem. The potential (p, k) is unchanged.

A slight modification of the procedure outlined above enables us to construct one of these Thünen assignments. An r_i satisfying (9) may not exist. Instead, we let r_i be the smallest number for which the right side of (9) does not exceed the left. If (10) obtains for this r_i, assign to q_i the proportion of α on the borderline that just fills out its allotment, and assign the rest to q_{i+1}. If the allotment of q_{i+1} is also exhausted, assign the rest to q_{i+2}, etc.

We continue to generalize and now abandon the assumption that $h(s)$ is the Euclidean distance from s to the nucleus. Instead, we let it be any bounded measurable function on S. Of course, $h(s)$ is interpreted as ideal distance, and our present generalization amounts to breaking the identification between geographical and economic distance. Thus we can incorporate irregularities of terrain and the transportation grid, tariffs not proportional to distance, zonal tariffs, etc.

Formally this generalization makes little difference because the theory of the allotment-assignment problem makes no assumption that $h(s)$ has any relation to a metric on S. We cannot assume that there is a unique feasible weight-falloff measure because the (10) phenomenon invalidates the uniqueness theorem. We can still assert the existence of such a measure, constructing it by the procedure outlined above. This measure still minimizes (8), and the potential (p, k) retains its properties.

But despite all this formal similarity, the *geographical appearance* of the resulting Thünen system can be radically altered. The land-use zones lie between contours of $h(s)$, as in (5). If $h(s)$ is Euclidean distance from the nucleus, these contours are concentric circles, and the familiar pattern emerges. With general $h(s)$, however, the zones can become quite irregular.

Let $\{s \mid h(s) = r_i\}$ be the "borderline" between zones occupied by land uses q_i and q_{i+1}. Then $q_1, \ldots, q_i$ occupy the open disc $\{s \mid h(s) < r_i\}$. (The borderline itself, if it has positive area, may be occupied by q_i, by q_{i+1}, or may be shared by them, etc.) The geographical shapes of the Thünen "rings" can then be gauged by examining the shapes of these open discs. This returns us to the discussion of Section 8.3. For the city-block metric (generated perhaps by a rectangular road grid) the open discs will be concentric diamonds, and the "rings" will be the regions between two diamonds at ideal distances r_{i-1} and r_i, $i = 1, \ldots, n$. For the distance function determined by a system of high-speed radial arteries converging on the nucleus, the open discs will have an amoeboid shape, projecting outward along the arteries. We would then expect high-intensity land uses to "sprawl" along the arteries, while sites away from the arteries, at similar geographical distances from the nucleus, would have lower intensity, more "rural" uses.

Limited-access transportation systems lead to ideal distances with disconnected open discs and a corresponding fragmentation of the Thünen rings. The resulting geographical pattern is intriguing. Consider, for example, a plane that is uniform except for a high-speed commuter railway connecting the nucleus with a series of isolated stops s_1, s_2, $s_3 \ldots$ in order of increasing distance. Away from the railway, land uses will be arranged in concentric circular rings in order of decreasing land-use weight q_1, q_2, $\ldots$. Suppose the first stop s_1 satisfies $r_9 < h(s_1) < r_{10}$, so that points off the railway at that distance are in the zone of q_{10} (s_1 itself is geographically far beyond that zone, but the railway shrinks the economic distance). The weight-falloff condition then calls for a *secondary sequence* of concentric rings, centered on s_1 and beginning with *land use* q_{10} rather than q_1. The second stop s_2 will be further out, say at ideal distance between r_{16} and r_{17}. Then another secondary sequence will spring up around s_2, starting at q_{17}, $q_{18}, \ldots$, and so on for $s_3, s_4, \ldots$.

The resulting pattern is a simple "urban hierarchy" organized in the form of a "metropolitan region." The "central" city grows up about the nucleus. "Satellite" cities grow up about the points of access to the transportation network. However, these cities lack the full range of land uses; the heaviest, most intensive, most "urban" uses are missing. The further out the satellite is, the more uses will be missing. The incidence matrix, indicating which land uses are present in which cities, is of the "central place" type in that if a given land use occurs in a given city, all lower type land uses are also present (in this case, "lower" means "lighter").

We could make the transportation system itself hierarchical, which would lead to the satellite cities themselves spawning satellites, etc. The central place pattern would still obtain. The interesting point is that this rather complicated scheme follows from simple and plausible assumptions concerning the transportation system, coupled with the weight-falloff condition, which itself follows from the minimization of transport costs (8).

To round off the discussion of non-Euclidean ideal distances, it should be mentioned that the "borderlines" $\{s \mid h(s) = r_i\}$ can themselves be broad geographic zones. That is, changes in geographic distance lead to no change in transport cost over a certain range. This can occur in a thoroughgoing regime of freight absorption or uniform charges over broad zones. This is quite common in postal and telephone rates, utility charges, retail deliveries, etc. Conversely, ideal distance can jump discontinuously, as at a tolling point or political border.

In all this the pattern of land-value densities depends on ideal distance and will therefore behave irregularly with respect to geographical distances. Thus land values will reach local peaks at points of access to the transportation system (highway interchanges, railway stops, etc.), will have ridges along radial arteries, etc.

We have devoted almost all our attention so far to modifications in Space and its associates α and h. Let us now turn to $\mathbf{Q}$. This has been assumed finite with unequal weights $w_1, \ldots, w_n$, with Σ_q being all subsets of $\mathbf{Q}$.

Now suppose that two different land uses have the same weight, say $w_1 = w_2$. Then uniqueness of a feasible weight-falloff assignment can in general no longer be guaranteed. Either q_1 or q_2 may occupy the innermost Thünen ring; more generally, any mixture of these over the innermost two zones that satisfies their allotments will do. (Correspondingly, some premise of the uniqueness theorem must fail. In this case it is the premise that Σ_q is the σ-field inversely induced by w; e.g., the singleton $\{q_1\}$ is not the inverse image of any real Borel set.)

Let us now assume that $\mathbf{Q}$ is infinite. This encompasses the case, for example, where continuous variation in the intensity of a land use is possible (recall that intensity variations in "one" land use are to be formally considered as *different* land uses.) We have an allotment measure β on $(\mathbf{Q}, \Sigma_q)$ and a measurable weight function $w:\mathbf{Q} \rightarrow$ reals, and we assume as usual that $\beta(\mathbf{Q}) = \alpha(S) < \infty$. We also assume that different land uses have different weights. It is then natural to assume that Σ_q is the σ-field inversely induced by w from the real Borel field. Also assume that $\{w(q) \mid q \in \mathbf{Q}\}$ is a real Borel set. Finally, assume that S, Σ_s, α, and h take the classical form, $h(s)$ being Euclidean distance to the nucleus (which is the center of circular disc S) and α being 2-dimensional Lebesgue measure.

Under these conditions, exactly one feasible assignment ν^o exists satisfying the measurable weight-falloff condition. Uniqueness follows at once from the uniqueness theorem. To prove existence (end of Section 8.5), we construct topologies $\mathfrak{J}_s$, $\mathfrak{J}_q$ on S and $\mathbf{Q}$ respectively, making S and $\mathbf{Q}$ Borel subsets of separable and topologically complete spaces, with respective Borel fields Σ_s and Σ_q, and making h and w semicontinuous. For $\mathfrak{J}_s$ we choose the usual topology on the plane, restricted to S. For $\mathfrak{J}_q$ we take all sets of the form $\{q \mid w(q) \in E\}$, E ranging over all open subsets of the real line with its usual topology. This collection is a topology over $\mathbf{Q}$, the topology *inversely induced* by w. One verifies that all the stated conditions are satisfied.

This unique Thünen assignment is intuitively a limiting case of the classical pattern. We still have circular symmetry of land-use assignments, but do not necessarily have broad rings devoted to a single land use. Instead there may be continuous variation of land uses as one moves outward—always going from heavier to lighter uses, of course. The potential (p, k) may still be constructed by the line integral formulas (5.83) and (5.84). Half-potential p is still convex (and decreasing, if w is positive); k is still concave and increasing. If p is differentiable at any given distance x, its slope is given by $-w$, where w is the weight of the land use located "at" distance x. The only novelty is that p and k need not have polygonal graphs, but may have continuously turning tangents over certain ranges.

As a final generalization, we can relax the factorability assumption (transport cost = ideal weight times ideal distance), which underlies the objective function (8). Recall that the allotment-assignment problem in its general form calls for the minimization of

$$\int_{S \times Q} f(h(s), w(q)) \, v(ds, dq) \qquad (8.6.11)$$

where f has positive (or nonnegative) cross-differences. Integral (8) is merely the special case in which $f(x, y) = xy$. Now the measurable weight-falloff condition characterizes the optimal solutions to (11) in general. Hence all the qualitative features of the Thünen system carry over the more general situation of minimizing (11). The one novelty is that the potential (p, k) may lose its convexity or concavity.

This concludes our discussion of the application of the allotment-assignment problem to the classical Thünen model and its generalizations. The variety of situations covered is already considerable. We now generalize still further, departing more or less radically from the classical interpretation.

ALLOCATION OF EFFORT AS A THÜNEN PROBLEM

In this book we have taken up two broad types of optimization problems: the allocation-of-effort problem of Chapter 5 and the transportation problem of Chapter 7 (of which the allotment-assignment problem is a special case). These problems are quite different in construction; but there is one kind of situation to which they both apply, each in its own way.

Consider the following search problem. One is prospecting an unexplored region for minerals and distributing search effort over Space so as to maximize expected return. A single location is the base of operations, say the railhead that connects one with civilization. Now, even if the prospect of finding minerals is uniform over the region, it seems reasonable that a person should search more intensively in the vicinity of the base than far away from it. For greater transport costs are incurred at distant points: the process of exploration itself is more costly, and so is the shipment of any minerals that are discovered.

This search problem is of the allocation-of-effort type of Chapter 5; yet the optimal solution appears to be a pattern that is symmetric about the base of operations such that intensity falls off with increasing distance from the base. This certainly mimics the Thünen pattern. The question then arises, Can we associate an allotment-assignment problem with the original allocation-of-effort problem such that their optimal solutions correspond in some way? This would provide insight into the phenomenon of the Thünen pattern arising in an apparently very different type of problem.

Before going into this, we give a few more examples. Consider a farmer faced with the problem of distributing fertilizer over his fields. The farmhouse provides the "base of operations." If there are no geographic irregularities, the distribution that maximizes total return would again appear to conform to the Thünen pattern, with more intensive fertilizer use on the nearer fields.

Again, consider a discriminating monopolist with a single factory located at s_o. Suppose that consumers are uniformly distributed over Space and all have identical demand functions for the commodity produced at s_o. Assume that the monopolist pays for transportation and that he can freely vary delivered price from one location to another—or, what is the same thing, that he can freely choose the density pattern of deliveries over Space, $\delta:S \rightarrow$ reals (δ is measured in units of, say, tons per year per acre). The profit-maximizing density would then appear to have the Thünen pattern in the sense that δ should decrease with increasing distance from s_o.

As a simple generalization of this case, drop the assumption that consumers are uniformly distributed. Then we still reach the same conclusion, provided $\delta(s)$ is interpreted as the density of deliveries *per person*, rather than *per acre*. A nonrigorous argument for the "intensity falloff" of δ goes as follows. Suppose there were sites s_1, s_2, with s_2 further from the factory than s_1 but with $\delta(s_2) > \delta(s_1)$. Then interchange these densities between one person at s_1 and one person at s_2. This leaves gross revenue intact but reduces transport costs, hence the original distribution was nonoptimal.

As a last example, let some public-service facility such as a police or postal station be located at s_o. This is to serve a certain hinterland and the question is, How shall the services be spatially distributed over this region? (Intensity of service could be measured, say, by frequency of police patrols or mail pickups.) With uniformly distributed population and with uniform benefits as a function of service intensity, it again seems reasonable that intensity of service should decline with distance from s_o, to maximize total benefits net of transportation costs.

We now proceed to the analysis. The allocation-of-effort problem is determined by a σ-finite measure space, (S, Σ_s, α), three measurable functions, $u:S \times$ reals $\rightarrow$ reals and $b, c:S \rightarrow$ extended reals, and two extended real numbers L_o and L^o. The problem is to choose a measurable function $\delta:S \rightarrow$ reals to maximize

$$\int u(s, \delta(s))\ \alpha(ds) \tag{8.6.12}$$

subject to the constraints

$$L_o \leq \int_S \delta\ d\alpha \leq L^o \tag{8.6.13}$$

$$b \leq \delta \leq c \tag{8.6.14}$$

Here (12) is an indefinite integral over S, and "maximization" is to be understood in the sense of standard ordering of pseudomeasures. (If (12) is well defined and finite as a definite integral for all feasible δ, this reduces to the ordinary size comparison of definite integrals.) It is also required that $\int_S \delta\ d\alpha$ be finite even if L_o or L^o is infinite. (For convenience we have changed notation somewhat from that of Chapter 5.)

We now identify a special subclass of the problems (12)–(14), one that embraces all the examples cited and for which the Thünen pattern emerges. The first special assumption is that b and c be *constants* (one or both may be infinite). The second special assumption is that

$$u(s, x) = -f(h(s), x) \tag{8.6.15}$$

for all $s \in S$, x real, for some measurable functions $h:S \rightarrow$ reals, $f:\text{reals}^2 \rightarrow$ reals, where f has positive cross-differences. There are a few more minor technical assumptions, but these are the two major ones. We now interpret them.

The first states that the limits on the range of permissible densities do not vary from point to point. This is plausible for all our examples. In fact it is reasonable to take $b = 0$, since negative densities are meaningless for them and no positive minimal density is required. (Exception: In the case of public services there may be an institutional requirement to reach some social minimum for all regions or persons. In this case, $b > 0$ but still constant.) As for the upper limit c, there may be either no such limit ($c = \infty$) or there may be some uniform "saturation level" of intensity for regions or persons.

As for the second assumption, the special form (15) is clearly reminiscent of the integrand in the allotment-assignment problem (11). (The minus sign is inserted because the objective in (12) is to maximize, whereas one minimizes in allotment-assignment.) We show that (15) is a plausible form for the objective in each of our examples. In evaluating a density pattern there are two factors to consider: the *gross returns* over Space accruing from that pattern and the *transportation cost* incurred by that pattern. In the exploration example, gross return would be the expected payoff from mineral discoveries (less searching costs other than transportation). For the farmer, gross return is profit on crops grown, less the cost of fertilizer. For the monopolist, it is revenue from sales. And for the public service facility, it is the social benefit from the service rendered.

In all these cases the gross return at a point will depend only on the intensity at that point, not on the location of the point per se. This results from our uniformity assumptions. The gross return is represented by a measurable function g:reals $\to$ reals. As for transportation cost, we make the simplest factorability assumption: transport cost incurred at point s is the product of intensity at s and the distance $h(s)$ of s from the base of operations. We then have

$$u(s, x) = g(x) - h(s) \cdot x \qquad (8.6.16)$$

The critical observation to make is that (16) is of the form (15) (with f having positive cross-differences) regardless of what the function g is. This follows at once from the fact that g does not depend on s. We may also discard the special transport cost assumption and assume more generally that transport cost is of the form $f_1(h(s), x)$ for some measurable function f_1 with positive cross-differences. Throwing in $g(x)$ does not alter this property, and (15) is still valid. Thus all the examples given seem to be encompassed by our special assumptions.

We now want to translate from the language of the allocation-of-effort problem to the language of the allotment-assignment problem. The latter speaks of land uses. We now let *each possible intensity level* be a land use. The set of land uses Q may be formally identified with the real line. (Actually, the interval $[b, c]$ would suffice.) This is a radically stripped-down version of the full-blown land-use concept. For the latter, the answer to the question, What's going on here?, would require at least the specification of two measures over $R \times T$, production and consumption. Here it requires just the specification of a real number, indicating intensity (intensity of search, sales, fertilizer use, etc.). The weight function w is simply taken to be the identity and may be ignored.

Choice in the allotment-assignment problem is over measures ν on $(S \times Q, \Sigma_s \times \Sigma_q)$; in the allocation problem, choice is over measurable functions δ:$S \to$ reals. To translate from the latter to the former, use the formula

$$\nu(G) = \alpha\{s \mid (s, \delta(s)) \in G\} \qquad (8.6.17)$$

for all $G \in \Sigma_s \times \Sigma_q$ (Σ_q is the Borel field on the real line $\mathbf{Q}$). Here α is the measure of (12)–(13) and has the interpretation of acreage, population, etc. That is, given an intensity distribution represented by δ, (17) shows how to express it as an assignment measure ν. For example, let G be the rectangle $E \times F$, and let α be areal measure. Then $\nu(E \times F)$ is the area in region E assigned to intensities among the numbers F, and this is just what the right side of (17) equals.

To see that ν defined by (17) is indeed a measure, note that $s \rightarrow (s, \delta(s))$ is a measurable mapping from S to $S \times \mathbf{Q}$. Equation (17) states that ν is the measure on $S \times \mathbf{Q}$ induced by this mapping from α on S. (There is one fine point here: the solutions to the allocation problem are not really the densities δ but the indefinite integrals $\int \delta \, d\alpha$ they represent. Two densities, δ_1 and δ_2, represent the same indefinite integral iff they are identical α-almost everywhere. Equation (17) would hardly be satisfactory if two equivalent δ's yielded different ν's. But they yield the same ν, as one may verify.)

In the allotment-assignment problem, ν is feasible iff it has the prescribed marginals α and β. As our notation indicates, we are taking over the measure space (S, Σ_s, α) of the allocation-of-effort problem bodily to be the left marginal space of the allotment-assignment problem. We now show that any assignment ν defined by (17) automatically satisfies the left marginal allotment-assignment constraint. In fact, for any $E \in \Sigma_s$, $\nu'(E) = \nu(E \times \mathbf{Q}) = \alpha(E)$ from (17), so that α is the left marginal of ν. We have not yet defined the allotment measure β on $(\mathbf{Q}, \Sigma_q)$; this will be taken up later.

Theorem: Let (S, Σ_s, α) be a nonatomic σ-finite measure space; let $h:S \rightarrow$ reals and $f:$reals$^2 \rightarrow$ reals be measurable, with f having positive cross-differences and $f(y, \cdot)$ being upper semicontinuous for each real y; let $b, c, L_o,$ and L^o be four extended real numbers.

Let δ^o be unsurpassed for the problem of minimizing

$$\int f(h(s), \delta(s)) \, \alpha(ds) \tag{8.6.18}$$

over the set of measurable functions $\delta:S \rightarrow$ reals satisfying

$$L_o \leq \int_S \delta \, d\alpha \leq L^o \tag{8.6.19}$$

the integral in (19) being well defined and finite, and

$$b \leq \delta \leq c \tag{8.6.20}$$

Then ν^o, defined by (17) for $\delta = \delta^o$, satisfies the (measurable) weight-falloff condition (with respect to h and w, the latter being the identity function).

Proof: The premises imply that an extended real number p_o and a set $E_o \in \Sigma_s$ exist such that $\alpha(E_o) = 0$, and for each $s \in S\backslash E_o$, $\delta^o(s)$ minimizes

$$f(h(s), x) + p_o x \tag{8.6.21}$$

over the real numbers x in the closed interval $[b, c]$ (p. 249).

Suppose first that p_o is finite, and let $s_1, s_2 \in S\backslash E_o$. Using the abbreviation f_{ij} for $f(h(s_i), \delta^o(s_j))$, $i = 1, 2, j = 1, 2$, we then have

$$f_{11} + p_o \delta^o(s_1) \leq f_{12} + p_o \delta^o(s_2)$$

$$f_{22} + p_o \delta^o(s_2) \leq f_{21} + p_o \delta^o(s_1)$$

These yield

$$f_{11} + f_{22} \leq f_{12} + f_{21} \tag{8.6.22}$$

It is impossible that both $h(s_1) < h(s_2)$ and $\delta^o(s_1) < \delta^o(s_2)$ be true, for in this case (22) would violate the positive cross-differences condition. Hence

$$h(s_1) < h(s_2) \quad \text{implies} \quad \delta^o(s_1) \geq \delta^o(s_2) \tag{8.6.23}$$

Now suppose $p_o = +\infty$. Minimization of (21) then means (by convention) that x is to be as small as possible, so that $\delta^o(s) = b$ for all $s \in S\backslash E_o$. Then (23) holds automatically. A similar conclusion follows for $p_o = -\infty$. We conclude that (23) is true for any $s_1, s_2 \in S\backslash E_o$.

We now go to the product space $(S \times Q, \Sigma_s \times \Sigma_q)$. Let G_1, G_2 be two measurable sets, with G_1 southwest of G_2; i.e., if $(s_i, q_i) \in G_i$, $i = 1, 2$, then $h(s_1) < h(s_2)$ and $q_1 < q_2$. (Remember that Q is the real line.) Define two new sets H_1, H_2 by $H_i = G_i\backslash(E_o \times Q)$, $i = 1, 2$. At least one of the sets H_1, H_2 must own no point of the form $(s, \delta^o(s))$, for otherwise (23) would be violated. Hence, from (17), $\nu^o(H_i) = \alpha(\phi) = 0$ must be true for at least one index $i = 1, 2$. We also have $\nu^o(E_o \times Q) = \alpha(E_o) = 0$ so that $\nu^o(G_i) \leq \nu^o(H_i) + \nu^o(E_o \times Q) = 0$ must be true for at least one index $i = 1, 2$. Thus ν^o satisfies measurable weight-falloff. ∎

This result ratifies our intuition concerning the nature of optimal solutions to the allocation problems discussed. They do indeed yield a Thünen system when translated by the natural formula (17) into the proper language. The assumptions made, in addition to the two already discussed, are that α be nonatomic and that $f(y, \cdot)$, as a function of its second argument alone, be upper semicontinuous for each real y.

We mention a few generalizations. As discussed in Section 5.4, the nonatomicity assumption on α could be dropped at the cost of assuming $f(h(s), \cdot)$ to be convex for all s in the atomic part of S. (If α is population measure, the atomic part may be thought of as the "cities"; if α is areal measure, it may safely be assumed to be nonatomic.) For the objective function of type (16), convexity of f is the same as concavity of g (= diminishing marginal returns to intensity of effort), which is not implausible but somewhat restrictive. Upper semicontinuity of f is the same as lower semicontinuity of g, which is so weak as to amount to no assumption at all from the practical point of view.

The assumption that b and c are constants can also be weakened to the following. For any s_1, $s_2 \in S$, if $h(s_1) < h(s_2)$, then $b(s_1) \geq b(s_2)$ and $c(s_1) \geq c(s_2)$. The preceding proof still goes through, with some minor complications whose discussion we omit.

It remains for us to complete the construction of the allotment-assignment problem to be derived from the allocation-of-effort problem (18)–(20). We make the additional assumption at this point that α is finite and define the allotment measure β on the real line $(\mathbf{Q}, \Sigma_q)$ by

$$\beta(F) = \alpha\{s \mid \delta^o(s) \in F\} \qquad (8.6.24)$$

for all Borel sets F. This is the measure induced on the real line from α by the function δ^o. (If α were infinite, one could not guarantee the σ-finiteness of β; in particular, β would not be σ-finite if δ^o were constant.)

Theorem: Assume all the premises of the preceding theorem, and in addition assume that f and α are bounded. Consider the following allotment-assignment problem on the space $(S \times \mathbf{Q}, \Sigma_s \times \Sigma_q)$. Minimize

$$\int_{S \times Q} f(h(s), q) \, \nu(ds, dq) \qquad (8.6.25)$$

over the set of measures ν on $S \times \mathbf{Q}$ that have left and right marginals α and β respectively; β is given by (24), where δ^o is unsurpassed for the problem (18)–(20).

Then the assignment ν^o, defined by (17) for $\delta = \delta^o$, is best for this problem.

Proof: First, we immediately verify that ν^o has α and β for its left and right marginals respectively, so that it is feasible. By the preceding theorem ν^o satisfies measurable weight-falloff; and this fact, together with the boundedness of f and α (hence β), implies that ν^o is best (p. 455). ∎

This artificially constructed allotment-assignment problem has the following intuitive meaning. Start with the density pattern δ^o, which is unsurpassed for the allocation-of-effort problem. Now consider any "reshuffling" of the pattern that preserves the distribution of density aggregated over all S; e.g., if $\alpha\{s \mid \delta^o(s) < 20\} = 100$, the reshuffled pattern will also have 100 acres with density under 20, though the actual region of low density may be different. These reshufflings remain feasible for the allocation-of-effort problem, hence none can surpass δ^o. Now in the associated allotment-assignment problem, the allotment measure β is precisely this density distribution, and the corresponding constraint assures that, in a sense, the feasible assignment ν are reshufflings of ν^o.

The boundedness of f and α insures that (18) is well defined and finite as a definite integral. In fact, it is equal to (25) if ν is derived from δ by (17). Hence the objective functions of the two problems coincide. The allotment-assignment problem is in a sense a subproblem of the original allocation-of-effort problem in that it imposes the extra allotment constraint that restricts comparison to reshufflings. This is the general role of the "artificial" allotment constraint.[25]

Thünen Systems without a Nucleus

Until now, the nucleus or "base of operations" has played a crucial role in the interpretation of the allotment-assignment problem, since the ideal distance $h(s)$ has been taken to be $h(s, s_N)$, the distance between location s and the nucleus s_N. We now show that a considerably more general interpretation is possible, one that need not single out any particular location for special treatment.

The idea is this. We have assumed that any point s must trade exclusively with nucleus s_N. Suppose instead that the pattern of trade of a point is given by a *distribution* ρ over Space. $\rho(S) = 1$ (i.e., ρ is formally a probability measure), and $\rho(F)$ is taken to be the fraction of total trade (exports plus imports, measured by ideal weight) of s, which terminates in region F for all measurable F. The "nuclear" case is precisely that in which ρ degenerates to a measure simply concentrated at s_N.

For example, there may be several nuclei (say $s_{N_1}, \ldots, s_{N_k}$), and trade is to be divided among them in the given proportions $\rho_1, \ldots, \rho_k$. Or ρ may be nonatomic, proportional perhaps to the distribution of population over Space. In any case, ρ is given as a condition of the problem.

We now generalize still further by allowing the spatial distribution of trade to depend on location. To represent this, we take ρ to be a *conditional* probability measure, with domain $S \times \Sigma_s$; i.e., for any $s \in S$, $\rho(s, \cdot)$ is a measure over S, with $\rho(s, S) = 1$, and for any $F \in \Sigma_s$, $\rho(\cdot, F)$ is a measurable function. The interpretation is: $\rho(s, F)$ is the fraction of total trade of location s that terminates in region F.

This allows a good deal of flexibility and realism to be incorporated

---

25. A more careful discussion would note that the constructed allotment-assignment problem introduces extraneous solutions; e.g., a 50-50 mixture of densities 10 and 30 is not the same as a density of 20. (Mixtures are meaningless for the allocation problem.) Hence the last theorem is stronger than it looks, since ν^o is optimal not only against assignments ν of type (17) but against "mixtures" as well.

into the conditions of the problem. We expect in a general way, for example, that locations tend to trade heavily with regions near themselves. One restriction is essential, however, for the analysis that follows. Though distribution may depend on location, it must not depend on the land use whose exports and imports are being distributed.

It remains to show the conditions under which this more general scheme still leads to an allotment-assignment problem. Feasibility conditions are the same. There is an areal measure α on (S, Σ_s) and an allotment measure β on (Q, Σ_q), which are the marginals of any feasible assignment ν on $(S \times Q, \Sigma_s \times \Sigma_q)$. As for the objective, we still postulate a weight function $w:Q \to$ reals. But the distance function $h:S \to$ reals seems to be missing, since there is no nucleus. Instead, we go back a step and assume that unit transport cost between any two locations can be expressed by a measurable function $g:S_1 \times S_2 \to$ reals (S_1 and S_2 are both identical to S; the subscripts are inserted for clarity). Function g need not obey any of the postulates for a metric except for symmetry: $g(s_1, s_2) = g(s_2, s_1)$.

The total transport cost incurred by an assignment ν is then

$$\int_{S_1 \times Q} \nu(ds_1, dq) \int_{S_2} \rho(s_1, ds_2) \, w(q)g(s_1, s_2) \qquad (8.6.26)$$

Integrating from right to left in (26), the integration over S_2 yields the transport cost (per ideal acre) incurred in the process of distributing the exports and imports of land use q at s_1, according to the spatial pattern $\rho(s_1, \cdot)$. But if we now define the function $h:S \to$ reals by

$$h(s) = \int_{S_2} g(s, s_2) \, \rho(s, ds_2) \qquad (8.6.27)$$

then h is measurable, and (26) is equal to

$$\int_{S \times Q} h(s)w(q) \, \nu(ds, dq) \qquad (8.6.28)$$

which is in allotment-assignment form.[26]

Thus our generalized interpretation has the same formal structure as before, provided we define "ideal distance" by the special rule (27). Let us check to see what happens when there *is* a nucleus s_N. In this case we have $\rho(s, F) = 1$ if $s_N \in F$, and $= 0$ if $s_N \in S\backslash F$, all $s \in S$, $F \in \Sigma_s$. The integral (27) then reduces to $g(s, s_N)$, and this is indeed our definition of $h(s)$ in the "nuclear" case.

The optimal solutions to the allotment-assignment problem with the

---

26. We have implicitly assumed that all integrals discussed are finite, to avoid pseudo-measure complications.

objective of minimizing (28) must of course satisfy the measurable weight-falloff condition with respect to (h, w). Thus the heavy land uses will be assigned to regions of low h-values. In the nuclear case this has the geometric interpretation that the heavy land uses crowd in about the nucleus. What interpretation offers itself in the general, nonnuclear case?

We discuss this under the assumption that ρ is independent of s; i.e., a fixed measure ρ over Space determines the distribution of exports and imports from any location with any land use. Also assume that g is a metric on S. In the nuclear case, the nucleus can be characterized as the site for which h is minimal ($h(s_N) = 0$, and h is otherwise positive). This suggests looking for a site s_o that minimizes $h(s)$ of (27). Indeed, the weight-falloff condition requires that heavy land uses crowd in about s_o just as they do about the nucleus when the latter exists.

The problem of finding a location s_o that minimizes (27) is basic in spatial economics. This is the *Weber problem* (to be discussed in detail in Chapter 9), and an optimal location is a *Weber point* or *median* of the distribution ρ. (Remember that ρ is independent of s; hence it may be thought of as simply a measure, not a conditional measure.) Thus land uses will tend to arrange themselves in a pattern that mimics the nuclear case, the median of ρ playing the role of "pseudonucleus." The difference is that the median may be otherwise just like any other point, with no tendency for transportation flows to concentrate on it. Also, the points on a borderline between successive land uses will not generally be equidistant from the median, as they would be from the nucleus.[27]

QUALITY COMPLEMENTARITY

It is a matter of everyday observation that rich people tend to live in better housing than poor people, that good students tend to go to good schools, that abler managers hold more responsible positions, and that "desirable" husbands tend to marry "desirable" wives. Do these cases have anything in common?

First, they refer to associations of two kinds of entities: people and housing, students and schools, workers and jobs, husbands and wives. Second, each of these two kinds of entities are ordered on a quality scale of some sort (by wealth, ability, aesthetic appeal, etc.). And third, out of all the possible ways that entities of various qualities could associate with each other, those of "high" quality on one scale tend to associate with those of "high" quality on the other, and similarly the "lows" associate with the "lows."

These characteristics establish a formal link with Thünen systems.

---

27. A borderline is a locus of constant h-value. These are "isodapanes" or "isovectures" of the Weber problem.

Here the two kinds of entities are locations and land uses, i.e., the points of S and Q respectively. The scaling of these entities is accomplished by the distance and weight functions h and w respectively. For vividness, think of heavy land uses as being of "high" quality, and similarly for "close," low-h locations. Site quality varies inversely with h; this inversion is needed to conform to the above usage. The high-high, low-low quality association is represented precisely by the weight-falloff condition that characterizes optimal land-use assignments ν.

We illustrate this last point by the housing example. Interpret S as the set of housing types and Q as the set of family types. An assignment of families to housing is represented by a measure ν over $S \times Q$; $\nu(E \times F)$ is, say, the number of acres of housing of types E occupied by families of types F. Let $w(q)$ be the wealth of family type q, and let $h(s)$ be an index of quality of housing type, low h corresponding to high quality (h stands for "humbleness"). Then the fact that wealthy families occupy high-quality housing is expressed by the measurable weight-falloff condition on ν: if G_1 is southwest of G_2 (so that $(s_i, q_i) \in G_i$, $i = 1, 2$, implies s_1 is of higher quality and q_1 of lower wealth than s_2, q_2 respectively), then either $\nu(G_1) = 0$ or $\nu(G_2) = 0$.

This all suggests that the analytical apparatus we have developed in this chapter may serve as an explanation of these rather diverse phenomena. This apparatus so far has run exclusively in terms of optimization (specifically, the minimization of cost in the allotment-assignment problem). Some of the situations mentioned may be interpreted directly as optimization problems. For example, a firm has an executive staff of varying ability and a set of positions of varying responsibility, and must fill the latter with the former to maximize profits. But more often we are talking about *social equilibrium* situations, where no single person decides the final outcome. Even here, it is often enlightening to express the equilibrium as the solution to an optimization problem, though no one is consciously trying to solve this problem. (In Section 8.7, we connect the "classical" Thünen equilibrium with the allotment-assignment problem in this way.)

We examine both these approaches for the situations mentioned. For the optimization approach it is convenient to choose a slightly altered form of the allotment-assignment problem, in which one *maximizes* rather than minimizes. Specifically, one is to maximize

$$\int g(-h(s), w(q)) \, \nu(ds, dq) \qquad (8.6.29)$$

over feasible assignments ν. Here $w(q)$ is the quality index of $q \in Q$, and $-h(s)$ is the quality index of $s \in S$. The minus sign is used to facilitate comparison with the original form of the problem.

If we define the function f:reals$^2 \to$ reals by

$$f(x, y) = -g(-x, y) \qquad (8.6.30)$$

we can see that the preference ordering determined by (29) is the same as that determined by minimizing

$$\int f(h(s), w(q))\, v(ds, dq) \qquad (8.6.31)$$

which is the original allotment-assignment objective function.

Now the critical feature of the allotment-assignment problem is that f in (31) should have *positive cross-differences*. By (30), this is true iff the function g has positive cross-differences.

What is the concrete significance of this property? One may speak of positive cross-differences of g in (29) as expressing *complementarity* between the two quality indexes w and $-h$. We argue below that any such complementarity associates high with high, and low with low, qualities of the two indexes. Becker uses essentially the same idea to explain assortative mating, the only formal difference being that he uses the positivity of the cross-*derivative* in place of positive cross-differences. This differentiability assumption is not needed. The exact relation is: let D_i stand for the partial derivative with respect to the ith argument, $i = 1, 2$, and consider the following condition.

$$D_2[D_1 g(x, y)] \geq 0 \qquad (8.6.32)$$

for all x, y. Now one may show, provided $D_2[D_1 g]$ exists and is finite, that g has nonnegative cross-differences iff (32) holds.[28] Besides the widespread tendency of people with similar traits to marry, Becker adduces the association of informed customers with honest shopkeepers, of able farmers with modern farms, and of able workers with able firms.[29]

Consider the job-assignment example. A firm has a managerial staff (graded by ability) and positions to be filled (graded by responsibility). Suppose $g(x, y)$ is the profit generated by a person of ability x placed in a position of responsibility y, total profit generated by an assignment v being given by (29). It is plausible that ability differentials show up more strongly, the more responsible the position to be filled. Thus brilliant A might outshine mediocre B in a leadership position but not do much better as a subaltern because the latter position does not afford much scope for A's talents. Letting $x_1 < x_2$ be two ability levels and $y_1 < y_2$ be two job responsibility levels, the assumption just stated is that

28. Also, under these conditions g has *positive* cross-differences iff (32) holds, *and* the inequality is strict on a set of points (x, y) that is dense in the plane. (To prove these statements, use the mean-value theorem.) Incidentally, the existence of $D_2[D_1 g]$ does not imply the differentiability or even the continuity of $g(x, \cdot)$, for any x.

29. G. S. Becker, A theory of marriage, in T. W. Schultz, ed., *Economics of the Family: Marriage, Children and Human Capital* (Univ. Chicago Press, Chicago, 1974), pp. 299–344, esp. p. 312. See also Ch. 9, note 28 below.

$$g(x_2, y_2) - g(x_1, y_2) > g(x_2, y_1) - g(x_1, y_1) \qquad (8.6.33)$$

But this is precisely the positive cross-differences condition on g!

Again, to take the school-assignment example, let $g(x, y)$, measured in dollars, be the "social benefit" from having a student of ability level x attend a school of quality level y. Condition (33) then states that the differential benefit in favor of the abler student is greater at the higher quality school. This again is not an implausible assumption. If we now imagine a coordinated school policy aimed at assigning students to maximize overall social benefit, it will have an allotment-assignment problem to solve. The optimal solution will then be, under rather general conditions, a weight-falloff measure, which means that the better students go to better schools.

To illustrate how a "weight-falloff" assignment might be a social equilibrium, consider the housing example. Again we make the unrealistic assumption that there is a single quality dimension along which housing types can be arrayed. Let s_1 and s_2 be two housing types, s_1 having the higher quality (say s_1 has a scenic view that s_2 lacks, or s_1 has central air conditioning, etc.). A given family is willing to pay a premium to occupy housing type s_1 rather than s_2. We now assume that the wealthier the family, the greater the premium it is willing to pay for the quality differential. This highly plausible assumption leads to the weight-falloff equilibrium.

The basic argument can be illustrated in the case where there are just two wealth levels, "rich" and "poor." Suppose rich families are willing to pay \$100 to occupy s_1 with its scenic view rather than s_2, while poor families are willing to pay just \$10 for this privilege. It cannot then happen that in equilibrium there are both rich families living in the lower quality housing s_2 and poor families living in the higher quality housing s_1. For if the rent differential exceeds \$10, the poor families in s_1 do better to switch to s_2; while if the rent differential is less than \$100, rich families in s_2 do better to switch to s_1. Since one of these cases must occur, somebody is out of equilibrium. We conclude that the rich occupy high-quality housing and the poor occupy low-quality housing in equilibrium.

This situation is associated in a natural way with the following "artificial" allotment-assignment problem. Let $x_1 > x_2$ and choose a function g:reals$^2 \to$ reals such that

$$g(x_1, y) - g(x_2, y) \qquad (8.6.34)$$

equals the premium families of wealth y are willing to pay to occupy housing of quality index x_1 rather than x_2. The assumption we made above is that the difference (34) increases with y for fixed x_1, x_2; this is the same as saying that g has positive cross-differences.

Let measure α on the set of housing types S and measure β on the set of family types Q be given by: $\alpha(E)$ = square feet of housing of types E and $\beta(F)$ = square feet of housing occupied by family types F in the above social equilibrium ν^o. Then ν^o will be the optimal solution to the problem of maximizing (29) (the integrand g satisfying (34)) over assignments with marginals α and β. Furthermore, the prices associated with the various housing qualities turn out to be a left half-potential for this problem.

Note that (34) does not determine g uniquely. Indeed, adding to g an arbitrary function that depends only on y will not affect (34). But this transformation does not alter the preference order determined by (29), hence it yields essentially the same allotment-assignment problem. One should be cautious in drawing normative conclusions from the fact that ν^o optimizes the problem just constructed: The market implicitly weights the preferences of different families, and this weighting need not coincide with that derived from some ethical principle.

We have run through the foregoing analysis rather rapidly because Section 8.7 will cover the same ground more elaborately in the context of Thünen systems proper.

A COMBINATORIAL APPLICATION

Given $2n$ real numbers, $x_1 < x_2 < \cdots < x_n$ and $y_1 < \cdots < y_n$, consider the problem of minimizing the sum

$$x_1 y_{\pi(1)} + x_2 y_{\pi(2)} + \cdots + x_n y_{\pi(n)} \qquad (8.6.35)$$

over all permutations π of $\{1, \ldots, n\}$. According to a theorem of Chebishev, the unique minimum occurs when the y's are taken in *reverse* order, matching y_n with x_1, etc. This is easily proved. For any other permutation π, there is an index j such that $y_{\pi(j)} < y_{\pi(j+1)}$. But then

$$x_j y_{\pi(j)} + x_{j+1} y_{\pi(j+1)} > x_j y_{\pi(j+1)} + x_{j+1} y_{\pi(j)} \qquad (8.6.36)$$

so switching these two y's reduces the sum (35). A finite number of these transpositions leads to the reversing permutation, which therefore minimizes.

This same argument yields the following generalization. Instead of (35) we minimize

$$f(x_1, y_{\pi(1)}) + \cdots + f(x_n, y_{\pi(n)}) \qquad (8.6.37)$$

Then if f has nonnegative cross-differences, the reversing permutation minimizes (37). If f has *positive* cross-differences, this minimizer is unique. (Inequality (36) is replaced by the cross-difference inequality. Note that (35) is the special case of (37) where f is the product function $f(x, y) = xy$.)

As suspected, there is an allotment-assignment problem lurking about. Indeed, let $S = Q = \{1, \ldots, n\}$, $\Sigma_s = \Sigma_q =$ all subsets, $\alpha =$

β = counting measure, $h(i) = x_i$, and $w(j) = y_j$ for $i, j = 1, \ldots, n$, and f be as in (37). The resulting allotment-assignment problem reads as follows.

Minimize the sum of the n^2 terms

$$f(x_i, y_j)\nu_{ij} \qquad (8.6.38)$$

(i, j ranging independently over $\{1, \ldots, n\}$), over all nonnegative (n, n) matrices (ν_{ij}) whose rows and columns all sum to 1 ("bistochastic matrices").

The values (37) are embedded among the values (38). Specifically, for feasible matrices consisting of just 0's and 1's ("permutation matrices"), (38) reduces to (37). If f has positive cross-differences, the unique optimal solution is the "weight-falloff" measure given by $\nu_{ij} = 1$ if $i + j = n + 1$, $\nu_{ij} = 0$ otherwise. This solution corresponds to the reversing permutation above.

The allotment-assignment problem just constructed is a special case of the *assignment* problem of ordinary linear programming. This in turn is a special case of the transportation problem. The positive cross-differences property of the objective function enables us to read off the optimal solution at sight.

8.7. THE THÜNEN SYSTEM AS A SOCIAL EQUILIBRIUM: FORMAL THEORY

A social equilibrium is a system involving several agents (with possibly conflicting preferences) such that no agent can take any action that improves the situation according to his own preferences. An example is the real-estate market of Chapter 6, in which agents acquire regions of Space (or Space-Time). Here each agent has a preference ordering over pairs consisting of the region acquired and the cost of acquiring it. Equilibrium consists of a pattern of real-estate values (represented by a measure over S or $S \times T$) and a partition of S (or $S \times T$) among the agents such that no agent can improve his position by switching to another region under the existing pattern of prices.

The equilibrium of this section is similar to that of the real-estate market but goes a step deeper. Namely, we assume that people acquire land to operate land uses on it. The decision problem facing each agent is accordingly more complicated. He must decide not only what land to acquire but what to do with it, i.e., what the land-use assignment is to be. The separate decisions of the various agents then result in a pattern of land uses over the whole system. We will show that under certain mild assumptions this overall land-use assignment is a Thünen system, in that it satisfies the measurable weight-falloff condition.

Now consider the formal model. We are given the measure space

(S, Σ_s, α), where S is physical Space and α is the areal measure on its σ-field Σ_s. The measurable function $h:S \to$ reals gives the distance of locations from the nucleus. Also given is the measurable space of land uses (Q, Σ_q), with the measurable weight function $w:Q \to$ reals. Area, distance, and weight are all "ideal" and may be considerably distorted from the corresponding physical magnitudes (recall Section 8.2). Specifically, they have the following properties. Let ν be a measure on $(S \times Q, \Sigma_s \times \Sigma_q)$ representing a certain land-use assignment: $\nu(E \times F)$ is the (ideal) acreage in region E devoted to land uses of types F. Then α measures the capacity of regions to accommodate land uses, in the sense that any feasible ν must satisfy

$$\nu(E \times Q) \leq \alpha(E) \tag{8.7.1}$$

for all regions E. Also, the transportation cost incurred in region E by assignment ν is

$$\int_{E \times Q} f(h(s), w(q))\, \nu(ds, dq) \tag{8.7.2}$$

Here $f:$reals$^2 \to$ reals is a given measurable function; one case we have mentioned often is the product function, $f(x, y) = xy$.

We now introduce a countable (possibly finite) number of agents labeled $n = 1, 2, \ldots$. At time zero when the system starts up, there is a big real-estate auction that leads to a measurable partition of Space, S, among the agents; S_n is the region falling under the control of agent n. Upon acquiring S_n, agent n chooses an assignment ν_n, which is a measure over $S_n \times Q$. The only constraint on ν_n is that it satisfy (1) for all sub-regions E of S_n. (Agent n will also have a budget constraint, but we suppose that this is reflected indirectly in his preference ordering to be discussed below and therefore need not be taken into account explicitly.) The several assignments ν_n on $S_n \times Q$, $n = 1, 2, \ldots$, then yield by direct summation an overall assignment ν on $S \times Q$.

Two kinds of costs are incurred by agent n. The first is transportation cost, which is given by (2) with $E = S_n$, $\nu = \nu_n$. The second is land cost. We suppose the real-estate market leads to a system of land values that is represented by a measure (or perhaps a signed measure) μ over Space. The net cost of land to agent n is

$$\mu(S_n) \tag{8.7.3}$$

(Actually $\mu(S_n)$ is the opportunity cost of land in the sense that even if agent n uses his own land and therefore pays no rent, he still loses the *opportunity* to rent or sell to someone else.) Total cost incurred by him is the sum of land cost and transportation cost.

We now come to the structure of agents' preferences. The idea is to make assumptions that are very weak yet lead to substantive conclusions. Let β_n be the right marginal of assignment ν_n; i.e., $\beta_n(F) = \nu_n(S_n \times F)$ for all $F \in \Sigma_q$. It is natural to call β_n the *allotment* corresponding to assignment ν_n. We now assume that the preference ordering of agent n satisfies the following condition: if two different actions yield the same allotment, the first is at least as preferred as the second iff the cost incurred under the first does not exceed the cost incurred under the second. Symbolically this may be written

$$(\beta, c_1) \succcurlyeq_n (\beta, c_2) \quad \text{iff} \quad c_1 \leq c_2 \tag{8.7.4}$$

Here c_i is the cost incurred under action i, $i = 1, 2$, β being the common allotment. No assumption is made concerning preferences among actions leading to different β's; these need not even be comparable. Also preferences may vary in an arbitrary way from agent to agent except that they all satisfy (4).

The rationale for (4) is worth examining in some detail. At first glance (4) appears to be so weak as to border on tautology. It states that, *ceteris paribus,* more money is preferred to less. This condition would appear to be universally satisfied except possibly for the small minority for whom poverty is a virtue, and even these have the option of throwing away excess money. In particular, (4) should be distinguished from the much stronger condition of profit maximization (or cost minimization in this case). Cost minimization entails indifference between two actions yielding the same cost, whereas (4) states nothing concerning actions that yield different allotments.

However, (4) does carry an implicit substantive assumption: agent n is indifferent to all aspects of his own land-use assignment other than allotment and cost. These other aspects include layout, shape of parcel, whether his land is in one piece or fragmented, etc. From one point of view this assumption is not implausible. Consider, for example, an agent contemplating two actions, both of which yield an identical allotment of one acre devoted to a certain residential land use, three acres devoted to his various business activities, two acres for recreation, etc. The agent's life-style will be the same in both cases, as will his income-expenditure pattern (except for transportation and land costs). The only difference lies in the spatial distribution of these activities, and why should this concern him? The only reason is that transportation plus land costs may vary with the spatial arrangement, and this factor is already taken into account in (4).

Thus the argument for (4) reduces, roughly, to the following: if one can do the same thing in region A or region B, one should be indifferent to location except for cost. This is fine except for one difficulty, and that is to make sure that all spatially varying costs are counted in. Now the transport cost formula (2) embodies the basic Thünen assumption that all trips involve the nucleus either as origin or destination (or are two-leg trips passing through the nucleus). In reality there are also "local" trips such as walking from room to room in one's house. The cost of these local trips will be higher in narrow or fragmented parcels than in "chunky" parcels. The postulated preference orders are unrealistic to the extent that these cost components are omitted. (A related omission is that of possible neighborhood effects among an agent's own land uses.)

Given the preference order of agent n on land-use assignments and regions, this induces an indirect preference order on regions alone; viz., $E_1 \succcurlyeq E_2$ by agent n iff for any land-use assignment on region E_2 there is an assignment on E_1 at least as preferred. This invites comparison with the preference orderings in the real-estate market of Chapter 6, which are also over regions. In general, the approach of this section does not lead to an additive utility function of the type discussed in Chapter 6.

We now gather up the strands of this discussion in the following definition.

Definition: A *social equilibrium* for the system (S, Σ_s, α), (Q, Σ_q), h, w, f, and preference orders $\succcurlyeq_n$, $n = 1, 2, \ldots$, consists of

  (i) a signed measure μ on (S, Σ_s) ("land values");
  (ii) a measurable partition (S_n), $n = 1, 2, \ldots$, of S;
  (iii) measures ν_n on $S_n \times Q$, $n = 1, 2, \ldots$;

such that the capacity constraint (1) is satisfied by the ν_n and for each agent n total cost is finite, and the pair

$$\left[\nu_n'', \int_{S_n \times Q} f(h(s), w(q)) \, \nu_n(ds, dq) + \mu(S_n) \right] \qquad (8.7.5)$$

cannot be surpassed in the agent's preference order by substitution of any other feasible region and assignment for S_n and ν_n.

Here the orderings $\succcurlyeq_n$, the partition elements S_n, and the measures ν_n all have the same countable index set $n = 1, 2, \ldots$; and $\nu_n'' = \beta_n$ is the right marginal of ν_n. In short, each agent finds that his choice (S_n, ν_n) yields an allotment-cost pair (5) that is unsurpassed in the set of options available.

Note that, unlike the situation in the allotment-assignment problem, the allotments β_n are not given in advance but are to be chosen by the agents. Also, μ is σ-finite, since land cost to each agent is finite.

Before launching into details, we mention an alternative (less realistic) way of defining social equilibrium for the system above. The definition just given requires *exclusive* control of land in the sense that Space is partitioned, and agent n alone decides on the land-use assignment in the region S_n that he acquires. Suppose we now drop this requirement and allow *joint* control of land. Agent n must choose merely a land-use assignment ν_n, which is now a measure on $S \times \mathbf{Q}$. The universe set for ν_n is the entire product space for each agent. There is no longer a need for agents to partition Space among themselves into proprietary regions.

To spell out this alternative model we must specify constraints and costs. Let ν be the sum of all individual assignments:

$$\nu = \nu_1 + \nu_2 + \nu_3 + \cdots \tag{8.7.6}$$

Then the areal capacity constraint (1) remains valid, with ν given by (6). That is, the area required by all agents together must not exceed the area available, so there is one global constraint. Transportation cost incurred by agent n is simply (2) with $E = S$ and $\nu = \nu_n$.

Land cost is a bit more complicated. Just as above, a land-value (signed) measure μ over Space arises in the real-estate market. This cost is then prorated among the agents in proportion to the area they require in various regions. To be precise, we have for each n, $\nu_n' \leq \nu' \leq \alpha$ from (6) and (1), where the prime denotes the left marginal. Hence ν_n' is absolutely continuous with respect to ν'; and since α (hence ν') is σ-finite, the Radon-Nikodym theorem asserts the existence of a function $g_n : S \rightarrow$ reals such that

$$\nu_n' = \int g_n \, d\nu' \tag{8.7.7}$$

We have $0 \leq g_n(s) \leq 1$ for ν'-almost-all locations s, and g_n represents the pro rata share of land of agent n. The land (opportunity) cost for agent n is then

$$\int_S g_n \, d\mu \tag{8.7.8}$$

Under exclusive control, (8) reduces to (3).[30]

Everything else is the same as in the "exclusive control" model. Preferences still satisfy (4), where β_n is the right marginal of ν_n. Social equilibrium is given by a land-value signed measure μ, and assignments (ν_n), $n = 1, 2, \ldots$, such that the capacity constraint (1) is satisfied and the allotment-cost combination yielded by ν_n is unsurpassed in the preference order of agent n for all n.

There is a simple relation between the "exclusive" and "nonexclusive" concepts of social equilibrium. Namely, the former may be thought of as the special case of the latter in which the additional requirement is imposed that the assignments of any two agents be *mutually singular* in the

---

30. In (7) g_n is unique only up to a region of ν'-measure zero. For (8) to be well defined, it must give the same value for any function g_n satisfying (7). It turns out that equilibrium μ must be absolutely continuous with respect to ν', which guarantees uniqueness in (8).

following strong sense: there is a measurable partition $(S_1, S_2, \ldots)$ of S such that $\nu_n[(S \backslash S_n) \times \mathbf{Q}] = 0$ for all n. To be precise, we identify these measures ν_n (which all have universe set $S \times \mathbf{Q}$) with their restrictions to $S_n \times \mathbf{Q}, n = 1, 2, \ldots$. These restricted measures are of the form used in the "exclusive control" model. It is now easy to check that the "nonexclusive" versions of areal constraints, transportation, and land costs reduce to the "exclusive" versions.

Given the land-use assignments $\nu_1, \nu_2, \ldots$, the overall pattern of land uses ν in the nonexclusive control model is given by summation (6). If the mutual singularity condition just mentioned obtains and we reinterpret ν_n to refer to the restriction of agent n's assignment to $S_n \times \mathbf{Q}$, then the same ν is obtained by *direct* summation:[31]

$$\nu = \nu_1 \oplus \nu_2 \oplus \cdots \tag{8.7.9}$$

The nonexclusive control model is in most respects less realistic than the exclusive control model. Joint control of land is relatively rare. Even rarer is a free market in partial shares of land: generally, a person must buy (or rent) a parcel completely or not at all. Finally, where such "fractional" markets do exist, it is unreasonable to ignore neighborhood effects from one agent's activities to another, since these activities are not merely "adjacent" but thoroughly "mixed" with each other. There are, however, some situations in which something resembling joint control is in effect: easements, joint usage of public facilities such as roads or parks, and perhaps land control by organizations.

In the following we restrict attention to the exclusive control model. For the record, however, we state that the nonexclusive control equilibrium can also be shown to have an overall land-use pattern satisfying the weight-falloff condition. The argument for this is similar to the one given below; it is even a bit simpler and does not need the "nonatomic" assumption (10). (For notational convenience we replace boldface $\mathbf{Q}$ by ordinary Q in the following formal discussion.)

Theorem: Given σ-finite measure space (S, Σ_s, α), measurable space (Q, Σ_q), measurable functions $h:S \to$ reals, $w:Q \to$ reals, $f:$reals$^2 \to$ reals, with f having positive cross-differences, and

$$\alpha\{s \mid h(s) = x\} = 0 \tag{8.7.10}$$

for all real x. Let $(\succeq_n), n = 1, 2, \ldots$, be a countable family of partial orders each satisfying (4).

Let there be a *social equilibrium* for this system consisting of signed

---

31. Recall that ordinary summation of measures, as in (6), is defined iff all summands are over the same measurable space, while *direct* summation, as in (9), is defined iff the universe sets of the summands are disjoint from each other.

measure μ on S, measurable partition (S_n), $n = 1, 2, \ldots$, of S, and feasible measures v_n on $S_n \times Q$, $n = 1, 2, \ldots$. Let $f(h(\cdot), w(\cdot))$ be half-bounded on each set $S_n \times Q$.

Then the overall assignment v, given by (9), satisfies the *(measurable) weight-falloff* condition (with respect to h, w).

Proof: First we show that each individual v_n, considered in isolation as a measure on $S_n \times Q$, must satisfy measurable weight-falloff. Indeed, consider any other measure $\tilde{v}_n$ on $S_n \times Q$ that has the same left and right marginals as v_n. Since $\tilde{v}_n$ satisfies the areal capacity constraint (1), it is feasible for agent n and therefore cannot surpass v_n in his preference ordering. Now v_n and $\tilde{v}_n$ have the same allotment; they also occupy the same region S_n, hence incur the same land cost. It follows that the transportation cost (2) incurred under v_n cannot exceed the transportation cost incurred under $\tilde{v}_n$.

Thus v_n is unsurpassed for the *allotment-assignment* problem on $S_n \times Q$ defined by its marginals and by f, h, and w. Since $f(h(\cdot), w(\cdot))$ is half-bounded on this set and f has positive cross-differences, it follows (p. 430) that v_n does indeed satisfy measurable weight-falloff.

Now suppose the overall assignment v violates measurable weight-falloff, so that there are measurable sets G_1, $G_2 \subseteq S \times Q$ of positive v-measure, with G_1 southwest of G_2. This means that S and Q can each be split into two measurable sets S^1, S^2 and Q^1, Q^2 respectively such that

$$h(s_1) < h(s_2) \tag{8.7.11}$$

$$w(q_1) < w(q_2) \tag{8.7.12}$$

for all $s_i \in S^i, q_i \in Q^i, i = 1, 2$, and such that

$$G_i \subseteq S^i \times Q^i \tag{8.7.13}$$

$i = 1, 2$.

Relation (13) implies that $S^1 \times Q^1$ and $S^2 \times Q^2$ both have positive v-measure. Since the number of agents is countable, there must be some agent n_1 such that $v[(S_{n_1} \cap S^1) \times Q^1]$ is positive. Furthermore, since $v' \leq \alpha$ is σ-finite, a region E_1 exists such that $v[(E_1 \cap S_{n_1} \cap S^1) \times Q^1]$ is positive and *finite*. Similarly, there is an agent n_2 and region E_2 such that $v[(E_2 \cap S_{n_2} \cap S^2) \times Q^2]$ is positive and finite.

The two agents n_1 and n_2 must be *distinct;* for if not, the measurable weight-falloff condition would be violated within one individual realm, contrary to our finding above.

Now consider the two functions g_i:reals $\to$ reals, $i = 1, 2$, given by

$$g_1(x_1) = v[(\{s \mid h(s) > x_1\} \cap E_1 \cap S_{n_1} \cap S^1) \times Q^1]$$

$$g_2(x_2) = v[(\{s \mid h(s) < x_2\} \cap E_2 \cap S_{n_2} \cap S^2) \times Q^2] \tag{8.7.14}$$

These functions are nonnegative but not identically zero; monotone (non-increasing for g_1 and nondecreasing for g_2); approach zero asymptotically; and are, because of (10), continuous. Hence numbers x_1^o, x_2^o exist such that

$$0 < g_1(x_1^o) = g_2(x_2^o) < \min\left[\lim_{x_1 \to -\infty} g_1(x_1), \lim_{x_2 \to \infty} g_2(x_2)\right] \qquad (8.7.15)$$

Now define the regions H^1, H^2 by

$$H^1 = \{s \mid h(s) > x_1^o\} \cap E_1 \cap S_{n_1} \cap S^1$$

$$H^2 = \{s \mid h(s) < x_2^o\} \cap E_2 \cap S_{n_2} \cap S^2 \qquad (8.7.16)$$

From (15) we then have

$$\infty > \nu(H^1 \times Q^1) = \nu(H^2 \times Q^2) > 0 \qquad (8.7.17)$$

Let c be the common value of $\nu(H^i \times Q^i)$. Next, define region J^1 by

$$J^1 = \{s \mid h(s) \leq x_1^o\} \cap E_1 \cap S_{n_1} \cap S^1 \qquad (8.7.18)$$

From the right-hand inequality in (15) it follows that $\nu(J^1 \times Q^1)$ is positive. Now the two sets $J^1 \times Q^1$ and $H^1 \times Q^2$ stand in a southwest-northeast relation, by (12), (16), and (18). Furthermore, both H^1 and J^1 are subsets of S_{n_1}. It follows that

$$\nu(H^1 \times Q^2) = 0 \qquad (8.7.19)$$

otherwise the measurable weight-falloff condition would be violated within the individual realm of agent n_1. A similar argument for agent n_2 shows that

$$\nu(H^2 \times Q^1) = 0 \qquad (8.7.20)$$

Relations (17), (19), and (20) are what is needed for the rest of this proof.

In what follows, the notation ν_G stands for the *truncation* of ν to the set G; i.e., for $G, K \in \Sigma_s \times \Sigma_q$ we have

$$\nu_G(K) = \nu(G \cap K) \qquad (8.7.21)$$

Now consider the following policy changes. Agent n_1, instead of taking S_{n_1} as his region of control, chooses $(S_{n_1} \backslash H^1) \cup H^2$. That is, he relinquishes control of region H^1 and acquires region H^2. Agent n_2 makes just the opposite switch, so that his region of control becomes $(S_{n_2} \backslash H^2) \cup H^1$. Also, the original pattern of land uses on $H^1 \cup H^2$ is altered. Specifically, the following signed measure is added to the original overall land-use assignment ν:

$$\Delta v = -v_{H^1 \times Q^1} - v_{H^2 \times Q^2}$$

$$+ \frac{1}{c} \left[(v_{H^2 \times Q^2})' \times (v_{H^1 \times Q^1})'' \right] + \frac{1}{c} \left[(v_{H^1 \times Q^1})' \times (v_{H^2 \times Q^2})'' \right] \qquad (8.7.22)$$

(To explain: The fourth term in (22), e.g., is the product measure of the *left* marginal of $v_{H^1 \times Q^1}$ by the *right* marginal of $v_{H^2 \times Q^2}$ divided by c, the common value in (17).) On H^1, because of (19), the first term of (22) knocks out the original assignment; the fourth term is what replaces it; similarly, on H^2, because of (20), the second term knocks out the original assignment, and the third term is what replaces it.

We first check this new assignment for feasibility. All measures in (22) are finite, and it is clear that $v + \Delta v$ is nonnegative. Using (17), we verify that the *left* marginal of Δv is identically zero. Hence the areal capacity constraint (1) remains satisfied for $v + \Delta v$. This establishes feasibility.

Next, we verify that the *right* marginal of the *first* plus *third* term of (22) is identically zero. These two terms give precisely the change from the original assignment of agent n_1. Thus the *allotment* attained by agent n_1 is exactly the same as with his original assignment. Since the original assignment of agent n_1 is unsurpassed in his preference ordering, it follows from (4) that the change in total cost must be nonnegative. Thus

$$\int_{S \times Q} f(h(\cdot), w(\cdot)) \, d \left[\frac{1}{c} ((v_{H^2 \times Q^2})' \times (v_{H^1 \times Q^1})'') - v_{H^1 \times Q^1} \right]$$

$$+ \mu(H^2) - \mu(H^1) \geq 0 \qquad (8.7.23)$$

The integral in (23) is the change in the transport cost incurred by agent n_1. The other two terms yield the change in his land cost, incurred by acquiring region H^2 and relinquishing H^1. (Finiteness of costs in the social equilibrium, together with the half-boundedness condition, insure that (23) is well defined.)

The same argument applied to agent n_2 yields (23) with the superscripts 1 and 2 interchanged. Adding (23) to the corresponding inequality for n_2, the land-value terms drop out, and we obtain

$$\int_{S \times Q} f(h(\cdot), w(\cdot)) \, d(\Delta v) \geq 0 \qquad (8.7.24)$$

But (24) is false. To show this we use the argument already employed in (5.17)–(5.19). The applicability of this argument follows from the observations that f has positive cross-differences, that $H^1 \times Q^1$ is southwest of $H^2 \times Q^2$ (from (11) and (12)), and that all measures in (22) have the same value on $S \times Q$, viz., c. Then Δv of (22) has the same form as v in

(5.11) except for the inessential factor c. Just let G_i in (5.10) be $H^i \times Q^i$ of the present argument, $i = 1, 2$. It follows that the integral in (24) must be negative. This contradiction shows that ν cannot violate measurable weight-falloff. ∎

Note that while this theorem severely constrains the pattern of land *uses*, it says nothing about land *users*. The individual regions S_n controlled by the various agents might have any irregular shapes, be fragmented, intermixed, etc. But the kinds of land uses at a site will depend (as far as their ideal weights are concerned) only on the ideal distance of that site, not on the agent who controls it. Smith will run a land use on one of his plots similar in weight to that run by Jones on an adjacent plot, and *dis*similar to the land use run by Smith on another of his plots at a different ideal distance from the nucleus.

Before going on, we mention a generalization of the preceding theorem that is of theoretical but little practical interest. The premise that land value μ must be bounded on each region S_n can be dropped from the definition of social equilibrium. Instead, μ can be any σ-finite signed measure on S (or even a pseudomeasure) without invalidating the conclusion that assignment ν satisfies the measurable weight-falloff condition.[32]

We now turn from the study of land *uses* to land *values*. Our aim is to show that land-value density is essentially a (left) *half-potential* for the allotment-assignment problem associated with the foregoing social equilibrium. But before doing so, we must face up to certain conceptual difficulties that we managed to evade in the preceding proof.

The first difficulty concerns the assumption of perfect competition embodied in our definition of social equilibrium. To be precise, we assume that agents can freely acquire or dispose of land at the fixed prices given by the signed measure μ. In reality, search and bargaining problems arise between the transacting agents, which are unlikely to be captured by an additive set function such as μ. (Transfer of land involves the displacement of one pattern of land uses by another on this land, hence changes in the allotments of the several agents; in general, the monetary compensations for these changes will not take a form that is additive over regions.) As a rule, the more agents there are, the more satisfactory the competitive assumption becomes. But even with a countably infinite number of agents the difficulty does not vanish, since these agents will still hold positive acreages.[33]

---

32. Condition (4) on preference orders must be correspondingly modified, since "cost" may no longer be representable as a real number. The natural generalization is to interpret $c_1 \leq c_2$ as referring to standard ordering of pseudomeasures. Cf. the real-estate market of Chapter 6. We omit detailed discussion of this point.

33. These difficulties motivate the "measure space of agents" approach, the number of agents being now uncountable and each literally having zero influence. But this creates new conceptual difficulties of its own. See Section 6.7.

The second difficulty involves a slightly paradoxical strengthening of the social equilibrium concept. We start with an analogy. Suppose one has a divisionalized firm, the various divisions trading with both the outside world and each other. A possible rule of operation is that each division offer the same prices both to outsiders and sister divisions and that each accepts the best offer, ignoring the affiliation status of trading partners. In short, the divisions act as if they were independent firms. Now consider agent n holding region S_n. Partitioning this region into $S_{n1}, \ldots, S_{nk}$, we may think of the land uses on each piece S_{ni} as being run by a "division" of agent n. We now make the "independent firms" assumption about the behavior of these divisions. To be precise, the "custodian" of region S_{ni} has the option of buying a piece of the territory of his alter ego, the "custodian" of S_{nj} ($j \neq i$), at the going market price and displacing the latter's land uses on the transferred region. The change in costs resulting from this transaction is to be computed just as if S_{nj} were controlled by an agent other than n; i.e., one takes account neither of the land uses ousted from S_{nj} nor of the compensation received by the custodian of this region.

This rule is rather awkward to justify if taken literally, and the correct interpretation seems to be as follows. The land-value measure μ plays a double role. Externally, between agents it functions as a market price; internally, within an agent's territory it functions as a "shadow" price, equilibrium under the above rule being a necessary condition for the optimal assignment of land uses. If there is just one agent in the entire system, μ plays a shadow role exclusively. As the number of agents rises, the shadow role shrinks and the "market" role expands, but the former does not disappear even for a countable infinity of agents. (It *would* disappear in the "measure space of agents" model.)

In the preceding proof we avoided these conceptual difficulties by centering the argument around a swap of territories that reduced combined transportation costs. In the following proof we are not so lucky. The special "independent firms" rule shows up in the proof below by allowing the argument to go through even if the agent "acquires" some of his own land.

We now give the details. The assignment ν on $S \times Q$ induces a measure λ on the plane via the mapping $(s, q) \rightarrow (h(s), w(q))$. Let λ and the function f:reals$^2 \rightarrow$ reals be given.

Definition: The measurable function p:reals $\rightarrow$ extended reals is a *left half-potential for* λ *almost everywhere* (with respect to f) iff there is a real Borel set F with $\lambda'(F) = 0$ such that $p(x)$ is finite for $x \notin F$ and

$$p(x_1) + f(x_1, y_2) \geq p(x_2) + f(x_2, y_2) \tag{8.7.25}$$

for all real numbers x_1, x_2, y_2 such that $x_1, x_2 \notin F$ and (x_2, y_2) is a point of support for λ.

Here λ' is the left marginal of λ. If $F = \phi$, this reduces to the ordinary definition of left half-potential, as in (5.67).

In (27) below, $\alpha - \nu'$ is the *areal excess capacity measure*. Since both ν' (the left marginal of ν) and α may be infinite measures, subtraction must be understood in the sense of Section 3.1.

Theorem: Let overall assignment measure ν on $S \times Q$, land value signed measure μ on S, and regions S_n, $n = 1, 2, \ldots$, constitute a *social equilibrium* for the system defined by (S, Σ_s, α), (Q, Σ_q), $h{:}S \rightarrow$ reals, $w{:}Q \rightarrow$ reals, $f{:}\text{reals}^2 \rightarrow$ reals, $\succcurlyeq_n$, $n = 1, 2, \ldots$, as in the preceding theorem.

Here α is σ-finite; h, w, and f are measurable; and, as always, ν satisfies (1), $\succcurlyeq_n$ satisfies (4), all n, and all agents have finite costs. In addition, assume that h and α satisfy (10), that f is continuous with positive cross-differences, that $f(x, y)$ is strictly increasing in x for all numbers y in the range of w, and that $f(h(\cdot), w(\cdot))$ is half-bounded on each set $S_n \times Q$. Then:

(i) There is a nonnegative, nonincreasing function $p{:}\text{reals} \rightarrow$ extended reals that is a left half-potential for λ almost everywhere (with respect to f)—here λ is the plane measure induced from ν by $(s, q) \rightarrow (h(s), w(q))$—and for which

$$\mu = \int (p \circ h) \, d\alpha \qquad (8.7.26)$$

(Hence $\mu \geq 0$.)

(ii) There is an extended real number x_o such that

$$(\alpha - \nu')\{s \mid h(s) \leq x_o\} = 0 \qquad (8.7.27)$$

$$\nu'\{s \mid h(s) \geq x_o\} = 0 \qquad (8.7.28)$$

$$\mu\{s \mid h(s) \geq x_o\} = 0 \qquad (8.7.29)$$

Proof: (i) First, we show that $\mu \geq 0$. Suppose $\mu(E)$ were negative for some region E. Since μ is σ-finite, there is a subregion $F \subseteq E$ for which $\mu(F)$ is negative and finite. Choose any agent n, and let him acquire region F and leave it vacant. Then his land cost falls, while his transport cost and his allotment remain the same; this contradicts the fact that he was in equilibrium to begin with. Hence $\mu \geq 0$. (Note that the agent may be "acquiring" land from himself, in accordance with our discussion above.)

Next we show that μ is absolutely continuous with respect to ν'. Let $\mu(E) > 0$ for some region E. Then $\mu(E \cap S_n) > 0$ for some n, since the sets S_n countably partition S. If $\nu'(E \cap S_n) = 0$, agent n could simply divest himself of $E \cap S_n$, reducing his land cost while leaving his transport cost and allotment unchanged. This contradicts the unsurpassedness of his initial position. Hence $\nu'(E \cap S_n) > 0$, so that $\nu'(E) > 0$. Thus $\mu \ll \nu'$.

Now $\nu' \leq \alpha$ is σ-finite. By the Radon-Nikodym theorem, it follows that a function $\tilde{p}:S \to$ reals exists such that

$$\mu = \int \tilde{p} \, d\nu' \qquad (8.7.30)$$

Since μ is a measure, $\tilde{p}$ may be chosen to be nonnegative.

Now consider the mapping from S to the plane given by $s \to (h(s), \tilde{p}(s))$. This induces a measure ρ on the plane from the measure ν' on S. We now show that ρ satisfies the weight-falloff condition. Suppose, on the contrary, that there are two measurable sets F_1 and F_2 in the plane, of positive ρ-measure with F_1 southwest of F_2. The inverse images of these sets are regions G_1 and G_2 respectively, of positive ν'-measure such that numbers x_o, z_o exist satisfying

$$h(s_1) \leq x_o \leq h(s_2) \qquad (8.7.31)$$

$$\tilde{p}(s_1) \leq z_o \leq \tilde{p}(s_2) \qquad (8.7.32)$$

for all $s_1 \in G_1$, $s_2 \in G_2$. Furthermore, at least one of the inequalities in (31) and at least one in (32) can be chosen strict (this follows from the definition of "southwest").

Now (10) holds with ν' in place of α. Using an argument similar to (14)–(17), we can find two subregions H_1 and H_2, of G_1 and G_2 respectively, such that H_2 is contained in the realm of some one agent: $H_2 \subseteq S_n$, say, such that μ is bounded on $H_1 \cup H_2$ and such that

$$\infty > \nu'(H_1) = \nu'(H_2) > 0 \qquad (8.7.33)$$

Let c be the common value in (33).

Consider now the following changes in the action of agent n. He relinquishes region H_2 and acquires H_1. The corresponding change in his assignment is

$$\Delta \nu = -\nu_{H_2 \times Q} + \frac{1}{c} [(\nu_{H_1 \times Q})' \times (\nu_{H_2 \times Q})''] \qquad (8.7.34)$$

The notation is as in (21) and (22). The first term in (34) knocks out his original assignment on H_2; the second term is the assignment he places on the newly acquired region H_1. (Displacement of the original assignment on H_1 does not appear in (34), since it does not appertain to agent n. As discussed above, this displacement is to be ignored even if $H_1 \cap S_n$ is nonempty.)

Let us test this for feasibility. It suffices to take regions $E \subseteq H_1$ and to check that $\Delta v(E \times Q) = v(E \times Q)v(H_2 \times Q)/c = v'(E)$ does not exceed $\alpha(E)$, as indeed it does not, since the original assignment was feasible.

The right marginal of Δv is identically zero, so that the allotment of agent n is unchanged. The change in land cost is given by

$$\mu(H_1) - \mu(H_2) = \int_{H_1} \tilde{p}\, dv' - \int_{H_2} \tilde{p}\, dv'$$

from (30). But

$$\int_{H_1} \tilde{p}\, dv' \leq z_o c \leq \int_{H_2} \tilde{p}\, dv' \qquad (8.7.35)$$

from (32) and (33). Furthermore, one of the inequalities in (32), hence in (35), is strict. It follows that land cost has been *reduced*.

The change in transport cost is given by

$$\int_{S \times Q} f(h(s), w(q))\, \Delta v(ds, dq) \qquad (8.7.36)$$

But

$$\int_{S \times Q} f(h(s), w(q))\, v_{H_2 \times Q}(ds, dq) \geq \int_{S \times Q} f(x_o, w(q))\, v_{H_2 \times Q}(ds, dq)$$

$$= \int_Q f(x_o, w(q))\, (v_{H_2 \times Q})''\, (dq)$$

$$\geq \int_{S \times Q} f(h(s), w(q)) \left[\frac{1}{c}\, (v_{H_1 \times Q})'\, (ds) \times (v_{H_2 \times Q})''\, (dq) \right] \qquad (8.7.37)$$

The first integral in (37), being within the original realm of agent n, is finite. Hence all the integrals in (37) are well defined. The first inequality in (37) arises from the right side of (31) and the fact that $f(\cdot, y)$ is increasing for y in the range of w; the equality arises from the induced integrals theorem, using the projection $(s, q) \to q$. The measure $(v_{H_2 \times Q})''$ is the right marginal of *both* measures in (34). Hence, passing over to the second measure in (34) and using the left side of (31) and the increasingness of $f(\cdot, y)$ once again, we obtain the last inequality of (37). Furthermore, at least one of the inequalities in (31), hence in (37), is strict.

But (36) equals the last integral of (37) minus the first integral of (37). Thus transport cost is also reduced. The change in *total* cost is then *negative*. This improvement contradicts the fact that agent n is in equilibrium.

Thus measure ρ does indeed satisfy the weight-falloff condition. Expressing this in topological form, no two points of support for ρ stand in a southwest-northeast relation.

Now let X be the set of real numbers x with the following property: There is exactly one number y such that (x, y) is a point of support for ρ. Define the function $p:X \rightarrow$ reals by letting $p(x)$ be the unique number for which $(x, p(x))$ supports ρ.

X is a Borel set. To show this, let

$$X_i = \{x \mid (x, y) \text{ supports } \rho \text{ for at least } i \text{ distinct numbers } y\}$$

$i = 1, 2$. Set X_1 is the left projection of the support of ρ. The support is a closed set, hence a countable union of bounded closed sets; X_1 itself is therefore a countable union of closed sets, hence Borel. Also, X_2 is a Borel set, since it is countable. To see this, associate with each $x \in X_2$ a rational number between y' and y'', where (x, y'), (x, y'') are distinct points of support for ρ. These rationals are distinct, by weight-falloff. Since the rationals are countable, so is X_2. But $X = X_1 \backslash X_2$, thus X is Borel.

We now show that X includes "almost all" the real numbers in the following sense. Let ρ' be the left marginal of ρ. Note that ρ' is also the measure on the real line induced by h from v' on S. For, letting E be a Borel set, we have

$$\rho'(E) = \rho(E \times \text{reals}) = v'\{s \mid h(s) \in E\} \tag{8.7.38}$$

We now assert that

$$\rho'(\text{reals}\backslash X) = 0 \tag{8.7.39}$$

First, $\rho'(\text{reals}\backslash X_1) = 0$, since ρ is zero on the complement of its support. Second, $\rho'(X_2) = 0$, since X_2 is countable and ρ' is zero on each singleton, by (10) and (38). This proves (39).

Now extend p from domain X to the entire real line as follows: $p(x)$ is the supremum of zero and the values $p(z)$ for $z \in X$, $z \geq x$. This extended function will also be denoted by p. The original p is nonnegative and nonincreasing, and one easily verifies that the extended p retains these properties.

Next we show that the composite function $p \circ h$ is equal to $\tilde{p}$, v'-almost everywhere on S. Indeed, consider the region H given by

$$\{s \mid h(s) \in X\} \cap \{s \mid (h(s), \tilde{p}(s)) \text{ supports } \rho\} \tag{8.7.40}$$

For any site $s \in H$, there is exactly one number y such that $(h(s), y)$ supports ρ, so that $p(h(s)) = y = \tilde{p}(s)$. Thus $\tilde{p}$ and $p \circ h$ coincide on H. The complement of the first region in (40) has v'-measure zero, by (38) and (39). The complement of the second region in (40) also has v'-measure zero, since ρ is zero on the complement of its support. Thus $\tilde{p} = p \circ h$ almost everywhere.

It follows from (30) that

$$\mu = \int (p \circ h) \, dv' \tag{8.7.41}$$

Now consider the mapping from $S \times Q$ to the plane, given by $(s, q) \to (h(s), w(q))$. This induces a measure λ on the plane from the measure v on $S \times Q$. Since v is a social equilibrium, the preceding theorem implies that it, hence also λ, satisfies weight-falloff. Expressed in topological form, this means that no two points of support for λ stand in a southwest-northeast relation.

Let Z be the set of real numbers z with the following property. There is *exactly one* number y such that (z, y) supports λ. Using the same arguments for Z and λ that we used above for X and ρ, we conclude that Z is a Borel set and that

$$\lambda'(\text{reals} \setminus Z) = 0 \tag{8.7.42}$$

where λ' is the left marginal of λ. We also have $\lambda' = \rho'$, since for Borel sets E, $\lambda'(E) = \lambda(E \times \text{reals}) = v\{(s, q) \mid h(s) \in E\} = v'\{s \mid h(s) \in E\}$. (cf. (38)).

We now verify that the function p is a left half-potential for λ almost everywhere. Specifically, we show that (25) holds for all $x_1, x_2 \in X \cap Z$, and all real y_2 such that (x_2, y_2) supports λ. Note that, from (39) and (42), the complement of $X \cap Z$ has λ'-measure zero, so the "almost everywhere" condition is met.

Suppose (25) were false, so that numbers $x_1, x_2 \in X \cap Z$, and y_2 real exist for which (x_2, y_2) supports λ, and

$$p(x_2) + f(x_2, y_2) > p(x_1) + f(x_1, y_2) \tag{8.7.43}$$

Suppose (which may not be the case) there were an increasing sequence of numbers $x^1 < x^2 < \cdots$ belonging to $X \cap Z$, with limit x_2. For each n let y^n be the number such that (x^n, y^n) supports λ. Then $y^1 \geq y^2 \geq \cdots$, and these numbers are bounded below by y_2. Hence $\lim y^n$ exists as $n \to \infty$ and is finite. Since the support of λ is closed, it follows that $(x_2, \lim y^n)$ belongs to it. Hence

$$\lim_{n \to \infty} y^n = y_2 \tag{8.7.44}$$

since there is exactly one point of the form (x_2, y) supporting λ. A similar argument establishes (44) in the case of a decreasing sequence $x^1 > x^2 > \cdots$ belonging to $X \cap Z$, with limit x_2.

Furthermore, the same argument, applied to the support of ρ instead of λ, yields

$$\lim_{n \to \infty} p(x^n) = p(x_2)$$

for any monotone sequence (x^n) belonging to $X \cap Z$ with limit x_2. This argument applies also to any number $x \in X \cap Z$, in particular to $x = x_1$. Thus p, restricted to $X \cap Z$, is *continuous*.

Now let $b > 0$ be the difference between the left and right sides of (43). Choose $\epsilon > 0$ so that all the following relations are satisfied for $i = 1$ and $i = 2$:

$$|p(x) - p(x_i)| < b/4, \text{ for all } x \in X \cap Z \text{ such that } |x - x_i| < \epsilon$$
$$(8.7.45)$$

$$|f(x, y) - f(x_i, y_2)| < b/4,$$
$$\text{for all } x, y \text{ such that } |x - x_i| < \epsilon \quad \text{and} \quad |y - y_2| < \epsilon \quad (8.7.46)$$

The existence of such an ϵ follows from the continuity of p at points x_1 and x_2, and the continuity of f at points (x_1, y_2) and (x_2, y_2). Next, choose a $\delta > 0$ so that

$$\lambda\{(x, y) \mid |x - x_2| < \delta, |y - y_2| \geq \epsilon\} = 0 \qquad (8.7.47)$$

The existence of such a δ follows from (44). For δ can be chosen so small that for any $x \in X \cap Z$ such that $|x - x_2| < \delta$, the y such that (x, y) supports λ is within ϵ of y_2. The set in (47) will therefore own no points of support, hence has measure zero.

Having chosen ϵ, then δ, consider the two regions G_i given by

$$\{s \mid |h(s) - x_i| < \min(\epsilon, \delta)\}$$

$i = 1, 2$. Both these regions have positive ν'-measure, since there are numbers y_1, y_2 such that (x_i, y_i) supports λ, $i = 1, 2$. (The following argument does not require that G_1, G_2 (or H_1, H_2 for that matter) be disjoint.)

Just as above, we can now find two subregions H_1 and H_2, with $H_i \subseteq G_i$, $i = 1, 2$, such that H_2 is contained in the realm of some one agent: $H_2 \subseteq S_n$, and such that (33) holds.

Continuing just as above, we contemplate the policy change for agent n in which he relinquishes region H_2 and acquires H_1. The corresponding change in his assignment is again given by $\Delta\nu$ of (34). As above, this change is feasible and leaves his allotment unchanged. The reduction in his land cost arising from the relinquishing of region H_2 is given by

$$\int_{H_2} (p \circ h) \, d\nu' = \int_{H_2 \times Q} p(h(s)) \, \nu(ds, dq) \qquad (8.7.48)$$

The left integral in (48) comes from (41). The equality arises from the induced integrals theorem.

Adding to this the reduction of transport cost on H_2, we find the reduction of *total* cost arising from the relinquishing of region H_2 to be

$$\int_{H_2 \times Q} [p(h(s)) + f(h(s), w(q))] \, v(ds, dq) \tag{8.7.49}$$

We now claim that the integral (49) exceeds cm, where c is the common value $v'(H_1) = v'(H_2)$, and m is the average of the two sides of (43). To see this, split the set $H_2 \times Q$ into two pieces $H_2 \times L$ and $H_2 \times (Q \backslash L)$, where

$$L = \{q \mid \mid w(q) - y_2 \mid < \epsilon\}$$

Now $v(H_2 \times (Q \backslash L)) = 0$, from (47). However, for all points of $H_2 \times L$ (except for a v-null set) the integrand in (49) exceeds m. (The exceptional points (s, q) are among those for which $h(s) \notin X \cap Z$. These have measure zero, by (39) and (42).) For (45) and (46) imply that the integrand value differs from the left side of (43) by less than $b/2$, hence it never gets halfway toward the right side of (43). It follows that (49) exceeds $m \cdot v(H_2 \times Q) = mc$.

Next consider the increase in costs arising from the acquisition of region H_1. For land cost this is

$$\int_{H_1} (p \circ h) \, dv' = \int_{H_1 \times Q} p(h(s)) \left[\frac{1}{c} (v_{H_1 \times Q})' (ds) \times (v_{H_2 \times Q})'' (dq) \right] \tag{8.7.50}$$

The left integral in (50) comes from (41). The equality arises from the induced integrals theorem on noting that the left marginal of the bracketed measure in (50) coincides with v' truncated to H_1.

Adding to this the increase in transport cost, we find the increase in *total* cost arising from this acquisition to be

$$\int_{H_1 \times Q} [p(h(s)) + f(h(s), w(q))] \left[\frac{1}{c} (v_{H_1 \times Q})' (ds) \times (v_{H_2 \times Q})'' (dq) \right] \tag{8.7.51}$$

The bracketed measure in (51) has value zero on $H_1 \times (Q \backslash L)$, from (47). However, for all points of $H_1 \times L$ (except for a null set as above) the integrand in (51) is *less than* m. For (45) and (46) imply that the integrand differs from the *right* side of (43) by less than $b/2$, hence it never gets halfway toward the *left* side. It follows that (51) is less than

$$m \left[\frac{1}{c} (v_{H_1 \times Q})'(H_1) \cdot (v_{H_2 \times Q})''(L) \right] = m \cdot \frac{1}{c} \cdot c \cdot c = mc$$

from (47), since $(v_{H_2 \times Q})''(L) = v(H_2 \times L) = v(H_2 \times Q) = c$.

Thus (49) exceeds (51). Hence the net increase in costs, which is (51) minus (49), is negative. Since his allotment remains the same, this contradicts the premise that agent n was initially in an unsurpassed position. Thus (43) must have been false, and we have established that p is an almost everywhere left half-potential for λ.

This completes part (i) of the theorem, except for the fact that ν' appears in (41) while α appears in (26). The proof of part (ii) will justify the last step of replacing ν' by α.

(ii) Let x_1 be the infimum of all numbers x satisfying

$$\nu'\{s \mid h(s) > x\} = 0 \qquad (8.7.52)$$

and let x_2 be the supremum of all numbers x satisfying

$$(\alpha - \nu')\{s \mid h(s) < x\} = 0 \qquad (8.7.53)$$

Three cases follow.

If $x_1 < x_2$, let x_o be any number satisfying $x_1 < x_o < x_2$. Equations (27) and (28) then follow at once, and (29) follows from (28), since, as shown above, μ is absolutely continuous with respect to ν'.

If $x_1 = x_2$, let x_o be their common value. Then (52) holds for $x = x_o$ because the set $\{s \mid h(s) > x_o\}$ is the limit of an increasing sequence of sets of ν'-measure zero. Similarly, (53) holds for $x = x_o$. Also, $\alpha - \nu'$ and ν' are both zero on the set $\{s \mid h(s) = x_o\}$, from (10). This again establishes (27) and (28), hence (29).

To finish the proof, we need only show that the last possibility $x_1 > x_2$ leads to a contradiction. Choose a number x_o such that $x_1 > x_o > x_2$. Then

$$\nu'\{s \mid h(s) > x_o\} > 0 \qquad (8.7.54)$$

from (52), and

$$(\alpha - \nu')\{s \mid h(s) < x_o\} > 0 \qquad (8.7.55)$$

from (53). From (55) an agent n exists for which

$$(\alpha - \nu')[S_n \cap \{s \mid h(s) < x_o\}] > 0 \qquad (8.7.56)$$

For the time being we also assume that

$$\nu'[S_n \cap \{s \mid h(s) > x_o\}] > 0 \qquad (8.7.57)$$

Using an argument similar to (14)–(17), we can find regions H_1 and H_2, contained in the sets of (56) and (57) respectively, such that $\infty > (\alpha - \nu')(H_1) \geq \nu'(H_2) > 0$.

Now consider the following change in policy of agent n. He relinquishes region H_2 and increases his assignment on H_1. Specifically, the change in his assignment $\Delta \nu$ is given by

$$-\nu_{H_2 \times Q} + [(\alpha - \nu')_{H_1} \times (\nu_{H_2 \times Q})'']/(\alpha - \nu')(H_1) \qquad (8.7.58)$$

The left measure in (58) knocks out the original assignment on H_2. Note that the original assignment on H_1 is *not* displaced, but *added to* by the right measure in (58). The excess capacity on H_1 allows the areal constraint (1) to remain inviolate, as one verifies. Hence the changed policy is feasible.

The right marginal of $\Delta \nu$ is identically zero, so that the allotment of agent n remains unchanged. The land cost is reduced, if anything, since he divests himself of H_2. (There is no change in land cost for H_1, since he controls this region to begin with.) Transportation cost must fall, by the argument of (37). (There is an obvious change of notation for the last integral of (37), and both inequalities there are now strict.) Hence total costs fall, and agent n has improved his position. This contradicts the fact that he is initially in equilibrium. Thus (57) must be false.

This same argument proves even more. Start with (56) and (57), with *any* number x substituted for x_o (the same x in both (56) and (57)). The same contradiction arises, so (56) and (57) cannot hold simultaneously for *any* x in place of x_o. It follows from this that a (possibly infinite) number x_{oo} exists such that

$$\nu'[S_n \cap \{s \mid h(s) \geq x_{oo}\}] = 0 \qquad (8.7.59)$$

$$(\alpha - \nu')[S_n \cap \{s \mid h(s) \leq x_{oo}\}] = 0 \qquad (8.7.60)$$

To see this, apply the argument at the beginning of part (ii) to the region S_n, rather than to all of S.

From (60) and (56) we obtain

$$(\alpha - \nu')[S_n \cap \{s \mid x_o > h(s) > x_{oo}\}] > 0 \qquad (8.7.61)$$

Because of (59), $\alpha - \nu'$ can be replaced by α in (61). Also,

$$\mu[S_n \cap \{s \mid x_o > h(s) > x_{oo}\}] = 0 \qquad (8.7.62)$$

from (59).

Now, from (54), an agent n^* exists for which

$$\nu'[S_{n^*} \cap \{s \mid h(s) > x_o\}] > 0 \qquad (8.7.63)$$

Arguing in the usual way, we can find regions H and H^*, contained in the sets of (61) and (63) respectively such that $\infty > \alpha(H) \geq \nu'(H^*) > 0$.

Consider the following policy changes by agent n^*. He relinquishes region H^* and acquires H, making the change of assignment $\Delta \nu$ given by

$$-\nu_{H^* \times Q} + [\alpha_H \times (\nu_{H^* \times Q})'']/\alpha(H)$$

One verifies the feasibility of this change and the fact that the allotment for agent n^* remains the same.

The change in land cost is nonpositive because H is free, by (62). The change in transport cost is negative, by the argument of (37) (with H^* in place of H_2, and obvious changes of notation in the last integral of (37); both inequalities in (37) are strict). Thus agent n^* has reduced his total costs and improved his position, contradicting the fact that n^* is initially in equilibrium.

This shows that $x_1 > x_2$ is false where x_1, x_2 are defined by (52) and (53). The proof of part (ii) is now complete.

Finally, we show that

$$\int_E (p \circ h) \, dv' = \int_E (p \circ h) \, d\alpha \qquad (8.7.64)$$

for all regions E. In conjunction with (41) this will prove (26). First, α and v' coincide on all subregions of $\{s \mid h(s) \le x_o\}$, from (27). Hence (64) is true for any E contained in this set. Second, for $E \subseteq \{s \mid h(s) > x_o\}$, the left side of (64) is zero, from (28). As for the right side, we first note that

$$\sup X \le x_o \qquad (8.7.65)$$

To prove (65), let $x \in X$. Then (x, y) supports ρ for some number y, which implies that

$$v'\{s \mid \; |h(s) - x| < \epsilon\} > 0 \qquad (8.7.66)$$

for all $\epsilon > 0$, from (38). If x exceeded x_o, (66) would contradict (28). This proves (65). But $p(x) = 0$ for all $x > \sup X$, so that $p(h(s)) = 0$ for all $s \in E$. Thus the right side of (64) is also zero, and (64) is again true. Finally, any region E splits into sets of these two types, so (64) is true in general. This completes the proof of (26). ∎

These results are worth contemplating. The number x_o is the "natural radius" of the Thünen system and corresponds to the classical "margin of cultivation." Beyond x_o land is permanently vacant and free, while land closer than x_o is filled to capacity (but "filled" with lower and lower density uses as one moves outward). Note that x_o can equal $+\infty$, indicating an infinite Thünen system with no boundary.

The natural radius is "almost" unique in the following sense: x_{oo} is another natural radius iff

$$\alpha\{s \mid h(s) \text{ is between } x_o \text{ and } x_{oo}\} = 0 \qquad (8.7.67)$$

(This statement follows from (27) and (28).)   In the classical Thünen situation, for example, where S is the plane, h is Euclidean distance from the nucleus, and α is ordinary area, (67) implies $x_{oo} = x_o$, so the natural radius is indeed unique.

The fact that p is an (almost everywhere) left half-potential yields much information about p (and thus about land values, via (26)), for the results of Section 8.5 apply. To see what is involved, define k:reals $\rightarrow$ extended reals by

$$k(y) = \inf\{p(x) + f(x,y) \mid x \in X \cap Z\} \qquad (8.7.68)$$

where X and Z are defined as in the preceding proof. This implies that

$$p(x) = \sup\{k(y) - f(x,y) \mid y \text{ real}\} \qquad (8.7.69)$$

for all $x \in X \cap Z$. To prove (69), first note that (68) implies

$$p(x) \geq k(y) - f(x,y) \qquad (8.7.70)$$

for all $x \in X \cap Z$, all real y. Let $x_2 \in X \cap Z$ be given, and let y_2 be the (unique) number such that (x_2, y_2) supports λ. From (25) it follows that the infimum in (68) for $y = y_2$ is attained at $x = x_2$. Hence (70) is satisfied with equality at (x_2, y_2), so that (69) is proved for $x = x_2$.

The pair (p, k) have most of the characteristics of a topological potential for λ (relative to f). But k, unlike p, has no simple intuitive interpretation. (One may strain a bit and interpret $k(y)$ as the maximum cost per unit area an agent operating a land use of weight y would tolerate.)

Now (69) may be applied as in Section 8.5. For example, if $f(\cdot, y)$ is a *concave* function for each y, then p, as the supremum of convex functions, is *convex*. More precisely, if $x_1, x_2, x_3 \in (X \cap Z)$ and $x_2 = \theta x_1 + (1 - \theta)x_3$, where θ lies between 0 and 1, then

$$p(x_2) \leq \theta p(x_1) + (1 - \theta)p(x_3) \qquad (8.7.71)$$

(Since (71) may not apply to all triples x_1, x_2, x_3, the stated condition might better be called "convexity almost everywhere.")

A Simplified Approach

The foregoing results concerning the structure of a Thünen social equilibrium, while rather striking, are also rather complicated to derive. At the same time, the underlying arguments are basically quite simple, though this fact is obscured by the requirements of rigor. It seems worthwhile, then, to present an alternative approach that is heuristic and informal. The sacrifice of rigor extends not only to the reasoning but even to the concepts themselves. Thus the notion of "running a land use *at* a location" is inherently vague and must be translated into measure-theoretic language to attain clarity. But to carry out all the clarifications needed would simply land us back in the complexities of the preceding proofs. Let us therefore plunge ahead boldly.[34]

Let there be a social equilibrium, with agent n_i running land use q_i at site s_i, $i = 1, 2$. The weight of land use q_i is $w(q_i) = y_i$, and the distance of site s_i from the nucleus is $h(s_i) = x_i$, $i = 1, 2$. We assume as usual that the transport cost incurred per acre at distance x for weight y is given by $f(x, y)$, where f has positive cross-differences. Also assume that there is a land-value density function $\tilde{p}(s)$.

Agent n_1 has the option of relinquishing an acre of land "at" site s_1, acquiring an acre of land "at" site s_2, and switching his land use q_1 from s_1 to s_2. This does not change his allotment; hence, since he is initially in equilibrium, this change cannot reduce his total costs. Thus

$$\tilde{p}(s_1) + f(x_1, y_1) \leq \tilde{p}(s_2) + f(x_2, y_1) \qquad (8.7.72)$$

The left side of (72) is approximately the total cost incurred by agent n_1 from the running of one acre of land use q_1 at site s_1. The right side is about what his total costs would be from running q_1 instead on an acre at s_2. A similar argument applies to agent n_2, who has the option of switching his land use q_2 from site s_2 to s_1. This yields

$$\tilde{p}(s_2) + f(x_2, y_2) \leq \tilde{p}(s_1) + f(x_1, y_2) \qquad (8.7.73)$$

(which is just (72) with subscripts interchanged).

These are the key relations. Adding (72) and (73), $\tilde{p}$ drops out and we obtain

$$f(x_1, y_1) + f(x_2, y_2) \leq f(x_1, y_2) + f(x_2, y_1) \qquad (8.7.74)$$

Now if $x_1 < x_2$, we cannot have $y_1 < y_2$, for otherwise (74) would contradict the fact that f has positive cross-differences; i.e., it cannot happen that a lighter land use is "at" a nearer site while a heavier use is "at" a more distant site. This may be taken as a heuristic characterization of the *weight-falloff condition*.

Next, take the special case where $x_1 = x_2$. Then (72) and (73) yield $\tilde{p}(s_1) \leq \tilde{p}(s_2) \leq \tilde{p}(s_1)$, so that $\tilde{p}(s_1) = \tilde{p}(s_2)$; i.e., whenever two sites have the same ideal distance, they have the same land-value density. It follows that $\tilde{p}$ may be written as a composite function, $p \circ h$. Then (73) reduces to

$$p(x_2) + f(x_2, y_2) \leq p(x_1) + f(x_1, y_2) \qquad (8.7.75)$$

This is a heuristic form of the condition (25) that p be a *left half-potential* for λ, the plane measure induced by the mapping $(s, q) \to (h(s), w(q))$ from the assignment ν.

---

34. The following development is essentially that of Faden, *Essays in Spatial Economics*, Ph.D. diss., Columbia Univ., 1967, pp. 170–88.

From (75) one can derive several properties of p from corresponding properties of f. The two most interesting are monotonicity and convexity.

Let $f(\cdot, y)$ be *strictly increasing* for each y, and choose $x_1 < x_2$. Assume there is a location s_2 at distance x_2 from the nucleus, with a land use q_2 in operation "at" s_2; let y_2 be the weight of q_2. Then (75) is satisfied for this x_1, x_2, y_2. Hence $p(x_1) - p(x_2) \geq f(x_2, y_2) - f(x_1, y_2) > 0$, so that p is *strictly decreasong*.

Let $f(\cdot, y)$ be a *concave* function for each y, and let x_1, x_2, x_3 satisfy $x_2 = \theta x_1 + (1 - \theta)x_3$, where θ is a number between 0 and 1. Assume as above that there is a location s_2 at distance x_2, with a land use q_2 of weight y_2 operating "at" s_2. Then (75) is satisfied for this x_1, x_2, y_2, and we also have

$$p(x_2) + f(x_2, y_2) \leq p(x_3) + f(x_3, y_2) \tag{8.7.76}$$

upon substituting x_3 for x_1 in (75). Now multiply (75) by θ, (76) by $(1 - \theta)$, and add. After rearrangement we obtain

$$\theta p(x_1) + (1 - \theta)p(x_3) - p(x_2)$$
$$\geq f(x_2, y_2) - \theta f(x_1, y_2) - (1 - \theta)f(x_3, y_2) \tag{8.7.77}$$

But the right side of (77) is nonnegative, since $f(\cdot, y_2)$ is concave. Hence p is a *convex* function. (This argument can be spruced up to provide an alternative rigorous derivation of the "almost everywhere convexity" property (71).)

Finally, we indicate how one goes from (75) to the important line integral representation of p (cf. (5.83)):

$$p(z_1) - p(z_2) = \int_{z_1}^{z_2} D_1 f(x, y)\, dx \tag{8.7.78}$$

Here $D_1 f(x, y)$ is the partial derivative of f with respect to its first argument, and the integral is taken along the line of support in the plane, connecting the points of support for λ. We may assume that $z_1 > z_2$. Choose a sequence $x_0 < x_1 < \cdots < x_n$, with $x_0 = z_1$, $x_n = z_2$, and let y_i be a number such that (x_i, y_i) is on the line of support, $i = 0, \ldots, n$. We then obtain

$$[f(x_i, y_i) - f(x_{i-1}, y_i)] \leq [p(x_{i-1}) - p(x_i)]$$
$$\leq [f(x_i, y_{i-1}) - f(x_{i-1}, y_{i-1})] \tag{8.7.79}$$

for $i = 1, \ldots, n$. The left inequality in (79) arises from substituting the triple of numbers (x_{i-1}, x_i, y_i) for (x_1, x_2, y_2) in (75); the right inequality in (79) arises from substituting the triple (x_i, x_{i-1}, y_{i-1}) for (x_1, x_2, y_2) in (75).

Now add the inequalities (79) over $i = 1, \ldots, n$. The p terms in the middle add to $p(x_0) - p(x_n)$, which is the left side of (78). Also, assuming continuity of $D_1 f$, both the left and right f-differences in (79) can be approximated by $(x_i - x_{i-1}) D_1 f(x_i, y_i)$. Adding this over i yields essentially a Riemann sum for the line integral in (78). A limit argument then establishes (78).

8.8. THE THÜNEN SYSTEM AS A SOCIAL EQUILIBRIUM: DISCUSSION

One striking fact about the model just developed is its great generality, or—what is the same thing—the weakness of the assumptions from which we start. We recapitulate the main ideas. First, area, distance, and weight are all "ideal," which allows great flexibility in incorporating geographic and institutional irregularities. Second, no assumption is made concerning the region controlled by any one agent. It may consist of many scattered parcels, near and far, of irregular shapes. Third, the only assumption concerning preferences is (7.4), which may be roughly stated as "other things being equal, lower costs are preferred to higher."

This generality of preferences is especially important for the real-estate market, such as we consider here. In industry analysis one deals with businessmen engaged in similar activities and has some basis for assuming similar motivations (typically assuming they all maximize income). But in the real-estate market we have religious, commercial, governmental, residential, etc., land users all participating. Thus philanthropists and robber barons, bureaucrats and businessmen, are all competing cheek-by-jowl in the same market, and one can hardly assume uniformity of preferences among these agents. (These remarks assume there are no zoning restrictions. With a zoning law the market splits to some extent into "noncompeting groups," but still with considerable diversity within these groups.)

The two other main assumptions are rather more restrictive. These are the absence of any constraint on land-use assignments other than the areal capacity constraint (7.1) and the characteristic form for transportation costs (7.2) (with f having positive cross-differences). In Section 8.9 we argue that these assumptions apply to a broader range of cases than might be expected at first glance.

We now examine the foregoing Thünen system from the point of view of *social welfare*. Is the social equilibrium at which the market arrives optimal in some sense? We know that the equilibrium assignment ν satisfies the measurable weight-falloff condition, and that it is therefore optimal for the allotment-assignment problem defined by its marginals

and the transport cost integral (7.2) (assuming certain other weak premises). That is, ν minimizes total transport costs over the set of all measures
on $S \times Q$ having left and right marginals ν' and ν'' respectively. This
has been vaguely recognized (but never really proved or stated clearly)
by several writers who claim that land uses arrange themselves to minimize the "friction of space."

But does this property imply social optimality? According to Alonso
it does not because the reduction of transport cost is just one desideratum,
which must be balanced against others such as freedom from congestion
and the quality of life.[35] This statement is correct. However, it is also
irrelevant for the question under discussion because ν minimizes transport
cost over a set of measures with the *same* right marginal ν'', and this
constraint has the effect of *holding constant* all the other desiderata.

To spell this out, let ν_n on universe set $S_n \times Q$ be the assignment of
agent n, $n = 1, 2, \ldots$. We then have $\nu'' = \nu_1'' + \nu_2'' + \cdots$, where ν_n'' is the
right marginal of ν_n. Consider now any alternative system of assignments
$\tilde{\nu}_n$, $n = 1, 2, \ldots$, yielding overall assignment $\tilde{\nu}$ by direct summation such
that these satisfy

$$\tilde{\nu}_n'' = \nu_n'' \qquad (8.8.1)$$

for all $n = 1, 2, \ldots$, and

$$\tilde{\nu}' = \nu' \qquad (8.8.2)$$

Adding (1) over all n, we obtain

$$\tilde{\nu}'' = \nu'' \qquad (8.8.3)$$

Then (2) and (3) imply that transport cost under $\tilde{\nu}$ is at least as large as
under ν. At the same time, (1) indicates that the mode of life of all agents
is the same under these two situations.

Since all other desiderata are held constant, it would seem that
minimization of transport cost is a *necessary* condition for social optimality. As in so many other contexts it appears that here too the competitive market solution has socially desirable features.

However, the same inefficiencies that crop up in competitive solutions
also arise here. Transport cost is the sum of all individual costs as personally assessed by each separate agent. Now in transportation particularly all sorts of costs are imposed on other agents, which are not assessed
against the traveler or shipper and would not be counted in his private
costs. These "external" costs include delays and crowding for other
travelers, increased risks of accident for them, pollution and noise, etc.[36]

---

35. Alonso, *Location and Land Use*, pp. 101–5. See note 11.
36. See, e.g., D. M. Winch, *The Economics of Highway Planning* (Univ. Toronto Press,
Toronto, 1963).

Second, the market implicitly weights the preferences of different agents by treating all dollars equally, no matter who spends them; one may decide to reject this weighting on ethical grounds. In either of these cases, the fact that the market minimizes "total transport costs" loses its normative significance.[37]

We now turn to the structural implications of the model. Since the Thünen social equilibrium is an optimal solution to the allotment-assignment problem, the extensive discussion of Section 8.6 applies to it. Thus we get Thünen "rings" of land uses of decreasing weight as we go outward from the nucleus. The "rings" tend to be elongated along major transport arteries, leading to "urban sprawl," and to form disconnected subcenters around points of access to limited-access transportation systems (highway interchanges, airports, railway stations). This latter process yields a "central place" hierarchy of centers and land uses,[38] the "higher-order" activities being those that are heavier.

Now consider the weight-falloff property more closely. First, note what it does not state. It says nothing about *how much* land is to be allotted to various uses, or even whether a given land use will appear in the system at all.[39] It only states that if two land uses do appear, the heavier will be closer to the nucleus (or, at worst, equidistant from it). Cities exhibit certain broad regularities in the ordering of their land uses by distance, and it is enlightening to compare these regularities with the predictions of the weight-falloff property.

Consider multiple-story structures. There is a general (but not perfect) tendency for land uses involving many stories to be heavier than land uses involving few stories. We would expect, then, the skyline of a city to get lower as one moves away from the nucleus (i.e., the CBD), as indeed it does.

The exceptions to this rule are no less enlightening than the uses that conform to it. The exceptions are there for most part thanks to political intervention that consciously sets itself against the natural tendency of the market—either by direct public ownership in the case of streets, parks, etc., or by laws forcing private owners to limit the coverage of their lots and the size and bulk of their buildings. The case of the transportation system is particularly interesting. There is some tendency for transporta-

---

37. The fact that total transport cost as given by (7.2) is a linear function of assignment v, hence of resource flows, indicates the absence of external effects. Thus the introduction of external effects would vitiate not only the ethical conclusions from the model but the model itself.

38. B. J. L. Berry and A. Pred, *Central Place Studies,* 2nd print. (Reg. Sci. Res. Inst., Philadelphia, 1965). (See pp. 463, 528).

39. To answer these questions, one needs the more elaborate models of the new urban economics. See note 15.

tion to conform to the general pattern, with elevated highways, subways, etc., in downtown areas; but the stacking of transportation surfaces is not carried as far as the stacking of other kinds of land uses. Actually, transportation is *not* a land use in the sense in which we are using the term in the Thünen model. Here "land use" is by definition sedentary. Thus there is no strong reason to expect that transportation would conform to the general weight-falloff pattern.

The analysis just given identifies Space with the surface of the earth and considers vertically stacked floors to be parts of one land use. An alternative view identifies Space with horizontal surfaces of support, including the separate floors of a multiple-story structure (see Section 8.4). On this view, each story supports a separate land use. This alternative view yields some additional predictions. The ideal distance of a site from the nucleus now includes not only the cost of surface transportation but also *vertical* transportation costs if the site is not at ground level. Thus ideal distance rises with each successively higher story (successively lower in the case of subterranean structures).

Applying the model to this situation, one expects a progressive lightening of land uses as one moves vertically away from ground level, together with a fall in the value of floor space—just as if one had moved horizontally further from the nucleus. (A rough analogy would be the climatic zones on a mountain at the equator; the zones passed through going upward approximate those passed through going poleward.)

Are these predictions borne out? In general, rental value falls as one goes from the ground floor to the second or third floors, and this pattern conforms to expectations. But (if there is an elevator) it levels out and perhaps rises for higher floors.[40] To put these facts in perspective, note the following points. First, successive floors are not perfect substitutes for each other, either technically (in terms of the weight of machinery, vibration, etc., they can sustain) or psychologically (higher floors have cleaner air, less street noise, commanding views, and the general quality of "upmanship" that leads people to climb mountains, sit on raised platforms, and wear elevator shoes). This lack of homogeneity violates one of the conditions of the model and helps explain the rise in rentals on upper floors.

Second, the "ideal" height of even the tallest existing skyscraper is rather small—just a few minutes of travel time. If it took, say, a half-hour to get to the top, we would begin to find some palpable differentiation among stories. Also, much of the incremental cost is incurred in the first floor or two (in the form of waiting time or stair climbing), which helps

---

40. R. Turvey, *The Economics of Real Property* (Allen & Unwin, London, 1957), pp. 16–18. The "Sheridan-Karkow formula" for the rental value of office space shows rent increasing with height; see *Buildings* (Dec. 1959): 30.

explain the rapid initial fall of rent. The well-known specialization of ground floors in retail trade seems to be due to their high visibility from the street, a factor that again violates one of the conditions of the model. This also contributes to the initial fall in rent.

Let us now turn to *land speculation*, i.e., the policy of delaying the onset of a land use to some point in the future. As discussed in Section 8.4, this delayed activity is a land use in its own right, the *forward time displacement* of the original activity. In general, a forward time-displaced land use is lighter than the original. Hence, if use q and its displacement q' both appear in a Thünen system, q' will locate further from the nucleus than q, by the weight-falloff condition.

We would expect, then, a general tendency for land uses to become more delayed in onset as we move outward from the nucleus. The initial fallow period, before any imports or exports arise from the site, should become longer with distance. This is observed, and is the phenomenon of *suburbanization*.

At the same time, certain exceptions to this rule are also implied by the weight-falloff condition. Let q_1 be a land use and q_1' a forward time displacement of q_1. While q_1' is lighter than q_1, it may well be heavier than another land use q_2, whose onset occurs before that of q_1'. In this case, if q_1' and q_2 appear in the system, q_1' will locate closer to the nucleus than q_2. There will then be an interval of time during which a nearer region (that occupied by q_1') will remain vacant while a further region (occupied by q_2) will be busily importing and exporting. This is the phenomenon of *leapfrogging*, which has been widely observed and commented on. Note that this does *not* constitute an "intensity reversal" contradicting the weight-falloff condition if "intensity" is correctly measured as an integral over the *entire* time horizon of the system, and not as a short-run indicator.[41]

The typical land uses of the central business district (front-office and professional activities, advertising, finance, department stores, theaters, night clubs, and hotels) should be the heaviest in the system, since they are the most central. Their weight arises from their eminent "stackability" into multiple stories and the high densities of people with relatively high-cost time involved in them.[42]

---

41. Cf. R. Sinclair, Von Thünen and urban sprawl, *Ann. Assoc. Am. Geogr.* 57(Mar. 1967):72–87; and E. M. Hoover, *The Location of Economic Activity* (McGraw-Hill, New York, 1948), pp. 170–71, for examples. An alternative explanation of leapfrogging is based on the presence of imperfect information in the real-estate market: see W. R. Thompson, *A Preface to Urban Economics* (Johns Hopkins Press, Baltimore, 1965), pp. 326–27.

42. Another centralizing factor is that these land uses are "communication-oriented" and generate heavy information flows among themselves. See E. M. Hoover and R. Vernon, *Anatomy of a Metropolis* (Doubleday Anchor, Garden City, N.Y., 1962), pp. 59 ff. But this factor lies outside the Thünen framework of this chapter.

At the other extreme, typically suburban land uses, such as golf courses and cemeteries, shade off finally into agriculture. These are all very light compared with most urban land uses, and the pattern again conforms to weight-falloff.

Manufacturing is found at all distances, and this reflects the heterogeneous character of this class of land uses. Suburban manufacturing is typically a one-story affair, with large areas devoted to parking, storage, and landscaped grounds, while central manufacturing tends to occur in lofts with a densely crowded labor force. (The typical designation of certain manufacturing activities as "light" or "heavy" has nothing to do with their ideal weights. The "heavy" activity of steelmaking, for example, is much lighter in ideal weight than the "light" activity of apparel manufacture; the much more centralized location of the latter activity bears this out.)

In general, the higher the average population density associated with a given land use, the heavier the weight of that land use, so that one expects a gradual diminution of crowding as one moves outward. (The CBD itself generally has a lower *residential* density than its environs. But in terms of the actual location of people (at work, shopping, etc.) it would probably have the highest density of all.)

However, certain exceptions are implied by the weight-falloff condition. Ideal weight depends not only on density but on trip-taking propensities and the cost of moving the people involved. The weight of a retirement colony, for example, would be reduced for both these reasons. Retired people tend to be "light" because their incomes are low, and their earned incomes are very low; furthermore, their trips to the center of town will be relatively infrequent. Thus we would expect retirement colonies to locate farther out than most other land uses having their population density.

Again, consider the influence of family size on location of residence. Extra children generate more local trips (to schools, playgrounds, etc.) but probably few if any extra trips to the center of town. At the same time, they increase the family's demand for floor space. Since local trips do not add to activity weight, the net result would seem to be that larger families tend to choose lower weight residential uses than smaller families. As a result, larger families should generally live farther from the center of town than smaller families. (This argument assumes that other things besides family size are equal—in particular, tastes and standard of living, as measured, say, by total income, or per-capita income, or something in between.)

We now turn to the relation of family income and residential distance from the nucleus. The question of whether the rich or the poor live closer to the nucleus reduces, by the weight-falloff principle, to the question of whether the rich or the poor choose the heavier residential land use. Income will affect residential weight in two ways: via the density at which people live and via the travel costs incurred per person.

The effect on density is well known. Low-income housing involves on the average more people per acre for several reasons. There are fewer square feet of floor space per person; a greater proportion of housing is in the form of multiple-story dwellings; and there is less open space per person in the form of lawns, playgrounds, parking facilities, etc. (There are, to be sure, high-rise luxury apartments, but these are exceptional.) The residential land uses of the poor thus tend to be heavier than those of the rich, hence closer to the nucleus.

The effect of income on travel cost is less clear. Travel cost resolves itself into the frequency with which one takes trips to the nucleus and the cost per unit distance of a trip. Trip frequency in turn depends on steadiness of employment, tastes in recreation and shopping, etc.; the influence of income on these factors is unclear. The cost of travel time rises with income, especially earned income. However, the rich are more likely to drive cars, ride taxis, etc., which compensate for these costs. While the overall effect is thus unclear, it may well be large, possibly large enough to overbalance the density effect. In this case the rich would live closer to the nucleus than the poor.[43]

This ambiguous conclusion is matched by ambiguous evidence. In developed countries the poor generally live closer to the center of town, but this is often reversed in the underdeveloped world and in past civilizations.[44]

The tendency for the poor to live closer to the center is abetted by the fact that cities grow outward, so that the older, more dilapidated buildings tend to be located toward the center. The poor will gravitate toward the lower quality housing stock (see p. 477) and thus inadvertently settle close to the nucleus.[45] This factor escapes the confines of the formal model of this section because it involves a change in control of land over time. The original occupants of central-city housing hand it over to poorer successors as it deteriorates. The formal model, however, contemplates a single real-estate auction at "time zero."

---

43. For further discussion, see Faden, *Essays,* pp. 179–82.

44. See G. Sjoberg, *The Preindustrial City* (Free Press, Glencoe, Ill., 1960), pp. 97–100; L. F. Schnore, On the Spatial Structure of Cities in the Two Americas, in P. M. Hauser and L. F. Schnore, eds., *The Study of Urbanization* (Wiley, New York, 1965), Ch. 10.

45. Examining urbanized areas in the 1960 U.S. Census, Schnore finds that the younger central cities tend to have higher status residents than their suburbs, while the older central cities have the opposite more usual pattern. This may reflect a housing quality effect, the younger cities not having had time to become too dilapidated toward the center. L. F. Schnore, The socio-economic status of cities and suburbs, *Am. Sociol. Rev.* 28(Feb. 1963): 76–85.

In the above discussion, it has been difficult to avoid making statements of one-way causality among variables. This is misleading, since we are really dealing with the simultaneous determination of equilibrium values. For example, the housing choices of rich vs. poor families (even assuming identical tastes) will depend on the structure of rental prices for the various qualities of housing, and these prices in turn depend on the actions of all land users in the system. For that matter, the level of family income itself is in part subject to choice, to be decided on jointly with the choice of housing quality, location, trip frequencies, etc.

Similarly, take the relation of car ownership to location. An automobile acts as a general "levitator," so that one expects car owners to make lighter residential choices and therefore locate farther out. This makes car ownership appear to be the exogenous causal variable. But one can also argue that families who decide to locate farther out find car ownership relatively advantageous and are therefore more likely to buy cars. Actually, the problem for each family is one of choosing the most preferred configuration of several variables jointly, and it is in general not correct to argue that the choice of one variable causes the choice of another. We do find a positive correlation between car ownership and distance from the center of town, as expected, but no clear-cut causal relation has so far emerged, which again is to be expected in view of the above comments.[46]

We now turn from the structure of land uses to land values. As usual, $f(x, y)$ is the transport cost per acre incurred by a land use of weight y at distance x from the nucleus, and $p(x)$ is the land value per acre at distance x (area, weight, and distance all being "ideal" quantities). We concluded above that if for each y the function $f(\cdot, y)$ is increasing and concave, the function p will be decreasing and convex. Is this premise realistic, and if so, is the conclusion borne out in real cities?

One difficulty that arises in trying to answer this question is that transport costs and land values are given in terms of physical, not ideal, acres and miles, so that a translation problem arises. For most urban land uses, one flat, well-drained parcel is about as suitable as another, and for these we may identify real and ideal area. For the sake of argument we also identify real and ideal distances. (One way to judge the realism of this assumption is to see how well the isochrones, the loci of points of equal travel time from the nucleus, approximate to concentric circles.) For fixed weight y, transport cost certainly increases with distance. Increasing congestion on approach to the center of town tends

---

46. J. F. Kain, Urban Travel Behavior, in L. F. Schnore and H. Fagin, eds., *Urban Research and Policy Planning* (Sage Publications, Beverly Hills, Calif., 1967), pp. 161–92, esp. pp. 173–75.

to make it concave as well. That is, the closer one is to the nucleus, the more costly it is (in terms of time, mental strain, etc.) to travel a given physical distance; this makes f concave in distance. As for the conclusion, the most striking manifestations of the decreasing convexity of p are the extreme heights to which land values per acre rise in the central business districts of large cities.

These results must be interpreted with some care. "Land value" refers to time zero, where all land is assumed to be uniform (except for varying distances from the nucleus). To be concrete, think of the original state as a vacant lot, with no capital improvements on it (except perhaps for drainage and leveling). Indeed, the formal model envisions just one omnibus real-estate auction at time zero, so that land values at other times are not even meaningful. In reality, of course, the real-estate market endures, with control of the same parcel passing from agent to agent. The question then arises, Must the cross-sectional distribution of land values at any future time conform to the Thünen pattern?

The answer is no. To take a not unrealistic example, suppose someone invests heavily in a suburban plot, while an "inner city" plot is allowed to become "blighted." The former can easily become more valuable than the latter, reversing the original order. Only if one can isolate a "pure site value" might the original Thünen pattern be retained. (Note that the "improvement value" of a site can be negative, so that clearance of the site—which restores its original vacant condition—improves its market value.)

Let us finally indicate some ways in which the assumptions of the Thünen model might be weakened in a realistic direction without invalidating its conclusions. We begin by introducing real-estate *taxation*. Suppose there is a *tax function* $t:S \times Q \rightarrow$ reals, where $t(s, q)$ is the tax liability incurred per acre by an agent running land use q at site s. More precisely, if agent n controls region S_n and chooses assignment ν_n over $S_n \times Q$, his total tax liability is

$$\int_{S_n \times Q} t(s, q)\ \nu_n(ds, dq) \qquad (8.8.4)$$

This tax liability must be reflected in the preferences of the agents participating in the system. The natural way to do this is by a simple reinterpretation of preference condition (7.4); viz., of two actions leading to the same allotment, the agent prefers the one incurring lower total costs, where costs include not only land costs (7.3) and transport costs (7.2) but also tax costs (4).

We do not attempt a fully rigorous discussion of the influence of this new factor but instead follow the informal simplified approach of (7.72)–

(7.75). Let there be a social equilibrium, with agent n_i running land use q_i at site s_i, $i = 1, 2$. The weight of q_i is y_i, and the distance of s_i from the nucleus is x_i, $i = 1, 2$. Then

$$t(s_1, q_1) + \tilde{p}(s_1) + f(x_1, y_1) \leq t(s_2, q_1) + \tilde{p}(s_2) + f(x_2, y_1)$$
$$(8.8.5)$$

where $\tilde{p}$ is the land-value density function.

The argument for (5) is essentially the same as for (7.72); viz., agent n_1 has the option of switching one acre of land use q_1 from site s_1 to s_2. This leaves his allotment unchanged; hence, since he is initially in equilibrium, this switch cannot decrease his costs. Thus the left side of (5), which is the tax plus land plus transport cost incurred from running an acre of q_1 at s_1, does not exceed the right side of (5), which is the total cost he incurs from running q_1 at s_2. Reversing the roles of agents n_1 and n_2, we find that (5) remains true if the subscripts 1 and 2 are interchanged throughout.

We now make the assumption that the tax function t is *separated;* i.e., we assume functions $g : S \rightarrow$ reals and $u : Q \rightarrow$ reals exist such that

$$t(s, q) = g(s) + u(q)$$
$$(8.8.6)$$

for all $s \in S$, $q \in Q$. How realistic is this assumption? First, consider a Henry George type tax on unimproved site value. This depends only on s, not on q, hence is of the form (6), with u identically zero. Next, consider a tax on improvements only. It is not implausible that the improvement value of a given land use q will not vary much from site to site; in this case the tax function will be more or less independent of s, and again be in the form (6), with g identically zero. (The principle of "treating equals equally" would tend to make such a tax independent of site in any case.) Actual real-estate taxes are generally in the form of a sum of the two types just mentioned, hence again in the form (6).

There are a number of circumstances, however, in which (6) is not realistic. One example is when different land uses are taxed at different rates—say when industrial land uses are taxed more heavily than residential. For in this case the rate of taxation of site value depends on the land use occupying the site, and there is interaction between s and q. Another such case arises when Space is partitioned into several different political jurisdictions, each with its own tax structure. Thus one jurisdiction may grant tax concessions to industry, while another may not; this leads to interaction between s and q. Even when the tax *structures* of different jurisdictions are similar and they differ only in tax *levels*, (6) is violated; for in this case s and q would combine multiplicatively rather than additively. Very roughly then, (6) seems a good approximation *within* a single political jurisdiction but not *between* them.

We now trace the implications of (6). Substituting in (5), we obtain

$$[g(s_1) + \tilde{p}(s_1)] + f(x_1, y_1) \leq [g(s_2) + \tilde{p}(s_2)] + f(x_2, y_1) \qquad (8.8.7)$$

Furthermore, if we first interchange subscripts in (5) and then substitute, we obtain (7) with *its* subscripts interchanged. Now (7) and its interchange differ from (7.72) and (7.73) respectively only in the fact that the function $\tilde{p}$ in (72)–(73) is replaced by $g + \tilde{p}$ in (7). The argument following (7.73) may now be repeated verbatim except for replacing $\tilde{p}$ by $g + \tilde{p}$. We conclude that the equilibrium assignment still satisfies the weight-falloff condition, even with the tax. Also, $g + \tilde{p}$ depends only on ideal distance h, so that a function p:reals $\rightarrow$ reals exists satisfying $g(s) + \tilde{p}(s) = p(h(s))$ for all $s \in S$. The function p has the properties that if $f(\cdot, y)$ is increasing for all y, then p is decreasing and if $f(\cdot, y)$ is concave for all y, then p is convex.

These results must be interpreted with care. The fact that weight-falloff is preserved does not mean that the tax has no behavioral effects. On the contrary, heavily taxed land uses (high u-values) will tend to occupy fewer acres and may disappear entirely, to be replaced by lightly taxed uses. But the *order* in which land uses are ranged, by increasing distance from the nucleus, will not change. Heavyweight uses will still be closer than lightweight uses.

It is no longer land values per se that have the simple regularity properties of the Thünen system, but only land values *plus* pure site taxes. Thus if one parcel is more heavily taxed than its neighbor, its equilibrium land value will be sufficiently below that of its neighbor so that the sum of value plus tax is about the same on both. This is essentially the classical conclusion that a pure site tax falls completely on the landlord, even if levied on the tenant. (This is a bit inaccurate: the tax has wealth redistribution effects that reverberate throughout the system.)

We conclude the discussion of taxes with an exercise. Show that if (6) is altered by adding the term $\theta(h(s), w(q))$ on the right, θ:reals$^2 \rightarrow$ reals being any function having nonnegative cross-differences, then all the conclusions above remain valid, with the single exception that in determining the properties of p the function $f + \theta$ should be used in place of f alone.

The remaining generalizations involve reinterpretations of the concepts "transport cost" and "activity weight" used in the formal model. The basic idea of the model, after all, is that people try to move their activities closer to the nucleus to economize on transport costs; "activity weight" measures the strength of this pull. Suppose there were some other factor (not necessarily having anything to do with transportation) that made it desirable to be closer to the nucleus. This factor would then operate as a "pseudo transport cost" and could be incorporated formally into the model *provided*—this is a strong assumption—its effects could be summarized in the ideal distance function *h* already in use for transportation. The "weights" of the various activities would be adjusted upward to reflect this new attractive force, the size of the adjustment depending on the impact of the factor on the particular activity in question. Similarly, there could be a factor that made it desirable to be *further from* the nucleus, and this would lead to a *downward* adjustment of activity weights. (One could even have mixed effects, some activities being attracted and others repelled from the nucleus because of some factor.) We discuss several examples of such factors.

Take *danger,* for instance. This comes in several forms. Danger from invasion and raiding parties has been important in the past, as the existence of walled cities testifies; it still is present today in "unpacified" countries. This danger is greatest at the periphery and diminishes as one moves toward the center of town. Hence it constitutes an attractive force and increases ideal weights. As for the differential effects, consider once again residential activities of rich vs. poor families. The rich will presumably be willing to pay more for a given increment of safety than the poor. Hence the residential activities of the rich gain more weight than those of the poor, and this creates a tendency for the rich to live closer to the center of town.[47]

Under modern conditions this source of danger is minor. Two other sources, however, are quite important: danger from criminals and danger from thermonuclear attack. These have just the opposite pattern, being least dangerous at the periphery and most dangerous toward the center. (For a partial explanation of this, see Section 5.8.) Hence they have a dispersive effect, reducing ideal weights. The rich should be more sensitive to these influences than the poor, accentuating the tendency for the rich to live farther from the center than the poor.

Most nontransport factors seem to follow this latter pattern, adding a centrifugal rather than centripetal force, reducing rather than increasing ideal weights. This applies to other aspects of the "urban syndrome," such as pollution. The fact that pollution decreases as one moves outward makes it a dispersive force. The rich, being willing to pay more than the poor for a given physical decrease in pollution intensity, will again tend to reside farther from the nucleus on this account.

A different kind of influence may be allowed for by relaxing the entrepôt assumption that all (nonlocal) trips must go through the nucleus. Suppose instead there are also *external trips,* in which one travels to the outside world by heading directly away from the nucleus. Examples are

---

47. Sjoberg mentions external danger as one factor making for centralized residences for the rich in the "preindustrial" city. Sjoberg, *Preindustrial City,* pp. 97–100.

furnished by certain types of outdoor recreation such as pleasure driving in the countryside. These trips exert a pull opposed to that of the nucleus, hence act as "levitators." The reduction in weight will be large for people with a strong taste for outdoor recreation or other activities involving external trips, and they will tend to reside at the periphery.

As a final example, suppose some factor of production can be obtained on terms that vary systematically by distance from the nucleus. Specifically, suppose that wages decline as one moves away from the nucleus.[48] Then any land use with positive labor cost can reduce it by moving outward. The wage gradient therefore acts as a levitator. The land uses most affected will be those that are most labor intensive; to be precise, the reduction in weight among land uses is proportional to their labor/land ratios.

8.9. THÜNEN SYSTEMS AND THE REAL WORLD

The purpose of this section is, first, to make a brief survey of some real-world Thünen systems and, second, to discuss briefly the origin and function of such systems and how they fit together in a hierarchy.

Whether a given real situation does indeed exemplify some theoretical concept is in general a matter of judgment. One rarely finds a pure case in which the assumptions of the theory are realized exactly. Instead, there is generally more or less "noise" that distorts the ideal conditions of the theory. If the "noise" is intense, the distortion may be so severe that the theory is useless for understanding what is going on. The question, then, is not whether the theory as interpreted is true or false, but whether it is a good or poor approximation to the situation in hand.

In the present case we consider a real-world pattern to be a Thünen system if it satisfies two conditions. First, there should be some indication of the concentric ring structure predicted by the weight-falloff condition. Second, there should be some indication that this structure arises via the central mechanism of the theory, viz., the attraction of the nucleus on the various land uses in operation.

It would be highly desirable to develop a series of indices of "goodness of fit" to ascertain in quantitative terms just how closely the following patterns *do* approximate to the ideal Thünen pattern. To do this, however, would take us far off course. In the absence of such indices the following comments may be helpful. The approximation to the ideal should be best at *intermediate* distances from the nucleus, but poor when very close to the nucleus and at the periphery of the system.

There are several reasons why the fit should be poor at the periphery.

---

48. Reasons for this decline are given in A. Lösch, *The Economics of Location,* trans. W. H. Woglom and W. F. Stolper (Yale Univ. Press, New Haven, 1954), p. 43, n. 10.

The pull of the nucleus is very weak here, so that this systematic effect is more easily overshadowed by random "noise." Furthermore, the pattern tends to be disrupted by external forces (such as the pull of nuclei of neighboring Thünen systems), which are strongest at the periphery. Near the nucleus, on the other hand, the fit is poor because of a scale effect. The nucleus, after all, is not really a geometrical point as the model demands, but a region that is small compared with the system as a whole (e.g., the inner central business district of a city). As distances become sufficiently small to be comparable in size to the nucleus itself, the direction of pull becomes uncertain and its strength attenuated, just as the pull of gravity *weakens* as one goes below the surface of the earth.

The distinction between local and nonlocal movement is important here. The latter refers to trips to and from the nucleus while the former refers to short-distance trips that do not affect the weight of land uses. Whether a trip is local depends on the particular Thünen system being investigated. Thus a trip to a neighborhood shopping center would be local in the context of the city as a whole, but nonlocal when the neighborhood itself is thought of as a miniature Thünen system. The distinction allows a great deal of flexibility in applying the apparently rigid requirements of the entrepôt model, that all trips must go to or come from the nucleus. The totality of local movements may far overshadow nonlocal movements; they undoubtedly do so for the larger Thünen systems. But nonlocal movements are focused on the nucleus, while local movements are diffuse in direction. This allows the former to exercise a systematic influence and lead to Thünen ring formation.

We now turn to real-world applications. (The following references are just a selection from a much larger number that could have been cited.) The bulk of our illustrations have been at the city level, with the CBD playing the role of nucleus. The applicability of the Thünen model to cities was noted by Isard (the germ of the idea can be traced back to Thünen himself).[49] Independently, the "concentric-circle" theory of city structure was promulgated by Burgess.[50] This was essentially an idealized description, stressing the (positive) correlation between distance from the center of town and socioeconomic status of residents. No satisfactory explanation of this pattern was given, but as we have seen, this relation can be incorporated into a modernized Thünen analysis.

---

49. W. Isard, *Location and Space-Economy* (M.I.T. Press, Cambridge, 1956), App. to Ch. 8. A notable pioneering effort in this direction, rich in descriptive detail, is that of R. M. Hurd, *Principles of City Land Values,* 1st ed. (The Record and Guide, New York, 1903).

50. E. W. Burgess, Urban Areas, in T. V. Smith and L. D. White, eds., *Chicago: An Experiment in Social Science Research* (Univ. Chicago Press, Chicago, 1929), pp. 113–38. The theory first appeared in 1925.

We pass over the plethora of other studies of internal city structure (by sociologists, geographers, economists, and others) and turn to the question of the existence of the nuclear attraction mechanism underlying the Thünen model. There are several possible approaches here. The most straightforward is to take a census of traffic flows. Many studies confirm the fact that a large fraction of all trips have the CBD as one terminus[51] (the fraction appears to be declining, however, as suburbanization continues). A related approach examines the structure of the transportation grid, noting the extent to which it is "radial" (focusing on the nucleus) vs. "peripheral" (bypassing the nucleus). The predominance of a radial transportation grid is both a symptom of strong nuclear attraction, since transport arteries tend to get built along routes of heavy traffic flow, and a further influence strengthening that attraction. The same can be said about scheduled common-carrier routes; e.g., bus routes in Queens, New York, run predominantly east-west, indicating that they serve as feeder lines for Manhattan commuter traffic.[52] Railroads in particular exhibit a strong radial pattern about large cities, extending like spokes of a wheel from the city hub into the hinterland.[53] Furthermore, the city itself, especially if large, tends to be situated at a point of high natural "nodality," having good access to navigable rivers, mountain passes, the ocean, etc.[54] (An explanation for this will be given in Section 9.4.) This tends to make regional traffic go through the city, making it a natural entrepôt and underpinning the assumptions of the Thünen model.

From the city level we can travel both upward and downward in scale. We first go downward. The Thünen model may be applied to communities within a city. E. Franklin Frazier found that Harlem in New York City exhibited a ring structure that reproduced in miniature the Burgess concentric-zone model.[55] At this level the attraction of the nucleus is likely to stem from the availability there of shopping and recreational facilities and from the fact that the nucleus provides a transportation gateway between the community and the rest of the city.

Farm villages serve as nuclei for the surrounding fields. The nearer fields tend to be cultivated more intensively than the more distant, as

---

51. For example, Illinois Department of Public Works and Buildings, *Chicago Area Transportation Study*, 3 vols. (Western Engraving and Embossing Co., Chicago, 1960-62).

52. D. Rogers, *110 Livingston Street* (Random House, New York, 1968), p. 44.

53. For example, N. Spulber, *The State and Economic Development in Eastern Europe* (Random House, New York, 1966), pp. 17-18 on Budapest; Lösch, *Economics of Location*, p. 129, n. 4, on Vienna and Prague.

54. L. F. Schnore and D. Varley, Some concomitants of metropolitan size, *Am. Sociol. Rev.* 20(Aug.1955):408-14, note the tendency for large cities to have water access.

55. E. F. Frazier, Negro Harlem: An ecological study, *Am. J. Sociol.* 43(July 1937): 72-88. He even pinpointed the nucleus, at the intersection of 7th Ave. and 135th St.

befits a Thünen system. Here the attraction of the nucleus stems from its residences and associated local facilities; it may also serve as a collection point for farm products. An individual farm with uniform terrain will tend to have a similar structure, the nucleus being the farmhouse. The problem of laying out the farm efficiently, in fact, can be formulated as an allotment-assignment problem (if the farmer makes a few simplifying assumptions), and the solution to this satisfies the weight-falloff condition.[56]

The Thünen model applies roughly to theaters, arenas, and other facilities for public exhibitions. Here the nucleus is the center of attention (stage, boxing ring, etc.), and "transport cost" is incurred in the form of deteriorating quality of the view as one moves farther from the center. "Land values" are the prices of seats, and these tend to rise as one moves closer to the center. "Land uses" are persons occupying seats, and their "weights" are given by their willingness to pay for a given improvement in quality of view. The "heavy" people will then gravitate to ringside, while the "light" people become groundlings.

On a desk or workbench a rational layout would find the more frequently used items closer to hand. One can give many more examples of such microscopic Thünen systems called into existence by nuclear attraction. We now, however, drop these small-scale systems and direct attention upward from the city level.

First, the influence of a city generally extends far beyond its political boundaries. This influence is reflected in the correlations that exist between distance from the city and such variables as population density, land values, intensity of farming, income, etc. This "metropolitan" system is merely the continuation outward of the city system; the latter (to which we have been devoting most attention) constitutes the inner rings of the complete system. When taking the metropolitan point of view, it is often convenient to think of the entire city as being the nucleus. This is the approach of Thünen himself and is implicit in some of the citations given above.

Above the city-metropolitan level lies what might be called the *subcontinental* level. The system may be mostly contained in one country, as in the United States, or may embrace several countries, as in western Europe. The nucleus is the "industrial heartland" of the region, which has the greatest concentration of population and capital goods and provides the major market of the system. In the United States this is the

56. On farms and farm villages as Thünen systems, see M. Chisholm, *Rural Settlement and Land Use* (Hutchinson Univ. Library, London, 1962), Ch. 4; Lösch, *Economics of Location*, p. 62, n. 45.

"northeastern industrial belt."[57] In Europe it is the Rhine-Ruhr complex. Land values, population density, and the "weight" of land uses all tend to decline as one moves away from the heartland. Also, per capita incomes tend to decline. This last fact establishes a link between subcontinental Thünen systems and the tendency for developing economics to develop a "dual" structure, with a high-income sector utilizing modern technology and a lagging low-income sector using traditional methods. Very roughly, the "modern" sector will occupy the inner rings of the system, while the "traditional" sector occupies the periphery. For example, a United Nations study points up a very general tendency for the low-income sector *within* each western European country to be that part farthest from the Rhine-Ruhr complex.[58]

One very suggestive aid for delineating Thünen systems at this macro-level is John Q. Stewart's concept of "potential" (not to be confused with the potentials in the measure-theoretic transportation problem of Section 7.5). For the most general definition start with a measure space (S, Σ, μ) on which is defined a metric, $h:S \times S \rightarrow$ reals, $h(s, \cdot)$ being measurable for each $s \in S$. Then the *(Stewart) potential for μ and h* is the function p with domain S given by

$$p(s) = \int_S \frac{\mu(dy)}{h(s, y)^k} \qquad (8.9.1)$$

where $k = 1$. S is usually a portion of the surface of the earth, and h is usually Euclidean distance (perhaps great-circle distance if the curvature of the earth over S is significant). Usually, μ is population or income measure (in which case p is called population or income potential respectively), but other measures such as employment or retail sales are also used.

The potential p is in effect a spatial moving average of measure μ and may be thought of as an index of "generalized closeness" to μ. In using p to delineate Thünen systems, one takes p to index "closeness" to the nucleus, so that the isopotential contour lines are taken to be loci of equal ideal distance. The justification for doing this is empirical. Potential has a high correlation with many variables one expects to be related to ideal distance in a Thünen system, and the contour maps of potential look reasonable in terms of where the Thünen rings should lie. Unfor-

57. See E. L. Ullman, Regional development and the geography of concentration, *Reg. Sci. Assoc. Pap.* 4(1958):179–98; C. D. Harris, The market as a factor in the localization of industry in the United States, *Ann. Assoc. Am. Geogr.* 44(Dec. 1954):315–48.

58. Economic Commission for Europe, *Economic Survey of Europe in 1954* (Geneva, 1955), Ch. 6. See also A. Melamid, Some applications of Thünen's model in regional analysis of economic growth, *Reg. Sci. Assoc. Pap.* 1(1955):L1–L5.

tunately, no good theoretical reason has been given as to why a function of the form (1) should have these properties.

There is a very crude argument for setting the exponent k in (1) equal to 1, as is almost always done. Setting k equal to 0 obviously makes p a constant, clear case of *over*smoothing. But (assuming S is the earth's surface and that μ has a continuous density function with respect to Euclidean area) we can show that, as k approaches 2 from below, p becomes proportional to the density function of μ; thus we are in effect reproducing μ, a clear case of *under*smoothing. The reasonable values for k thus lie between 0 and 2, and 1 seems a good compromise. This argument is weak, and it would be interesting to determine whether modifications of (1) give better fits.

Taking S to be the United States (excluding Alaska and Hawaii), one finds that the maps of income and population potential are similar. The national peak is in the New York City area, and the contours are roughly concentric about this point clear out to the Rocky Mountains. West of the Rockies, a much lower secondary system has emerged, centered on California. The primary system has persisted in main outline for more than a century.[59] This may be taken as an estimate of the subcontinental Thünen structure for the United States.

At the highest level of all, we treat the *entire world* as a Thünen system. Here the nucleus is the "North Atlantic Heartland" of northeastern United States and western Europe. One way of assessing the structure of the system is by examining international trade and travel, the bulk of which originates or terminates in this region. At the world level it is reasonable to assume that most "nonlocal" trips cross national borders and are picked up in the international transaction accounts. Another approach uses potentials. Warntz has constructed a world income-potential map (using great-circle distances).[60] This shows twin peaks on the two sides of the North Atlantic, the world peak being in the New York City area. The contours are for the most part roughly concentric about the Heartland, with a much lower secondary system beginning to emerge centered on Japan.

This picture of a world Thünen system evokes some familiar echoes.

---

59. For maps and correlations with various social phenomena, see J. Q. Stewart, Empirical mathematical rules concerning the distribution and equilibrium of population, *Geogr. Rev.* 37(July 1947):461–85; J. Q. Stewart, Demographic gravitation: Evidence and applications, *Sociometry* 11(Feb.–May 1948):31–58; J. Q. Stewart and W. Warntz, Macrogeography and social science, *Geogr. Rev.* 48(Apr. 1958):167–84; W. Warntz, *Macrogeography and Income Fronts* (Reg. Sci. Res. Inst., Philadelphia, 1965). Harris, The market as a factor in the localization of industry, also uses various potentials for market delineation.

60. Warntz, *Macrogeography,* cover and p. 92. His map of world *population*-potential, p. 111, has peaks in China and India, as one might expect, and conforms much more poorly to the world trade and travel pattern.

The period of high colonialism (say 1880–1914) has often been thought of in similar terms, especially by Marxian analysts. Here the mother countries of western Europe struggle for colonies that function as sources of raw materials and as outlets for manufactured goods and investments; the mother countries are the centers of managerial and financial control. As a result, trade becomes strongly polarized into the radial Thünen pattern.

This system transcends the Thünen model we have been discussing, since it involves military and political factors operating outside the market system. Nonetheless, there is reason to believe that a similar pattern would have evolved even under a single world free-market system. "Colonial" relations did arise in the U.S. subcontinental Thünen system, between eastern merchants and Wall Street on the one hand, and western farmers and southern planters on the other. And on the metropolitan level one finds similar relations between city banks and hinterland farmers mortgaged to them.

In this same vein we recall Marshall Lin Piao's analogy in which the western world corresponds to the "cities" and the underdeveloped world to the "countryside."[61] In effect he is comparing the world Thünen system with the Chinese subcontinental Thünen system, with the thought that the laws (of political struggle) that operate on one level should operate on another.

Consider the concept of *ideal distance* in the context of the world Thünen system. We have argued (Sections 8.2 and 8.3) that economic distance based on unit transport cost is the proper concept to use in Thünen models; this led us to expect certain characteristic distortions in the Thünen rings, such as elongations along major transport arteries and isolated pieces around points of access to the transportation grid. These same arguments apply to Thünen systems at any level.

At the world level, considerable differences still remain between ideal and physical (i.e., great-circle) distances, although the advent of air and automotive transportation, as well as radio communication, has tended to make ideal distances conform more closely to physical distances than used to be the case. The ideal/physical distance ratio tends to be relatively low across bodies of water and to be relatively high across mountains, deserts, and regions characterized by very cold climates, endemic diseases, and unsettled political conditions (pirates, bandits, guerrillas). Ideal distance takes a discontinuous jump across national boundaries because of tariff and customs barriers, exchange controls, travel restrictions, etc.

Let us concentrate on the implications of land vs. water travel. The

---

61. Quoted in Barbara Ward, *Spaceship Earth* (Columbia Univ. Press, New York, 1966), p. 130.

advantage of water was very great in the prerailroad nineteenth century, when the shortest route from New York to California lay around Cape Horn. As a result the oceans of the world played a role in the world Thünen system similar to that which high-speed radial highways play in a city-metropolitan system. One consequence was that the *interiors* of continents tended to be farther in ideal distance from the North Atlantic Heartland than their coastal regions. The coast of Australia was closer to London than the interior of Africa was.

We would then expect the interiors of continents to have lighter land uses than their coastal regions, by the weight-falloff condition. A rough index of land-use weight is population density. And, indeed, an estimated 2/3 of the human population live within 300 miles of the sea; 56% of the population live at altitudes below 200 meters, on just 28% of the world's land area.[62] (Other factors besides mere distance from the sea keep continental interiors relatively empty, however, e.g., dryness, rough terrain, and cold.)

Furthermore, continents are internally differentiated in terms of sea access. In particular, land along navigable rivers will be relatively close to the world nucleus in ideal terms and should carry heavier uses than less accessible land—a sort of continental "urban sprawl" functionally similar to that spreading along intercity highways.

The various potentials (1) should presumably be calculated in terms of ideal distances h rather than physical distances, since this will yield a more accurate index of "generalized accessibility." Using ideal distances would strengthen the impression that we are dealing with a single world center rather than an American–West European bipole, since the North Atlantic Ocean gap would shrink.

This completes our survey of real-world Thünen systems. The discussion has been impressionistic, intended to establish a framework around which more substantial quantitative work can be organized.

We now turn to the discussion of Thünen systems over Time rather than over Space. We describe a number of idealized historical patterns, into one or more of which most real-world Thünen systems should approximately fit.

Our basic model does have dynamic features, since the concept of "land use" allows for nonstationary import-export patterns over Time. Certain dynamic phenomena, such as land speculation and suburbanization, have already been discussed and partially explained in terms of the model. Nonetheless, many aspects of the development of real-world

---

62. G. T. Trewartha, *A Geography of Population: World Patterns* (Wiley, New York, 1969), p. 79.

Thünen systems fit into the model poorly or not at all. These include transportation construction, the origin and changing functions of the nucleus, and population movements and the consequent transfer of control over land.

Actually, many aspects of population redistribution can be incorporated into our model without undue strain. Births, deaths, aging, and other changes of state are accounted for in the *capital structure* of the various land uses in operation. (We have been concentrating almost exclusively on the export-import component of land uses in this chapter, and ignoring the capital component because the latter makes no *direct* contribution to land-use weight.) Commuting has been discussed at length. This leaves *migration*.

First, we make a point of definition. There is no sharp distinction between migration and commuting. The ultimate data we deal with here are contained in a person's *itinerary*, the function giving his location at each moment of his existence. If this function is roughly periodic over a certain interval, we speak of "commuting." If it changes more or less permanently, we speak of "migration." There are various intermediate cases.[63]

In any case, both migration and commuting are accounted for by examining a person's trips. We distinguish, as above, between local and nonlocal trips—local trips being relatively short distance and not touching the nucleus, nonlocal trips being those going to, from, or through the nucleus. The distinction depends on the system of reference. Thus a rural-to-urban migration trip would be considered nonlocal in the context of the destination city Thünen system, but local in the context of the world Thünen system. Nonlocal trips contribute to the weight of the land use at which they originate or terminate; local trips do not. But even for nonlocal trips, migration (as the term is commonly understood) probably makes just a minor contribution to weight; for it is nonrepetitive, and the cumulative impact of commuting trips and goods shipments will tend to swamp it.

The significance of migration for the evolution of Thünen systems lies, rather, in its influence on the capital structure of land uses. (This influence is the net result of migration, births, deaths, aging, etc. Construction and mining (including improvements, maintenance, scrappage, and demolitions) play a similar role in the capital structure of nonhuman resources.) When a land use enters into a more intensive phase of exporting and importing, this will in general be accompanied by a rise in resource density on the site: more people, plant, equipment, inventories per acre. We may expect net in-migration at the beginning of this phase.

---

63. For more detailed discussion of these concepts, see Faden, *Essays,* pp. 20–23.

Similarly, a phase of reduced intensity should be accompanied by net out-migration.

Consider suburbanization, for example. This was explained in terms of the tendency for lighter land uses to have longer initial periods of vacancy. By the weight-falloff condition, this implies a general tendency for land more distant from the nucleus to remain vacant longer than land close in. In-migration should occur about the time this initial idle period comes to an end. Thus we may expect a ring of intensive in-migration, the ring itself expanding away from the nucleus over time. The migrants themselves may originate from points closer to the nucleus, from points farther out, or as "immigrants" from outside the Thünen system.

The predominant direction of flow of migration provides an important principle for classifying Thünen systems. *Condensation* systems are those in which the main flow of migration is inward, toward the nucleus. *Dispersion* systems are characterized by outward migration, away from the nucleus. One rarely deals with a "pure" case; e.g., at the city level, suburbanization is mainly a dispersion process, the old urbanites moving out; but at the same time the condensation process of rural to urban migration still goes on. Furthermore, the same system might be dispersive during one epoch of its history and condensive during another. Nonetheless, the distinction is useful, and we now discuss the conditions that can be expected to yield one or the other of these processes.

A dispersion system is likely to arise when the surrounding countryside is vacant or occupied by a sparse aboriginal population living at a low technological level. The United States subcontinental Thünen system furnishes one example; this was peopled by "westward expansion," which is also "outward dispersion" from the eastern nucleus. A second example is the eastward expansion of Russia into Siberia (this may perhaps be thought of as dispersion into the eastern periphery of the western European subcontinental system). In the U.S. case an additional force making for dispersion was the great tide of migration from Europe that entered at the nucleus (so that, in a sense, this was dispersion into the *western* periphery of the western European system).[64]

A condensation system is likely to arise, however, when the surrounding countryside is occupied by a dense initial population. (Incidentally, this assumption does not invalidate the model used in this chapter. It is required only that initial conditions be *uniform* over Space, not necessarily

---

64. For descriptions see H. S. Perloff, E. S. Dunn, Jr., E. E. Lampard, and R. F. Muth, *Regions, Resources, and Economic Growth* (Johns Hopkins Press, Baltimore, 1960); R. A. Billington, *Westward Expansion,* 3rd ed. (Macmillan, New York, 1967); G. A. Lensen, ed. *Russia's Eastward Expansion* (Prentice-Hall, Englewood Cliffs, N.J., 1964). In *The Great Frontier* (Houghton Mifflin, Boston, 1952), Walter Prescott Webb treats the expansion of Europe (the Metropolis) since 1500 into the newly discovered lands (the Frontier) as the central theme of modern history. Curiously, though Russia is part of the Metropolis, Siberia is not part of the Frontier (map, p. 10).

that there be vacancy everywhere.) The major type is rural-to-urban migration. Condensation also characterizes "dual" economies, with people in the traditional sector migrating into the small modern sector.

The spatial distribution of income will in general differ in these two kinds of Thünen system. There is a general tendency for migrants to move from regions of low per capita incomes to regions of high per capita incomes. Hence a condensation system should be characterized by higher incomes toward the center, and a dispersion system by higher incomes toward the periphery. These expectations (especially the latter) are by no means certain. For migration is selective, and it is entirely possible that the migrants improve their position even if average income (including income of nonmigrants) is lower at destination than at origin. In particular, the "pioneers" who migrate in a dispersion system must do without the amenities of civilization, and tend to be drawn from the lower income strata.[65] Thus even a dispersion system might have declining income with increasing distance from the nucleus.

These two kinds of Thünen systems appear to be exemplified in two well-known economic development models. The model of W. Arthur Lewis[66] leads to a condensation system. Here labor in the traditional sector of an underdeveloped economy is available in completely elastic supply to the modern sector, which creates a ceiling on wages in the latter sector. Expansion of the modern sector pulls in surplus labor from the countryside. The model of Frederick Jackson Turner,[67] on the other hand, leads to a dispersion system. This is the mirror image of the Lewis model, for Turner postulates an unlimited supply of *land* at the frontier of settlement; this creates a *floor* ("safety valve") on wages at the center, for any tendency of wages to fall will lead to out-migration to the periphery.

Neither condensation nor dispersion requires long-distance migration. The same population redistribution can be accomplished by a series of short-distance moves (short in comparison to the radius of the Thünen system). This step-by-step movement is in fact very common. People move, leaving a vacancy into which people behind them eventually move, who in turn leave a vacancy, etc. (Example: The "filtering" process by which housing gets handed down from higher to lower income families, accompanying movement into the suburbs.) Or, instead of

65. But see I. Bowman, *The Pioneer Fringe* (American Geogr. Soc. Spec. Publ. 13, New York, 1931), on the changing character of modern pioneering.

66. W. A. Lewis, Economic development with unlimited supplies of labour, *Manchester School* 22(May 1954):139–92. For more recent work see J. C. H. Fei and G. Ranis, *Development of the Labor Surplus Economy* (Irwin, Homewood, Ill., 1964); and A. C. Kelley, J. G. Williamson, and R. J. Cheetham, *Dualistic Economic Development: Theory and History* (Univ. Chicago Press, Chicago, 1972).

67. F. J. Turner, *The Frontier in American History* (Holt, New York, 1921). For more recent work, much of it critical, see G. R. Taylor, ed., *The Turner Thesis* (Heath, Boston, 1956). Webb, *Great Frontier,* applies this thesis to the entire western world.

a series of "pulls" there may be "pushes." People move, which leads to crowding and conflict at their destination, and the driving out of people ahead of them; these in turn push out others, etc. (The European *Völkerwanderung* was partly of this type.)

We now turn to the question of origins of Thünen systems. In the condensation case there appear to be (at least) two ways in which these systems originate. One is the passing of a certain *threshold* of concentration. This is a compound of density of population, size of per capita income, and mutual accessibility. (The peak regional income potential provides a rough index of concentration.) At the threshold the market is just wide enough to make certain centrally located enterprises economically viable. These come into existence and start a snowballing process, involving the founding of linked enterprises; deepening division of labor; construction of plant, utilities, transportation arteries, and communication links; in-movement of population; innovations; etc.[68] The second way is for some autonomous localized innovation to occur, such as the discovery of a mineral deposit or the opening of trade relations with the outside world. The center itself tends to be founded at a point of high general accessibility, either by virtue of natural "nodality" (river confluence, good harbor, etc.) or because of previous transportation construction. And the very act of founding the center leads to further radial transportation construction that artificially increases the advantages of the site.

Dispersion systems typically originate through an incursion from the outside world (say first by explorers, missionaries or soldiers, then traders, then settlers). The nucleus is then likely to be near the original port of entry to the region.

Once the snowballing process begins, the nucleus exerts an attractive force on the entire region, which tends to pull land uses into the Thünen ring pattern. The nucleus acts as trade center, gateway to the outer world, source of specialized goods and services, and major employment center.

As the system matures, a number of forces arise to retard growth at the center, perhaps even to halt and reverse it. Vacant land becomes scarce; traffic congestion grows. The capital plant becomes aged and obsolescent; pollution rises with density of population. As the world develops, the original location of the center may become less advantageous; mineral deposits are played out; harbors become silted. At the same time, concentration in various parts of the hinterland reaches a threshold of its own, and new competing centers arise. (These are likely to be located near points of access to the transportation grid feeding the original center, e.g., railway stations, highway interchanges, and airports.)

---

68. See Thompson, *Preface to Urban Economics,* Ch. 1.

These processes go on simultaneously in Thünen systems of all levels. We now change our point of view and study the interrelations among these systems. The following idealized model appears to be a fair approximation to many real-world situations. Thünen systems form a *hierarchy* in the sense that for each system there exists a higher level system to which it stands in a certain subordinate relation. (The exception is the world Thünen system, which stands at the apex of the hierarchy.) Let the system occupying region S_1 with nucleus s_1 be subordinate to the system occupying region S_2 with nucleus s_2. This means, first, that S_1 is a subregion of S_2; second, trips between points of S_1 and nucleus s_2 go through nucleus s_1. That is, s_1 functions as a *gateway* between its own hinterland and the superior nucleus s_2.

This gateway function comes in several diverse forms. Nucleus s_1 may be the site of a major railway station or highway intake point, so that long-range traffic funnels through it. Out-of-region telephone calls may go through a local exchange at s_1. Wholesalers or manufacturers at s_2 may ship to retailers at s_1, who in turn sell to customers in the hinterland S_1. Conversely, s_1 may serve as an assembly point for goods produced in S_1, these goods then being shipped in bulk to s_2. Mail addressed to points in S_1 may be shipped from s_2 to a local post office at s_1, and thence distributed by letter carriers; outgoing mail follows the opposite route. Businesses and government agencies may have regional offices at s_2 and local offices at s_1; complaints, information, orders, merchandise, tax forms, etc., which flow between hinterland residents and these organizations, may then be channeled through these local offices.

Now let the superior Thünen system (S_2, s_2) itself be subordinate to the higher level system (S_3, s_3). Trips, shipments, and messages between points in the hinterland S_1 and nucleus s_3 will then go through *both* intermediate centers s_1 and s_3. The picture that emerges is that of a chain of command, in which messages from one unit to another go first up the chain to the lowest unit superior to both of them, then down the chain to the receiving unit. Approximations to this structure are found in the circulation of mail through the postal system, messages through the telephone system, and checks through the bank clearing system, as well as in the administrative structure of complex organizations.

Trips between two sites in S_1 are to be considered local when analyzing the system (S_2, s_2), because such trips are short-circuited through the nucleus s_1 and nucleus s_2 exerts no extra pull on the land uses generating them. Similarly, for the system (S_3, s_3) the much wider collection of trips between sites in S_2 are to be considered local. The same land use thus gets lighter as it is referred to higher and higher level systems.

The general conclusions of Thünen analysis (the weight-falloff condition, the declining convex structure of land values, etc.) apply at each level of the hierarchy. As one goes up the chain of systems, a broader picture emerges; but more detail is lost because more trips are relegated to the "local" category and ignored. Thus an analysis at several levels is needed to get the full picture.

What cannot be explained by Thünen analysis alone is the hierarchical structure itself. We list a few of the factors involved. There are, first, bulk economies in transportation and communication. Thus one builds a small number of channels used collectively, rather than a separate channel between every possible pair of sites. This creates artificial nodes around which centers can grow. A distributional apparatus arises for collecting, storing, assembling, shipping in bulk, and disassembling into retail lots. Second, various enterprises exhibit scale economies, which when combined with distributional costs lead to a diversity of preferred spacings for different industries. (This will be discussed further in Section 9.6.)

The preceding discussion of hierarchy touches central-place theory at many points.[69] The central-place and Thünen approaches are complementary in that the latter concentrates on the structure of land uses in the field, while the former concentrates on the activities at the various nuclei. One major theme in central-place theory is that a total ordering (allowing indifference) of both activities and centers exists such that nth level activities are present only in nth and higher level centers. We make two brief comments about the relations between these approaches. First, while the Thünen analysis is a social equilibrium based on the interactions of agents choosing most preferred actions, the central-place analysis does not rest on this foundation. Instead, it is an idealized description; and it is not at all clear, for the most part, how the patterns it postulates arise from the interactions of rational agents. Second, one of the few exceptions to this last generalization arises from the Thünen analysis itself. We have seen (p. 463) that with a limited-access transportation grid, the weight-fall-off condition leads to the "central-place" pattern of activities and centers. The centers form at the access points; the higher order centers are those closer to the nucleus in ideal distance; the higher order activities are the heavier ones. While this model alone is inadequate, it suggests that a modified Thünen framework, incorporating such hierarchy-producing assumptions as bulk economies in transportation, might prove to be an adequate theoretical underpinning for the central-place model.

---

69. The pioneering work is W. Christaller, *Central Places in Southern Germany,* trans. C. W. Baskin (Prentice-Hall, Englewood Cliffs, N.J., 1966), publ. 1933. Much of the literature is reviewed in Berry and Pred, *Central Place Studies* (the second printing has a supplement through 1964). B. J. L. Berry, *Geography of Market Centers and Retail Distribution* (Prentice-Hall, Englewood Cliffs, N.J., 1967), is a good recent work in this tradition. An interesting verbal synthesis of the ideas of Thünen and Christaller, yielding a hierarchical arrangement similar to the one described here, may be found in E. von Böventer, Towards a unified theory of spatial economic structure, *Reg. Sci. Assoc. Pap.* 10(1962): pp. 184–86.

Finally, a word about the dynamics of the Thünen hierarchy. One index of the "importance" of a particular Thünen level is the ratio of nonlocal to local trips, for this measures the degree to which the nucleus participates in the overall functioning of the system. The data for this index are generally not available, but one can often make do with a cruder index presumably correlated with the original. For example, the "importance" of the world Thünen system can be estimated by the ratio of the value of total international trade to gross world product. This ratio rose from, say, 0.03 in 1800 to 0.33 in 1913 but declined somewhat thereafter.[70] This illustrates one very long-term trend: the rise in importance of the higher level centers at the expense of the lower level. This trend has certainly not been steady, but if we examine human history in terms of millenia rather than centuries, it is palpable enough.[71] It is associated with other trends that have been much commented on, such as the deepening division of labor, the rise of national states and incipient world organizations, and the shift from agricultural to blue collar to white collar work, especially of the information-processing variety. One basic causal factor in all this is doubtless the reduction of the real cost of transportation and communication, especially for long distances, but the detailed causal interrelations are still obscure.[72]

---

70. S. Kuznets, Quantitative aspects of the economic growth of nations. X. Level and structure of foreign trade: long-term trends, *Econ. Dev. Cul. Change* 15, no. 2, Pt. II (Jan. 1967):pp. 3–7.

71. The central theme of N. S. B. Gras, *An Introduction to Economic History* (Harper and Brothers, New York, 1922), is the successive rise to dominance of larger and larger centers and associated hinterlands, i.e., of Thünen systems, from villages to world metropolises.

72. The kind of dynamic Thünen analysis we have been discussing has much in common with the literature on "growth poles" or "growth centers." See, e.g., J. Friedmann, *Regional Development Policy* (MIT Press, Cambridge, 1966); J. R. Boudeville, *Problems of Regional Economic Planning* (Edinburgh Univ. Press, Edinburgh, 1966). A very incisive analysis is to be found in A. O. Hirschman, *The Strategy of Economic Development* (Yale Univ. Press, New Haven, 1958), Ch. 10.

9

Some General Problems of Location

In this final chapter we treat a number of topics of location theory from the measure-theoretic point of view. The major omission will be Thünen theory, which was discussed extensively in Chapter 8. We begin with an attempt to set up a comprehensive framework into which these various topics can be fitted. The discussion here will be informal for the most part.

9.1. SPATIAL PREFERENCES

One convenient approach to the problems of location is by classification of the preference orderings that enter into them. In normative problems this involves classification of the objectives to be optimized. For example, how should plants be arranged to minimize the total cost of production plus distribution? Or how should residences and industry be arranged to minimize overall pollution costs? Or again, how should population be distributed to facilitate social contact and the exchange of information? Or, in the case of mutually hostile groups, how shall these be arranged to minimize friction? Production costs, pollution, social contact, and conflicts are examples of the many objectives that arise in location problems.

In positive problems involving social interaction, each of the participants will have a preference ordering, and each tries to optimize over the set of options believed available to himself. Thus a person may have objectives concerning income, career, marriage and family, recreation, and environment in which to live, and on the basis of these makes decisions concerning what to buy and sell, where to work and live, when to move, etc. These preferences may vary considerably from person to person. There are preferences for urban over rural surroundings, and conversely; for the presence or absence of children; for the presence or absence of one or another ethnic group; etc. The problem, then, is: given this spectrum of preference orderings, and given the pattern of options (as determined by resource ownership, the legal system, technology, etc.), how will these people sort themselves out, what activities will they engage in, and how will they run the enterprises they control?

Ideally, for any two possible worlds a preference ordering states whether the first is preferred, indifferent, or dispreferred to the second. (A "possible world" is given by a measure over the set of all possible histories, as discussed in Chapter 2. If uncertainty is to be incorporated, a still more complicated construct is needed—the preference ordering is over the set of all probability measures on the set of possible worlds. See Section 2.8). The problem is to find relatively simple ways to represent, at least approximately, common preferences such as those discussed above.

Our approach is to examine separately various aspects of preference (such as pollution, social contacts, consumption, or services) and develop plausible utility functions for each. Since an agent will be concerned with several of these aspects, these various utility functions must be amalgamated into one function representing overall preferences. The simplest way of doing this is just to add them up (after appropriate scaling to reflect the relative importance the agent attaches to the various aspects of preference). We are not particularly concerned with this amalgamation problem, however, but rather with the form of the special utility functions for each aspect of preference.

All the utility functions to be examined are in the form of integrals $\int_A f \, d\mu$, where μ is constructed from various physical stock or flow measures, while f is a "pricing" or "weighting" function. One example with which we have already dealt extensively is the representation of transportation cost; here A is the space of origin-destination pairs, μ is the shipment-flow measure, and f is the unit transport cost function.

Consider the representation of pollution costs, for example. Take a person at a particular point in Space-Time (s', t'). The release of resources into circulation at various times and places will impinge on (s', t'). In Section 4.4 we discussed a simple model for this physical circulation; here we are interested in the evaluation of these flows. We assume this may be approximated by the integral

$$\int_{R \times S \times T} f(r, s, t) \, \mu(dr, ds, dt) \qquad (9.1.1)$$

Here μ is a measure on $(R \times S \times T, \Sigma_r \times \Sigma_s \times \Sigma_t)$, which we may interpret as follows: $\mu(E \times F \times G)$ is the mass of resources of types E released into circulation in region F in period G, and $f(r, s, t)$ is the cost per unit mass of resource type r released at location s at instant t to the given person at (s', t'). This cost function may represent the valuation of the pollutee himself (how much he is willing to pay to avoid contact), or it may perhaps represent the valuation of a committee of health experts, etc. Function f may take on both positive and negative values— negative referring to things the person prefers to encounter (say flowers, perfume, or music).

Integral (1) refers to a single point in a person's itinerary. To find lifetime costs we must integrate (1) over T. Thus let t_o, t^o be the moment of a person's birth and death respectively, and let $g:[t_o, t^o] \to S$ be his itinerary, giving his location at each moment of his existence. Lifetime pollution is then approximated by

$$\int_{[t_o, t^o]} d\tau \int_{R \times S \times T} f(r, s, t, g(\tau), \tau) \, \mu(dr, ds, dt) \qquad (9.1.2)$$

where we have added two arguments to f that were suppressed in (1), viz., the space-time location of the pollutee. Time discounting may be built directly into f. (Integral (2) is likely to be a bad approximation if there are important interactions among pollutants or other nonlinearities. We have also sidestepped the problem of what to do when a person's values themselves change over his lifetime.)

There is a similar formulation in terms of activities rather than resources. We postulate a function $f : S \times Q \times S \times Q \to$ reals with the interpretation that $f(s_1, q_1, s_2, q_2)$ is the cost imposed upon activity q_2 located at s_2 by activity q_1 located at s_1, per unit level of q_1 per unit level of q_2. (Negative values of f correspond to "positive external effects" from (s_1, q_1) to (s_2, q_2), and vice versa.) The total cost imposed upon unit level of q_2 at s_2 is then

$$\int_{S_1 \times Q_1} f(s_1, q_1, s_2, q_2) \, \mu(ds_1, dq_1) \qquad (9.1.3)$$

Here μ is the assignment measure on $(S \times Q, \Sigma_s \times \Sigma_q)$: $\mu(E \times F)$ is the total activity of types F going on in region E. (The unit of measurement of activity levels was taken to be "acres" in Chapter 8; here we leave it arbitrary.)

So far we have considered the cost imposed upon one person or activity. From a normative point of view it is of interest to consider the *total* cost imposed upon all persons or activities together. In (3) this involves a further integration over the remaining two variables S_2 and Q_2. Total cost equals

$$\int_{S_2 \times Q_2} \mu(ds_2, dq_2) \int_{S_1 \times Q_1} f(s_1, q_1, s_2, q_2) \mu(ds_1, dq_1)$$

By Fubini's theorem, this can be written as a single integral with respect to a product measure:[1]

$$\int_{(S \times Q) \times (S \times Q)} f \ d(\mu \times \mu) \qquad (9.1.4)$$

The form of (4) is quite interesting. It is a special case of the following:

$$\int_{A \times A} f(a_1, a_2) \ [\mu(da_1) \times \mu(da_2)] \qquad (9.1.5)$$

where μ is a measure on space (A, Σ), and $f:A \times A \rightarrow$ reals is measurable with respect to $\Sigma \times \Sigma$. Integral (5) will be called a *quadratic form* in the measure μ. Indeed, if Σ is a finite σ-field, one easily verifies that (5) is a quadratic form in the ordinary algebraic sense of that term, in the n-tuple $(m_1, \ldots, m_n)$ of values that μ assumes on the partition generating Σ.

This quadratic form may be taken as a general representation of the costs associated with a measure μ when these costs are generated by the interaction of *pairs* of "entities" from some realm A. In the pollution example these entities were activities at sites, $A = S \times Q$, and the interaction was that of polluter and pollutee. But this paradigm extends to a surprisingly wide range of phenomena that have nothing to do with pollution.

For example, let μ be the spatial distribution of people at a certain point in time. To be precise, let H be the set of all people types (H is a subset of R), and let μ be the measure on $H \times S$ for which $\mu(E \times F)$ is the number of people of types E living in region F, for all measurable $E \subseteq H, F \subseteq S$. We suppose there is a function $f:H \times S \times H \times S \rightarrow$ reals, where $f(r_1, s_1, r_2, s_2)$ is the cost imposed upon a person of type r_2 located at s_2 by the presence of a person of type r_1 located at s_1. This f may take on both positive values (absence of r_1 preferred by r_2 to presence of r_1) and negative values (presence preferred to absence).

These costs may stem from the direct liking or disliking of r_1 by r_2, or they may stem from consequences that r_2 believes (perhaps erroneously) to be associated with the presence of r_1—such as loss of prestige, poorer schools, exposure to crime or disease, slovenliness, welfare burdens.

---

1. We have discussed Fubini's theorem only in connection with nonnegative integrands f. For general measurable f, write $f = f^+ - f^-$ and apply Fubini's theorem separately to f^+ and f^-. If in at least one of these two cases the integrals involved are finite, we may subtract the equalities and conclude that Fubini's theorem continues to hold for f (we ignore here a minor complication involving null sets); μ is assumed to be abcont.

Whatever the source of these preferences, the consequences for the behavior of r_2 will be the same.[2]

The total cost of these people interactions is given by

$$\int_{H \times S \times H \times S} f \, d(\mu \times \mu) \qquad (9.1.6)$$

This is formally identical to the pollution cost formula (4), with H in place of $\mathbf{Q}$. (Some religions recognize the phenomenon of "ritual pollution";[3] this can be represented in the function f of (6) and is indeed formally indistinguishable from physical pollution.)

By specifying the cost function f in various ways, we can capture a number of commonly recognized preference patterns. We first make a simplification. It is plausible to assume that locations s_1 and s_2 enter the cost function f only in terms of the distance between them. To be precise, we assume that f may be written in the form

$$f(r_1, s_1, r_2, s_2) = g(r_1, r_2, h(s_1, s_2)) \qquad (9.1.7)$$

Here $h:S \times S \to$ reals has the interpretation that $h(s_1, s_2)$ is the distance from s_1 to s_2. This can be ordinary physical distance, but more generally it is "ideal" distance, not necessarily satisfying any properties of a metric. For example, if the prevailing wind blows from s_1 to s_2, then $h(s_1, s_2)$ may be small compared to $h(s_2, s_1)$. The function $g:H \times H \times$ reals $\to$ reals appearing in (7) has the interpretation that $g(r_1, r_2, x)$ is the cost imposed on a person of type r_2 by a person of type r_1 at distance x from r_2.

The limit of $g(r_1, r_2, x)$ as $x \to \infty$ represents the "absolute" preference of r_2 for the existence or nonexistence of r_1. In general, one expects $g(r_1, r_2, x)$ to decline in absolute value as x increases, since "neighborhood effects" become attenuated with distance. But the decline may not be monotonic. Two very general motives are embodied in the variation of g with distance: "sociability" on the one hand, and dislike of crowding on the other. The latter motive is reflected in a sharp peaking of positive cost at very small distances. Sociability is a complex motive that may be further analyzed into gregariousness per se, sexual and familial motives, desire for religious communion, for exchange of goods and services, for information acquisition, etc. It is reflected in a negative cost component that increases in absolute value as distance declines. Together, these components will usually yield a function $g(r_1, r_2, x)$, which for given r_1, r_2 will attain a mini-

---

2. Cf. G. S. Becker, *The Economics of Discrimination*, 2nd ed. (Univ. Chicago Press, Chicago, 1971), p. 16. It may be important to distinguish these different sources of preference for normative, policy-making purposes.

3. See M. Douglas, *Purity and Danger: An Analysis of Concepts of Pollution and Taboo* (Praeger, New York, 1966).

mum at some positive "optimal" distance x.[4] Observers have noted characteristic national differences in these preferred distances.[5]

The foregoing discussion indicates that the objective function (6) may be a plausible representation of *distance* preferences. However, (6) is not adequate for the related but distinct component of *density* preferences. This refers to such things as preferences concerning total population or rural vs. urban living. (The latter in turn may stem from more ultimate preferences concerning the clean air, unhurried pace, and closeness to nature of rural living vs. the variety, culture, and bright lights of the city.) That (6) is not adequate to represent density preferences is clear from the observation that multiplication of μ by a constant c multiplies (6) by c^2. Thus there is no way to represent the preference for an intermediate density level over a higher or lower one.

There are various ways to remedy this defect. The simplest is to convert the homogeneous quadratic form (6) into a nonhomogeneous form by adding a term linear in μ, converting (6) into

$$\int_{H \times S \times H \times S} f \, d(\mu \times \mu) + \int_{H \times S} f' \, d\mu \qquad (9.1.8)$$

for some given function $f':H \times S \to$ reals. (Conversely, (8) indicates a general method for introducing nonlinearities into linear objective functions. In the transportation problem of Chapters 7 and 8, for example, we studiously avoided taking account of congestion phenomena. This could be done by adding a term of the form (6) (with $A \times B$ in place of $H \times S$) to the objective function, obtaining (8). The "kernel" function f should be positive definite. This converts the measure-theoretic linear program into a measure-theoretic *quadratic* program.) Density preferences vary greatly, from Daniel Boone at one extreme to Samuel Johnson on the other.[6]

We now turn to the question, How does the cost function $g(r_1, r_2, x)$ vary with r_1 and r_2? Liking or disliking can be directed toward other persons *qua* individuals (relatives, lovers, friends, enemies) or *qua* members of groups (ethnic, religious, occupational, or income groups; women; children; etc.). Both these situations can be handled formally: the first by making distinctions among people types that are sufficiently fine for the

4. To take an analogy from Schopenhauer, a group of porcupines will huddle together closely for warmth, but not *too* closely. A. Schopenhauer, Studies in Pessimism, p. 100, in *The Complete Essays of Schopenhauer,* trans. T. B. Saunders (Willey Book Co., New York, 1942).

5. For the general study of "proxemics" or "personal space" preferences, see E. T. Hall, *The Hidden Dimension* (Doubleday, Garden City, N.Y., 1966).

6. "No, Sir, when a man is tired of London, he is tired of life; for there is in London all that life can afford." *Boswell's Life of Johnson* (Oxford Univ. Press, London, 1953), p. 859.

particular individuals in question to be distinguishable from all others; the second by taking $g(r_1, r_2, x)$ constant over $r_1 \in E$, where E is, say, the set of people types comprising a certain ethnic group, for a given r_2, x, indicating that the attitude of r_2 is the same toward all members of group E. The same distinction can be made with reference to r_2, the person whose preferences are represented by $g(r_1, r_2, x)$. Attitudes common to a group E are reflected in the fact that $g(r_1, r_2, x)$ is constant over $r_2 \in E$ for given r_1, x; while attitudes idiosyncratic to some individual can be represented by making sufficiently fine distinctions.

For simplicity assume that H can be partitioned into a finite number of sets $E_1, \ldots, E_n$ such that

$$g(r_1, r_2, x) = g(r_1', r_2', x) \tag{9.1.9}$$

whenever r_1 and r_1' both belong to the same set E_i, and r_2 and r_2' both belong to the same set E_j. We may then simply speak of the attitude of group E_j toward group E_i as reflected in the cost function g in (9). (The terminology "group E_j" is convenient but slightly inaccurate; we should say "the set of people who, at time t, are of types in E_j''.)

The costs that arise from a group's association with *itself* may be positive or negative. Perhaps the most common case is where cost is negative, becoming greater in absolute value as distance shrinks (down to some point at which congestion becomes serious); i.e., there is mutual attraction: "birds of a feather flock together." But mutual repulsion also occurs. Borrowing a term from ethology, this phenomenon may be called "territoriality."[7] An example would be competitors for some economic or political niche: "this town's not big enough for both of us." (However, the tendency for competing firms to spread out spatially can arise as an indirect result of responses to local price variations. We need not invoke any direct repulsive forces to explain this phenomenon; cf. Section 9.6.)

The costs of association between different groups may again be positive or negative. One common pattern is where attraction depends on the degree of similarity of groups, "similarity" being a composite of "distances" on several socioeconomic dimensions: race, religion, language, occupation, income. Similar groups may find themselves mutually attracted (an extension of the "birds-of-a-feather" phenomenon), while repulsion is greater, the greater the "dissimilarity" or "social distance" between groups (the phenomenon of xenophobia). However, some dissimilarities are attractive, the most notable being a difference in sex. Finally, the pattern need not be symmetric: group E_i may be attractive to

---

7. C. R. Carpenter, Territoriality: A Review of Concepts and Problems, in A. Roe and G. G. Simpson, eds., *Behavior and Evolution* (Yale Univ. Press, New Haven, 1958), pp. 224–50.

E_j while E_j is repulsive to E_i; e.g., a low-status group E_j may try to live near a high-status group E_i, while E_i flees from E_j.

One important modification of the above ideas is where the continuous scale of distances x collapses into a dichotomy, which we may call "association" and "nonassociation." For example, the cost that r_1 imposes on r_2 may depend only on whether they live in the same town, not on the physical distance between them. Here association means living in the same town. One might instead have "house," "neighborhood," or "country" in place of town. There is also a nonspatial interpretation; association can mean belonging to the same *organization,* which might be a family, a firm, a union, a political party, etc. That is, r_1 imposes a cost (or benefit) on r_2 iff they both belong to the organization in question.[8]

We now turn to quite a different situation that again leads to a cost function of the form (4), the *layout* problem. Consider an industrial process consisting of n activities $q_1, \ldots, q_n$. A total (ideal) weight of w_{ij} is to flow from activity q_i to activity q_j. There are m possible sites $s_1, \ldots, s_m$ at which activities may be located. The (ideal) distance from s_k to s_l is given by h_{kl}. Thus if q_i is located at s_k and q_j at s_l, this pair generates a transportation flow that costs $w_{ij} h_{kl}$ (plus $w_{ji} h_{lk}$ in the opposite direction). In general, there may be fractions of an activity assigned to a site. If x_{ik} is the fraction of activity q_i assigned to s_k, the total transport cost incurred with this layout is

$$w_{ij} h_{kl} x_{ik} x_{jl} \qquad (9.1.10)$$

quadruply summed over $i, j = 1, \ldots, n$ and $k, l = 1, \ldots, m$. A typical problem is then to minimize (10) (perhaps augmented by terms linear in x), subject to all activities being allocated, to capacity constraints at each site, and perhaps to x_{ik} constrained to equal 0 or 1.[9]

The restriction to a finite number of activities and sites is somewhat artificial. The measure-theoretic generalization is similar to the allotment-assignment problem. One is to find an assignment measure μ on $(S \times Q, \Sigma_s \times \Sigma_q)$ to minimize

$$\int_{(S_1 \times Q_1) \times (S_2 \times Q_2)} w(q_1, q_2) h(s_1, s_2) [\mu(ds_1, dq_1) \times \mu(ds_2, dq_2)] \qquad (9.1.11)$$

Here S_1 and S_2 are replicas of S, Q_1 and Q_2 are replicas of Q, $w(q_1, q_2)$ is the mass that is to flow from activity q_1 to q_2, per unit level of each;

8. The well-known Bogardus social distance scale is based on the degree of association to which one would admit someone: marriage, residential proximity, citizenship, etc. E. S. Bogardus, Measuring social distance, *J. Appl. Sociol.* 9(1925):299–308.

9. T. C. Koopmans and M. J. Beckmann, Assignment problems and the location of economic activities, *Econometrica* 25(Jan. 1957):53–76. Much literature has appeared since this early article. For recent work see R. L. Francis and J. A. White, *Facility Layout and Location: An Analytical Approach* (Prentice-Hall, Englewood Cliffs, N.J., 1974), esp. Ch. 8.

$h(s_1, s_2)$ is the distance from s_1 to s_2. The constraints include the specification of μ'', the right marginal of μ (this is the allotment that defines the industrial process in question); and also areal capacity constraints, perhaps integer-value constraints, etc. When Σ_s and Σ_q are both finite, (11) reduces to (10). Clearly (11) is a special case of the quadratic form (4), in which the integrand f has the factored form $w \cdot h$.

Consider the special case of (11) in which $h(s_1, s_2) = 1$ if $s_1 = s_2$, and $= 0$ if $s_1 \neq s_2$. This is the formal representation of the "association" vs. "nonassociation" dichotomy mentioned above. We spell this out, reinterpreting $w(q_1, q_2)$ as an interaction cost between activities q_1 and q_2, which comes into effect iff the activities are associated—i.e., in this case, colocated at the same site. In the Reiter-Sherman model a term linear in μ is tacked on to the quadratic, and plants must be assigned to sites ("cities") in all-or-none fashion, so that we have a quadratic integer program.[10] Without the linear term it does not matter to which particular cities the various plants are assigned, but only how they are associated with each other. We have in this case a problem of optimal *partitioning*, which is conveniently reformulated in abstract terms as follows. Given measure space (A, Σ, μ) and measurable function $f: A \times A \to$ reals, choose a measurable partition $\{A_1, \ldots, A_n\}$ of A to minimize

$$\int_{A_1 \times A_1} f \, d(\mu \times \mu) + \cdots + \int_{A_n \times A_n} f \, d(\mu \times \mu)$$

For example, A may be the space of activities $\mathbf{Q}$ and f the function w as above; μ is the given allotment measure (written μ'' above); A_1 is the set of activities colocated at site s_1, etc. Or A may be physical Space, and the problem may be one of optimal partitioning into administrative or electoral districts. Or A may be some realm of objects such as biological specimens that have to be classified.[11]

We now take a different (and somewhat more traditional) point of view. Think of the general location problem in terms of *facilities* and of *transportation flows* among them. Facilities come in many forms: factories, stores, warehouses, residences, places of entertainment, schools, churches, hospitals, police stations, etc. It is sometimes useful to think of

10. S. Reiter and G. R. Sherman, Allocating indivisible resources affording external economies or diseconomies, *Int. Econ. Rev.* 3(Jan. 1962):108–35; also S. Reiter, Choosing an investment program among interdependent projects, *Rev. Econ. Stud.* 30(Feb. 1963):32–36.

11. This is just one approach to the classification problem, which has an enormous literature stretching over most fields of human knowledge. As examples we may note R. R. Sokal and P. H. A. Sneath, *Principles of Numerical Taxonomy* (Freeman, San Francisco, 1963), Ch. 7; B. J. L. Berry, Grouping and Regionalizing, in *Quantitative Geography*, Pt. I, pp. 219–51, W. L. Garrison and D. F. Marble, eds. (Northwestern Univ. Stud. Geogr. 13, Evanston, 1967); W. D. Fisher, *Clustering and Aggregation in Economics* (Johns Hopkins Press, Baltimore, 1968).

neighborhoods or even entire cities as single facilities. Facilities are places at which various resources congregate to participate in activities. As a rule, a facility will have a more or less permanent capital plant of structures, equipment, and inventories specialized for the particular activities carried on there; but in limiting cases the facility provides little more than a meeting place for people to assemble and interact. Thus auditoriums or even street corners or open fields would be facilities from this point of view.

In the course of participating in activities at facilities, resources are created, destroyed, and transformed (as indicated by their itineraries and transmutation paths). Transportation effects a rearrangement of the resource "bundles" at the various facilities into other bundles at other facilities. The itinerary of any one person takes the form of a sequence of alternative "flights" and "perchings." The latter represent sojourns at various facilities for various time intervals; the former represent intervals of transit from one facility to another. It is convenient to distinguish *routine* from *nonroutine* patterns of movement in a given itinerary. The simplest case of routine movement is that of a *periodic* itinerary; i.e., there is a number $x > 0$ such that $g(t) = g(t + x)$ for all t in a certain time interval. Here $g(t)$ is one's location at instant t, and x is, say, 24 hours. More generally, routine can be interpreted in a statistical sense, characterized by the average frequency with which one visits a given site, the average fraction of one's time spent at that site, etc., provided these averages settle down approximately to constants over some long time interval.

Routine movements may be called *commuting*, using that term to refer not only to the shuttling between home and work but to all trips between places visited routinely. Nonroutine movements may be called *migration*, and we may distinguish *partial* migration, in which one changes some but not all of the places one visits routinely (e.g., changing place of employment while keeping the same residence), from *full* migration, in which one changes to a disjoint set of sites visited routinely (e.g., long-distance movement involving a complete turnover of one's places of residence, work, recreation, etc.).[12]

What are the costs (or benefits) arising in this "facilities" point of view? These fall into two broad categories, those associated with the transport flows between facilities and those associated with the facilities themselves. Transport costs have already been discussed and are much simpler in structure than facility costs (a reflection of the simplicity of structure of physical Space S or Space-Time $S \times T$, compared with the space of activities Q). We turn to the latter.

---

12. For further discussion of these concepts, see A. M. Faden, *Essays in Spatial Economics,* Ph.D. diss., Columbia Univ., 1967, pp 20–23. Chapter 1 of this work is devoted to the study of commuting, as defined above.

Consider the preferences of a single agent over different commuting patterns. (We do not discuss migration as defined above.) Aspects of preference will include the frequency of visitation at various sites, the length of sojourn, the activities at these sites, and the agent's role in them.[13] These preferences in turn reflect a blend of underlying motives, among which may be listed: changes in the agent's wealth resulting from participation in activities, including monetary gains and losses (e.g., admission charges, service fees, wages), and changes in the value of his property, including the "human capital" embodied in the agent himself; association with other participating agents; the enjoyment or suffering of the agent when participating in these activities.

The agent will, for several reasons, be concerned about what goes on in a wider class of facilities than those he actually visits: (i) What goes on elsewhere affects the value of the *option* he has to visit them.[14] (ii) He may have financial interests in other facilities, or their operation may indirectly affect his wealth (e.g., by influencing prices). (iii) He may have general values concerning what types of activities should or should not exist. (This last aspect of preference might be represented by an integral such as (3).)

We now compare this "facilities" approach with the "pollution" approach discussed above. To illustrate the essential differences, consider an industry consisting of a number of identical plants scattered over the landscape, each charging an identical mill price for its product. Suppose a given agent is affected by this industry in two ways: he buys its product and suffers its pollution. The cost of the latter is obtained by a weighted *summation* over the plants. The cost of the former, however, depends only on distance to the *nearest* plant. (If price varied from plant to plant, cost would still depend on the single most advantageously located plant.) There is an extra degree of freedom for the agent in a facility model in that he may choose to interact separately with the various parts of his environment, whereas interaction is fixed by relative locations in pollution models. (In more elaborate pollution models one can introduce options of creating *barriers* to various kinds of interaction, options that are in a certain sense dual to the facility options of choosing transport flows.)

---

13. We mention just one application of an analysis of this sort. Assuming that facilities benefit is given by $B(x_1, \ldots, x_n)$, where x_i is visitation frequency to facility i and B is a function subject to rather mild conditions, and assuming that the agent chooses $x_1, \ldots, x_n$ to maximize B minus transport cost, we can deduce a form of the "gravity" law of spatial interaction. See Faden, *Essays,* pp. 42–50. A different approach to this mysterious empirical regularity is found in J. H. Niedercorn and B. V. Bechdolt, Jr., An economic derivation of the 'gravity law' of spatial interaction, *J. Reg. Sci.* 9(Aug. 1969):273–82.

14. This point is stressed in various writings of A. G. Hart, T. C. Koopmans, and B. A. Weisbrod.

9.2. SOME INFORMAL MODELS

The rest of this chapter is devoted to spelling out the implications of some of these preference patterns under suitable constraints. In the following sections we develop a number of formal models. Most of these are facility models, and indeed transportation costs enter as an essential ingredient in all. To balance the picture, this section will present an informal nonrigorous discussion of some pollution type models. We have divided the subject this way because we found the facility models much easier to formalize than the pollution models. Presumably one reason is that the facility models build on a long tradition of theoretical work in location theory (W. Launhardt, A. Weber, A. Lösch, and others), while the "pollution" models lack this background.

Consider first the results stemming from attractions and repulsions among people as represented in a cost function of type $g(r_1, r_2, x)$ (cf. (1.7)). We assume that attraction or repulsion intensifies as distance decreases (formally, $|g(r_1, r_2, x)|$ is a decreasing function of x for given r_1, r_2) except for very small distances at which congestion effects may overcome attraction. With this assumption, people try to reduce their distance from other distant attractive people and to increase their distance from other repulsive people.

The distance between oneself and others can be affected either by changing one's own location or by inducing others to change theirs (or inducing them to remain where they are). Actions of both types are important. We first examine the effects of people changing their own locations.

If all people were mutually attracted and no other aspects of preference existed, they would pile up in a gigantic ant hill. Taking account of the inevitable congestion effects, equilibrium would occur at some intermediate spacing, where the congestion gradient just balances the attraction gradient (cf. Schopenhauer's porcupines, footnote 4). This may be taken as a very crude model of "big-city" life. (Recall that attraction among people stems not merely from sociability but also from the economic advantages that result from proximity; the latter are probably more important in the explanation of the growth of large cities.) Conversely, if all people were mutually repelled and again no other forces were in operation, they would tend to spread uniformly over the available surface.

Now suppose the birds-of-a-feather pattern prevails. Specifically, suppose people can be partitioned into N types such that there is mutual attraction within each type that is stronger than any attraction between types. Then the people of each type would indeed tend to flock together into N communities or neighborhoods. If there is also mutual attraction between types, these communities will tend to settle near each other to form a "supercommunity." If there is mutual repulsion, the N communities will isolate themselves. If there are both attractions and repulsions between types, we may get several isolated supercommunities. More complicated patterns can arise if attraction is not transitive. Suppose groups A and B and also B and C are mutually attracted, while A and C are repelled. An equilibrium might involve A and C settling on opposite sides of B.[15]

How do these preference patterns work themselves out in the context of areal capacity limits, leading to land scarcity and a real-estate market? Start with an initial nonequilibrium distribution of population. Suppose there is a disc having a relatively high initial concentration of group E, which exhibits high mutual birds-of-a-feather attraction. Members of group E will then be willing to pay more for locations near this disc than nonmembers because the proximity of their compatriots makes these locations desirable for members but not nonmembers. The initial concentration will then tend to snowball as more group E members outbid nonmembers and move in. The other groups are acting in a similar manner, so that a process of mutual segregation occurs.

This account is oversimplified in several respects. If there are several initial concentrations of group E, the larger concentrations may gain at the expense of the smaller, eventually with all group E members ending up near one of the original concentrations, the others being abandoned. If there are positive migration costs that must be balanced against gains of relocation, the segregation process may not go to completion.

Consider next the internal structure of one of these segregated neighborhoods. The *centrality-peripherality* index of a location s_o is given by

$$\int_S g(h(s, s_o))\, \mu(ds) \tag{9.2.1}$$

Here μ gives the spatial distribution of the group in question: $\mu(F)$ is the number of group members in region F; $h(s, s_o)$ is the distance from location s to s_o, and $g(x)$ is the benefit to a group member from another member at distance x. (Function g in (1) is the same as g in (1.7) with the first two arguments set equal to the type of the group in question and with sign reversed to make attraction read positive. Note that Stewart's potential is a special case of (1), with $g(x) = 1/x$. See (8.9.1).)

Integral (1) is an index of "generalized closeness" to the group. Points of higher centrality are more desirable for group members, so that land values will tend to be higher and densities greater than at more

---

15. A more comprehensive theory might include the tendency for such "unbalanced" or "dissonant" taste patterns to disappear. Cf. F. Heider, *The Psychology of Interpersonal Relations* (Wiley, New York, 1958); L. Festinger, *A Theory of Cognitive Dissonance* (Stanford Univ. Press, Stanford, Calif., 1957). This idea goes back at least to Sumner's "strain toward consistency in the mores." William Graham Sumner, *Folkways* (Ginn, Boston, 1906), Ch.1, Sects. 5, 45.

peripheral points. The whole pattern is not unlike that of a Thünen system, although the latter arises from a different structure of preferences and constraints. Peripheral land values may be depressed not only because these locations are "further" from the group in general but also because they tend to be "closer" to adjacent neighborhoods of different, perhaps repulsive, groups.[16]

Finally, suppose there are variations within the group in the intensity of birds-of-a-feather preferences. Let s_1 be a central location and s_2 a peripheral one, and let r_1 be a person willing to pay a greater differential than person r_2 for the privilege of living at s_1 rather than s_2. Then, in general, r_1 will outbid r_2 for s_1, and vice versa for s_2. The upshot is that the more central locations tend to be occupied by people with more intense group proximity preferences (where intensity is measured by monetary differential bids). The argument here is essentially the same as that in the competitive Thünen system.

We now look more closely at relations *between* groups. Suppose there are two groups M and F, each M member being attracted by F members and vice versa. The obvious example is males and females, but the same situation occurs more generally between resource types that are *complementary* in the sense that they customarily participate together in desired activities: pens and ink, needles and thread, locks and keys, cans and can openers, left and right shoes, even capital and labor in general. (In the case of inanimate resources, the attraction is of course a reflection of the desires of the people who control their movements. Note that the male-female pair is itself a special case of complementary resource types.) The tendency here will be for the spatial distributions of M and F to become identical; i.e., the proportion of all M to be found in any region will be about the same as the proportion of all F in that region. This same process tends to concentrate the total mass.

Mutual repulsion between groups causes them to separate, unless other forces are present leading them to maintain "antagonistic proximity." There is an interesting middle case, where A is attracted by B but B is repelled by A. This "approach-avoidance" situation is not uncommon. One case already mentioned is when B is a high-prestige group and A a low-prestige group—A tries to get close to B so that the prestige "rubs off" on A; B tries to avoid A so that its prestige is not "diluted" by A. This process commonly occurs in Resource space as well as in physical Space. That is, A not only "imitates" B's location, but imitates B in other traits as well: in clothing, life-style, model of speech. Then B may avoid A by "moving" to other less easily imitated states, e.g., by consuming conspicuously at a level beyond A's means if the prestige differential is based

---

16. For more on the effects of ethnic borderlines on land values see M. J. Bailey, Note on the economics of residential zoning and urban renewal, *Land Econ.* 35(Aug. 1959):288–92.

on wealth. If the imitation is successful, the prestige base does indeed become diluted, as when colleges become diploma mills or army medals become as easy to get as Boy Scout badges. Another case arises when A has a parasitic or predatory relation to B. For example, A may be a group of criminals who prey upon a group of victims B.

In an approach-avoidance situation, if neither group has any control over the movements of the other, equilibrium occurs when both groups are about equally spaced throughout the available region. For if, say, B has a higher than average concentration in some disc, members of A outside the disc will rush into it, causing the members of B in the disc to rush out; a similar argument applies to *lower* than average concentrations of A.

It is interesting to compare this with the Criminal-Victim model of Section 5.8. There the approach-avoidance preferences are expressed in terms of *densities* rather than distances, and the conclusion is that both criminals and victims will be distributed at uniform density in equilibrium (a conclusion similar in tone, if not in substance, to the one based on distances). The argument for this conclusion is essentially the one just given, dressed up in formal language to meet the requirements of rigor.[17]

An amusing case related to this approach-avoidance scheme arises from different cultural definitions of proper distance in, say, conversation. Of two people, the one preferring the smaller distance will keep advancing, and the one preferring the large distance will keep retreating as they converse. Similar anomalies arise in preferred seating distances, furniture arrangement, etc. These situations are grist for the mill of "proxemics" (footnote 5). Many approach-avoidance situations arise at this individual level; one example is unrequited love.

Let us now turn to actions that influence the *other* fellow's location rather than one's own. Preferences are still expressed by $g(r_1, r_2, x)$, the cost that a person of type r_1 imposes on a person of type r_2 when the distance from r_1 to r_2 is x. In the following analysis, r_2 influences x by inducing r_1 to move (or refrain from moving). The possible cases may be classified by three different dichotomies: (i) whether r_1 is attractive or repulsive to r_2, (ii) whether the action is to induce r_1 to move to a

---

17. The reader may have trouble distinguishing a preference for "low density" from a preference for "wide spacing," say, or distinguishing "uniform density" from "equal spacing." These concepts are, in fact, completely different, the first arising from an areal measure and applying to absolutely continuous measures, the second arising from a metric and applying to integer-valued measures. The confusion arises from the special connection between Euclidean area and Euclidean metric: with "ideal" areas of "ideal" distances the distinction is clear. There is a general method for constructing measures from metrics that is not discussed in this book and may prove important in future applications. The derivation of Euclidean area (= Lebesgue 2-dimensional measure) from Euclidean distance is a special case. See C. A. Rogers, *Hausdorff Measures* (Cambridge Univ. Press, Cambridge, 1970), pp. 50–53.

better location (nearer r_2 in the case of attraction, further in the case of repulsion) or not to move to a worse location, and (iii) whether the inducement to r_1 is voluntary or involuntary. (Here "voluntary" is taken to mean that r_1 is offered a reward for compliance with the wishes of r_2; "involuntary" is taken to mean that r_1 is threatened with punishment for noncompliance or is physically compelled to comply. This dichotomy is not really strict, since the carrot and the stick can be combined in a single offer.) By combining these dichotomies we get $2^3 = 8$ possible cases. We shall illustrate them.

We first consider voluntary inducements. Actions that induce attractive people to move closer to oneself are very common. Much advertising falls in this category: to attract customers, to recruit workers, to entice industries into a community or immigrants into a country. Next consider actions that induce attractive people to stay when they are tending to leave. Big-city establishments have a variety of programs to stem the tide of suburbanization of their middle-class residents and consequent erosion of their tax base and political and economic support, urban renewal being the most important.[18] Examples of actions that induce repulsive people to move away voluntarily are harder to find. The action of a duopolist who pays off his rival to get out of town would fall in this category. An example of an action to keep repulsive people at bay would perhaps be an agreement among oligopolists not to poach in each other's territory. Possibly some support for poverty and civil rights programs stems from the feeling that they will stave off future incursions into one's own neighborhood. If so, this support falls in the present category.

We now turn to involuntary inducements. Actions that compel attractive people to move closer to oneself are uncommon. The operation of the fugitive slave law is an illustration in that it returned escapees to an exploitative relation with their owners, an association the latter found attractive. An example of actions that keep attractive people from moving away are exit barriers imposed on potential emigrants by certain authoritarian countries—either on everyone or on persons of special value to the state (e.g., skilled workers).

Now consider actions that compel repulsive people to move away from oneself—deportation, exile, and expulsion of people from a territory, the latter typically in connection with nation formation and expansion. ("Nationalism" can be analyzed into a complex of motives of the sort we have been considering: birds-of-a-feather attraction toward compatriots and xenophobic revulsion toward others. The new ingredient is attachment to one's "sacred territory," with the irredentist demand for

---

18. For discussion of these programs see E. P. Wolf and C. N. Lebeaux, Class and Race in the Changing City, in L. F. Schnore and H. Fagin, eds., *Urban Research and Policy Planning* (Sage Publ., Beverly Hills, Calif., 1967), pp. 99–129.

exclusive control and perhaps occupation by compatriots.) At the city level such programs as urban renewal and highway construction often have the indirect effect of making the city less attractive to the poor and racial minorities, so that these programs fall in the present category from the point of view of city dwellers who find these groups repulsive. Imprisonment also falls in this category.

Finally, consider actions that compel repulsive people to keep their distance. This is a rich category. The institution of private property itself (in particular, private property in land) enables people to keep others at bay by prohibiting trespass. Then there are entry barriers, which may take the form of fences, locks, and other physical appurtenances but also may take institutional forms such as immigration restrictions, exclusive clubs, difficult licensing requirements, etc. An interlocking system of institutional actions serves to keep the poor, racial minorities, and other "undesirables" out of certain neighborhoods, e.g., large-lot zoning, legal harassment, unavailability of mortgages, restrictive covenants, and finally violence.[19]

The following comment will round out this analysis. All the actions considered so far refer to the efforts of r_2 to reduce the cost $g(r_1, r_2, x)$ by changing distance x. It may also be possible to change r_1, i.e., to alter the state of other people so that they become more attractive (or less repulsive) to oneself. Consider the poverty program. From the point of view of an affluent supporter, it may be construed as an action to change the state of his less fortunate neighbors in a direction that reduces the cost to him of crime, welfare payments, bad conscience, etc. An extreme instance of change of state is to nonexistence: in plain language, r_2 kills r_1. (Execution is a common alternative to exile or imprisonment.) Person r_2 may also wish to change his *own* state to one that finds other people and the world in general more pleasant (e.g., by religious practices or drugs).

The erection of barriers involved in several of the action categories discussed above has an interesting double effect. It tends to prevent movement of people to locations less desired by the agent, and also it alters the ideal distance function itself. Without moving, two people on opposite sides of a new barrier are now farther apart in ideal terms. Since the x in $g(r_1, r_2, x)$ refers to ideal distance, the level of preference of r_2 can be affected over and above any constraint on the movements of r_1. It may be, for example, that "good fences make good neighbors," not because of the physical barriers they interpose, but because of the sense of distance they convey. Any number of illustrations may be found in the little symbolic acts of everyday life: labeling possessions, putting up "do not disturb" signs, and other steps to insure privacy. Similarly, the improve-

---

19. M. Grodzins, *The Metropolitan Area as a Racial Problem* (Univ. Pittsburgh Press, Pittsburgh, 1958). U.S. Commission on Civil Rights, *Report: Housing* (USGPO, Washington, D.C., 1961).

ment of transportation links has the double effect of facilitating movement and reducing the sense of distance. Destruction of barriers or transport links have effects opposite to their erection, of course.

Consider next the effects of a differential distribution of spacing preferences in the population, say the most preferred distance to one's nearest neighbor. If these are the only aspects of preference in operation and enough space is available, then people will sort themselves out—with people of similar spacing preferences settling relatively near each other, spacing in each separate community being the common most preferred spacing of the people in that community. Thus "urbanites" will settle together with relatively small distances between neighbors, and "ruralites" will settle together with relatively large spacing between neighbors. Note that, although people of similar preferences settle together, this does not stem from any birds-of-a-feather preference pattern. Indeed, people are assumed to be indifferent to the preferences per se of their neighbors.

A similar result obtains if preferences are in terms of most preferred *densities* rather than spacing. People of similar density preferences will tend to settle together at the common density most preferred by them. In Euclidean space, density preferences are not very different from spacing preferences, but in more general spaces the two are independent (footnote 17).

An interesting variant of these patterns is where each person has a preference ordering over the size of the community he lives in. People again sort themselves out, but now in terms of size clusters rather than spacing or density. If the number of people most preferring to live in a community of size N is an exact multiple of N for each $N = 1, 2, \ldots$, all preferences can be optimized simultaneously.

We now turn to a class of models that again lead to the sorting out of people of similar preferences, although they are quite different in structure from those discussed above. The following are facility models, and the sorting out occurs by people migrating to the vicinity of facilities they wish to use heavily.

We begin with the well-known Tiebout model.[20] Here there are a large number of communities, each run by a political entrepreneur, say a city manager. Each of these offers a fixed bundle of public services with a scheme for financing them. People are fully mobile and knowledgeable and migrate to the community whose services-financing bundle best suits their tastes. People with a strong taste for public services will move to communities in which the level of services (and the accompanying finan-

---

20. C. M. Tiebout, A pure theory of local expenditures, *J. Polit. Econ.* 64(Oct. 1956): 416–24. For recent work see E. S. Mills and W. E. Oates, eds., *Fiscal Zoning and Land Use Controls* (Heath, Lexington, Mass., 1975). These essays stress control of mobility via zoning, local taxes, and subsidies and are therefore relevant to the topic discussed on pp. 543–46.

cial burden) is high, and vice versa for those with a weak taste for such services. Families with many children will tend to migrate to communities with a strong educational program. People with no children will tend to avoid such communities to escape the tax burden. The poor will tend to move to communities with generous welfare programs; the affluent will tend to avoid them.

Suppose community A is more efficient than community B in the sense that A provides a larger bundle of services for the same financial burden or the same bundle of services for a lesser burden. Then people will tend to migrate from B to A. In this way efficiency is "rewarded" by attaining a larger constituency. (The inefficiency of B may stem from several sources: managerial incompetence, corruption, poorer city layout.)

We make several comments to place this model in a more general perspective. The same logic applies to "private" as well as to "public" services. We need not discuss here the vexing question of how best to define the concept of public service. Whether we take this to mean a service provided via the political process, one that provides joint benefits for technological reasons, one from which people cannot be "excluded," or one that enters simultaneously into several people's preference orders (these concepts are neither exclusive nor exhaustive), the same logic applies. The only exception is the limiting case of a service whose benefit requires no access (i.e., benefit is independent of one's location), for such a service provides no incentive to migrate. Few if any services approach this limiting case.[21] All other cases impel people to move toward the source of the service and thus promote the sorting-out process discussed above.

Consider now a "private" (or at worst "semipublic") service, religion. Suppose various denominations build churches at various places. People will then tend to move toward a church of their choice. Thus each church will tend to be surrounded by its particular patrons, just as each "city hall" of the Tiebout model becomes surrounded by clients for the particular bundle of services it provides. (Agnostics will move away from all churches, as the land around them is bid up in value.) Similarly, the opening of various places of entertainment will sort out the various clienteles by taste—say highbrows to the opera, lowbrows to the bowling alley. A distribution of stores—some selling luxuries, some necessities— tends to sort people out by income level. A distribution of stores selling "soul" products tends to sort out people by ethnic group.

Nature itself provides a basis for this sorting-out process by furnishing a diversity of environments. Insofar as some people prefer hilly country and some flat, some prefer the seashore and some the interior, some

---

21. On this point see J. Margolis, A comment on the pure theory of public expenditure, *Rev. Econ. Stat.* 37(Nov. 1955):347–49.

prefer sharp seasonal contrasts and other uniformity, the same kind of selective migration will occur. Sometimes man inadvertently accomplishes the same thing. Thus, since pollution tends to be more severe in urban areas, we would expect that rural-to-urban migration is selective of those who are relatively less averse to pollution.[22]

A second point in connection with these models is that there is generally a feedback effect. The distribution of people affects the distribution of facilities as well as vice versa. Consider the Tiebout model, where we now drop the assumption that the services-financing bundle for any community is fixed. Instead, we assume more realistically that this bundle responds to the preferences of the residents of the community. We now claim that this feedback process leads in the direction of greater *differentiation* among communities in the bundles they offer. Suppose two cities A and B start out with fairly similar programs—say A has a slightly stronger educational program, with taxes a bit higher, than B. Thus there is a weak tendency for families with many children to move to A from B, and families with no children from A to B. But this movement changes the residency composition at A in a direction more favorable to educational expenditure, at B in a direction less favorable. Voter influence will then tend to widen the initial difference between the two programs. This in turn strengthens the tendency to selective migration, so we have a snowball effect ("positive feedback").[23]

The same differentiating effect arises from feedback in the case of private facilities. When coreligionists settle in a relatively compact region, this encourages more and bigger churches of that denomination to be built there, which in turn encourages further in-migration of parishioners (and out-migration of others as land values rise). The ingathering of an income, occupational, or ethnic group encourages the founding of facilities catering to the special tastes of that group, which in turn encourages further in-migration and sorting out.

This differentiation process arises in nonspatial contexts as well. A magazine with a good sports section may acquire a clientele especially interested in sports, which encourages it to expand this section, attracting still more buyers with this taste; it may end up specializing completely in

---

22. According to J. H. Dales, *Pollution Property and Prices* (Univ. Toronto Press, Toronto, 1968), p. 25.

23. We could easily write a system of differential equations to represent this process. A similar phenomenon arises in the "pollution" models discussed above where, for example, if the black population in a neighborhood comes to exceed a certain critical percentage, a cumulative process "tips" the neighborhood toward complete segregation. See T. C. Schelling, The Process of Residential Segregation: Neighborhood Tipping, in A. H. Pascal, ed., *Racial Discrimination in Economic Life* (Heath, Lexington, Mass., 1972) pp. 157–84. E. P. Wolf, The tipping-point in racially changing neighborhoods, *J. Am. Inst. Planners* 29(Aug. 1963):217–22, is skeptical concerning the realism of such models.

sports. A religious sect with a particular slant on doctrine or ritual may acquire adherents pushing it still farther in this direction; a political party with a mild ideological slant may attract zealots pushing it to a more extreme position.[24] Other forces pull in the opposite direction: the Hotelling argument just footnoted, mass advertising, the setting of national standards. Nonetheless, powerful tendencies militate against the uniformization of culture.[25]

Another consequence of this feedback phenomenon is the clustering of facilities catering to a common clientele. Suppose facility A attracts patrons of type B, leading them to concentrate near A. This concentration may induce facility A', which is also heavily patronized by people of type B, to be built near A. The proximity of A and A' does not arise from any direct economic linkage between them, but from the fact that while both are striving to be close to B, they necessarily get close to each other. The same phenomenon of *indirect linkage* can arise via several intermediate direct linkages.

One example of this is the clustering of facilities by the typical income level of their clients. Second-hand stores, pawnbrokers, "numbers" runners, brothels, and low-quality housing show similar locational patterns, since these are facilities catering to low-income clients. Similarly, trades of a particular type will often cluster. A shoe store may find it advantageous to open near other shoe stores because these other stores provide the "external economy" of attracting a pool of potential shoe buyers.

The contrast between these facility models and the pollution-attraction models discussed above should be stressed. Similar people congregate in facilities models, not because they are attracted to each other, but because they are attracted to a common group of facilities. Suppose that of three people A, B, and C the first two have a strong taste for government services, while C prefers low taxes; at the same time B and C are mutual friends, while between A and B and between A and C there is indifference. The facility models stress the tendency for A and B to settle

---

24. This line of argument should be compared with that of Harold Hotelling, who argues that competition for adherents should drive parties, sects, firms, etc., to become more, not less, similar to each other. H. Hotelling, Stability in competition, *Econ. J.* 39(Mar. 1929):41–57. The issues involved, embracing the whole theory of the division of labor, remain one of the great unexplored areas of economics, despite the auspicious start of Adam Smith.

25. J. Keats, *The Insolent Chariots* (Lippincott, Philadelphia, 1958), Ch. 7, claims that mobility induced by the automobile leads to just such a dead uniformity. Glenn and Simmons found that differences in attitudes among the four U.S. census regions are, if anything, widening over time, as judged by age differences in questionnaire responses. They attribute this to differentials in urbanization and its concomitants, however, rather than to differential migration. N. D. Glenn and J. L. Simmons, Are regional cultural differences diminishing? *Public Opin. Quart.* 31(Summer 1967):176–93.

near each other, while the preceding models stress the tendency of B and C to settle near each other.[26]

These observations raise some doubts as to the adequacy of the critique of Tiebout's model by Samuelson.[27] In arguing that Tiebout's model does not produce an arrangement that is optimal in any ethical sense, Samuelson adduces an analogous "free market" in marriages, in which the more desirable boys pair off with the more desirable girls.[28] But this is a mutual-attraction model and the analogy is false; according to Tiebout people are attracted to facilities, not to each other. The facilities themselves, unlike spouses, have no satisfaction levels.

We now change our perspective and see how the ideas of this section might be harnessed to explain a real-world phenomenon: the formation of *urban neighborhoods*, with particular reference to differentiation by income. The discussion here is "synthetic" in that we adduce whatever motives seem plausible, in contrast to the preceding "analytic" approach in which one examines some simple motive in isolation and sees where it leads. Consequently, the next few paragraphs are even more impressionistic than the preceding ones.

We start from the Thünen framework of Chapter 8. Here the desire to get close to the center of town to save on transport costs leads to a competition for space and increasing land costs as one approaches the center. This leads to the formation of "rings" of land uses concentric with the center of town. People with similar tastes and incomes tend to locate at about the same (ideal) distance x from the center. If the metric is Euclidean, each "ring" will be a geometrical annulus, the region between two concentric circles.

Next suppose that facilities catering to this taste-income stratum will be attracted to its vicinity and so settle at about distance x from the center. (A more refined argument would note that Thünen forces operate on facility locations also, so that the characteristic distance might be rather different from x.) If there are many identical facilities of a certain type, no problem arises: they space themselves more or less uniformly

---

26. In his article Tiebout makes a passing reference to the fact that the consumer-voter desires association with "nice" people, and thereby introduces a note foreign to facilities models. Tiebout, Pure theory of local expenditures, p. 418, n. 12. This merely underlines the fact that his major argument places no reliance whatsoever on motives of this sort.

27. P. A. Samuelson, Aspects of public expenditure theories, *Rev. Econ. Statist.* 40(Nov. 1958):332–38. The critique occurs on pp. 337–38.

28. A truly free marriage market with negotiable dowries (possibly negative) would yield this outcome only if the "income" produced by spouse qualities has nonnegative cross-differences. See pp. 476–77. Becker notes the tendency for quality-quality association in marriage to be strengthened by institutional barriers to negotiability. G. S. Becker, A theory of marriage, in T. W. Schultz, ed., *Economics of the Family* (Univ. Chicago Press, Chicago, 1974), p. 321.

around the annulus. But if cost and demand are such that there is "room" for just one or two such facilities, the Thünen ring structure cannot be an equilibrium. People will tend to crowd round the facilities, breaking up the annulus into a few blobs of more compact shape. Feedback from this movement to the location of facilities accentuates the tendency for the ring structure to disintegrate.[29] Suppose now that birds-of-a-feather attraction exists among the members of this taste-income stratum. This again tends to pull the Thünen ring into a more compact shape.

There is a characteristic interaction between these forces and the Thünen forces. Suppose we start with a pure Thünen equilibrium and suddenly introduce mutual attraction among the members of a certain stratum. To get closer to each other, this group will choose smaller residences. This means that the ideal weight of their residential land uses will *rise;* hence, by the weight-falloff condition, this group will relocate *closer* to the center of town. Similarly, if mutual repulsion suddenly develops, the group will choose larger residences to keep each other at bay. This makes their chosen land uses lighter, so they locate farther out than before. Thus we may say that (holding other things such as income, family size, ethnic affiliation, etc., constant) more sociable people will live closer to the center of town than less sociable people.

These arguments have been framed in the context of a Euclidean metric. Suppose instead there is a system of high-speed arterial roads converging on the center of town. The Thünen "rings" are then no longer annuli but extend in starfish fashion into the hinterland along each artery (p. 505).

Now introduce facilities or mutual attraction. The shape of a neighborhood resulting from the operation of these forces would appear not to be circular but elongated along the arterials. That is, the "starfish" tendency of the Thünen rings becomes accentuated, and the resulting neighborhoods take on a "sectoral" pattern along the arterials, as in the idealized descriptions of Hoyt.[30] However, suppose the road pattern stresses high-speed *circumferential* routes (such as Route 128 around Boston). Neighborhoods under mutual attraction will then tend to elongate along *these* routes. But the original Thünen rings are already elongated in this way. We conclude that the presence of low-cost circumferential routes tends to *moderate* the disruption of the Thünen annular pattern caused by facilities or neighborhood attraction, while the presence of low-cost arterial routes tends to *accentuate* the disruption.

---

29. William Alonso has noted that the annulus is a rather unwieldy shape for a single individual's domain. The argument just given notes the same thing for neighborhoods. We have examined Alonso's proposed solution above in Section 6.8. See W. Alonso, *Location and Land Use* (Harvard Univ. Press, Cambridge, 1964), App. B.

30. H. Hoyt, *The Structure and Growth of Residential Neighborhoods in American Cities* (FHA, Washington, D.C., 1939).

We mention one more phenomenon that applies to the segregation of neighborhoods by income level. The fact that person A is not in general indifferent to the quality of housing maintained by his neighbor B has been noted by Davis and Whinston.[31] They go on to argue that this external effect leads to general undermaintenance in the sense that everyone would simultaneously be in a more preferred position if maintenance expenditures were increased a bit on all housing. However, a more subtle effect also follows from this external effect. A wealthy man is in general willing to spend more money for the same quality improvement than a poor man. Thus he will spend more on a house, more to maintain it, *and* more to live next door to a well-maintained house than one poorly maintained. Thus rich men not only choose to live *in* higher quality housing but choose to live *near* such housing as well. But by this very token, a rich man will tend to live near other rich men. Likewise, the poor will live near each other, not because they are attracted to each other, but because they cannot afford not to do so. We conclude that this housing neighborhood effect leads to segregation by income.[32]

Finally, we mention one important dimension that has been ignored in the preceding models (indeed, throughout this book), viz., *changes* in preferences that arise from one's experiences. There is a definitional problem here. If one sits down to a meal hungry and gets up satiated, it is possible to say *either* that one's taste for food has been (temporarily) downgraded by the experience of eating *or* that preferences are over the various possible *time patterns* of eating and fasting, and that the hunger-satiation cycle does not signify a change of tastes but merely a carrying out of the most preferred time pattern. From this second point of view, the term "change of preference" is to be reserved for changes in the preference ordering of the time patterns themselves. In either case time enters in an essential way.

Thus instead of considering mere attraction or repulsion of one person by another as we did above, we may consider preferences over the fraction of one's time θ spent in the presence of that other person. ("Attraction" may be taken to mean that $\theta = 1$ is preferred to $\theta = 0$; Clearly some intermediate θ-level might be preferred to either extreme.) For preferences of this sort, "equilibrium" cannot be adequately expressed in terms of mere spatial arrangement. Itineraries must be specified, and there is a synchronization problem to assure that each pair of itineraries are copresent the proper fraction of the time.

---

31. O. A. Davis and A. B. Whinston, Economics of urban renewal, *Law Contemp. Probl.* 26(Winter, 1961):105–17.

32. This argument amounts to an extension of the quality-quality association model of Section 8.6 from two factors to N factors (the participants in the housing market). There is, indeed, a corresponding extension of the allotment-assignment model from two factors (the left and right marginal spaces) to N factors, which will not be discussed in this book.

In some cases even the fractional time exposure θ is not a sufficient description of preferred time patterns. Rather, the duration of uninterrupted association is crucial. In activities such as eating a meal, seeing a movie, or having an operation, the completion of the activity is of importance, and this requires a sufficiently long time interval. "Equilibrium" here involves an even more delicate synchronization problem among itineraries.

Among the dynamic taste patterns occurring in connection with the use of facilities, we may note fatigue patterns and deprivation patterns. A *fatigue* pattern is one in which continued use of a facility (or the goods and services associated with it) diminishes one's desire for further use, at least temporarily. ("Strength of desire" might be measured objectively by the money one is willing to put up for, say, one minute's use of the facility.) This includes satiation of appetites, the demand for novelty and adventure, the need to "get away from it all."

A *deprivation* pattern is one in which prolonged abstention from the use of a facility leads to gradually increasing strength of desire for its use. This may result, for example, from the depletion of inventories obtained at the last use of the facility. (From this point of view, periodic shopping exemplifies a deprivation pattern.) Or it may result from the accumulation of waste inventories that require periodic use of waste-disposal facilities.

A periodic time pattern of activities may be thought of as resulting from an interplay of fatigue and deprivation patterns; the former regulating the duration of each facility use, the latter regulating the duration of the interval between successive uses of the same facility.

These same patterns occur in the rhythm of interpersonal attraction and repulsion. The palling of attraction with continued exposure— "familiarity breeds contempt"—illustrates a fatigue pattern, while "absence makes the heart grow fonder" illustrates a deprivation pattern.

We now go to the more radical concept of preference change in which one "migrates" from one time pattern to another rather than merely "commuting" within the same time pattern. In this "irreversible" change one may distinguish two dynamic processes that are in a sense opposite to fatigue and deprivation respectively. *Extinction* refers to a process whereby prolonged abstention from the use of a facility eventually reduces the strength of desire for its use. The interpersonal equivalent is the process whereby a long period of noninteraction leads to decay of attraction—"out of sight, out of mind".[33] This is the opposite of a deprivation pattern effect, but the two are not incompatible. It may happen that with prolonged lack of interaction, strength of desire first rises but eventually falls.

---

33. Hence the advice of Dr. Johnson: A man, Sir, should keep his friendship *in constant repair. Boswell's Life of Johnson,* p. 214.

Habituation is the process whereby the use of a facility increases the desire to use that facility. This is the opposite of a fatigue pattern effect, but again the two are not incompatible. Each incident of use may end in fatigue, while an increasing fraction of one's time is devoted to this facility because of habituation.

The following parable will illustrate these processes. Two countries, Eastland and Westland, have populations whose tastes are at first identical. Rice is relatively cheap in Eastland, wheat relatively cheap in Westland. As a result, the rice/wheat consumption ratio is higher in Eastland than in Westland. Habituation and extinction lead to taste changes in these countries such that the taste for rice relative to wheat becomes stronger in Eastland than in Westland.[34] Thus even if prices became equal in the two countries at some future date, the Eastlanders would still consume more rice relative to wheat than the Westlanders. The interpersonal equivalent of habituation is the process whereby interacting people tend to become friendly.[35]

Introducing preference changes into the preceding models leads to complications. Consider the birds-of-a-feather model in which several groups sort themselves out by mutual attraction. Not only will mutually attractive people move toward each other but, by habituation, people who are close to each other tend to become attracted. In general, then, the final equilibrium distribution of people by location and sentiment (if it exists) will not be independent of the initial distribution or of transitory disturbances. Complications of this sort must eventually be faced in any adequate model of national development, for example. For it appears that the present partition of the world into nation-states (the basis of powerful sentiments of attraction and repulsion) has arisen in part from "accidental" historical circumstances such as military cease-fire lines.[36]

9.3. INTERPLANT AND INTERINDUSTRY LOCATION

Consider the following layout problem. There are m "plants" whose location is already fixed, and one is to choose the location of an additional n "plants." The total flow of goods from plant i to plant j is given as

---

34. That is, wheat-rice indifference curves become more closely parallel to the wheat axis in Eastland than in Westland. Note that the resulting shifts in tastes tend to equalize price ratios. There is an interesting analogy here with the Heckscher-Ohlin theory of international trade.

35. See G. C. Homans, *The Human Group* (Harcourt, Brace, New York, 1950).

36. According to Pieter Geyl, part of the present frontier between Holland and Belgium traces back to the cease-fire line between the armies of Maurice of Orange and the Duke of Parma about 1590. This line corresponded neither to the religious nor the linguistic frontier of the time. P. Geyl, *Debates with Historians* (Meridian Books, World, Cleveland, 1958), pp. 206–9. The frontiers between the two Koreas, Israel and the Arab states, and India and Pakistan are possible contemporary instances of the same process in the making.

w_{ij}, $i, j = 1, \ldots, m + n$. The cost of transporting unit weight from location s' to s'' is given as $h(s', s'')$. Plants are numbered so that the first m have fixed locations, say $s_1, \ldots, s_m$. The problem is to choose $s_{m+1}, \ldots, s_{m+n}$ so as to minimize the total transport costs incurred on the flows between all pairs of plants. This total cost is the sum of the $(m + n)^2$ terms $w_{ij} h(s_i, s_j)$, where i, j run independently from 1 to $m + n$.

There are many interpretations of this model. Clearly, a plant can be any entity that may be thought of as located at a single geometrical point. In practice, plants will be entities whose size is very small compared with the spatial realm of the problem. Thus in broad regional planning the plants may be entire cities, the problem being where to locate n new cities advantageously in relation to m existing cities. Or on the same scale, a plant may be a complex of interrelated industries, the problem being where to locate n new complexes in relation to m existing ones. Or plants may refer to departments in a factory or a university, rooms in a house, facilities in a city, etc.

Consider next the meaning of the distinction between the fixed locations of plants $1, \ldots, m$ and the variable locations of plants $m + 1$, $\ldots, m + n$. One obvious interpretation is that the m plants are already in existence, making the cost of moving them prohibitive (the plant may be fixed by nature, e.g., a mineral deposit). Another interpretation is that we are dealing with a *subproblem*. That is, although some or all of these m plants are also not located yet, we specify their locations provisionally and see what these imply for the location of the remaining plants and for total costs. Now $s_1, \ldots, s_m$ are parameters and might themselves be subject to choice in a second problem. (The role of subproblems will be discussed more thoroughly in Section 9.4.)

The numbers w_{ij} and the function h are "ideal" weights and distances respectively and need not have any close resemblance to the corresponding physical magnitudes. Weight w_{ij} may reflect such things as the shipment of goods, the movement of people, the flow of mail or telephone calls from plant i to plant j. Furthermore, it will reflect a cumulative weighting of flows over time, the weights properly adjusted to take account of discounting and changes in unit transport cost over time. All this has been discussed extensively in connection with the concept of land use (Sections 8.2 and 8.4).

A problem does arise, however, as to the nature of the "givenness" of w_{ij} and h. The same dual interpretation as given above for $s_1, \ldots, s_m$ applies here. On the one hand, the givenness of h may reflect the fact that the transportation system is fixed, and the givenness of w_{ij} may reflect technological rigidities that force a fixed flow between each pair of plants regardless of location. On the other hand, some or all of the w_{ij} and h may be interpreted parametrically, as being themselves subject to choice in a further problem. Thus even though the w_{ij} may themselves be subject to choice, we may wish to know the optimal location pattern entailed by some (or any) particular choice of these numbers. In this case we are dealing again with a subproblem.

So far we have just been referring to plants. But a similar problem can be formulated in terms of *industries*. That is, given the location of *m* industries and given that (ideal) weight w_{ij} is to flow from industry *i* to industry *j*, find the location of *n* other industries to minimize overall transport costs. The difference is that while the location of a "plant" is specified by a single point, the location of an "industry" is specified by a *measure* over Space. This raises the additional question, Given measures μ_i and μ_j for the locations of industries *i* and *j*, just what *is* the transport cost incurred by the flow from *i* to *j*? Clearly the industry problem involves a considerably deeper conceptual apparatus than the plant problem. Therefore we delay consideration of it until the results of the plant problem are in; the latter involves no measure theory. (It is all the more remarkable, then, that the main result for the industry problem follows from the main result for the plant problem.)

We impose no constraints other than the givenness of $s_1, \ldots, s_m$, w_{ij}, and *h*. That is, any point of Space is eligible to be a plant location (this involves very little loss of generality, as we shall see). In particular, it is allowable for several plants to colocate at the *same* point.

This colocation is in fact the very essence of the following results, which we refer to collectively as *coincidence theorems*. A coincidence theorem is one that states there is an optimal location pattern in which two or more plant (or industry) locations coincide. The significance of coincidence theorems arises from the insight they provide into the phenomenon of *agglomeration,* the mysterious but pervasive tendency for resources to pile up into cities, industrial districts, etc.[37]

INTERPLANT LOCATION

We start with a simple case. Space is the real line with the ordinary Euclidean metric, $h(s', s'') = |s' - s''|$.

Lemma: Given real numbers $s_1, \ldots, s_m$, and w_{ij}, $i, j = 1, \ldots, m + n$, the *w*'s being nonnegative, consider the problem of minimizing

$$\sum_{i=1}^{m+n} \sum_{j=1}^{m+n} w_{ij} |s_i - s_j| \tag{9.3.1}$$

over real numbers $s_{m+1}, \ldots, s_{m+n}$. An optimal solution $s^o_{m+1}, \ldots, s^o_{m+n}$

---

37. There are of course other approaches to the agglomeration problem, such as the Thünen apparatus of Chapter 8.

exists in which each of the s_{m+i}^o, $i = 1, \ldots, n$, coincides with one of the s_i, $i = 1, \ldots, m$.

Proof: Say that plants $i_1, \ldots, i_k$ form a *cluster* at x iff $s_{i_1} = s_{i_2} = \cdots = s_{i_k} = x$, and no other plant is located at x. Now suppose there is an arrangement of plants in which some s_{m+i} does not coincide with any of the fixed positions $s_1, \ldots, s_m$. Consider the effect on (1) arising from moving the *cluster* to which plant $m + i$ belongs; i.e., the locations of all plants in this cluster are changed simultaneously so that they remain coincident. So long as this cluster does not cross the location of another cluster, it is easily seen that each term in (1)—hence (1) as a whole— is *linear* in x, the cluster location. Hence the cluster may be moved in at least one direction (left or right) without increasing (1).

If the cluster in question lies between two others, move it so that (1) does not increase until it comes into coincidence with one of its neighboring clusters. If the cluster in question is the rightmost of all, move it left into coincidence with its neighbor. This again cannot increase (1), since the cluster is simultaneously getting closer to all others. (Here is where $w_{ij} \geq 0$ comes in.) Similarly, if the cluster in question is the leftmost, move it right into coincidence with its neighbor, which does not increase (1).

We have shown that in any arrangement for which some s_{m+i} does not coincide with any $s_1, \ldots, s_m$ there is another arrangement with (1) at least as small and having *one less cluster*. The argument can be repeated for *this* arrangement, and we see by induction on the number of clusters that an arrangement exists in which each s_{m+i} coincides with some $s_1, \ldots, s_m$, with (1) for this arrangement being no greater than (1) for the original.

But the number of these special arrangements is finite. In fact, s_{m+i} can take on at most m values, so the number does not exceed m^n. Hence at least one of these arrangements must be optimal. ■

We now want to generalize this result from the real line to an *arbitrary* metric space. But a price must be paid for this generality: the restriction to $m = 2$. (The case $m = 1$ is trivial, since all plants can colocate with the single fixed plant at a total cost of zero.) It will be convenient to change notation slightly, letting s_o and s_{oo} be the given locations, while $s_1, \ldots, s_n$ are to be found. In view of its application to the interindustry problem, it will also be convenient to generalize a bit further and prove the result for semimetric spaces.

Definition: $h{:}A \times A \rightarrow$ reals is a *semimetric* on A iff

(i) $h(a, a) = 0$ for all $a \in A$;
(ii) $h(a_1, a_2) = h(a_2, a_1)$, all $a_1, a_2 \in A$ (symmetry);

(iii) $h(a_1, a_2) + h(a_2, a_3) \geq h(a_1, a_3)$, all $a_1, a_2, a_3 \in A$ (triangle inequality).

As an exercise, the reader may verify that (ii) and (iii) imply $h \geq 0$. Thus a semimetric differs from a metric only in that the former allows distinct points to be at distance zero. We are now ready for the generalized coincidence theorem.

Theorem: Let $h{:}S \times S \rightarrow$ reals be a semimetric, let points $s_o, s_{oo} \in S$ be given, and let numbers $w_{ij} \geq 0$ be given, $i, j = o, oo, 1, 2, \ldots, n$. Consider the problem of minimizing

$$C = \sum_i \sum_j w_{ij} h(s_i, s_j) \qquad (9.3.2)$$

over points $s_1, \ldots, s_n$ of S (i and j run independently over the set $\{o, oo, 1, 2, \ldots, n\}$ in this summation).

An optimal solution $s_1^o, \ldots, s_n^o$ exists in which each s_i^o coincides either with s_o or with s_{oo}.

Proof: For $j = o, oo, 1, 2, \ldots, n$, let

$$x_j = h(s_o, s_j) - h(s_j, s_{oo}) \qquad (9.3.3)$$

We have

$$x_i - x_j = \left[h(s_o, s_i) - h(s_o, s_j) \right] + \left[h(s_j, s_{oo}) - h(s_i, s_{oo}) \right]$$

$$\leq h(s_j, s_i) + h(s_j, s_i)$$

by the triangle inequality. Hence

$$x_i - x_j \leq 2h(s_j, s_i) \qquad (9.3.4)$$

Reversing the roles of i and j and combining with (4), we obtain

$$| x_i - x_j | \leq 2h(s_i, s_j) \qquad (9.3.5)$$

for all $i, j = o, oo, 1, 2, \ldots, n$. Relations (5) and (2) imply

$$2C \geq \sum_i \sum_j w_{ij} | x_i - x_j | \qquad (9.3.6)$$

(summations over $\{o, oo, 1, 2, \ldots, n\}$).

Now consider the problem on the *real line* of minimizing the right-hand side of (6) over numbers $x_1, \ldots, x_n$, when x_o and x_{oo} are given. By the lemma above, we know an optimal solution exists to this problem, in which each x_i^o coincides either with x_o or with x_{oo}. Let P_o be the set of indices i for which $x_i^o = x_o$, and let P_{oo} be the set of indices i for which

$x_i^o = x_{oo}$ (note that $o \in P_o$, $oo \in P_{oo}$). Substituting these values into the right side of (6) yields

$$| x_o - x_{oo} | \sum_{i \in P_o} \sum_{j \in P_{oo}} (w_{ij} + w_{ji}) \tag{9.3.7}$$

To see this, note first that $| x_i - x_j | = 0$ if i and j are both in P_o or both in P_{oo}, and that $| x_i - x_j | = | x_o - x_{oo} |$ if $i \in P_o, j \in P_{oo}$ or vice versa. Since (7) minimizes the right side of (6), it follows that (7) cannot exceed $2C$.

Going back to the definition of x_o and x_{oo} from (3), we find that $x_o = -h(s_o, s_{oo})$, $x_{oo} = h(s_o, s_{oo})$. Hence $| x_o - x_{oo} | = 2h(s_o, s_{oo})$. Substituting this in (7), it follows that

$$C \geq h(s_o, s_{oo}) \sum_{i \in P_o} \sum_{j \in P_{oo}} (w_{ij} + w_{ji}) \tag{9.3.8}$$

Thus we have established a lower bound for the possible values of (2). We now show that this lower bound can be attained. Consider the following location pattern in S:

$$s_i^o = s_o \text{ if } i \in P_o, \qquad s_i^o = s_{oo} \text{ if } i \in P_{oo} \tag{9.3.9}$$

Substituting these values in (2), we find that the right side of (8) is indeed attained. Hence the arrangement (9) is optimal. This arrangement is of the stipulated form. ∎

In the preceding argument, the symmetry of h played an essential role. This is not a bad assumption for distances, but sometimes it is not realistic (one-way streets, tariffs, upwind vs. downwind). Therefore, the possibility of dropping this assumption is of some interest. The following result shows that symmetry of h can be dropped, *if* symmetry of w is substituted in its place; i.e., $w_{ij} = w_{ji}$. This is much less plausible in general than symmetry of h, but there are certain conditions that tend to bring it about. First, if the movement is mainly of people rather than goods, the prevalence of commuting cycles tends to equalize flows in opposite directions. Second, $w_{ij} \neq w_{ji}$ tends to create a backflow of underutilized transportation equipment in the less frequented direction, and this in turn encourages backhaul traffic.[38] Note that dropping the symmetry assumption means that h need no longer be nonnegative.

Theorem: Let $h : S \times S \to$ reals satisfy the triangle inequality, with $h(s, s) = 0$, all $s \in S$; let s_o, s_{oo} be given points of S; and let w_{ij} be non-

---

38. Some examples are cited in E. M. Hoover, *The Location of Economic Activity* (McGraw-Hill, New York, 1948), p. 22.

negative numbers satisfying $w_{ij} = w_{ji}$ for $i,j = o, oo, 1, 2, \ldots, n$. Then the problem of minimizing (2) over locations $s_1, \ldots, s_n$ again has an optimal solution in which s_i^o coincides either with s_o or with s_{oo}, all $i = 1, \ldots, n$.

Proof: Define the function $f: S \times S \to$ reals by

$$f(s', s'') = [h(s', s'') + h(s'', s')]/2 \qquad (9.3.10)$$

One easily verifies that f is a semimetric. Furthermore, for any $i, j = o, oo, 1, 2, \ldots, n$ we have

$$w_{ij}h(s_i, s_j) + w_{ji}h(s_j, s_i) = w_{ij}[h(s_i, s_j) + h(s_j, s_i)] = 2w_{ij}f(s_i, s_j)$$

$$= w_{ij}f(s_i, s_j) + w_{ji}f(s_j, s_i) \qquad (9.3.11)$$

The first equality in (11) arises from the symmetry of w; the second arises from (10), and the last arises from the symmetry of w and of f. Equation (11) implies that (2) is unaltered in value if h is replaced by f. By the preceding theorem, an optimal solution, with s_i^o coinciding with s_o or s_{oo}, exists to the problem of minimizing (2) with f in place of h. By the equality just established, the solution minimizes (2) itself. ∎

What happens if symmetry is abandoned both for h and for w? The following counterexample shows that coincidence can break down. Space consists of three points s_o, s_{oo}, and s'. There are three plants, o, oo, and 1. Plants o and oo have locations fixed at s_o, s_{oo} respectively; the location of plant 1 is to be found. Weight $w_{o,1} = w_{oo,1} = 1$; all other w's are zero. $h(s_o, s_o) = h(s_{oo}, s_{oo}) = h(s', s') = 0$; $h(s_o, s') = h(s_{oo}, s') = 1$; $h(s', s_o) = h(s', s_{oo}) = h(s_o, s_{oo}) = h(s_{oo}, s_o) = 3$. One easily checks that h obeys the triangle inequality, although of course it is not symmetric. If plant 1 locates at s_o or at s_{oo}, the total cost incurred (2) is 3 units; if it locates at s', the total cost is 2 units. Thus s' is the unique optimal location and coincides with neither s_o nor s_{oo}.

Now consider the possibility of certain constraints being imposed on the pattern of plant locations; e.g., certain regions may be prohibited to certain plants ("zoning" restrictions). Or certain pairs of plants may not be allowed to approach too closely to each other. The possibilities are very rich, but the following observation shows that the preceding results still hold under quite general circumstances. Suppose the imposed constraints still allow any location pattern $s_1, \ldots, s_n$ in which each s_i coincides with some fixed plant location. Then one of these patterns must be optimal if there are two fixed plants. This is an application of the principle of "independence of irrelevant alternatives": if a_o is optimal in the set of alternatives A, and $a_o \in B \subseteq A$, then a_o is optimal in the set of alternatives B.

Furthermore, suppose certain subsets of plants, say $\{i_1, \ldots, i_k\}$, $\{j_1, \ldots, j_l\}$, etc., are constrained to colocate (perhaps because there are strong economies arising from their coincidence). Then the preceding results still apply. For we can redefine the problem by taking each of these *clusters* to be a single plant, the flow from cluster P_1 to cluster P_2 being the sum of w_{ij} over $i \in P_1$, $j \in P_2$. The preceding conclusions apply to this new problem, hence there is an optimal solution with each cluster located at s_o or s_{oo}. The same is therefore true for each plant.

Our main result has been for *two* plants with fixed locations. This cannot be relaxed even on the Euclidean plane, for there is a well-known case with *three* fixed plants in which the optimal location of a fourth variable plant coincides with none of them. Namely, let the three plants be placed at the corners of an equilateral triangle, and let each ship one unit to a fourth plant, all other flows being zero. Then the optimal location is at the centroid of the triangle (a result known to Fermat in the seventeenth century).

In view of this limitation one may raise a question as to the practical significance of the results we have attained. We delay discussion of this point until after some results for the interindustry problem are in. However, there are two other lines of argument, independent of each other and not as rigorously developed as the present approach, both of which indicate that these "noncoincident" solutions should be rare. The first stems from Palander and was elaborated by Hoover.[39] This stresses the peculiarities of the transportation system that tend to make such "intermediate" locations inferior: terminal costs and long-haul economies in general and the limited number of routes and junctions.

The second line of argument stems from Leon Moses.[40] He notes that the inflow-outflow pattern from a plant is in general not fixed but subject to choice, and the choice depends on the prices at which these can be bought or sold. Suppose the plant in question locates closer to a second plant. The saving in transport cost makes flows to or from that plant more attractively priced, so the plant in question will tend to buy and/or sell more to the linked plant. But this substitution increases the attraction of the second plant, which tends to make the plant in question move still closer to it. Thus "intermediate" locations tend to be un-

---

39. Hoover, *Location of Economic Activity*, pp. 30–31, 40–42, 301–2; E. M. Hoover, *An Introduction to Regional Economics* (Knopf, New York, 1971), pp. 48–54.

40. L. N. Moses, Location and the theory of production, *Quart. J. Econ.* 72(May 1958):259–72. We discuss this and related work by Sakashita and Alonso further in Section 9.4 in connection with subproblems.

stable; small displacements generate further displacements in the same direction.

In terms of our plant location model, these conclusions would take the following form. Realistic parameter values are such as to make coincidences occur even if the number of "fixed" plants exceeds two. The Palander-Hoover argument makes h responsible for this behavior; the Moses argument makes w_{ij} responsible.

INTERINDUSTRY LOCATION

We are now ready for the interindustry problem. As noted, the natural representation of an "industry" is as a measure over Space. We begin by placing a σ-field Σ_s on S; Σ_s will be the domain of all measures considered. Let $\mu_i:\Sigma_s \rightarrow$ reals be the distribution of industry i. It will be convenient here to stipulate that

$$\mu_i(S) = 1 \qquad (9.3.12)$$

for all industries i. Thus μ_i is formally a probability measure. But to avoid the misleading connotations of the word "probability" we refer to (12) as making μ_i a *normalized* measure.

For any region E, $\mu_i(E)$ may then be interpreted as the *fraction* of industry i located in E. We take this to mean the following. If mass w_{ij} is to flow from industry i to industry j, then $w_{ij}\mu_i(E)$ will originate in region E, and $w_{ij}\mu_j(F)$ will terminate in region F for all $E, F \in \Sigma_s$. (We say "originate in E," not "be exported from E" because some of this mass may be shipped to industry j in the *same* region E. For the same reason, "terminate in F" is correct, not "be imported into F." Furthermore, these numbers exclude transhipments, if any.)

There is a unit transport cost function $h:S \times S \rightarrow$ reals that is assumed to be measurable with respect to $(S \times S, \Sigma_s \times \Sigma_s)$. Let λ be a shipment measure on $(S \times S, \Sigma_s \times \Sigma_s)$; i.e., $\lambda(E \times F)$ is the total mass shipped from points of region E to points of region F. The total cost incurred by λ is then

$$\int_{S \times S} h \, d\lambda \qquad (9.3.13)$$

The format of the interindustry problem is similar to that of the interplant problem. Since we only consider the case of two fixed-location industries, we specialize notation to this case. Thus we are given the distributions μ_o and μ_{oo} of industries o and oo. We are also given w_{ij}, the total mass to be shipped from industry i to industry j for $i, j = o, oo$, $1, 2, \ldots, n$, as well as the cost function h. The problem is to choose

$\mu_1, \ldots, \mu_n$, the distributions of the remaining industries, so as to minimize overall transport costs.

Let $T(\mu_i, \mu_j)$ be the transport cost incurred in moving one unit of mass from distribution μ_i to distribution μ_j. Then total transport cost is given by

$$\sum_i \sum_j w_{ij} T(\mu_i, \mu_j)$$

where the summations run independently over $\{o, oo, 1, 2, \ldots, n\}$.

But what is the relation between $T(\mu_i, \mu_j)$ and h? Let λ be a shipment measure that shifts mass from μ_i to μ_j. The cost incurred is given by (13). Many measures λ accomplish this shift, and $T(\mu_i, \mu_j)$ is precisely the minimum (or rather, the infimum) of the costs incurred over all such feasible λ. That is, there is a subproblem of finding efficient transport flows, which is buried in the interindustry problem and whose solution is embodied in the function $T(\mu_i, \mu_j)$.

We specify this subproblem. This may be formulated in (at least) two different ways: as a *transportation* problem (Sections 7.1–7.5) or as a *transshipment* problem (Sections 7.6–7.11). We choose the latter for two reasons. First, the transshipment formulation is more plausible in the present context. It offers a wider range of shipment options; and if some of these reduce costs further, why should they not be taken? Second, not only is the transshipment formulation more realistic, it is mathematically much simpler in the present context. (Indeed, it is not at all clear that the coincidence theorem proved below is even true if $T(\mu_i, \mu_j)$ is interpreted as the value of a *transportation* problem.)

Using the transshipment formulation, we have

$$T(\mu_i, \mu_j) = \inf \int_{S \times S} h \, d\lambda \qquad (9.3.14)$$

the infimum taken over all bounded measures λ on $(S \times S, \Sigma_s \times \Sigma_s)$ that satisfy

$$\lambda' - \lambda'' = \mu_i - \mu_j \qquad (9.3.15)$$

(λ', λ'' are the left and right marginals of λ respectively).

Recall the meaning of (15). For any region E, $\mu_i(E)$ is the fraction of mass originating in E, and $\mu_j(E)$ is the fraction of mass terminating in E. Hence $\mu_i(E) - \mu_j(E)$ is the net surplus in E, and this is the amount that must be exported (or imported, if negative) from E to have a feasible shift of mass from μ_i to μ_j. On the other hand, $\lambda'(E) = \lambda(E \times S)$ is the gross shipments *from* points of E, and $\lambda''(E) = \lambda(S \times E)$ is the gross shipments *to* points of E. Hence $\lambda'(E) - \lambda''(E)$ is precisely the net exports from region E. Flow λ is feasible exactly when these two differences are equal for all E; this is the statement of (15). (The *transportation* problem formulation requires that $\lambda' = \mu_i$ and $\lambda'' = \mu_j$; these imply (15), but not vice versa.)

A few more comments on (14) and (15) follow. Some restriction must be placed on h to insure that (14) is well defined for all feasible λ. (A more general formulation, in which (14) is interpreted as a pseudomeasure, has the drawback that the infimum might not exist.) As for (15), recall that μ_i and μ_j are both normalized, so that $\mu_i(S) = \mu_j(S)$ (λ need not be normalized, though it *is* bounded). This equality implies that a *feasible* solution exists (see (17) below). However, there is no guarantee that an *optimal* λ exists. There may be no feasible λ that actually attains the cost $T(\mu_i, \mu_j)$. But the function T is still well defined.

Equations (14) and (15) define T as a function on the domain **M** × **M**, where **M** is the set of all *normalized* measures on (S, Σ_s). The following result expresses its basic properties.

Theorem: Let h be nonnegative, bounded, and measurable. Then T is nonnegative and bounded, $T(\mu, \mu) = 0$ for all $\mu \in$ **M**, and T obeys the triangle inequality:

$$T(\mu_1, \mu_2) + T(\mu_2, \mu_3) \geq T(\mu_1, \mu_3) \tag{9.3.16}$$

If in addition h is symmetric, then T is a *semimetric* on **M**.

Proof: Since $h \geq 0$, $\int_{S \times S} h \, d\lambda \geq 0$ for any feasible λ. Hence $T \geq 0$. For the pair (μ, μ), $\lambda = 0$ is feasible, yielding a cost of 0. Hence $T(\mu, \mu) = 0$ for all $\mu \in$ **M**.

To show that T is bounded, choose μ, $\nu \in$ **M**, and let λ be the product measure: $\lambda = \mu \times \nu$. Flow λ is a feasible shipment measure for the pair (μ, ν). For

$$\lambda'(E) - \lambda''(E) = \mu(E)\nu(S) - \mu(S)\nu(E) = \mu(E) - \nu(E) \tag{9.3.17}$$

for all regions E, so (15) is satisfied. (The first equality in (17) arises from the definition of product measure, the second from the fact that μ and ν are normalized.) Also, $\lambda(S \times S) = \mu(S)\nu(S) = 1$. It follows that the cost for this λ does not exceed k, where k is any upper bound for h. This proves that T is bounded above, hence bounded.

Now we come to the triangle inequality. Let λ_{12} be feasible for the pair (μ_1, μ_2), and let λ_{23} be feasible for (μ_2, μ_3). Thus

$$\lambda'_{ij} - \lambda''_{ij} = \mu_i - \mu_j \tag{9.3.18}$$

for $ij = 1, 2$, and $ij = 2, 3$, from (15). Adding the two equalities (18), we obtain

$$\lambda' - \lambda'' = \mu_1 - \mu_2 + \mu_2 - \mu_3 = \mu_1 - \mu_3 \tag{9.3.19}$$

where we have written λ for the sum $\lambda_{12} + \lambda_{23}$. Equation (19) asserts that λ is feasible for the pair (μ_1, μ_3). Hence

$$\int_{S \times S} h \, d\lambda_{12} + \int_{S \times S} h \, d\lambda_{23} = \int_{S \times S} h \, d\lambda \geq T(\mu_1, \mu_3) \qquad (9.3.20)$$

Taking the infimum on the left side of (20) over all λ_{12} feasible for (μ_1, μ_2) and over all λ_{23} feasible for (μ_2, μ_3), we obtain (16).

Finally, assume in addition that h is symmetric. To show that T is a semimetric, it remains only to prove that T is symmetric:

$$T(\mu, \nu) = T(\nu, \mu) \qquad (9.3.21)$$

all $\mu, \nu \in \mathbf{M}$. Let λ be a feasible shipment for (μ, ν) and consider its transpose λ^*, defined by $\lambda^*(E) = \lambda\{(s_1, s_2) \mid (s_2, s_1) \in E\}$, all $E \in \Sigma_s \times \Sigma_s$. (This is well defined; see Section 7.11.) Since the right marginal of λ^* equals the left marginal of λ and vice versa, it follows that λ^* is feasible for the pair (ν, μ). Also,

$$\int_{S \times S} h(s_1, s_2) \, \lambda^*(ds_1, ds_2) = \int_{S \times S} h(s_2, s_1) \, \lambda(ds_1, ds_2)$$

$$= \int_{S \times S} h(s_1, s_2) \, \lambda(ds_1, ds_2) \qquad (9.3.22)$$

(The first equality in (22) arises from the induced integrals theorem via the mapping $(s_1, s_2) \rightarrow (s_2, s_1)$ from $S \times S$ to itself. The second equality arises from the symmetry of h.) Equation (22) states that the costs incurred under λ and λ^* are equal. Thus for any shipment feasible for (μ, ν), there is a shipment feasible for (ν, μ) that does as well in terms of costs. Since the opposite is also true, equality (21) is established. ∎

We did *not* have to assume that h satisfies the triangle inequality. This is important, and it is worthwhile to discuss the empirical significance of this property. There are two ways of interpreting h. The first takes $h(s_1, s_2)$ to be the cost of moving unit mass over a *direct route* linking s_1 to s_2. There is no reason why such an h should obey the triangle inequality. There may be good direct connections between s_1 and s_2 and also between s_2 and s_3, but only poor direct connections between s_1 and s_3; and we then might well have $h(s_1, s_3) > h(s_1, s_2) + h(s_2, s_3)$. On the other hand, h may itself arise from the solution to a preliminary optimization problem; viz., $h(s_1, s_2)$ is the minimal cost over all routes, direct and indirect. This second h may reasonably be expected to satisfy the triangle inequality, for $h(s_1, s_2) + h(s_2, s_3)$ is the minimum cost over all routes from s_1 to s_3 passing through s_2. All these routes and others are available for $s_1 \rightarrow s_3$ routing, so $h(s_1, s_3)$ should be at least as small as this sum. The first interpretation of h is more fundamental in the sense that the second h is the result of processing the raw data of the first.

Now the transhipment formulation is designed to work directly on this raw data. Transhipment optimization automatically solves this preliminary optimal routing problem. That is why T satisfies the triangle inequality (16) even though it was not assumed for h. The transportation problem formulation, however, by its nature does not allow indirect routing. It is not surprising then that the triangle inequality (16) may not hold if T is the infimum for the transportation formulation.[41]

It is appropriate that T be called the *Kantorovitch (semi)metric* on **M**, after the man who first suggested such a function.[42]

After all these preliminaries, the coincidence theorem for the interindustry problem follows with surprising rapidity.

Theorem: Given measurable space (S, Σ_s), let $h:S \times S \rightarrow$ reals be measurable, bounded, nonnegative, and symmetric. Let two normalized measures μ_o and μ_{oo} on Σ_s be given, and let numbers $w_{ij} \geq 0$ be given, $i, j, = o, oo, 1, 2, \ldots, n$. Consider the problem of minimizing

$$\sum_i \sum_j w_{ij} T(\mu_i, \mu_j) \tag{9.3.23}$$

over normalized measures $\mu_1, \ldots, \mu_n$ on Σ_s (i and j run independently over the set $\{o, oo, 1, 2, \ldots, n\}$ in this summation, and T is defined by (14) and (15)).

An optimal solution $\mu_1^o, \ldots, \mu_n^o$ exists in which each μ_i^o coincides with either μ_o or μ_{oo}.

Proof: By the preceding theorem, T is a semimetric on **M**, the space of normalized measures over (S, Σ_s). Comparing (23) with (2), we see that this inter*industry* problem on S is formally identical with the inter*plant* problem on **M**, with T playing the role of h. Hence an optimal solution $\mu_1^o, \ldots, \mu_n^o$ exists in which each "point" $\mu_i^o \in$ **M** coincides with one of the "points" μ_o or $\mu_{oo} \in$ **M**. This completes the proof! ∎

We also have the following theorem.

---

41. Here is a counterexample: S has three points; h is symmetric, $h(s, s) = 0$, all $s \in S, h(s_1, s_2) = h(s_2, s_3) = 1, h(s_1, s_3) = 3$. Let μ_i be concentrated on $\{s_i\}$, $i = 1, 2, 3$, and show that (16) is false.

42. L. Kantorovitch, On the translocation of masses, *Comptes Rendus (Doklady) de l'Académie des Sciences de l'URSS,* 37, no. 7–8 (1942):199–201. Recall, however, that in this early article Kantorovitch did not clearly distinguish between transhipment and transportation. Indeed, his problem is stated in transportation form.

Theorem: Let the premises be the same as above with two exceptions: h need not be symmetric, but w is symmetric: $w_{ij} = w_{ji}$ for all $i, j = o, oo, 1, 2, \ldots, n$. Then again there is an optimal solution $\mu_1^o, \ldots, \mu_n^o$ to the problem of minimizing (23) in which each μ_i^o coincides with either μ_o or with μ_{oo}.

Proof: It is still true that $T(\mu, \mu) = 0$, all $\mu \in \mathbf{M}$, and that T satisfies the triangle inequality (16). T need not be symmetric, but we can still apply the inter*plant* theorem with w symmetric to the space $\mathbf{M}$. Hence the same conclusion follows as above. ■

One point should be noted regarding the meaning of "optimality" in these two theorems. As we have remarked, there is no guarantee that a shipment measure λ exists from μ_i to μ_j that actually attains the cost minimum $T(\mu_i, \mu_j)$. However, we can say that for any $\epsilon > 0$, no matter how small, there is a feasible λ whose cost comes within ϵ of $T(\mu_i, \mu_j)$. Optimality must be understood in this ϵ-sense. If we wish to avoid even this quibble, it suffices to assume that an optimal shipment measure exists in the two cases μ_o to μ_{oo} and μ_{oo} to μ_o (since these are the only origin-destination pairs that occur in the optimal location pattern). If h is symmetric, it suffices to assume the existence of an optimal shipment measure in just one of these cases, since if, say, λ^o is optimal for shipping from μ_o to μ_{oo}, its transpose λ^{o*} is optimal for shipping in the opposite direction. (This follows from (22).)

Let us discuss some practical implications of these results. We first note that the concepts discussed clarify the notion of "orientation." The location literature is full of remarks such as "industry x is market oriented," "materials oriented," or "labor oriented." Market orientation, for example, has a clear meaning if the "market" is a single point. The industry locates at or close to this point to reduce shipping costs on its output. But what does it mean if the market consists of hundreds of cities scattered over the landscape? The following explication suggests itself.

Let both the industry and the market be represented by (normalized) measures over Space. The meaning of this representation for the industry has been discussed. To say that the market has the distribution μ means that $\mu(E)$ is the fraction of total sales of the industry's product that occurs in region E for all $E \in \Sigma_s$. (If the industry makes several products, a separate market measure may be defined for each product.) *Market orientation* means that the industry distribution ν is close to the market distribution μ, "closeness" being taken in the sense of *Kantorovitch distance:* $T(\nu, \mu)$ is small.

The rationale for this definition is that $T(\nu, \mu)$ is precisely the "cost of shipping goods to market" (per unit weight). Market-oriented behavior on the part of the industry is then the choosing of ν to make

$T(\nu, \mu)$ small. The extreme case is where $\nu = \mu$; i.e., the fraction of industry output in any region equals the fraction of industry sales in that region. This avoids transport cost on final product completely.

This argument assumes that the industry acts to reduce overall transport cost as if it were controlled by a single person. Presumably an efficiently functioning market for transportation services would tend to bring about this situation. However, market imperfections (such as lack of information, distortive tariffs, basing-point systems, territorial divisions, and other collusive practices) reduce the power of the "invisible hand," and thereby reduce the usefulness of the definition based on Kantorovitch distance.

We may summarize the preceding results by saying that if there are just two industries whose location is fixed, all other industries in the complex considered will be "oriented" to one of these two. The major weakness of these results is the restriction to just two fixed industries. This weakness can be partially overcome, however.

First, if the premises are "approximately" fulfilled, we may expect that a location pattern $\mu_1^o, \ldots, \mu_n^o$ exists of the stipulated kind that is "close" to being optimal. Let us make this statement more precise. We consider two kinds of approximate fulfillment of the premises.

Suppose more than two industries with fixed locations exist but that several industries have identical distributions, so that just two distinct distributions μ_o and μ_{oo} exist. Then we could think of all industries with distribution μ_o as constituting a single composite industry, and similarly for the cluster with common distribution μ_{oo}. The flow weights w_{ij} for these composite industries are the sums of the weights of the component industries. The coincidence theorems are applicable, and we conclude that an optimal distribution pattern $\mu_1^o, \ldots, \mu_n^o$ exists, with each μ_i coinciding with either μ_o or μ_{oo}.

Now suppose these conditions are approximately fulfilled in the sense that the fixed industries fall into two clusters, the industries in each cluster having distributions "close" to each other. Here closeness is measured by the Kantorovitch semimetric. Let μ_o be any distribution from the first cluster, and μ_{oo} any from the second. Then there should be a distribution pattern $\mu_1^o, \ldots, \mu_n^o$, each μ_i^o coinciding with either μ_o or μ_{oo}, which incurs a total cost not far from the minimum attainable. This informal argument could be made rigorous by showing that the minimal total cost attainable is a *continuous* function of the distributions of the fixed industries (continuity defined in terms of the Kantorovitch semimetric on **M**).

The second kind of approximate fulfillment we consider involves the structure of the w_{ij} matrix. Suppose there are more than two fixed industries, but also for all but two of these industries the w_{ij} weights linking them to all other industries are "very small." Then again there should be a "near optimal" solution in which every nonfixed industry coincides in distribution with one of the two remaining fixed industries. Again, these concepts and the argument could be made rigorous by showing that minimal attainable cost is a continuous function of the numbers w_{ij}.

As an illustration, suppose there is an initial tendency for industries to fall into two broad categories, market-oriented and resource-oriented. The "ideal" market-oriented industry has a distribution coinciding with the spatial distribution of consumption: $\mu(E)$ is the fraction of total consumption done by residents of region E. (One may want to make a correction for government spending by adding in consumption by military bases, space centers, etc.) The "ideal" resource-oriented industry is a little more difficult to pin down. Perhaps it may be taken as proportional to the rate of extraction of natural resources (in tons per year) from various regions, including in this total water and farm products as well as minerals, forest products, fishery products, etc. That is, this distribution is a mass-weighted average of the distribution of all extractive industries.

The point is that—under certain assumptions—once this dichotomous pattern comes into existence, the interindustry coincidence theorem indicates that it tends to perpetuate itself. For any new industry finds that its optimal location pattern will roughly coincide *either* with the ideal market-oriented distribution *or* with the ideal resource-oriented distribution. The assumptions required are fairly stringent, however, including such things as the neglect of rents, climate, and other nontransportation aspects of industrial location and the nonexistence of major geographical discoveries.[43]

SIMPLY CONCENTRATED INDUSTRIES

In the preceding theory we made use of the fact that an industry distribution could be considered a "plant" in the space $\mathbf{M}$ of normalized measures. Conversely, we now note that plants may be identified with a special type of industry distribution. Namely, we identify the plant located at point s_o with the normalized measure μ *simply concentrated* at s_o; i.e., $\mu(E) = 1$ if $s_o \in E$; $\mu(E) = 0$ if $s_o \notin E$ for all regions E.

We return to the general interindustry problem. Given fixed $\mu_1, \ldots, \mu_m$, we are to choose $\mu_{m+1}, \ldots, \mu_{m+k}$ to minimize

$$\sum_i \sum_j w_{ij} T(\mu_i, \mu_j) \tag{9.3.24}$$

where i, j run over $\{1, \ldots, m + k\}$. In the development above we re-

---

43. To judge how good an approximation this dichotomous scheme represents, we would need extensive data concerning industries in terms of their *Kantorovitch distances*, rather than in terms of "location quotients," "segregation indices," etc. Unfortunately, except for some sporadic solutions to the transhipment problem, such data do not exist.

These ideas may cast some light on the tendency for bipolar configurations to arise periodically in politics and warfare.

stricted ourselves to the special case $m = 2$. Now we specialize in a different way by assuming each of the fixed measures $\mu_1, \ldots, \mu_m$ is simply concentrated (there is no restriction on m). Under certain additional conditions our conclusion will be that an optimal solution $\mu^o_{m+1}, \ldots, \mu^o_{m+k}$ exists in which each μ^o_i is simply concentrated.

This is not a coincidence theorem. To grasp its significance, start from the inter*plant* location problem in which one is given locations $s_1, \ldots, s_m$ and must choose locations $s_{m+1}, \ldots, s_{m+k}$ to minimize transport costs. Identify a location with the normalized measure simply concentrated at that point. Now suppose that, instead of being restricted to simply concentrated measures, one is given the much wider option of choosing *any* normalized measures $\mu_{m+1}, \ldots, \mu_{m+k}$ for the distribution of the remaining industries. The gist of the following theorem is that this wider range of choice is of no benefit because an optimal solution already exists among the k-tuples of simply concentrated measures. Economists will recognize the similarity in tone of this result to the Arrow–Georgescu–Roegen–Koopmans–Samuelson substitution theorem for input-output systems, which states that under certain conditions varying bundles of final demand are optimally met by producing each good with inputs in fixed proportions, even though one has the option of varying these proportions.[44]

We shall be working in n-dimensional space for some $n = 1, 2, \ldots$. If A is a Borel set in n-space, (A, Σ) will refer to the measurable space formed by Σ, the Borel field of n-space restricted to subsets of A. We need the following concept. (A set is *convex* iff it contains the line segment between any two of its points.)

Definition: Let A be a convex Borel set in n-space, and let μ be a normalized measure on (A, Σ). The *centroid* of μ is the n-tuple whose ith component is $\int_A x_i \, \mu(dx_1, \ldots, dx_n), i = 1, \ldots, n$.

The centroid (if it exists) is a point of A.

Definition: $f{:}A \rightarrow$ reals is a *convex function* iff A is a convex set and $f(\theta x + (1 - \theta)y) \leq \theta f(x) + (1 - \theta)f(y)$, all $x, y \in A$, all θ between zero and one.

We may now state *Jensen's inequality*.

Lemma: Let A be a convex Borel set in n-space, and let μ be a nor-

44. T. C. Koopmans, ed., *Activity Analysis of Production and Allocation* (Wiley, New York, 1951), Ch. 7–10.

malized measure on (A, Σ) whose centroid $\bar{a}$ exists. Let $f{:}A \rightarrow$ reals be a convex measurable function. Then the integral of f is well defined (possibly infinite), and $\int_A f \, d\mu \geq f(\bar{a})$.

We now come to the main result. The terms "closed" and "continuous" refer to the usual topology of n-space relativized to subsets.

Theorem: Let S be a bounded, closed, convex subset of n-space for some $n = 1, 2, \ldots$. Let $h{:}S \times S \rightarrow$ reals be continuous and convex, let it satisfy the triangle inequality, and let $h(s, s) = 0$, all $s \in S$.

Let normalized simply concentrated measures $\mu_1, \ldots, \mu_m$ on (S, Σ_s) be given as well as numbers $w_{ij} \geq 0$, $i, j = 1, \ldots, m + k$, and consider the problem of minimizing (24) over normalized measures $\mu_{m+1}, \ldots, \mu_{m+k}$ on (S, Σ_s). (In (24) T is the value of the transhipment problem (14)–(15) determined by the function h.)

Then an optimal solution $\mu^o_{m+1}, \ldots, \mu^o_{m+k}$ exists in which each μ^o_i is *simply concentrated*.

Proof: Let μ_i, μ_j be normalized measures on (S, Σ_s). We first show that the transhipment problem (14)–(15) has an optimal solution λ^o whose left and right marginals coincide with μ_i, μ_j respectively. To see this, let $\mathfrak{J}$ be the usual topology of n-space relativized to the subset S; $\mathfrak{J}$ is separable and topologically complete; Σ_s is the σ-field generated by $\mathfrak{J}$; and h is continuous and also bounded (boundedness of h follows from closed boundedness of S and continuity of h). These conditions, with $\mu(S_i) = \mu(S_j) < \infty$, imply the existence of an optimal solution λ^o to the *transportation* problem on $S \times S$, with source, sink measures μ_i, μ_j respectively. The results of Section 7.10 show that this λ^o is also optimal for the corresponding *transhipment* problem.

This λ^o is a normalized measure on $(S \times S, \Sigma_s \times \Sigma_s)$. Now $S \times S$ is a bounded closed convex subset of $2n$-space. Boundedness of $S \times S$ implies that the centroid of λ^o exists. We write this centroid as an ordered pair $(\bar{s}_i, \bar{s}_j)$, where $\bar{s}_i$ and $\bar{s}_j$ are each n-tuples. In fact, one easily checks that $\bar{s}_i$ and $\bar{s}_j$ are precisely the centroids of the left and right marginals of λ^o, which are μ_i, μ_j respectively.

We now apply Jensen's inequality to λ^o and h, obtaining

$$\int_{S \times S} h \, d\lambda^o \geq h(\bar{s}_i, \bar{s}_j) \tag{9.3.25}$$

But the left side of (25) equals $T(\mu_i, \mu_j)$, since λ^o is optimal for transhipment. Thus the unit transhipment cost for the pair (μ_i, μ_j) is bounded below by the distance from the centroid of μ_i to the centroid of μ_j.

Consider now the effect of replacing the pair (μ_i, μ_j) by $(\bar{\mu}_i, \bar{\mu}_j)$, where $\bar{\mu}_i$ is the normalized measure that is simply concentrated at centroid $\bar{s}_i$, and similarly for $\bar{\mu}_j$ at $\bar{s}_j$. The normalized measure $\bar{\lambda}$ on $S \times S$ that is simply concentrated at $(\bar{s}_i, \bar{s}_j)$ is feasible for this latter pair and incurs the cost $h(\bar{s}_i, \bar{s}_j)$. Then (25) implies that the transhipment cost for the latter pair does not exceed the cost for the original pair.

Now consider any feasible solution $\mu_{m+1}, \ldots, \mu_{m+k}$. Replace each μ_i by $\bar{\mu}_i$, the normalized measure concentrated at the centroid of μ_i. Do the same for the fixed measures μ_i, $i = 1, \ldots, m$. No term in (24) is increased by this replacement, hence the same is true of the total cost (24) itself. Furthermore, each fixed measure μ_i, $i = 1, \ldots, m$, is simply concentrated to begin with. Since the centroid of a simply concentrated measure is its point of concentration, the replacement process yields the original μ_i back again; thus feasibility is maintained.

We have proved that for any feasible k-tuple of industry distributions $\mu_{m+1}, \ldots, \mu_{m+k}$ there is another feasible k-tuple $\bar{\mu}_{m+1}, \ldots, \bar{\mu}_{m+k}$ of simply concentrated distributions that does at least as well. It remains to show only that within the latter class of k-tuples there is a best one. We may identify these k-tuples with k-tuples $s_{m+1}, \ldots, s_{m+k}$ of points of S. Similarly, identify the fixed measures μ_i with their points of concentration. For this arrangement, total cost incurred is

$$\sum_{i=1}^{m+k} \sum_{j=1}^{m+k} w_{ij} h(s_i, s_j) \qquad (9.3.26)$$

Function h is continuous, hence (26) (as a function of the variable k-tuple of locations) is continuous with domain S^k. The latter set is closed and bounded, hence compact. Therefore a k-tuple minimizing (26) exists. The corresponding k-tuple of simply concentrated measures is optimal. ∎

As for applications, this theorem implies the same sort of self-perpetuating process we found for the coincidence theorem. That is, if the industries linked to a not-yet-located industry are all simply concentrated, then this remaining industry should also concentrate in a single plant.

The conditions on h are satisfied by the Euclidean metric, the city-block metric, and a wide variety of other plausible ideal distance functions (in particular, by any metric derived from a norm).

We finish by investigating the special case in which there is just *one* industry to be located. This case is of interest in itself in connection with the Weber problem (see Section 9.4). It also enables us to lighten considerably the load of assumptions needed in the preceding theorem. In particular, all references to convexity, n-space, and topology may be dropped.

For convenience we change notation as follows. Let w_i^{in} be the mass to be shipped *from* fixed industry i *to* the variable industry, and let w_i^{out} be the mass to be shipped *to* fixed industry i *from* the variable industry. Letting μ_i be the given distribution of industry i, and ν the distribution of the variable industry, the total transport cost incurred then takes the form

$$C(\nu) = \sum_{i=1}^{m} \left[w_i^{in} \, T(\mu_i, \nu) + w_i^{out} \, T(\nu, \mu_i) \right] \qquad (9.3.27)$$

As usual, T is the transhipment cost (14)–(15). The problem is to minimize (27) over ν.

Theorem: Let (S, Σ_s) be a measurable space, with all singleton sets measurable. Let $\mu_1, \ldots, \mu_m$ be normalized measures on S, simply concentrated at the points $s_1, \ldots, s_m$ respectively. Let $h: S \times S \to$ reals be bounded measurable, let h satisfy the triangle inequality, and let $h(s_i, s_i) = 0$, $i = 1, \ldots, m$. Let numbers w_i^{in}, w_i^{out}, $i = 1, \ldots, m$, be non-negative.

Let point s_o minimize the expression

$$f(s) = \sum_{i=1}^{m} \left[w_i^{in} h(s_i, s) + w_i^{out} h(s, s_i) \right] \qquad (9.3.28)$$

over all $s \in S$. Then the normalized measure ν^o that is simply concentrated at point s_o minimizes (27) over the set of all normalized measures ν on (S, Σ_s).

Proof: Consider the transhipment problem (14)–(15) for the pair (μ_i, ν), where ν is normalized and μ_i is the given distribution of industry i. Since the latter is simply concentrated and $\mu_i(A) = \nu(A) < \infty$, and h has the properties assumed above, we may invoke a theorem of Section 7.10 and conclude that an optimal solution λ^o exists having μ_i and ν as its left and right marginals respectively. This λ^o is unique, being given by

$$\lambda^o(G) = \nu\{s \mid (s_i, s) \in G\} \qquad (9.3.29)$$

for all $G \in \Sigma_s \times \Sigma_s$ (s_i is the point of concentration of μ_i).

Since λ^o is optimal, we have

$$T(\mu_i, \nu) = \int_{S \times S} h \, d\lambda^o = \int_{\{s_i\} \times S} h \, d\lambda^o = \int_{S} h(s_i, s) \, \nu(ds) \qquad (9.3.30)$$

(The last equality in (30) arises from the induced integrals theorem under the mapping $(s_i, s) \to s$, using (29).)

A similar argument for the pair (ν, μ_i) yields

$$T(\nu, \mu_i) = \int_S h(s, s_i) \, \nu(ds) \qquad (9.3.31)$$

It follows from (30) and (31) that

$$C(\nu) = \int_S f \, d\nu \qquad (9.3.32)$$

where C and f are defined by (27) and (28) respectively. Now f is minimized at s_o. Hence

$$C(\nu) \geq f(s_o) \qquad (9.3.33)$$

from (32).

Now consider $\nu = \nu^o$, the normalized measure simply concentrated at s_o. From (30) and (31) we obtain $T(\mu_i, \nu^o) = h(s_i, s_o)$, $T(\nu^o, \mu_i) = h(s_o, s_i)$. Hence $C(\nu^o) = f(s_o)$. This with (33) proves that ν^o minimizes (27). ■

Thus under rather general conditions we may conclude the following. If each of the industries linked with a given not-yet-located industry is concentrated in a single plant, then the industry in question should concentrate in a single plant to minimize overall transport costs.

9.4. THE WEBER PROBLEM

Let Σ_s again be the σ-field of Space S, and let $h{:}S \times S \to$ reals be measurable. (Actually, for (1) it suffices to assume that $h(x, \cdot){:}S \to$ reals is measurable for each $x \in S$.) As above, we interpret $h(x, y)$ as the cost of transporting unit (ideal) mass from location x to location y. Let μ be a measure on (S, Σ_s). The *Weber problem* is the problem of finding a point x_o that minimizes

$$\int_S h(x, s) \, \mu(ds) \qquad (9.4.1)$$

over all $x \in S$.

As a first interpretation, suppose a plant is to be located to minimize the overall transport cost incurred on its shipments. The pattern of shipments to be made is given by measure μ over S: $\mu(E)$ is the total mass to be shipped to region E, for all $E \in \Sigma_s$. Then the integral (1) expresses this total cost, and the optimum location problem coincides with this Weber problem.

This formulation suffers from the apparent defect of neglecting transport cost incurred on shipments *into* the plant. This can be remedied by supposing that we are given two measures μ^{out} and μ^{in}; $\mu^{\text{out}}(E)$ is the total mass to be shipped *from* the plant *to* region E, while $\mu^{in}(E)$ is the total mass to be shipped *to* the plant *from* region E, for all $E \in \Sigma_s$. Total transport cost incurred is then given by

$$\int_S h(x, s) \, \mu^{out}(ds) \; + \; \int_S h(s, x) \, \mu^{in}(ds) \qquad (9.4.2)$$

Suppose now that h is symmetric, so that $h(s, x) = h(x, s)$. Then (2) reduces to $\int_S h(x, s) \left[\mu^{out} + \mu^{in}\right] (ds)$. But this has the same form as (1) if we define $\mu = \mu^{out} + \mu^{in}$. In other words, if h is symmetric, there is a second more useful interpretation of (1); viz., $\mu(E)$ is the total shipments in both directions between the plant and region E, all $E \in \Sigma_s$, and (1) then gives the total cost incurred on imports *plus* exports of the plant.

There is yet a third interpretation. Suppose the patterns of in-shipments and out-shipments *coincide:* $\mu^{in} = \mu^{out}$. Let μ be this common measure. Then (2) reduces to

$$\int_S [h(x, s) + h(s, x)] \, \mu(ds) \qquad (9.4.3)$$

Letting $h'(x, s) = h(x, s) + h(s, x)$, we see that (3) reduces to (1) with h' in place of h; $h'(x, s)$ is simply the unit cost of a *round trip* between points x and s. Thus in this special coincidence situation, (1) again yields total transport cost if we interpret h to be unit round trip cost. This coincidence condition is not implausible if the "shipments" of the plant consist largely of people engaged in commuting.

The Weber problem arises in a remarkable variety of contexts, usually as a subproblem. We have mentioned the plant-location interpretation, in which μ refers to the spatial distribution of inflows and outflows. The "plant" may, however, not be a factory but a facility serving some population. Consider the problem of locating a high school for a certain district, for example. One criterion might be to minimize the total travel costs incurred by the students in going to and from school (cost may include factors such as travel time, risk of accident, transport mode). Then, if we take for μ the population distribution of students by residence and take for h the ideal distance function incorporating these cost features, the stated criterion leads again to a Weber problem. Or consider the problem of locating a communitywide facility such as a municipal airport, sewage-treatment plant, courthouse, or city hall. As a criterion of location, one might minimize the overall difficulty of access to the facility by its potential users. "Difficulty" may be measured by the integral of (ideal) distance with respect to the distribution of population in the community, which again yields a Weber problem. (For the sewage-treatment plant, a better index would be the integral of unit pumping costs with respect to the distribution of sewage "production"; this is still a Weber problem.)

At a higher level, the "plant" to be located may be an entire city; e.g., consider the problem of locating a new capital for a country. One might minimize the difficulty of access to the capital by inhabitants of the nation, measured by the integral of distance with respect to the distribution of the national population.

For another example, consider a single person or family contemplating where to reside. We suppose that all other commuting sites (workplaces, shopping places, schools, churches) have already been selected and that trip frequencies to these sites have already been decided on (e.g., five round trips a week to the workplace, one round trip a week to church). The family might then locate its residence to minimize overall transport costs. This may again be expressed as a Weber problem. The integrand is ideal distance, and the measure μ is concentrated at the finite number of commuting sites. The mass at site s_i is the trip frequency to that site, adjusted for the composition of family members visiting there, with their different time costs, etc.

Uncertainty can be incorporated into these problems. In locating a fire station, for example, the relevant measure is μ, where $\mu(E)$ is the *expected* rate of outbreak of fires per unit time. Similar considerations apply to the location of police stations and hospitals, where the distribution of demand for the services of these facilities cannot be known with certainty. (Strictly speaking, of course, this is true for any real-world problem, but in some problems the gain in realism may be outweighed by the extra complications introduced by an explicit treatment of uncertainty.)

The Weber criterion (1) is plausible only if differential costs due to transportation dominate other sources of cost differentials as location varies. The latter include variations in land-value densities, geographic irregularities such as climatic or topographic variations, and differential availability and quality of already existing structures. This raises a question as to the adequacy of the Weber problem formulation for locating facilities within a city, since variations in land values are quite marked in urban regions. However, a modified Weber formulation incorporating some results from the theory of Thünen systems can be used here (see below).

How does the Weber problem relate to the interplant and interindustry location problems of the preceding section? None of these subsumes any of the others. The interplant problem has several plants to locate, while the Weber problem has just one; on the other hand, the Weber problem allows a general fixed measure μ, while the interplant problem is restricted to a finite scattering of fixed plants. We can, however, formulate a problem that simultaneously generalizes both the Weber problem and the interplant problem.

Given k fixed measures $\mu_1, \ldots, \mu_k$ over Space (S, Σ_s), given ideal distance function $h:S \times S \rightarrow$ reals, and given numbers w_{ij}, $i, j = 1, \ldots, k$, minimize

$$\sum_{i=1}^{k} \int_{S} h(x_i, s) \, \mu_i(ds) + \sum_{i=1}^{k} \sum_{j=1}^{k} w_{ij} h(x_i, x_j) \qquad (9.4.4)$$

over all k-tuples of locations $x_1, \ldots, x_k \in S$. This may be interpreted as follows: k plants are to be located to minimize overall transport costs. The latter come in two categories. First, each plant i incurs costs in shipping to a distribution μ_i that is fixed in Space. These costs are represented by the integrals in (4). Second, there are flows w_{ij} between each pair of these plants. These costs are represented by the double summation in (4).

The problem (4) will be called the *multiplant Weber problem*. If $w_{ij} = 0$ for all i, j, this reduces to k separate ordinary Weber problems. If each of the measures μ_i is concentrated at a finite number of points, this reduces to the ordinary interplant location problem.[45]

Note the following relation between multiplant Weber problems and inter*industry* location problems. If each fixed industry is simply concentrated (i.e., if it is actually a single plant), we know from the last part of Section 9.3 that under fairly general circumstances an optimal solution exists to the interindustry problem in which each variable industry is simply concentrated. The restriction to simply concentrated industries yields a subproblem that is precisely the multiplant Weber problem, and we are then guaranteed that an optimal solution for this subproblem is optimal for the original interindustry problem.

The interplant and interindustry problems are subject to the same limitations as the Weber problem. The criterion of transport cost minimization is plausible only if differential transport costs dominate other differentials (in land-value density, production costs).

In recent years, however, the relevance of the Weber formulation has been questioned from an entirely different point of view. It is claimed that even if land values and production possibilities are uniform, the Weber criterion does not accurately mirror the relative advantages of different locations because it neglects the options for input-output substitutions and neglects spatial market structure.[46] These criticisms have yielded several valuable insights but have been unduly severe on the Weber for-

---

45. This statement is slightly inaccurate. If h is not symmetric we need integrals for both in- and out-shipments, as in (2), to attain a full generalization of the interplant problem. Special cases of this problem are treated in Francis and White, *Facility Layout and Location*, Ch. 5.

46. L. N. Moses, Location and the theory of production, *Quart. J. Econ.* 72(May 1958): 259–72. N. Sakashita, Production function, demand function and location theory of the firm, *Reg. Sci. Assoc. Pap.* 20(1968):109–22; W. Alonso, A reformulation of classical location theory and its relation to rent theory, *Reg. Sci. Assoc. Pap.* 19(1967):23–44, beginning at p. 34.

mulation. The following remarks aim at attaining a certain perspective on the issues.

The key to understanding the role of the Weber problem is to view it as a subproblem. We examine this concept in full generality. A *problem* may be defined as a set of options A, with an ordering $\geq$ on the elements of A representing the relative desirability of these options. A *solution* to this problem is any element $a_o \in A$ that is optimal for this ordering. (If the ordering $\geq$ is merely partial, "optimal" means "unsurpassed.") A *subproblem* of the problem $(A, \geq)$ is a pair $(B, \geq')$, where B is a subset of A, and $\geq'$ is $\geq$ restricted to $B \times B$. We immediately have the principle of "independence of irrelevant alternatives": if a_o is a solution to $(A, \geq)$ and $a_o \in B$, then a_o is a solution to $(B, \geq')$. Furthermore, if $\mathcal{B}$ is a collection of subsets of A that covers A, then any solution a_o to the problem $(A, \geq)$ will be a solution to the problem $(B, \geq')$ for some $B \in \mathcal{B}$. (This follows at once from the preceding statement, since a_o must belong to some member of $\mathcal{B}$.)

The point of these observations is that the subproblem $(B, \geq')$ is often much simpler than the original and that, by a clever choice of the covering $\mathcal{B}$, one can often derive important *necessary* conditions for a solution. Two examples will illustrate. Let A be an open subset of n-space, and consider the problem of maximizing the differentiable function $f{:}A \to$ reals. As is well known, it is necessary for all partial derivatives of f to vanish at any maximizing point $(x_1^o, \ldots, x_n^o)$. That the ith partial must vanish results from the fact that any solution satisfies the 1-dimensional subproblem in which all coordinates but the ith are held fixed. The second example is Bellman's principle of optimality.[47] Here one deals with multistage decision problems, and the principle states that any solution must remain optimal for the subproblem specified by fixing the state of the system at any stage.

We now return to the general location problem. Suppose one must choose not only where to locate a plant but also such things as what in-put-output combinations to use, where and how much to buy and sell, what pricing policies to use. The question then arises, Can the set of options A be covered by a collection $\mathcal{B}$ such that the subproblem determined by B is a Weber problem for each $B \in \mathcal{B}$? If so, then any solution to the original problem also solves a Weber problem, yielding an important and useful necessary condition.

The answer to this question is, Sometimes yes, sometimes no. It is negative in an example adduced by Alonso (pp. 34–36 of the article cited), that of f.o.b. pricing by a monopolist, because the spatial distribution of shipments μ varies with the location of the plant and cannot be taken as

---

47. R. Bellman, *Dynamic Programming* (Princeton Univ. Press, Princeton, 1957), p. 83.

given as required by the Weber formulation. However, for a monopolist who can discriminate among locations, Weber *is* a valid subproblem (Alonso, pp. 36–37). For let the pricing function $p:S \rightarrow$ reals sustain a shipment pattern μ over Space. The monopolist now has the option of holding the function p fixed as he varies his plant location, absorbing any changes in transport cost incurred. For the set of alternatives characterized by this fixed p, all nontransport costs and revenues are fixed as location varies. Thus profit maximization is equivalent to transport-cost minimization within this set, and we do indeed have a Weber subproblem.

To reach this conclusion, we do not have to assume that the agent in question is a profit maximizer. The Weber problem arises as a subproblem in the location of churches, government facilities, private residences, and other cases where income is certainly not the sole factor motivating the decision maker. By taking a subproblem in which everything is fixed except location and assuming such conditions as geographic uniformity, we are in effect holding constant the level of attainment of all nonpecuniary aspects of motivation. To generate a Weber subproblem, we need to assume only that, *ceteris paribus,* more money is preferred to less—a much weaker condition than profit maximization.[48]

We could list a wide variety of situations in which the Weber problem arises as a valid subproblem but give only a few additional examples. The fact that input and output levels can be varied causes no difficulties: consider the covering $\mathcal{B}$ where each $B \in \mathcal{B}$ is the set of options corresponding to a specific given bundle of input-output levels and a specific spatial distribution of shipments. The Weber subproblem also arises when shipments are between plants controlled by the same agent. For example, each variable plant in the interplant location problem individually solves the Weber problem determined by the distribution of all other plants to which it is linked. Finally, Lösch has observed that the Weber problem arises under a basing-point system, where each plant is allocated a sales quota distribution μ under fixed prices and may locate freely.[49]

The Weber problem stems from a long historical tradition, from Alfred Weber himself (1909) back through Launhardt, Steiner, and Fermat. The great bulk of this work deals with what is, from our point of view, a rather special case: S is the plane, h is Euclidean distance, and μ is concentrated at a finite number of points (usually three). This case is still yielding interesting results, as witness the recent work of Kuhn.[50] Kuhn's

---

48. Cf. the more detailed discussion of this point above (p. 481). It was shown there that this very weak motivational assumption was all that was needed to insure the formation of a Thünen system.

49. A. Lösch, *The Economics of Location,* trans. W. H. Woglom (Yale Univ. Press, New Haven, 1954), p. 164, n. 53.

50. H. W. Kuhn, On a Pair of Dual Nonlinear Programs, in J. Abadie, ed., *Nonlinear Programming* (Wiley, New York, 1967), pp 37–54.

programs are closely related to the equilibrium-of-forces condition in the Varignon frame, an analog device for solving the traditional case of the Weber problem. We do not discuss this further, except to state that the Varignon frame can be extended to solve the *multiplant* Weber problem (add interplant attractive forces proportional to the w_{ij}). The Varignon frame, original and extended, works (in principle) in Euclidean space of any dimension, although it is not very practical even in two dimensions.[51]

One more aspect of the Weber problem is worth mentioning. This is the concept of nodality, which is quite important but hard to formalize. Suppose locations are connected by a limited number of routes forming a network. This may arise from natural irregularities (rivers, ports, mountain passes, valleys, oases, islands), from transportation construction (road, rail, pipeline, airport), or from institutional features (closed borders with limited points of entry, designated air routes). A *node* is a location at which at least three arcs meet. Consider a Weber problem on this system of arcs; specifically, consider the effect on transport cost as the "plant" moves along arcs in the vicinity of a node s_o. Each arc a_i emanating from s_o taps a "hinterland" S_i: the set of locations for which the optimal route to s_o lies along arc a_i. Now, as the plant moves closer to s_o along arc a_i, it moves farther from all locations in S_i, but *closer* to the points of S_j for all $j \neq i$ (since, as a rule, the optimal route to these points goes through s_o). Thus, unless there is an i for which the mass $\mu(S_i)$ outweighs all the other masses $\mu(S_j)$ together, the node s_o is a *local minimizer* of transportation cost.

This informal argument indicates a tendency for the least-cost point to lie at a node; this is called *nodality*. When one takes a sequence of Weber problems over time, this tendency is self-reinforcing; for the attractiveness of a node is enhanced by shipment linkages to the plants located at the node as solutions to preceding Weber problems. This tendency for nodal location is also reinforced by transportation construction; for connecting routes tend to be built between existing agglomerations, and this enhances the relative locational advantages of the points connected.

To round out the picture, nodality must be supplemented by ideas from the theory of Thünen systems. For activities have positive space requirements, and this makes it impossible for them to agglomerate literally at single points. Instead, they will spread out around these points,

---

51. For the classical Weber problem, see C. J. Friedrich, ed., *Alfred Weber's Theory of the Location of Industries* (Univ. Chicago Press, Chicago, 1928); E. M. Hoover, *Location Theory and the Shoe and Leather Industries* (Harvard Univ. Press, Cambridge, 1937), Ch. 3, 5; W. Isard, *Location and Space-Economy* (MIT Press, Cambridge, 1956), Ch. 5; H. W. Kuhn and R. E. Kuenne, An efficient algorithm for the numerical solution of the generalized Weber problem in spatial economics, *J. Reg. Sci.* 4, no. 2(Winter 1962): pp. 21–33; and Francis and White, *Facility Layout and Location*, Ch. 4.

forming Thünen systems. In other words, the nuclei of Thünen systems, which are simply assumed to be given in Thünen theory, tend to be the nodes of the transportation grid that grow through agglomeration for the reasons outlined above.[52]

FORMAL THEORY

We now come to the formal theory. The first step is to place sufficient restrictions on the measure μ and distance function h so that the Weber criterion

$$u(x) = \int_S h(x, s)\, \mu(ds) \qquad (9.4.5)$$

is well defined. We assume that the measure μ is bounded. In fact, it is convenient to assume that μ is *normalized:* $\mu(S) = 1$. This involves no real loss of generality, and yields the interpretation that $\mu(E)$ is the fraction of total shipments of the plant that terminate in region E, for all $E \in \Sigma_s$.

However, we shall (usually) *not* assume that h is bounded. This raises a problem as to the meaning of (5), for the definite integral there need not be well defined. As before, we interpret (5) as a *pseudomeasure,*[53] expressed as usual in the form of an *in*definite integral. Of two locations x_1 and x_2, x_1 is taken to be preferred or indifferent to x_2 ($x_1 \succcurlyeq x_2$) iff $u(x_2) \succcurlyeq u(x_1)$, where $\succcurlyeq$ is to be understood in the sense of *standard ordering* of pseudomeasures. The reason for the reversal of subscripts is that we are minimizing: the larger pseudomeasure corresponds to the less preferred option.

This pseudomeasure interpretation is ideally suited for the Weber problem. It allows us to deal with the case of unbounded h with almost no extra complications whatever, since the special form of (5) reduces standard ordering of pseudomeasures to a very simple criterion, by the standard integral theorem (3.3.15). Specifically, we have the following.

Theorem: For the problem of minimizing (5), in the sense of (reverse) standard ordering of pseudomeasures, we have $x_1 \succcurlyeq x_2$ iff

---

52. These remarks should be compared with the discussion of the hierarchy of Thünen systems and "generalized accessibility"—as indexed, say, by Stewart's potential—of Section 8.9. A point displaying "nodality" is likely to be a point at which "accessibility" attains a local maximum, so that the two approaches are not necessarily incompatible. On nodality see H. J. Mackinder, *Britain and the British Seas* (Appleton, New York, 1902); and the discussion of "interposts" in R. G. Hawtrey, *The Economic Problem* (Longmans, Green, London, 1926), Ch. 7, 9. Hoover, *Location of Economic Activity,* pp. 41, 301–2, has an interesting hypothetical example.

53. Readers who do not feel at home with pseudomeasures are free to add the extra premise that h is bounded. Everything that follows remains valid under this restrictive assumption.

$$\int_S [h(x_2, s) - h(x_1, s)] \, \mu(ds) \geq 0 \qquad (9.4.6)$$

That is, $x_1 \geqslant x_2$ iff, first, the *definite* integral (6) is well defined (if not, x_1 and x_2 are not comparable); and, second, its value is nonnegative. If (5) is well defined and finite for both x_1 and x_2, this criterion reduces to the ordinary size comparison of the two integrals (5). But (6) may hold even if (5) is not well defined, or if it is infinite, for one or both of x_1 and x_2.

If h obeys the triangle inequality (as it usually will) things are even simpler, for we then have $h(x_2, x_1) \geq h(x_2, s) - h(x_1, s) \geq -h(x_1, x_2)$ for all $s, x_1, x_2 \in S$. This proves that for all $x_1, x_2 \in S$ the integrand in (6) is bounded. Since μ is also bounded, the definite integral (6) is always well defined. Hence x_1 and x_2 are *comparable*: if h obeys the triangle inequality, the partial order determined by standard order is *complete*. It follows that the distinction between *best* and *unsurpassed* solutions to the Weber problem disappears in this case.

The significance of these ideas comes through clearly in the special case where Space is the real line. We need the following familiar concept.

Definition: Let μ be a normalized measure on the real line; a number x is a *median* of μ iff $\mu\{y \mid y < x\} \leq \frac{1}{2} \leq \mu\{y \mid y \leq x\}$.

It is well known that a median always exists and that the set of medians is a bounded closed interval (usually a single point).

Consider now the Weber problem on the real line where h is the ordinary metric $h(x, y) = |x - y|$. We are to minimize

$$\int_{-\infty}^{\infty} |x - s| \, \mu(ds) \qquad (9.4.7)$$

over real numbers x; i.e., find the point about which the first absolute moment attains its minimum. It is well known that $x = x_o$ minimizes (7) iff x_o is a median, *provided* that (7) is finite for all x. However, if (7) is infinite for some x, it is easily shown to be infinite for all x, and the connection with the median obviously disappears. We now show that if (7) is reinterpreted as a *pseudomeasure,* then this connection is reestablished for *any* normalized μ. (The limits $\pm\infty$ are left off in (8) to indicate the indefinite integral.)

Theorem: Let μ be a normalized measure on the real line. Number x_o minimizes

$$\int |x - s| \, \mu(ds) \qquad (9.4.8)$$

over real numbers x (in the sense of reverse standard ordering of pseudo-measures) iff x_o is a median of μ.

Proof: By the preceding theorem, x_o minimizes (8) in the pseudo-measure sense iff

$$\int_{-\infty}^{\infty} [|x - s| - |x_o - s|] \, \mu(ds) \geq 0 \tag{9.4.9}$$

for all real x, where (9) is an ordinary definite integral on the real line. Since the integrand is bounded, this is well defined. Given $x \neq x_o$, let

$$I_1 = \{s \mid s \leq \min(x, x_o)\}, \qquad I_3 = \{s \mid s \geq \max(x, x_o)\}$$

and let $I_2 = \text{reals} \setminus (I_1 \cup I_3)$, which is the set of numbers lying strictly between x and x_o. Splitting the domain of integration in (9) into these three intervals, we find that

$$\int_{I_1} = (x - x_o)\mu(I_1), \qquad \int_{I_3} = (x_o - x)\mu(I_3) \tag{9.4.10}$$

while

$$\int_{I_2} \geq -|x_o - x| \, \mu(I_2) \tag{9.4.11}$$

Inequality (11) follows from the fact that

$$|x - s| - |x_o - s| \geq -|x_o - x| \tag{9.4.12}$$

Now let x_o be a median, and consider two cases. First, if $x > x_o$, then by (10) and (11) the integral (9) is at least as large as

$$(x - x_o)\mu(I_1) + (x_o - x)\mu(I_2 \cup I_3) = (x - x_o)[2\mu(I_1) - 1]$$

$$= (x - x_o)[2\mu\{s \mid s \leq x_o\} - 1] \geq 0 \tag{9.4.13}$$

(The inequality arises from the median property.) Second, if $x < x_o$, then the integral (9) is at least as large as

$$(x - x_o)\mu(I_1 \cup I_2) + (x_o - x)\mu(I_3) = (x_o - x)[2\mu(I_3) - 1]$$

$$= (x_o - x)[2\mu\{s \mid s \geq x_o\} - 1] \geq 0 \tag{9.4.14}$$

(The inequality again arises from the median property.) This proves the inequality (9) in all cases. Hence x_o minimizes (8).

For the converse let x_o again be a median, and let x not be a median. Then inequality (13) or (14) is strict, or else $\mu(I_2) > 0$. Furthermore, for $s \in I_2$ the inequality (12) is strict. These last two facts imply that inequality (11) is strict, so in all cases (13) or (14) in turn implies that the integral (9) is strictly positive.

Now reverse roles and assume that x_o is *not* a median. For x choose any median (there always is one). The argument just given shows that integral (9) is strictly *negative*. Hence x_o does not minimize (8) in the pseudomeasure sense. ∎

We now return to the general Weber problem determined by normalized measure μ on (S, Σ_s) and measurable function $h{:}S \times S \rightarrow$ reals. We refer to any optimal solution of this problem as a *median* of μ (with respect to h). The theorem just proved shows that this coincides with the usual concept of median on the real line (with respect to the usual real metric), so that this definition may be viewed as a natural extension of the median concept to abstract measure spaces.

Certain general statements can be made about the *location* and *existence* of medians. The basic idea is that if a certain bounded region E contains more than half the total mass of measure μ, then the median (if it exists at all) cannot lie beyond a certain distance from E. Intuitively, we can think of each "particle" of mass as exerting a pull, and the median as the point of equilibrium of forces. If the point lies too far from E, all the mass of E pulls it in approximately the same "direction." Being more than half the total, this pull outweighs all other pulls together; hence such a point cannot be in equilibrium. At any rate, this argument seems valid in Euclidean space. We find it holds, remarkably enough, in a general metric space.

Once this bounding of the possible location of a median is established, we can prove the existence of a median from a few extra mild conditions. The following two theorems formalize these arguments.

Theorem: Let μ be a normalized measure on (S, Σ_s), and let $h{:}S \times S \rightarrow$ reals be measurable and symmetric and satisfy the triangle inequality. Let $r \geq 0$ be real, $s_o \in S$, and let E be the closed disc with center s_o, radius $r{:} E = \{s \mid h(s_o, s) \leq r\}$. Let $\mu(E) = w$ be greater than $\frac{1}{2}$.

Then no point x outside the closed disc F with center s_o and radius

$$2rw/(2w - 1) \tag{9.4.15}$$

can be a median for the Weber problem of minimizing

$$\int h(x, s)\, \mu(ds) \tag{9.4.16}$$

over $x \in S$.

Proof: Let x lie outside the closed disc F, so that

$$h(s_o, x) > 2rw/(2w - 1) \tag{9.4.17}$$

We show that point s_o is superior to x, so that the latter is no median. We must prove that

$$\int_S [h(x,s) - h(s_o,s)] \, \mu(ds) > 0 \qquad (9.4.18)$$

First, $h(x,s) - h(s_o,s) \geq -h(s_o,x)$, by the triangle inequality, so that

$$\int_{S\backslash E} [h(x,s) - h(s_o,s)] \, \mu(ds) \geq -h(s_o,x)\mu(S\backslash E) = -h(s_o,x)(1-w) \qquad (9.4.19)$$

Second, for points $s \in E$ we have

$$h(x,s) \geq h(x,s_o) - h(s,s_o) = h(s_o,x) - h(s_o,s)$$

by the triangle inequality and symmetry. Also $h(s_o,s) \leq r$. Hence

$$h(x,s) - h(s_o,s) \geq h(s_o,x) - 2r \qquad (9.4.20)$$

so that

$$\int_E [h(x,s) - h(s_o,s)] \, \mu(ds) \geq [h(s_o,x) - 2r]w \qquad (9.4.21)$$

Adding (19) and (21), the integral in (18) must be at least as large as $[h(s_o,x) - 2r]w - h(s_o,x)(1-w) = h(s_o,x)(2w-1) - 2rw$. But this is positive, by (17). ■

We make two comments. First, this theorem remains valid if instead of *closed* discs E and F we have *open* discs with the same respective centers and radii. The proof is virtually the same as above, with $\geq$ replacing $>$ in (17), the opposite replacement in (20) and (21), and some minor changes in wording.

Second, if we specialize to a vector space S with Euclidean metric h, then a much sharper bound for the possible location of a median can be obtained. Namely, the result above is true with the radius (15) replaced by

$$rw(2w - 1)^{-1/2} \qquad (9.4.22)$$

We omit the proof of this statement. As an example, suppose that 51% of the total mass lies in a certain disc of radius 1. Then, according to formula (15) the median (if it exists at all) must lie in the concentric disc of radius 51 ($= 1.02/0.02$). However, if h is known to be a Euclidean metric, by (22) the median must lie in the concentric disc of radius $0.51 \, (0.02)^{-1/2}$, which is less than 4.

As an exercise the reader might show that there actually is a case, with $r > 0$ and $w = 1$, in which a median exists on the rim of the disc F with radius $2rw/(2w - 1) = 2r$. In this sense the preceding theorem is the best possible, at least for $w = 1$. The bound (22) is also the best possible in the Euclidean case for any $w > \frac{1}{2}$.

We now take up the question of existence of a median. The following theorem has a premise involving compactness. Recall that a set E in a metric space is *compact* iff every sequence in E has a subsequence converging to a point in E. The same definition applies to semimetrics. In the proof we make use of the well-known theorem that a continuous function on a nonempty compact domain has a minimizer.

Theorem: Let μ be a normalized measure on (S, Σ_s), and let measurable $h: S \times S \to$ reals be a metric (or even a semimetric) such that every closed disc is compact. Then a median exists for the Weber problem (16).

Proof: Choose a point $s_o \in S$, and define the function $f: S \to$ reals by

$$f(x) = \int_S [h(x, s) - h(s_o, s)]\, \mu(ds)$$

(By the triangle inequality this is indeed well defined and finite.) It can be seen that a point x_o is a median iff x_o minimizes f. Hence it suffices to show that f has a minimizer.

First we show that f is continuous. For $x', x'' \in S$ we have

$$f(x') - f(x'') = \int_S [h(x', s) - h(x'', s)]\, \mu(ds) \qquad (9.4.23)$$

Hence $|f(x') - f(x'')| \le h(x', x'')$, since the integrand in (23) is bounded in absolute value by $h(x', x'')$. Then for $\epsilon > 0$ any two points within distance ϵ have their f values differing by at most ϵ. This proves that f is (uniformly) continuous.

Next consider the sequence $\mu\{s \mid h(s_o, s) \le n\}$, $n = 1, 2, \ldots$. This approaches $\mu(S) = 1$ as a limit. Hence an integer n_o exists for which $\mu\{s \mid h(s_o, s) \le n_o\} > \frac{1}{2}$. By the preceding theorem, there is a closed disc F with center s_o such that s_o is superior to every point $x \notin F$ for the Weber problem (16). Thus

$$f(s_o) < f(x) \qquad (9.4.24)$$

for all $x \in S \backslash F$. Furthermore, F is compact, so that a point $x_o \in F$ exists that is a minimizer of f restricted to F. By (24), x_o minimizes f over all of S, hence x_o is a median. $\blacksquare$

The compactness premise in this theorem is valid for quite a variety of metrics, perhaps for all that are found in practice. In particular it holds in n-space for the Euclidean metric and the city-block metric, in fact for any metric derived from a norm. Thus we are assured of the existence of a median under quite general conditions.

This same line of reasoning can be used to prove a best solution exists to the *multiplant* Weber problem under the above compactness premise.

The remainder of this section will be devoted to obtaining qualitative results for the Weber problem by using notions of *symmetry* and *convexity*. We finish by showing how these ideas may be applied in the context of a Thünen system to obtain some surprising qualitative results.

We assume that the Weber criterion

$$u(x) = \int_S h(x,s)\, \mu(ds) \qquad (9.4.25)$$

is well defined and finite as a definite integral for all $x \in S$. Thus we need no longer worry about pseudomeasures.

Consider a measurable transformation $g{:}S \to S$ that leaves both the measure μ and the unit cost function h *invariant;* i.e.,

$$\mu(E) = \mu\{s \mid g(s) \in E\} \qquad (9.4.26)$$

for all regions E, and

$$h(g(x), g(y)) = h(x, y) \qquad (9.4.27)$$

for all $x, y \in S$.

Theorem: Let measurable $g{:}S \to S$ satisfy (26) and (27). Then

$$u(x) = u(g(x)) \qquad (9.4.28)$$

for all $x \in S$.

Proof: $u(x) = \displaystyle\int_S h(x,s)\, \mu(ds) = \int_S h(g(x), g(s))\, \mu(ds)$

$$= \int_S h(g(x), s)\, \mu(ds) = u(g(x))$$

Here the second equality arises from (27), while the third equality arises from the induced integrals theorem under the mapping g, since by (26) the measure induced from μ by g is μ itself. ∎

Thus the value of the Weber criterion is itself invariant under g. We refer to any g satisfying (26) and (27) as a *symmetry* for the Weber problem.

For the next result we assume that Space is n-space for some $n = 1, 2, \ldots,$ or more generally, a convex Borel subset of n-space.

Theorem: Let $h(\cdot, s){:}S \to$ reals be a convex function for all $s \in S$. Then $u{:}S \to$ reals given by (25) is convex.

Proof: Let $z = \theta x + (1 - \theta)y$, where $x, y \in S$ and θ is a number between 0 and 1. Then

$$u(z) = \int_S h(z,s)\, \mu(ds) \leq \int_S [\theta h(x,s) + (1 - \theta)h(y,s)]\, \mu(ds)$$

$$= \theta u(x) + (1 - \theta)u(y) \quad \blacksquare \tag{9.4.29}$$

The fact that u is a convex function implies that the set of medians for the Weber problem is convex. Let x and y be medians, and let z be a point between them. Then $u(x) = u(y) \geq u(z)$. But since x and y minimize u, this last must be an equality; thus z also minimizes u; i.e., z is a median.

These results extend to the *multiplant* Weber problem (4). Here there are k plant locations $x_1, \ldots, x_k$ to be found, and we may write $u(x_1, \ldots, x_k)$ for the value of the objective function (4) for this k-tuple. Let $g:S \to S$ again satisfy (26) and (27). Then

$$u(x_1, \ldots, x_k) = u(g(x_1), \ldots, g(x_k)) \tag{9.4.30}$$

Equation (30) follows from the fact that each of the $k + k^2$ terms in the multiplant Weber cost function is invariant under transformation by g. For the k integral terms this follows at once from (28). For the k^2 interplant flow terms this is an immediate consequence of (27). So much for symmetry.

As for convexity, let Space be a convex Borel set in n-space. Each plant location is an n-tuple of real numbers, and the k-tuple of plant locations may be thought of as a single point in a kn-dimensional "phase space." We now assume that the function h is convex. (This is stronger than the previous assumption that $h(\cdot, s)$ is convex for each $s \in S$.) Then $u(x_1, \ldots, x_k)$ is a *convex function*. This follows from the fact that each of the $k + k^2$ terms in the multiplant Weber cost function (4) is itself a convex function. (A sum of convex functions is convex.) For the k integral terms this is an immediate consequence of (29). For the k^2 interplant flow terms this follows at once from the convexity of h and the nonnegativity of the weights w_{ij}.

We apply these symmetry and convexity results to some concrete Weber problems. Consider the very simple problem in which Space is the Euclidean plane and μ is the uniform distribution over some circular disc, which we may assume to be centered at the origin $(0, 0)$. What is the median? It *seems* to be the origin, but how does one prove this? A brute-force method is to differentiate under the Weber integral and set the resulting expression equal to zero. But this yields a mess even in the present simple case; besides, the Weber value u may not always be differentiable, so the method itself is dubious.

Instead, use the following approach. First, the existence theorem guarantees that there is a median (x_o, y_o). Next consider the reflection mapping $(x, y) \to (-x, -y)$. One easily verifies that this is a symmetry. Hence $u(-x_o, -y_o) = u(x_o, y_o)$, by (28). This proves that the point

$(-x_o, -y_o)$ is also a median. Finally, the Euclidean metric is convex, hence u is convex, hence the set of medians is convex. But $(0, 0)$ lies between (x_o, y_o) and $(-x_o, -y_o)$. Hence $(0, 0)$ is indeed a median, and we are finished! (With a little more effort one could show that this is the only median, but we shall not take up questions of uniqueness.)

The method just used applies well beyond the special case discussed. The origin will be a median for the Weber problem (16) if h and μ satisfy the following conditions: μ is symmetric about the origin, in the sense of being invariant under reflection through the origin; h is convex; and $h(-x, -y) = h(x, y)$. These conditions on h are satisfied not only by the Euclidean metric but by the city-block metric, indeed by any metric derived from a norm, among others.

We now give a very useful sufficient condition for the invariance of μ. Going back for a moment to a general measure space S with transformation $g: S \to S$, suppose μ may be expressed as an indefinite integral

$$\mu = \int \delta \, d\lambda \tag{9.4.31}$$

where measure λ is invariant under g, and also where

$$\delta(s) = \delta(g(s)) \tag{9.4.32}$$

for all $s \in S$. It then follows from the induced integrals theorem that μ is invariant under g. For the special case of the plane, let λ be 2-dimensional Lebesgue measure; λ is invariant under reflection through the origin. Indeed, λ is invariant under any linear transformation with Jacobian ± 1 (this includes the other transformations discussed below). For any such g, choose any nonnegative measurable function δ satisfying (32). This yields a measure μ via (31) that is invariant under g.

For the circular disc discussed above we have δ a positive constant inside the disc, and $\delta = 0$ outside the disc. This satisfies (32) under the reflection $(x, y) \to (-x, -y)$. The same holds for a uniform distribution on any plane lamina that has a point of symmetry with respect to reflections through that point. Thus the point of symmetry of an annulus, a parallelogram, or an even-sided regular polygon is the median for the corresponding distributions, under the Euclidean metric, the city-block metric, or indeed any convex h invariant under reflection through the point of symmetry.

Similar arguments apply to reflections through a straight line. (The reflection of point P through line L is the point P' such that L is the perpendicular bisector of segment PP'.) As an application, let us show that for a uniform distribution μ over an equilateral triangular lamina, the centroid of the triangle is a median, under the Euclidean metric. This triangle has no point of symmetry, but it has three lines of symmetry L_1, L_2, L_3. Let g be the reflection through L_1, say. Lebesgue measure and the Euclidean metric are invariant under g. Also, (32) holds, by symmetry. Thus μ is invariant under g, so that $u(P) = u(g(P))$, where $u(P)$ is the value of the Weber integral at point P. We know that a median P_0 exists. By reflecting P_0 repeatedly through the lines L_1, L_2, L_3, we get up to five additional points $P_1, \ldots, P_5$, all of which must be medians. Finally, the centroid belongs to every convex set containing $\{P_0, \ldots, P_5\}$; since the set of medians is convex, the centroid is a median.   Q.E.D.

(Caution: For the city-block metric, the centroid of the equilateral triangle is usually not a median. Where does the preceding argument break down?) These arguments extend easily from 2 to n dimensions, where we also may wish to consider reflections through higher dimensional linear varieties.[54]

THE WEBER PROBLEM IN A THÜNEN SYSTEM

We now apply these concepts to Thünen systems. Recall some salient facts about such systems. Space (S, Σ_s) is furnished with an *areal measure* α; intuitively, $\alpha(E)$ is the capacity of region E for holding land uses. A special location s_N, the *nucleus,* serves as origin or destination for all traffic flows. There is a measurable function $g:S \to$ reals, $g(s)$ being the *ideal distance* between point s and the nucleus.

From the elaborate theory built upon these foundations in Chapter 8, we use only the fact that land uses tend to be arranged symmetrically around the nucleus—roughly speaking, points at equal distances carry land uses of equal ideal weight. With this in mind we frame the following definition.

Definition: Given α and g as above, μ on (S, Σ_s) is a *Thünen measure* iff it can be expressed as an indefinite integral

$$\mu = \int (f \circ g) \, d\alpha \qquad (9.4.33)$$

for some measurable function f:reals $\to$ reals.

That is, μ is a Thünen measure iff it has a density function that has the same value at points equidistant from the nucleus. We may expect most structural features of a Thünen system (such as the distribution of population or of capital goods by value) to have a distribution of the form (33), at least approximately.

To introduce Weberian ideas, we want to consider distances between any two points in the system, not just between the nucleus and other points. Accordingly, let $h:S \times S \to$ reals be given, where $h(x, y)$ is as usual the ideal distance from x to y. We assume that h is measurable and

---

54. For an earlier discussion of symmetry and convexity in the location of medians see Faden, *Essays,* pp. 101–3.

symmetric: $h(x, y) = h(y, x)$. The function g above is obtained from h by holding one of its arguments fixed at s_N: $g(x) = h(s_N, x)$.

Consider now the problem of locating a plant that is to make shipments whose distribution is a Thünen measure μ. The median for μ will usually be a point near the nucleus. If rent differentials are negligible for the plant, the median may be taken as the optimal location. If not—and rent differentials are generally important in the Thünen context—a reasonable location objective is to minimize the *sum* of rent and transport costs incurred (both discounted to the present). This has the effect of pushing the optimal location further from the nucleus than the median is.

This cross between Weber and Thünen introduces a certain conceptual difficulty. On the one hand, the plant is treated as if located at a single point. On the other hand, it incurs rent and therefore occupies a region of positive area. This is a contradiction (unless we have a strange areal measure assigning positive values to single points). The only fully adequate way to resolve this paradox is to treat the plant as a spatially extended entity, as we did with the regions controlled by individual agents in Section 8.7. But this approach entails great complications, and we do not follow it here. Instead, we leave the paradox unresolved, so that the following arguments must be construed as merely heuristic.

Is it realistic to assume that the shipment distribution will be a Thünen measure? Doubts on this score arise when the plant is located far off center; for then points at equal distances from the nucleus tend to be at quite *un*equal distances from the plant. Would this not lead to unequal demand densities along the same Thünen ring? This doubtless happens, but there are many cases where (33) would seem to be a good approximation. First, this is true where a uniform pricing policy is in effect: stores making free deliveries, or uniform charges for telephone service, mail, gas, water, or sewage, independent of location within the system. Second, (33) seems reasonable for government services that are "plant-initiated" rather than "consumer-initiated," e.g., routine police patrolling, building inspections. Third, (33) seems reasonable for goods and services having inelastic demands.

We now specialize to the case where S is the plane, and α is 2-dimensional Lebesgue measure. Also let h be the Euclidean metric. Without loss of generality, let the nucleus be at the origin: $s_N = (0,0)$. We observe that the nucleus is a median for any Thünen measure. For α is invariant under reflection through the nucleus: $(x, y) \rightarrow (-x, -y)$. The Thünen rings are circles, and these are also invariant; hence the density $f \circ g$ is invariant under reflection. Thus μ is invariant. Finally, h is invariant and convex besides, which proves that s_N is a median by an argument given above.

But the nucleus is also the point at which rent density hits its peak. The sum of rent and transport costs will usually be minimized at some distance away from the nucleus.

The interesting result arises when we consider the *multiplant Weber problem* in this Thünen context. Recall that in the multiplant Weber problem (4), transport costs incurred fall into two categories. There is a cost for shipments between each plant and the rest of the world of the form

$$\int_S h(x_i, s)\, \mu_i(ds) \qquad (9.4.34)$$

for plant i. Here x_i is the location of plant i that is to be chosen, and μ_i is a given spatial distribution, as in the ordinary Weber problem. And there are the interplant shipment costs of the form

$$w_{ij} h(x_i, x_j) \qquad (9.4.35)$$

for the pair plant i, plant j. Here w_{ij} is a given nonnegative number. Total costs, then, are the sum of *three* parts: (34) summed over all plants, plus (35) summed over all pairs of plants, *plus* the rentals incurred on all plants.

We assume that each of the measures μ_i is a Thünen measure. *If rents are negligible, an optimal solution is for all plants to locate at the nucleus;* for this arrangement simultaneously minimizes all terms in (34) and (35), since the nucleus is a median for any Thünen measure.

Assume that an optimal solution exists. We now show that an optimal solution exists of a very special form: *all plants are on the same ray from the origin.*

To prove this, let (x_i, y_i) be the cartesian coordinates of plant i in an optimal arrangement ($i = 1, \ldots, k$ if there are k plants in all). Consider the new arrangement in which plant i is now located at $((x_i^2 + y_i^2)^{1/2}, 0)$ for all i; i.e., each plant is now located on the positive X-axis at the same Euclidean distance from the nucleus as before.

Consider the effect on the three parts of total costs. Shipment costs (34) are the same for each plant. To see this, note that the shift of the location of plant i can be effected by a rotation of Space around the nucleus. This transformation leaves α invariant and leaves the density function of μ_i invariant; hence it leaves μ_i itself invariant. Also, h is invariant. By a theorem above, it follows that Weberian costs (34) are invariant.

Interplant shipment costs (35) are, if anything, lower than before; for plant i and plant j are rotated on two concentric circles. A pair of points on these respective circles are at minimum distance iff they lie on the same ray from the origin, which is precisely how they do lie in the new arrangement.

Finally, rentals are the same as before for each plant, since rentals depend only on Euclidean distance from the nucleus, which is unchanged. Thus total costs are no larger than before. Since the original arrangement is optimal, the new arrangement must be also.

This concludes the argument. The heuristic part is in the preceding paragraph, where we implicitly assume that the rental incurred by a plant depends only on the rent-density of its location.

Clearly, there is nothing special about the X-axis. The plants could have been rotated to any other ray emanating from the nucleus. This entire argument generalizes easily from 2- to n-dimensional space (with α being n-dimensional Lebesgue measure, h the Euclidean metric in n-space, etc.), and we again conclude that there is an optimal arrangement for the multiplant Weber problem in which all plants lie along the same ray emanating from the nucleus (provided an optimal arrangement exists at all).

Can we also draw the converse conclusion that any arrangement in which all plants do *not* lie on a ray through the nucleus is *not* optimal? The answer is yes, provided the multiplant Weber problem does not reduce to two independent subproblems. To make this notion precise, say that plant i and plant j are *directly linked* iff $w_{ij} + w_{ji} > 0$. Say that plant i and plant j are *linked* iff there is a sequence of plants, beginning with i and ending with j, such that each adjacent pair of plants in the sequence are directly linked. Finally, say that the multiplant Weber problem is *irreducible* iff every two plants are linked.

Then, if the problem is irreducible, any optimal arrangement must have all plants lying on some ray emanating from the nucleus. To prove this, take an arrangement in which plants i_o and j_o are not on the same ray. Taking a sequence of directly linked plants beginning with i_o and ending with j_o, there must be some adjacent pair of plants in this sequence that are not on the same ray, say i_1 and j_1. Now perform the rotation shift to a single ray as above. No cost term is increased by this shift. Furthermore, the sum of the two terms $w_{i_1 j_1} h(x_{i_1}, x_{j_1}) + w_{j_1 i_1} h(x_{j_1}, x_{i_1})$ is strictly less than before, since h declines and one of the w's is positive. Hence the original arrangement cannot be optimal. This concludes the proof.

If the multiplant Weber problem is reducible, then the set of plants can be partitioned, so that any two plants within the same partition set are linked and no two plants in different partition sets are linked. Hence, in an optimal arrangement all plants in the same partition set must lie on a single ray, but the rays for different partition sets may be different and can be chosen arbitrarily.

We now drop the assumption that the metric is Euclidean, and assume instead that it is *city-block*. Thus (in n-space) the distance between n-tuples $(x_1, \ldots, x_n)$ and $(y_1, \ldots, y_n)$ is given by $|x_1 - y_1| + \cdots + |x_n - y_n|$. (We work in n-space because the following arguments are more transparent with this generality; $n = 2$ is the only case of possibly practical importance.)

First, we make a general observation on the Weber problem under the city-block metric. This metric has the happy property of decomposing the Weber problem (even the multiplant Weber problem) in n-dimensional space into n separate 1-dimensional problems, which are much easier to solve. To see this, consider the two kinds of transport costs in the multiplant problem.

For the external shipments of plant i the cost is

$$\int_S [\,|\,x_{i1} - s_1\,| + \cdots + |\,x_{in} - s_n\,|\,]\, \mu_i(ds_1, \ldots, ds_n)$$

$$= \int_{-\infty}^{\infty} |\,x_{i1} - s_1\,|\, \mu_i^1(ds_1) + \cdots + \int_{-\infty}^{\infty} |\,x_{in} - s_n\,|\, \mu_i^n(ds_n) \qquad (9.4.36)$$

when the plant is located at point $(x_{i1}, \ldots, x_{in})$; cf. (34). Here the given shipment distribution μ_i can be any bounded measure, not necessarily a Thünen measure. In (36) the measures $\mu_i^1, \ldots, \mu_i^n$ are the n marginals of μ on the 1-dimensional component subspaces of n-space S. Equation (36) itself follows from the induced integrals theorem: split the original integral into a sum of n integrals in the obvious way, then project the l-th integral onto the l-th component subspace, $l = 1, \ldots, n$.

For the interplant shipment from plant i to plant j the cost is

$$w_{ij} \,[\,|\,x_{i1} - x_{j1}\,| + \cdots + |\,x_{in} - x_{jn}\,|\,] \qquad (9.5.37)$$

when plants i and j are located at $(x_{i1}, \ldots, x_{in})$ and $(x_{j1}, \ldots, x_{jn})$ respectively; cf. (35). If there are k plants in all, (36) and (37) together contain $(k + k^2)n$ terms, the sum of which is to be minimized over the kn unknowns x_{il} = the l-th component of the location of plant i.

Group the kn unknowns into n sets of k variables each, viz., $\{x_{1l}, \ldots, x_{kl}\}$, $l = 1, \ldots, n$. The $(k + k^2)n$ terms can be grouped into n sets of $k + k^2$ terms each such that the l-th set of terms contains only variables from the l-th set of unknowns. Thus the original problem falls apart into n separate minimization problems. The prescription for solution is this: Given the problem determined by measures $\mu_1, \ldots, \mu_k$ and weights w_{ij}, $i, j = 1, \ldots, k$, solve the 1-dimensional problem determined by the marginals $\mu_1^l, \ldots, \mu_k^l$ and the same weights w_{ij} (and the usual metric on the real line) for each $l = 1, \ldots, n$. Let the optimal solution be the k-tuple $(x_{1l}^o, \ldots, x_{kl}^o)$, for each l. Then the optimal location for plant i in the original problem is the n-tuple $(x_{i1}^o, \ldots, x_{in}^o)$, for $i = 1, \ldots, k$.

Let us now return to the Thünen context: S is n-space, and each measure μ_i in the multiplant Weber problem is a Thünen measure of the form (33). In (33) $g(s)$ is the city-block distance of point s from the nucleus s_N, which is at the origin $(0, \ldots, 0)$; α is n-dimensional Lebesgue measure. The problem again is to find an arrangement of plants minimizing the sum of total transport costs *and* total plant rentals. Total transport costs are given by the sum of all terms in (36) and (37). We assume that an optimal solution exists.

In the Euclidean case we concluded there was an optimal arrangement in which all plants located on the same ray from the origin; any ray would do. In the *city-block* case we claim again that an optimal arrangement exists in which all plants locate on the same ray; but we claim this only for *certain* rays. Specifically, we claim that *an optimal arrangement exists on the ray passing through the point* $(\pm 1, \pm 1, \ldots, \pm 1)$. Here any of the 2^n possible sign patterns may be chosen, and any of the 2^n corresponding rays holds an optimal arrangement. In 2-space these are the four rays emanating from the origin at 45° to the X and Y axes in the northeast, northwest, southeast, and southwest directions respectively.

We prove this for the ray passing through the point $(1, 1, \ldots, 1)$; the slight modifications needed for the other eligible rays should be obvious. Let $(x_{i1}^o, \ldots, x_{in}^o)$, $i = 1, \ldots, k$, be an optimal arrangement of plants, and consider the new arrangement in which plant i is located at

$$(1, 1, \ldots, 1)[\,|x_{i1}^o| + \cdots + |x_{in}^o|\,]/n \qquad (9.4.38)$$

for $i = 1, \ldots, k$; i.e., each plant is shifted to the special ray while remaining at the same city-block distance from the nucleus as before. Let us compute the effect on each component of total cost.

All rental costs remain the same, since rent density depends only on ideal distance from the nucleus, which is unchanged. (This is the same heuristic argument used in the Euclidean case.)

As for interplant transport costs (37), they decline, or at worst remain the same, because the distances between each pair of plants are no greater than before. This results from the following theorem. Let 0 be the origin, and s_1, s_2 two points such that $h(s_i, 0) = c_i > 0$, $i = 1, 2$; then $h(s_1, s_2)$ attains its minimum whenever s_1 and s_2 lie on the same ray from the origin. This statement is true not only for the city-block metric but for any metric h derived from a norm $\| \cdot \|$. The proof runs as follows:

$$h(s_1, s_2) = \|s_1 - s_2\| \geq |\,\|s_1\| - \|s_2\|\,|$$
$$= |h(s_1, 0) - h(s_2, 0)| = |c_1 - c_2| \qquad (9.4.39)$$

Relation (39) establishes a lower bound for $h(s_1, s_2)$. On the other hand, if s_1 and s_2 lie on the same ray, then

$$\|s_1 - s_2\| = \|s_1 - s_1(c_2/c_1)\| = |1 - (c_2/c_1)| \cdot \|s_1\| = |c_1 - c_2|$$

so that this lower bound is attained, and the proof is finished. Since plants are shifted onto the same ray while maintaining their respective distances from the origin, it follows that interplant distances do not rise.

Finally, we come to the external shipment costs (36). The argument here is a bit more complicated. We consider two types of transformations of Space into itself. The first is of the form

$$(x_1, \ldots, x_n) \to (x_1, \ldots, x_{l-1}, -x_l, x_{l+1}, \ldots, x_n) \qquad (9.4.40)$$

for some $l = 1, \ldots, n$; i.e., the sign of coordinate x_l is reversed, every-thing else remaining the same (reflection through the hyperplane $x_l = 0$). The second is a cyclic permutation of coordinates:

$$(x_1, \ldots, x_n) \rightarrow (x_n, x_1, x_2, \ldots, x_{n-1}) \qquad (9.4.41)$$

Both these transformation leave Lebesgue measure α invariant, since they have Jacobians equal to ± 1. Both transformations also preserve city-block distances from the nucleus. It follows that both transformations leave any *Thünen* measure μ_i invariant. Also, both transformations leave city-block distances in general invariant. It follows that the total external shipment costs incurred by plant i, (36), are invariant under transforma-tion of its location by either (40) or (41).

Now, starting from the original location $(x_{i1}^o, \ldots, x_{in}^o)$ of plant i, apply the transformation (40) once for each *negative* coordinate x_l. It follows that

$$u(x_{i1}^o, \ldots, x_{in}^o) = u(\,|\,x_{i1}^o\,|\,, \ldots, |\,x_{in}^o\,|\,) \qquad (9.4.42)$$

where $u(x_1, \ldots, x_n)$ is the value of the Weber cost (36) when plant i is located at point $(x_1, \ldots, x_n)$. Next, apply (41) n times to the point $(\,|\,x_{i1}^o\,|\,, \ldots, |\,x_{in}^o\,|\,)$. The resulting n points all have the same u-value (42). Furthermore, the *mean* of these n points is precisely the point (38) to which plant i is shifted. But u is a convex function, since the city-block metric is convex. Hence the u-value at point (38) is less than or equal to the common u-value (42). This proves that external shipment costs do not rise.

Thus the shift to arrangement (38) causes no rise in any cost category. Since the original arrangement was optimal, so is (38). This concludes the argument for the city-block case.

On the city-block plane, for example, the multiplant Weber problem will have an optimal solution with all plants strung out along the 45° line heading northeast (or northwest, southeast, or southwest).

Do these results have any exemplifications in the real world? This is not easy to answer. As for the special "diagonal direction" solution in the city-block case, this seems to be rather sensitive to small deviations from "city-blockness," so that we would expect the specific directional effect to be obliterated, even when the transportation grid approximates the ideal rectangular form. But the predicted tendency for linked plants to be strung out along *some* ray emanating from the nucleus appears to be more robust.

This would show up, for example, in a tendency for cities to grow persistently in one (or a few) directions, rather than to expand uniformly about their nuclei—a directional bias that would show up even in the absence of natural irregularities in geography. Another possible exempli-fication might be the tendency noted by Lösch for some cities to exhibit alternating sparsely and densely settled corridors, a tendency he explained by a rather dubious theory of superimposed hexagonal industry networks.[55]

Two points should be kept in mind in any such suggested interpretation. First, any tendency for plants to string themselves out in rays would stimulate the construction of radial transportation arteries along these rays. This reinforces the original tendency but also undermines the metric structure upon which the theory expounded above rests. Conversely, the existence of radial arteries furnishes an independent reason for the "stringing-out" phenomenon to occur.

Second, a widespread tendency for plants to string themselves out in a Thünen system undermines the Thünen ring structure itself and makes it less plausible that the distribution of shipments from any newly located plant will be a Thünen measure. In this sense the theory is "self-undermining," just as, say, a Keynesian equilibrium involving a nonzero net investment rate is "self-undermining."

9.5. MARKET REGIONS

We now turn our attention from the location of facilities to the pattern of *shipments*. The setup is the same as in the transportation problem of Chapter 7. There is an origin measure and a destination measure, and the problem is to find a flow measure (on the product space, origin × destination) that transfers the mass from origin to destination in an efficient manner. A special assumption distinguishes this section: we assume that at least one of the two measures, origin or destination, is *concentrated at a countable (possibly a finite) number of points*. Just as in the allotment-assignment problem where a different special assumption was made, this yields results that are not available for the general transportation problem.

There are many real-world illustrations of this model. We consider only cases in which the origin and destination spaces are the same—usually a portion of physical Space S but sometimes a portion of Space-Time $S \times T$.

Consider a multiplant industry that is to ship its product to satisfy a spatial demand distribution. The latter is the destination measure. Each plant has a certain given production level, which may be interpreted as a capacity, as a quota assigned by a cartel or central manager, as a point on a supply schedule, or in other ways. The origin measure is the spatial output distribution for the industry, and this will be concentrated at the finite number of locations occupied by the plants. The problem is to find

---

55. Lösch, *Economics of Location,* pp. 125 and 129, n. 13.

a pattern of shipment from each plant such that together they satisfy demand, such that plant production levels are not exceeded, and such that overall transportation cost is minimized over the set of shipment patterns satisfying these feasibility conditions.

An example of the opposite flow pattern is furnished by milk distribution to large cities.[56] Here the origin measure is the spatial distribution of milk production, while the destination measure is urban milk consumption. If we think of cities as being located at single points (which is reasonable from a broad regional point of view), the destination measure is concentrated at the finite number of city locations.

Let us refer to these two setups as demand systems and supply systems respectively. Formally, a *demand system* is a transportation problem in which the origin measure is concentrated at a countable number of points; a *supply system* is one in which the destination measure has this property. (These conditions may hold simultaneously. Indeed, if both origin and destination measures are concentrated at a *finite* number of points, things reduce to the ordinary transportation problem of linear programming.)

There is another very common situation that is both a demand and a supply system. This is the case in which resources flow in both directions. Most commonly, this takes the form of round trips by "commuting" resources such as people or vehicles. Consider the problem of assigning children to schools. Here there are two relevant measures: the spatial distribution of children by residence and the spatial distribution of school capacity. If schools are thought of as located at single points, the latter measure is concentrated at these points.

Suppose one wants to minimize overall transport cost of school trips, subject to meeting the school capacity and compulsory education constraints. This factors into two separate problems. The optimal flow pattern *from* home *to* school comes from the *supply* system in which residential distribution of children is the origin measure and school capacity distribution is the destination measure. The optimal flow pattern *to* home *from* school comes from the *demand* system in which the roles of these measures are interchanged.

This procedure has a flaw, however. Solving the "ingress" and "egress" problems separately may result in children being reshuffled to different homes from the ones they left. Since children are not interchangeable, this is unacceptable. The difficulty can be overcome by an artifice. In place of the unit cost function $h(x, y)$ substitute the *round-trip* cost $h(x, y) + h(y, x)$ and find the optimal flow pattern λ for the home-to-school supply system with this cost function. It follows from the symmetry of round trip costs that the optimal flow pattern for the school-to-home demand system is precisely the transpose (i.e., the reversal) of λ, so all children can indeed return to the homes from which they started. (The proof of this statement for the transportation problem, with equality constraints and identical origin and destination spaces, is virtually the same as the corresponding proof for the transhipment problem p. 566.)

This same supply-demand system approach applies to urban facilities in general—not only to schools but to the assignment of hospitals, court-houses, libraries, fire stations. Still another class of applications arises in the organization of hierarchical structures. Given the scattering of higher order and lower order units, which of the latter should be con-nected to or controlled by which of the former to minimize overall costs of connection, communication, or control? Examples would be the divisional hierarchy in a firm, the regional chain of command in the military, the clearing apparatus in the banking system, the postal and telephone systems. In all these cases the facilities or higher order units will have capacity limits on their ability to serve or control the population of lower order units, clients, etc.[57]

All the examples so far have been in physical Space. An example in Space-Time is provided by the marketing of a crop with a sharply defined harvest time. We may think of this crop as making an appearance in the system at isolated points (s, t) of Space-Time, where s is, say, the location of a grain elevator and t is a moment at which the harvest arrives at s. The unit transport cost function has domain $S \times T \times S \times T$, $h(s_1, t_1, s_2, t_2)$ being the cost of transporting unit mass from site s_1 at in-stant t_1 to site s_2 at instant t_2. Here "transportation" is used in the wide sense that includes storage (= transportation through time), and cost includes storage cost. The destination measure is the Space-Time pattern of consumption μ, a measure on $S \times T$: $\mu(E \times F)$ is the mass consumed in region E in period F. The origin measure is concentrated at the count-able number of points (s, t) at which the crop appears in the system. Thus we again have a demand system.

Going a step further, we might consider a system operating in Time alone. An example would be the model just discussed with all references to Space suppressed. The unit cost function has domain $T \times T$, $h(t_1, t_2)$ being the cost of storing unit mass from moment t_1 to t_2.[58]

---

57. M. Yeates has formulated the school assignment problem as a transportation problem; P. R. Gould and T. R. Leinbach have done the same for hospital services. For references and discussion see P. Haggett and R. J. Chorley, *Network Analysis in Geography* (St. Martin's Press, New York, 1969), pp. 208–10. Boulding uses the market region model to explain spatial distribution of political sovereignty, plants being "home base" points of maximal power. K. E. Boulding, *Conflict and Defense* (Harper and Row, New York, 1962), pp. 78–79, 230–45. Similarly for the distribution of competing ideologies, Boulding, p. 280. Distribution of "culture areas" may be explained similarly; see C. Wissler, *Man and Culture* (Crowell, New York, 1923) and A. L. Kroeber, *Anthropology*, 2nd ed. (Harcourt, Brace and World, New York, 1948).

58. P. A. Samuelson, Intertemporal price equilibrium: A prologue to the theory of speculation, *Weltwirtschaftliches Archiv* 79(1957):181–227.

We comment on these models involving Time. First, the value of h at points for which $t_2 < t_1$ is, strictly speaking, infinite, since one cannot go backward in time. In the transportation problem, however, the cost function must be finite. This difficulty can probably be overcome by replacing h by a function assigning very high finite values in place of $+\infty$, sufficiently high to discourage any flows that go backward in time. Second, to have a bona fide demand system, the appearance of mass at the points of origin (s, t) must be instantaneous. For suppose, say, a crop arrives steadily at some grain elevator over a positive time interval; then the origin measure is no longer concentrated at a countable number of points: an enduring point in Space corresponds to a "line" in Space-Time, which has an uncountable number of points.

We now sketch the main lines of inquiry. First, there is a characteristic property of optimal flow patterns in demand or supply systems. Just as allotment-assignment problems yielded solutions satisfying the *weight-falloff condition* (the fundamental property of Thünen systems), the transportation problem in a demand or supply system yields solutions in which the flow is "partitioned" into *market regions*. Exact definitions will come later; a rough description of the meaning of this term will suffice for now. In a demand system, this means that the space is partitioned, with one partition set or "market region" for each point of concentration of the origin measure. The demand in each market region is then satisfied exclusively by shipments from its corresponding point of concentration. The situation for supply systems is similar, except that the flow is *from* the market region *to* the corresponding point of concentration of the *destination* measure. Geographically, market regions tend to be relatively "close" to their corresponding points of concentration.

In the case of the multiplant industry shipping its product, each plant will have a territory surrounding it to which it ships exclusively. In the case of milk marketing to cities, each city will have a "milkshed," the dairy farms in which ship exclusively to that city. For the school system, each school will have a neighborhood "school district" from which it draws its student body exclusively.[59] For the seasonal crop marketing example, the "market regions" will be chunks of Space-Time, each following its corresponding point of concentration.

One of our aims is to pin down rigorously the extent to which this "market region" property does indeed characterize optimal flows in demand or supply systems. Our second main objective is to investigate the role of *prices* in these systems. We know from Chapter 7 that, under rather general conditions, an optimal solution λ to the transportation problem having origin space A and destination space B determines a pair

---

59. "Busing" is the policy of departing from this transport-cost-minimizing assignment to attain other social objectives.

of functions (p, q), $p:A \rightarrow$ reals, $q:B \rightarrow$ reals (called a *potential*), having a certain relation to λ. It turns out that p and q act *as if* they were origin and destination *price fields,* in the sense that the flow λ could arise from the behavior of agents participating in competitive markets and confronted with these prices. (For this reason the dual values arising in ordinary linear programming are often referred to as shadow-prices.)

The opposite approach is also of interest. That is, given competitive markets and prices in the overall framework of a demand or supply system, does the resulting flow pattern partition into market regions? Is it optimal in some sense?

We have been discussing market systems in terms of the transportation problem. The question arises, Why not use the *transhipment* formulation instead? Let μ' and μ'' be the origin and destination measures, and let λ' and λ'' be the left and right marginals of shipment measure λ on $(A \times A, \Sigma \times \Sigma)$ respectively. Then for the transportation problem we have the constraints on λ:

$$\lambda' = \mu', \qquad \lambda'' = \mu'' \tag{9.5.1}$$

while for the transhipment problem we have the weaker constraints

$$\lambda' - \lambda'' = \mu' - \mu'' \tag{9.5.2}$$

(We are choosing the equality-constrained variants of both problems and assuming μ', μ'' are bounded in (2).) Condition (2) seems preferable at first glance, since (1) imposes artificial restrictions that do not really constrain a shipper. The trouble is that the transhipment formulation (2) obliterates the special feature that characterizes market systems; i.e., μ' or μ'' is concentrated on a countable set, a feature that disappears in their difference $\mu' - \mu''$. Measures μ' and μ'' enter only through their difference in (2), whereas they retain their separate identities in (1). Thus we are forced to rely on the transportation problem format if we want any chance of developing a special theory for market systems.

In applying the results we obtain, one should check that there is no advantage in transhipping, so that it suffices to consider only the translocations allowed by (1). Section 7.10 gives conditions that guarantee this, the main one being that the unit cost function h satisfies the triangle inequality. Similarly, the fact that the origin and destination spaces are identical is irrelevant from the perspective of the transportation problem but imposes certain additional constraints in reality. (Example: A potential for the transportation problem is a pair of functions (p, q) on the origin and destination spaces respectively; we should have $p = q$ if the spaces are identical.) These considerations are, however, "metatheoretic": in the theory of market systems per se, the triangle inequality, for example, plays no role and will not be mentioned again.

Let us now get down to business. The first task is to define the concept of "market region" ("demand region" if we are dealing with a demand system, "supply region" if we are dealing with a supply system). Two entirely different concepts are needed—a *potential* concept indicating where the shipments of a plant *might* go to (or come from) under competition, and an *effective* concept indicating where the shipments of a plant actually *do* go to (or come from).

We begin with the "potential" concept for a demand system. We are given a countable number of points $a_1, a_2, \ldots$ in a space A, together with a unit transport cost function $h: A \times A \to$ reals and a family of *mill prices* $p_1, p_2, \ldots$, one for each plant (plant i is located at point a_i). Consider the situation of a potential buyer located at point $b \in A$ and contemplating which plant to purchase from. If he purchases from plant i, the unit cost (counting in both mill price and transport cost) will be

$$p_i + h(a_i, b) \qquad (9.5.3)$$

We suppose that the buyer, if he makes any purchases at all, will buy from a plant (or plants) i for which (3) is a minimum. This motivates the following definition.

Definition: The *potential demand region* of plant i is

$$P_i = \{b \mid p_i + h(a_i, b) \le p_j + h(a_j, b) \quad \text{for all} \quad j = 1, 2, \ldots\} \qquad (9.5.4)$$

That is, for point $b \in A$ form the numbers $p_j + h(a_j, b)$, one for each plant, and find the smallest (if there is a smallest). Then b belongs to the potential demand region of whichever plant attains this smallest number.

If a potential buyer is *not* located in P_i, he will *not* purchase from plant i, since the cost to him is less from some other plant. The converse of this statement is false: a buyer may well be located in P_i without purchasing from plant i—because either he purchases from another equally costly plant or perhaps chooses not to purchase at all.

We note some properties of potential demand regions. If the number of plants is finite, the regions cover the entire space A. For, given b, the numbers $p_j + h(a_j, b)$ have at least one smallest among them, so b belongs to at least one P_i. But P_1, P_2, ... need not be a packing; there may be overlaps where two or more plants are equally lowest priced. Call these common points *border points* between two demand regions. As an example, let A be the plane and let h be Euclidean distance. Border points b between P_i and P_j must satisfy

$$p_i + h(a_i, b) = p_j + h(a_j, b) \qquad (9.5.5)$$

The locus of points b satisfying (5) is either empty or a hyperbola (or in limiting cases a straight line or a ray). Thus on the Euclidean plane, potential demand regions are bordered by hyperbolic arcs (and/or line segments).

Specializing still further leads us to a case of considerable importance. Consider the Euclidean plane where (i) the price at all plants is the same: $p_i = p_j$, all i, j and (ii) each plant has six nearest neighboring plants at distance $c > 0$, these forming the corners of a regular hexagon. Condition (i) assures that each point of the plane belongs to the potential demand region of the plant to which it is closest. (The resulting demand regions must then be convex polygons, often called *Dirichlet regions, Voronoi polygons,* or *Thiessen polygons.*) Condition (ii) is the *honeycomb lattice* arrangement of plants. It implies that the number of plants is infinite. In this case the potential demand regions are regular hexagons, all of side length $c/\sqrt{3}$, centered on their respective plants; they cover the plane and are nonoverlapping except along their edges. This is the basic structure on which Christaller and Lösch build their theories.

We now turn to the "potential" concept for *supply* systems. We again have plants at $a_1, a_2, \ldots$, with mill prices $p_1, p_2, \ldots$; but these plants are now recipients rather than donors of shipments, and transport cost must be *subtracted* from mill price to find the net revenue of the seller. Consider the situation of a potential seller located at point b. If he sells to plant i, his net revenue will be

$$p_i - h(b, a_i) \tag{9.5.6}$$

If he sells at all, it will be to a plant (or plants) for which (6) is a maximum. This suggests the following concept.

Definition: The *potential supply region* of plant i is

$$P_i = \{b \mid p_i - h(b, a_i) \geq p_j - h(b, a_j) \quad \text{for all} \quad j = 1, 2, \ldots\}$$

That is, for point $b \in A$ form the numbers $p_j - h(b, a_j)$, one for each plant, and find the largest (if there is a largest). Then b belongs to the potential supply region of whichever plant attains this largest number.

For every statement we have made about potential demand regions there is a corresponding statement about potential supply regions. Thus if a potential seller is *not* located in P_i, he will *not* sell to plant i; but the converse is not necessarily true. If the number of plants is finite, the potential supply regions cover space A, but these regions will in general overlap. Dirichlet regions and, in particular, the regular hexagons of a honeycomb lattice can be constructed in terms of supply regions instead of demand regions.

The system of potential demand or supply regions tells us relatively little about the actual resource flow pattern. For a simple illustration, take a demand system on the Euclidean plane and let just *one* plant a_o exist. There is just one potential demand region, viz., the entire plane. But the actual flow will typically be to a bounded region in the vicinity of a_o because transport costs become prohibitively expensive for distant points. Thus we must distinguish the region to which the flow from the plant *might* go (its *potential* demand region), from the subregion to which the flow *actually does* go (its *effective* demand region).

To introduce the "effective" concepts, we need some technical constructions. (From here on we consider only demand systems. A completely parallel theory exists for supply systems. To derive it, reverse the roles of the origin and destination spaces in what follows, "transpose" h, and substitute $-p_i$ for p_i.) We again have a space A and a countable set of points $A' = \{a_1, a_2, \ldots\}$. Demand systems have been discussed so far in terms of a transportation problem on the product space $A \times A$. Now it will be more convenient to let the product space be $A' \times A$; i.e., the origin space is the set of plant locations. (This involves no real loss of generality, since by definition of a demand system there is no resource flow out of the complementary set $A \setminus A'$.)

Sets A and A' are provided with σ-fields Σ and Σ' respectively. We now make the additional specification that Σ' is *discrete;* i.e., Σ' is the class of all subsets of A'. Since A' is countable, this is equivalent to assuming that $\{a_i\} \in \Sigma'$ for each point a_i. (Since $A' \subseteq A$, one might think that Σ' should be the restriction of Σ to A'. However, there is no need to make such an assumption; indeed, the fact that A' is a subset of A is completely irrelevant for the theory that follows.)

Now let λ be a shipment measure, i.e., a measure on the product space $(A' \times A, \Sigma' \times \Sigma)$. The discreteness of Σ' enables us to decompose λ into the separate shipments out of each plant.

Definition: λ_i, the *shipment distribution* from plant i, is the measure on (A, Σ) given by $\lambda_i(E) = \lambda[\{a_i\} \times E]$, all $E \in \Sigma$.

(Note that λ_i is a measure on A, while λ itself is a measure on the product space $A' \times A$.) The sum of all the measures λ_i, $i = 1, 2, \ldots$, is none other than the right marginal λ'' of λ.

The effective demand region of plant i is, intuitively, the set of points "at" which measure λ_i is located. Any point b carrying positive measure should clearly be in this region. But in general other points should also be included; if not, we would have to say that a nonatomic measure is not located anywhere. The concept appears to depend unavoidably on topological ideas. Accordingly, we introduce a topology $\mathfrak{J}$ on A in addition to its measure structure.

Definition: The *effective demand region* of plant i (notation: E_i) is the support of measure λ_i (relative to topology $\mathfrak{J}$).

Recall that for any measure space (A, Σ, μ) with topology $\mathfrak{J}$, point $a_o \in A$ *supports* μ iff every measurable neighborhood of a_o carries positive

μ-measure. The *support* of μ is the set of all such points and is always a closed set. As an example, let A be the plane, let $\Im$ be the usual plane topology (the one generated by the set of all open circular discs), and let Σ be the usual 2-dimensional Borel field. Then point b supports λ_i iff every circular disc of positive radius centered at b has positive λ_i-measure. The effective demand region of plant i is then the set of all such points. This is an eminently reasonable explication of the notion of the "location" of the shipments of plant i and, indeed, seems to be the one implicitly used by all authors who deal with the concept on the plane.

We also must clarify the concept, for the shipment measure λ, of mass flowing "from" plant i "to" point b. Again it is too restrictive to confine this to the case where the singleton pair $\{(a_i, b)\}$ has positive λ-measure, for no pairs would be eligible if λ were nonatomic. Instead, we say that *plant i ships to b* iff the pair (a_i, b) supports λ. Since λ is defined on the product space $A' \times A$, we need a topology on this space to give meaning to the concept of support. We use the product topology $\Im' \times \Im$ for this purpose, where $\Im$ on A has already been discussed and $\Im'$ is the *discrete* topology on A'; i.e., $\Im'$ is the class of all subsets of A' (thus $\Im' = \Sigma'$).

The following result shows that the two concepts just defined are related in an intuitively appealing way.

Lemma: Point b belongs to effective market region E_i iff plant i ships to b.

Proof: Let $b \notin E_i$. Then sets $G \in \Im$, $F \in \Sigma$ exist such that $b \in G \subseteq F$ and $\lambda_i(F) = 0$. Then

$$(a_i, b) \in [\{a_i\} \times G] \subseteq [\{a_i\} \times F] \tag{9.5.7}$$

The two bracketed sets in (7) belong respectively to $\Im' \times \Im$ and $\Sigma' \times \Sigma$, since $\Im'$, Σ' are discrete. Also $\lambda[\{a_i\} \times F] = \lambda_i(F) = 0$ Hence (a_i, b) does not support λ; i.e., plant i does not ship to b.

Conversely, let a_i not ship to b. Then sets $H \in \Im' \times \Im$, $K \in \Sigma' \times \Sigma$ exist such that $(a_i, b) \in H \subseteq K$ and $\lambda(K) = 0$. Then

$$b \in \{a \mid (a_i, a) \in H\} \subseteq \{a \mid (a_i, a) \in K\} \tag{9.5.8}$$

The two sets in (8) belong respectively to $\Im$ and Σ. Also

$$\lambda_i \{a \mid (a_i, a) \in K\} \leq \lambda(K) = 0$$

Hence b does not support λ_i; i.e., $b \notin E_i$. ∎

Potential and effective market regions are entirely different concepts, the former defined in terms of mill prices p_i and unit cost function h, the latter in terms of the support of a shipment measure λ. Yet in some cases they coincide, which has led to the unfortunate custom in location theory of bracketing them both under the same term (e.g., "market area") and rarely distinguishing explicitly between them.[60] This has led to confusion, for a number of standard results in the literature are valid for just one of these two concepts. For example, we mentioned above that, on the Euclidean plane, potential market regions will be bounded by hyperbolic arcs (or by straight-line segments in special cases).[61] This statement is in general *false* for *effective* market regions. With population uniform in tastes and income, effective market regions are likely to be bounded in part by circular arcs (at the distance from the plant at which transport cost just chokes off demand). If the population distribution has blank spots (bodies of water, deserts, etc.), the effective market regions may well have holes or indentations in them, a contingency that cannot arise for potential regions.

A curious case arises in the Löschian system, where there is a postulate that all market regions together exhaust space.[62] This postulate is ambiguous, since Lösch never distinguishes between the "potential" and "effective" region concepts. If we interpret the postulate to refer to potential regions, then it is true, since the potential market regions cover space (in the case of the finite number of plants assumed by Lösch). However, if we interpret it to refer to *effective* regions (as do Mills and Lav), then the proposition is sometimes *false,* as these authors were the first to point out.[63]

We next turn to the question of the relation between potential and effective demand regions P_i and E_i. Of course, without further assumptions there need not be any particular relation between them. Some connection is needed between p_i, $i = 1, 2, \ldots,$ and h on the one hand and the support of λ on the other.

We have already touched on one possible connection arising from cost-minimizing behavior on the part of participants in the demand system, viz., that no one located outside the potential demand region of plant i will purchase from plant i. There are several ways to explicate this notion. One of them is

$$E_i \subseteq P_i \qquad (9.5.9)$$

60. An explicit distinction between them is drawn in Faden, *Essays,* pp. 70–71, 141. A closely related distinction between "virtual" and "actual" production is made by R. E. Kuenne, *The Theory of General Economic Equilibrium* (Princeton Univ. Press, Princeton, N.J., 1963), p. 441. In fact Kuenne uses the term "potential market area" for what we call "potential demand region."

61. F. A. Fetter, The economic law of market areas, *Quart. J. Econ.* 38(May 1924): 520–29.

62. Lösch, *Economics of Location,* p. 95.

63. E. S. Mills and M. R. Lav, A model of market areas with free entry, *J. Polit. Econ.* 72(June 1964):278–88. We discuss this paper in Section 9.6.

for all $i = 1, 2, \ldots$. Condition (9) may be rationalized as follows. If $b \in E_i$, then plant i is shipping to b, so that someone "at" b is purchasing from plant i; by the cost-minimizing principle, $b \in P_i$.

Another connection may be formulated in terms of prices and transport costs rather than sets. Namely, for all plants i and all points $b \in A$,

$$\text{if plant } i \text{ ships to } b, \quad \text{then} \quad p_i + h(a_i, b) \le p_j + h(a_j, b) \quad (9.5.10)$$

for all plants $j = 1, 2, \ldots$. Condition (10) may also be rationalized in terms of cost-minimizing behavior. If someone at b purchases from plant i, the total unit cost (including mill price and transportation) from plant i will not exceed the same from any other plant.

Theorem: (9) is true iff (10) is true.

Proof: Assume (9), and let plant i ship to b. By the lemma above, $b \in E_i$; hence $b \in P_i$, which yields the inequality in (10) by definition of potential demand region.

Conversely, let (10) be true, and let $b \in E_i$. Again by the lemma above, plant i ships to b; hence the inequality in (10) is true for all plants j, so that $b \in P_i$. ■

Thus the different-looking conditions (9) and (10) are actually logically equivalent. We refer to either as constituting the *topological competitive rule* for the demand system. (The reason for this terminology will soon become clear.)

Both forms are suggestive; (9) gives a geometrical picture of the demand system, while (10) indicates that the mill prices p_i (which may be thought of as a function $p:A' \to$ reals) fulfill the conditions for a *left half-potential* for the transportation problem of minimizing total transport costs. This suggests a connection between mill prices and duality theory for the transportation problem, a suggestion that will be explored below.

There is yet another important way of expressing the behavior of purchasing from the least expensive plant. We first make a few observations concerning measurability. Consider the unit transport cost function $h:A' \times A \to$ reals (note that we are restricting the domain of h from $A \times A$ to $A' \times A$ just as we did with the rest of the demand system). It is sometimes useful to think of h as a family of functions $h_i:A \to$ reals, $i = 1, 2, \ldots$, one for each plant, defined by $h_i(b) = h(a_i, b)$ for all plants i, all points b. Thus $h_i(b)$ is the unit transport cost from plant i to point b.

It now follows readily from the facts that A' is countable and Σ' is discrete that h is measurable iff h_i is measurable for all $i = 1, 2, \ldots$. (Measurability of h refers to the product σ-field $\Sigma' \times \Sigma$; measurability of h_i refers to Σ. Similarly, it follows from the fact that $\mathfrak{I}'$ is discrete that h

is continuous iff h_i is continuous for all $i = 1, 2, \ldots$; continuity refers to the topologies $\mathfrak{J}' \times \mathfrak{J}$ and $\mathfrak{J}$ respectively.)

Lemma: If h is measurable, then $P_i \in \Sigma$ for all plants $i = 1, 2, \ldots$.

Proof: For all plants $i, j = 1, 2, \ldots$, the functions h_i, h_j are measurable; hence the same is true of their difference $g_{ij} = h_i - h_j$. It follows that the sets $F_{ij} = \{b \mid g_{ij}(b) \leq p_j - p_i\}$ belong to Σ. For fixed i, P_i is simply the intersection of F_{ij} over all $j = 1, 2, \ldots$. As the intersection of a countable number of measurable sets, P_i itself is measurable. ∎

This result insures that condition (11) to be discussed now is well defined.

Assume that h is measurable, and consider again the behavioral condition that no one located outside P_i will purchase from plant i. Perhaps the simplest explication of this condition, involving no topological concepts, is

$$\lambda_i(A \backslash P_i) = 0 \qquad (9.5.11)$$

for all plants $i = 1, 2, \ldots$. Condition (11) states literally that plant i does not ship to the complement of its potential demand region. We refer to (11) as constituting the *measure-theoretic competitive rule.*

We now have two competing competitive rules: (11) on the one hand and (9)–(10) on the other. Both involve a shipment measure λ on space $(A' \times A, \Sigma' \times \Sigma)$, together with a (measurable) transport cost function $h:A' \times A \to$ reals and system of mill prices $p:A' \to$ reals. In addition (9)–(10) involves a topology $\mathfrak{J}$ on A (as well as the discrete topology $\mathfrak{J}'$ on A').

It is of interest to find conditions under which one of these competitive rules implies the other. The more important implication runs from the topological to the measure-theoretic rule. Recall that a topology $\mathfrak{J}$ has the *strong Lindelöf property* iff every collection G of open sets has a countable subcollection whose union is $\cup \mathsf{G}$. This condition is fairly easy to fulfill; e.g., it holds for the usual topology of n-space for any finite n and for any subspace thereof.

Theorem: Let topology $\mathfrak{J}$ on A have the strong Lindelöf property. Then the topological competitive rule (9)–(10) implies the measure-theoretic competitive rule (11).

Proof: By (9), $E_i \subseteq P_i$, all $i = 1, 2, \ldots$. Let $b \in A \backslash P_i$; then $b \in A \backslash E_i$. Since b does not support λ_i, sets $G_b \in \mathfrak{J}$ and $F_b \in \Sigma$ exist such that $b \in G_b \subseteq F_b$ and $\lambda_i(F_b) = 0$. Let G be the collection of the sets G_b for all b not belonging to P_i. By the strong Lindelöf property there is a countable subcollection $\mathcal{G}' = \{G_{b_1}, G_{b_2}, \ldots\}$ such that $\cup \mathcal{G} = \cup \mathcal{G}'$. Collection $\mathcal{G}$ covers $A \backslash P_i$, so that

$$A \backslash P_i \subseteq (\cup \mathcal{G}) = (G_{b_1} \cup G_{b_2} \cup \cdots) \subseteq (F_{b_1} \cup F_{b_2} \cup \cdots)$$

Hence $\lambda_i(A \backslash P_i) \leq \lambda_i(F_{b_1}) + \lambda_i(F_{b_2}) + \cdots = 0$. Thus $\lambda_i(A \backslash P_i) = 0$ for all $i = 1, 2, \ldots$, which is (11). ∎

In the opposite direction we have the following.

Theorem: Let $A \backslash P_i \in \mathfrak{J}$ for all $i = 1, 2, \ldots$ (i.e., each P_i is a $\mathfrak{J}$-closed set). Then the measure-theoretic competitive rule (11) implies the topological competitive rule (9)–(10).

Proof: Let $b \in A \backslash P_i$. By assumption, $A \backslash P_i \in \mathfrak{J}$, and since h is measurable, $A \backslash P_i \in \Sigma$ also. Finally, $\lambda_i(A \backslash P_i) = 0$ by (11), so that b has a measurable neighborhood of λ_i-measure zero. Thus $b \in A \backslash E_i$. This yields (9). ∎

A sufficient condition for each P_i to be closed is that h be continuous. (This may be proved along the lines of the preceding lemma.) Thus under fairly general conditions the two competitive rules are actually equivalent.

The relations between topological and measure-theoretic competitive rules established by these last two theorems bear a striking analogy to the relations between topological and measure potentials in the transportation problem (Section 7.5).

We now want to explore the relations between competitive rules as discussed above and *optimality* of shipment flows in the sense of transport cost minimization. Specifically, let λ^o be a given measure on $(A' \times A, \Sigma' \times \Sigma)$ and consider the transportation problem: minimize

$$\int_{A' \times A} h \, d\lambda \qquad (9.5.12)$$

over measures λ on $(A' \times A, \Sigma' \times \Sigma)$ satisfying

$$\lambda' = \lambda^{o\prime}, \qquad \lambda'' = \lambda^{o\prime\prime} \qquad (9.5.13)$$

(Here λ', $\lambda^{o\prime}$ are the left marginals and λ'', $\lambda^{o\prime\prime}$ are the right marginals of these respective measures. We assume the marginals of λ^o are σ-finite. If (12) is not well defined and finite, we interpret it as an indefinite integral in the pseudomeasure sense and take "minimize" to refer to (reverse) standard ordering of pseudomeasures.)

We first show that (under certain boundedness conditions) a measure λ^o for which a system of mill prices, $p:A' \to$ reals, exists such that competitive rule (11) is satisfied is, in fact, optimal for the problem (12)–(13).

Here we make the usual assumptions: $A' = \{a_1, a_2, \ldots\}$ is countable, Σ' is discrete, and h is measurable. No topological concepts are used.

Theorem: Let λ^o be a bounded measure on $(A' \times A, \Sigma' \times \Sigma)$, and let $h{:}A' \times A \to$ reals and $p{:}A' \to$ reals be bounded functions such that the resulting system of potential demand regions P_i satisfies the measure-theoretic competitive rule:

$$\lambda_i^o(A \setminus P_i) = 0 \qquad (9.5.14)$$

all $i = 1, 2, \ldots$. Then λ^o is *best* for the problem of minimizing (12) over measures λ satisfying (13).

Proof: Define $q{:}A \to$ reals by

$$q(b) = \inf\{p_i + h_i(b)\} \qquad (9.5.15)$$

the infimum taken over all plants i. We show that the pair (p, q) is a *measure-potential* for λ^o, which implies that λ^o is best for (12)–(13) (Section 7.3).

We must verify four conditions. The first is measurability: q is the infimum of the countable collection of measurable functions $p_i + h_i(\cdot)$, $i = 1, 2, \ldots$, hence is itself measurable. The measurability of p is an automatic consequence of the discreteness of Σ'.

Second, we must show that the integrals

$$\int_{A'} p \, d\lambda^{o\prime}, \qquad \int_A q \, d\lambda^{o\prime\prime} \qquad (9.5.16)$$

are well defined and finite. λ^o and p are bounded; q is bounded, since p and h are bounded (we may ignore the trivial case $A' = \phi$). This and measurability insure that these integrals are well defined and finite.

Third,

$$q(b) - p_i \le h_i(b) \qquad (9.5.17)$$

for all plants i and points $b \in A$. Inequality (17) follows at once from (15).

Finally, we must show that

$$\lambda^o(G) = 0 \qquad (9.5.18)$$

where $G = \{(a_i, b) \mid q(b) - p(a_i) < h(a_i, b)\}$.

Suppose $b \in P_i$. It follows from the definitions of q and of potential demand region that $q(b) = p(a_i) + h(a_i, b)$ so that $(a_i, b) \notin G$. Hence G is contained in the union of all sets of the form

$$\{a_i\} \times (A \setminus P_i) \qquad (9.5.19)$$

one for each plant i. Each set (19) has λ^o-measure zero, by (14), and this implies (18).

Hence (p, q) is indeed a measure-potential for λ^o. ∎

This theorem gives a useful sufficient condition for optimality of a shipment measure in a demand system. But the boundedness assumptions that have to be made on h and λ^o are sometimes irksome. For example, if A is the Euclidean plane, the Euclidean metric h is certainly unbounded. Furthermore, suppose the plants form a honeycomb lattice over the plane, with each plant shipping exclusively to its Dirichlet region (the regular hexagon of points closer to the given plant than to any other). To be specific, suppose the destination measure is proportional to Lebesgue measure within each hexagon. Then the shipment measure λ is also unbounded. Thus the preceding theorem does not apply. Yet at the same time one feels intuitively that the shipment pattern specified must be optimal in some sense for the problem of supplying the plane uniformly from a honeycomb lattice of plants. Can any result along these lines be established?

It can. We first need a more general concept of Dirichlet region. We again have a countable set of points $A' = \{a_1, a_2, \ldots\}$ in a space A and a unit cost function $h{:}A' \times A \to$ reals. No price concepts are needed.

Definition: The *Dirichlet demand region* of plant i is

$$D_i = \{b \mid h_i(b) \leq h_j(b), \text{ for all } j = 1, 2, \ldots\}$$

Comparing this with the definition of potential demand region (4), we see that D_i is simply P_i in the special case where all prices $p_1, p_2, \ldots$ are identical. (Dirichlet *supply* regions may be defined by specializing potential supply regions in the same way.) Point b belongs to D_i iff the cost of unit shipment from plant i to b is no greater than this cost from any other plant.

For example, on the Euclidean plane Dirichlet demand regions are closed convex sets. For the honeycomb lattice arrangement of plants they are the usual congruent regular hexagons centered on their respective plants, so that the case mentioned above falls under the following theorem.

Theorem: Let λ^o be a measure on $(A' \times A, \Sigma' \times \Sigma)$ having σ-finite marginals; let $h{:}A' \times A \to$ reals be measurable and bounded below, with the property that

$$\lambda_i^o(A \backslash D_i) = 0 \qquad\qquad (9.5.20)$$

for all plants $i = 1, 2, \ldots$. (Here D_i is the Dirichlet demand region of plant i, as determined by h.)

Then λ^o is *unsurpassed* (in the sense of reverse standard ordering of pseudomeasures) for the transportation problem of minimizing (12) subject to (13).

Proof: We use the result (p. 342) that if λ^o has a measure potential (p,q) in the *wide sense,* with at least one of the two integrals (16) well defined and finite, then λ^o is unsurpassed for the problem (12)–(13).

We must show the existence, then, of functions $p:A' \rightarrow$ reals and $q:A \rightarrow$ reals such that all the conditions in the preceding proof are fulfilled, except that in (16) only one of the integrals needs to be well defined and finite, instead of both.

We define p to be *identically zero* and then follow the preceding proof exactly for the special case $p_i = 0$, all i. Function q defined by (15) is finite, since h is bounded below. Inequality (17) is again a simple consequence of (15). The first integral in (16) is well defined and finite, since p is zero. (We cannot say anything about the second integral, but this is not needed.) As for (18), we need merely to observe that when p is zero the regions P_i are the same as D_i. The argument goes through with this replacement, establishing (18). ∎

Thus the usual honeycomb lattice shipment pattern is optimal if one interprets "optimal" to mean unsurpassed in the sense of reverse standard ordering of pseudomeasures. But the theorem goes well beyond this special case. It establishes the unsurpassedness of any shipment pattern in which no plant ships outside its own Dirichlet demand region (this is (20)). The other conditions needed are very weak. Thus h will usually be nonnegative, which automatically makes it bounded below (and this could be weakened even further, since it is used only to make q finite). For λ^o to have σ-finite marginals, it suffices that no Dirichlet demand region receive infinite mass.

Comparing this result with the previous one, we see that most of the boundedness conditions have been dropped, but at a cost. First, the result holds only for a system of Dirichlet demand regions rather than for potential demand regions in general ((20) is a special case of the competitive rule (14)). Second, we can infer only the unsurpassedness of λ^o, a weaker conclusion than bestness.

We now take up the question of implications in the opposite direction. Given an optimal shipment measure λ^o (optimal in the sense that it redistributes mass from its left marginal distribution to its right marginal distribution with minimal transportation cost incurred), must there exist a system of mill prices $p:A' \rightarrow$ reals such that some competitive rule is satisfied? The answer is yes, under certain conditions. Specifically, the competitive rule satisfied by λ^o and p is the topological one (9)–(10).

The usual assumptions are in effect: $A' = \{a_1, a_2, \ldots\}$ is the countable set of points from which all shipments originate, and topology $\mathfrak{I}'$ and σ-field Σ' on A' are both discrete.

Theorem: Let topology $\mathfrak{I}$ and σ-field Σ be defined on space A, with $\mathfrak{I} \subseteq \Sigma$; let $h{:}A' \times A \to$ reals be bounded and continuous (with respect to product topology $\mathfrak{I}' \times \mathfrak{I}$);[64] let λ^o be a measure on $(A' \times A, \Sigma' \times \Sigma)$, with σ-finite marginals, which is *unsurpassed* for the transportation problem of minimizing (12) subject to (13).

Then a function $p{:}A' \to$ reals exists such that the topological competitive rule (9)–(10) is satisfied: if plant i ships to b, then

$$p_i + h_i(b) \le p_j + h_j(b) \tag{9.5.21}$$

for all plants $j = 1, 2, \ldots$ (as usual we have written p_j for $p(a_j)$, all $a_j \in A'$).

Proof: Since $\mathfrak{I}' \subseteq \Sigma'$; $\mathfrak{I} \subseteq \Sigma$; h is bounded, continuous, and measurable; and λ^o is unsurpassed, a topological potential (p, q) exists for λ^o (p. 365). That is, the functions $p{:}A' \to$ reals, $q{:}A \to$ reals satisfy

$$q(b) \le p_j + h_j(b) \tag{9.5.22}$$

for all plants j and all points b; furthermore, (22) is satisfied with equality if the pair (a_j, b) supports λ^o.

Now let plant i ship to b; in other words, (a_i, b) supports λ^o. Hence $p_i + h_i(b) = q(b) \le p_j + h_j(b)$ for all plants j, which yields (21).[65] ∎

A result of this sort might be applied in a number of ways. First by providing a necessary condition for optimality (under the stated assumptions), it furnishes a method for locating optimal shipments and testing for optimality. Second, it provides a clue for designing a system that will sustain a desired shipment pattern λ^o. That is, suppose a social planner for some reason wanted the shipment pattern λ^o to occur and he had con-

---

64. The continuity of h implies its measurability. For h continuous implies h_i continuous for all plants i, which implies h_i measurable (since $\mathfrak{I} \subseteq \Sigma$), which implies h measurable (since A' is countable and Σ' is discrete).

65. McIlroy has outlined the derivation of a related result in a special case. Namely, he adduces the existence of numbers v_i (the negatives of our mill prices p_i) such that, in our terminology, the measure-theoretic competitive rule (14) obtains. (Since he deals with the plane, whose topology has the strong Lindelöf property, this follows from our conclusion (21).) His paper also inspired the idea we have used of working with mill prices p_i only, rather than with a price field defined over the entire space A. M. D. McIlroy, On Placement of Central Offices, Bell Labs internal memo., Holmdel, N.J., May 12, 1961. McIlroy's work was drawn to my attention by Dr. W. W. Hardgrave. An abstract appears in R. L. Graves and P. Wolfe, eds., *Recent Advances in Mathematical Programming* (McGraw-Hill, New York, 1963), pp. 335–36.

trol of mill prices p_i (e.g., he might own all the plants). By choosing these appropriately, he *might* induce a competitively organized market to ship in the desired pattern. (Actually, further controls would be needed to insure that the right destination measure is attained.)

Our final investigation explores more deeply along these lines. Generally, it is not the translocation λ^o per se that one aims at attaining but rather the origin and destination measures, with *some* efficient redistribution pattern from the first to the second measure. Consider the following problem.

We are given a measure μ on (A, Σ), bounded but otherwise completely arbitrary. We are also given a countable number of plants, at $a_1, a_2, \ldots$, with respective capacities $m_1, m_2, \ldots$, the sum of all the capacities being $\mu(A)$. Finally, we have the usual unit transport cost function $h: A' \times A \rightarrow$ reals, where $A' = \{a_1, a_2, \ldots\}$. Question: Does a system of mill prices $p_1, p_2, \ldots$ exist, together with a shipment measure λ having the specified marginals ($\lambda'\{a_i\} = m_i$, all i, and $\lambda'' = \mu$) such that no plant ships outside its potential demand region as determined by the mill prices?

This may be rephrased in more behavioristic terms. Suppose a completely inelastic demand pattern is specified by μ: $\mu(E)$ is the total mass that will be purchased in region E. Potential buyers will purchase from the least expensive plant, counting in both mill price and transport cost. (If several plants are tied for least expensive, the buyer may distribute his purchases over them in any manner.) Does a system of mill prices (and a way of distributing purchases in case of ties) exist such that (i) demand μ is satisfied and (ii) the capacity m_i of each plant i is just exhausted?

The possibility of ties for least expensive source (or, equivalently, the overlapping of potential demand regions) is what leads to complications. In some cases these complications can be avoided. Call a measure μ *segregating* iff for any two distinct plants i and j and for any number c, we have

$$\mu\{b \mid h_i(b) - h_j(b) = c\} = 0 \tag{9.5.23}$$

For example, on the Euclidean plane the sets in (23) are either hyperbolic arcs or parts of straight lines, all of which have 2-dimensional Lebesgue measure zero. Hence any μ that is absolutely continuous with respect to Lebesgue measure is segregating. If μ is segregating, there is a unique least expensive plant almost everywhere, and ties may be ignored.

But in other cases it is precisely the possibility of ties that allows a shipment measure λ satisfying specifications to exist. To take an extreme case, suppose μ is simply concentrated at b_o. Then *all* plants must ship to this single point, which requires that b_o belong to *all* potential demand regions. This can be accomplished by choosing mill prices so that $p_i + h_i(b_o)$ is the same for all plants i, resulting in a universal tie at b_o.

We first discuss a subproblem. Suppose mill prices have been selected, so that a system of potential market regions $P_1, P_2, \ldots$ has been determined. Under what conditions does a measure λ on $(A' \times A, \Sigma' \times \Sigma)$ exist, meeting specifications? There are three specifications:

$$\lambda'' = \mu \tag{9.5.24}$$

(the right marginal equals the given destination distribution μ),

$$\lambda_i(A) = \lambda'\{a_i\} = m_i \tag{9.5.25}$$

for all plants i (plant i exports its given capacity m_i), and

$$\lambda_i(A \backslash P_i) = 0 \tag{9.5.26}$$

for all plants i (no plant ships outside its potential demand region—the measure-theoretic competitive rule).

The following result confines itself to the case of a *finite* number of plants. No use is made of the fact that the sets P_i are potential demand regions; they can be any measurable sets. (In particular, they need not cover Space as potential demand regions must.)

Theorem: Let (A, Σ, μ) be a bounded measure space; let $A' = \{a_1, \ldots, a_k\}$ be furnished with the discrete σ-field Σ'; let $m_1, \ldots, m_k$ be nonnegative numbers such that

$$m_1 + \cdots + m_k = \mu(A) \tag{9.5.27}$$

and let $P_1, \ldots, P_k \in \Sigma$. Then a measure λ exists satisfying (24), (25), and (26) iff the following 2^k inequalities hold, one for each subset $\{i_1, \ldots, i_r\}$ of $\{1, \ldots, k\}$:

$$m_{i_1} + m_{i_2} + \cdots + m_{i_r} \leq \mu(P_{i_1} \cup \cdots \cup P_{i_r}) \tag{9.5.28}$$

Proof: Let λ satisfy (24), (25), and (26). For each i we have $m_i = \lambda_i(A) = \lambda_i(P_i)$, by (25) and (26). Hence, letting $Q = P_{i_1} \cup \cdots \cup P_{i_r}$, we have

$$m_{i_1} + \cdots + m_{i_r} = \lambda_{i_1}(P_{i_1}) + \cdots + \lambda_{i_r}(P_{i_r})$$

$$\leq \lambda_{i_1}(Q) + \cdots + \lambda_{i_r}(Q) \leq \lambda''(Q) = \mu(Q)$$

by (24), which proves (28).

To prove the converse, we use network flow theory. Assume all inequalities (28) hold, and consider the following capacitated graph consisting of $2 + k + 2^k$ nodes and $k + 2^k + k \cdot 2^{k-1}$ directed arcs: two nodes, O ("origin") and D ("destination"), k nodes labeled by the points

$a_1, \ldots, a_k$, and 2^k nodes labeled by the subsets $\{i_1, \ldots, i_r\}$ of $\{1, \ldots, k\}$. There is an arc from O to each a_i, of capacity m_i; an arc from each $\{i_1, \ldots, i_r\}$ to D of capacity $\mu(Q_1 \cap \cdots \cap Q_k)$, where

$$Q_i = P_i \quad \text{if } i \in \{i_1, \ldots, i_r\}, \qquad Q_i = A \backslash P_i \quad \text{if } i \notin \{i_1, \ldots, i_r\} \qquad (9.5.29)$$

and, finally, an arc of infinite capacity from a_i to $\{i_1, \ldots, i_r\}$ iff $i \in \{i_1, \ldots, i_r\}$.

A *flow* is an assignment of numbers ≥ 0 to arcs such that (i) no number exceeds the capacity of its arc, and (ii) for all nodes except O and D, the sum of the numbers assigned to arcs terminating at the given node equals the sum of the numbers assigned to arcs originating at the given node. The *size of a flow* is the sum of the numbers assigned to arcs originating at O, and also equals the sum of the numbers assigned to arcs terminating at D. A *cut* is a partition of the nodes into two sets C_O and C_D such that $O \in C_O$ and $D \in C_D$. The *size of a cut* is the sum of the capacities of the arcs terminating at nodes in C_D and originating at nodes in C_O.

The *min-cut-max-flow theorem* states that the minimum over the sizes of all possible cuts, *if finite*, equals the maximum over the sizes of all possible flows (under this condition such a maximal-sized flow exists).[66]

We shall prove the existence of a flow of size $\mu(A)$. From this a λ satisfying (24), (25), and (26) can be constructed. Let us first prove this last statement. Let $x_{i,\{i_1, \ldots, i_r\}}$ be the number assigned to the arc going from node a_i to node $\{i_1, \ldots, i_r\}$ (if there is no such arc set this to zero). Choose a subset $\{i_1, \ldots, i_r\}$, and let E be a measurable subset of $Q_1 \cap \cdots \cap Q_k$, where the Q's are defined by (29). Now let

$$\lambda_i(E) = x_{i,\{i_1, \ldots, i_r\}}[\mu(E)/\mu(Q_1 \cap \cdots \cap Q_k)] \qquad (9.5.30)$$

(If $\mu(Q_1 \cap \cdots \cap Q_k) = 0$, set $\lambda_i(E) = 0$.) The sets $(Q_1 \cap \cdots \cap Q_k)$ partition A into 2^k pieces (some may be empty). To define λ_i for any $F \in \Sigma$, intersect F with each of these pieces, apply (30), and sum the 2^k numbers obtained. Clearly, λ_i is a measure on (A, Σ), and the λ_i's together determine a measure λ on $(A' \times A, \Sigma' \times \Sigma)$. We show that this measure works.

To verify (24), note that the number assigned to the arc from $\{i_1, \ldots, i_r\}$ to D must equal its capacity $\mu(Q_1 \cap \cdots \cap Q_k)$. For the sum of the capacities of all such arcs is $\mu(A)$, the size of the flow, so each such arc must be saturated. Hence the sum of the numbers assigned to arcs *terminating* at $\{i_1, \ldots, i_r\}$ must be $\mu(Q_1 \cap \cdots \cap Q_k)$. Now add (30)

---

66. L. R. Ford, Jr., and D. R. Fulkerson, *Flows in Networks* (Princeton Univ. Press, Princeton, 1962), Ch. 1.

over $i = 1, \ldots, k$, holding the set $\{i_1, \ldots, i_r\}$ fixed. By the argument just given, the sum of the x's is $\mu(Q_1 \cap \cdots \cap Q_k)$. Hence we obtain

$$\lambda''(E) = \lambda_1(E) + \cdots + \lambda_k(E) = \mu(E) \qquad (9.5.31)$$

(This is obviously also valid if $\mu(Q_1 \cap \cdots \cap Q_k) = 0$.) By summation, (31) continues to hold for any $F \in \Sigma$, which yields (24).

To verify (25), note that the number assigned to the arc from O to a_i must equal its capacity m_i. The argument is the same as before. The sum of the capacities of all such arcs is $\mu(A)$, by (27), so each such arc must be saturated. Hence the sum of the numbers assigned to arcs *originating* at a_i must be m_i. By (30) we have

$$x_{i,\{i_1,\ldots,i_r\}} = \lambda_i(Q_1 \cap \cdots \cap Q_k) \qquad (9.5.32)$$

Summing (32) over all subsets $\{i_1, \ldots, i_r\}$ of $\{1, \ldots, k\}$, we get m_i on the left by the argument just given, and we get $\lambda_i(A)$ on the right. This is (25).

To verify (26), note that $A \backslash P_i$ is the union of all sets of the form $(Q_1 \cap \cdots \cap Q_k)$ for which $Q_i = A \backslash P_i$. For each such set

$$\lambda_i(Q_1 \cap \cdots \cap Q_k) = 0$$

since there is no arc from a_i to the node labeled $\{i_1, \ldots, i_r\}$ corresponding to such a set $Q_1 \cap \cdots \cap Q_k$. This yields (26). Thus λ works.

It remains only to prove that a flow of size $\mu(A)$ exists. The cut for which $C_D = \{D\}$ has size $\mu(A)$, the sum of capacities of all arcs from $\{i_1, \ldots, i_r\}$ to D. From the min-cut-max-flow theorem we only need to prove that every other cut has a size $\geq \mu(A)$.

Consider the cut in which C_O consists of O, nodes $a_{i_1}, \ldots, a_{i_r}$, and the set of nodes $\mathfrak{N}$, each *member* of $\mathfrak{N}$ being labeled by a *subset* of $\{1, \ldots, k\}$. The arcs from C_O-nodes to C_D-nodes fall into three categories. (i) Arcs from O to a_i, where i ranges over the *complement* of $\{i_1, \ldots, i_r\}$. The sum of the capacities of these arcs is

$$\mu(A) - (m_{i_1} + \cdots + m_{i_r}) \qquad (9.5.33)$$

from (27). (ii) Arcs from a_i to nodes labeled by subsets of $\{1, \ldots, k\}$. Even a single such arc yields a cut size of infinity. Hence for a finite cut size there must be no such arcs, which requires that each subset of $\{1, \ldots, k\}$ that meets $\{i_1, \ldots, i_r\}$ must belong to $\mathfrak{N}$. (iii) Arcs from members of $\mathfrak{N}$ to D. To make the cut size as small as possible, $\mathfrak{N}$ should consist *only* of the nodes whose labels meet $\{i_1, \ldots, i_r\}$. A typical arc of this sort has a capacity $\mu(Q_1 \cap \cdots \cap Q_k)$, where at least one of the following conditions holds:

$$Q_{i_1} = P_{i_1}, \quad \text{or } Q_{i_2} = P_{i_2}, \quad \text{or} \ldots, \quad \text{or } Q_{i_r} = P_{i_r} \qquad (9.5.34)$$

The sum of $\mu(Q_1 \cap \cdots \cap Q_k)$ over all $(Q_1 \cap \cdots \cap Q_k)$ satisfying (34) is precisely

$$\mu(P_{i_1} \cup \cdots \cup P_{i_r}) \tag{9.5.35}$$

The size of the cut in question, then, is (33) plus (35), and the cut of minimal size is certainly included among these. But it follows from (28) that the sum of (33) and (35) is $\geq \mu(A)$. Hence $\mu(A)$ is the minimal cut size, so a flow of size $\mu(A)$ indeed exists. ∎

If the sets P_i intersect only on μ-null sets, things are much simpler. For we then have $\mu(P_{i_1} \cup \cdots \cup P_{i_r}) = \mu(P_{i_1}) + \cdots + \mu(P_{i_r})$, and the 2^k inequalities in (28) reduce to k equalities

$$m_i = \mu(P_i) \tag{9.5.36}$$

$i = 1, \ldots, k$, in view of (27). In contrast to the preceding proof, it is then easy to show that (36) is necessary and sufficient for the existence of λ satisfying (24), (25), and (26).

Let us now return to the main problem, which is to find mill prices p_i and λ such that λ satisfies (24), (25), and (26) for the system of potential demand regions P_i determined by the p_i. The preceding theorem enables us to reduce this to a problem involving mill prices alone; viz., find mill prices so that the resulting system of potential demand regions satisfies the inequalities (28). For then we are assured of the existence of a feasible λ.

This approach can be carried through in certain special cases. However, the deepest results are not obtained by this route but by finding λ first and then finding p. The idea is to find a λ optimal for the transportation problem and then use preceding results to show the existence of prices such that all stipulations are satisfied. This alternative approach works not only for the finite number of plants of the preceding theorem but also for a countable infinity of plants. The major limitations of scope of the following results are, as usual, boundedness restrictions on μ and on h. It is curious that topological assumptions are needed even though the original problem does not use topological concepts. These assumptions are fairly easily satisfied. For example, if A is any Borel subset of n-space for any finite n, the usual topology and σ-field on A satisfy them. As usual, $\mathfrak{I}'$ and Σ' are the discrete topology and σ-field on countable $A' = \{a_1, a_2, \ldots\}$.

Theorem: Let (A, Σ, μ) be a bounded measure space; let $\mathfrak{I}$ be a topology on A that is separable and topologically complete such that Σ is its Borel field (more generally, A, $\mathfrak{I}$, and Σ may be the restriction to a Borel subset of such a space). Let $h{:}A' \times A \to$ reals be bounded and continuous (hence measurable). Let numbers $m_1, m_2, \ldots \geq 0$ be given, one for each a_i, such that

$$m_1 + m_2 + \cdots = \mu(A) \tag{9.5.37}$$

Then a function $p{:}A' \to$ reals and a measure λ^o on $(A' \times A, \Sigma' \times \Sigma)$ exist such that (24), (25), and (26) are satisfied for the system of potential demand regions P_i determined by p.

Proof: The assignment $\nu\{a_i\} = m_i$ uniquely determines a measure ν on (A', Σ'). Consider the transportation problem: minimize

$$\int_{A' \times A} h \, d\lambda \tag{9.5.38}$$

over measures λ on $(A' \times A, \Sigma' \times \Sigma)$ satisfying

$$\lambda' = \nu, \qquad \lambda'' = \mu \tag{9.5.39}$$

(λ', λ'' are the left and right marginals of λ respectively).

We will apply the theorem on p. 349 to prove the existence of an optimal solution to this problem. We must show that the appropriate premises of this theorem are satisfied. The discrete topology $\mathfrak{J}'$ on countable space A' is easily shown to be separable and topologically complete, and discrete Σ' is its Borel field. The same is true of $\mathfrak{J}$ and Σ on A, by explicit assumption. (More generally, A, $\mathfrak{J}$, and Σ are restricted to a Borel subset, which still suffices (p. 354).) Furthermore, $\nu(A') = \mu(A)$, by (37); μ is bounded; and h is bounded continuous. Applying the theorem, we conclude that an optimal λ^o does indeed exist.

Equations (25) and (24) are simply the feasibility conditions (39); hence these are satisfied by λ^o. It remains to show the existence of a system of mill prices $p{:}A' \to$ reals such that (26) is satisfied.

To do this we invoke a preceding theorem (p. 614). Noting again that h is bounded continuous, that $\mathfrak{J} \subseteq \Sigma$ (since Σ is the Borel field of $\mathfrak{J}$), and that λ^o is unsurpassed for (38)–(39), it follows that a function $p{:}A' \to$ reals exists such that the topological competitive rule (9)–(10) is satisfied.

Finally, $\mathfrak{J}$ is separable metrizable, hence it has the strong Lindelöf property. It follows (p. 609) that the measure-theoretic competitive rule (11) is satisfied. This is the same as (26). ∎

This is a rather powerful result. To get a feel for it the reader might check what it says in certain simple special cases, such as when A is a line-segment or a loop, when μ is concentrated at just two points, or when there are just three plants. Even in these cases the existence of a suitable λ^o and p is far from obvious. Another corollary is that, under the conditions stated, for k plants k prices exist such that the resulting potential demand regions satisfy the 2^k inequalities (28)—a statement that seems difficult to verify directly. (Incidentally, there are just $k - 1$ degrees of freedom in choosing prices, since adding a constant to all prices leaves the potential demand regions P_i invariant; this follows at once from the definition of P_i.)

In summary, given any (completely inelastic) demand pattern μ and any system of plant capacities $m_1, m_2, \ldots$ satisfying (37), a system of mill prices exists that will sustain a competitive shipment pattern satisfying the demand and capacity constraints under rather general conditions. Furthermore, the resulting shipment pattern minimizes transport costs over the class of shipments satisfying these constraints.

9.6. SERVICE SYSTEMS

INTRODUCTION: EXAMPLES

Consider a typical demand system of the kind studied in the preceding section, e.g., a number of plants manufacturing a certain commodity that is sold off to customers located in the surrounding countryside. In our analysis both the location and output of plants and the distribution of demand over Space were taken as given. Only the shipment pattern from the first to the second distribution was open to choice.

We now consider one or both of these distributions themselves to be unknown. That is, we are to find the locations and outputs of the plants themselves and/or the distribution of spatial demand. Finding the shipping pattern of the preceding section is a subproblem of this extended problem.

We first take up this problem from a normative point of view. In the preceding section we took as our objective the minimization of *transportation costs*. If the location and output of plants is to be determined, we will also take *production costs* into consideration. Assume there is a function of the form $C(s, x)$, giving the total cost of producing an output of $x \geq 0$ in a plant located at point s. Total production cost is then a sum of terms $C(s_i, x_i)$, one for each plant i.[67] The problem then becomes one of minimizing total costs, production plus transportation.

Next, if the distribution of demand μ is to be determined, we suppose there is a function giving the *benefits* accruing from each such μ. The objective now is to maximize net benefits, i.e., benefits minus total costs.

We shall work with one particular kind of benefit function. Let A be the space in which the system is located (A may be physical Space, or Space-Time, etc.). Space A is provided with a σ-field Σ on which an *areal*

---

67. This statement is strictly correct only if the sum is finite. For infinite total production costs, e.g., in the case of a lattice on the endless plane, we must introduce pseudo-measure concepts as discussed below.

measure α is defined (α may be ordinary surface area or volume if A is physical Space, 4-dimensional volume if A is Space-Time, etc.). We consider only demand distributions μ that have a density δ with respect to areal measure:

$$\mu = \int \delta \, d\alpha \qquad (9.6.1)$$

We suppose there is a benefit function $b:A \times \text{reals} \rightarrow \text{reals}$ such that the total benefit accruing from μ is given by

$$\int_A b(a, \delta(a)) \; \alpha(da) \qquad (9.6.2)$$

Thus benefit per unit area depends on location and deliveries per unit area, and the total benefit is obtained by integration.[68]

There is an alternative interpretation of the function b that is of some interest. Suppose the system is run by a profit-maximizing monopolist, one who can discriminate perfectly over space A. He will choose locations, outputs, and shipment and delivery patterns to maximize total *revenue* minus total costs. Total revenue in turn is given by an integral of the form (2), where $b(a, x)$ is now the revenue per unit area arising at point a from a delivery of quantity x per unit area. Thus the same model may be used for both social optimization and monopolistic profit maximization.[69]

We now switch to a *social equilibrium* point of view in which the behavior of the system arises from the interactions of several agents with diverse powers and preferences. The possibilities here are very rich, for there may be variations in the number of agents, the portions of the system they control, and the specific options open to them (e.g., whether they have the power to enter or exit from the system, relocate, charge admission, discriminate in pricing, etc.) as well as variations in preferences.

It is convenient to think of the participating agents as being divided into two groups, "firms" and "clients." The "firms" are the agents who run the discrete side of the market, the system of plants. They are not necessarily business firms in the ordinary sense but may be government agencies, nonprofit organizations, private cooperatives, or clubs. The "clients" are spread out over space and interact with the plants, perhaps in the form of visits by clients to plants, by plant personnel to clients, or by the flow of resources or messages between them. Production costs are typically incurred by the firms, while benefits accrue to the clients, perhaps in exchange for payments to firms. Transportation costs may be incurred

---

68. The setup in (1)–(2) is the same as the objective function in the allocation-of-effort model of Chapter 5.

69. For still other interpretations see Section 8.6.

by either side, depending on who initiates trips or shipments between plants and clients.

What is the relation between "firms" and "plants?" The former is the decision-making unit and the latter is the technological unit; there need not, in principle, be any particular relation between them. The simplest assumption is that there is a 1–1 correspondence between plants and firms, with each firm controlling exactly one plant. This is the assumption of Lösch, for example.[70] Alternatively, each firm may control several plants, as in the case of branch plants, chain stores, denominational churches, etc. The limiting case is where one firm controls *all* the plants in the system. This is the monopolistic case mentioned above. It occurs in "company towns," or, more often, in the case of regulated utilities; e.g., all phone booths in a town will typically be controlled by the same company. The most important examples, however, are those in which the monopolist is a government agency, controlling, say, all the schools, courts, libraries, post offices, or fire stations in a town.

The opposite is also possible. Control of a single plant may be shared among several firms. Consider a broad regional model of a highly urbanized industry, where several firms are located in each city. The production costs incurred by a firm in a city will in general depend on the output of the other firms in that city as well as on its own output, for there are "external economies" in the form of access to common pools of labor, customers, and auxiliary services, along with the possibility of informational exchange, common research and marketing facilities, etc.[71] To take the facilities of each firm as a separate plant would violate one of our assumptions, viz., that total production costs are the sum of production cost at each plant $C(s, x)$, which depends only on its own output x. This difficulty can be surmounted by considering that all the facilities of the firms in a single city constitute just *one* big plant (assuming there are no significant intercity interactions in production costs). From a locational point of view this involves treating a city as if it were a single geographical point, an idealization that is reasonable in broad regional studies.

In some systems the firms and the clients are not independent decision-making entities. Rather, the latter are "stockholders" of a sort in the former, in the sense that firm preferences are an indirect reflection of client preferences, and firm costs and revenues accrue to clients (over and

70. Lösch, *Economics of Location,* Ch. 1, 2. This is also the assumption in most models of spatial competition, e.g., Hotelling, Stability in competition.

71. A. Marshall, *Principles of Economics,* 8th ed. (Macmillan, London, 1920), Book 4, Ch. 10, 11, 13; E. M. Hoover and R. Vernon, *Anatomy of a Metropolis* (Doubleday, Garden City, N.Y., 1962), pp. 45–49, 58–69; R. M. Lichtenberg, *One Tenth of a Nation* (Harvard Univ. Press, Cambridge, 1960).

above any direct user charge for the firm's services). This is the case when the firms are government agencies, since operating deficits are generally financed by taxes on clients. Most government-run systems involve *nonlocal* stockholder relations in that the deficits incurred by, say, a school are financed not only by taxpayers using that school but also by taxpayers throughout the school system. A *local* stockholder system, on the other hand, is one in which the control, costs, and revenues of a given plant accrue only to clients in the market region of that plant. An example would be a system of churches organized on a congregational rather than a hierarchical basis, in which each congregation controls and finances the services of the particular church it attends. Another example would be a system of neighborhood swimming pools, each owned by the group of families who use it.

The preceding discussion has been in terms of the *pattern of control* of service systems. We now discuss them from the point of view of the *physical flows* involved in their normal operations. Each plant has a market region within which the clients making use of its goods or services are located. To obtain this use, resources (human and/or nonhuman) located at the plant must be brought into physical contact with resources at the client's location. This point of contact may be either at the plant or at the client's location. In the first case resources must move from the client to the plant; in the latter case the movement is in the opposite direction. In the case of services (as opposed to goods production) there is an additional return movement after processing is completed, so that we are really dealing with a round-trip rather than a one-way movement. Recall the discussion of services in Section 2.7, which concluded that the essential feature was that the resources of several different owners are brought together. Thus if *A* performs a service for *B*, some resources (human or nonhuman) of *A* and *B* are brought together. This may take the form of *B*'s *renting A*'s resources for a time, or the form of *bailment* of *B*'s resources into *A*'s hands for a time. The study of service systems focuses on the actual spatial movements involved in these arrangements.

To classify service systems from this point of view, we may ignore flows that are common to all, or almost all, of them. Thus almost all service systems will have a labor force engaged in routine commuting, e.g., converging on the respective plants in the morning and dispersing in the evening. But only in some cases will employees move about in the field on service calls or patrols. Almost all service systems will require utility inflows of water, electricity, etc., and will generate outflows of garbage and sewage. (The exceptions are the respective utility systems themselves, which provide self-services in their specialty.)

We also ignore flows of a purely "chauffeuring" character, in which a person takes a trip solely to transport resources involved in the operation of the system (e.g., deliveries, routine shopping trips).

In the following table a cross ($\times$) in row i column j indicates that the type of flow named in the column head occurs frequently in the service system named in the row stub; a blank indicates that the type of flow named is small, rare, or nonexistent. The flow types are:

- A. flow of goods (i.e., nonhuman resources) from plant to clients
- B. flow of goods from clients to plant
- C. visits (i.e., round trips) at the plant by clients
- D. visits at the plant by goods
- E. visits to clients by plant goods
- F. visits to clients by plant employees:
    - F_1. visits in response to client requests
    - F_2. visits on the initiative of the plant manager

Service System Type	A	B	C	D	E	F_1	F_2
manufacturing	$\times$	$\times$					
retail stores	$\times$	$\times$					
warehouses	$\times$	$\times$		$\times$			
repair shops, laundries				$\times$			
beauty shops			$\times$			$\times$ [72]	
schools			$\times$				
churches			$\times$				
courts (civil actions)			$\times$				
restaurants		$\times$	$\times$			$\times$ [72]	
hospitals			$\times$			$\times$	
police stations			$\times$			$\times$	$\times$
fire stations					$\times$	$\times$	
inspections—health, welfare, etc.						$\times$	$\times$
home repairs, construction	$\times$				$\times$	$\times$	
libraries (lending)			$\times$		$\times$		
libraries (reference)			$\times$				
auto rentals					$\times$		
museums			$\times$				
banks	$\times$	$\times$	$\times$	$\times$			
TV broadcasting	$\times$						
street maintenance	$\times$					$\times$	$\times$
political campaign headquarters							$\times$

Consider a few examples. Banks engage in check clearing (A and B); people visit them to negotiate loans (C), and they store valuables (D).

---

72. See No place like. . . , *Newsweek,* Mar. 14, 1966, pp. 95–96.

Police arrest people (C), respond to requests for aid (F_1), and also patrol on their own initiative (F_2). (Note that visits "to" clients involve proximity rather than physical contact in this last case. Clients receive benefits from proximity; cf. Section 5.8. The same holds for street maintenance and some inspections.) Firemen respond to calls (F_1) and also bring equipment into the field to process fires (E).

This list of service system types is far from exhaustive and is meant only to illustrate the variety of forms encompassed under this rubric. The list of flow categories also omits a few minor types (e.g., outflow of the newborn from hospitals; inflow of lifers to prisons). From the table one may discern some major subtypes of service systems; e.g., the "people-processing" institutions are characterized by a × in column C.

Service systems may be classified in other ways besides control patterns and physical flows. Teitz[73] mentions the following categories. (i) Distributors vs. collectors (already encompassed above); (ii) point vs. network patterns—i.e., systems that do not or do have a specialized distribution grid of pipes or wires respectively (most systems fall in the first class, which may be further subdivided into systems using the public roads, those using the "natural highways" of air or water, and those using electromagnetic transmission); and (iii) whether or not the territories of different plants are administratively delineated.

The great variety of interpretations we have adduced for the service system model raises a question of goodness of fit. Just how much distortion is involved in squeezing the characteristics of a real system into a mold that, after all, was designed (by Lösch) just for manufacturing? A full answer would require a special investigation of each type of system, for each has individual quirks not shared by the others. We limit discussion to a few general remarks and examples.

The essential features of the model are (i) function $C(s, x)$ giving the *production cost* incurred by a plant with output x located at point s; (ii) unit transport cost function $h(s_1, s_2)$, with total *transport costs* incurred by a shipment measure λ given by the usual transportation problem integral (the marginals of λ are given by the distribution of plant outputs and the distribution μ of demand respectively); and (iii) *benefits or revenues* associated with μ given by (1)–(2).

Note that every actual service system involves flows of a multiplicity of resources, while the model allows just one resource type to be produced, shipped, and consumed. If all the resources are processed and shipped in fixed proportions, any could be used as an index. If not, there is an aggregation problem and some distortion is inevitable.

The meaning of plant "output" is clear for manufacturing, but

---

73. M. B. Teitz, Toward a theory of urban public facility location, *Reg. Sci. Assoc. Pap.* 21(1968):35–51, esp. pp. 39–40.

problematical for some of the other systems listed. For retail stores or repair shops it may be taken as an index of the physical volume of sales. But what is the output of, say, a church? One possibility is to take some objective index of usage such as parishioner-hours per week or perhaps parishioner-visits per week, possibly weighted by person and point of time. (These may be taken as proxies for some less observable output variables, depending on one's theological views.) The same indices are available for the other people-processing institutions, together with other indices such as "illnesses cured" or "gains in educational achievement scores," etc.

For police systems one possible output index is man-hours per week of patrolling. The situation here is typical of systems having "mobile sub-plants" that roam around their market regions performing services for the clients with whom they come in contact. These mobile subplants include police patrol cars, library bookmobiles, garbage trucks, taxicabs, and ice-cream peddlers. In all these cases "subplant-hours per week" may be a reasonable index of output.

The crucial question, however, is not whether one can find an output index but whether this index can be combined with cost and benefit/revenue functions to fit the model in a reasonable way.

First consider transportation cost. For most systems this is reason-ably straightforward, provided one allows for the costs of time, risk of accident, etc. In television transmission, however, while there are no transportation costs in the usual sense of the term, there is a reduction in strength and clarity of the signal with increasing distance. This is a special form of the "friction of space" and formally should be included as a cost of transportation. Investments in booster stations and cables are entirely analogous to roadway construction. Both reduce the friction of space and require more than the simple transportation problem integral to express the costs incurred. Again consider the fire-fighting system. It incurs trans-portation costs in the usual sense in getting to fires; but these costs are swamped by the extra expected fire damage arising from the delay in getting to the fire. This is a special kind of time cost and should be in-cluded among the costs of transportation if one is interested in social optimization. (The same kind of cost arises in any system that responds to emergency calls, e.g., rescue squads, ambulances, riot police.) Finally, for systems having mobile subplants there are difficult problems of optimal routing and scheduling of these subplants, and for network systems there are problems of optimal timing, placement, and capacity in the construc-tion and maintenance of the transmission grid; these problems again transcend the framework of the present model.

We now turn to the "production" costs and benefits/revenues side of the model. The major difficulties that arise here in fitting the model reside in variations in the *quality* of service. Here we are not referring to the fact that costs and benefits are in general nonlinear functions of output and consumption (this can be captured in nonlinear C and b functions) but to the fact that these costs and benefits depend on variables they are not allowed to depend on in the model.

Consider the matter of *congestion,* especially in connection with people-processing institutions. The greater the intensity of usage, the lower the quality of service in all of them, e.g., museums, schools, concert halls, swimming pools, parks. More accurately, in some activities there may be a phase in which quality of service rises with intensity, owing to factors such as sociability, mutual aid, exchange of information, and coordinated roles; e.g., this occurs in education, religion, team sports, and dances. But past a certain degree of crowding, quality must deteriorate with rising intensity of use. In any case, quality varies with "output." But this means that benefits or revenues will depend not only on location s and "delivery" density δ but also on the total output x of the plant; no provision is made for this dependence in the model.

Quality varies for a different reason in what may be called *connective* systems. Consider the post office, for example. Here the "plants" are the local postal stations; "output" may be taken as the volume of mail handled; quality of service involves such things as speed and reliability of delivery. Inevitably, quality is influenced by congestion; but over and above this, quality depends (in the case of nonlocal mail) not only on conditions at one's own plant but on conditions at other plants, as well as on their location and spacing. Flows *between* plants of the system are an essential feature of its operation, so that one cannot look at a single plant and its hinterland in isolation to assess costs and benefits. The model makes no provision for this interaction. Other connective systems include banks in their clearing-house role and telephones; the highway grid itself may be thought of as a connective system with access points playing the role of plants.[74]

As a final example of the difficulties that may arise, again consider the police. We count as benefits only the crime-preventive effects of police work, ignoring its other functions. As argued in Section 5.8, crime density depends on the density of police, population (or perhaps movable wealth), and potential criminals. If all other distributions are fixed, then benefits might be expressed as $b(s, \delta)$, δ being the density of police. The trouble is that the distribution of criminals is not fixed but will itself respond to police deployment, so that we have a game situation of the kind analyzed in Section 5.8. Thus crime density at a point depends not only on police

---

74. Connective systems especially are often organized in hierarchical form; this marks still a further departure from the service system model. For an approach to such hierarchies in terms of Thünen systems see Section 8.9.

density there but on the distribution of police throughout the system. The model makes no provision for this effect.

Feedback effects of this sort are present in most service systems. Establishment of a plant creates an incentive for clients to locate closer to the plant to reduce transportation costs, except in those cases where clients do not bear such costs, e.g., free deliveries, blanket pricing. For any one service system such effects are probably slight. To take them into account would require a theory combining the Thünen and service system models, and we do not attempt to develop such a theory.[75]

We conclude that one should apply the service system model with caution. Nonetheless it still seems to have a wide range of applications, at least as a first approximation. It also has an extensive literature, not all of which is correct. Thus it is well worth investigating and we now turn to this task.

SYSTEMS WITH ONE PLANT

We first take the case in which the entire system is controlled by a single agent who optimizes according to some preference ordering. This can be social optimization in some sense, profit maximization, etc. We also assume to begin with that there is *just one plant*. This is of interest in itself and is also relevant for the multiplant case; for if decisions involving all plants except one have been made, the problem of optimizing with respect to the last plant becomes a one-plant subproblem of the original.

In the Weber problem the output and distribution of the product are given, and one must find the optimal location of the plant. Here we take just the opposite tack, assuming that plant location is given and that output and distribution of the product are to be found. (The most general problem has both these and location as unknowns. The present problem and the Weber problem are both subproblems of this.)

Formally, we are given a (σ-finite) measure space (A, Σ, α); $\alpha(E)$ may be interpreted variously as the area of region E, the population of region E, or perhaps the income of region E, depending on the particular service system under discussion. In any case we refer to α as areal measure. There is also a measurable function $h:A \to$ reals, $h(a)$ being interpreted as the unit transport cost from the plant (at its given location) to point $a \in A$. Also there is a measurable function $b:A \times$ reals $\to$ reals, $b(a, x)$

---

75. For further discussion on costs and benefits in various service systems see W. S. Vickrey, General and Specific Financing of Urban Services, in H. G. Schaller, ed., *Public Expenditure Decisions in the Urban Community* (Resources For the Future, Inc., Washington, D.C., 1963), pp. 62–90.; A. W. Drake, R. L. Keeney, and P. M. Morse, eds., *Analysis of Public Systems* (M.I.T. Press, Cambridge, 1972).

being interpreted as benefit (or revenue) density at point $a \in A$ arising from a delivery density of x at that point (all densities are with respect to measure α). Finally, there is a function C:reals $\rightarrow$ reals, $C(x)$ being the total production cost of output x.

The problem is to find a measure μ over (A, Σ), $\mu(E)$ being interpreted as the total deliveries to region E. We consider only measures μ that are *finite* and absolutely continuous with respect to α. (This last restriction is intuitively plausible, since it merely states that no positive deliveries are to be made to any region E having zero area or zero population, income, etc.: $\alpha(E) = 0$.) This is equivalent to requiring that μ be expressible as a finite indefinite integral

$$\mu = \int \delta \, d\alpha \qquad (9.6.3)$$

for some measurable nonnegative function δ:$a \rightarrow$ reals. For most purposes it is convenient to think of density δ rather than μ as the unknown to be found.[76]

The objective is to maximize total benefit (or revenue) minus total transportation and production costs. Expressed in terms of δ, this is

$$\int_A [b(a, \delta(a)) - h(a)\delta(a)] \, \alpha(da) - C\left[\int_A \delta \, d\alpha\right] \qquad (9.6.4)$$

To explain: The last term in (4) is total production costs. Output is $\mu(A)$, expressed as an integral via (3). The left integral in (4) breaks into two pieces. The first is the integral of benefit (or revenue) density $b(a, \delta(a))$ and yields total benefit (or revenue). The second is the integral of $h\delta$, and this yields total transport costs. To see this, express transport costs in terms of μ: $\int_A h \, d\mu$, and apply (3) to convert to an expression in δ. (We assume for now that b, h, and α are such that the integral in (4) is well defined and finite for all feasible δ; this restriction will be relaxed later.)

As mentioned above, the interpretation of $b(a, \delta)$ as *benefit* density is appropriate for social optimization, as when the plant is run by a government agency that is noncorrupt and efficient. The interpretation of $b(a, \delta)$ as *revenue* density is appropriate for a profit-maximizing discriminating monopolist.

How does one characterize optimal solutions δ^o to (4)? It is convenient to take still another subproblem and assume that total output is given:

$$\int_A \delta \, d\alpha = L \qquad (9.6.5)$$

---

76. Recall, however, that two density functions δ_1 and δ_2, which are equal α-almost everywhere, yield the same measure μ in (3) and are therefore not to be considered as distinct solutions.

The last term in (4) is then fixed, and the problem reduces to maximizing the left integral in (4). This should look familiar—it is just a special case of the *allocation-of-effort* problem of Chapter 5! We use the theory of that chapter to characterize optimal solutions to the problem in hand.

First, recall a few facts. Since α is σ-finite, the universe set A may be split into two measurable pieces A_o and $A \backslash A_o$, such that α restricted to A_o is nonatomic, and α restricted to $A \backslash A_o$ is σ-atomic. As an example, suppose α represents population distribution over the plane, idealized so that cities are represented as single points. Then $A \backslash A_o$ may be taken as the set of all points a for which $\mu\{a\} > 0$ (i.e., as the set of city-points), A_o itself being the complementary "rural" region. If α itself is nonatomic, $A_o = A$ and the following theorem simplifies. (All references to the concavity of the benefit function may be dropped.)

Recall that a function f:reals $\rightarrow$ reals is *concave* iff $f(\theta x + (1 - \theta)y) \geq \theta f(x) + (1 - \theta)f(y)$ for all x, y real and all θ between 0 and 1. Applied to $b(a, \cdot)$, concavity is an expression of the law of diminishing marginal returns: each successive dose of resource applied yields a lower (or at best an equal) increment of benefit or revenue than the preceding one.

Concavity is a strong assumption though not an implausible one, and it is desirable to replace it by weaker conditions if possible. On the nonatomic region A_o, concavity can be replaced by lower semicontinuity, which is so weak a condition as to be negligible from an empirical point of view. (For definition, see pp. 230–31.)

The following result specializes the theorem of p. 249 to the problem in hand, using the notation of the present section.

Theorem: Let (A, Σ, α) be a σ-finite measure space, with nonatomic part A_o. Let h:$A \rightarrow$ reals and b:$A \times$ reals $\rightarrow$ reals be measurable functions, with $b(a, \cdot)$:reals $\rightarrow$ reals lower semicontinuous for all $a \in A_o$ and concave for all $a \in A \backslash A_o$. Let L be a real number ≥ 0.

Let δ^o be optimal for the problem of maximizing the left integral in (4) over the set of all nonnegative measurable functions δ:$A \rightarrow$ reals satisfying (5).

Then an extended real number p^o exists such that, for α-almost-all $a \in A, \delta^o(a)$ maximizes

$$b(a, x) - (h(a) + p^o)x \tag{9.6.6}$$

over the set of real numbers $x \geq 0$.

Proof: The objective function $f(a, x)$ of Section 5.4 is here $b(a, x) - h(a)x$. It suffices to assume concavity, etc., for $b(a, \cdot)$, since the remainder of the integrand is linear in x. The lower and upper bound functions b and c of that section are identically 0 and ∞ respectively. ∎

Let us dispose of the possibility that p^o is infinite. Recall the convention of Section 5.4 that if $p^o = -\infty$, then maximization of (6) is taken to mean that $\delta^o(a)$ must be as large as possible. But this means that $\delta^o = +\infty$ almost everywhere, which is not feasible (except in the trivial case $\alpha = 0$). Similarly, $p^o = +\infty$ means that $\delta^o(a)$ must be as small as possible, so that $\delta^o = 0$ almost everywhere. This is the identically zero distribution and is feasible iff $L = 0$; i.e., iff plant output is zero. (The option of not producing at all should not be overlooked. It may well be optimal.) We conclude that if output L is positive, then the number p^o of the theorem must be finite.

Number p^o turns out to have a far-reaching economic interpretation. It is the *marginal net benefit* (or *revenue*) arising from output in a sense to be discussed below. Let us return to the wider problem in which output L is itself a variable to be chosen, rather than a given value. Suppose δ^o is an optimal solution for this wider problem, the optimal output level being L^o. The preceding theorem still applies to δ^o, since this density must remain optimal for the subproblem in which output L is fixed at L^o. To solve this wider problem, we need to know how *net benefits,* which is the integral in (4) evaluated at an optimal density function, vary with output. We assume that an optimal δ^o exists for each output level in a certain interval. (The premises of the preceding theorem concerning $b(a, x)$ do not guarantee the existence of optimal solutions; see Section 5.6. Rather than complicate things further, we simply assume this existence outright.)

Now define

$$B(L) = \int_A [b(a, \delta_L(a)) - h(a)\delta_L(a)] \; \alpha(da) \qquad (9.6.7)$$

Here δ_L is the optimal density associated with output level L; $B(L)$ is the net benefits (total benefits minus total transportation costs) that can be obtained from output L.

Just as the optimal density δ^o may vary with parameter L, so may the number p^o appearing in (6). We write p_i for the number associated with L_i. Our aim is to characterize the structure of the function B and its relation to the numbers p_i.

Theorem: Make the assumptions of the preceding theorem and also assume that an optimal density δ_i exists for each number L_i in the closed interval $[L_o, L^o]$, where $0 \le L_o < L^o < \infty$. Let the function B, with domain $[L_o, L^o]$, be defined by (7).

Then for any two numbers L_1, L_2 in this interval we have

$$B(L_1) - p_1 L_1 \ge B(L_2) - p_1 L_2 \qquad (9.6.8)$$

where p_1 is the number p^o associated with the parameter $L = L_1$ in the preceding theorem.

Proof: If $p_1 = +\infty$, then $L_1 = 0$, and (8) is obviously correct. We may assume then that p_1 is finite. We have

$$f(a, \delta_1(a)) - p_1\delta_1(a) \geq f(a, \delta_2(a)) - p_1\delta_2(a) \qquad (9.6.9)$$

for α-almost-all $a \in A$, by (6), where we have used the abbreviation

$$f(a, x) = b(a, x) - h(a)x \qquad (9.6.10)$$

in (9). Integrating (9) with respect to α over A yields (8). ∎

Inequality (8) is a very strong result. It states that B has a linear support function at each point of its domain (except possibly the left endpoint if $L_o = 0$). That is, the linear function

$$g(x) = p_1 x + B(L_1) - p_1 L_1 \qquad (9.6.11)$$

which equals the value of B at L_1, is at least as large as B at all other points of its domain. This implies that B is a concave function.[77] Furthermore, if B is differentiable at point L (and it must be differentiable at all but a countable number of points of its domain), the function g of (11) is *tangent* to B at L. This means that p_L equals the derivative of B at L (where p_L is the shadow price associated with output L):

$$p_L = DB(L) \qquad (9.6.12)$$

whenever the latter exists. This is the sense in which p^o may be interpreted as the *marginal net benefit* arising from output.

But even if B is not differentiable, it always possesses left and right derivatives (denoted D^- and D^+ respectively) at each interior point of its domain, and for these we have

$$D^-B(L) \geq p_L \geq D^+B(L) \qquad (9.6.13)$$

Finally, let $L_o \leq L_1 < L_2 \leq L^o$; then $p_1 \geq p_2$. That is, the marginal net benefit associated with higher output is less than (or at best equal to) that associated with lower output. All these results follow from (8) by standard arguments.

Now, finally, to maximize (4)! The production cost function C comes in at last. One simple procedure is in two stages. First, choose L^* to maximize

$$B(L) - C(L) \qquad (9.6.14)$$

over $L \geq 0$. Second, choose δ^o to optimize the allocation-of-effort problem, discussed above, with the parameter L in (5) set equal to the L^* found in stage one. An alternative procedure uses p as the search variable. For a

---

77. If α is nonatomic, no concavity assumptions at all are being made. Nonetheless, the conclusion is still valid: B is concave.

trial value of p, use (6) to compute δ; substitute δ in (7) to find net benefits and in (5) to find L; and finally substitute in (14) to find benefits net of *all* costs. Then iterate, depending on which direction seems to increase (14).

There is a related approach that the entire market can carry out in a decentralized manner. As a preliminary, we must show that (13) is a sufficient, as well as necessary, condition for a number p to serve as the parameter in (6) determining an optimal density for output L.

Theorem: Make the assumptions of the preceding theorem, and also assume that $b(a, \cdot)$ is continuous for all $a \in A$. Let $L_o < L_1 < L^o$, where $[L_o, L^o]$ is an interval of outputs for which optimal densities exist, and let δ_1 be an optimal density for L_1.

Then a number p satisfies (6) for $\delta^o = \delta_1$ iff it satisfies (13) for $L = L_1$.

Proof: If p satisfies (6), it satisfies (8), and we have noted that (13) then follows by standard convexity arguments.

Conversely, let p satisfy (13). If B is differentiable at L_1, there is just one number satisfying (13), viz., the derivative $DB(L_1)$. Since a p_1 satisfying (6) must exist, $p = p_1$, and the proof is complete in this case.

If B is not differentiable at L, it has a "corner," and $D^-B(L_1) > D^+B(L_1)$. Let $p \neq p_1$ be any number *strictly* between these limits. Thus

$$D^-B(L_1) > p > D^+B(L_1) \tag{9.6.15}$$

We first demonstrate that a measurable $\delta : A \to$ reals exists such that $\delta(a)$ maximizes

$$f(a, x) - px \tag{9.6.16}$$

over $x \geq 0$ for almost all $a \in A$ (f is as in (10)). Let δ^o be an optimal density for output L^o, and let p^o be a number satisfying (6) for δ^o. Since $b(a, \cdot)$, hence $f(a, \cdot)$ and hence (16) is continuous in x, the set of maximizers of (16) over the closed bounded interval $[0, \delta^o(a)]$ is nonempty and closed. Hence there exists a *largest* such maximizer. Call it $\delta(a)$. Let $y > \delta^o(a)$. Then

$$f(a, \delta(a)) - p\delta(a) \geq f(a, \delta^o(a)) - p\delta^o(a)$$

$$\geq f(a, \delta^o(a)) - p^o\delta^o(a) - (p - p^o)y$$

$$\geq f(a, y) - p^o y - (p - p^o)y = f(a, y) - py, \tag{9.6.17}$$

for almost all $a \in A$. The first inequality in (17) follows from $\delta(a)$ maximizing (16) over $[0, \delta^o(a)]$. The middle inequality follows from $(p - p^o)(y - \delta^o(a)) \geq 0$, which in turn results from

$$p > D^+B(L_1) \geq D^-B(L^o) \geq p^o \tag{9.6.18}$$

a consequence of (15), the fact that $L_1 < L^o$ and that B is concave, and (13) for p^o and L^o respectively. The last inequality in (17) arises from the fact that $\delta^o(a)$ maximizes $f(a, x) - p^o x$ over *all* $x \geq 0$, and holds for almost all $a \in A$. Relation (17) shows that $\delta(a)$ maximizes (16) over *all* $x \geq 0$, for almost all $a \in A$.

That the resulting function $\delta : A \to$ reals is *measurable* follows from the argument on pp. 255–56.

Next we have the two relations

$$f(a, \delta_1(a)) - p_1\delta_1(a) \geq f(a, \delta(a)) - p_1\delta(a) \tag{9.6.19}$$

$$f(a, \delta(a)) - p\delta(a) \geq f(a, \delta_1(a)) - p\delta_1(a) \tag{9.6.20}$$

holding for almost all $a \in A$. Inequality (19) follows from (6) for δ_1 and p_1, while (20) follows from $\delta(a)$ maximizing (16). Adding and canceling, we obtain

$$(p - p_1)(\delta(a) - \delta_1(a)) \leq 0 \tag{9.6.21}$$

almost everywhere.

Since $p \neq p_1$, it follows from (21) that $\delta - \delta_1$ never takes opposite signs, except for a null set. Hence there are two possibilities: if $\delta = \delta_1$ almost everywhere, then p satisfies (6) for δ_1, and the proof is finished. If $\delta \neq \delta_1$ on a set of positive α-measure, then

$$L = \int_A \delta \, d\alpha \neq \int_A \delta_1 \, d\alpha = L_1 \tag{9.6.22}$$

Inequality (22) leads to a contradiction. For take any numbers L_2, L_3 satisfying

$$L_o < L_2 < L_1 < L_3 < L^o \tag{9.6.23}$$

Let p_i, δ_i be associated with $L_i, i = 2, 3$. The argument of (18) yields

$$p_2 > p > p_3 \tag{9.6.24}$$

Applying the argument of (19)–(20) with subscript 2, and then 3, in place of 1 yields

$$(p - p_i)(\delta(a) - \delta_i(a)) \leq 0 \tag{9.6.25}$$

$i = 2, 3$ for almost all $a \in A$. From (24) and (25) we obtain

$$\delta_2 \leq \delta \leq \delta_3 \tag{9.6.26}$$

almost everywhere. Integration of (26) yields $L_2 \leq L \leq L_3$. But this holds for any L_2, L_3 satisfying (23). It follows that $L = L_1$, contradicting (22). We conclude that $\delta = \delta_1$ almost everywhere. Thus p satisfies (6).

This concludes the proof for all p strictly inside the interval (15). Finally, take a sequence of such p's approaching $D^+B(L_1)$, and take another sequence approaching $D^-B(L_1)$. Noting that (16) is continuous in p and that $\delta_1(a)$ maximizes (16) for all p's in the sequence, we conclude that $\delta_1(a)$ still maximizes when the endpoint values $D^\pm B(L_1)$ are substituted for p in (16). ∎

Now consider what happens if the plant controller, rather than fixing the entire distribution μ of output, simply sets a *price p* at the plant, allowing the various clients to take whatever quantities they wish at that price. How much will they choose? Suppose a client situated at point $a \in A$ derives a *benefit* $b'(a, x)$ from consumption of quantity x. He incurs transportation costs of $xh(a)$ along with purchasing costs of xp. Thus the net benefit to him is

$$b'(a, x) - (h(a) + p)x \qquad (9.6.27)$$

We assume that the client chooses a quantity x that maximizes this expression.

Comparing (27) with (6), we see that they have the same form, except that the functions b and b' might be different. Thus if $b = b'$, the clients, in maximizing (27) for each $a \in A$, are fulfilling a necessary and sufficient condition for the resulting distribution to be optimal, given the total amount demanded.

Two questions arise. First, does $b = b'$? This of course depends on the preferences of the plant manager. From the point of view of a classical liberal, the identity of b and b' is a value axiom. This assumption is fortunate for applications, since it spares the plant manager the task of actually ascertaining what the function b' is. A simple pricing policy will achieve the optimum no matter what b' is (provided only that the concavity, etc., conditions are fulfilled). However, an interventionist would point to the divergence between private and social benefits, the effects on income distribution, etc., that would lead a market solution of the type contemplated to be suboptimal. And for a profit-maximizing monopolist we definitely have $b \neq b'$, since $b(a, x)$ is total revenue while $b'(a, x)$ includes "consumers' surplus." For the rest of this discussion we assume $b = b'$.

The second question arises from the fact that x in (27) is a *quantity* while x in (6) is a *density* with respect to α. Just what is the relation between α and the population of decision-making clients? The simplest solution conceptually is the following. The space A itself is the set of clients, Σ = all subsets of A, and α is counting measure—so that each client gets a weight of one, the integral $\int_A \delta \, d\alpha$ reduces to a sum $\delta(a_1) + \delta(a_2) + \cdots$, and "density" is the same as quantity. This approach can accommodate a finite or even countably infinite number of clients.[78] On the other hand there are technical advantages in taking α to be nonatomic. Assumptions can be weakened considerably, concavity being relaxed to lower semicontinuity. This is an idealization, however, since the individual decision makers are absorbed in an amorphous mass.[79] We have a "continuous" approximation to a "discrete" reality.

We still must answer the question, At what level should the mill price be set? The answer (assuming $b = b'$ and C differentiable) is that it should be set to equal *marginal cost* (the principle of marginal-cost pricing). We show that this rule will at least maintain an optimum. Suppose that output $L^o > 0$ maximizes $B(L) - C(L)$, and that p^o is the marginal cost at L^o: $p^o = DC(L^o)$. It is easy to see that we must have

$$D^- B(L^o) \geq p^o \geq D^+ B(L^o) \tag{9.6.28}$$

For if the left inequality is violated, $B(L) - C(L)$ could be increased by reducing L; if the right inequality is violated, $B(L) - C(L)$ could be increased by increasing L. (If $DB(L^o)$ exists, then (28) is just the familiar marginal cost equals marginal benefit (or revenue) condition.)

Now (28) is the same as (13). Hence, by the theorem just proved, if δ^o is an optimal density for L^o, then $\delta^o(a)$ will maximize (6) for almost all $a \in A$, with this p^o in (6). In turn, this means that an individual client at $a \in A$, when confronted with the mill price p^o, will find that $\delta^o(a)$ is a quantity that maximizes his net benefits (27). Thus marginal-cost pricing sustains an equilibrium at the optimal solution.

Getting to the optimum is a different matter. One simple scheme is a sequence (L_t, p_t), $t = 1, 2, \ldots$, where p_t is the marginal cost at L_t and L_{t+1} is the quantity demanded under p_t. Whether this converges to an optimal solution depends on the starting point and the nature of the function C. We do not discuss this further, except to note that the option of no production at all ($L = 0$) should be investigated separately, especially in the common case where there is a jump discontinuity "setup cost" at 0.

Thus we have a justification for extending the marginal cost pricing rule to a general measure-theoretic context. The interpretation of "marginal cost" for a particular service system may not be completely straightforward. For example, in community swimming pools and some other

---

78. It may also be combined with a "continuous" Space as we have done in the analysis of competitive Thünen equilibrium, Section 8.7, and market equilibrium, Section 6.6.

79. Taking α nonatomic is precisely the Aumann "continuum-of-traders" approach. See Section 6.7. Location theorists have worked with uniform distributions of consumers, etc., for a long time, so it may be said that they have anticipated Aumann by several decades! Of course it may also be said that Aumann was the first to realize what he was doing (and to draw deep conclusions from this realization).

people-processing systems, marginal cost may be taken as the extra congestion at the "plant" caused by the presence of one more participant, and the "mill price" (= admission fee) should reflect this.

Pseudomeasure Treatment of the One-Plant Case

We shall go on to the general, multiplant case. Here we must at last face up to the possibility that costs and/or benefits may be infinite. This is certainly true for the classical model of plants on the endless plane in a lattice arrangement. We introduce pseudomeasures to handle this contingency.

Let us first quickly run through the one-plant analysis again from this more general point of view. We now drop the restriction that δ must make the integral (4) well defined and finite. We do this, not because infinite costs or benefits are very likely in the one-plant case, but because most of the conceptual issues arising from the introduction of pseudomeasures are already present in this case and the generalization is easy.

The objective remains to maximize the algebraic sum of benefits, transport costs, and production costs, as in (4). However, we now let these be *pseudomeasures* rather than real numbers. For these three components to be addible, the pseudomeasures must be over the *same* space. Total benefits are given by $\int b(a, \delta(a)) \, \alpha(da)$ where this is now an *indefinite* integral over space A.

As for transportation costs, in Chapter 7 and Section 9.5 this was expressed as an integral on a product space such as $A' \times A$. But in (4) this collapsed to an integral over A, the destination space alone. It seems natural to follow this lead and think of transport cost as an indefinite integral over A:

$$\int h(a)\delta(a) \, \alpha(da) \qquad (9.6.29)$$

Note the meaning of (29). All costs incurred in moving a shipment from the plant located at a_o, say, to site b are to be attributed to site b. (If total costs incurred are finite, it does not matter how they are distributed over A; but if they are infinite, it may.)

Finally, there is production cost $C(L)$. This again must be thought of as a pseudomeasure over A, and again there is a "natural" way to do this; viz., think of production cost as an ordinary *measure, simply concentrated* at the plant site a_o, and of mass $C(L)$.

With these reinterpretations of the terms of (4), every feasible density δ has a utility value $U(\delta)$ that is a pseudomeasure over A. Preference ordering among densities is reflected in *standard ordering* among their utilities: $\delta_1 \succeq \delta_2$ iff $U(\delta_1) \succeq U(\delta_2)$. Recall that when two pseudomeasures are expressed as indefinite integrals with respect to the same measure, there is a very simple criterion for the standard ordering relation between them. Namely,

$$[\int g_1 \, d\mu \geqslant \int g_2 \, d\mu] \, \Big| \, \text{iff} \int_A [g_1 - g_2] \, d\mu \geq 0 \qquad (9.6.30)$$

(the standard integral theorem). That is, to check whether $\int g_1 \, d\mu$ is at least as large as $\int g_2 \, d\mu$ (in the sense of standard order), evaluate the *definite* integral $\int_A [g_1 - g_2] \, d\mu$. The relation holds iff this definite integral is well defined and, furthermore, nonnegative (it may equal $+\infty$).

Now the revised utility function (4) would be of this special form but for the production cost term $C(L)$. Even with this term there is a simple criterion of the (30) type.

Theorem: Let δ_1, δ_2 be two feasible density functions over (A, Σ), and let them be evaluated by (4) as a pseudomeasure. Then $\delta_1 \geqslant \delta_2$ (in the sense of standard ordering) iff

$$\int_A [f(a, \delta_1(a)) - f(a, \delta_2(a))] \, \alpha(da) \geq C(L_1) - C(L_2) \qquad (9.6.31)$$

Here f is the net benefit density function: $f(a, x) = b(a, x) - xh(a)$, and L_i is the output generated by δ_i, $i = 1, 2$.

Proof: Let $\nu = \alpha + \mu_{a_o}$, where μ_{a_o} is the measure concentrating a mass of *one* at the plant site a_o. Measure ν is σ-finite, and both α and μ_{a_o} are absolutely continuous with respect to ν. Hence they both have densities with respect to ν by the Radon-Nikodym theorem, so that *all* terms in (4), including production cost, may be expressed as indefinite integrals with respect to measure ν. Now (30) may be applied. In the resulting definite integral the production cost terms may be placed under separate integral signs, since they have finite integrals. Simplifying the resulting expression we obtain (31). Details are left as an exercise. ∎

Thus, to check whether $\delta_1 \geqslant \delta_2$, evaluate the definite integral in (31). The relation holds iff this definite integral is well defined and, furthermore, is at least as large as the difference in production costs. Note that if $f(a, \delta_1(a))$ and $f(a, \delta_2(a))$ both have finite integrals, then (31) is merely a restatement of the ordinary objective function (4). However, (31) may apply even if these integrals are infinite or not well defined. Thus our pseudomeasure approach constitutes a bona fide generalization of the case where all benefits and costs are finite.

Since standard ordering is in general not complete, we must distinguish two meanings of the term "optimal." Feasible δ^o is a *best* solution iff $\delta^o \gtrsim \delta$ for all feasible δ, while δ^o is merely *unsurpassed* iff there is no feasible δ such that $\delta > \delta^o$.

With these preliminaries disposed of, we now return to the one-plant case. We still assume that $b(a, \cdot)$ is lower semicontinuous on A_o and concave on $A \backslash A_o$. In fact, we make the same assumptions as above, except that b, h, and α are no longer restricted so as to make the objective function well defined and finite as a definite integral.

First, take the subproblem in which output L is given (L real, non-negative). Then the right side of inequality (31) is zero, and we are left with an allocation-of-effort problem of the Chapter 5 type. Let δ^o be *unsurpassed* for this problem. Then we obtain (6), the same conclusion as above. An extended real number p^o exists such that $\delta^o(a)$ maximizes

$$f(a, x) - p^o x \tag{9.6.32}$$

over $x \geq 0$, for α-almost-all $a \in A$. Furthermore, this conclusion suffices for δ^o to be *best* (p. 249), so that the concepts of best and unsurpassed solutions coincide for this particular problem.

The definition of $B(L)$ given by (7) is no longer convenient, since it may not be finite. This difficulty is easily overcome. Assume as above that an optimal—specifically, an unsurpassed—density δ_i exists for each L_i in a closed interval $[L_o, L^o]$, where $0 < L_o < L^o < \infty$ (note that $L_o > 0$). Let L_i, L_j be two numbers in this interval. Using (32), we obtain

$$p_j(\delta_i(a) - \delta_j(a)) \geq [f(a, \delta_i(a)) - f(a, \delta_j(a))] \geq p_i(\delta_i(a) - \delta_j(a)) \tag{9.6.33}$$

for almost all a, where p_i, p_j are numbers p^o satisfying (32) for δ_i, δ_j, respectively (since L_i, $L_j > 0$, p_i, p_j cannot be infinite).

Now the left and right expressions in (33) integrate with respect to α to $p_j(L_i - L_j)$ and $p_i(L_i - L_j)$ respectively. Since these are finite, it follows that the middle expression in (33) must have a definite integral that is well defined and finite.

Choose a fixed arbitrary L_3 in the interval $[L_o, L^o]$ with corresponding optimal δ_3 and define function B on this interval by

$$B(L_i) = \int_A [f(a, \delta_i(a)) - f(a, \delta_3(a))] \, \alpha(da) \tag{9.6.34}$$

By the argument just given, B is well defined and finite. Furthermore,

$$B(L_1) - B(L_2) = \int_A [f(a, \delta_1(a)) - f(a, \delta_2(a))] \, \alpha(da)$$

$$\geq \int_A p_1(\delta_1 - \delta_2) \, d\alpha = p_1(L_1 - L_2) \tag{9.6.35}$$

(The first equality in (35) arises from subtracting integrands in (34); this is valid since the integrals are finite. The inequality arises from (33).) But (35) yields precisely the conclusion (8) above. Thus B has a linear support function at each point of its domain. We conclude as above that B is concave, that (12) and (13) hold, etc.

Finally, let δ_1, δ_2 be unsurpassed for L_1, L_2. Note that criterion (31) may be rewritten: $\delta_1 \succcurlyeq \delta_2$ iff $B(L_1) - C(L_1) \geq B(L_2) - C(L_2)$. It follows that if a given L^* maximizes $B(L) - C(L)$, and δ^o is an unsurpassed density for this output, then δ^o is best for the overall maximization problem with unrestricted output. (Recall that "best" and "unsurpassed" were found to be equivalent for the fixed output problem discussed.) Thus B defined by (34) has exactly the crucial properties of B defined by (7), and the entire preceding analysis is preserved. In particular, the entire discussion revolving around marginal-cost pricing remains valid. We conclude that the introduction of pseudomeasures leads to complications that are minor at worst.[80]

Systems with Many Plants

We now formulate the *multiplant service system model*. We are given the overall measure space (A, Σ, α), α σ-finite; the benefit-density function $b:A \times$ reals $\rightarrow$ reals and the unit transport cost function $h:A \times A \rightarrow$ reals, both measurable; and the production cost function $C:A \times$ reals $\rightarrow$ reals, $C(a, x)$ being the total cost of producing output x at plant site a.

There are a countable (possible finite) number of plants, plant i being located at $a_i, i = 1, 2, \ldots$. We abbreviate $h(a_i, a)$ by $h_i(a)$, and $C(a_i, x)$ by $C_i(x), i = 1, 2, \ldots$. The problem is to choose locations and density functions $\delta_i:A \rightarrow$ reals, one pair for each plant. Density δ_i represents the distribution of the output of plant i over space A. We are to maximize

$$\int \left[b(a, \delta(a)) - h_1(a)\,\delta_1(a) - h_2(a)\,\delta_2(a) - \cdots \right] \alpha(da)$$
$$- C_1(L_1) - C_2(L_2) - \cdots \qquad (9.6.36)$$

where

$$\delta = \delta_1 + \delta_2 + \cdots \qquad (9.6.37)$$

is the total density function, and

$$L_i = \int_A \delta_i \, d\alpha \qquad (9.6.38)$$

---

80. We comment briefly on the zero-output case, which was not discussed above. If $L_i = 0$, the possibility that $p_i = +\infty$ cannot be eliminated; if so, the right-hand inequality in (33) disappears. $B(0)$ is still well defined by (34) but might take on the value $-\infty$. Apart from this the preceding discussion remains valid for $L = 0$.

is the output of plant i, $i = 1, 2, \ldots$. Expression (36) starts with an indefinite integral over A, interpreted as a pseudomeasure. The integrand is composed of the benefit function minus the sum of transportation costs incurred by each plant (the integral of $h_i \delta_i$ is the transport cost for plant i). The relevant density for assessing benefits is the *total* density (37) from all plants together. The sum of all production costs is then subtracted. As in the one-plant case, the production cost for each plant i is thought of as a measure on (A, Σ) of mass $C_i(L_i)$ concentrated at the plant site a_i.

There are a number of feasibility constraints to be imposed. All these are of a technical nature, designed to insure that expression (36) is well defined as a pseudomeasure. First, each δ_i must be measurable and non-negative. Second, δ must be finite (to guarantee that the undefined expression $b(a, \infty)$ does not occur). Third, each L_i must be finite (infinite production at a plant is not defined).

Fourth, for all $a \in A$, the series

$$h_1(a)\,\delta_1(a) + h_2(a)\,\delta_2(a) + \cdots \tag{9.6.39}$$

must be *absolutely convergent* (i.e., the sum of the absolute values of the terms in (39) must be finite). The reason for this condition is that an integrand must be finite for the indefinite integral to be well defined as a pseudomeasure. Anything less than absolute convergence would make the sum depend on the order of summation, which seems counterintuitive.

Fifth, and finally, $\{a_i\} \in \Sigma$ for all $i = 1, 2, \ldots$. This guarantees that the sum of all production costs (each thought of as a simply concentrated measure) is σ-finite.

How restrictive are these conditions? One may guess that no intuitively plausible solutions would be eliminated by them. In any case, to remove them would involve going beyond even pseudomeasures, so we keep them. Conditions one, two, and four could just as well be required only *almost* everywhere.

These are the only overall conditions we impose. We now find it useful to consider the *subproblem* in which all plant locations are given in advance and all plant outputs are also fixed. Thus we have a sequence (finite or infinite) of distinct points a_1, a_2, ... and a corresponding sequence of nonnegative real numbers L_1, L_2, $\ldots$. Equation (38) is imposed as an additional constraint for each i. The production cost terms are then predetermined constants and may be dropped from (36) without disturbing the ordering.

How does one characterize optimal solutions to this multiplant service system problem? To begin with we take note of two further subproblems contained in this one. First, suppose that the total density δ is given. The benefit term $b(a, \delta(a))$ is then independent of the particular

$\delta_1, \delta_2, \ldots$ chosen and may be dropped from (36); the objective then reduces to the minimization of overall transport costs. The resulting subproblem is almost the same as the demand region problem of the preceding section, differing only in that (36) values are pseudomeasures over A, while the demand region objective function values are pseudomeasures over $A' \times A$, A' being the set of plant sites.

A second subproblem arises when δ_i is given for all plants i except one, say $i = i_o$. This fixes $h_i \delta_i$ for all $i \neq i_o$, and these terms may be dropped from (36). The problem of optimizing δ_{i_o} is then precisely of the one-plant form that we have already considered. By varying i_o, we get as many subproblems as there are plants.

Now, under fairly general conditions, an optimal solution to the demand region problem has associated with it a system of "shadow prices," $p_1, p_2, \ldots$, one for each plant, as discussed in the preceding section. Also, an optimal solution to the one-plant service system problem has associated with it a shadow price p', as discussed above, so that we get a second sequence $p_1', p_2', \ldots, p_i'$ corresponding to the plant i subproblem. Let the sequence $(\delta_1^o, \delta_2^o, \ldots)$ be optimal for the multiplant service system problem. Then it is optimal for the subproblems, so that there will be two shadow price sequences (p_i), (p_i'), $i = 1, 2, \ldots$, associated with it. It turns out that (again under fairly general conditions) there is *one* sequence that fulfills both roles simultaneously; i.e., we can choose $p_i = p_i'$ for all i. Conversely, if there is a single sequence serving as shadow prices *simultaneously* for both subproblems, then the corresponding solution is optimal for the multiplant service system problem. The following two theorems make these statements precise.

Let us first recall some concepts. Given a sequence $p_1, p_2, \ldots$, one number for each plant, the *potential demand region* of plant i is the set

$$P_i = \{a \mid p_i + h_i(a) \leq p_j + h_j(a) \quad \text{for all} \quad j = 1, 2, \ldots\} \qquad (9.6.40)$$

(We allow some or all p_j's to be infinite here.)

Let λ_i be the measure giving the distribution of the output of plant i over space A. Shipment pattern $(\lambda_1, \lambda_2, \ldots)$ obeys the *measure-theoretic competitive rule* iff

$$\lambda_i(A \backslash P_i) = 0 \qquad (9.6.41)$$

for all plants i; i.e., no plant ships outside its own potential demand region. In terms of densities δ_i, (41) reads:

$$\int_{A \backslash P_i} \delta_i \, d\alpha = 0 \qquad (9.6.42)$$

The integrand in (42) is nonnegative. The integral is then zero iff the set on which δ_i is positive has measure zero; i.e., the measure-theoretic competitive rule is equivalent to the statement:

$$\alpha[\{a \mid \delta_i(a) > 0\}\backslash P_i] = 0 \tag{9.6.43}$$

for all plants i.

Given a sequence of densities $\delta_1^o, \delta_2^o, \ldots$, one for each plant, define the functions $b_i: A \times \text{reals} \rightarrow \text{reals}$ by

$$b_i(a, x) = b(a, \delta^o(a) - \delta_i^o(a) + x) \tag{9.6.44}$$

In the one-plant subproblem in which δ_j^o is held constant for all j except $j = i$, the appropriate benefit function is b_i of (44). The reason is that if $\delta_i(a)$ is set equal to x, the density of *total* shipments to point a is x *plus* $\delta^o(a) - \delta_i^o(a)$, the sum of the fixed densities; hence the resulting benefit density is precisely $b_i(a, x)$. This explains the appearance of the functions b_i in the following theorem.

Theorem: Let $(\delta_1^o, \delta_2^o, \ldots)$ be unsurpassed for the multiplant service system problem (including the constraints (38) and fixed plant locations). Let $b(a, \cdot): \text{reals} \rightarrow \text{reals}$ be lower semicontinuous for all $a \in A_o$ (the nonatomic set of measure α) and concave for all $a \in A\backslash A_o$. Also let $b(a, \cdot)$ be *differentiable* at the point $\delta^o(a)$, for all $a \in A$ such that $\delta^o(a) > 0$.

Then extended real numbers $p_1, p_2, \ldots$ exist, one for each plant, such that

(i) $\delta_i^o(a)$ maximizes

$$b_i(a, x) - (h_i(a) + p_i)x \tag{9.6.45}$$

over $x \geq 0$, for almost all $a \in A, i = 1, 2, \ldots$;

(ii) *and* the resulting set of potential demand regions satisfies the measure-theoretic competitive rule.

Proof: Since $(\delta_1^o, \delta_2^o, \ldots)$ is unsurpassed, each separate δ_i^o must be unsurpassed for the one-plant subproblem resulting from fixing δ_j^o for all $j \neq i$. Also, each function b_i inherits the concavity or lower semicontinuity properties of b. Part (i) of the conclusion now follows from the one-plant theory above.

Thus for each plant i there exists an extended real p_i and a null set E_i such that $\delta_i^o(a)$ maximizes (45) for all $a \in A\backslash E_i$. Let $E = E_1 \cup E_2 \cup \cdots$. As a countable union of null sets, E itself is null.

Let $a' \in \{a \mid \delta^o(a) > 0\}\backslash E$. Since $\delta^o(a') \geq \delta_i^o(a') > 0$, it follows that $b(a', \cdot)$ is differentiable at the point $\delta^o(a')$. Hence $b_j(a', \cdot)$ is differentiable at the point $\delta_j^o(a')$ for all j, these derivatives all being equal to $Db(a', \delta^o(a'))$. We then have

$$h_i(a') + p_i = Db_i(a', \delta_i^o(a')) = Db(a', \delta^o(a'))$$
$$= Db_j(a', \delta_j^o(a')) \leq h_j(a') + p_j \qquad (9.6.46)$$

for all plants j. The first equality arises from the fact that $\delta_i^o(a') > 0$ maximizes (45) at a differentiable point, so the derivative of (45) must be zero there. The other equalities arise from the definitions of b_i and b_j. The inequality in (46) arises from the fact that $\delta_j^o(a')$ maximizes (45) with j in place of i, (45) again being differentiable at that point; hence the derivative of (45) must be *nonpositive* there. (Note that the existence of a' implies that p_i is finite; (46) obviously is still valid for any $p_j = +\infty$.)

Relation (46) implies that a' belongs to the potential demand region of plant i. Thus

$$[\{a \mid \delta_i^o(a) > 0\}\backslash E] \subseteq P_i \qquad (9.6.47)$$

for all plants i. Relation (47) in turn implies that the set in (43) is contained in E. Since E is null, (43) itself follows. But (43) is the same as (41), which is what we had to prove. ■

The interpretation of infinite p_i values is as in the one-plant case: $p_i = -\infty$ is ruled out if $\alpha(A) > 0$, while $p_i = +\infty$ is possible only if plant i is shut down ($L_i = 0$).

For the following converse theorem we need a condition on α. Recall that α is said to be *segregating* iff for any two distinct points a_i and a_j and for any number c we have $\alpha\{a \mid h_i(a) - h_j(a) = c\} = 0$. (For example, Lebesgue measure on the Euclidean plane is segregating.) The effect of this property is to make the system of potential demand regions P_1, P_2, ..., nonoverlapping almost everywhere (provided not all p_i's are infinite).

Theorem: Let $(\delta_1^o, \delta_2^o, \ldots)$ be feasible for the multiplant service system problem (including the constraints (38) and fixed plant locations). Let extended real numbers $p_1, p_2, \ldots$ exist, one for each plant, satisfying the conclusions of the preceding theorem. Also let

$$L_1 |p_1| + L_2 |p_2| + \cdots < \infty \qquad (9.6.48)$$

(remember that $0 \cdot \infty = 0$), and let α be segregating. Then $(\delta_1^o, \delta_2^o, \ldots)$ is *best* for this problem.

Proof: If any p_i is infinite, then $L_i = 0$, so that the only feasible density is $\delta_i = 0$ almost everywhere. We get an equivalent problem by simply eliminating such plants from the system. Thus we may assume that all p_i's are finite.

Let H_i be a null set such that $\delta_i^o(a)$ maximizes (45) for all $a \in A\backslash H_i$, $i = 1, 2, \ldots$, let

$$F_{ij} = \{a \mid h_i(a) - h_j(a) = p_j - p_i\} \tag{9.6.49}$$

where (i, j) runs over all pairs of *distinct* plants, and let

$$G_i = \{a \mid \delta_i^o(a) > 0\} \backslash P_i \tag{9.6.50}$$

$i = 1, 2, \ldots$. Let G be the union of all the H_i, all the F_{ij}, and all the G_i. This is a countable union; also, the F_{ij} are null sets since α is segregating, and the G_i are null sets by the competitive rule (43). Hence G itself is a null set.

Now let $(\delta_1^{oo}, \delta_2^{oo}, \ldots)$ be another feasible solution. We must show that $[(\delta_i^o)_{i=1,2,\ldots}] \gtrdot [(\delta_i^{oo})_{i=1,2,\ldots}]$ (standard order). Since production costs are fixed by (38), this condition is equivalent to

$$\int_A [f(a, \delta_1^o(a), \delta_2^o(a), \ldots) - f(a, \delta_1^{oo}(a), \delta_2^{oo}(a), \ldots)] \, \alpha(da) \geq 0 \tag{9.6.51}$$

where $f(a, x_1, x_2, \ldots) = b(a, x_1 + x_2 + \cdots) - x_1 h_1(a) - x_2 h_2(a) - \cdots$. (Cf. (31); f is well defined for any feasible $(\delta_1, \delta_2, \ldots)$.)

Now let point a' belong to $P_i \backslash G$. If a' belonged to P_j for some $j \neq i$, it would also belong to F_{ij}, contradicting $F_{ij} \subseteq G$. Hence $a' \notin P_j$ for all $j \neq i$. If $\delta_j^o(a') > 0$, then $a' \in G_j$, contradicting $G_j \subseteq G$. We conclude that $\delta_j^o(a') = 0$ for all $j \neq i$. Hence $\delta^o(a') = \delta_i^o(a')$, and the two functions $b(a', \cdot)$ and $b_i(a', \cdot)$ coincide.

Since $\delta_i^o(a')$ maximizes (45), we obtain

$$[b(a', \delta_i^o(a')) - \delta_i^o(a')(h_i(a') + p_i)]$$
$$\geq [b(a', \delta^{oo}(a')) - \delta^{oo}(a')(h_i(a') + p_i)] \tag{9.6.52}$$

Now,

$$\delta^{oo}(a')(h_i(a') + p_i) = [\delta_1^{oo}(a')(h_i(a') + p_i) + \delta_2^{oo}(a')(h_i(a') + p_i) + \cdots]$$
$$\leq [\delta_1^{oo}(a')(h_1(a') + p_1) + \delta_2^{oo}(a')(h_2(a') + p_2) + \cdots] \tag{9.6.53}$$

The equality in (53) is simply the expansion of $\delta^{oo}(a')$ by (37). The inequality arises from the fact that a' belongs to the demand region P_i. On the right side of (53) the sum of the negative terms is finite; hence the series converges, possibly to $+\infty$.

The series $\delta_1^{oo}(a')h_1(a') + \delta_2^{oo}(a')h_2(a') + \cdots$ is absolutely convergent, since δ^{oo} is feasible. This allows us validly to transfer these terms bodily from one side of an inequality to the other. Combining (52) and (53) and carrying out this transfer, we obtain after some further rearrangement

$$[b(a', \delta_i^o(a')) - \delta_i^o(a')h_i(a')] - [b(a', \delta^{oo}(a')) - \delta_1^{oo}(a')h_1(a') - \cdots]$$
$$\geq \delta_i^o(a')p_i - [\delta^{oo}(a')p_1 + \delta_2^{oo}(a')p_2 + \cdots] \tag{9.6.54}$$

Finally, recalling that $\delta_j^o(a') = 0$ for all $j \neq i$, we see that on the left side of (54) the first bracketed expression equals $f(a', \delta_1^o(a'), \delta_2^o(a'), \ldots)$, while the second bracketed expression equals $f(a', \delta_1^{oo}(a'), \delta_2^{oo}(a'), \ldots)$. The right side of (54) may also be rewritten, and we obtain

$$f(a', \delta_1^o(a'), \ldots) - f(a', \delta_1^{oo}(a'), \ldots)$$

$$\geq [(\delta_1^o(a') - \delta_1^{oo}(a'))p_1 + (\delta_2^o(a') - \delta_2^{oo}(a'))p_2 + \cdots] \qquad (9.6.55)$$

The right side of (54), hence of (55), converges.

Inequality (55) was proved for any $a' \in P_i \backslash G$. However, we notice that the specific index i does not appear in (55); hence (55) is valid for any point a' of the set

$$(P_1 \cup P_2 \cup \cdots) \backslash G \qquad (9.6.56)$$

Next, suppose $a' \in A \backslash (G \cup P_1 \cup P_2 \cup \cdots)$. If $\delta_j^o(a') > 0$, then $a' \in G_j \cup P_j$ from (50), a contradiction. Hence $\delta_j^o(a') = 0$ for *all* plants j. Inequality (52) remains valid for all i, since $a' \in A \backslash H_i$, and from these inequalities we obtain

$$b(a', 0) \geq b(a', \delta^{oo}(a')) - \delta^{oo}(a')g(a') \qquad (9.6.57)$$

where

$$g(a') = \inf_i (h_i(a') + p_i) \qquad (9.6.58)$$

Relation (53) is no longer valid as it stands, but it becomes revalidated if $(h_i(a') + p_i)$ is replaced by $g(a')$. This follows at once from (58). (Note that either $\delta^{oo}(a') = 0$ or $g(a')$ is finite, from (57), hence the right side of (53) still converges.) With this change, the rest of the preceding argument remains valid, so (55) is established also on the set

$$A \backslash (G \cup P_1 \cup P_2 \cup \cdots)$$

Combined with (56), the validity of (55) is established on $A \backslash G$. Thus (55) is true for *almost all a'*.

Now the left side of (55) is precisely the integrand in (51). Hence, if it can be shown that the definite integral of the *right* side of (55) is well defined and equal to *zero*, this will establish (51) and complete the proof.

Define the function $\theta: A \to$ extended reals by

$$\theta(a) = (\delta_1^o(a) + \delta_1^{oo}(a)) | p_1 | + (\delta_2^o(a) + \delta_2^{oo}(a)) | p_2 | + \cdots$$

The typical term on the right side of (55) is

$$(\delta_i^o(a) - \delta_i^{oo}(a))p_i \qquad (9.6.59)$$

The absolute value of the sum of the terms (59) for $i = 1, \ldots, n$ does not exceed $\theta(a)$ for any $a \in A$ and any finite n. Furthermore,

$$\int_A \theta \, d\alpha = 2L_1 | p_1 | + 2L_2 | p_2 | + \cdots < \infty$$

by monotone convergence and (48). Hence we may apply the dominated convergence theorem and conclude that the integral of (55), right side, is the limit of the integrals of the partial sums. But the integral of a typical term (59) is $(L_i - L_i)p_i = 0$. Hence the integral of each partial sum is zero, and so is the entire integral. ∎

Let us examine this pair of theorems. The former, giving *necessary* conditions for an optimum, uses premises that are scarcely more stringent than in the one-plant case. In the latter, which gives *sufficient* conditions for an optimum, the assumption that α is segregating is usually satisfied. Assumption (48) is automatically satisfied if the number of plants is finite, but it fails, for example, in the case of an infinite lattice repetition of plants on the Euclidean plane unless prices are zero.

Of the two theorems, the one giving necessary conditions is the more useful because we are dealing with a subproblem. Optimality in a subproblem does not guarantee optimality in the original problem. However, suppose $(\delta_1^o, \delta_2^o, \ldots)$ is unsurpassed for the *general* multiplant problem, with variable plant outputs $L_1, L_2, \ldots$. Then it clearly remains unsurpassed for the subproblem in which outputs are fixed at $L_1^o, L_2^o, \ldots$, the levels generated by $(\delta_1^o, \delta_2^o, \ldots)$. Let benefit function b satisfy the premises of the "necessity" theorem. It follows that numbers $p_1, p_2, \ldots$ exist that function simultaneously as shadow prices in all one-plant subproblems and in the demand region subproblem of the general multiplant service system problem.

Let us now allow output levels to vary. We saw in the one-plant problem that under certain conditions an optimal solution could be sustained by a free market system if the plant controller used a *marginal-cost-pricing* rule. The conditions were that the production cost function was differentiable and that benefits $b'(a, \cdot)$ as seen by a client at point a coincided with benefits $b(a, \cdot)$ in the objective function. Does this situation continue to hold in the multiplant case?

It does, provided $b = b'$ again, all cost functions C_i are differentiable, α is segregating, and the premises of the "necessity" theorem hold. For suppose $(\delta_1^o, \delta_2^o, \ldots)$ is an unsurpassed solution (with positive output at each plant), and let $p_1, p_2, \ldots$ be the marginal costs: $p_i = DC_i(L_i^o)$, $i = 1, 2, \ldots$. Each δ_i^o is unsurpassed for the one-plant subproblem in which δ_j^o is fixed, all $j \neq i$. The analysis of the one-plant subproblem showed that p_i functioned as a "shadow-price," so that (45) is maximized by $\delta_i^o(a)$ almost everywhere. The proof of the "necessity" theorem then shows that for the same marginal costs $p_1, p_2, \ldots$, the resulting set of potential demand regions obeys the measure-theoretic competitive rule. Let G be the union of the sets H_i, F_{ij} of (49) and G_i of (50), as in the preceding proof. Then G is a null set: $\alpha(G) = 0$.

Consider a client located at point $a' \notin G$. He must decide which plant or plants to buy from. These will be the plants i for which transport cost plus mill price is a minimum. Under marginal-cost pricing the criterion is $h_i(a') + p_i$, so that the client will buy only from plants in whose potential demand regions he is located (how his purchases are distributed among these plants is a matter of indifference).

There are two cases. First, let a' belong to some potential demand region, say $a' \in P_i$. Since $a' \notin G$, it follows that a' belongs to no other demand region P_j, so that $\delta_j^o(a') = 0$ for all $j \neq i$; hence $\delta_i^o(a') = \delta^o(a')$. Also $\delta_i^o(a')$ maximizes (45), which here is the same as saying that $\delta^o(a')$ maximizes

$$b(a', x) - x(h_i(a') + p_i) \qquad (9.6.60)$$

over $x \geq 0$. Now the client at a' will buy exclusively from plant i, since $h_j(a') + p_j$ has a unique minimizer at $j = i$. The amount he will buy maximizes his benefits minus transport costs and plant charges; i.e., it maximizes (60).

Second, suppose a' belongs to no potential demand region. Since $a' \notin G_i$, it follows that $\delta_i^o(a') = 0$ for all i. Hence $x = 0$ maximizes (45) for all i, which implies

$$b(a', 0) \geq b(a', x) - xg(a') \qquad (9.6.61)$$

for all $x \geq 0$, $g(a')$ defined as in (58). From (61) we deduce that a client at a' had best buy nothing at all. For suppose he buys x_i from plant i, $i = 1, 2, \ldots$. Let $x = x_1 + x_2 + \cdots$. His net benefit is

$$b(a', x) - x_1[h_1(a') + p_1] - \cdots$$
$$\leq [b(a', x) - (x_1 + x_2 + \cdots)g(a')] \leq b(a', 0) \qquad (9.6.62)$$

(The first inequality in (62) arises from (58); the second is from (61).)

Thus we have shown that, except for a null set G, if mill prices are set equal to marginal costs at the respective optimal outputs L_i^o, then the unsurpassed solution $(\delta_1^o, \delta_2^o, \ldots)$ can be sustained by individual clients each maximizing his own net benefit in a free market system.

One might try to attain an optimal solution by successive trial values of prices, seeing how market demand varies and adjusting until price equals marginal cost. This procedure has all the vicissitudes of the one-plant case plus the additional problem of interaction. Changing p_i will affect demand not only at plant i but at neighboring plants, as clients are shifted between demand regions.

SHIPMENT PATTERNS AND MILLS-LAV ARRANGEMENTS

Our discussion thus far has run solely in terms of *potential* demand regions. It is also of interest to investigate the system of *effective* demand regions. Recall that E_i, the effective demand region of plant i, is intuitively the region to which plant i actually ships. This is completely different from the potential demand region concept, which refers to the region where plant i has lowest mill price plus unit transport cost.

Formally, we are given a topology $\mathfrak{I}$ on space A, over and above its measure structure. The *effective demand region* of plant i, E_i, is then the support of λ_i (with respect to $\mathfrak{I}$), where λ_i is the distribution of shipments from plant i over A. In terms of the density function δ_i we have $\lambda_i = \int \delta_i \, d\alpha$.

In general, very little can be said about the shape of E_i without further assumptions. Irregularities in the benefit function $b(a, x)$ can lead to all sorts of irregular holes and indentations in E_i. We now assume that *the benefit function does not depend on location;* i.e., $b(a, x)$ is independent of its first argument. Let us rewrite the benefit function as b:reals $\to$ reals, $b(x)$ being the benefit density arising from a delivery density of x.

Consider the one-plant case on the Euclidean plane. If the benefit function is uniform over Space, it seems intuitive that an optimal delivery pattern would have an effective demand region that is a circular disc centered at the plant site. With a more general transport cost function h_i, one expects again that deliveries will be made to the more accessible points, so that perhaps the effective demand region would be a set of the form

$$\{a \mid h_i(a) \leq c\} \tag{9.6.63}$$

for some number c. What if there are many plants? From the theory of market regions we know that under fairly general conditions we have $E_i \subseteq P_i$ (the topological competitive rule). This suggests that E_i might be the *intersection* of P_i, the potential demand region, and a region of the form (63).

It turns out that these ideas are essentially correct, though there are a few complications and some extra mild assumptions seem to be needed. Of these assumptions only one needs discussion.

Definition: Given measure space (A, Σ, α) and topology $\mathfrak{I}$, α is *ubiquitous* iff support $\alpha = A$.

For example, with the usual topology of n-space, Lebesgue measure is ubiquitous; the normal distribution is ubiquitous on the real line.[81] In practice the ubiquity condition on α will usually be satisfied. Even when it is not, one can usually redefine the problem so that it *becomes* satisfied:

---

81. The notion of a ubiquitous resource in location theory seems to involve this concept, strengthened by the condition of availability everywhere on the same terms.

Suppose there is a set F of α-measure zero, such that $F \cup (\text{support } \alpha) = A$. (The existence of such an F is guaranteed if $\mathfrak{J}$ has the strong Lindelöf property.) No positive deliveries can be made to F; hence we have essentially the same problem if Space is restricted to $A \backslash F$, restricting Σ, α, b, and all h_i to this set and taking the relative topology $\mathfrak{J}'$. But with these changes, α is now ubiquitous on $A \backslash F$. Thus the assumption is mild indeed.

The conclusion of the following theorem uses the topological concepts of "closure" and "interior." Given $\mathfrak{J}$ on A, the *interior* of a set $P \subseteq A$ is the union of all open sets contained in P; we denote this $\text{Int}(P)$. For example, if P is a hexagon on the plane with its usual topology, $\text{Int}(P)$ is obtained simply by removing all boundary points from P. The *closure* of a set $H \subseteq A$ is the intersection of all closed sets containing H; we denote this $\text{Cl}(H)$. For example, if H is the circular disc $\{(x, y) \mid x^2 + y^2 < r^2\}, r > 0$, then $\text{Cl}(H)$ is obtained by replacing $<$ by $\leq$.

The concept of "essential supremum" is used in the proof. Recall that the *essential supremum* of a function $h:G \to$ reals, abbreviated ess sup $(h \mid G)$, is $\inf_F \sup\{h(a) \mid a \in G \backslash F\}$, the infimum taken over all α-null sets F.

Theorem: Given (A, Σ, α), $b:A \times$ reals $\to$ reals, $h_i:A \to$ reals, and $L_i, i = 1, 2, \ldots$, let $(\delta_1, \delta_2, \ldots)$ be feasible densities for the resulting multiplant service system problem.

Let $b(a, x) = b(x)$ be uniform over space, let α be segregating, and let extended real numbers $p_1, p_2, \ldots$ exist such that $\delta_i(a)$ maximizes (45) for almost all $a \in A$, $i = 1, 2, \ldots$, and such that the resulting system of potential demand regions $P_1, P_2, \ldots$ satisfy the measure-theoretic competitive rule. Also, let there be a topology $\mathfrak{J}$ on A such that h_i is continuous (as well as measurable), $i = 1, 2, \ldots$, and such that α is ubiquitous.

Then for each *effective* demand region E_i there is an extended real number c_i such that

$$\text{Cl}[\{a \mid h_i(a) < c_i\} \cap \text{Int}(P_i)] \subseteq E_i \subseteq [\{a \mid h_i(a) \leq c_i\} \cap P_i] \qquad (9.6.64)$$

Proof: Define c_i by

$$c_i = \text{ess sup}[h_i \mid \{a \mid \delta_i(a) > 0\}] \qquad (9.6.65)$$

$i = 1, 2, \ldots$.

We first prove the right-hand inclusion in (64). By the theory of market regions (p. 610), P_i is closed, since each h_j is continuous; this and the measure-theoretic competitive rule imply the topological competitive rule; i.e.,

$$E_i \subseteq P_i \qquad (9.6.66)$$

Next suppose a point $a_o \in E_i$ existed with $h_i(a_o) > c_i$. The set $\{a \mid h_i(a) > c_i\}$ is open, hence it is a measurable neighborhood of a_o. Since a_o supports λ_i, we must have

$$\lambda_i\{a \mid h_i(a) > c_i\} > 0 \qquad (9.6.67)$$

Translated into density form, (67) reads

$$\alpha\{a \mid h_i(a) > c_i \quad \text{and} \quad \delta_i(a) > 0\} > 0 \qquad (9.6.68)$$

(Compare the translation of (41) into (43).)

Inequality (68) implies $c_i < +\infty$. Let $x_1, x_2, \ldots$ be a decreasing sequence of numbers with limit c_i. Consider the sets obtained by replacing c_i successively with each x_n in set (68). This is an increasing sequence of sets whose limit is set (68). By the continuity theorem for measures, an $x > c_i$ exists such that (68) remains true when c_i is replaced by x. But this is incompatible with the definition of c_i, (65). This contradiction proves that

$$E_i \subseteq \{a \mid h_i(a_o) \leq c_i\} \qquad (9.6.69)$$

Inclusions (66) and (69) together yield the right inclusion in (64).

We shall establish that

$$[\{a \mid h_i(a) < c_i\} \cap \text{Int}(P_i)] \subseteq E_i \qquad (9.6.70)$$

Set E_i, being a support, must be closed. Hence (70) implies the left inclusion in (64). Thus the establishment of (70) will complete the proof.

We argue by contradiction. If (70) is false, there is a point $a_o \in (\text{Int}(P_i))\backslash E_i$ with $h_i(a_o) < c_i$. Since a_o does not support λ_i, sets $G \in \mathfrak{I}$ and $K \in \Sigma$ exist such that $a_o \in G \subseteq K$ and

$$\lambda_i(K) = 0 \qquad (9.6.71)$$

Translated into density form, (71) reads

$$\alpha[K \cap \{a \mid \delta_i(a) > 0\}] = 0 \qquad (9.6.72)$$

Also,

$$a_o \in [\{a \mid h_i(a) < c_i\} \cap G \cap \text{Int}(P_i)] \subseteq [\{a \mid h_i(a) < c_i\} \cap K \cap P_i] \qquad (9.6.73)$$

The smaller bracketed set in (73) is open, and the larger is measurable. Also a_o supports α, since α is ubiquitous. It follows that

$$\alpha[\{a \mid h_i(a) < c_i\} \cap K \cap P_i] > 0 \qquad (9.6.74)$$

For extended real numbers x, define the sets

$$J_x = \{a \mid h_i(a) < x \quad \text{and} \quad \delta_i(a) = 0\} \cap K \cap P_i \qquad (9.6.75)$$

The bracketed set in (74) is contained in the union of J_{c_i} and the bracketed set in (72). It follows that $\alpha(J_{c_i}) > 0$.

Inequality (74) implies that $c_i > -\infty$; hence there is an increasing sequence of numbers x_1, x_2,... with limit c_i (note that c_i may equal $+\infty$). An argument similar to the one following (68) shows that an $x_o < c_i$ exists such that

$$\alpha(J_{x_o}) > 0 \qquad (9.6.76)$$

We must also have

$$\alpha \{a \mid h_i(a) > x_o \text{ and } \delta_i(a) > 0\} > 0 \qquad (9.6.77)$$

For if (77) were false, the ess sup in (65) would be less than c_i.

Consider the set

$$J_{x_o}\backslash[H_i \cup \bigcup \{F_{ij} \mid j = 1, 2, \dots, j \neq i\} \cup \bigcup \{G_j \mid j = 1, 2, \dots\}] \quad (9.6.78)$$

where H_i is a null set such that $\delta_i(a)$ maximizes (45) for all $A\backslash H_i$, F_{ij} is given by (49), and G_j by (50). (F_{ij} is not well defined if $p_i = p_j = \infty$; but if $p_i = \infty$, then $\delta_i(a) = 0$ almost everywhere, so $c_i = -\infty$, contradicting (74); hence p_i must be finite.) Since α is segregating, F_{ij} is null for all $j \neq i$; G_j is null for all j, by the measure-theoretic competitive rule. Hence from (76) the set (78) still has positive α-measure, so it is not empty.

Let a' be a point of set (78). We claim that $\delta_i(a')$ maximizes

$$b(x) - x[h_i(a') + p_i] \qquad (9.6.79)$$

over real $x \geq 0$. First, $\delta_i(a')$ maximizes

$$b(\delta(a') - \delta_i(a') + x) - x[h_i(a') + p_i] \qquad (9.6.80)$$

over real $x \geq 0$, by (45), since $a' \notin H_i$. Furthermore, $a' \in P_i$, by (75); hence $a' \notin P_j$ for all $j \neq i$, since $a' \notin F_{ij}$; hence $\delta_j(a') = 0$ for all $j \neq i$, since $a' \notin G_j$. But this means that (80) is the same as (79).

We now repeat the entire argument beginning at (78) with J_{x_o} replaced by the set in (77). The corresponding set in (78) is again nonempty. Let a'' belong to this set. We claim that $\delta_i(a'')$ maximizes

$$b(x) - x[h_i(a'') + p_i] \qquad (9.6.81)$$

over real $x \geq 0$. For $\delta_i(a'')$ maximizes (80) with a'' in place of a', since $a'' \notin H_i$. Also $\delta_i(a'') > 0$ from (77), and $a'' \notin G_i$; hence $a'' \in P_i$. The remainder of the argument is the same as above.

We now show that the respective maximizations of (79) and (81) yield a contradiction. Letting $\delta_i(a'') = y > 0$, and noting that $\delta_i(a') = 0$, by (75), we obtain $b(0) \geq b(y) - y[h_i(a') + p_i]$ from (79), and $b(y) - y[h_i(a'') + p_i] \geq b(0)$ from (81). Adding these inequalities and simplifying, we obtain $h_i(a'') \leq h_i(a')$. But, in fact, $h_i(a'') > x_o > h_i(a')$, by (77) and (75). This contradiction establishes (70) and completes the proof. ∎

The conclusion of this theorem, (64), brackets the effective demand region E_i between lower and upper bounds in the sense of set containment. In most cases of interest these bounds coincide. In these cases we can say unambiguously that E_i is the intersection of the potential demand region P_i and a disc of the form (63). Note that nothing in the premises of this theorem is said directly about the optimality of the shipment pattern $(\delta_1, \delta_2, \ldots)$; rather, the premise is framed in terms of the existence of shadow prices $p_1, p_2, \ldots$. This form is very convenient for applications.

As an example, we take the famous honeycomb lattice arrangement of plants on the Euclidean plane (α is plane Lebesgue measure and $\mathfrak{J}$ the usual plane topology). Recall that each plant has six nearest neighboring plants (at distance r, say), these forming the corners of a regular hexagon. We suppose the benefit function is uniform over space and that a shipment pattern $(\delta_1, \delta_2, \ldots)$ exists having real numbers $p_1, p_2, \ldots$ that are shadow prices in the sense of the theorem. We further suppose these numbers p_i are all equal. Under these conditions the potential demand regions P_i are (closed) regular hexagons centered at their respective plant sites, the distance from plant i to an edge of P_i being $r/2$, and the distance from plant i to a corner of P_i being $r/\sqrt{3}$.

All the premises of the preceding theorem are now satisfied, so we obtain (64). The lower and upper bounds in (64) coincide, so we can definitely say that each effective demand region is the intersection of a regular hexagon and a concentric circular disc. Furthermore, it is not difficult to show that the numbers c_i of (65) must all have a common value c; i.e., the radii of the discs are equal, so that the regions E_i are congruent to each other. There are four cases.

1. $c = -\infty$. This is the no-production case, and all effective demand regions are empty.
2. $0 < c \le r/2$ (the case $-\infty < c \le 0$ is impossible). Here the disc is contained in the hexagon, so that the effective demand regions are circular discs.
3. $r/2 < c < r/\sqrt{3}$. Here neither the disc nor the hexagon contains the other. The effective demand regions are twelve-sided "polygons," having six straight and six curved sides alternating with each other.
4. $c = r/\sqrt{3}$. Here the hexagon is contained in the disc, so that the effective demand regions coincide with the potential demand regions: $E_i = P_i, i = 1, 2, \ldots$. (The case $c > r/\sqrt{3}$ is impossible, by the measure-theoretic competitive rule.)

This example brings us close to the work of Mills and Lav[82] who, however, approach the subject in terms of a specific institutional frame-

82. E. S. Mills and M. R. Lav, A model of market areas with free entry, *J. Polit. Econ.* 72(June 1964):278–88.

work (viz., monopolistic competition), with each firm a profit maximizer and having just one plant. We trace the implications of this framework.

Assume for the present that the number and locations of all plants are fixed (not necessarily in a honeycomb lattice pattern). Pricing is on an f.o.b. basis; i.e., the price to a client purchasing from plant i is the mill price p_i plus the unit transport cost. We are still on the Euclidean plane, and α, $\mathfrak{I}$, and h_i retain the meanings they had above; b is still uniform over Space. We also make the weak assumption that the potential demand regions cover Space, so that each client finds at least one plant for which the combined mill price and transport cost $p_i + h_i(a)$ is minimized. Each client will purchase from the least expensive outlet, which means he will buy only from plants in whose potential demand regions he is located. For all but a null set of locations the choice of plant is uniquely determined, since α is segregating. Furthermore, the quantity purchased by a client will be a maximizer of (45). Thus we may expect that the mill prices $p_1, p_2, \ldots$ will fulfill the shadow-price conditions for the resulting shipment pattern.

This argument is identical with that used in showing how a decentralized pricing system can sustain an optimal shipment pattern. The only difference here is that the plant manager will generally set mill price above marginal production cost rather than equal to it.

But the foregoing argument, along with the other standard assumptions, shows that the premises of the preceding theorem are fulfilled. Thus again we have (64). The upper and lower bounding sets in (64) will generally coincide (except in some rather artificial special cases). In general, then, E_i will be the intersection of a circular disc with P_i; P_i in turn is the intersection of hyperbolic regions and half-planes. In the special case in which firms have their plants arranged in a honeycomb lattice and all mill prices are identical, the four cases outlined above are the only possibilities. Thus the speculations of Mills and Lav concerning effective market regions in the shape of $6n$-polygons seem to be off the track.[83]

The most important accomplishment of Mills and Lav is the demonstration that there are situations where, in equilibrium, effective demand regions do *not* coincide with potential demand regions, contrary to an impression left by Lösch.[84] A single counterexample is sufficient to refute an invalid generalization. But it is also useful to examine the underlying principles involved. Accordingly, we restate the Mills-Lav argument in a

---

83. This same conclusion is reached in Faden, *Essays,* p. 147.

84. Lösch postulates that the entire space is served. Since he never clearly distinguishes potential from effective regions, it is possible that Lösch is merely stating that *potential* demand regions cover Space, rather than making the (incorrect) statement that *effective* demand regions cover Space. Lösch, *Economics of Location,* p. 95.

general service system context. We start by focusing attention on just one plant, located at a_o. Our assumptions are as follows.

1. The production-cost function at the plant site $C(x)$ is nonnegative and $C(0) = 0$.
2. The benefit-density function $b(a, x)$ is uniform over space, so it may be written $b(x)$; b is continuous, and strictly concave on the nonnegative real numbers.
3. Let $\delta(a, p)$ be the maximizer of

$$b(x) - x[h(a) + p] \qquad (9.6.82)$$

over $x \geq 0$. Here $h(a)$ is the unit transport cost to a from plant site a_o. From the assumptions on b there is at most one such maximizer; in fact, either such a maximizer exists or (82) increases indefinitely with x. In the latter case we set $\delta(a, p) = +\infty$ by convention.

By pp. 255–56, $\delta(\cdot, p)$ is measurable for all p. Our third assumption is

$$\int_A \delta(a, 0) \; \alpha(da) < \infty \qquad (9.6.83)$$

Inequality (83) implies that the anomalous region on which $\delta(a, 0) = +\infty$ must be an α-null set: demand is finite almost everywhere. The same must be true for any positive p, since $\delta(a, p)$ is nonincreasing in p.

4. Our fourth and last assumption is that if there were just a single plant at a_o and no other plants in the system, then a price p_o exists generating positive demand, at which (economic) profits are zero, and such that profits are negative at any other price.

Before we begin drawing conclusions from these assumptions let us examine their economic meaning. Essentially no assumption is imposed on production costs C. The strict concavity of b is the principle of diminishing marginal benefits. (If b is quadratic, we get linear demand functions, the case examined by Mills and Lav.) Continuity of b merely assures that b takes no "jump" at $x = 0$. Inequality (83) states that at mill price zero total demand is finite.

Assumptions 1–3 are all weak and plausible; the fourth is the one with teeth in it and provides the nub of the argument. The idea is that, even with no competitors, the firm just barely manages to survive (in the sense of not exiting from the industry). If a competitor now tears a piece out of the firm's demand region and thereby shrinks its demand, this is the last straw and the firm leaves the industry.

It is easy to find functions satisfying all these assumptions, since C is virtually unrestricted. We now show how the argument just sketched can be justified.

The critical price p_o in assumption 4 must be nonnegative, for negative p_o implies negative revenue, hence negative profit. It must be finite, since demand is positive. Let F be a region satisfying

$$\alpha[F \cap \{a \mid \delta(a, p_o) > 0\}] > 0 \qquad (9.6.84)$$

We will show that if the firm is denied the opportunity to sell in region F (say by competitive encroachment), it cannot avoid making losses at any positive production level.

From the assumptions on b it is not difficult to show that $\delta(a, p)$ is continuous in p, wherever $\delta(a, p)$ is finite. From this it follows that the demand function $f(p) = \int_A \delta(a, p) \, \alpha(da)$ (defined for $p \geq 0$) is continuous. For let $p_1, p_2, \ldots$ be a monotone sequence converging to p. For each $a \in A$, $\delta(a, p_n)$, $n = 1, 2, \ldots$, is a monotone sequence converging to $\delta(a, p)$. Inequality (83) allows us to apply the monotone convergence theorem and conclude that $f(p_n) \to f(p)$.

Furthermore, for each $a \in A$, $\delta(a, p)$ converges monotonically to zero as p goes to $+\infty$. By monotone convergence again, it follows that $f(p) \to 0$ as $p \to +\infty$.

Now suppose that after the loss of region F, a price p' existed generating a demand $q' > 0$, for which profit is nonnegative:

$$p'q' - C(q') \geq 0 \qquad (9.6.85)$$

Obviously $p' \geq 0$. We clearly have $q' \leq f(p')$, since $f(p')$ is the demand over the whole space including region F. Since f is continuous and goes to zero, $p'' \geq p'$ must exist such that $f(p'') = q'$. Hence

$$p''f(p'') - C(f(p'')) \geq p'q' - C(q') \qquad (9.6.86)$$

Inequalities (85) and (86) imply that the firm makes nonnegative profits when it sets the mill price to p'' and has the entire space, including region F, at its disposal. By assumption 4 there is just one price p_o at which this is possible. Hence $p'' = p_o$, and the left side of (86) equals zero. It follows that (86) is actually an equality. This is possible iff $p' = p''$, since the quality sold is positive.

But this yields a contradiction. For (84) implies that q', the quantity sold at price $p' = p_o$ after region F has been deleted, must be *less* than $f(p_o) = f(p'')$, the quantity sold at price p_o before F has been deleted. Hence the firm cannot avoid losses at positive production levels if F is deleted.[85]

The rest of the argument is purely geometrical. Go to the special case of the Euclidean plane and assume that production costs C (as well as benefits b) are uniform over Space. There are b and C functions such

---

85. A less elaborate version of this argument may be found in Faden, *Essays*, pp. 141–43.

that the firm just manages to survive in the absence of any competitors. The effective demand region will be a closed circular disc centered on the plant site (say of radius r), by (64). Now enter other firms, each operating a single plant. Suppose two such plants were closer than distance $2r$ to each other. Then at least one of the firms will have a region F satisfying (84) deleted from its demand region. By the argument just given it cannot then avoid losses and will leave the industry. Thus the only possible equilibrium situation is one in which the distance between any two plants is $\geq 2r$. It is then obvious that the effective demand regions cannot cover the plane. This concludes the Mills-Lav argument. (This argument assumes that r is finite. If $r = +\infty$, a firm needs the entire plane to survive, and there is no room for even one additional firm.)

A number of comments on this argument are called for. First, assumption 4, that the firm just manages to survive at one mill price p_o, could be weakened slightly, to allow a modicum of encroachment on the circular demand region by other firms, but not enough to fill in all the gaps in the plane. If the arrangement of plants is a honeycomb lattice, the effective demand regions will have the form of case 3, twelve-sided "polygons" with alternating straight and curved sides.

Second, a similar argument could be made for discriminating monopolists, rather than the f.o.b.-pricing firms discussed above. For, in general, the effective demand region of a lone discriminating monopolist on the plane will again be a circular disc. Benefit and production cost functions exist such that the monopolist just survives, and arguments similar to those above show that plants must be far enough apart to avoid encroaching on each other's effective demand regions, which again means that the plane will not be covered.

Finally, it is quite instructive to see what happens if, instead of the Euclidean metric on the plane, we have the *city-block* metric. For a single plant on the plane, the effective demand region will be a *diamond* rather than a circular disc, by (64). The diamond of radius r centered at (x_o, y_o) is $\{(x, y) \mid |x - x_o| + |y - y_o| \leq r\}$. As above, we can find functions b and C such that the firm just manages to survive without competitive encroachment. But the geometrical part of the argument breaks down, since diamonds, unlike circles, *can* be fitted together to cover the plane. As an example, consider the arrangement in which plants occupy all points of the form (m, n), where m, n range over the integers $0, \pm 1, \pm 2, \ldots$, and $m + n$ is an even number.[86] One easily verifies that the set of diamonds of radius one centered at these plant sites covers the plane.

---

86. A *diamond* lattice is defined as any lattice obtained from this one by the affine transformation $(x, y) \rightarrow (ax + b, ay + b)$, a, b real, $a \neq 0$. The present argument applies to any diamond lattice.

Furthermore, these diamonds overlap only along their border lines. Since the borders have area zero, this is not an encroachment in the sense of (84). Thus the Mills-Lav argument breaks down for the city-block metric.

VARIABLE PLANT LOCATIONS AND NUMBERS

It will be noticed that the arguments above rely only on the option of firms to *leave* the industry; nothing is said about the option of new firms to *enter* the industry. The possibility of entry imposes further conditions on any equilibrium; there must be no point at which a new plant can insinuate itself and make positive profits. Lösch believed that new firms would crowd into the industry until profits of all firms were driven to zero.[87] However, later writers noted that firms could indeed be spread out sufficiently to make positive profits, but not so far apart as to allow a niche where another firm could enter with positive profits.[88] Thus the free entry condition does *not* imply that equilibrium profits must be zero, even with identical firms and uniform cost and benefit functions.

A closely related Löschian principle is rather interesting; viz., *the number of firms is maximized* (subject to no firm making losses).[89] The heuristic argument for this condition again turns on the free entry option (and a limitless supply of potential entrepreneurs). The principle is at best an approximate characterization of equilibrium, and one easily finds situations that violate it. The Beckmann example cited above does so. In general, just as in packing a trunk, careful coordination of the locations of plants is needed to maximize numbers; helter-skelter entry generally leads to a submaximal result.

Nonetheless, even as an approximating·principle it commands attention because of its simplicity, generality, and extremal form. No normative significance should be attached to the fact that something is being maximized; this is merely a convenient descriptive device.[90]

To implement the Löschian principle, we must face up to an aspect of the service system problem that we have studiously avoided to this point and must allow locations to vary. (Up to now we have only considered subproblems in which locations of all plants are given.) It turns out that

---

87. Lösch, *Economics of Location,* pp. 95 and 109 n. 1, following E. H. Chamberlin, *The Theory of Monopolistic Competition,* 1st ed. (Harvard Univ. Press, Cambridge, 1933).

88. B. J. L. Berry and W. L. Garrison, A note on central place theory and the range of a good, *Econ. Geog.* 34(1958):304–11, and more formally in M. Beckmann, *Location Theory* (Random House, New York, 1968), p. 44.

89. Lösch, *Economics of Location,* pp. 8, 95, 259. Lösch does not state the parenthetical condition explicitly, but it is obviously needed.

90. As a champion of small business, Lösch finds virtue in this principle on sociological grounds. *Economics of Location,* pp. 66–67. By the usual economic criteria, it will generally be true that the resulting equilibrium has too many firms, each producing too little at too high a price.

the variable location problem is quite difficult. Furthermore, several alleged "results" widely accepted in the location literature turn out to be erroneous. Finally, serious conceptual difficulties arise concerning the very meaning of "maximization" in the Löschian principle or in social optimization.

Our discussion is devoted mainly to clarifying these conceptual issues. Difficulty arises when the number of plants is infinite—an important theoretical case since it includes the lattices, honeycomb and other. Our basic method for deciding when one arrangement is superior to another is to interpret the resulting values in the objective function (e.g., (36) for social optimization) as *pseudomeasures* and then apply *standard ordering* to them. This approach, which has served so well throughout this book, proves to be inadequate when plant locations are allowed to vary in the service system problem.

To see where the trouble lies, let space A be the plane (although the following discussion is valid in fairly general spaces). Consider the social optimization objective function (36) and ignore everything except the *production* cost terms, which may be written

$$C(a_1', L_1) + C(a_2', L_2) + \cdots \tag{9.6.87}$$

Here $C(a_i', L_i)$ is the cost incurred in producing output L_i at plant site a_i'. For simplicity assume that C is a nonnegative function. Expression (36), including the part (87), is to be interpreted as a pseudomeasure over space A. The natural way of doing this for (87) is to think of each $C(a_i', L_i)$ as representing a measure over A, *concentrated at the point* a_i', the production cost being the total mass. Then (87) is the sum of this countable collection of measures. (As mentioned above, the assumption that each singleton $\{a_i'\}$ is measurable insures that (87) is σ-finite. Since plant locations may now vary, we must assume that all singleton sets $\{a\}$ are measurable, as is true for the Borel field on the plane.)

Let $A' = \{a_1', a_2', \ldots\}$, the set of plant sites in (87). Consider now the effect of a shift to a new set of plant sites $A'' = \{a_1'', a_2'', \ldots\}$ *disjoint* from the old set. (Since there are just a countable number of points to avoid, this can be accomplished simply by a rigid translation of the plane.) Suppose that (87), thought of as an ordinary infinite series, goes to infinity and that the same is true for the new set of plant sites A'' (so that the two resulting measures are infinite, though σ-finite).

Then the original and the shifted arrangements are *not comparable* under standard ordering of pseudomeasures. To see this, recall that $\mu \succeq \nu$ (standard order) iff

$$(\mu - \nu)^+(A) \geq (\mu - \nu)^-(A) < \infty \tag{9.6.88}$$

If μ is the cost measure (87) and ν is the same after the shift, then

$\mu(A') = \nu(A'') = \infty$, and $\mu(A'') = \nu(A') = 0$. Thus μ and ν are infinite and mutually singular when restricted to $A' \cup A''$. The upper and lower variations in (88) are then both infinite, and this precludes comparability. Nor would the inclusion of the benefit and transport cost terms of (36) alter this conclusion, for $A' \cup A''$, being countable, has area zero, so that these other terms are zero on this set.

The same situation arises if one uses the Löschian principle as a criterion instead of social optimization. Assume each firm has just one plant. How does one compare the number of firms under two arrangements, when this number is infinite in both cases? The natural thing is to think of the "number of firms" as a measure over space. Specifically, let each firm correspond to a measure of mass *one* concentrated at its plant site; the number of firms is simply the sum of these measures. But this leads to just a special case of the preceding construction, viz., the case in which each $C(a_i, L_i)$ of (87) has the value *one*. (We are also maximizing the number of firms here and minimizing costs above; but this reversal does not affect the comparability argument just given.)

We conclude that standard ordering is inadequate for the variable location problem because it refuses to make a comparison between many pairs of arrangements that *do* seem comparable to intuition. To take a concrete example, apply the Löschian principle to two honeycomb lattice arrangements of plants, nearest neighbors being distance x apart in the first lattice and distance $y < x$ apart in the second. Intuition suggests that the second lattice has "more firms" than the first because plants are "more densely packed" into space. But these arrangements are not comparable under standard ordering if the two sets of plant sites are disjoint.

What we need is a more complete ordering among pseudomeasures than standard ordering allows. At the same time, if ψ_1 and ψ_2 are comparable under standard order, intuition suggests that the indicated preference should be preserved. That is, if $\psi_1 > \psi_2$ (standard order), then we should have $\psi_1 >' \psi_2$ for the more complete ordering $>'$ that replaces standard order, and similarly $\psi_1 \sim \psi_2$ should imply $\psi_1 \sim' \psi_2$.

Chapter 3 defined a large class of orderings that have precisely these properties: the *extended orderings* of pseudomeasures. We recall the basic concepts, referring to Chapter 3 for details and justifications. An extended ordering is determined by an *extension class,* which is a class of measurable sets $\mathcal{G}$ having the following two properties. (i) For every $G_1, G_2 \in \mathcal{G}$, there is a $G_3 \in \mathcal{G}$ such that $G_3 \supseteq (G_1 \cup G_2)$, and (ii) there is a countable subcollection of $\mathcal{G}$-sets whose union is the universe set A.

Given the extension class, the extended ordering $\geqslant$ is defined as follows. First define the *restricted domain* Σ_ψ of pseudomeasure ψ as the class of all measurable sets E for which $\psi^+(E)$ and $\psi^-(E)$ are not both infinite. On Σ_ψ define $\psi(E)$ as $\psi^+(E) - \psi^-(E)$.

Definition: $\psi_1 \succcurlyeq \psi_2$ (extended ordering with respect to extension class $\mathcal{G}$) iff

(i) $\mathcal{G} \subseteq \Sigma_\psi$, where Σ_ψ is the restricted domain of $\psi = \psi_1 - \psi_2$;
(ii) for all $\epsilon > 0$ there is a set $G \in \mathcal{G}$ such that for all $H \supseteq G$, $H \in \mathcal{G}$, we have $\psi(H) > -\epsilon$.

There still remains the problem of which of the many extended orderings to choose. This is a matter of intuitive appeal. Choose an ordering and see whether the preferences it expresses accord with intuition. On the plane we choose the ordering determined by the class of all *closed circular discs,* i.e., the class of all sets of the form

$$\{a \mid d(a_o, a) \leq r\} \qquad (9.6.89)$$

for all points a_o and real numbers $r \geq 0$, d being the Euclidean distance. One immediately verifies that this is indeed an extension class. For any two discs there is obviously a third disc containing both of them; also, for fixed a_o the discs (89) with $r = 1, 2, \ldots$ are a countable covering of the plane. We refer to this extended ordering as the *circles criterion.*[91]

As discussed in Chapter 3, the circles criterion is a natural generalization of the "overtaking" criterion on the nonnegative real line. The overtaking criterion in turn is an extension of the Ramsey method for comparing improper integrals in one dimension; the Ramsey method is just a special application of standard ordering (see pp. 162–63, 179–81). Thus the extension to the circles criterion is the 2-dimensional analog of the extension from Ramsey ordering to "overtaking."[92]

Given two arbitrary pseudomeasures on the plane, the task of determining their comparability under the circles criterion is not all that easy. It is therefore useful to find some easily verified sufficient conditions for comparability. The following theorem gives a condition that is satisfied by a wide variety of pairs of pseudomeasures, including those arising from the lattice arrangements common in location theory.

The *diameter* of a set E in a metric space is given by

$$\text{diam}(E) = \sup\{d(a, b) \mid a, b \in E\} \qquad (9.6.90)$$

---

91. The class of *open* discs, and the class of *open* and *closed* discs together, both yield the same pseudomeasure ordering as (89). See Chapter 3, note 23. Alternatively, other normed metrics such as the city-block metric could be used. It seems unlikely that any two "reasonable" pseudomeasures on the plane would be ordered differently by these alternatives. In any case, the following theorem is valid for all these alternative versions.

92. Other criteria have been proposed for comparing arrangements in two- and higher-dimensional spaces. See L. Fejes Tóth, *Lagerungen in der Ebene auf der Kugel und im Raum,* 2nd ed. (Springer, Berlin, 1972), pp. 55–56, and C. A. Rogers, *Packing and Covering* (Cambridge Univ. Press, 1964), p. 22. These criteria do not seem as appealing as those deriving from pseudomeasure theory.

A set is *bounded* iff it has finite diameter. In the following, one key assumption is that the plane is partitioned into *uniformly bounded* sets; i.e., there is a fixed real number k such that none of the sets has a diameter exceeding k. The distances referred to are the Euclidean distances.

Theorem: Let (A, Σ, α) be 2-dimensional Lebesgue measure ("area") on the Borel field of the plane. Let μ, ν be two σ-finite measures on the plane, with ν finite on every bounded measurable set. Let $\{A_1, A_2, \ldots\}$ and $\{B_1, B_2, \ldots\}$ be two countable measurable partitions of the plane such that

(i) each A_n and each B_n has positive finite area;
(ii) there is a real number k with $\operatorname{diam}(A_n) \le k$, $\operatorname{diam}(B_n) \le k$ for all $n = 1, 2, \ldots$;
(iii) numbers x and y exist such that

$$\mu(A_n)/\alpha(A_n) \ge x > y \ge \nu(B_n)/\alpha(B_n) \tag{9.6.91}$$

for all $n = 1, 2, \ldots$ (x may equal $+\infty$).

Then $\mu > \nu$ on the circles criterion.

Proof: Suppose there were a closed circular disc F_o and a *positive* number c such that for all closed circular discs $F \supseteq F_o$, we have

$$\mu(F) - \nu(F) \ge c \tag{9.6.92}$$

Then $\mu > \nu$ (circles criterion). To prove this, we first show that $\mu \ge \nu$. Let (ψ^+, ψ^-) be the Jordan form of pseudomeasure $\psi = (\mu, \nu)$. By the equivalence theorem for pseudomeasures,

$$\psi^+ + \nu = \psi^- + \mu \tag{9.6.93}$$

It now follows that

$$\psi(F) = \psi^+(F) - \psi^-(F) = \mu(F) - \nu(F) \ge c \tag{9.6.94}$$

for any disc $F \supseteq F_o$. The inequality in (94) is simply (92). Since F is bounded, $\nu(F)$ is finite; hence $\psi^-(F)$ is finite, by the minimizing property of the Jordan form. This justifies transposing the terms of (93) at F to obtain the second equality in (94). The first equality in (94) is true by definition, and incidentally we have shown that $F \in \Sigma_\psi$.

Condition $\mu \ge \nu$ follows at once from (94). For, given any $\epsilon > 0$, choose F_o; then for any closed circular disc $F \supseteq F_o$ we have $\psi(F) \ge c > -\epsilon$.

Furthermore, $\nu \not\ge \mu$. To see this, choose positive $\epsilon < c$, and for any closed circular disc H choose a closed circular disc $F \supseteq (H \cup F_o)$. Relation (94) applies to F, so $(\nu - \mu)(F) = -\psi(F) \le -c < -\epsilon$, so that we cannot have $(\nu - \mu)(F) > -\epsilon$.

Thus given (92), we have proved that $\mu > \nu$. To complete the overall proof, we show the existence of a disc F_o and number c such that (92) is satisfied for discs $F \supseteq F_o$.

To this end, let a_o be any fixed point of the plane, and let r be any real number $\geq k$. Let E, F, G be three closed circular discs all centered at a_o, with radii $r - k$, r, and $r + k$ respectively. Let H be the union of all sets A_n *contained* in F, and let J be the union of all sets B_n *not disjoint* from F. We claim that

$$E \subseteq H \subseteq F \subseteq J \subseteq G \qquad (9.6.95)$$

To prove the first inclusion, let $a \in E$; a must belong to some A_n, say $a \in A_m$. For any $b \in A_m$ we have $d(a, b) \leq k$; hence $d(a_o, b) \leq d(a_o, a) + d(a, b) \leq (r - k) + k = r$; hence $A_m \subseteq F$, so that $A_m \subseteq H$. This proves $E \subseteq H$.

Inclusion $H \subseteq F$ is clear. Let $a \in F$; a belongs to, say, B_m, which is therefore not disjoint from F; hence $B_m \subseteq J$. This proves $F \subseteq J$.

Finally, let $a \in J$, so that, say, $a \in B_m$ where B_m meets F. Thus if $b \in B_m \cap F$, we have $d(a, b) \leq k$; hence $d(a_o, a) \leq d(a_o, b) + d(b, a) \leq r + k$, so that $a \in G$. This completes the proof of (95).

Now,

$$\mu(A_n) \geq x\alpha(A_n) \qquad (9.6.96)$$

for all A_n, from (91). This implies

$$\mu(F) \geq \mu(H) \geq x\alpha(H) \geq x\alpha(E) \qquad (9.6.97)$$

The left and right inequalities in (97) come from (95). Add inequalities (96) over all A_n contained in H; since these partition H, we obtain the middle inequality of (97). Similar reasoning yields

$$\nu(F) \leq \nu(J) \leq y\alpha(J) \leq y\alpha(G) \qquad (9.6.98)$$

We then obtain

$$\mu(F) - \nu(F) \geq x\alpha(E) - y\alpha(G) = x\pi(r - k)^2 - y\pi(r + k)^2 \qquad (9.6.99)$$

The inequality in (99) results from subtracting (98) from (97). The equality arises from the observation that $\alpha(E)$ and $\alpha(G)$ are simply the ordinary areas of circular discs of radii $(r - k)$ and $(r + k)$ respectively.

We are now ready to choose F_o and $c > 0$ satisfying (92). If $x = \infty$, let F_o be any closed circular disc of radius $r > k$. Any disc $F \supseteq F_o$ must also have a radius $r > k$. Hence $\mu(F) - \nu(F) = \infty$, by (99), so (92) is satisfied by any $c > 0$.

If $x < \infty$, let F_o be any closed circular disc of radius

$$r > 2k(x + y)/(x - y) \qquad (9.6.100)$$

Any disc $F \supseteq F_o$ must also have a radius r satisfying (100). The right side of (99) may be rearranged to read

$$\pi r[r(x - y) - 2k(x + y)] + \pi k^2(x - y) \qquad (9.6.101)$$

The bracketed component in (101) is obviously positive if r satisfies (100). Hence (92) is satisfied for all closed circular discs $F \supset F_o$ if we choose $c = \pi k^2(x - y) > 0$. ∎

A number of comments are called for. Note that the theorem involves the comparison of two *measures* μ and ν. If, more generally, one is called upon to compare two *pseudo*measures ψ_1 and ψ_2, one must first express their difference $\psi_1 - \psi_2 = \psi$ as the difference of two measures in order to apply this theorem. One way of doing this is by writing $\psi = \psi^+ - \psi^-$, but other decompositions may be more convenient. A very convenient technique is to add a multiple of α to both ψ_1 and ψ_2. This method is available only if $\psi_i + x\alpha$ is a measure for sufficiently large real $x, i = 1, 2$, a condition usually fulfilled in practice.

The use of the circles criterion imposes no restriction on the unit transport cost function h; in particular, h need not be Euclidean, or even a metric.

We now show that this theorem is not vacuous, in the sense that measures exist that are not comparable under standard ordering but satisfy the premises of the theorem and are therefore comparable under the circles criterion. Consider two arrangements of plants in a diamond lattice, the diagonals of the diamonds being r_1 and $r_2 > r_1$ in length. We use the Löschian principle to evaluate these arrangements, so that each plant site carries a mass of *one*. We have noted that arrangements of this type are not comparable under standard ordering if none of the plant sites in the two arrangements coincide.

Let the partition elements $A_1, A_2, \ldots$ and $B_1, B_2, \ldots$ be the diamonds themselves in the respective lattices. (Border lines may be assigned to one or another adjacent region in any arbitrary manner.) Then $\mu(A_n) = \nu(B_n) = 1$ for all n, since each diamond contains just one plant. Also, $\alpha(A_n) = r_1^2/2$ and $\alpha(B_n) = r_2^2/2$ for all n, the areas of the respective diamonds. Hence $\mu(A_n)/\alpha(A_n) \geq 2/r_1^2 > 2/r_2^2 \geq \nu(B_n)/\alpha(B_n)$ for all n, so (91) is satisfied. The other premises are obviously also satisfied, so we may conclude that $\mu > \nu$ on the circles criterion. This accords with intuition; the "more densely" packed arrangement has "more" firms.

This example illustrates an important special case of the preceding theorem, where the arrangements being compared are "repetitive." Suppose for each arrangement Space can be partitioned into uniformly bounded regions such that "net benefits," b_i are the same for each respective region, $i = 1, 2$, and also the areas α_i are the same for the respective regions, $i = 1, 2$ (α_1, α_2 positive finite, b_1, b_2 finite). Then the first arrangement is preferable to the second on the circles criterion if

$$b_1/\alpha_1 > b_2/\alpha_2 \qquad (9.6.102)$$

This statement follows at once from the theorem if $b_1, b_2 \geq 0$. If this is not so, choose an x sufficiently large to make both $b_1 + x\alpha_1$ and $b_2 + x\alpha_2$ nonnegative, and then add $x\alpha$ to the two pseudomeasures being compared. (This amounts to translating them simultaneously in pseudomeasure space and leaves their comparability unaltered under the circles criterion, or any other vector partial order for that matter.) Under these new evaluations, b_i is changed to $b_i + x\alpha_i$, $i = 1, 2$; α_i is unaltered. Now (102) obviously implies $(b_1 + x\alpha_1)/\alpha_1 > (b_2 + x\alpha_2)/\alpha_2$, so that the preceding theorem again yields preferability of the first arrangement.

Intuitively, (102) says that the preferable arrangement is the one with higher "average density of benefits." When comparing repetitive arrangements, use of this criterion is almost inevitable; indeed it has been used by Isard, Mills and Lav, and perhaps others. It is therefore comforting to know that it can be deduced from the circles criterion. The trouble with (102) is that it is so specialized. Very few arrangements are sufficiently regular for anything like an average density of benefits to exist. That is why the service system literature has scarcely gone beyond the comparison of arrangements in which all potential market regions are congruent 3-, 4-, or 6-gons. This is not good enough. An arrangement alleged to be optimal must test its mettle against *all* feasible alternatives, not just those that are simple and regular. The circles criterion offers a general principle by which such a program could be carried out.[93]

We finish by examining the claims to optimality of the honeycomb lattice arrangement of plants. Here we are taking the Euclidean plane, assuming that the production cost function C is the same at all locations and that demand density is uniform over the plane at constant level $q > 0$. (This last condition may be thought of either as stating a subproblem of the general social optimization or Löschian problem or as arising from a benefit function of the form $b(x) = 0$ for $x < q$, $b(x) = m$ for $x \geq q$, where m is a "very large" number.)

The claim that a honeycomb lattice arrangement of plants (or cities) is efficient in some sense may have been first made by Christaller, and it was then greatly elaborated on by Lösch.[94] As mentioned above, the honeycomb lattice has been compared in the literature with only a very small number of alternative arrangements. Despite this, Lösch claimed

---

93. Of course the circles criterion may itself prove inadequate—perhaps because like standard ordering it refuses to make a comparison where intuition demands one. Where we should go in this circumstance is not clear. The alternative approaches of Fejes Tóth and Rogers should be kept in mind.

94. W. Christaller, *Central Places in Southern Germany,* trans C. W. Baskin (Prentice-Hall, Englewood Cliffs, N.J., 1966), p. 63 (German original, 1933); Lösch, *Economics of Location,* pp. 110–14.

the honeycomb lattice was indeed optimal, and this claim has been un-
critically repeated by later writers.[95]

An attempt to deduce the optimality of the honeycomb lattice from
packing theory has been made by Dacey.[96] A typical result from packing
theory is: given that no two plants may be closer than some fixed distance
r, the "densest packing" of plants (in a sense that can be made precise) is
attained with a honeycomb lattice arrangement. The trouble with this approach is that it introduces an irrelevant constraint. Firms are not constrained to avoid each other but to avoid losses. They can and will locate
"on top of each other" if profits can be increased by doing so (the "clustering of trades" phenomenon). This "minimal distance" constraint is
also irrelevant for the social optimization problem. Thus the packing approach does not accomplish anything.

Actually, the honeycomb lattice arrangement of plants is *not* always
optimal, at least for the social optimization version of the service system
problem. That is, for a certain class of production cost functions, there
is a nonhoneycomb arrangement surpassing any honeycomb arrangement
on the circles criterion. This superior alternative is "repetitive" in the
sense discussed above, and (102) obtains. The alternative has a higher
average density of benefits than any honeycomb lattice arrangement and
therefore surpasses them all on the circles criterion. We just outline the
major findings here, referring to the original article for proofs and discussion.[97]

As stated, the problem is set on the Euclidean plane with fixed constant demand density q. Since demand is fixed, the benefit terms of the
objective function are fixed and may be dropped. Social optimization
reduces to total cost minimization. And since we are dealing only with
"repetitive" arrangements, this in turn reduces to the minimization of
total cost per unit area, which is (102).

The key to the result is the form of the production cost function C.
We assume that

$$C(x) = ax^2 - bx^{3/2} + cx \qquad (9.6.103)$$

Here a, b, c are positive constants. Let these constants satisfy

$$32ac > 9b^2 \qquad (9.6.104)$$

---

95. For example, B. J. L. Berry and A. Pred, *Central Place Studies* (Reg. Sci. Res.
Inst., Philadelphia, 1961), p. 4, and J. Serck-Hanssen, The Optimal Number of Factories
in a Spatial Market, 269–81 in H. C. Bos, ed., *Towards Balanced International Growth*
(Wiley, New York, 1969), at p. 275, n. 6.

96. M. F. Dacey, The geometry of central place theory, *Geografiska Annaler,* 47B
(1965):111–24, at p. 113. On packing theory consult the works of Fejes Tóth and Rogers
cited above.

97. A. M. Faden, Inefficiency of the regular hexagon in industrial location, *Geogr.
Anal.* 1(Oct. 1969):321–28.

Actually, (104) is not needed for what follows but is inserted only to guarantee that marginal cost is positive. It should be emphasized that (103) and (104) determine a cost function that is not at all freakish or pathological; it is a perfect textbook example, with U-shaped average cost curve.

The desired result obtains if the parameter b lies in a certain interval. Specifically,

$$(8 - \log 27)2^{-1/2}3^{-7/4} > bq^{1/2}t^{-1} > (4 + \log 27)2^{-3/2}3^{-7/4} \qquad (9.6.105)$$

Here t is the cost of transporting unit mass unit distance; the left and right limits are about 0.486 and 0.377 respectively.

If (105) is satisfied, an arrangement exists that surpasses any honeycomb lattice arrangement. We indicate briefly the nature of this superior alternative and why it works. The right inequality in (105) guarantees the existence of a best one in the class of honeycomb lattices. The left inequality then guarantees that, in this best-sized hexagon, total costs can be reduced even further if demand at points near the corners of the hexagon is supplied locally. That is, new plants should be founded near the corners, producing on a small scale for their immediate vicinity and replacing shipments from the big plant at the center. This new arrangement is superior to the best honeycomb lattice, hence it is superior to any honeycomb lattice.[98]

Note we do not claim that *this* new arrangement is optimal. Some other pattern, perhaps a very irregular one, may do even better. Note also, that the counterexample above depends essentially on the fact that discs in the Euclidean metric (i.e., circles) cannot be fitted together to cover the plane (a fact also responsible for the Mills-Lav result). Thus the argument does not carry over to the diamond lattice in a city-block metric. It is at least possible then that the diamond lattice is unsurpassed, or even best, in this metric.

We end our discussion at this point. In particular, we will not go on to discuss the central place scheme of a hierarchy of service systems. We have indeed found in Chapter 8 that a city-activity hierarchy arises from a Thünen system with a limited-access transportation network; but we are referring here to the geometrical aspect of the scheme, with its rather dubious structure of nested honeycomb lattices. In view of the preceding discussion, it is perhaps prudent not to build too high when the foundation itself is shaky.

---

98. In a sententious though obscure footnote, Isard mentioned the possibility that the honeycomb lattice could be improved in this way. However, he did not demonstrate that such a possibility was realizable and seems to have misstated the needed conditions on the cost function. Isard, *Location and Space-Economy*, p. 242, n. 39.

Title Index

Author and Name Index

Subject Index

Abcont measures, 83–84
  conditional, 87
  Fubini's theorem, 85–86, 88
  Hahn decomposability, 132–33
  induced, 333n
  singularity of net flows, 381
  transportation feasibility, 337
Ability-responsibility association, 476
Absence makes the heart grow fonder, 554
Abstinence and taste changes, 554
Accessibility, generalized
  measured by Stewart potential, 519, 522, 526
  nodality, 581–82
  Thünen entrepôt systems, 526, 581–82
Activities, 12, 16–17, 108–13, 200. *See also* Land uses
  allowable, 199
  disposal, 223
  everyday, 117–20
  exemplified, 112–13
  finite spaces, 202–3
  fission and fusion, 118
  land uses vs., 414
  linkage. *See* Linkage, activities
  location, 109–11, 199
  measured in areal units, 414
  participants, 116
  scattering, 118
  sedentary, 109, 111, 199
  segregating, 118
  steady state, 110–11
  stock and flow components, 109–10, 200, 202, 413. *See also* Consumption; Production
  types, 111–13
  Weberian, 110n. *See also* Weber problem
Activity analysis model, 199–204
  no indivisibility, 419

  no neighborhood effects, 205–8, 419, 482
  no scale effects, 419
Addition, measures, 95, 139, 210–11
Additive
  countably, 30, 34, 36, 79, 93–94
  finitely, 30, 190
  set of measures, 210
  uncountably, 35–36
  utility functions, 299–300
Admission fees, 638
Adventure, 554
Advertising, 120, 545, 550
Africa, interior, 522
Agglomeration, 557
Aggregation of a measure, 41, 55
  limited information processing capacity, 122
  transportation problem, 331
Aggregation problem, service systems, 626
Agnostics, 548
Agriculture
  allotment restrictions, 423n
  form of mining, 119
  light land use, 508
  self-sustaining process, 119
  shift away, 529
  Thünen systems, 466, 517–18
Air pollution, 195–98, 535
Aleksandrov's theorem, 348, 352
Algorithms. *See* Computational schemes
Allocation-of-effort problem, 221–24
  anomalous atom case, 242
  constraints, 222–24
  examples, 219, 465–67
  feasibility, 251–53
  generality, 220
  imperfect markets, 291
  investment, 222–24
  no shadow price, 228–29, 242